BIOMECHANICS

A QUALITATIVE APPROACH FOR STUDYING HUMAN MOVEMENT

ELLEN KREIGHBAUM
Montana State University

KATHARINE M. BARTHELS
California Polytechnic State University

Illustrations by the authors

BIOMECHANICS

THIRD EDITION

Macmillan Publishing Company
New York

Editor: Robert Miller
Production Supervision: Kirsten Stigberg/Publication Services, Inc.
Text Design: Sheree Goodman
Cover Designer: Sheree Goodman
Illustrations: Katherine M. Barthels and Ellen Kreighbaum

This book was set in Palatino by Publication Services, Inc.,
and printed and bound by R.R. Donnelley & Sons.
The cover was printed by Phoenix Color Corp.

Printed in the United States of America

Macmillan Publishing Company
866 Third Avenue, New York, New York 10022

Collier Macmillan Canada, Inc.

Library of Congress Cataloging-in-Publication Data

Kreighbaum, Ellen.
Biomechanics : a qualitative approach for studying human movement
/ Ellen Kreighbaum, Katharine M. Barthels.—3rd ed.
p. cm.
Includes bibliographical references.
ISBN 0-02-366310-3
1. Human mechanics. 2. Biomechanics. I. Barthels, Katharine M.
II. Title.
QP303.K72 1990
612'.76—dc20 89-12667
CIP

Printing: 1 2 3 4 5 6 7 Year: 0 1 2 3 4 5 6

The authors dedicate this text to their parents, who guided, influenced,
and instilled attitudes necessary to pursue such a task;
to professional colleagues, who inspired and encouraged their questioning;
and to their students, whose comments and criticisms were an integral
part of the text development, and who provided the main reason for its writing.

Foreword

Authoring a textbook is like becoming a parent. It is a mixture of joy, a measure of heartache, considerable pride, and lots of work! It begins in rapture, sometimes planned, often not. If the timing and conditions are right, conception takes place, followed by a finite period of gestation when the myriad parts develop and are integrated into a complete functioning organism capable of life on its own. At birth (or publication of the first edition, whichever side of the simile you choose to adopt), the organism is, at the same time, complete and incomplete. It has the basic framework and all the organs necessary for survival, but it is yet an infant.

The first revision can be likened to adolescence when the awkwardness of rapid growth occurs and must be accomodated. The second revision signals the arrival of early adulthood where self-identity has been achieved and responsible contribution becomes the accepted calling.

Ellen Kreighbaum and Kathy Barthels, in the creation of this textbook, have been good parents. Their basic conception was magnificent, that of producing a source book for students, teachers, and researchers which emphasizes the qualitative aspect of studying human movement. It fulfilled a need in the field of movement analysis that had been largely ignored. The addition of concept modules and enhanced understanding sections in the first revision (second edition) provided necessary growth, development, and strength. This, the second revision (third edition) of the textbook, marks its status as a fully developed body of knowledge. Its uniqueness remains the authors' dedication to presenting a way of analyzing movement that is purposely qualitative. New to this edition are requisite refinements, knowledge updates, the dedication of a separate chapter for biomechanical analysis, and the addition of a new chapter on instruments that facilitate the analysis process.

Just as an adult retains certain identifiable features of his or her childhood, the same is true of this treatise on biomechanics. The authors have remained true to their "original intent and perspective," that of "promoting an understanding of biomechanical principles that govern the effectiveness of human movement skills" within a "conceptual" and "qualitative" framework. Three foundational features of the original work remain clearly evident in this new edition. They are:

1. The importance of biomechanical principles. Principles are accepted basic rules that guide action. They attempt to explain why the body moves in certain ways, why it obeys the laws of physics/mechanics. They provide insight and allow generalizing to take place in the application of knowledge.
2. The belief that knowledge of such principles can improve performance. The insight gained from understanding biomechanical principles speeds motor skill learning. The principles provide goals to be achieved in the performance. They offer a template against which performance can be compared.
3. The importance of a conceptual/qualitative approach in learning these principles. How one learns new concepts frequently is as important as what those new concepts are. The emphasis here is always on getting the "idea" or the "big picture" first, then filling in the details. It's not that the details are unimportant. They are essential. It's just that, in the initial stages of understanding a complex body of knowledge, too many details can sometimes impede progress. In skilled motor performance, it's the old "paralysis by analysis" phenomenon.

New features are also evident in this mature organism. Of particular interest are the following:

1. A separate chapter for biomechanical analysis which brings to life the application of the principles of this science.
2. The addition of a chapter on instruments that facilitate the analysis process.

It is an honor for me to present this complete work to you. I do it confidently and proudly as one announcing the graduation of a prized student. I have watched her grow and have even played a small part in her development. She has helped many others along the way, including me. She now takes her place as a competent and complete contributor in the ever improving and expanding field of biomechanical analysis.

Orwyn Sampson

Biomechanics Instructor
United States Air Force Academy

Preface

The revision of this text was undertaken for the following reasons: (1) to present an expanded and more thorough analysis system for explicit use by the teacher and coach, (2) to update material and expand the references and suggested readings sections following each chapter, and (3) to modify organization and content in response to readers' comments and suggestions.

The third edition of this text is in keeping with the original intent and perspective of the authors. Thus, it is focused on promoting an understanding of the biomechanical principles that govern the effectiveness of human movement skills. Furthermore, the focus is on presenting appropriate biomechanical information within conceptual frameworks that are based on the common overall performance objective of those skills being analyzed. In keeping with the AAHPERD Kinesiology Academy Guidelines and Standards, the conceptual, qualitative approach has been maintained to facilitate the student's grasp of the ideas presented.

In particular, the analysis section has been expanded greatly and has been developed into a separate chapter. Chapter 9, Observing and Analyzing Performance, presents a complete format for analysis. The teacher or coach is led through the initial steps of identifying the overall performance objective of the skill, the mechanical principles governing the accomplishment of that objective, and the identification of the critical features of performance that the coach can observe. Using this format, the practitioner should be able to identify and correct misplaced, unnecessary, or detrimental aspects of performance. The analysis chapter is placed after the modules and chapters dealing with anatomy or basic mechanics and before the chapters dealing with movement performances.

The text has remained qualitative as much as possible. Where necessary, equations are used with numerical examples so that the student can obtain a feel for the relative importance of the concept being presented. For those who use quantitative material in their classes, these relationships are presented in the appendixes for convenient reference. Furthermore, the third edition includes a chapter on biomechanics instrumentation in which the tools for quantitative analysis are described thoroughly. For those who incorporate data collection in their laboratory experiences, the chapter should serve as a valuable overview of those instruments used currently in biomechanics laboratories and at data collection sites.

The organization of the text into modules and chapters has been maintained from the second edition. The modules present basic biomechanical (anatomical or mechanical) material necessary for the understanding of groups of mechanically similar skills that follow in the chapters. For example, basic musculoskeletal information in Modules A and B and basic mechanical concepts of forces and torques in Modules C, D, and E are presented before the application of these concepts to the upper extremity, lower extremity, and trunk in Chapters 3–6. The authors firmly believe in the necessity of synthesizing anatomy and mechanics and applying them to the biological structure. Thus, we have not consented to organizing the text into an anatomy section followed by a mechanics section and finally an application section as many textbooks are prone to do. While this may be more difficult for some initially, the final result will encourage a unification of the bio- and the mechanics and, we hope, will minimize the dichotomy of the two, which has been prevalent in the past.

The module and chapter organization also facilitates ease of use for instructors who have many different approaches to teaching kinesiology or biomechanics. One may begin a course at any chapter or module. The modules and chapters that contain prerequisite information to this starting point are listed at the beginning of that chapter or module. Thus, the instructor need assign only those prerequisites to present the material in the chosen order. With this organizational scheme, virtually any order of presentation is possible. An anatomically based course could use Modules A–E and Chapters 1–7. A mechanically based course could use Modules C–K and Chapters 1 and 8–17.

The Understanding sections have been maintained and revised, and expanded where appropriate. These sections can be used for assignments, as a springboard for class discussion, or for laboratory work. The answers and points of discussion for the questions in the Understanding sections are given in the *Instructor's Manual*. Also included is useful information and direction for the instructor such as sources of biomechanics information from the literature, from professional organizations, and from professional meetings.

A further source of information is in the expanded and detailed lists of References and Suggested Readings, which follow each module and chapter. While some of these sources may exceed the level of understanding for some, they are important and useful for becoming familiar with the leading edge of the frontiers of biomechanics. They may also be important sources for reports and assignments.

Finally, the illustrations and photographs have been modified in some cases and expanded in others. In this edition, as in the last two editions, the authors have created their own illustrations and photographs. The authors believe that the application of a concept can be enhanced with the use of a quality illustration, and thus the illustrations should be used as a source of information as well as an example of concepts presented in the text.

Contributions from several individuals to the production of this edition are appreciated. We wish to thank sincerely the following people:

Brigitte Kohler, who did the final inking of the authors' original illustrations in the first edition, most of which are included in this edition

Bob Schwarzkopf, Jeannie Zumwalt, Don Jensen, and Rozan Pitcher, who served as subjects for some of the illustrations or for photographs

Carol Sanford, Marge Burgess, and Raeann Magyar, who assisted in various ways to complete the third edition

Orwyn Samson, who graciously consented to write the foreword to the third edition and whose comments and suggestions have been used to revise and improve the text.

E.K.
K.B.

Contents

CHAPTER **6**

BIOMECHANICAL RELATIONSHIPS IN THE TRUNK **255**

CHAPTER **7**

APPLICATION OF BIOMECHANICS TO NEUROMUSCULAR FITNESS ACTIVITIES **275**

CHAPTER **8**

BODY BALANCE AND STABILITY CONTROL **310**

CONCEPT MODULE **F**

LINEAR MOVEMENT RESPONSES TO APPLIED FORCES **334**

CHAPTER 1

The Study and Analysis of Human Movement

PREREQUISITES

None

CHAPTER CONTENTS

The primary objective of a course of study in physical education is to help the student understand the nature and function of human movement in sport, dance, physical recreation, and adapted movement activities (Figure 1.1). The competent professional should be well versed in the body of knowledge or subject matter of this specialized field, which demands an understanding of numerous subdisciplines.

FIGURE 1.1
Biomechanics for the teacher or coach.

1.1 Kinesiology and Biomechanics: Areas of Study

Within the broad definition of kinesiology, the physical educator or coach may study human movement from a variety of perspectives—psychological, philosophical, sociological, or physiological. In an undergraduate physical education curriculum, **kinesiology** is the title traditionally given to a course concerning the anatomical and mechanical bases of human movement related specifically to sport, dance, and adaptive activities. The combination of knowledge of the biological and material properties of the skeletal, articular, and neuromuscular systems and of the laws and principles set forth in mechanical physics provides the foundation for the development of a new name for this subdiscipline—biomechanics. Generally, **biomechanics** is defined as that area of study wherein the knowledge and methods of mechanics are applied to the structure and function of the living human system. Biomechanics is not an area reserved for study by the physical educator or coach, however. Figure 1.2 presents an overview of a variety of movement-related fields and occupations in which biomechanics information is applied.

Biomechanical knowledge is used in several diverse disciplines that include biology, physiology, medicine, and mechanical physics. Many professionals—engineers, designers, physical therapists, oral and orthopedic surgeons, cardiologists, and aerospace engineers—make practical applications of biomechanics. A biomedical engineer may study the biological and material properties of the human body and their mechanical aspects to understand the flow of blood within the arteries. The knowledge gained may then be applied to help reduce or solve circulatory problems. A human factors designer may study the physical characteristics of the human body and the mechanical aspects of its movements that are necessary in using a given work space. Such knowledge would then enable the designer to determine an efficient height for a worktable or to provide sufficient space between one work area and another. A teacher or coach studies the physical characteristics of the human body and the principles of mechanical physics so that guidance concerning mechanically effective performances can be given to the learner.

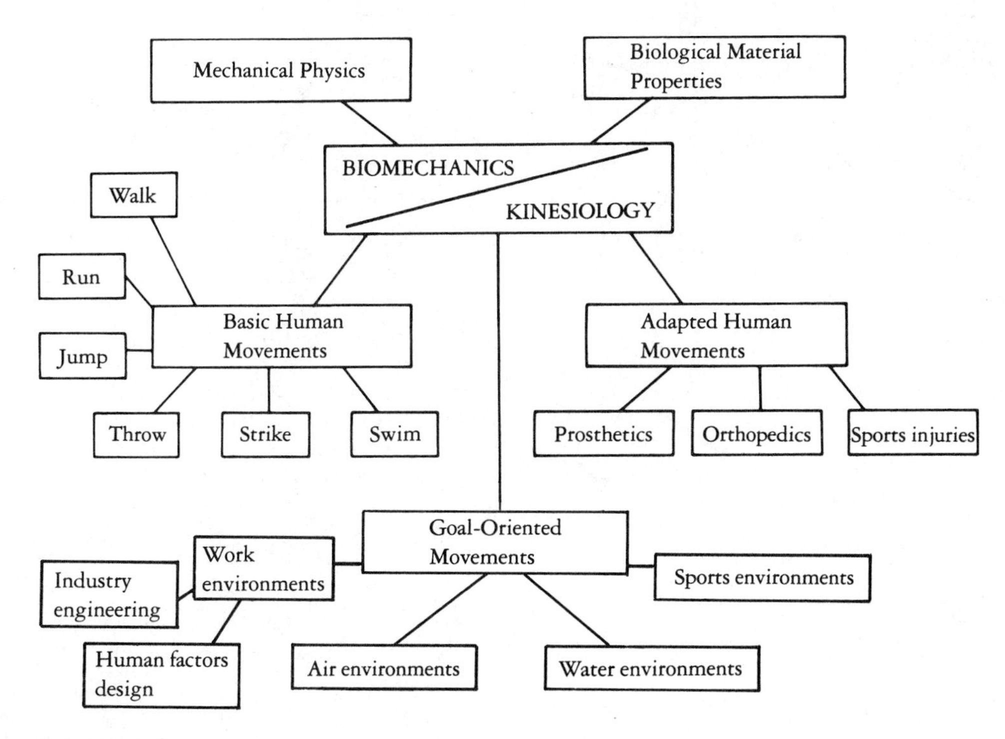

FIGURE 1.2
Application of biomechanics to movement-related fields.

Physical Characteristics of the Body

Knowing and applying knowledge of the physical characteristics of the human body is important when attempting to analyze its movement qualities. Information about the growth and structure of the bones, joints, and muscles can be used to determine appropriate or inappropriate movement activities for a variety of age groups.

The area of study concerned with measures of the body's physical characteristics is **anthropometrics**. Data gathered from research on human anthropometrics are presented in Appendix III. These measures of human body size, shape, and composition include measures of height, weight, and volume. Similar measures may be taken on each segment of the body—its length, weight, and volume. Anthropometric data are usually classified into two or more groups (e.g., by gender, age, body shape, or athletic

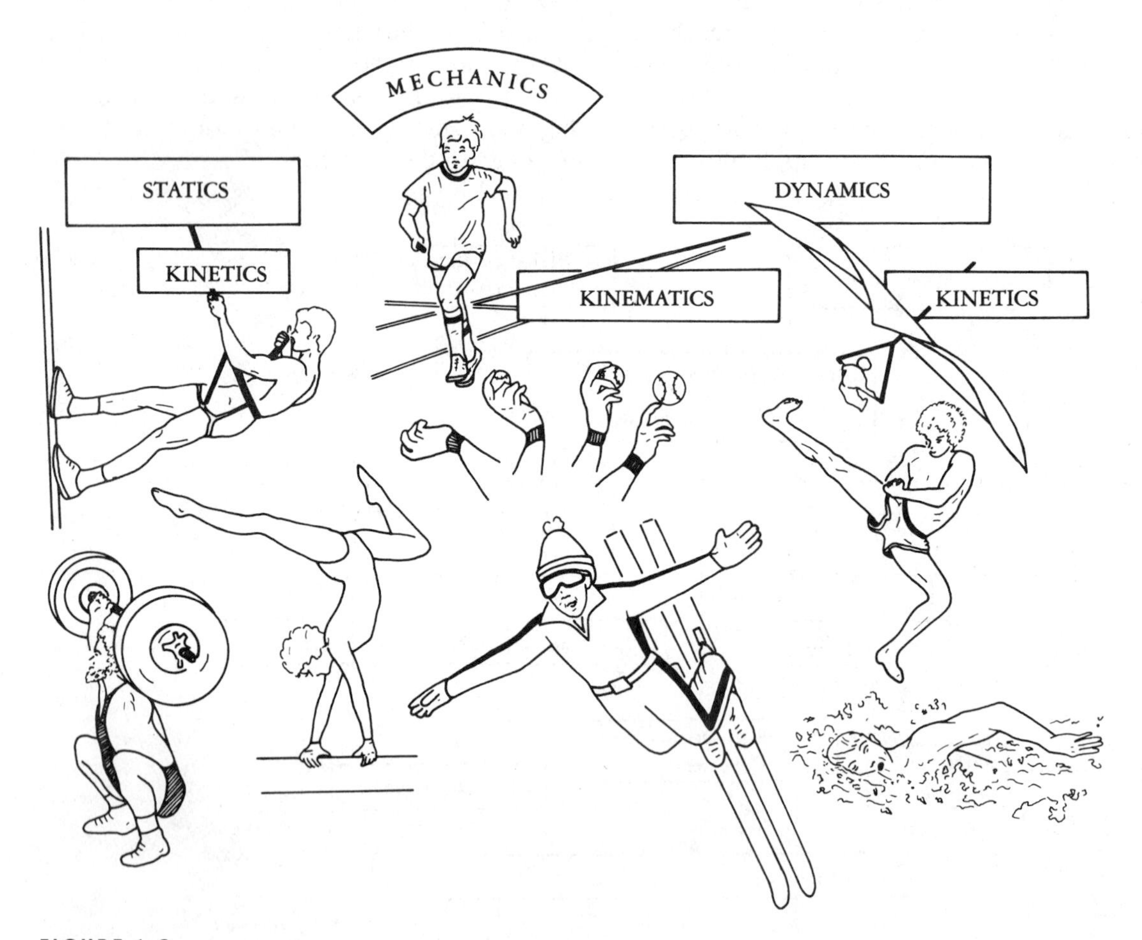

FIGURE 1.3
Areas of study in biomechanics.

participation) and then compared with the anthropometric measures of other groups.

Mechanical Characteristics of Human Movement

The area of study concerned with the mechanical aspects of any system is called **mechanics** and may be divided into two categories: **statics**, the study of factors associated with nonmoving systems, and **dynamics**, the study of factors associated with systems in motion. These relationships are shown in Figure 1.3.

One may attempt to remain motionless, but doing so is almost impossible. Even while standing, the body sways from side to side and forward and backward. Neverthless, a static analysis may be made of the human body in near-static activities such as while standing, balancing, and doing certain resistance exercises. In most sport situations, however, dynamic analysis is required because total body or segmental movement takes place. The area of dynamics is further divided into *kinematics*, the study of the time and space factors of motion of a system, and *kinetics*, the study of the forces acting on a body that influence its movement.

1.2 Approaches for Studying Movement

Two approaches are used for studying the biological and mechanical aspects of human movement—a quantitative approach and a qualitative approach. The **quantitative approach** involves describing a movement of the body or its parts in numerical terms. Such quantification of movement characteristics helps eliminate subjective description since it relies only on numerical data obtained from the use of instrumentation. Numbers obtained by measuring or counting can describe the physical situation. The observer then may use this quantification to explain or describe the actual situation. Usually, it is not economical or a wise use of time for a teacher or coach to analyze the movements of each of the performers quantitatively; the instrumentation can be expensive, often is not readily available, and if available, is difficult to transport to the field, swimming pool or gynmasium. The techniques of quantitative analysis can be quite time-consuming. An understanding of the information derived from a quantitative analysis should be within the grasp of most teachers and coaches so that the current scientific literature on a given sport or activity can be incorporated. An introduction to quantitative data gathering is given at the end of this text in Chapter 17.

The **qualitative approach** attempts to describe movement in nonnumerical terms (i.e., without measuring or counting any part of the performance). The impressions gained from a qualitative analysis may be substantiated with quantitative data, and many hypotheses for research projects are formulated in such a manner. The qualitative evaluations of a performance should be based on the teacher's ability to recognize the critical features of the skill. Subjective conclusions based on qualitative analyses can be rejected or affirmed on the basis of quantitative study.

Both qualitative and quantitative analyses provide valuable information about a performance; however, a qualitative assessment is the predominant method used by the teacher or coach in analyzing the movements of performers.

In most teaching and coaching situations,

analysis of movement is based on visual observation. The usual method employed in everyday situations is a qualitative analysis based on observation and memory of a performance so that subsequent evaluation can be made. Videotapes and films provide visual records that may be viewed and reviewed to identify more variables than can be recognized by a single observation and memory alone.

Although films and videotapes can provide records from which measurements may be taken for a detailed quantitative analysis, great care must be taken to prepare and conduct the filming so that valid measurements may be obtained. Often this is not possible unless special arrangements are made. Therefore, mechanical analyses based on films and videotapes are usually qualitative rather than quantitative.

Videotapes and films enhance the learning process because they provide valuable visual feedback to the performer. The teacher or coach should be able to perform a visual qualitative analysis if film or videotape is being used, so that an immediate comment can be given the performer while the movement experience is still fresh. The teacher's knowledge of biomechanics becomes increasingly important as the refinements become more minute. Without videotape or film or some other form of analytic tool, small refinements could not be observed at all, because of the limitations of human perceptual abilities. A more detailed discussion of approaches to qualitative analysis is located in Chapter 9.

Efficiency and Effectiveness of Performance

If one is to consider the efficiency of movement, one must incorporate the concepts of work and energy. An efficient movement or skill performance is one in which a given amount of work (the skill) is done with a minimum of energy expenditure. Efficient movement is a definite advantage to performers who want to do as much work as possible before their energy runs out, such as marathon runners or swimmers. A more efficient mover lasts longer given the same pace, or performs at a faster pace with less energy expended than does a less efficient mover. For most sport activities, however, the primary concern is not with teaching how to save energy to keep going. Biomechanists are concerned more with the **effectiveness** of a performance, that is, with determining the most appropriate movements to help the performer successfully accomplish the overall performance objective of the skill. The degree of work effort required by the skill or the amount of energy expended in performing it are not as important. For example, a 100-m run can be performed more *efficiently* by a runner who runs at half the speed of a more *effective* runner who arrives at the finish line first.

Maximize, Minimize, or Optimize?

In the 100-m run, the performer's purpose is to run with maximum speed for a short distance. Therefore, the effort put forth (energy output) in the running movements should be **maximized**. If the race were 5-km, however, a maximal rate of energy expenditure would be too great physiologically to maintain for the entire distance, and the runner probably would turn into a walker to finish. Nor would the athlete be effective (successful) in the mile race if the running speed were **minimized** to conserve energy. Instead, the runner would need to **optimize** the running speed (and rate of energy expenditure) to accomplish the purpose of the 5-km run. The optimal speed would be the specific race pace that is the fastest the runner could maintain for 5-km. The runner would learn this pace through practice. Running speed is but one example of biomechanical factors or quantities that frequently are *optimized* rather than *maximized* or *minimized*.

Lafortune and Cavanagh (1983) have shown that for cycling, a low correlation exists between efficiency measures and the mechanical effectiveness of the movements used to pedal. This work is a good example of *optimizing the efficiency* to *maximize the effectiveness*. Generally, an effective movement is optimally efficient; however, a maximally efficient movement is not necessarily biomechanically effective.

Understanding the Study of Human Movement

1. Determine whether each of the following activities require a static or dynamic analysis (a) a cartwheel, (b) a handstand, (c) a sprint starting position, (d) a hammer throw, (e) the en garde position in fencing, (f) a football line stance, (g) a chip shot in golf, (h) a basketball free throw.

2. A list of hypothetical articles related to movement follows. Which articles examine the kinematics of a movement, and which concern the kinetics of a movement?

- **a.** "The Timing of the Segments During a Golf Swing"
- **b.** "The Strength of the Erector Spinae Group During the Lifting Action in Man"
- **c.** "The Flight of the Discus"
- **d.** "Accuracy in the Tennis Forehand Drive"
- **e.** "Muscular Force Involved in Baseball Batting"
- **f.** "The Effects of Variation of Speed and Direction of Object Flight on Catching Tasks"
- **g.** "Ground Reaction Forces in the Golf Swing"
- **h.** "Force Production in the Clean and Jerk"
- **i.** "Mechanical Analysis of the Initial Velocity in the Standing Broad Jump"
- **j.** "Comparisons of the Force Production of Good and Poor Long Jumpers"
- **k.** "Effects of Stride Length and Stride Frequency on Running Velocity"

3. The following statements are from analyses of movement activities. Indicate which are qualitative and which are quantitative.

- **a.** The greater the angle of takeoff, the greater the vertical velocity component.
- **b.** The angle of takeoff was found to be 23 degrees.
- **c.** The style of high jump was more effective for jumper A than for jumper B.
- **d.** Jumper A jumped 30 cm higher than jumper B.
- **e.** Raising the arms over the head will raise the center of gravity within the body.
- **f.** An object swung in an arc and released from its circular path will travel in a new path tangent to the arc at the point of release.

4. Browse through recent issues of a sport-specific magazine such as *Ski, Golf, Tennis, Runner's World, The International Gymnast*, or *Track and Field Quarterly*. Find and list 10 phrases or statements that concern the biomechanical, anatomical, or mechanical bases of a movement or a piece of sport equipment.

5. Describe three activities in which efficiency should be reduced in favor of effectiveness.

6. Describe an activity in which force is optimized, but not maximized, for the most effective performance.

References and Suggested Readings

Arnold, R. K. (1978). Optimizing skill learning: Moving to match the environment. *JOPER, 49*(9), 84–86.

Asmussen, E. (1976). Movement of man and study of man in motion: A scanning review of the development of biomechanics. In P. V. Komi (Ed.), *Biomechanics V-A: Proceedings of the Fifth International Symposium on Biomechanics* (pp. 23–40). Baltimore: University Park Press.

Atwater, A. E. (1980). Kinesiology/biomechanics: Perspectives and trends. *Research Quarterly for Exercise and Sport, 51*(1), 193–218.

Bates, B. T. (1977). Scientific basis of human movement. *JOPER, 48*(8) 68–74.

Cavanagh, P. R. (1986). The cutting edge in physical education and exercise science research. *The Academy Papers, 20,* 115–119. Champaign, IL: Human Kinetics.

Dyson, G. (1976). Coaching philosophy. *Swimming Technique, 13*(2), 44–49.

Dyson, G. (1977). *The mechanics of athletics* (pp. 11–13). New York: Holmes & Meier.

Hatze, H. (1974). The meaning of the term biomechanics. *Journal of Biomechanics, 7*(2), 189–190.

Hays, J. (1979). Back to basics in physical education and dance. *JOPER, 50*(5), 33–35.

Lafortune, M. A., & Cavanagh, P. R. (1983). Effectiveness and efficiency during bicycle riding. In H. Matsui & K. Kobayashi (Eds.), *Biomechanics VIII-B: Proceedings of the Eighth International Congress of Biomechanics* (pp. 928–936). Champaign, IL: Human Kinetics.

Malina, R. M. (1975). Anthropometric correlates of strength and motor performances. In J. H. Wilmore and J. F. Keogh (Eds.), *Exercise and sports science reviews* (Vol. 3, pp. 249–272). New York: Academic Press.

Norman, R. W. (1975). Biomechanics for the community coach. *JOPER, 46*(3), 49–52.

CONCEPT MODULE A

The Skeletal System and its Articulations

PREREQUISITES

None

CONCEPT MODULE CONTENTS

Four anatomical disciplines directly relate to human body movement. They are (1) osteology, the study of the skeletal system; (2) arthrology, the study of the articulations; (3) myology, the study of the muscles; and (4) neurology, the study of the nervous system. Not all the information relating to these systems is directly applicable to the biomechanics of movement; however, those characteristics and properties of the skeletal and articular systems that do concern human movement are presented in the following section.

A.1 The Skeletal System

Structurally, the human skeleton is divided into two parts, the **axial skeleton**, which includes the skull, thorax, pelvis, and vertebral column, and the **appen-**

dicular skeleton, which includes the upper and lower extremities. Figures A.1a and b illustrate the anterior and posterior views of the human skeleton.

The axial skeleton consists of the skull (29 bones), the thorax (25 bones), and the vertebral column (26 separate bones). In addition to the 12 pairs of ribs, the sternum is the only other bone in the thorax. The sternum is a fused unit of three parts: the manubrium, the body, and the xyphoid process. Although the three separate sections of the sternum are fused, knowing their locations is important for identifying muscle attachments and the mechanics of muscle pull. The vertebral column (Figure A.1c) consists of five regions: the cervical spine (7 bones), the thoracic spine (12 bones), the lumbar spine (5 bones), the sacrum (5 bones fused into a single unit), and the coccyx (4 bones).

The appendicular skeleton consists of two basic divisions, the upper right and left extremities (32 bones each) and the lower right and left extremities (31 bones each). The upper extremity begins with the shoulder girdle complex and ends with the fingertips. The lower extremity begins with the pelvic girdle and ends with the toes. For the most part, the individual carpals (bones in the hand) and tarsals (bones in the foot) are not of primary functional significance in the biomechanics of sport, although two tarsals do play an important function in weight bearing and should be identified: the talus (directly below the tibia) and the calcaneus (the great bone of the heel, directly below the talus).

The skeleton has several important functions: (1) to protect vital organs within the body, (2) to support soft tissues, (3) to serve as the factory for making red blood cells, (4) to serve as a reservoir for minerals (calcium and phosphate), (5) to provide the attachments for skeletal muscles, and (6) to act as a system of machines to receive muscle torques and to make movement possible. Of these services, the last two are of particular concern to biomechanists and kinesiologists.

Classification of Bones

One method of classifying bones is by shape. Over time, bone tissue follows what is known as Wolff's Law: the shape of a bone to some extent determines its main function, and conversely, the function may alter its shape. Long bones are the predominant bones in the extremities and serve a particular function in providing levers for movement. The long bones in the lower extremities are generally larger and stronger than those in the upper extremities, for they must repeatedly bear the weight of the body during locomotion, and therefore, are subjected to repeated stress. The long bones in the upper extremities are generally smaller and lighter since they are used mainly in reaching, grasping, and throwing rather than weight bearing. Some smaller bones may be subjected to repeated stress, such as the humerus of a baseball pitcher or the metatarsals of runners.

The short bones generally appear as chunks of bone and are found predominantly in the hands and feet. They are compact and, in addition to allowing movement, serve collectively to provide elasticity, flexibility, and shock

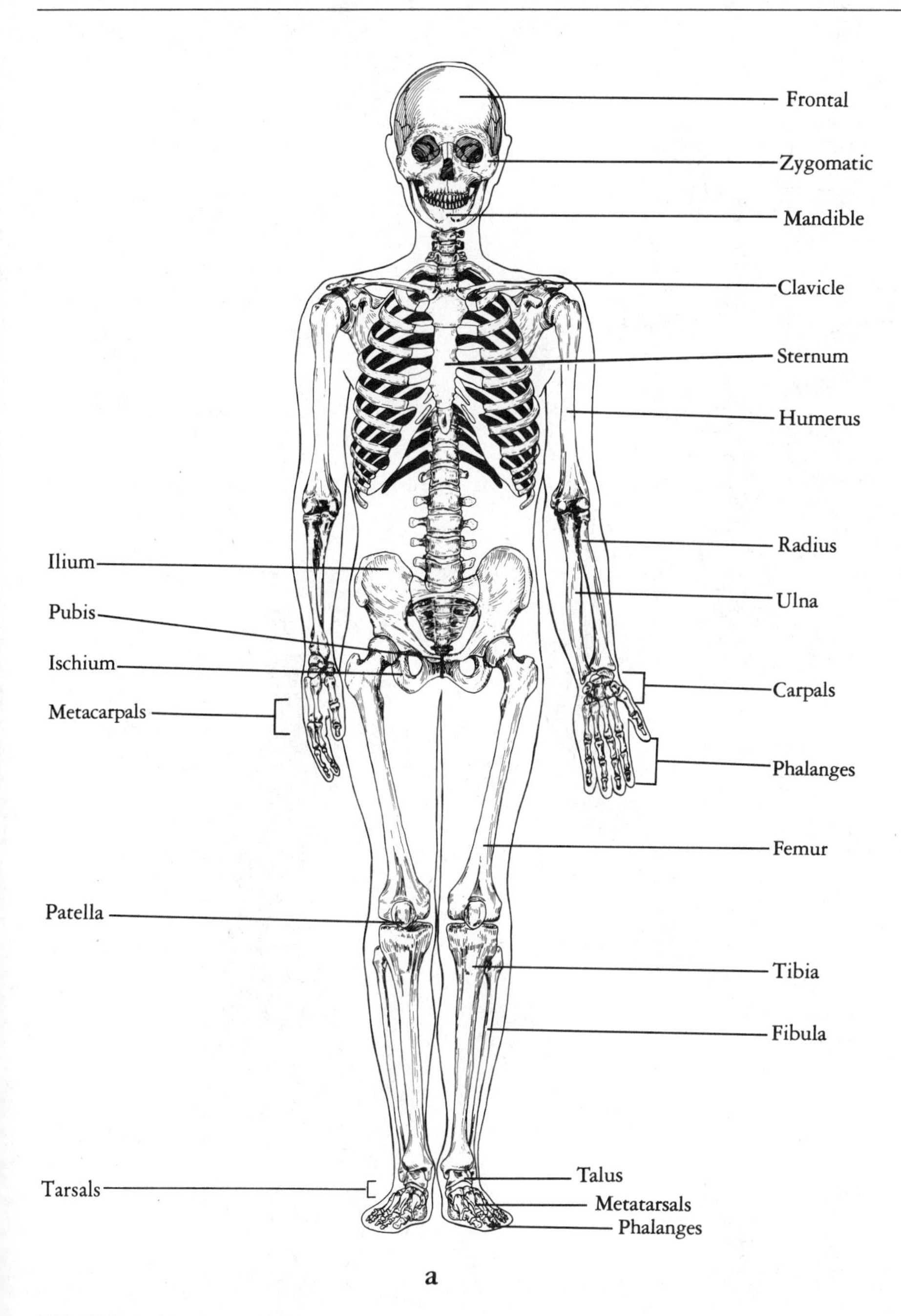

FIGURE A.1

(a) Anterior (front or ventral) aspect of the human skeleton. (b) Posterior (back or dorsal) aspect of the human skeleton. (c) Side view of the vertebral column.

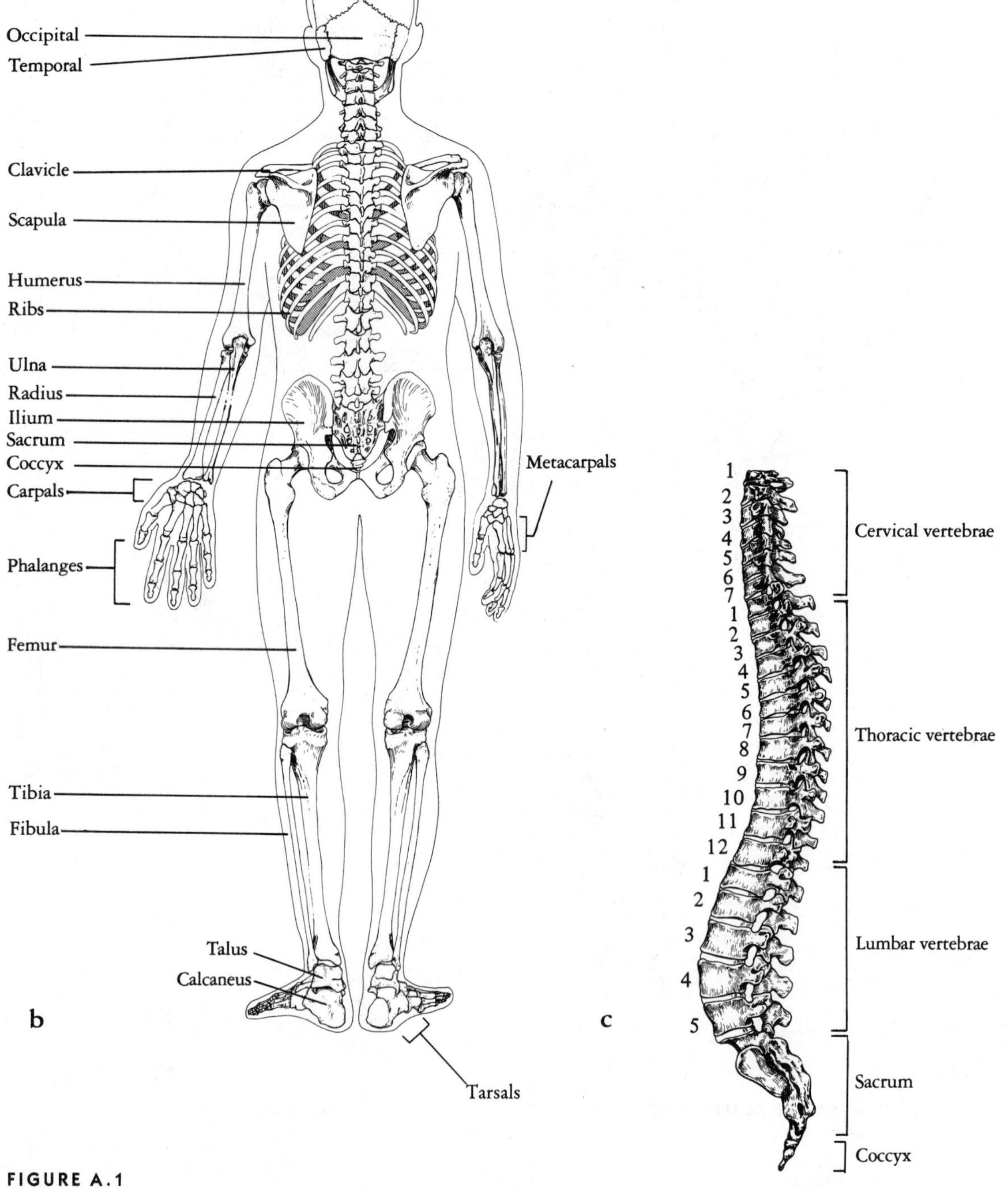
Parietal
Occipital
Temporal
Clavicle
Scapula
Humerus
Ribs
Ulna
Radius
Ilium
Sacrum
Coccyx
Carpals
Phalanges
Femur
Tibia
Fibula
Talus
Calcaneus
b
Metacarpals
Tarsals
c
1
2
3
4
5
6
7
Cervical vertebrae
1
2
3
4
5
6
7
8
9
10
11
12
Thoracic vertebrae
1
2
3
4
5
Lumbar vertebrae
Sacrum
Coccyx

FIGURE A.1
(continued)

absorption. The flat bones, such as those in the skull, pelvis, and scapula, provide a protective covering for structures lying below them and also serve as locations for muscular attachments. Irregular bones, such as the vertebrae, protect the spinal cord and branching nerves. The spinous and transverse processes of the vertebrae serve as "handles" for muscular attachments and as levers for muscle pull, and the bodies of the vertebrae partially absorb the shock of landing in walking, running, or jumping.

Gross Structure of Bone

The knowledge of some basic structural properties of bone is useful for understanding structural functional relationships. Figure A.2 illustrates some of the important gross features of a typical long bone, flat bone, and irregular bone.

If a long bone is in the extremities, the end of the bone closer to the trunk is the **proximal** end, and the end farther away from the trunk is the **distal** end. The **lateral** side of the bone is that side away from the midline of the body, while that side closest to the midline is the **medial** side.[1] If a bone is part of

[1] When anatomical position is assumed with palms facing forward, as in the left upper extremity in Figure A.1.

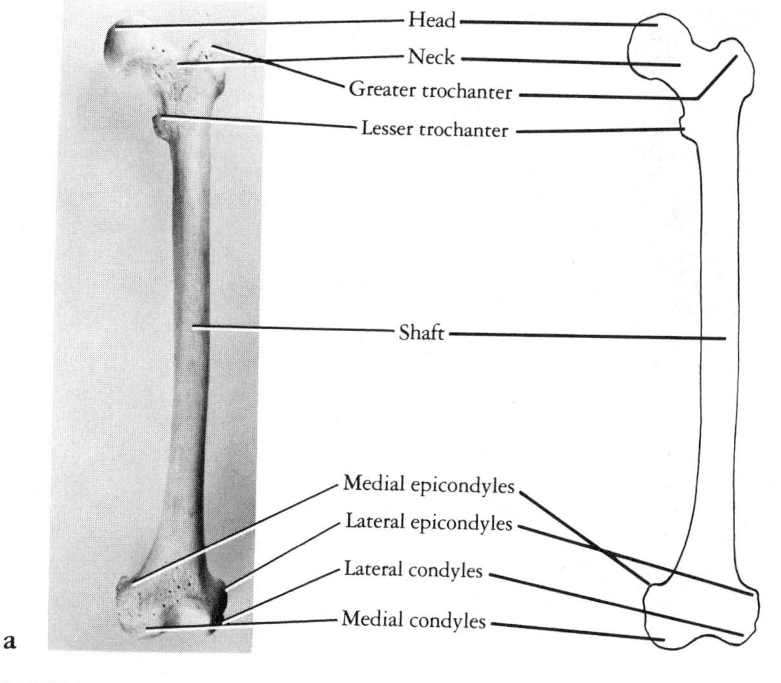

FIGURE A.2

Important parts of a long bone: (a) femur, (b) radius and ulna. (c) Parts of a flat bone (scapula). (d and e) Parts of an irregular bone (vertebrae).

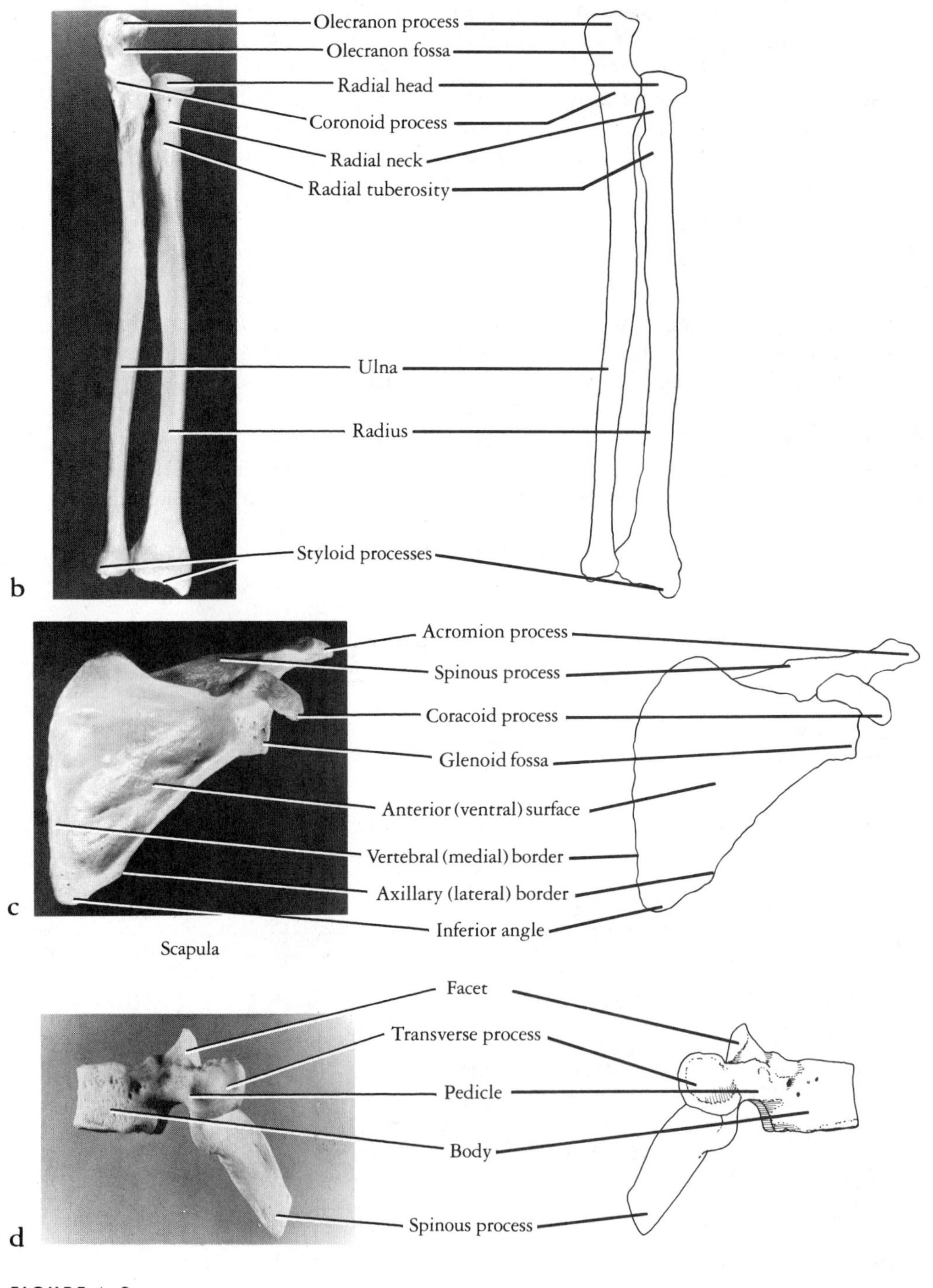
Olecranon process
Olecranon fossa
Radial head
Coronoid process
Radial neck
Radial tuberosity
Ulna
Radius
Styloid processes
b
Acromion process
Spinous process
Coracoid process
Glenoid fossa
Anterior (ventral) surface
Vertebral (medial) border
Axillary (lateral) border
Inferior angle
c
Scapula
Facet
Transverse process
Pedicle
Body
Spinous process
d

FIGURE A.2
(continued)

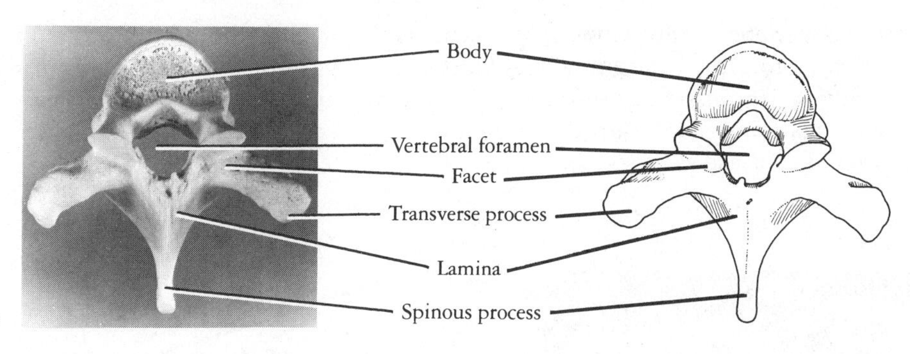

FIGURE A.2
(continued)

the axial skeleton, the border closest to the head is the **superior** border of the bone; that closest to the feet is the **inferior** border. Although all bones do not possess each of these characteristics, knowledge of the following landmarks is useful in identifying parts of the skeleton:

Condyle (Figure A.2a) — A rounded process of a bone that articulates with another bone.

Epicondyle (Figure A.2a) — A small condyle.

Facet (Figure A.2d, e) — A small, fairly flat, smooth surface of a bone, generally an articular surface.

Foramen (Figure A.2e) — A hole in a bone through which nerves or vessels pass.

Fossa (Figure A.2b, c) — A shallow, dish-shaped section of a bone that provides a space for an articulation with another bone or serves as a muscle attachment.

Process (Figure A.2b, c, d, e) — A bony prominence.

Tuberosity (Figure A.2b) — A raised section of bone to which a ligament, tendon, or muscle attaches; usually created or enlarged by the stress of the muscle's pull on that bone during growth.

Architecture and Development of Long Bone

Of special interest to developmental biomechanists is the role of forces in changing the composition, size, and structure of bones in the skeletal system. Although the primary determinant of the architecture and the material properties is genetic, adaptations may occur as a result of the amount and kind of stresses to which the bone is subjected and the nutritional situation of the individual. Traumatic acute or repetitive overuse injuries are influenced by the type of material making up the bone and the structure of the part of the bone subjected to these stresses.

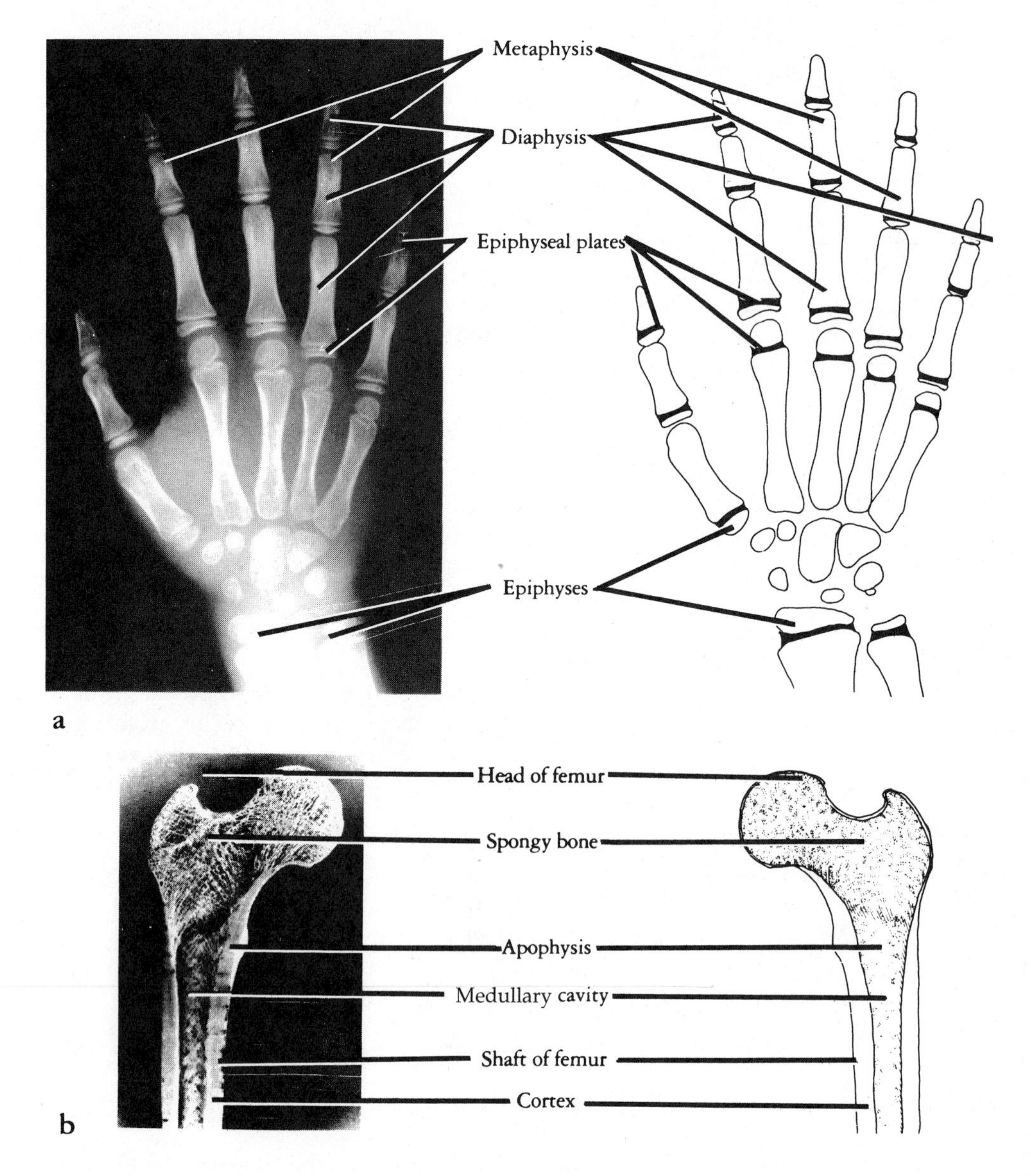

FIGURE A.3

Long bone growth and structural characteristics.

The structure of a typical long bone is illustrated in Figures A.3a and b. The articular surfaces of long bones, those parts that form joints with adjacent bones, are located at the distal and proximal ends of the bones. The shaft of the long bone is called the **diaphysis**. The diaphysis consists of an outer shell called the **cortex**. The outer side of the bone is covered by a thin membrane called the **periosteum**; the inside of the cortex is covered by a similar membrane called the **endosteum**. The cortex surrounds the **medullary** cavity, which contains the bone marrow. On either end of the diaphysis is an area filled with spongy, or **trabecular**, bone. This area is called the **metaphysis**. Bounding the proximal and distal ends of the metaphysis is a line, or plate, of cartilage that cuts across the end sections of the bone. This plate of cartilage is called the **epiphyseal** plate and separates the metaphysis from the **epiphysis**. The epiphyses also contain spongy bone, or trabeculae, and form the proximal and distal ends of the long bone.

A long bone grows in length by depositing new bone cells (osteoblasts) along the epiphyseal plates on the metaphysis side. In this manner, the bone continues to lengthen until internal maturational or nutritional signals indicate a stopping time. The completion or closing time of the epiphyseal plates, called **ossification**, varies with different bones and also varies for the same bones between the sexes. In females, ossification generally occurs up to 4 years earlier than in males.

At the location of muscle or muscle tendon insertions to the bone, a raised section of bone develops, which is called an **apophysis**. Similar to the epiphysis in structure, the apophysis is separated from the cortex by an apophyseal plate, along which osteoblasts are placed. The tibial tuberosity formed at the attachment of the patellar tendon (quadriceps) and the radial tuberosity, where the biceps tendon is attached to the radius, are primary examples. With continued use of a muscle, a number of new bone cells are laid down, creating a raised portion of bone at that location.

Changes in Bony Structure and Integrity

Bone tissue is adaptable. The changes in structure with changes in function were first identified by Wolff in 1892. Over the years, Wolff's Law has become the foundation on which sport scientists support the importance of exercise in the development and maintenance of integrity of bone tissue. Changes in bone tissue are reflected in increased or decreased diameter of the medullary canal, increased or decreased width of the cortical bone, changes in the mineral content of the tissue, and changes in the length, girth, or density of the bone tissue.

Physical stress on bone tissue can be either **compression** (pressing together), **tension** (pulling apart), **torsion** (twisting), or **shear** (tearing across). The mechanical properties of bone consist of hardness, elasticity, stress (force)–strain (deformation) relationships, energy absorption, and resistance to fatigue. Compression of either the epiphyseal or apophyseal plates before closure retards bone growth. Tension on the bone, such as that produced by muscle pull, activates bone cell growth, and shearing forces tend to separate the plate from the newly calcified bone. The quality of these properties possessed by bone appears to depend on one or more of the following factors: age, nutritional status, the type of bone tissue, hormonal status, and the length and magnitude of stress placed on the bone.

Recently, much research effort has been applied to identifying the factors that cause changes in bone tissue, specifically, changes during the growing years, the adult years, and the later years.

Changes During the Growing Years

Before ossification of the bones, particular injuries may occur to the areas surrounding the epiphyseal and apophyseal plates. In addition, the articular cartilage surrounding the ends of the long bones may also undergo changes in mechanical properties with stress (Micheli, 1986). Further, in growing athletes, an imbalance in the strength of musculoskeletal tissues results in the system injuring itself as a result of one tissue being stronger than an interfacing tissue. Micheli (1986) reports that the "growth factor" between the bone length and muscle tendon unit may harm the involved structures. For example, the rapid growth of the length of the femur and the tibia can place painful stress on the patellar tendon and the tibial tuberosity, because these structures have difficulty accommodating quickly to the increased distance that they must span.

During these times of rapid growth, young athletes should not be placed in intense or lengthy training programs, and special attention should be given to appropriate stretching exercises. The increased tension at the tendon attachment to the bone may result in microtears in the tendon, the apophysis or bony epiphysis. This condition is called Osgood-Schlatter's disease and is most frequently seen in males in their middle teen years. The condition is particularly aggravated by quick stops, in which the foot is planted and the deceleration is accommodated by the flexing knee. Landings in basketball, gymnastics, jumping events, and volleyball are events that place the athlete at high risk for developing Osgood-Schlatter's disease. For a thorough biomechanical discussion of avulsions of the proximal tibia see Mirbey et al., 1988.

A second situation that may occur is the imbalance in strength between the muscle tissue itself and the strength of its attachment to the bone. If placed in an intense strength training program during the rapid growth years, and particularly when the introduction of the hormone testosterone enhances the strength of muscles in males, the athlete's muscle tissue may become so strong that it may pull itself away from its attachment. The situation is particularly prevalent in strength-dominated events such as wrestling and competitive weight lifting. Coaches should be particularly cautious of intense strength-training programs for youths from 14 to 22 years.

If a child is active in a wide variety of activities, the child's bones grow and strengthen to accommodate stresses related to those activities. Similarly, if a child is subjected to a single activity repeatedly, such as throwing a baseball or playing tennis, the utilized bones grow and develop in a slightly different configuration than the less-stressed bones on the opposite side. A muscle's pull on a bone stimulates bone growth at the apophyseal plate, and the shape and strength of the bone–muscle junction accommodates itself to that type of activity. This adaptation provides a physiological reason (aside from the philosophical argument) for starting children early in activities that require specialized body movements, such as gymnastics and ballet. The inherent disadvantage to this type of program is that repeated stresses of isolated body parts that are integral to a sport may prove injurious to young musculoskeletal structures. For example, the impact of a tennis ball on a racket provides a sometimes overwhelming stress on an immature arm, and practicing baseball pitches repeatedly or pitching an entire game may be more than an immature arm can handle safely. When subjected to stress in this way, the bone may show signs of wear: small portions of the joint cartilage may chip off, the epiphysis may pull off the bone entirely, or the epiphyseal plate may close at an early age and thus arrest growth in the bone. From the standpoint of developmental kinesiology, these factors have serious implications for intensive youth sport programs.

Although impact and repeat stress do not always result in permanent injury, their effects may be severe in children before closure of the epiphyseal plates in the bone. Researchers have found that separation "of the medial epicondyle of the humerus is not uncommon prior to epiphyseal closure" in Little League pitchers (Woods, Tullos, and King, 1973). Epiphyseal injuries have been reported in adolescent weight lifters (Ryan and Salciccioli, 1976).

With the increased intensity and frequency of gymnastics participation, young gymnasts are subjected to greater musculoskeletal injury. In addition, female gymnasts are becoming more and more skillful at younger and younger ages. Gymnasts may suffer from the types of injuries described previously, such as avulsion fractures of the apophysis and osteochondritis; overuse injuries are becoming more prevalent. Stress fractures of the radius in gymnasts (Roy et al., 1985) have become more frequent with increased difficulty of the skills being performed. Possible factors are the use of upper extremity weight bearing on softer mats, the addition of twisting vaults and dismounts from upper extremity weight-bearing positions, twisting moves that involve pronation and supination of the radioulnar joint when the hand is fixed, and increasing numbers of repetitions in practice.

The most frequent overuse injuries in gymnasts have been the wrist, low back, shin, and foot (Caine and Lindler, 1985). Jackson (1976) reported a four-times higher incidence of low-back pathologies in young female gymnasts than in the rest of the population. It is becoming clear that coaches must consider the age of the athletes and the vulnerability of their musculoskeletal systems.

The large size of a child is not an indicator of bone maturation; in fact, many times the opposite is true. The child who is overweight or excessively large may be among the last to mature physically. Considerable evidence confirms that exercise and nontraumatic stress on the bones of children and adults enhance the size and strength of the bones. In growing animals, a low-intensity training either does not affect or stimulates bone growth in length and girth, but high-intensity training inhibits length and girth development. Bone density, on the other hand, is increased by high-intensity training but is not influenced by low-intensity training (Booth and Gould, 1975).

Changes in Bone During the Adult Years

Changes in bone tissue during the adult years, after complete ossification of the bones, involve altering the density of the bone tissue so as to change the strength of the tissue (Smith and Gilligan, 1987). Through studies of disabled adults and of adults undergoing reduced or zero-gravity environments, it has been determined that lack of use within the Earth's atmosphere or bone tissue subjected to reduced gravitational stress results in bone loss, demineralization, and consequently, reduced strength.

To determine the effects of exercise on bone density, adult athletes have been studied and compared to inactive peers, and adult athletes have been studied to identify bone composition differences between their dominant and nondominant limbs. Generally, bones subjected to greater stresses over time developed increases in strength as a result of greater diameters, densities, cortical widths, and calcium concentrations. These changes, however, may be influenced by initial weight, body fat, exercise intensity, and hormonal influences in females. As is becoming evident as a result of studying osteoporosis and its influence on bones, estrogen plays an important part in females' ability to maintain bone mineral concentration. Although we think of estrogen depletion as a condition in older women, we are learning that amenorrheic athletes with estrogen reduction may be located in any age group. The reduction of estrogen formation in these athletes has a profound effect on bone constitution and thus affects women in young and middle adulthood (Montoye, 1987).

Changes in Bone Tissue in Older Adults

After the ages of 30 years in males and 40 years in females, cortical diameter, mineralization, and density of bones have been shown to decrease (Montoye, 1987). Hayes (1986) reports that tensile strength and elasticity of bones decrease approximately 2% per decade from age 20 to age 90 years. Athletic participation has been shown to reduce the magnitude of these losses. Older adult males participating in athletic competition have been shown to have dominant limbs with greater cortical width and bone mineralization compared to the nondominant limbs. In females, the reduction in estrogen production after menopause has been associated with reduced bone integrity and resulting osteoporosis. Limited results so far have demonstrated that exercise stress on bones may help control the bone loss during the older adult years (Smith and Gilligan, 1987).

A.2 Articulations

The human skeleton is a system consisting of bones joined together to form segments or links. These links are movable and provide for the attachment of muscles to produce movement. The junction of two bones is called an **articulation**.

Classification of Articulations

An articulation may be classified according to the amount of movement it allows. Figure A.4 illustrates the types of articulations.

A **synarthrodial** articulation is the juncture between two bones that are considered immovable because they do not display any particular movement. Synarthrodial joints do allow slight movement if responding to an applied force and thus do have the ability to absorb shock. The sutures between the bones of the skull are examples of synarthrodial articulations. **Amphiarthrodial** articulations are slightly movable because a fibrocartilagenous disc (between the vertebrae) or a ligament or membrane links the two bones (scapula to the clavicle). Because of the connective tissue, the two bones may slightly shift in their relative positions. These types of joints also absorb shock by providing some "give" to the forces that tend to separate or force together the two bones.

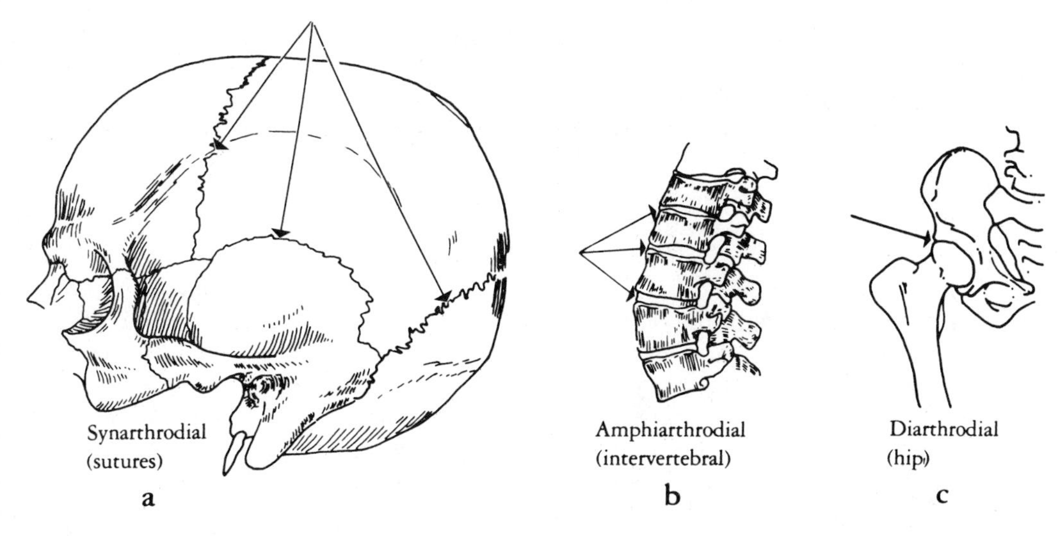

FIGURE A.4

Examples of the three types of articulations: (a) synarthordial (skull), (b) amphiarthrodial (vertebrae) and (c) diarthrodial (hip).

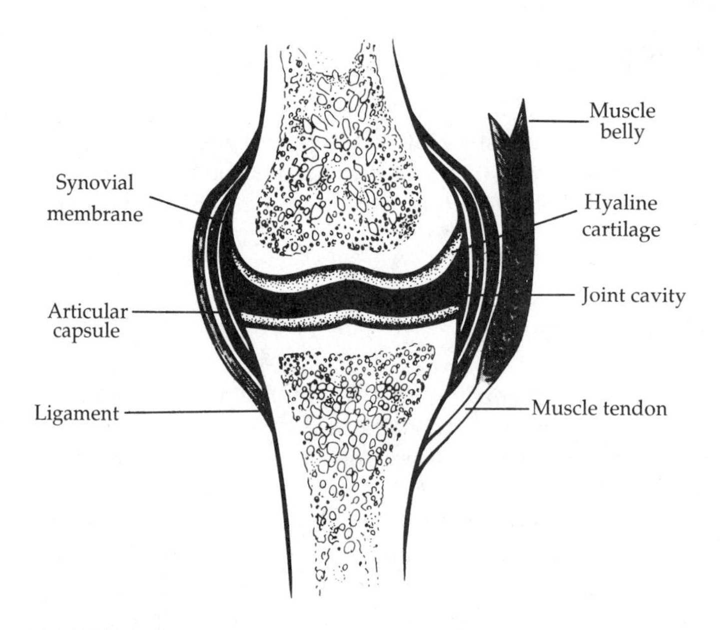

FIGURE A.5

Typical diarthrodial articulation. (Adapted from Kreighbaum, E. (1987). Anatomy and kinesiology. In N. Van Gelder (Ed.), *Aerobic dance-exercise instructor manual* (p. 43). San Diego: IDEA Foundation.)

The most common type of articulation encountered in the study of human movement is the freely movable joint named the **diarthrodial** articulation. There are several types of diarthrodial articulations named for the types of movements they allow. Figure A.5 illustrates the structures comprising a typical diarthrodial articulation.

Hyaline cartilage covers the adjoining ends of the bones. Such cartilage is the smooth, elastic substance similar to that which is found on the ends of chicken bones. Its main functions are to absorb shock and provide a relatively frictionless surface over which the bones may move. When degeneration of articulations occurs, the hyaline cartilage may crack, chip, or wear away from the end of the bone, resulting in painful movement as a result of uncovered bone endings or chips or spurs of cartilage becoming lodged within the joint.

The **joint cavity** provides a space for the movement of the two bones and houses the **synovial fluid**, a fairly viscous (thick) material that lubricates the hyaline cartilage. There is some indication that the viscosity of the fluid may change with activity; this has been implicated as being part of the "warm-up" effects that occur during the initial stages of movement. The **synovial membrane** surrounds the joint cavity but does not cover the hyaline cartilage. It is filled with nerve endings and is quite sensitive to changes within the capsule or pressure from foreign objects such as bone spurs. On the outside of the synovial membrane is the **articular capsule**. The capsule surrounds the articulation like a collar and helps hold the bones together. **Ligaments** are the connective tissues that connect the two bones and are on the outside of the capsule. **Tendons**, which are also on the outside of the capsule, are connective tissues that in many cases attach the muscles to the bone. Not all muscles have tendinous attachments. Some muscle fibers attach directly to the periosteum (covering) of the bone itself. A tendinous attachment serves two functions: (1) it concentrates the force of the muscle's pull into a small area on the bone, and (2) it allows a greater freedom in changing the angle at which the muscle pulls on the bone. The importance of this second factor will become more evident in the discussion on muscle mechanics. The fibrous attachments of the muscles are generally found on the origin (the least movable end) of the bone and are associated with large muscles, which may develop great force and pull on relatively large bones such as the pelvis or the scapula.

Biomechanics of Articular Connective Tissue

As with bone tissue, the mechanical properties of connective tissue have been the focus of research to determine what factors enhance strength and flexibility. The stress (force)–strain (deformation) relationship in connective tissue is important for determining the tissue's **elastic limit**. When a tissue is stretched, it will generally return to its original length. This property is called **elasticity**. If a tissue is stretched beyond its elastic limit, it will not return to its original length but will remain lengthened. This property of remaining in a deformed state is called **plasticity**. The point at which the tissue will not return to its original shape is its elastic limit. Some evidence is available to show that the elastic limit of a tendon or ligament can be enhanced by exercise and training and, of course, can be reduced by aging and inactivity (Woo, 1986).

Generally, the stretching of a ligament with a given force is greater than the stretching of a tendon. Woo (1986) summarizes the reported strain in ligaments as being 12%–50%, whereas tendon strain was from 9% to 30%. One must consider other factors besides the isolated mechanical properties of the tendon or ligament itself, specifically, the strength of the musculotendinous junctions (MTJs) or the bone–ligament junctions. In a study of the effects of warmup on musculoskeletal injury, Safran et al. (1988) found that although isometrically warmed muscle displayed greater elasticity when failure occurred, it occurred consistently at the MTJ.

One of the factors that influences the integrity of the interface between the connective tissue and the bone is age. Particularly relevant is the age before epiphyseal closure. Woo (1986) reported results of studies of rabbit medial collateral ligaments (MCL) in animals of varying ages. The younger the animals, the less stress the MCL–tibial attachment could take before avulsion (detachment). The smallest force produced tibial avulsion in animals in which epiphyseal closure had not yet occurred.

Rehabilitation of ligaments and tendons are influenced by the age of the person and by whether the injured structure was immobilized or not and for how long. While immobilized connective tissue may recover completely, the ligament–bone complex does not recover completely. Furthermore, immobilization has a much more profound effect on reducing the biomechanical properties of connective tissues than exercise has on enhancing them (Woo, 1986). This is why a bone fracture can be much less serious than a severe ligament sprain; a bone normally heals to a state that is stronger than the original, whereas a ligament does not.

Surgery may be the only method by which original integrity of connective tissue may be reinstated. With the development of synthetic ligaments and grafts, additional choices are being added to the list of rehabilitative protocols. The biomechanical properties of connective tissue are important to consider before choosing a rehabilitation method.

The final consideration in discussing the biomechanical properties of connective tissue is the influence of steroids. We know that local injections of steroids into tendons to reduce inflammation can also alter the material properties of the tissue, reduce the strength of the tendons, and predispose them to complete rupture (Unverferth and Olix, 1973). Although steroids may enhance the development of musclar strength, their influences on the connective tissue–bone complex have not been substantiated. The steroid-influenced athlete may be at greater risk in experiencing an avulsion of an apophysis as a result of the strength of the muscles involved—a situation similar to the adolescent athlete participating in a concentrated strength-development program.

Stability of Articulations

The **stability** of an articulation is defined as the ability of the skeletal framework to absorb shock and withstand motion without injury to the joints and their surrounding tissues, such as dislocation of the joint, sprain of the ligaments, or **strain** of the muscle tissue.[2] Three sources of stability for an articulation are: (1) a strong bony arrangement at the joint, whereby one bone "fits into" or around another, as at the elbow or hip; (2) a strong ligamentous arrangement, whereby the ligaments surrounding the joint are of sufficient quantity and quality to be able to resist dislocating forces, as is provided in the ligaments of the hip joint; and (3) strong muscular arrangements, whereby the muscles surrounding the joint and their lines of force during tension tend to pull the two bones together, as at the shoulder joint.

Two factors determine the possibility of a strong articulation: first, whether the articulation has the structures to provide the stability, and, second, how

[2] In mechanics, the word **stress** denotes a force, whereas strain denotes the deformation of a material under stress. In this text, the anatomical definition will be used. That is, **strain** refers to deformation or the tearing of muscular tissue, whereas **sprain** refers to the deformation or tearing of ligamentous tissue.

strong the available structures are. For instance, an articulation may have few ligaments on the anterior side; thus, the anterior side has a weak ligamentous arrangement. If there are ligaments on the anterior side, however, these ligaments may be strong or weak, depending on their condition. The stability of any articulation depends on the strength of the bony, ligamentous, and muscular factors. If one of these is weak, the entire articulation is somewhat weak. A person cannot do much to change the stability of the bony structure of an articulation, but evidence suggests that one can increase the strength of the ligaments through exercise, to help them resist any dislocating forces, and one can undergo strength-training programs for the muscles so that they may better maintain the integrity of the joint. Particularly in the case of articulations such as the knee and the shoulder, where the bony arrangements provide minimal stability, the muscles surrounding these joints must be strong to stabilize joints against dislocating forces and to help hold the two bones together.

Mobility of Articulations

The **mobility** of an articulation is defined as the degree to which an articulation is allowed to move before being restricted by surrounding tissues. Mobility is determined by the articulation's **range of motion** (ROM), which is the total amount of movement (usually measured in degrees) through which the articulation's segments may pass. This quality is commonly called **flexibility**. The ROM of an articulation depends on the shape of the articulated parts of the bone and the tightness of the ligaments and muscles surrounding it. Figure A.6 illustrates limits to the ROM at the elbow.

Figure A.6a shows an elbow hyperextended to a point where the olecranon process has contacted the olecranon fossa and hence has stopped its motion. If movement is forced beyond this point, bruising or breaking of the bone will occur. Usually, an articulation is prevented from reaching the point of bony contact by the ligaments or muscle tendons that cross the joint and become taut before bone contact is made. Figure A.6b illustrates the end of the elbow extension that is due to musculotendinous tightness. Flexibility exercises are designed to increase the ROM of joints by increasing the "stretchability" of the ligaments and tendons. Finally, an articulation is inhibited in its ROM by the compressing of two adjacent tissue structures, as shown in Figure A.6c. In this case, the elbow is flexed to the point where the tissues of the forearm contact the biceps muscle on the anterior side of the humerus. The larger the tissue bulk, or the more adipose tissue (fat), the sooner the tissues contact each other and limit motion.

Thus an articulation's ROM is the total range through which the segments can move before being stopped by bony, ligamentous, or muscular structures. Note that the stability and mobility of a joint are not mutually exclusive; that is, an articulation may have good stability and still maintain its mobility. (Consider a strong but flexible gymnast.) In fact, in joints that have a high degree of

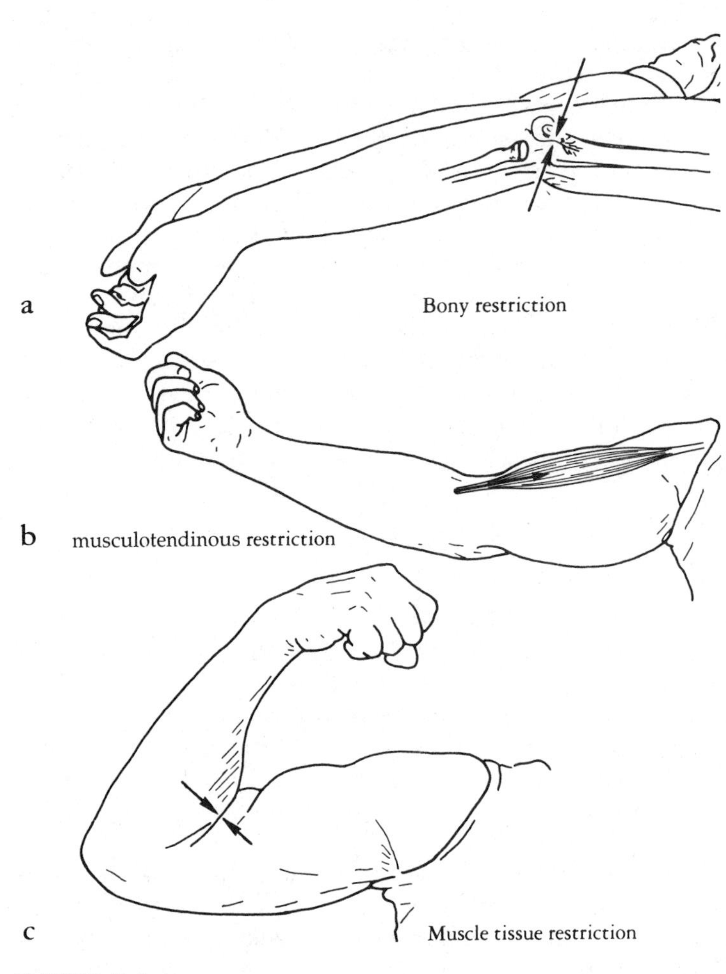

FIGURE A.6

The elbow joint is restricted in its range of motion by (a) bony tissue, (b) tendinous tissue, and (c) muscle tissue.

mobility such as the knee joint, muscle strength is an important component of the joint's reinforcement.

The term *muscle-bound* is used commonly to describe a person with large muscular tissue development that limits ROM because the adjacent tissues touch. A certain amount of muscle bulk develops when a person continually engages in intensive muscle-strengthening exercises through a relatively small ROM and does not bother to stretch the muscles sufficiently. The articulation becomes bound to a relatively limited ROM because of ligamentous, tendinous, and tissue bulk limitations. The concept that weight training causes a muscle-bound condition is a myth, for muscle bulk develops only with a high-resistance, low-repetition exercise program that does not incorporate any stretching following the workout. Females generally do not develop the muscle bulk that males do with similar training programs. This is due primarily to differences in the levels of the hormone testosterone (President's Council on Physical Fitness and Sports, 1975).

No evidence suggests that a flexible joint (one with a large ROM) is less strong than an inflexible one. The added ROM merely allows a flexible articulation to move safely into positions that an inflexible one cannot assume. Flexibility, then, is an important factor in the prevention of injuries and in efficient skill performance, but to satisfy these purposes, flexibility must be accompanied by ligamentous and muscular stability surrounding an articulation. The gymnast who is strong enough to perform on the rings also must have flexibility in the shoulder joints to allow the body to swing down to a long hang from a handstand. The movement requires not only strength to guide and control it but also enough flexibility in the surrounding tissues to prevent muscle strains and ligament sprains when the articulation is brought forcefully to the end of its range.

Most of us have met or heard about a person who is described as double-jointed and is able to move the body parts in ways not normally possible to others. Excluding some pathological deformity, the term is a misnomer, for no double joints exist. The condition is seen in articulations that are weak ligamentously and muscularly and thus allow excessive ROMs. An extended elbow that allows the joint angle to move beyond 180 degrees and knees that hyperextend beyond the normal straightening are typical examples of the lack of ligamentous and muscular stability in those joints. Although double-jointedness is used by some as an excuse for not being able to perform certain exercises, it is not the condition of exaggerated ROM that inhibits performance; it is the lack of muscular strength and tissue support in the joint that increases the required amount of force necessary to execute the exercise. The hyperextended position is not an efficient position from which to move because of the muscles' angles of pull. A discussion of this concept is in Chapter 3, Biomechanics of the Musculoskeletal System.

Understanding the Skeletal-Articular Structures

1. Using the skeleton shown in Figure A.7, identify by name the bones important to human movement in physical education and athletics. (Exclude individual bones of the skull, the hand, and the bones distal to the calcaneus in the foot.)
2. Study all the articulations of the bones in Figure A.7. Label these by name on the drawing. There are 13 articulations, counting the vertebral column as one. List which bones articulate together to make each articulation.

3. Using a manual goniometer or a protractor, roughly measure the ROM of elbow flexion, knee flexion, and hip flexion with the knee straight and with the knee flexed. Measure two females and two males. What are the differences between the ROMs of the males and females? Can you explain the differences among individuals within the group?
4. Measure and list the general ROMs for the following movements:
   a. *Hip:* Flexion–extension, hyperextension, mediolateral rotation
   b. *Ankle:* dorsiflexion and plantar flexion
   c. *Elbow:* flexion–extension

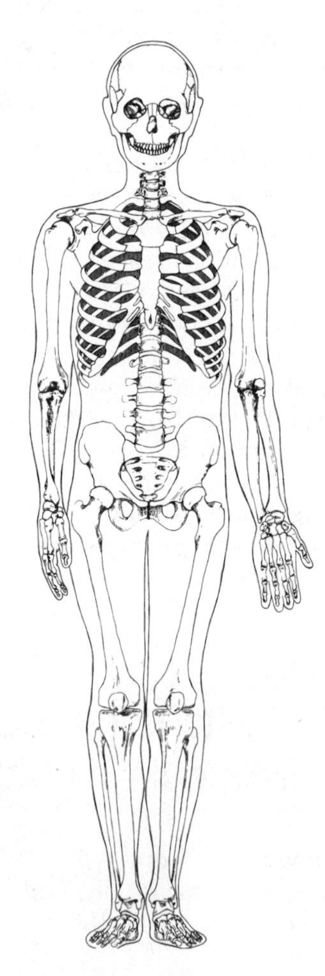

FIGURE A.7
The human skeleton.

d. *Radioulnar joint:* pronation–supination

e. *Shoulder joint:* mediolateral rotation

5. Are there differences between individuals in any of the above movements? Why?
6. Locate a person who can hyperextend the knees. Measure the ROM of the knee flexion and extension motion. What stops the flexion motion? What stops the extension motion? Locate a person who cannot hyperextend the knees. Answer the same questions as above and compare and discuss the differences.

References and Suggested Readings

Acuff, R. V. (1975, November). Advanced skill attainment and skeletal development. *Gymnast*, p. 47.

Adams, J. E. (1968). Bone injuries in very young athletes. *Clinical Orthopedics and Related Research, 58,* 129–140.

American Academy of Pediatrics. (1983). Weight training and weight lifting: Information for the pediatrician. *The Physician and Sportsmedicine, 11*(3), 157–161.

Benton, J. W. (1982). Epiphyseal fracture in sports. *The Physician and Sportsmedicine, 10* (11), 63–71.

Booth, F. W., & Gould, E. W. (1975). Effects of training and disuse on connective tissue. In J. H. Wilmore & J. F. Keogh (Eds.), *Exercise and Sports Science Reviews* (Vol. 3, pp. 95–103). Santa Barbara, CA: Journal Publishing Affiliates.

Bowers, K. D. (1981). Patellar tendon avulsion as a complication of Osgood-Schlatter's disease. *American Journal of Sports Medicine, 9*(6), 356–359.

Cage, J. B. (1983). Intracondylar fracture of the femur in an adolescent athlete. *The Physician and Sportsmedicine, 11*(6), 115–118.

Caine, D. J., & Lindler, K. J. (1985). Overuse injuries of growing bones: The young female gymnast at risk? *Physican and Sportsmedicine, 13,* 51–64.

Caine, D. J., & Lindler, K. J. (1984). Growth plate injury: A threat to young distance runners? *Physician and Sportsmedicine, 12,* 118–123.

Carter, J. E. L. (1981). Somatotypes of female athletes. In J. Barnes, M. Hebbelinck, & A. Venerando (Eds.), *The female athlete* (Vol. 15, pp. 85–116). Basel: S. Karger.

Chantraine, A. (1985). Knee joint in soccer players: Osteoarthritis and axis deviation. *Medicine and Science in Sports and Exercise, 17,* 434–439.

deVries, H. A. (1980). *Physiology of exercise* (pp. 403–404). Dubuque, IA: William C. Brown Co.

Dominguez, R. H. (1978). Shoulder pain in age group swimmers. In B. Eriksson & B. Furberg (Eds.), *Swimming Medicine IV: Proceedings of the Fourth International Congress on Swimming Medicine* (pp. 105–109). Baltimore: University Park Press.

Goldberg, V., & Aadalen, R. (1978). Distal tibial epiphyseal injuries: The role of athletics in 53 cases. *American Journal of Sports Medicine, 6*(5), 255–262.

Gugenheim, J. J., et al. (1976). Little League survey: The Houston study. *American Journal of Sports Medicine, 4*(5), 189–200.

Hayes, W. C. (1986). Bone mechanics: From tissue mechanical properties to an assessment of structural behavior. In Y. C. Fung (Ed.), *Frontiers in biomechanics* (pp. 196–209). New York: Schmid-Schonbein.

Howard, M. H., & Piha, R. J. (1965). Fractures of the apophyses in adolescent athletes. *Journal of the American Medical Association, 192*, 842–844.

Hunter, L. Y., & O'Connor, G. A. (1980). Traction apophysitis of the olecranon. *American Journal of Sports Medicine, 8*(1), 51–52.

Ishikawa, H., et al. (1988). Osteochondritis dissecans of the shoulder in a tennis player. *The American Journal of Sports Medicine, 16*, 547–550.

Jackson, D. (1976, December). Low back pain in the young gymnast. *International Gymnast*, p. 50.

Komor, A. J. (1985). The identification of force constraint functions for human joints under dynamic conditions. In D. A. Winter, R. W. Norman, R. P. Wells, K. C. Hayes, and A. E. Patla (Eds.), *Biomechanics IX-A* (pp. 166–170). Champaign, IL: Human Kinetics.

Kozar, B., & Lord, R. M. (1983). Overuse injury in the young athlete: Reasons for concern. *The Physician and Sportsmedicine, 11*, 116–122.

Krahenbuhl, G. S. (1968). Influence of exercise on bone growth and metabolism. In D. H. Clarke (Ed.), *Kinesiology Review* (pp. 43–48) Reston, VA: AAPHERD.

Kujala, U. M., Kvist, M., & Heinonen, O. (1985). Osgood-Schlatter's disease in adolescent athletes. *The American Journal of Sports Medicine, 13*, 236–240.

Larson, R. L. (1976). Little League survey: The Eugene study. *American Journal of Sports Medicine, 4*(5), 201–209.

Larson, R. L., & McMahan, R. O. (1966). The epiphyses and the childhood athlete. *Journal of the American Medical Association, 196*, 607–612.

Marino, M. (1983). Sports medicine in action: A look at swimming. *Muscle and Bone, 3*(3), 3–10.

McCarroll, J. R., Rettig, A. C., & Shelbourne, K. D. (1988). Anterior cruciate ligament injuries in the young athlete with open physes. *The American Journal of Sports Medicine, 16*, 44–47.

Metzmaker, J. N., & Pappas, A. M. (1985). Avulsion fractures of the pelvis. *The American Journal of Sports Medicine, 13*, 349–356.

Micheli, L. J. (1986). Pediatric and adolescent sports injuries: Recent trends. In K. B. Pandolf (Ed.), *Exercise and Sport Sciences Review* (pp. 359–374). New York: Macmillan.

Micheli, L., & Jupiter, J. (1978). Osteid osteoma as a cause of knee pain in the young athlete: A case study. *American Journal of Sports Medicine, 6*(4), 199–208.

Mirbey, J., Besancenot, J., Chambers, R. T., Durey, A., & Vichard, P. (1988). Avulsion fractures of the tibial tuberosity in the adolescent athlete. *The American Journal of Sports Medicine, 16*, 336–340.

Montoye, H. J. (1987). Better bones and biodynamics. *Research Quarterly, 58*, 334-347.

Morehouse, L. E., & Rasch, P. A. (1963). *Sports medicine for trainers* (2nd ed.). Philadelphia: W. B. Saunders.

Morey, E., & Baylink, D. (1978). Inhibition of bone formation during space flight. *Science, 201*(22), 1138–1141.

Oyster, N., Morton, M., & Linnell, S. (1984). Physical activity and osteoporosis in postmenopausal women. *Medicine and Science in Sports and Exercise, 16*, 44–50.

Pappas, A. M. (1983). Epiphyseal injuries in sports. *The Physician and Sportsmedicine, 11* (6), 140–148.

Peterson, C. A., & Peterson, H. A. (1972). Analysis of the incidence of injuries to the epiphyseal growth plate. *The Journal of Trauma, 12*, 275–281.

Pettrone, F. A. (Ed.) (1954). *American Academy of Orthopaedic Surgeons Symposium on Upper Extremity Injuries in Athletes*. St. Louis: Mosby.

President's Council on Physical Fitness and Sports. (1975). Joint and body range of motion, *Physical Fitness and Research Digest, 5*(4), 17.

Pruett, D. M., & Lopez, R. (1981, October). Preventing injuries in young ballet dancers. *JOPERD, 52*, 26–29.

Push, J. (1985). Mechanical aspects of osteoarthritis. In D. A. Winter, R. W. Norman, R. P. Wells, K. C. Hayes, & A. E. Patla (Eds.), *Biomechanics IX-A* (pp. 133–134). Champaign, IL: Human Kinetics.

Rarick, G. L. (1980, September). Motor development: Its growing base. *JOPERD, 51*, 26–27, 56–61.

Rians, C. B., et al. (1987). Strength training for prepubescent males: Is it safe? *The American Journal of Sports Medicine, 15*, 483–489.

Roy, S., Caine, D., & Singer, K. M. (1985). Stress changes of the distal radial epiphysis in young gymnasts. *The American Journal of Sports Medicine, 13*, 301–307.

Ryan, J. R., & Salciccioli, G. G. (1976). Fractures of the distal radial epiphysis in adolescent weight lifters. *American Journal of Sports Medicine, 4*(1), 26–27.

Safran, M. R., Garrett, W. E., Seaber, A. B., Glisson, R. R., & Ribbeck, B. M. (1988). The role of warmup in muscular injury prevention. *The American Journal of Sports Medicine, 16*, 123–129.

Salter, R. B., & Harris, W. R. (1963). Injuries involving the epiphyseal plate. *Journal of Bone and Joint Surgery, 45A*, 587–622.

Schwab, S. A. (1977). Epiphyseal injuries in the growing athlete. *Canadian Medical Association Journal, 117*, 626–630.

Shaffer, T. E. (1980). The uniqueness of the young athlete: Introductory rewards. *American Journal of Sports Medicine, 8*(5), 370–371.

Smith, E. L., & Gilligan, C. (1987). Effects of inactivity and exercise on bone. *The Physician and Sportsmedicine, 15*, 91–97.

Smith, E. L., Reddan, W., & Smith, P. E. (1981). Physical activity and calcium modalities for bone mineral increase in aged women. *Medicine and Science in Sports and Exercise, 13*(1), 60–64.

Stamford, B. A. (1988). Exercise and the elderly. In K. B. Pandolf (Ed.), *Exercise and Sport Science Reviews* (pp. 341–379). New York: Macmillan.

Stanitski, C. L. (1982). Low back pain in young athletes. *The Physician and Sportsmedicine, 10*(10), 77–91.

Stone, M. H. (1988). Implications for connective tissue and bone alterations resulting from resistance exercise training. *Medicine and Science in Sports and Exercise, 20*, S162–S168.

Torg, J. (1973, June). Little League: The theft of a carefree youth. *The Physician and Sportsmedicine, 1*(6), 72–78.

Unverferth, L., & Olix, M. L. (1973). The effect of local steroid injections on tendon. *Journal of Sports Medicine, 1*(4), 31–37.

Weltman, A., et al. (1986). The effects of hydraulic resistance strength training in prepubertal males. *Medicine and Science in Sports and Exercise, 18*, 629–638.

Whalen, R. T., Carter, D. R., & Steel, C. R. (1988). Influence of physical activity on the regulation of bone density. *Journal of Biomechanics, 21*, 825–838.

Wilkins, K. E. (1980). The uniqueness of the young athlete: Musculoskeletal injuries. *American Journal of Sports Medicine, 8*(5), 377–382.

Wolff, J. (1892). *Gesetz der Transformation der Knochen*. Berlin: Aug. Hirschwald.

Woo, S. (1986). Biomechanics of tendons and ligaments. In Y. C. Fung (Ed.), *Frontiers in biomechanics* (pp. 180–195). New York: Schmid-Schonbein.

Woods, G. W., Tullos, H., & King, J. (1973). The throwing arm: Elbow joint injuries. *Journal of Sports Medicine, 1*(4), 43–47.

Yang, K. H., Tzeng, C. R., & King, A. I. (1985). Response of the articular facet joint to axial loads. In D. A. Winter, R. W. Norman, R. P. Wells, K. C. Hayes, & A. E. Patla (Eds.), *Biomechanics IX-A* (pp. 149–153). Champaign, IL: Human Kinetics.

CONCEPT MODULE B

The Link System and Its Movements

PREREQUISITES

Module A

CONCEPT MODULE CONTENTS

B.1 The Link System and Its Movements

As we have seen from studying the skeleton and its articulations, the body is composed of segments linked together at their articulations. Thus, we say that the body is basically a system consisting of movable segments or links. The skeleton may be divided into minute divisions for precise description in which each bone, no matter how small, is considered a separate unit; or it may be divided into the largest possible units for simplicity (e.g., the head, neck, and trunk; the upper extremity; and the lower extremity). The more numerous the identified links, the more precise the description of what the body is doing; however, the description may become too intricate for practical purposes. The fewer the links, the simpler the description. If too few are included, however, identification of many important movements is impossible. Thus we must compromise. For our purposes, the body is divided into 11 functional segments, or links: head and neck, thorax and thoracic vertebrae, lumbar vertebrae, pelvis, thigh, leg, foot, shoulder girdle, arm, forearm, and hand. Technically, some of these segments are made up of more than one link (e.g., the shoulder girdle, the vertebral column, the hand, and the foot). The links are shown in Figure B.1.

Anatomical Position

The upright position, in which all joints are extended and the palms are facing forward, is called the anatomical position of the body. The skeleton in Figure

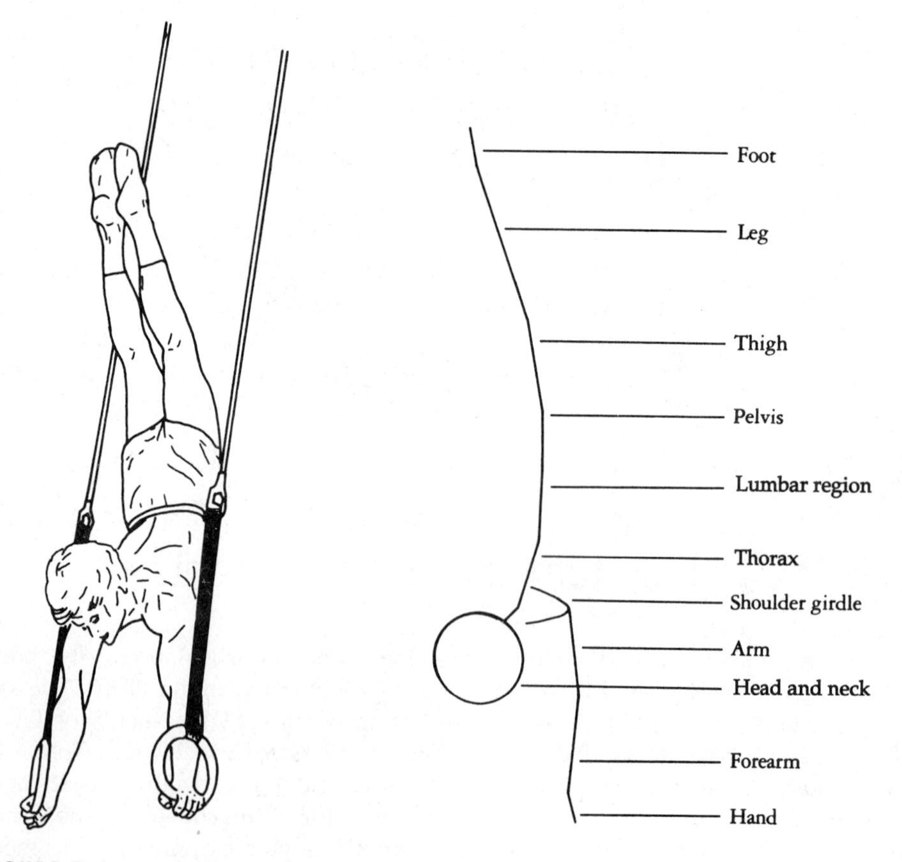

FIGURE B.1

The segmental links, used in describing human movement, identified for a body on the rings.

A.1a closely illustrates the anatomical position. (Both palms should be facing forward in anatomical position.)

Although this position is a convenient one to begin movement description, it is not a practical position to use when studying human movement because most movements of the body in sport or daily activities are not initiated from the anatomical position. Starting positions in sport activities begin from an infinite number of positions, and the movement patterns proceeding from these positions also involve complicated combinations. By using the anatomical position as a frame of reference, however, it is easy to describe and define the movements and positions of muscles, bones, and ligaments.

Body and Segmental Planes and Axes

A concept that is useful in defining segmental and body movements is the concept of the **plane of motion** created by a movement and an **axis of rotation** around which the movement occurs. If one were to punch a hole in the center of a piece of cardboard and insert a pencil through the hole, one would have a representation of a plane (cardboard) and an axis (pencil). If a dot were drawn on the cardboard, it would circle around the pencil when the cardboard was rotated around the pencil.

The pencil represents an axis *around* which movement occurs. The cardboard represents a plane of movement *along* which movement occurs. Thus, the dot on the cardboard moves *around* the axis (pencil) and *along* the plane (cardboard). If the pencil axis were inserted through a human body from the right side to the left side, the body could turn around the axis and move along the plane. Note that the axis and the plane are always perpendicular to each other; that is, they form a 90-degree angle. Figure B.2 illustrates the three planes of the body.

When an axis runs in a side-to-side direction relative to the body, as in this example, it is called a **mediolateral** (ML) **axis**. (The ML axis is sometimes referred to as the frontal axis or one of the two transverse axes.) The rotation of the body or any segment around an ML axis is along an invisible plane that is perpendicular to the axis. This plane, which divides the body or a segment into right and left sections, is called a **sagittal plane**. Movement of the total body around an ML axis and along a sagittal plane is shown in Figure B.3a.

When the axis and plane are used to describe total body movement, as in the previous example, they are designated as the **body's axis** and **plane**. In cases where the body rotates free of support, the body's plane and axis of movement pass through its center of gravity (CG). The planes and axes that pass through the body's CG divide the body into equally distributed mass halves (right and left, top and bottom, front and back), and they are called **principal** or **cardinal** planes and axes. In the example in Figure B.3a, the body is moving along a principal sagittal plane and around a principal ML axis. When the body is supported by the ground or another surface, the planes and axes of total body movement are not necessarily through the body's CG. For example, a gymnast

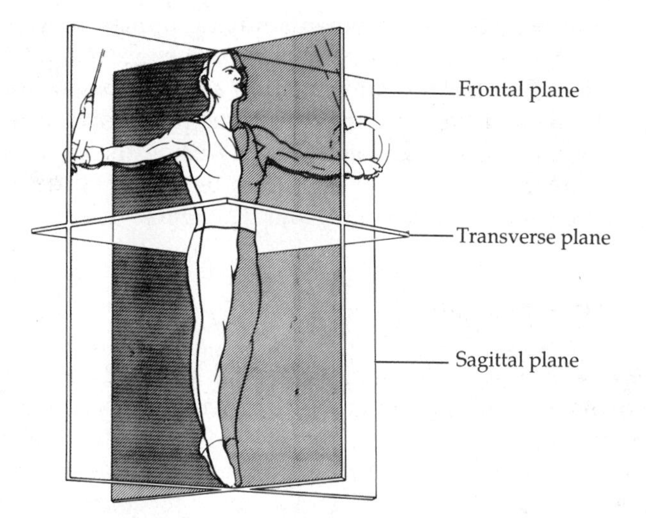

FIGURE B.2

The three planes of the body shown on a gymnast. (Adapted from Kreighbaum, E. (1987). Anatomy and kinesiology. In N. Van Gelder (Ed.), *Aerobic dance-exercise instructor manual* (p. 50). San Diego: IDEA Foundation.)

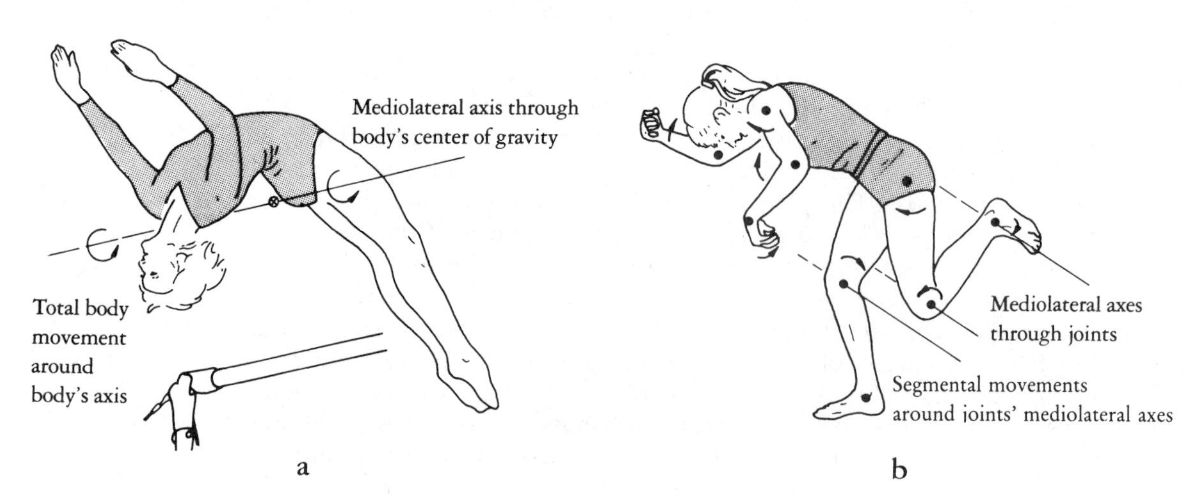

FIGURE B.3

(a) Movements of the total body around the body's mediolateral axis and along a sagittal plane. (b) Segmental movements around a mediolateral axis and along a sagittal plane.

swinging around the high bar moves along a sagittal plane and around an ML axis located at the bar, not through the body's CG. The axis in this illustration is not a principal axis.

A body segment that moves along a sagittal plane and around an ML axis is called a **segmental plane** and **axis**. The plane and axis have the same orientation as the body's plane and axis of movement but are located through the joint in which the movement occurs. Example movements are lifting the arm or the leg forward from the anatomical position. Shown in Figure B.3b are segmental movements occurring around ML axes and along sagittal planes. The location of these segmental axes is through the respective joints that are represented by dots.

Other examples of segmental movements around an ML axis and along a sagittal plane include pulling the foot up toward the leg, flexing the elbow joint, and bending over to touch the floor.

If the total body is turning in the air, as in an aerial cartwheel as shown in Figure B.4, the body is turning around an axis that runs through the body

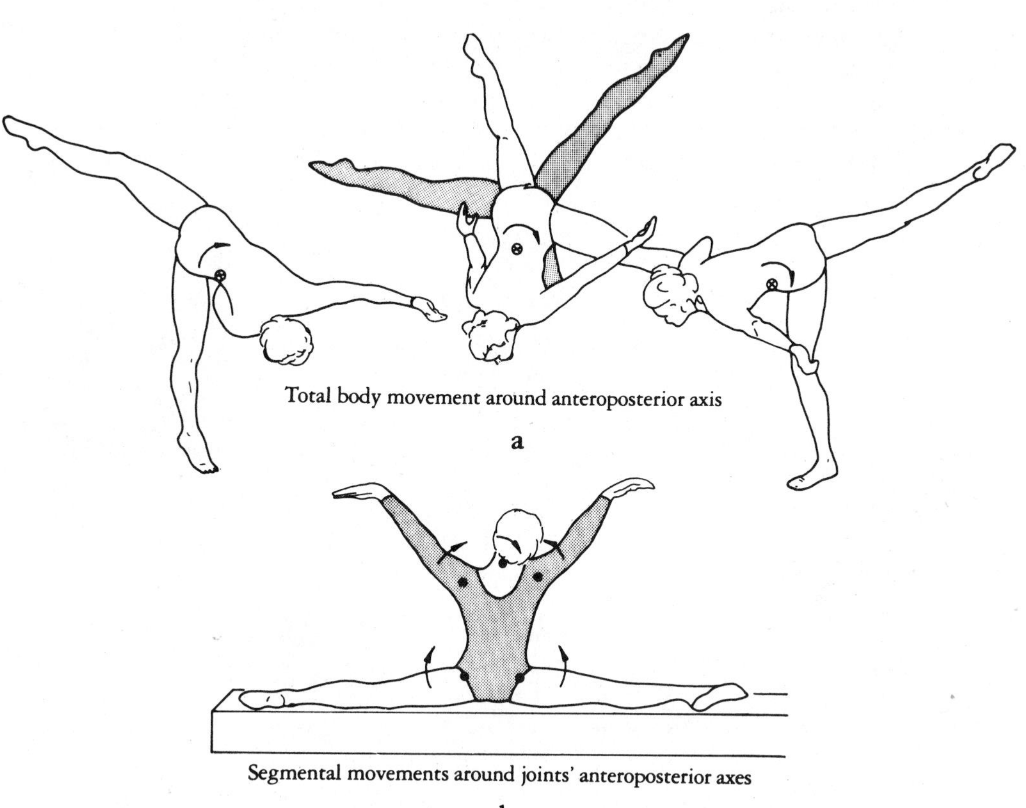

FIGURE B.4

Motions of the body (a) and of the arms and legs and head (b), when moving around the anteroposterior axis and the frontal plane.

from front to back, and the body is moving along a plane that passes from side to side (Figure B.4a). In this case, the plane is called a frontal plane because it divides the body into front and back parts. The axis around which the body or its segments turn while moving in a frontal plane is called the body's **anteroposterior** (AP) axis or sagittal axis. The AP axis passes through the body or an articulation from the front to the back. Figure B.4b illustrates the motion of the arm and leg when moving around an AP axis and along a frontal plane of the shoulder and hip, respectively.

Other examples of moving segments around the AP axis and along a frontal plane include flexing the trunk or the head to the side and moving the hand toward the thumb side or little-finger side of the forearm.

The third possibility for movements is represented in Figure B.5. The total body is performing a twisting movement around an axis running from the head end to the foot end of the body. The motion of the body is along a plane that is perpendicular to the axis and that divides the body into upper and lower parts. The axis is called the body's **longitudinal axis**, and the plane, the body's **transverse plane** (Figure B.5a).

The segmental movements shown in Figure B.5b are around a longitudinal axis and along a transverse plane. The segmental longitudinal axes are longitudinal relative to the bones and segments that are moving and not necessarily longitudinal relative to the total body. The plane and axis are mutually perpendicular and pass through the joint centers where the movements are taking

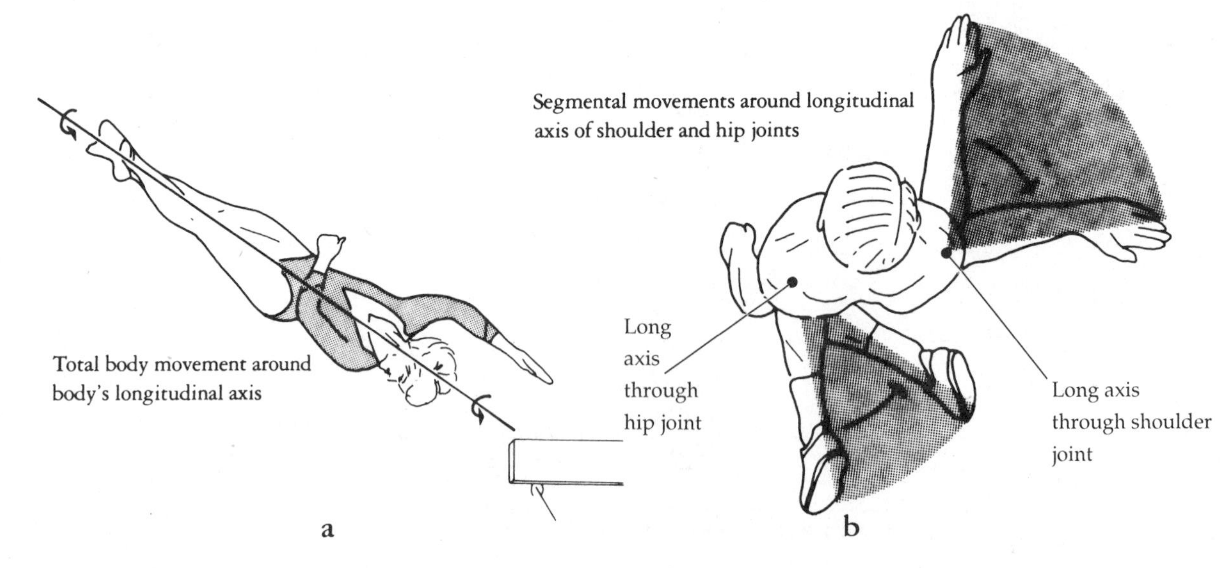

FIGURE B.5

(a) Total body movement around the body's longitudinal axis and along the body's transverse plane. (b) Segmental movements around a segmental longitudinal axis and along a segmental transverse plane.

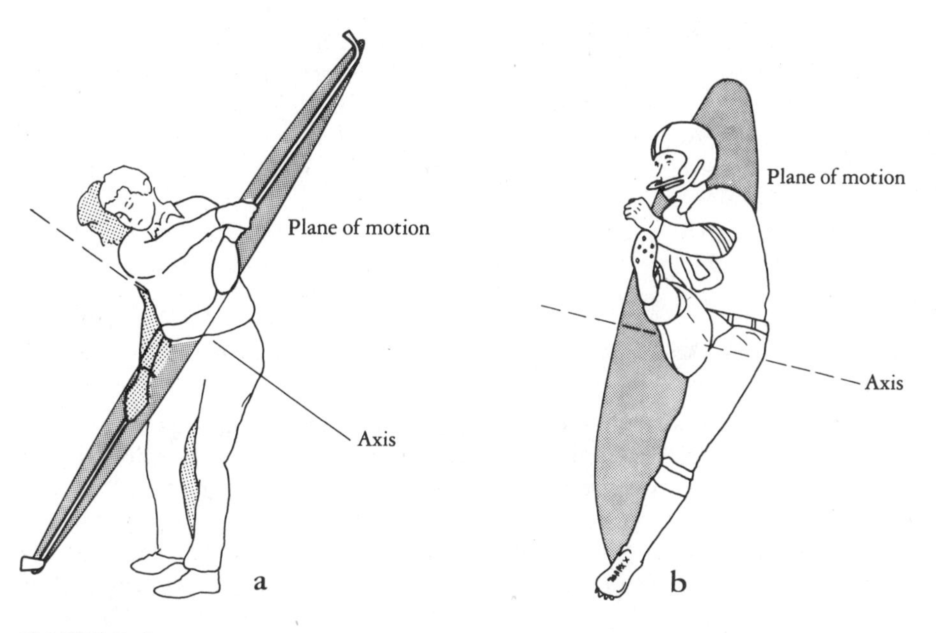

FIGURE B.6
(a) A golfer during the swing of the club, showing the plane and axis of rotation of the left arm. (b) A football punt, showing plane and axis of motion of the lower extremity.

place, in this case, the respective hip and shoulder joints. Other movements of the body segments around longitudinal axes and along transverse planes include rotating the head to the side and rotating the forearms in order to turn the hands.

In daily activities, sports, and games, the human body rarely uses the anatomical position, and movement occurs infrequently along only one of the three principal planes. Most movement occurs diagonally among all the principal planes. Aside from isolated movements such as a forearm curl, movements of the body begin and end from positions that require rather complicated descriptions. Nevertheless, all movements of the body segments occur around axes and along the planes perpendicular to the axes.

Since all possible axes and planes do not have individual names, those other than the ones described previously are called **diagonal** or oblique. The diagonal axis and the diagonal plane can be imagined in space, and the plane along which a movement occurs is always perpendicular to the axis around which it occurs. Figure B.6a shows the diagonal plane of motion and axis of rotation of a golfer in the midst of a swing.

To locate the axis and plane for the golfer's upper extremity movement, we can start by demonstrating the plane of that movement, or the plane along which the arm and club travel during the swing. The plane is diagonal; that is, it does not correspond to any single one of the three body planes just described. The plane is diagonal to the ground, and the axis is perpendicular to the plane.

A second example of a diagonal movement is the leg motion in a football punt (Figure B.6b). As the leg is brought up and forward, it moves across the front of the body slightly, thus producing a diagonal movement. The axis for the motion of the thigh is through the hip joint, and the plane of motion follows along the motion of the thigh and is diagonal to the three body planes previously described.

Terminology for Describing Segmental Movements

As we have seen, any skeletal link (segment) moves around an axis that passes through a joint and along a plane. Movements are named according to the articulation where the movement occurs and not for the segment doing the moving (e.g., elbow flexion rather than forearm flexion). Most movements of the segments occurring at the articulation can be categorized according to the plane of movement and the joint's axis of rotation. The unique movements of some joints are designated by special terms. Although the movement terms are somewhat the same from joint to joint, they usually must be memorized. As an aid to memorization, Tables B.1 through B.3 list the joint movements occurring around similar axes and along similar planes. With the exception of the transverse (horizontal) motions at the shoulder and hip joints, the movements may be initiated from the anatomical position.

Mediolateral Axis–Sagittal Plane Movements

Table B.1 lists the articulations and their movements that occur around an ML axis and along a sagittal plane. Movements illustrating these joint motions around an ML axis and along a sagittal plane are shown in Figure B.7a–h.

TABLE B.1

SEGMENTAL MOVEMENTS OCCURRING AROUND A MEDIOLATERAL AXIS AND ALONG A SAGITTAL PLANE

Movement Names	Articulations
Flexion, extension, and hyperextension	Hip, knee, shoulder, elbow, wrist, phalangeal joints, and vertebral column
Dorsiflexion and plantar flexion	Ankle joint
Transverse abduction (extension) and transverse adduction (flexion)[a]	Shoulder and hip joints

[a] Transverse abduction (extension) and adduction (flexion) movements occur around the ML axis and along the sagittal plane of the humerus and femur if moved into the transverse *body* plane by abduction of the shoulder or hip joints.

TABLE B.2

SEGMENTAL MOVEMENTS OCCURRING AROUND AN ANTEROPOSTERIOR AXIS AND ALONG A FRONTAL PLANE

Movement Names	Articulations
Lateral flexion	Vertebral column
Abduction and adduction	Shoulder, hip, and metatarsophalangeal joints
Radial and ulnar flexion (deviation)	Wrist joint
Elevation and depression	Sternoclavicular joint
Upward and downward rotation	Acromioclavicular joint
Transverse abduction (extension) and adduction (flexion)[a]	Shoulder and hip joints

[a] Transverse abduction and adduction movements occur around the AP axis and along the frontal plane of the humerus and femur if moved into the transverse *body* plane by flexion of the shoulder or hip joints.

TABLE B.3

SEGMENTAL MOVEMENTS OCCURRING AROUND A LONGITUDINAL AXIS AND ALONG A TRANSVERSE PLANE

Movement Names	Articulations
Medial and lateral rotation	Shoulder and hip joints; knee joint at 90-degree flexion
Pronation and supination	Radioulnar joint
Transverse (lateral) rotation	Vertebral column
Protraction and retraction[a]	Sternoclavicular joint[b]
Inversion and eversion	Subtalar joint (talocalcaneal)

[a] Protraction and retraction of the shoulder girdle are sometimes called abduction and adduction, respectively, when referring to the movement of the scapula relative to the vertebral column.
[b] Axis and plane are named relative to the sternum, not to the clavicle.

Anteroposterior Axis–Frontal Plane Movements

Table B.2 lists the articulations and their movements that occur around an AP axis and along a frontal plane. Movements illustrating these joint motions around an AP axis and along a frontal plane are shown in Figure B.8a–f.

Longitudinal Axis–Transverse Plane Movements

Table B.3 lists the articulations and their movements that occur around a longitudinal axis and along a transverse plane. Movements illustrating these joint motions around a longitudinal axis and along a transverse plane are shown in Figure B.9a–f.

In free space, the entire body may move around an axis and along a plane. The body's axis and plane associated with a given total body movement remain the same no matter what orientation the body segments assume. For instance,

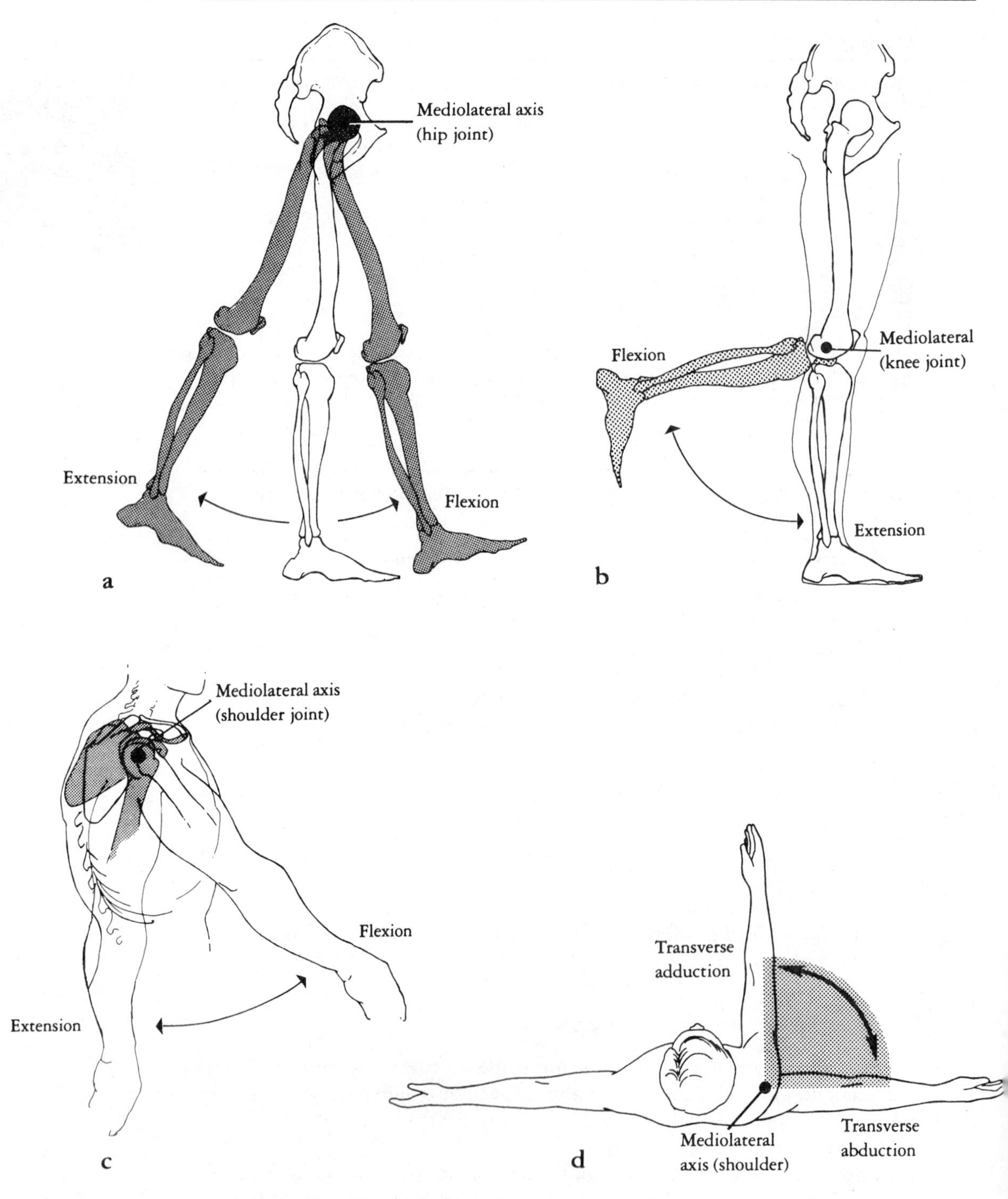

FIGURE B.7

Segmental movements illustrating joint motion around a mediolateral axis and along a sagittal plane.

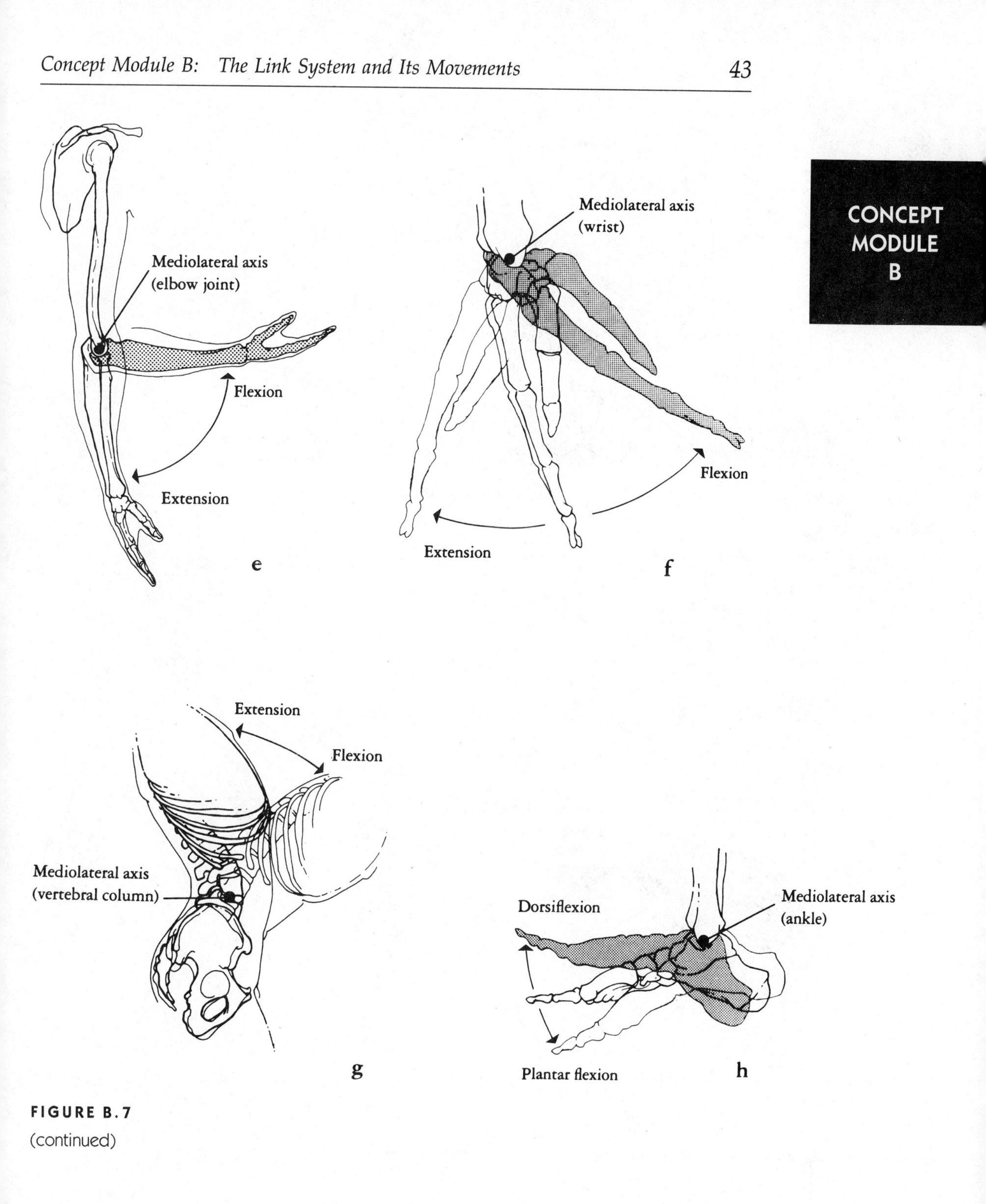

FIGURE B.7
(continued)

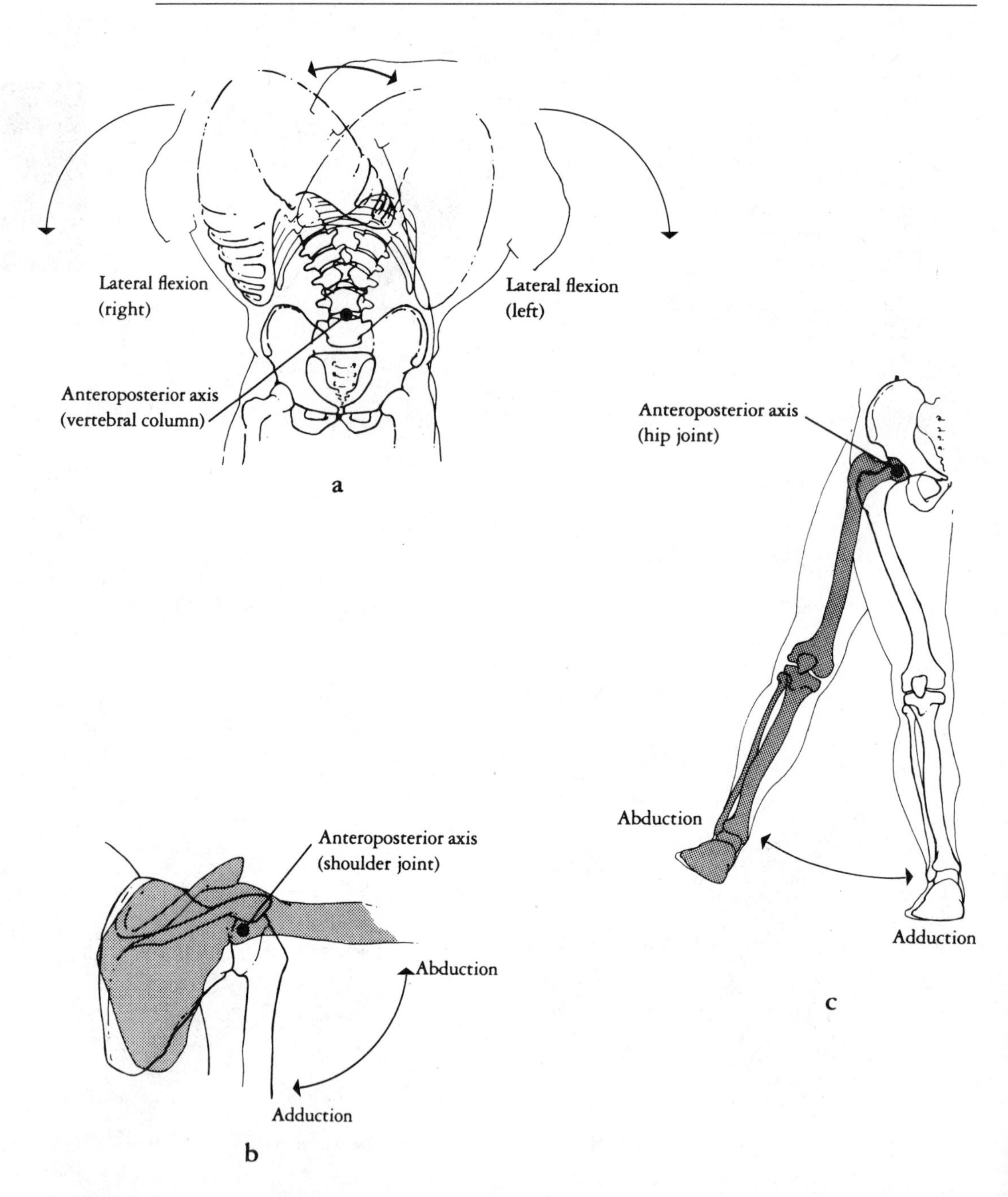

FIGURE B.8

Segmental movements illustrating joint motion around an anteroposterior axis and along a frontal plane.

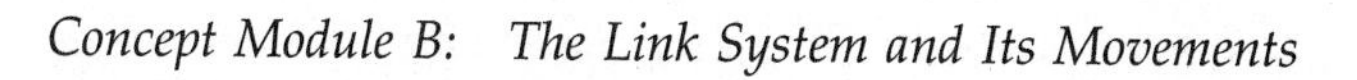

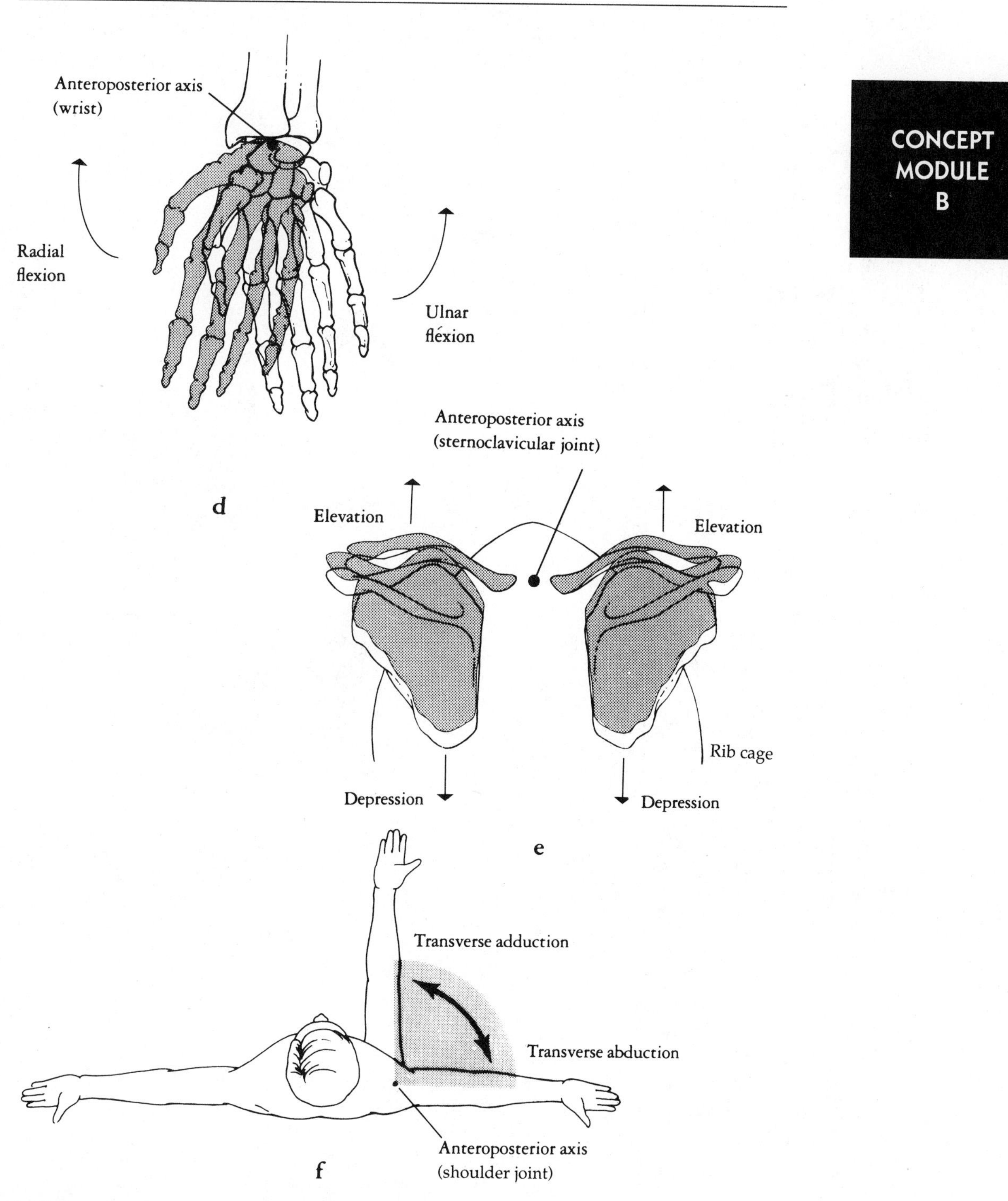

CONCEPT MODULE B

FIGURE B.8
(continued)

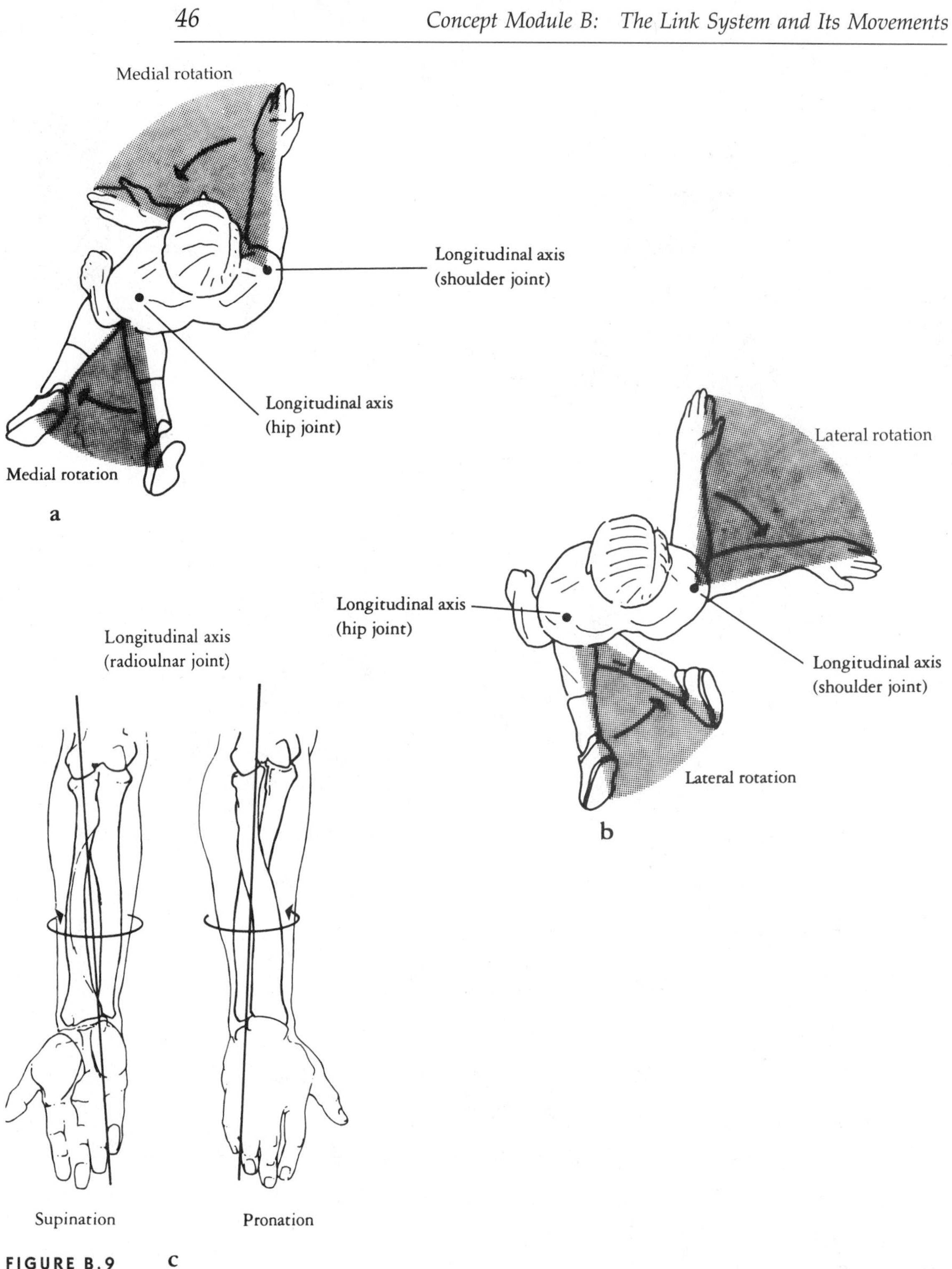

FIGURE B.9

Segmental movements illustrating joint motion around a longitudinal axis and along a transverse plane.

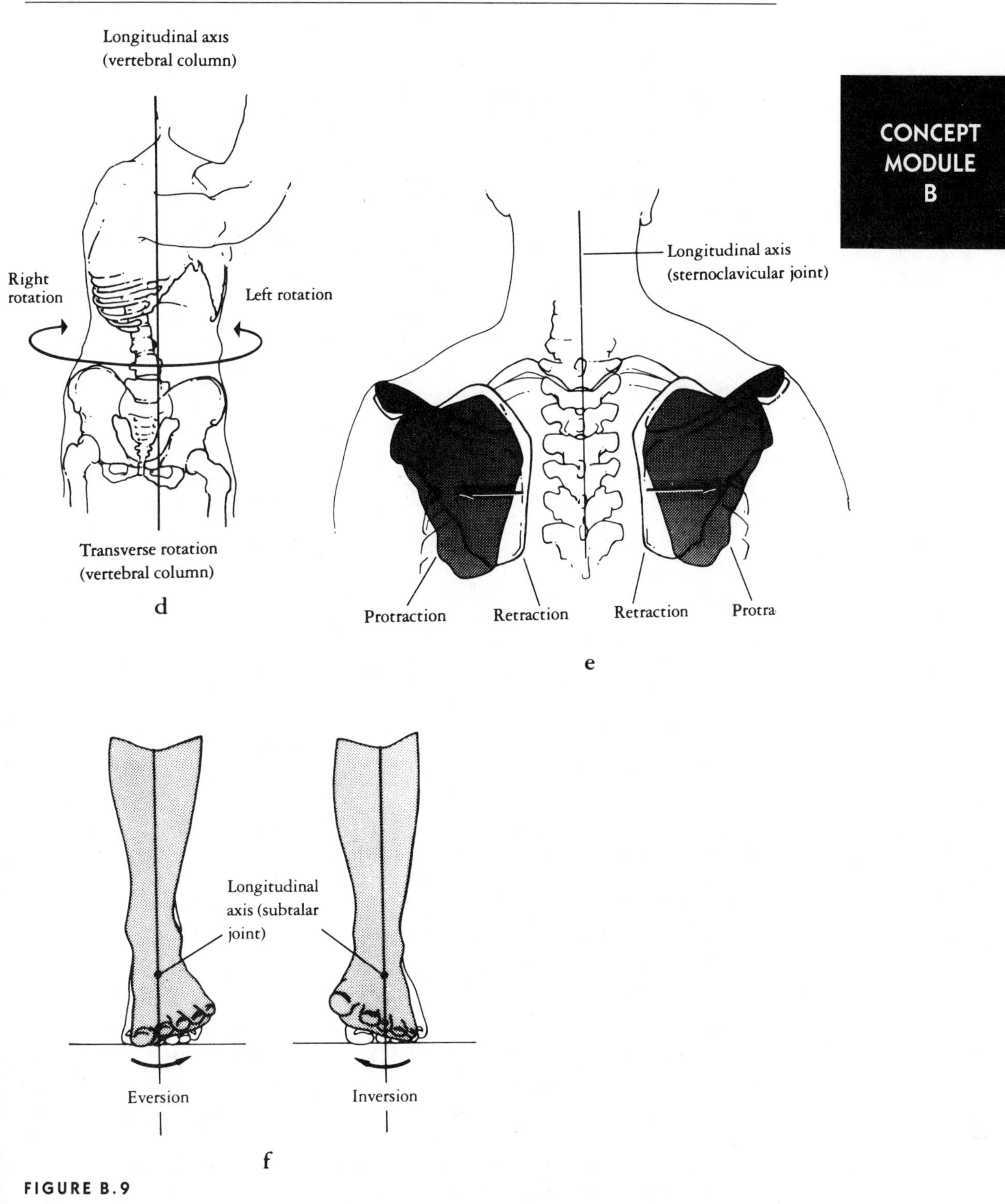

Longitudinal axis
(vertebral column)
Right
rotation
Left rotation
Transverse rotation
(vertebral column)
d
Longitudinal axis
(sternoclavicular joint)
Protraction
Retraction
Retraction
Protra
e
Longitudinal
axis (subtalar
joint)
Eversion
Inversion
f
CONCEPT
MODULE
B

FIGURE B.9
(continued)

FIGURE B.10

A body turning in space in four different body positions around similar parallel axes: (a) tuck, (b) pike, (c) layout, and (d) back arch.

Figure B.10 shows a body turning around its ML axis in tuck, pike, layout, and back-arch positions. Similarly, with segmental motion, the joint's axis and plane remain the same no matter what orientation the body part assumes or what position the body is in. For instance, the flexion of the elbow occurs around the elbow's ML axis, whether it is flexed from anatomical position or flexed when the arm is abducted at the shoulder joint.

Understanding the Body's Link System and Its Movements

1. Refer to Figure A.7. For each articulation labeled, list all the anatomical movements possible. Demonstrate each. Name the plane and axis used for each of the movements listed.

2. Demonstrate the anatomical position and movements along each of the three anatomical planes and around each of three axes of the body.
3. Experiment with the articulations of the body previously named. Determine the number of axes around which the more distal segment moves. (Since there are only three axes, the number will be zero, one, two, or three: nonaxial, uniaxial, biaxial, or triaxial, respectively.)
4. For the body illustrated in Figure A.7, name the segments in the body's link system most important to the biomechanist.
5. List all the articulations that are triaxial, that are biaxial, and that are uniaxial.
6. For the following movements, list the axis and plane used during the force phase of the movement: (a) the breathing movement for the front crawl (neck), (b) the vertical jump (knee), (c) the elementary backstroke (shoulder joint), (d) the forehand drive in tennis (shoulder joint), (e) the placekick (knee), (f) turning a door handle (radioulnar joint).
7. With the help of a partner, perform an easy run, a sit-up, a jumping jack, and a push-up. Describe the articulations being used, the axis and plane of movement of each segment being moved, and the degrees of motion of each segment.
8. With the help of a partner, perform the following movements: (a) flex the knee 90 degrees, (b) abduct the hip 30 degrees, (c) transverse abduct the shoulder 60 degrees, (d) medially rotate the shoulder joint 20 degrees, (e) laterally rotate the hip 30 degrees, (f) pronate the radioulnar joint 180 degrees, (g) medially rotate the knee 10 degrees, (h) laterally rotate the knee 20 degrees, (i) invert the subtalar joint as far as possible, (j) flex the elbow 90 degrees and then flex the fingers as far as possible and move the wrist in radial and ulnar flexion.
9. From the anatomical position, rotate the shoulder. (To be sure the movement is from the shoulder joint, flex the elbow.) How far does it rotate medially? laterally? What is the ROM of the pronation and supination of the radioulnar joint? At what joint do these movements occur? Explain the difference between the rotation of the shoulder joint and the rotation of the radioulnar joint. When the elbow is extended, what is the total ROM of the hand rotation together with the shoulder and forearm rotations? Can you think of any sport examples using these structures and motions in combination?

B.2 Anthropometric Characteristics of the Body

Many factors contribute to an athlete's success or effectiveness in exercise and sport performance. Among these factors are motivation, nutrition, interest, and opportunity for practice. Moreover, an athlete's size, weight, proportions, and physique also contribute to performance effectiveness in many sports and exercises. For example, a professional basketball player is typically thought of as being larger than average, and a female gymnast as being smaller than average. **Anthropometrics** is a subdiscipline concerning the measurement of size, shape, and proportions of the human body and its segments. The area of study concerned with the physical measurements of the human body as they relate to physical development, exercise, and sport performance is called **kinanthropometrics** (Hebbelinck and Ross, 1974).

Because the gathering of anthropometric data is often time consuming, studies using human measurements do not employ the exact measures of single subjects but, rather, use average data gathered from a population of similar subjects. Subjects are most frequently grouped by gender and by athletic event. The most gross generalizations are those comparing all athletes to all nonathletes. Many investigations compare anthropometric measures within gender and between sport groups, such as between female high jumpers and female sprinters or between male weight lifters and male wrestlers. Care must be taken, however, when considering *average values* from grouped data because misleading stereotypes of a group's physical characteristics may be developed from averages or norms. One should keep in mind that averages represent nothing more than a general impression of a group of individuals, each of whom actually may be quite different from the average. In fact, wide variations frequently exist among individuals measured. Consequently, it is important to note the standard deviations of those groups.

A comparison between the heights of males and females is an example of the misuse of group means. To believe that all males are taller than all females because the mean height for males is greater than the mean height for females is obviously erroneous. Similar stereotypical descriptions of weight, body fat, muscular endurance, strength, and other physical characteristics also may be misused. Physical characteristics of individuals within a population range from the greatest to the least on a continuum, and the division of a population into two or more groups usually results in considerable overlap of the groups. The heights of males and females are estimated to overlap by 60%; that is, according to heights, the upper 60% of females coincide with the lower 60% of the males.

A second consideration when using anthropometric data is the unclarity of how much of any measure depends on genetic factors and how much depends on an individual's training activity over several years. Although training does not affect a person's height, it does affect a person's weight. Whenever weight of the total body or of its segments is considered, the data are influenced by the individual's training state. deVries (1980) discusses a heavyweight class U.S. weight lifter who gave up his weight-lifting career and took up long-distance running. After 3 years, he had lost 64 lb of body weight, and his total snatch and clean-and-jerk lift weight had decreased by 196 lb. When his lift was measured per unit of body weight, however, he was stronger after 3 years of running than when he had been weight lifting competitively. He had lifted 3.23 lb per pound of body weight competitively but lifted 3.30 lb per pound of body weight as a distance runner. Other measures affected by training are circumferences of segments, lean body weight, and segmental proportions.

A third factor relating to anthropometric data concerns the biomechanical advantage an individual may have as a result of a given anthropometric measure. Any biomechanical advantage must be considered in light of other characteristics of the athlete, such as physiological capacities. In addition, certain anthropometric measures are more important for success in some events than others. For example, weight is more important to success in a football defensive lineman than in a javelin thrower. Within a given event, some anthropometric measures contribute more biomechanically to an individual's success than do other anthropometric measures. For example, the height of a person's center of gravity is more important than the person's arm length for success in the high jump.

Height and Weight

The most commonly encountered anthropometric measures, and those most easily gathered, are the height and weight of the body. The mean height and weight values for male and female athletes who participated in the 1968 Olympics are presented in Figure B.11.

Expected differences occur among heights and weights of athletes in different events. Male basketball players are the tallest, and weight throwers, the heaviest. Female gymnasts and divers are among the shortest and lightest. Hebbelinck and Ross (1974) report that in sports in which greater height has a biomechanical advantage, the mean height of athletes tends to increase over the years. Because of the selection process, the mean height of basketball players, rowers, and weight throwers as a group is increasing, whereas the mean height of athletes in events in which height does not give an advantage, such as gymnastics, is remaining stable. The height–weight relationship is important in wrestling since weight limitations exist within classifications. Although a long-limbed tall wrestler has an advantage in reaching and in increasing the size of his base for increased stability, a short opponent is likely to be a stronger wrestler. The short wrestler's weight consists of more muscle mass than the tall wrestler's weight, which consists of more skeletal weight.

Ponderal Index

The **ponderal index** (PI) is a measure of stature, or stoutness. It is determined by a particular ratio of the height and weight of an individual. One equation used to calculate the PI is the following:

EQUATION B.1

$$PI = 10^3 \times \sqrt[3]{\frac{W}{H}}$$

where W is body weight and H is body height.

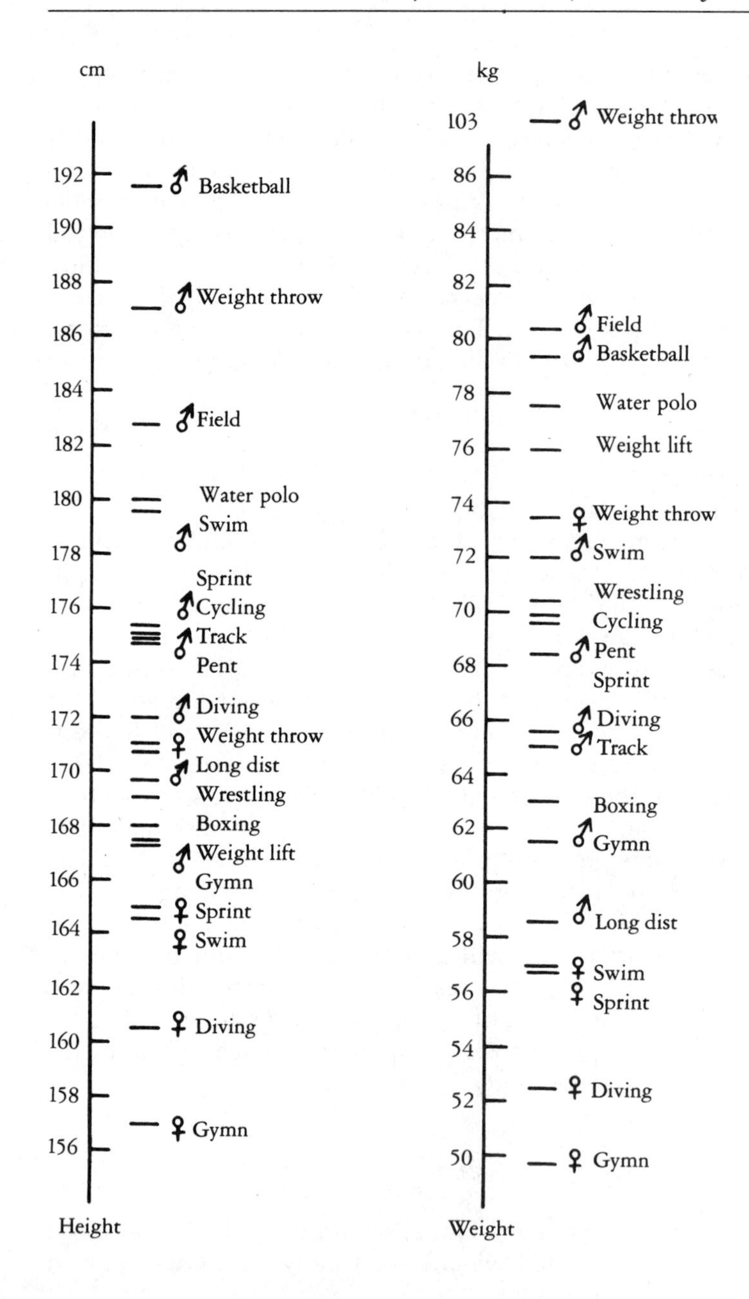

FIGURE B.11

Height and weight anthropometric parameters of 1968 Olympic athletes. (Adapted from Hebbelinck and Ross, 1974.)

According to the equation, two people of equal weight and height have equal stature, or PI. Given equal weight, a tall person has a smaller stature, or PI, than a short person. On the other hand, if two people are the same height, their PIs vary according to their weight differences. The person weighing the most has the larger PI. The PIs of elite athletes do not vary considerably between most events. Table B.4 presents data on several groups of Olympic athletes from the 1928, 1960, 1964, 1968, and 1972 games (Hebbelinck and Ross, 1974).

The distance runners and the basketball players tend to have the smallest PIs, whereas the throwers tend to have the greatest. Surprisingly, the marathon runners have a slightly higher PI than the 5,000-m and 10,000-m runners. As one would expect, the marathoners are lighter, but they are also shorter. Thus, the marathoners are more short than they are light as compared to the distance runners, and consequently their PI is slightly greater.

Before basketball players were encouraged to weight train, they were tall, lean, and had relatively little body mass. Their PI was small because they were relatively light. Since weight training has been incorporated into basketball conditioning programs, players have increased their muscle mass and have become heavier. Although as a group they have also become taller, their PI has increased because their average weight has increased more than their average height. PI does not distinguish between fat weight and muscular weight. A large and a small PI are shown for two people in Figure B.12a and b.

Somatotype

Related to an athlete's measure of stature is the measure of body type, or configuration. Although qualitative assessment of body types was used by the ancient Greeks, the best-known classification system, **somatotyping**, was named and developed by Sheldon in the 1940s. Somatotyping uses three classifications: the endomorph, the mesomorph, and the ectomorph. The **ectomorph** is described as being tall and thin; the **endomorph** as being short and fat; and the **mesomorph** as being stout and muscular. The mesomorph is what is commonly termed an "athletic build" (Morehouse and Rasch, 1963). Stereotypic bodies representing these somatotypes are shown in Figure B.13.

Since Sheldon's work, the somatotyping process has undergone many refinements. Because no definite delineations exist among the types, a person can have a somatotype consisting of differing degrees of all three components. The most commonly used rating method is the Heath-Carter somatotype method. The rating is a quantitative assessment of the amount of all three components. The endomorphic component is the relative fatness of the body; the mesomorphic component is a musculoskeletal rating per unit of body height; and the ectomorphic component is the "linearity" of the body. In a study using the Heath-Carter method, athletes were rated higher on mesomorphy and lower on endomorphy than nonathletes. Athletes from different sports varied in their degrees of endomorphy and ectomorphy, but all the athletes had a high mesomorphy component.

TABLE B.4

HEIGHT (cm), WEIGHT (kg), AND PONDERAL INDEX FOR OLYMPIC COMPETITORS IN SELECTED OLYMPIC GAMES[a]

Event	Amsterdam 1928	Rome 1960	Tokyo 1964	Mexico 1968	Munich 1972
100 m, 200 m, 4 × 100 m					
Height	172.7	177.7	176.0	175.4	176.0
Weight	64.7	70.8	70.8	68.4	69.8
Ponderal index	23.25	23.28	23.51	23.32	23.39
800 m and 1,500 m					
Height	174.4	180.5	177.0	177.3	177.2
Weight	66.7	68.9	65.8	65.0	67.2
Ponderal index	23.25	22.71	22.81	22.68	22.94
5,000 m and 10,000 m					
Height	167.7	172.2	173.6	171.9	172.4
Weight	60.3	60.7	60.3	59.8	61.3
Ponderal index	23.11	22.82	22.84	22.75	22.87
Marathon					
Height	166.1	168.3	170.3	168.7	170.0
Weight	59.6	60.2	60.8	56.6	60.2
Ponderal index	23.52	23.29	23.09	22.76	23.05
Jumpers					
Height	178.9	179.9	181.4	182.8	181.6
Weight	69.1	71.9	73.2	73.2	73.8
Ponderal index	22.94	23.06	23.06	22.88	23.10
Throwers					
Height	180.09	191.3	187.3	186.1	187.4
Weight	88.6	101.7	101.4	102.3	107.3
Ponderal index	24.64	24.40	24.90	25.13	25.36
Javelin					
Height	180.9	182.2	183.0	179.5	181.4
Weight	88.6	87.3	83.4	76.5	87.5
Ponderal index	24.64	24.35	23.87	23.65	24.47
Basketball					
Height		189.7	189.4	189.1	192.6
Weight		79.5	84.3	79.7	85.5
Ponderal index		22.67	23.15	22.76	22.87
Rowers					
Height	181.1		186.0	185.1	185.5
Weight	76.9		82.2	82.2	84.5
Ponderal index	23.48		23.38	23.49	23.66
Gymnasts					
Height	169.6		167.2	167.4	168.0
Weight	61.8		63.3	61.6	64.1
Ponderal index	23.31		23.84	23.59	23.82

[a]Adapted from Hebbelinck and Ross, 1974.

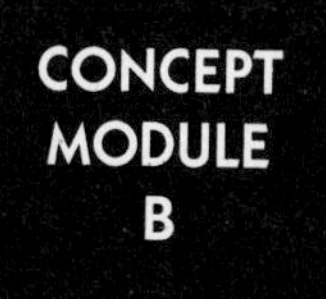

a b

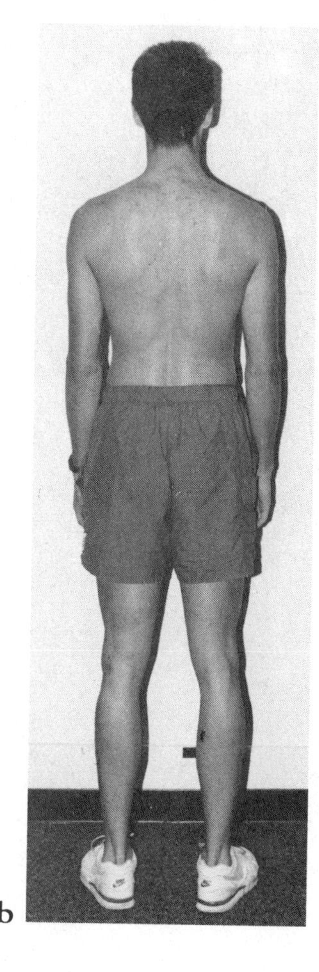

FIGURE B.12
(a) Large and (b) small ponderal indices.

Body Proportions

Absolute measures—the height of the body and the length of it's segments—are probably more important than the sizes of the segments relative to each other. Basketball, high jumping, throwing events, and volleyball are sports in which the absolute height of the body is important to success; the taller, the better. Given two performers of equal height, however, measurements can reveal quite different **proportions**, or relative lengths, of the segments making up that height. One person may have long lower extremities and a relatively short trunk; the other person may have a long trunk and short lower extremities; yet a third may have equally proportioned extremities and trunk. All would be the same height when standing. When seated, however, the person with the

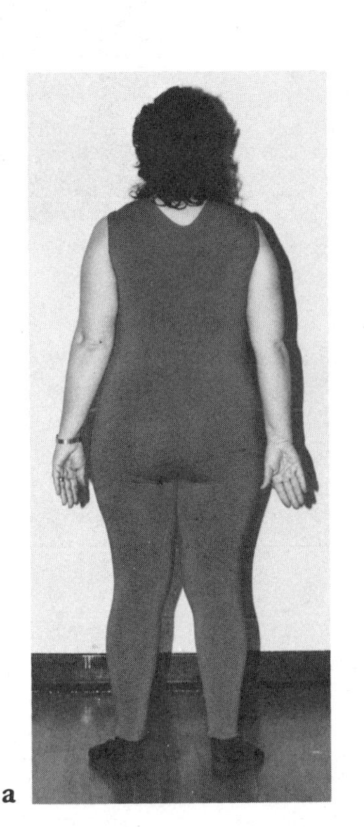

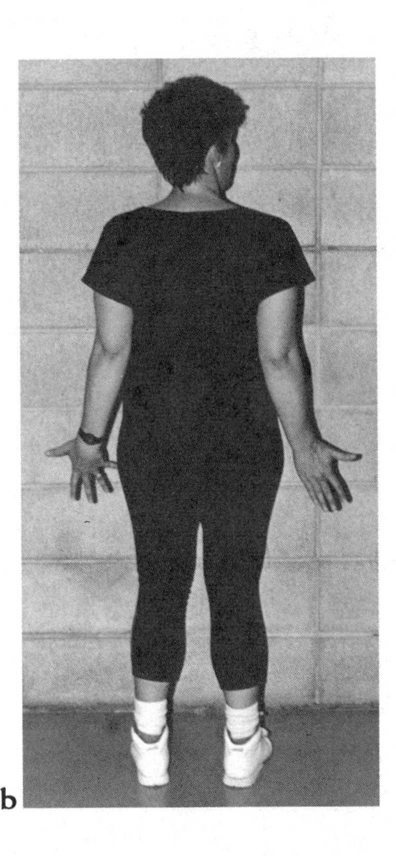

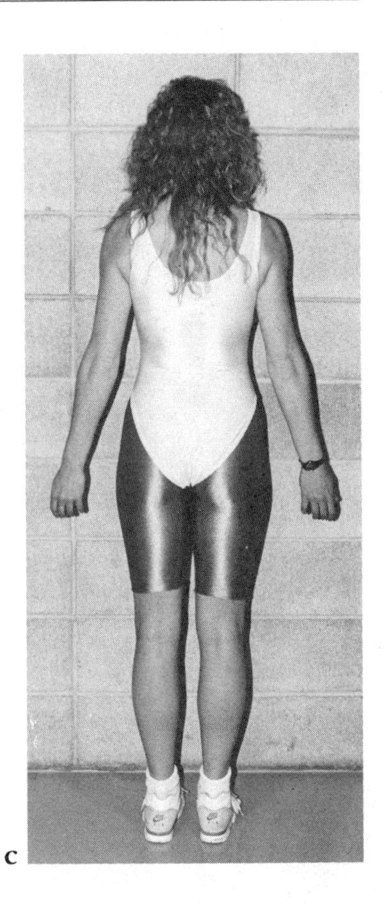

FIGURE B.13

Typical somatotypes: (a) endomorph, (b) mesomorph, and (c) ectomorph.

longest trunk would be taller than the person with the shortest trunk. If these same individuals were seated against a wall so that their legs were extended out from that wall, their leg lengths would reflect the opposite difference of their seated height. These differences can be expressed as a percentage of total body height. Figure B.14 illustrates proportional differences in athletes of similar heights.

Arm Lengths

Long upper extremity segments are advantageous to athletes in sports that include throwing and striking skills because a long lever affords greater ROM than a short lever. A long-armed boxer, wrestler, and martial artist have a

a c b

FIGURE B.14

Proportional differences of athletes of similar heights: (a) standing heights, (b) seated heights, and (c) lower extremity lengths.

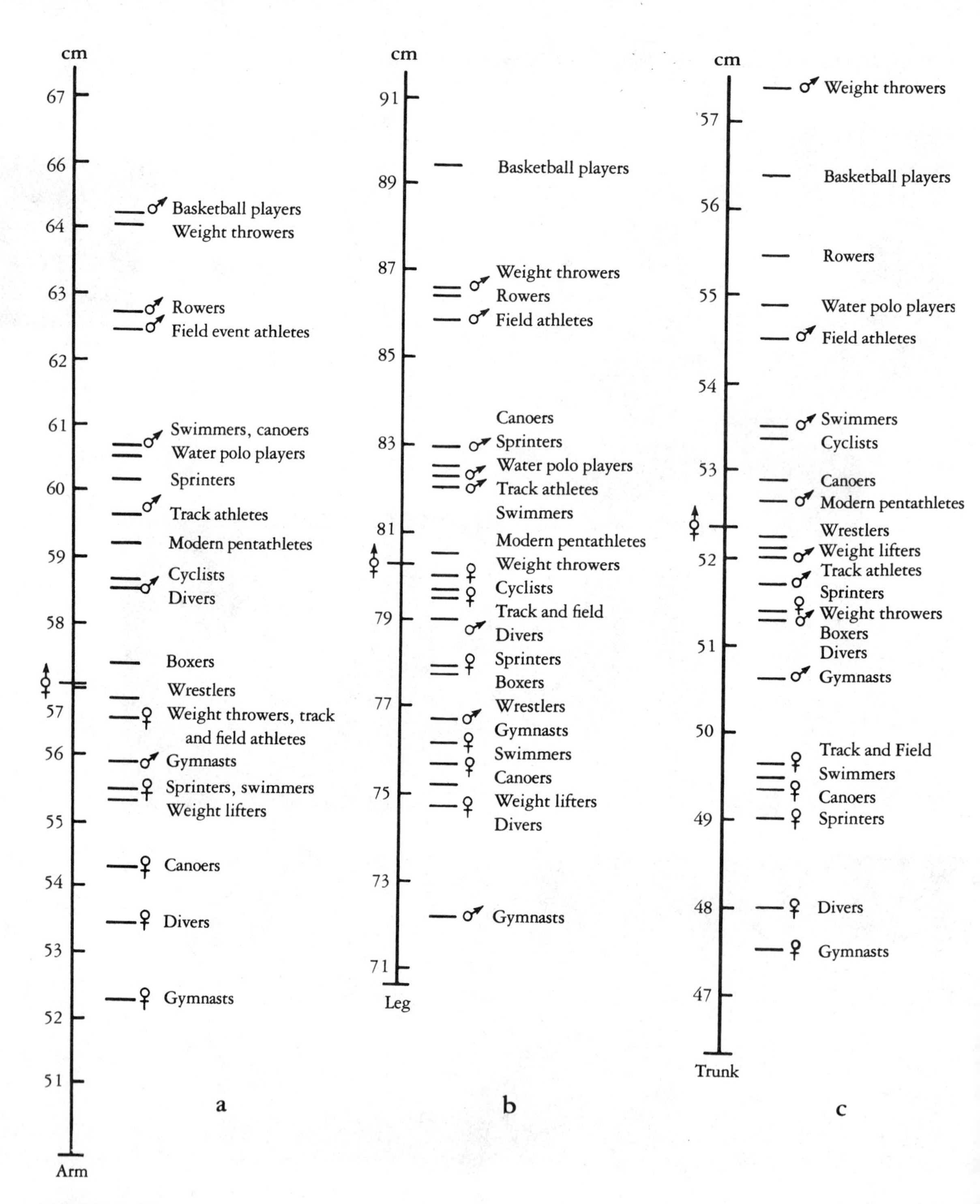

FIGURE B.15

Selected anthropometric parameters of 1968 Olympic athletes: (a) arm, (b) leg, and (c) trunk lengths (adapted from Hebbelinck and Ross, 1974.)

distinct advantage in reach over a short-armed opponent. From Figure B.15, we can see that basketball players, weight throwers, and rowers have the greatest arm lengths. The shortest arm lengths are found in gymnasts, divers, canoe paddlers, and weight lifters.

Leg Lengths

Proportional differences may give certain athletes a biomechanical advantage in their sport. Athletes with the greatest leg lengths are reported to be basketball players, weight throwers, rowers, and field event athletes. The shortest leg lengths occur in gymnasts, divers, wrestlers and boxers.

The **crural index**, the ratio of the length of the lower leg to that of the thigh, is used to describe the proportions of the leg. A ratio greater than one indicates a leg longer than a thigh; a ratio less than one indicates a leg shorter than a thigh.

A study of the proportions of animals of different species indicates that the fast runners and jumpers are those that have long distal segments, such as the horse, the kangaroo, and the cat. The thigh segments of these animals are short, ending before the segment leaves the pelvis or flank area. The leg segment is longer than the thigh; the foot segment is the longest, and the toes are also quite long. The animals described as the best jumpers have a leg considerably longer than the thigh segment and therefore a large crural index.

Burke and Brush (1979) reported that a select group of teenage female distance runners had on the average a longer leg length than thigh length compared with teenage female nonathletes. These authors speculated that a short thigh allows the runner to bring her leg through fast and with ease.

Long lower extremities are a disadvantage to the gymnast who must lift, hold, and swing those segments. Long legs offer greater resistance than short legs, and thus are often prohibitive to performing some gymnastic skills effectively. One should always keep this in mind when teaching gymnastics to nonathletes, for the tall, long-legged person will surely be at a disadvantage.

Trunk Lengths

Trunk lengths follow the trends set for arm and leg length measures. Weight throwers have longer trunks than basketball players; however, both of these groups have greater trunk lengths than other athletes. Again, gymnasts, divers, and boxers have the shortest trunks.

Implications of Anthropometrics in Teaching and Coaching

The anthropometric characteristics of individuals are often used as a rationale for success or lack of success in a particular activity. Anthropometric characteristics are certainly considered by coaches in selecting athletes for particular sports. (What basketball coach has not tried to recruit the tallest kid in the class?) Although a person's physical characteristics may provide an advantage, the effectiveness of movement patterns used to perform skills are also determined by these factors: the extent to which mechanical principles are used, motor ability, and motivation. For instance, individuals with long body segments experience more difficulty in performing many gymnastics skills than individuals of short stature; however, the quality (effectiveness) of a gymnastics performance also depends on the use and application of mechanical principles and the performer's motor ability, muscular strength, and endurance characteristics.

Physical characteristics may give a mechanical advantage in the performance of certain skills, especially in those events in which height or weight or both are primary factors for success. Therefore, learners should not be subjected to evaluative tools (skills tests) in which successful performance in part depends on the physical characteristics of the performer. For example, vertical jumping ability should not be assessed by measuring the height attained from the floor. Rather, it should be assessed by measuring the distance from an initial reach height to jump reach height since this method eliminates the influence of differing physical statures on the score.

Understanding Anthropometric Measures

1. Record your height, weight, and the segmental lengths of your head and neck, trunk, upper arm, forearm, hand, thigh, and leg. Calculate each segment's percentage of your total body height by dividing the length of each segment by your total height. Compare your percentage lengths with those given for your gender in Appendix III, Table III.2. Compare and discuss your findings with others in the class. Make some generalizations for tall, short, and average-height people. Are the percentages the same regardless of height?
2. Calculate your PI using Equation B.1. In Table B.4, find the sport group that you most closely match. Discuss your findings with others in the class. Are the matches between the statures and sport groups plausible?
3. How could a person's PI contribute biomechanically to success in movement activities?
4. Calculate your crural index by dividing your leg length by your thigh length. Of what importance might this be to performance?

5. According to the data reported in Figures B.11 and B.15, to which sport groups do you closely compare in height, weight, arm length, leg length, and trunk length? List these sport groups. Are they different or generally the same? How do you explain the results?
6. Draw a person of average stature. Imagining that the person you drew has a PI of 25, draw a person with a PI of 21 and a person with a PI of 28.

References and Suggested Readings

Arnold, J. A., Brown, B., Micheli, R. P., & Coker, T. (1980). Anatomic and physiologic characteristics to predict football ability. *American Journal of Sports Medicine, 8*(2), 119–122.

Burke, E. J., & Brush, F. C. (1979). Physiological and anthropometric assessment of successful teenage female distance runners. *Research Quarterly, 50*(2), 180–187.

Carter, J. E. L. (1981). Somatotypes of female athletes. In J. Borms, M. Hebbelinck, and A. Venerando (Eds.), *The female athlete: A socio-psychological and kinanthropometric approach* (pp. 85–116). Basel: Karger.

Carter, J. E. L. (1985). Morphological factors limiting human performance. In D. H. Clarke and H. M. Eckert (Eds.), *The American academy papers* (pp. 106–117). Champaign, IL: Human Kinetics.

deVries, H. A. (1980). *Physiology of exercise* (pp. 403–404). Dubuque, IA: Wm. C. Brown Co.

Hebbelinck, M., & Ross, W. D. (1974). Kinanthropometry and biomechanics. In R. C. Nelson & C. A. Morehouse (Eds.), *Biomechanics IV: Proceedings of the Fourth International Seminar on Biomechanics* (pp. 537–552). Baltimore: University Park Press.

Jensen, R. K. (1981). The effect of a 12-month growth period on the body moments of inertia of children. *Medicine and Science in Sports and Exercise, 13*(4), 238–242.

Katch, V. L., Katch, F. I., Moffatt, R., & Gittleson, M. (1980). Muscular development and lean body weight in body builders and weight lifters. *Medicine and Science in Sports and Exercise, 12*(5), 340–344.

Moffatt, R. J., Surina, B., Golden, B., & Ayres, N. (1984). Body composition and physiological characteristics of female high school gymnasts. *Research Quarterly for Exercise and Sport, 55*, 80–84.

Morehouse, L. E. & Rasch, P. A. (1963). *Sports medicine for trainers* (2nd. ed.). Philadelphia: W. B. Saunders.

Morrison, W. G., Frank, H. F., & Greg, R. (1979). An analysis of anthropometric data from female olympic rowers. In J. Terauds (Ed.), *Science in Sports: Proceedings of the International Congress of Sport Sciences* (pp. 185–192). San Diego: Academic Publishers.

Morrow, J. R., Jackson, A. S., Hosler, W. W., & Kachwick, J. K. (1979). The importance of strength, speed, and body size for team success in women's intercollegiate volleyball. *Research Quarterly, 50*(3), 429–437.

Reilly, T., Watkins, J., & Borms, J. (Eds.), (1986). *Kinanthropometry III, Proceedings of the VIII Commonwealth International Conference on Sport, Physical Education, Dance, Recreation and Health*. New York: E & FN Spon.

Shields, C. L., Whitney, F. E., & Zomar, V. D. (1984). Exercise performance of professional football players. *The American Journal of Sports Medicine, 12*, 455–459.

Slaughter, M. H., & Lobman, T. G. (1980). An objective method for measurement of musculo-skeletal size to characterize body physique with application to the athletic population. *Medicine and Science in Sports and Exercise, 12*(3), 170–174.

Spence, D. W., Dinch, J. G., Fred, H. L., & Coleman, A. E. (1980). Descriptive profiles of highly skilled women volleyball players. *Medicine and Science in Sports and Exercise, 12*(4), 299–302.

Ward, T., & Groppel, J. L. (1980). Sport implement selection: Can it be based upon anthropometric indicators? *Motor Skills: Theory Into Practice, 4*(2), 103–110.

Wilmore, J. H. (1983). Body composition in sport and exercise: directions for future research. *Medicine and Science in Sports and Exercise, 15*(1), 21–31.

CHAPTER 2

Neuromuscular Aspects of Movement

PREREQUISITES

Concept Modules A, B
Chapter 1

CHAPTER CONTENTS

2.1 Functional Aspects of the Muscular System

Muscle Properties

Among the several types of muscle tissue in the body, the skeletal (striated) muscles are of concern to us because they are the force-producing agents that cause or control the movements of our body segments. Skeletal muscle tissue has four functional properties, described by Gowitzke and Milner as *irritability, contractibility, distensibility,* and *elasticity*. Because skeletal muscle tissue is irritable, it responds to stimulation by contracting, or producing tension to pull its ends (attached to bones) closer together. The cessation of contraction is *relaxation*. Contraction takes time to develop to maximum value, and relaxation takes time to occur completely as contraction diminishes. Muscle can be distended, or stretched, within limits, to allow for nonrestricted joint movement caused by an antagonist (opposite) muscle or some other force. The fourth property of muscle and its connective tissue network is its elasticity, which gives it the ability to recoil from stretch.

Four aspects of muscle function that are of significance in physical education and athletics are flexibility, strength, power, and muscular endurance. **Flexibility** refers to the state of the muscle's length, which restricts or allows freedom of joint movement. **Strength** is the maximum amount of force that can be exerted by a body segment by means of muscular tension. Muscular **power** refers to the rate at which muscular force is applied to move a load, or the rate at which physical work is done. Normally, if a large muscular force is used to move a load at a fast rate, it is called a powerful movement. Muscular **endurance** is the ability of the muscles to exert force repeatedly or constantly.

These four muscle qualities are specific to each muscle group in the body; that is, each quality exists to some degree in a given muscle group, depending on the routine requirements placed on the muscle group.

More detail relevant to training these qualities is presented in Chapter 7.

Musculoskeletal Relationships

The skeletal muscle system is a voluntary system; that is, it needs an intact nerve supply to contract, and does so mostly as a result of conscious direction from the brain. The location and organization of the skeletal muscle tissue relative to the bones and articulations dictate the skeletal system's movements. All skeletal muscles have at least two attachments, each one being on a different bone so that the muscle crosses at least one articulation. Although labeling the attachments of a muscle is of little value in biomechanics, they may be identified by saying that their origin is usually the more proximal attachment, is usually the more fleshy attachment to the periosteum of the bone, and is usually on the more stable or stationary bone. The insertion of a muscle is normally considered to be the distal attachment, is usually tendinous, and is attached to the more movable bone. These characteristics, however, are only generalizations. If one flexes the hip to punt a soccer ball, the more movable attachment for the muscles acting across the hip is on the thigh and leg. On the other hand, if one performs sit-ups from a supported straight-leg position, then the trunk and pelvis become the more movable attachment for those same muscles.

The important characteristics of a muscle or muscle group are the *location of the attachment* on the bone, the *angle* found at the attachment, and the *number of articulations crossed* by the muscle between attachments.

Functional Muscle Groups

Frequently it is convenient and adequate to refer to the action of *groups* of individual muscles rather than trying to name each one that is or might be acting. The group name is based on the joint movement that the muscles are *able* to cause. For example, the muscles that are able to cause elbow extension belong to the functional muscle group called elbow extensors. Note that the elbow extensors *can* cause elbow extension, but that does not mean that every time elbow extension occurs, the elbow extensors necessarily cause it. After flexing your elbow 90 degrees (caused by the elbow flexors), the lowering of your arm is the joint movement of elbow extension, but it is caused by gravity, *not* the elbow extensors.

In summary, a muscle group name reflects the name of the joint movement that the group *can* but not necessarily *does* cause.

Classification of Muscles

Anatomists classify muscles or muscle groups in several ways, and muscles are named for any number of factors.

> The names of muscles have been derived from: (a) their situation, as the Brachialis, Pectoralis, Supraspinatus; (b) their direction, as the Rectus, Obliquus and Transversus Abdominis; (c) their action, as Flexors, Extensors; (d) their shape, as the Deltoideus, Trapezius, Rhomboideus; (e) the number of divisions, as the Biceps, Triceps, Quadriceps; (f) their points of attachment, as the Sternocleidomastoideus, Omohyoideus. (*Gray's Anatomy*, 1966, p. 381)

From the perspective of the biomechanist, however, more meaningful classifications exist, and they are discussed in the sections that follow.

Muscle Classification by Fiber Structure

Of interest to some biomechanists is the fiber structure of the muscles. Two basic types of fiber structures exist in skeletal muscle: fusiform and pennate. As shown in Figure 2.1a, **fusiform** muscle fibers run longitudinally to, and somewhat parallel with, the muscle's long axis. Although a single fiber does not run the entire length of the muscle, the many fibers that make up the muscle are aligned at zero angle with the muscle's long axis. The pennate muscles (Figure 2.1b–d) have tendons that run along the muscle's longitudinal axis, and their fibers run obliquely to this tendon. There are three basic **pennate** structures: the unipennate (Figure 2.1b), in which one set of parallel fibers runs obliquely to a single tendon; the bipennate (Figure 2.1c), in which two sets of parallel fibers run obliquely to a single tendon; and the multipennate (Figure 2.1d), in which there is more than one branchlike muscle off of the main tendon with associated pennate fibers to each branch.

Mechanically, there are advantages associated with each type of muscle structure. In muscles with longitudinal structure, the sarcomeres are arranged in series and thus can shorten a greater effective distance. These fibers can therefore pull the bones through a greater ROM. Structurally, however, the longitudinal muscle has fewer fibers per unit area and therefore does not produce the force of a pennate. The pennate structure allows for a greater number of fibers within a smaller area because the sarcomeres are arranged in parallel. The parallel structure allows for more short fibers than fewer long ones. The pennate structure produces less ROM than the longitudinal structure for a given shortening distance of the muscle fibers, however. For example, suppose a muscle fiber is aligned 20 degrees with the muscle's tendon. One can determine

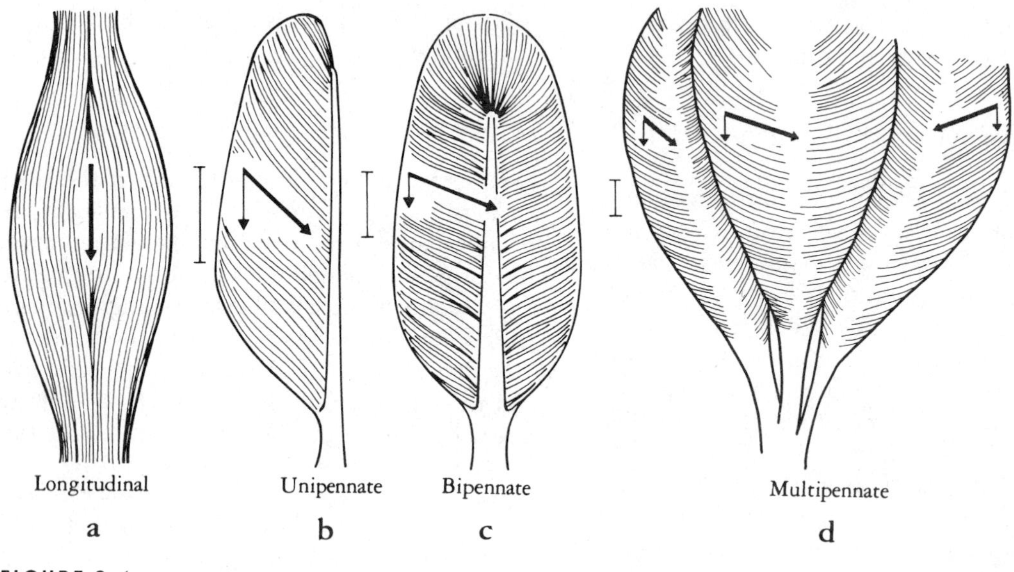

FIGURE 2.1
The various forms of gross muscle fiber structure have different force and range of motion (ROM) advantages.

a shortening component parallel with the long axis of the muscle. The greater the angle of the muscle fiber to the long axis of the muscle, the smaller the effective shortening component. In a given muscle, the more sets of oblique fibers, the greater the force available, but the lesser the ROM produced by that muscle. Thus, the multipennate structure allows for greater force but less ROM than the bipennate; the bipennate, more force but less ROM than the unipennate; and the unipennate, more force but less ROM than the longitudinal.

Muscle Classification by Relationship to Articulations

An important functional classification of muscles and muscle groups is by the number of articulations crossed by the muscle. A muscle can affect the movement of each articulation that it crosses. The simplest is the **uniarticulate** (one-joint) muscle. Although the muscle may function to produce, help produce, or restrict movements of the articulation it crosses, it can affect and be affected by only that articulation. When the muscle shortens, it can produce motion in that single articulation; similarly, when the uniarticulate muscle is stretched, it is stretched to accommodate only the ROM of that single articulation.

The **biarticulate** or **multiarticulate** muscle crosses two or more articulations. When it shortens, it tends to pull the two bony attachments toward the center of the muscle and therefore tends to produce motion in all the articulations that it crosses. Thus, the two-joint muscle is considered more efficient than the uniarticulate muscle because it can produce motion in more than one joint at the same time. Most of the muscles in the lower extremities that are used for locomotion are two-joint muscles and can efficiently cause motion in two or more of the articulations used for ambulation (Wells, 1988). Two-joint muscles have an important disadvantage, however. The two-joint muscle fibers cannot exert enough tension to shorten sufficiently to cause full ROM in both joints at the same time. This phenomenon has been called **active insufficiency** (O'Connell and Gardner, 1972). Similarly, it is very difficult for a two-joint muscle to be stretched enough to allow for a full ROM in both joints at the same time. This phenomenon has been called **passive insufficiency** (O'Connell and Gardner, 1972). Figure 2.2 illustrates the insufficiency concept of two-joint muscles.

An example of a two-joint muscle's insufficiency is that of the hamstring muscle group on the posterior side of the thigh. The muscles that constitute the hamstring group function to extend the hip and flex the knee. If both of these motions are to occur simultaneously, the hamstring group is not capable of shortening enough to produce full hyperextension in the hip while also producing full flexion at the knee (active insufficiency). The hamstrings can easily flex the knee fully if the hip is flexed. Similarly, the hamstring group is not capable of being stretched enough to allow full hip flexion and full knee extension, as when toe touches are performed from the long sitting position (passive insufficiency). The hamstrings do allow full extension of the knee,

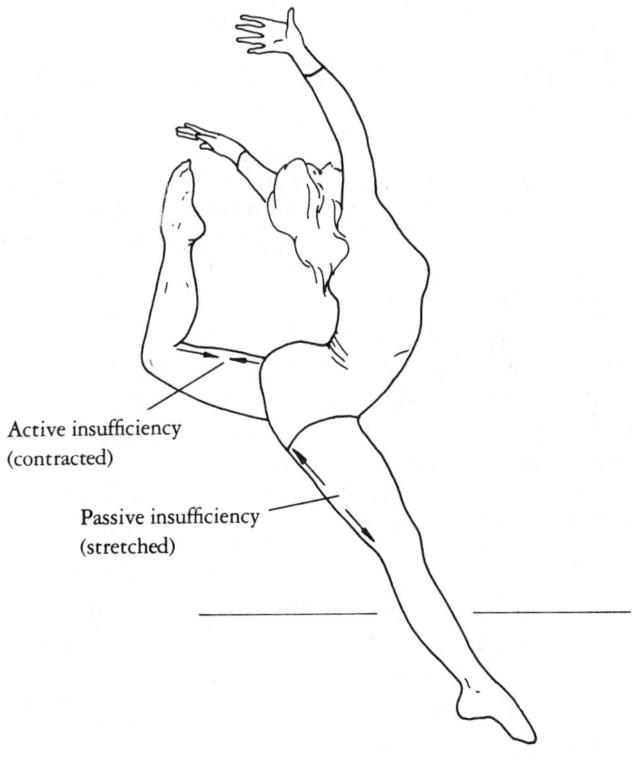

Active insufficiency
(contracted)

Passive insufficiency
(stretched)

FIGURE 2.2

Insufficiency of two-joint muscles to produce or allow total range of motion (ROM) over both joints. Active insufficiency of the hamstrings during simultaneous hip extension and knee flexion. Passive insufficiency of the hamstrings during hip flexion and knee extension.

however, if the hip is also extended, because the muscles are stretched at the knee end but are not stretched at the hip end. Wells (1988) reports a 7%–29% reduced mechanical work cost by using two-joint muscles rather than one-joint muscles in walking.

To summarize, the multijoint muscles are more efficient in that they can produce motion in two or more joints with a single contraction; however, they are insufficient to produce or allow full range of joint motion in both joints at the same time.

A second concept relating muscle positions to the articulations is the **agonist–antagonist** relationship. Those muscles that cause or help cause motion are called the agonists (movers). Those muscles that can perform the movement opposite to the movement being done are called the antagonists. When the agonist muscle group contracts to produce motion at an articulation, the antagonists relax to allow full ROM to occur. This phenomenon is called **reciprocal inhibition**. Because muscles can only pull on bony levers and cannot push, the opposition provided by the agonist–antagonist relationship enables the body to perform the variety of movements of which it is capable.

2.2 Types and Functions of Muscular Tension

Thus far we have mentioned only a single function of muscles in the body: to produce the force necessary to move a bone. Unfortunately, the movement function is so overemphasized that one tends to forget or to be unaware of the other important roles that muscles play in movement. Single muscles, movements, and functions should not be isolated, since muscles may change functions from one movement to another. Second, remember that muscular tension (contraction) is not the only force that produces movements of the body and its parts. Other external forces, produced by objects, other people, and gravity, may affect the function of a muscle as it contracts. One of two situations occurs when tension is produced in a muscle: a segment moves or it does not move. If the segment moves in the direction of applied muscular tension or opposite to the direction of the applied muscular tension, the tension is called **dynamic tension** or **dynamic contraction**. If a muscle produces tension or force against an opposing force or resistance, and the segment does not move, the tension is called **static tension** or **static contraction.**[1]

The three functions of muscular contractions are (1) to act as a motive force to move a body part (**concentric tension**), (2) to act as a resistive force to resist the movement of a body part (**eccentric tension**), and (3) to act to stabilize or fixate a body part (**isometric**, or **static, tension**). Any other function can easily fit into one of these three categories. Note that concentric and eccentric tension are types of dynamic tension.

Moving a Body Segment

If a muscle contracts and shortens sufficiently to cause movement of the articulation that it

[1] Traditionally, dynamic tension has been called *isotonic*. Isotonic means "same tension," which is not necessarily the case for a muscle that is changing its length. Static tension traditionally has been called *isometric*, which means "same measure"; that is, the muscle does not change its length.

crosses—that is, causes one or both attachments to move closer to the middle of the muscle—then that muscle's function is to move a body part(s). That muscle is said to be in a **concentric** type of dynamic contraction, during which the muscle becomes shorter (concentric meaning "moving to the center"). Concentric tension of a muscle is what anatomy texts refer to when they list the movements caused by that muscle. The hamstrings flex the knee. The elbow flexors function as the movers when the body is *pulled up* to a chinning bar (see Figure C.2a, p. 95). The elbow extensors function as movers during the *upward* phase of a push-up (see Figure D.2a, p. 112).

Resisting the Movement of a Body Segment

Muscle forces are not the only forces that may move a body segment. If another force is doing the moving, the muscles that perform the opposite function to the body movement being performed (antagonists) may be under tension to resist the movement. For instance, if one is lifting a barbell overhead, muscles are required to move body parts and the barbell against gravity to the up position, and thus the muscles function as movers. On the way down, however, gravity moves the body parts and the barbell. To control that downward movement, a certain amount of resistance must be applied upward, *against* gravity. The resistance is applied by tension of the muscles that could produce the *upward* motion. If we consider the muscles that extend the elbow on the upward movement, we can see that they function to extend the elbow to lift the barbell. On the way down, gravity is pulling the barbell, and the elbow is *flexing*; however, the *elbow extensors* control the descent by providing a resistive force. The elbow extensor muscles are being lengthened as the elbow is flexing, but the extensor muscle fibers must exert tension to provide the resistance and, thus, the control for the movement. This lengthening tension of a muscle group (e.g., the elbow extensors) is called **eccentric tension** (*eccentric* meaning "away from the center"). Muscles that function in this manner control the speed or direction of movement by resisting movement caused by another force (which may include another muscle's force). The segmental motion and muscular tension exhibited by the loser in an arm-wrestling match provides a good example of eccentric tension. The resistive (eccentric) functioning of the elbow flexors against the force of gravity is shown in the downward movement in a pull-up. The elbow extensors function as the resistance during the downward phase of a push-up.

Stabilizing or Fixating a Body Segment

Muscle tension that does not provide enough force to cause a movement and does not give way to another force attempting to cause movement of a body part is termed **static tension**, or more commonly, **isometric contraction**. There is muscular tension, but no discernible movement occurs between body parts. The muscles may attempt to shorten to move the body part, but, because of resistance provided by an external object, person, or another muscle, the body part does not move. Examples of static tension include the muscular tension that develops when you attempt to abduct your arm against a wall, the muscular tension that develops when you arm wrestle with someone of equal strength, and the muscular tension that develops in an antagonistic muscle when a two-joint muscle functions to produce motion in a single joint. In the last case, the antagonistic muscle must counteract the motion in the second joint of the two-joint muscle to allow movement in only the first

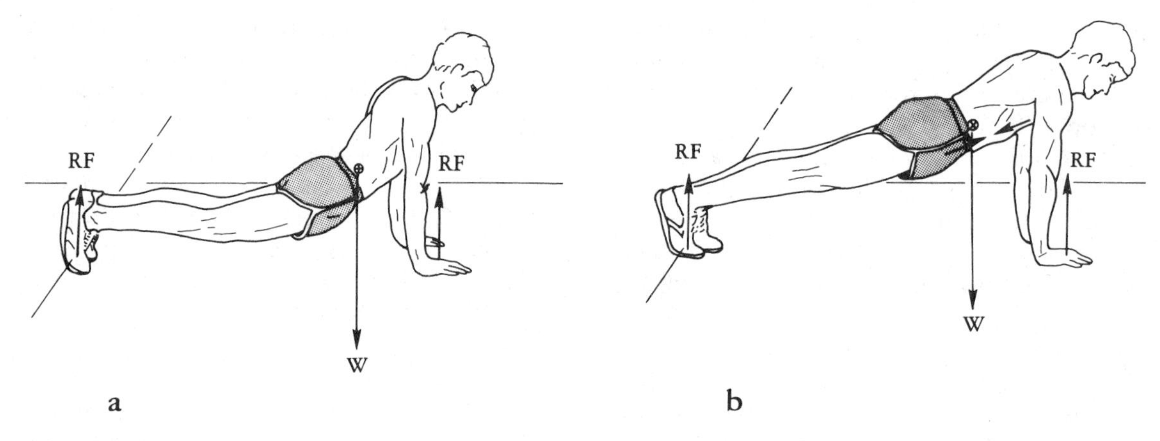

FIGURE 2.3

The vertebral column flexors perform a stabilizing function during a pushup. (a) Vertebral column hyperextension. (b) The flexors stabilze the vertebral column in a neutral position.

joint. The stabilization role as described in the last example may be used when a force acts to produce movement of a body part and that movement is not wanted. For example, while performing a push-up, one expects to work the shoulder and arm muscles. Gravity is also pulling on the trunk, however, so that without muscles to counteract the pull of gravity by fixating or stabilizing the trunk, the vertebral column would hyperextend. Figure 2.3 illustrates the fixating or stabilizing function of the trunk flexors (abdominals) during a push-up.

Sometimes the classification of muscular functions includes a function called **synergy**. *Synergy* means "working together" and is used to describe two muscles working together or helping each other to produce a single movement. For example, if a two-joint muscle is functioning to produce movement at one joint, the second joint must be stabilized by the contraction of an antagonist. This relationship is called true synergy.

The second type of synergy is seen when two muscles have a common mover function. Each muscle also has a second function. The second functions of these two muscles are antagonistic to each other. This relationship is sometimes called helping synergy, because the two muscles help each other perform the common movement while their antagonistic movements are canceled.

Although synergy is an important concept, it is not a real function of a muscle, but rather a description of how muscles work together. Certainly, one must always remember that muscles must work cooperatively to produce any quality of movement—forceful, smooth, or well-timed.

Understanding the Characteristics of the Muscular System and Muscular Tension

1. Using illustrations from a muscle anatomy textbook, name the fiber structures of the following muscles: (a) deltoid, (b) trapezius, (c) subscapularis, (d) triceps, (e) semimembranosus, (f) biceps femoris, (g) rectus femoris, (h) adductor

magnus, (i) tibialis posterior, (j) gastrocnemius.

2. What is the advantage of a fusiform muscle fiber structure in movement as compared with a pennate structure?

3. What advantage does a pennate fiber structure have over a fusiform structure?

4. Differentiate between active and passive insufficiency in two-joint muscles.

5. With the elbow flexed 90 degrees and the wrist in anatomical position, flex the fingers. With the elbow in the same position but with the wrist flexed 90 degrees, repeat the finger flexion. What is the difference in the ROM that the fingers display in the two situations? What is the name of this phenomenon and the reason for its occurrence?

6. Flex the shoulder joint 90 degrees, and then flex the elbow as far as possible. Repeat the elbow flexion with the shoulder joint flexed 180 degrees. Is there any discomfort? Why should one expect the second task to be more difficult? What is the name for this phenomenon?

7. Flex the hip joint with the knee flexed. Flex the hip joint with the knee extended. Note the ROM of each of these movements. Why is there a difference? What is the name for this phenomenon?

8. Stand on one foot, and flex the hip and the knee of the free leg as far as possible. Stand on one foot, hyperextend the hip, and flex the knee as far as possible. Is there any difference in the two exercises? What is the difference called, and why would you expect a difference?

9. If the muscle is the motive force for a segment's rotation, what type of muscle contraction is taking place?

10. What type of muscular tension serves to stabilize a body segment to prevent it from moving under the influence of another force?

11. If gravity is the motive force for vertebral flexion, what muscle group contracts to resist or control the speed of this movement, and what type of tension is this?

12. To what muscle *group* does a muscle belong if it can help produce wrist flexion? What is the name of the muscle group that can cause knee extension? hip abduction? shoulder transverse abduction? flexion of the vertebral column? medial rotation of the hip?

13. To what muscle group does a muscle belong if it *resists* knee flexion? if it *resists* knee extension? if it *resists* ankle dorsiflexion? if it *resists* hip abduction? if it *resists* radioulnar supination?

14. Name four pairs of agonist–antagonist muscle *groups* acting across joints you select.

15. In Figure 2.4a–k, a person is performing a series of exercises. Identify for the initial movement and the return movement:

    a. Each articulation where the movement is occurring

    b. Each joint movement occurring

    c. The motive force causing each joint movement

    d. The resistive force acting against each joint movement occurring

    e. The functional muscle group used and the type of contraction used during each joint movement

    f. The body parts or joints needing stabilization during the whole exercise

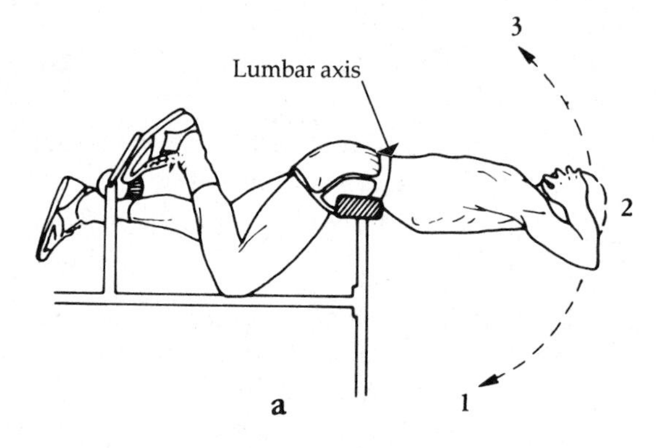

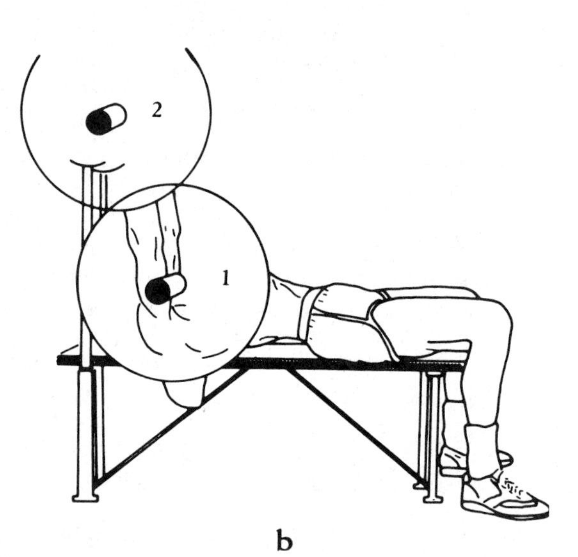

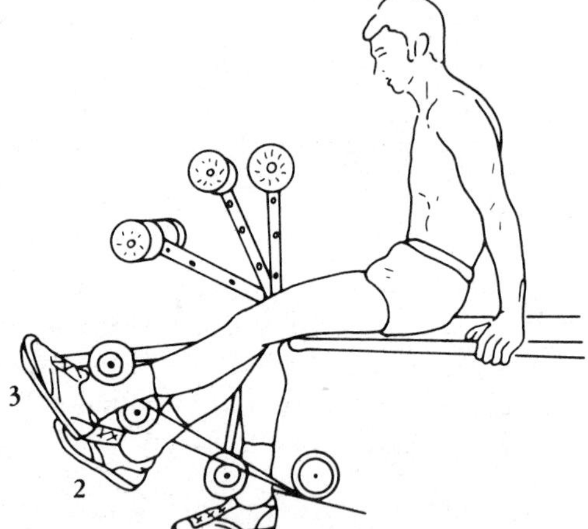

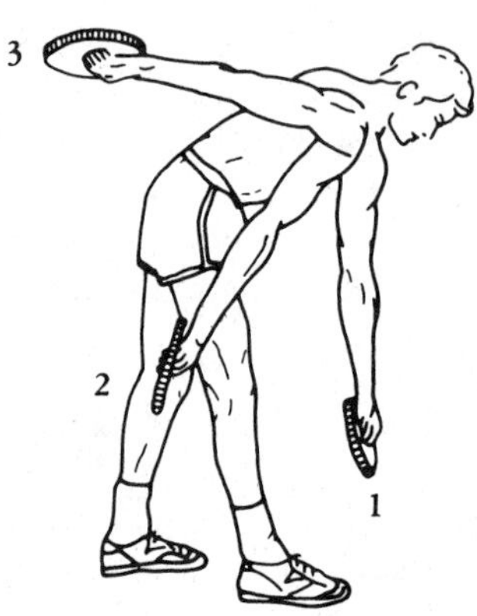

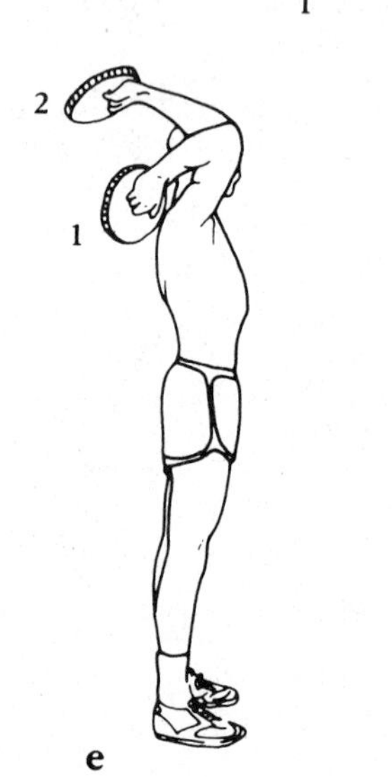

2
1
f

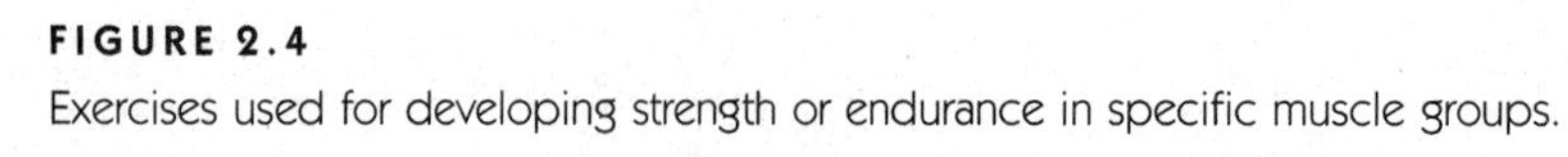

FIGURE 2.4

Exercises used for developing strength or endurance in specific muscle groups.

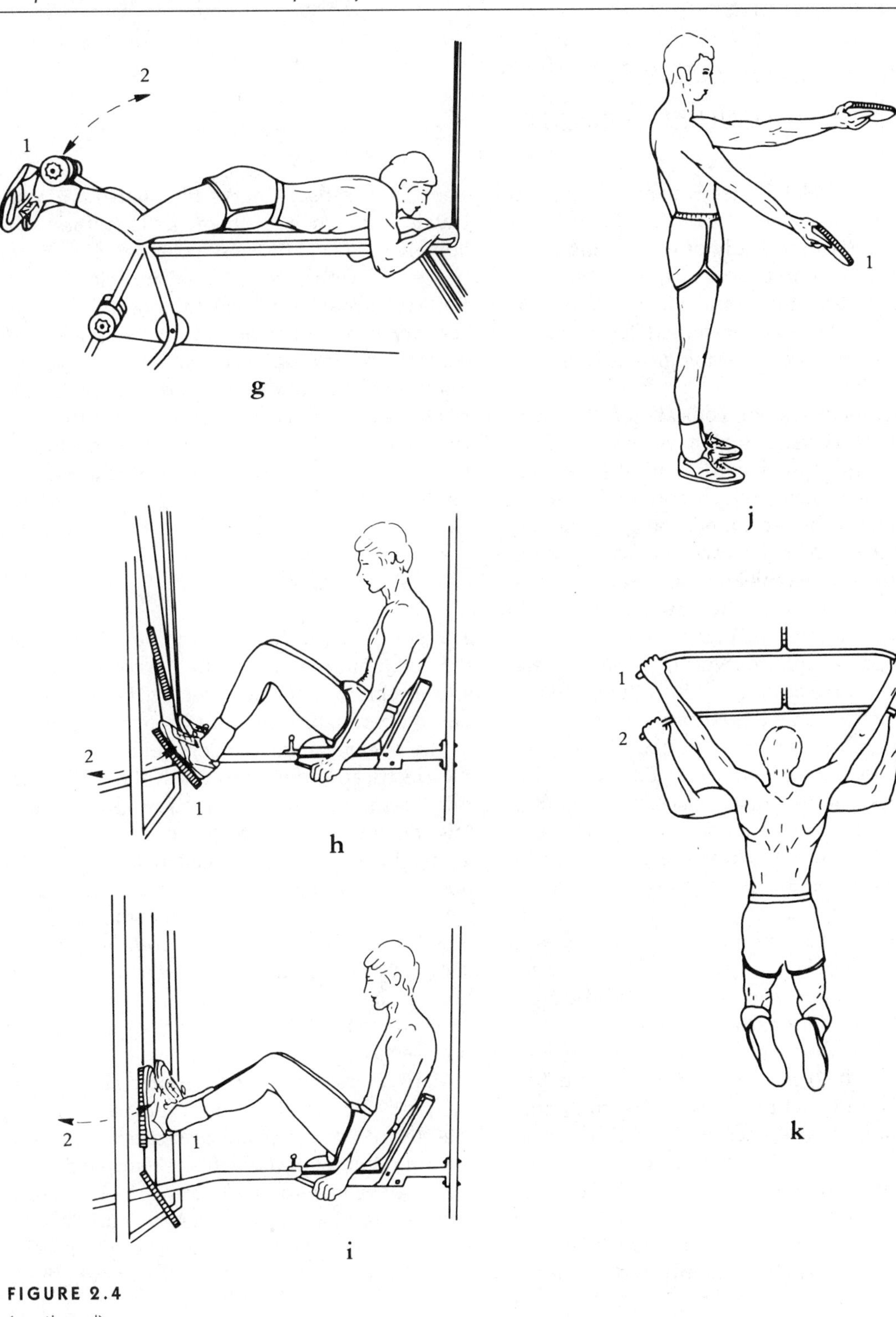

FIGURE 2.4
(continued)

2.3 Functional Aspects of the Neuromuscular System

The musculoskeletal mechanism is set into action and generally controlled by the nervous system. A thorough study of the nervous system is far too complex to be undertaken here; however, several important concepts of neuromuscular functions are examined because they aid understanding of kinesiological and biomechanical factors.

The nervous system consists of the central nervous system, which is composed of the brain and spinal cord, and the peripheral nervous system, which consists of nerve branches from the central nervous system that are arranged in pairs for each side of the body. Twelve pairs of cranial nerves and 31 pairs of spinal nerves are divided as follows: cervical, 8 pairs; thoracic, 12 pairs; lumbar, 5 pairs; sacral, 5 pairs; and coccygeal, 1 pair. These spinal nerves are associated with the vertebral column sections mentioned previously.

The spinal nerves are regrouped into functional sections called plexuses. Further dividing of these nerves takes place peripherally, and thus, the nerves radiate out into areas of the body and its extremities. The names and functions of the plexuses are: the *cervical plexus*, which innervates the muscles of the neck and consists of cervical spinal nerves 1 through 4; the *brachial plexus*, which innervates the shoulder, shoulder joint, and upper limb and consists of cervical spinal nerves 5 through 8 and thoracic spinal nerve 1; the *lumbar plexus*, which innervates the hip and the anterior and medial portions of the thigh and consists of lumbar spinal nerves 1 through 4; and the *sacral plexus*, which innervates the posterior thigh and the entire leg and foot and consists of lumbar spinal nerve 5 and sacral spinal nerves 1 through 3. These plexuses are considered to comprise mainly motor nerve fibers, but sensory nerve fibers also are associated with their pathways. Therefore, the spinal nerve pathways provide the motor nerves necessary for skeletal muscle contraction to produce movements and also the sensory nerves that provide the pathways for sensations back to the central nervous system. The mechanisms for voluntary muscular contractions, the sensory receptors, and the neuromuscular reflexes are important in kinesiological, biomechanical, and physical therapy applications.

The Motor Unit

The spinal nerves continue to subdivide until they come to the end of the nervous pathway and integrate with the muscle. The last single nerve fiber (axon) puts out branches to attach to the skeletal muscle fibers and thus allows the electrical impulses to reach the site where they can stimulate the chemical reaction that initiates muscular contraction.

The last single nerve cell body, its nerve fiber (axon) and its branches, and their associated muscle fibers make up the smallest functional unit of neuromuscular contraction, and is called a **motor unit**. Generally, motor units can be classified as slow or fast and possess different characteristics that are discussed in a later section.

Many motor units exist within a single muscle, and the number and type vary among muscles according to muscle size and function. The number of muscle fibers supplied by a motor neuron varies according to the precision required of the particular muscle. For example, those muscles that do small, precise movements, such as those in the fingers, have fewer muscle fibers associated with their many motor units than do the more powerful but less precise muscles of the thigh. Even in the simplest task, many motor units, with each of their respective associated muscle fibers, are required to be active.

The muscle fibers in a given motor unit are dispersed among a number of different bundles rather than being located in one small region. This means that any particular section of a whole muscle consists of fibers from different motor units. When an impulse is transmitted to muscle fibers from their common motor neuron, not all the muscle fibers in that motor unit contract. The stimulus does not reach the thresholds of some "lazy fibers." They can be activated on summation of stimuli, however (Gowitzke and Milner, 1988). The motor unit, therefore, does not exhibit an all-or-none response; only those fibers whose threshold is reached can respond together. The total muscle's tension is determined in part by the total number of fibers activated at any given time.

2.4 Factors Influencing the Effectiveness of Muscular Tension

Recruitment

A single motor unit with all its muscle fibers contracts if the stimulus is adequate for each of the fibers. To produce more force, more motor units must be used. An increase in the number of motor units and thus muscle fibers stimulated to produce a stronger contraction is called **recruitment**. Recruitment is the main muscular response used to produce greater muscle tension.

Firing Frequency and Synchronization[2]

Increasing the **frequency** of the nervous stimulus also produces increased muscular tension. Also, if the contractions are synchronized, additional tension results. The control of **synchronization** of motor unit contractions plays a major role in the level of skill of a performer. A skilled performer has developed the ability to synchronize the motor unit firing at the appropriate instants to produce well-timed, forceful movement, whereas an unskilled performer may have somewhat erratic motor unit activity and thus may lose coordinated movement as well as some tension production by the muscle being used.

[2] For comprehensive coverage of microscopic substructure and biochemical properties of contraction, the reader is referred to Gowitzke and Milner (1988).

Muscle Temperature

Deep muscle temperature affects the speed of muscular contraction and relaxation. Cooler muscles require more relaxation time than warmer muscles, and incomplete relaxation may cause resistance to contracting antagonist muscles (deVries, 1980). The value of muscular contraction warm-up exercise before demanding activity is significant. Heating the muscles through warm-up muscular contraction (active warm-up) is more effective than externally heating the area for faster contraction and relaxation (passive warm-up).

Muscle Fiber Type

Two fiber types, slow twitch (ST) and fast twitch (FT), were identified in the early stages of fiber type studies. Now, three fiber types (associated with their motor unit types) have been identified and classified according to various characteristics (Burke and Edgerton, 1975; deVries, 1980). Type I fibers previously called ST are now called slow, oxidative (SO). They have a low strength of contraction, low anaerobic capacity, and are small in size, but they have high capillary density and are highly resistant to fatigue.

The FT fibers are now differentiated into two types. Type IIa fibers are fast, oxidative, and glycolytic (FOG). They are the largest fibers, and they have a large capillary density for good blood supply. Their speed of contraction is fast, and their contraction strength is great. They have medium aerobic and anaerobic capacities, but they are more fatigable than SO fibers.

Type IIb fibers, the second type of FT fibers, are fast and glycolytic (FG). These fibers are the most fatigable, having low aerobic capacity and small capillary supply. On the other hand, they are strong and large, and they have a high anaerobic capacity.

All muscle fibers within one motor unit are of the same type, but within any given muscle is a mixture of motor unit (fiber) types. The percentage of fiber types varies among all the muscles in an individual. The fiber-type makeup of any one muscle in an individual is generally believed to be established early in life and to be unchangeable. Metabolic characteristics of fiber types, however, can be changed with training (Gollnick et al., 1972). For example, the aerobic capacity and glycogen content of the muscle can be improved with training.

Wood et al. (1983) indicate that changes in the motor neuron and changes at the neuromuscular junction occur with training. Such neural changes with training may account for the changes seen in FOG and FG fibers. Distance running, for example, seems to produce a shift from FG fibers to FOG fibers, and following weight training, from FOG to FG fibers (deVries, 1980).

Selective Recruitment of Fiber Types

The recruitment of fiber types usually occurs in a preferential manner. SO fibers are recruited during light-to-moderate ongoing activity. The FOG and then the FG fibers are recruited during exercise demanding greater muscular tension or high power or speed of movement or a combination of these (Campbell et al., 1979).

Gans (1982) reports results of studies on fatiguing activity that show a "standard" order of recruitment of fiber types. First recruited are SO fibers, then FOG, and lastly, FG. This same order of fiber recruitment has been found in fatigued downhill skiers (Eriksson, 1976).

The percentage of each fiber type, or the predominance of one type, also varies among individuals, but not between the sexes. Occasionally, muscle biopsies are used to determine the percentage of one fiber type over another in the same muscle of different individuals. An explanation for an athlete's success might be based on a certain fiber predominance in the muscles used in an activity. For example, individuals who have a predominance of one type or another may have an advantage in muscular endurance types of activities (SO fibers) or in quick, forceful activities (FOG, FG fibers).

Because fiber-typing has been related to predictions of success in endurance or in speed events, incentive exists to develop a simple test to predict fiber-type predominance so that individuals can be placed in those activities to which they may be best suited. A muscle biopsy is *not* such a simple test.

In a study of muscle fiber composition in women and their performance capacities, Campbell et al. (1979) found that predominance of one fiber type over another (determined by biopsies) had no relationship to performance before and after training; that is, of the four tests administered—before and after training—none were useful in discriminating between those with high and low percentages of FT fibers. These tests, which were not predictive, included: a maximal oxygen uptake on a cycle ergometer, a high-anaerobic-load power test on a cycle, a low-anaerobic-load power test, and a Sargent jump.

Improvement of all subjects occurred with training, but no significant differences in improvement were noted that would favor a high- or low-percentage FT group in terms of benefiting more from the training.

The complexity of contraction patterns and fatigue level during exercise may obscure any relationship that may exist between muscle fiber composition and *objective* measures of performance. Studies have not yet found a repeatable correlation. Efforts continue, however, because identification of "success" variables would be rewarding to coaches and athletes.

Elastic Energy: Storage and Recoil

When a muscle is in a tensed state and is then forcibly and quickly stretched immediately before shortening, the elastic recoil provides a significantly large initial force to move the segment in addition to the force provided by the muscle's contractile elements.

The stiffness, or the contracted state, of the muscle is necessary for the storage of elastic energy upon external stretching. The tension developed in individual fibers within a muscle is transmitted to the connective tissue network that envelopes the fibers to form bundles of fibers, which in turn are incorporated into larger and larger bundles. These connective tissues blend with the strong tendon tissues that then attach to the bones. The connective tissue network, which interfaces the contractile muscle fibers with the tendons and bones, has an elastic component that recoils following its stretch. This **parallel elastic component** allows the shortening of the muscle fibers to be smoothly transmitted to the bones. It also brings the muscle back to its original length following stretching.

Another source of elasticity is in the cross-bridges between the actin and myosin filament of the myofibrils. It is called the **series elastic component** and plays a major role in the storing of elastic energy when an active muscle is stretched (Komi, 1984b). The best use of the series elastic component is made when the muscle is quickly stretched and then shortened. Quick stretching followed by eccentric then concentric tension in a muscle group is a natural pattern of human muscular function in movement and is called the "stretch-shortening cycle" (Komi, 1984b). It occurs in running, jumping and all other activities in which muscle groups are suddenly stretched by impact or other external cause. When the sudden stretching occurs, a stretch reflex causes eccentric then concentric tension of the stretched muscles. This increases the muscle stiffness so that the elastic component can store more elastic energy. The action resulting from the elastic recoil plus the concentric tension is a much stronger, more powerful muscle shortening than could be achieved by concentric tension alone. The stretch-shortening cycle also saves energy by taking advantage of the elastic recoil that was made possible by an outside quick-stretching force. The reduction in effort is apparent, for example, when a fast walker breaks into a jog: Elastic energy is stored in the plantar flexors of the ankle and is then used for the pushoff in the bouncing gait.

Vertical jumps following quick landings

from approach steps are noted as being higher than those without the approach and landing. Vertical jumps following landing from a 0.4 m height (depth jump) have been found to be higher than those made from a semisquat with or without a countermovement, or dip (Asmussen and Bonde-Petersen, 1976, cited by Cavagna, 1977). Jumping from greater heights than 0.4 m, however, does not seem to produce as favorable a result. An explanation for this might be that upon landing, a deeper, slower flexion is required for shock absorption, and therefore the elastic energy may be dissipated during this process, rather than used immediately for leg extension. Generally, the smaller the amplitude of joint movement during the stretch and the quicker the stretch, the greater the speed of recoil and the greater the power developed. When movement takes place against a given mass, the effect of stretching is the generation of greater power to accelerate the mass. For example, the mass could be the body itself, as in jumping, or a segment with an implement.

Ikegami et al. (1981) report that a javelin thrower's kinetic energy (energy resulting from the running approach) appears to be stored as elastic energy in the early part of the throwing motion when the body is arched and the javelin is back. This elastic energy is then released to accelerate the javelin.

Prestretches or windups of segmental movements such as in throwing, striking, and kicking patterns can therefore augment the muscle contractile force that is to come. Shock absorption (stretching at foot plant) in running can contribute to greater force of push-off (Cavagna, Dusman, and Margaria, 1968). In sprint running, the power developed at each push-off was found to be increased due to the contribution of the muscle's recoil from the quick stretch applied at contact (0.15 to 0.1 sec).

Moreover, indications are that SO and FT fibers have different elastic properties (Komi and Bosco, 1978), with the SO fibers being reported to have the greater elasticity (Gowitzke and Milner, 1988). If this were so, individuals varying in fiber-type composition in a given muscle group would also tend to vary in their ability to store and use elastic energy in activities using prestretch along with motive muscular contraction force.

Length–Tension Relationship

A factor that influences the force of muscle contraction is its length at the time it is stimulated. The **length–tension relationship** states that the contractile tension the muscle is able to produce increases with the length of the muscle and is maximum when the muscle is at its resting length, that is, the length midway between the shortest and longest position. The contractile tension produced decreases as the muscle becomes shorter, or as it is stretched beyond resting length. In normal (intact) muscles, greater total tension is produced when the muscle is in an elongated position, greater than resting length. The increased tension seen with stretch, however, is not contractile muscle tension; it is due to the contribution of the elastic component in the tissues. Figure 2.5 shows a simplified length–tension curve for contractile tension and for the increase in total tension due to the elastic component.

If a muscle is *overly* stretched, that is, forced beyond the length to which it normally can be stretched in the body, the increase in elastic tension cannot make up for the decrease in contractile tension that occurs progressively from resting length. Just how much stretch is optimum depends on the muscle. Gowitzke and Milner (1988) report that parallel-fibered muscles produce maximum total tension near resting length, whereas other fiber arrangements can be stretched farther for more tension. In general, Gowitzke and Milner indicate that the greatest total tension can

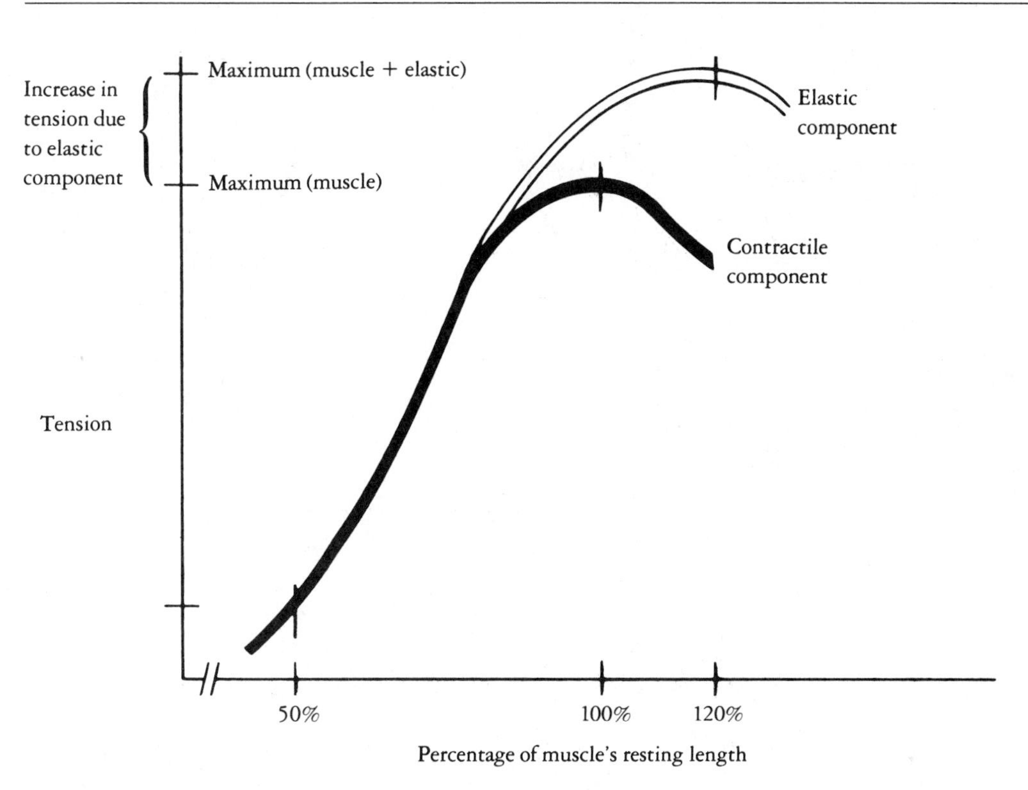

FIGURE 2.5
The length–tension relationship in skeletal muscle. The increase in total tension is due to the elastic component.

be produced at between 120% and 130% of resting length.

An example of a muscle group in a fully extended position to gain the maximum effect of the length–tension relationship is the leg and foot relationship at the ankle when the feet are placed in starting blocks (Figure 2.6a). Another example is the thigh, leg, and foot position at the hip, knee, and ankle when the starting position for swimming is assumed (Figure 2.6b). Both of these positions place the ankle plantar flexors at a length that is greater than their resting lengths, thus increasing the possible tension developed upon contraction. The placement of the segment in a position to gain maximum tension due to the length–tension relationship should not be confused with the dynamic situation in which the segment is moved into a position that causes a fast stretch on the muscles crossing a joint, as in a backswing.

Minimum tension at the shortest length usually occurs at the end of the range of motion of the joint across which the muscle has shortened. Gowitzke and Milner (1988) report that at lengths shorter than 50% of resting length, no contractile tension is able to be generated.

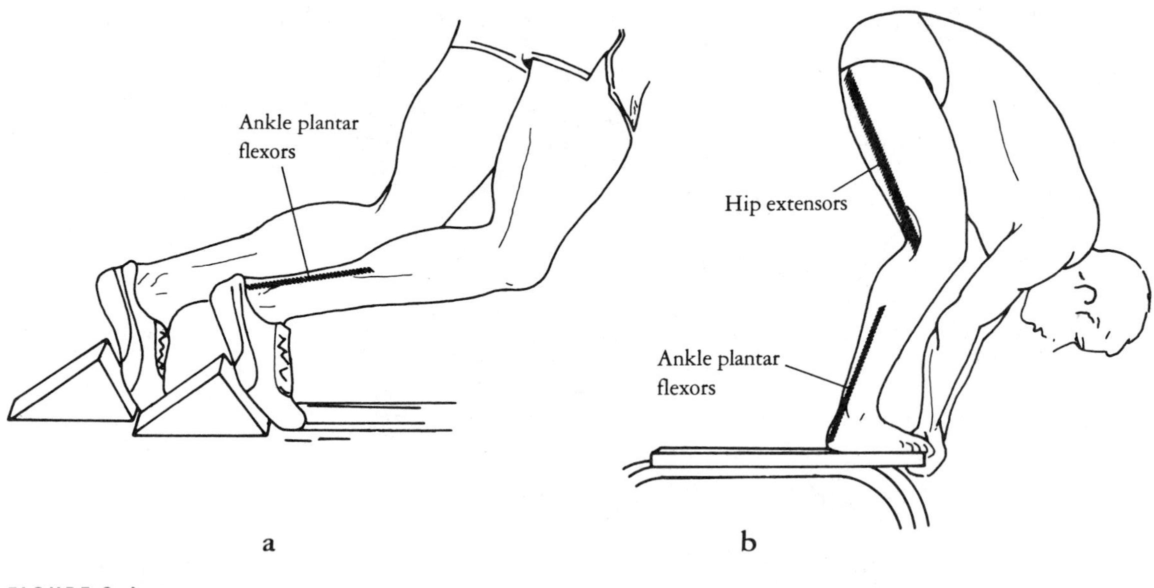

FIGURE 2.6

Placing muscles on stretch before contraction enables them to contract more forcefully.

Force–Velocity Relationship

The relationship that exists between the velocity (speed) of muscle shortening and the tension it is able to generate may be an important constraint, or limiting factor, in the performance of rapid movements. Laboratory experiments with excised muscles indicate that the tension a muscle is able to exert decreases as the speed of muscle shortening increases. In intact muscles within a given individual, the muscles that move light segments shorten faster than the muscles that move heavier segments (Gans, 1982). According to the *force–velocity relationship,* the force that a muscle can exert is inversely related to the speed of shortening. That is, if the velocity of shortening is slow, the tension that can be developed is large, and if the shortening velocity is fast, the tension developed is small. Figure 2.7 shows a force–velocity curve that includes concentric, isometric, and eccentric contractions. Note that the shortening velocity is zero for isometric (static) contraction; this occurs when the load is great enough so that the muscle cannot shorten. Such static or isometric loads are used often as measures of a person's maximum strength, since using zero velocity is a convenient way to standardize the tests. As the resistance is lightened, the muscle can shorten, and the velocity of muscle contraction increases. The shortening velocity is maximum when the load (tension developed) is zero. The maximum speed with zero load depends on the speed of contraction for the contractile units in the particular muscle examined (SO, FOG, and FG fiber types).

An investigation of this phenomenon noted that although our understanding of this process is incomplete, some type of neural regulatory mechanism may be involved in limiting the maximum voluntary tension levels, possibly as a safety mechanism for preventing anatomical injury (Perrine and Edgerton,

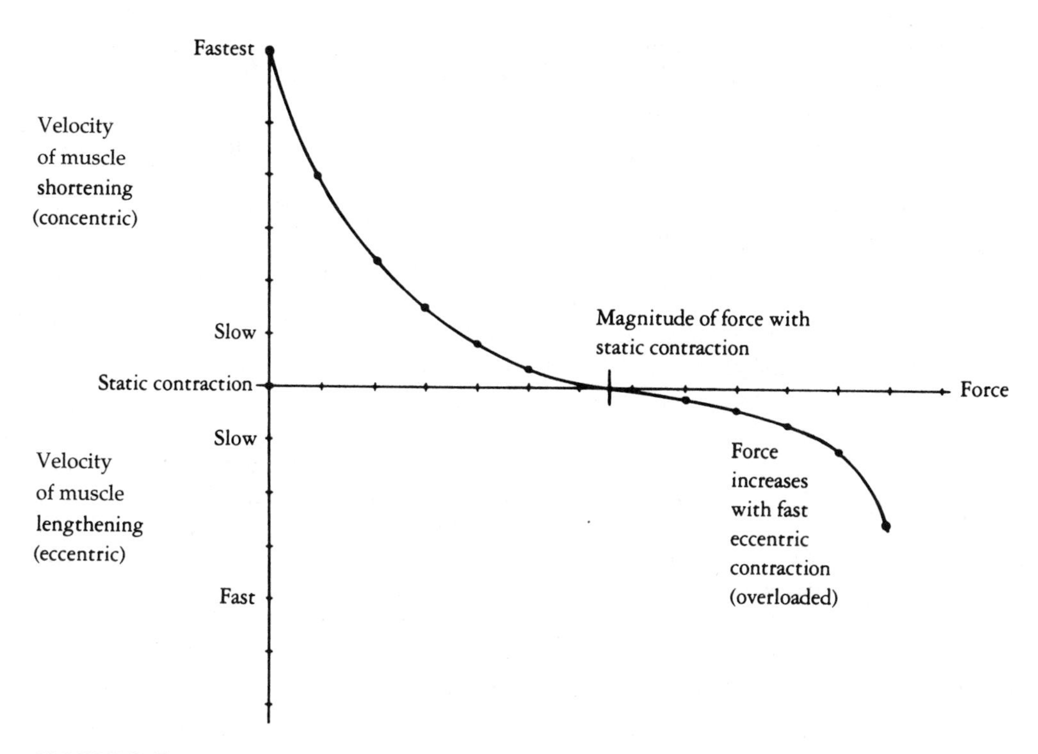

FIGURE 2.7
Force-velocity curve for an isolated muscle. The maximum contractile force is least for fast concentric tension and greatest for fast eccentric tension when overloaded.

1978). The possible limitation set by the force–velocity relationship is noteworthy in performances in which the speed of muscular contractions must cause a segment to keep up with or catch up to an object that is already moving to apply an accelerating force to that object. An example is striking a racquetball in a forward direction as it is moving forward from a rebound off the back wall. If the segments cannot move faster than the ball, there can be no impact force to accelerate the ball, that is, to force the ball to move faster than it is already moving.

Another example is trying to apply large forces to rapidly turning bicycle pedals. An interesting feature of the force–velocity relationship can be seen on the curve in Figure 2.7. When the muscle is "superloaded," thereby causing eccentric contraction against the moving load, the tension able to be generated increases with lengthening velocity, but to some limiting value. To generate muscle tension that is greater than the maximum isometric tension, the load must be greater than the isometric load that cannot be moved by the muscle; otherwise a decrease in tension is observed by the muscle simply acting with a submaximal controlling tension to control the speed of the load. Eccentric strength training, using supermaximal loads, has been incorporated into resistance-training programs of athletes, although such work is reported to generate the greatest amount of muscle soreness, which is probably due to some tissue damage.

Muscle power and *muscle strength* are terms often used incorrectly to refer to the same muscular ability. Strength refers only to the force production of a muscle. *Power* is the product of force and the speed with which that force is applied to move something (Power = Force × Velocity). For example, if you lifted your book 1 m high, and it took you 1 sec to do it, the power of your effort would be twice as great as it would be if you were to do it in 2 sec. If a force is applied to a resistance and no movement occurs, the power production is zero.

Muscle Power

The force–velocity relationship is an important consideration in the production of muscle power. Because an inverse relationship exists between the amount of force produced and the velocity of application, a trade-off between these two variables must occur. Because power equals force applied times the velocity with which that force is applied, some optimum speed of muscle shortening could be expected to produce the most powerful contraction. Such an optimum speed has been found to be approximately 25%–30% of the fastest speed of contraction (Astrand and Rodahl, 1977). At this rate, the force of contraction is approximately 30% of the maximal isometric force.

A simplified power-velocity curve is shown in Figure 2.8. It illustrates how power increases with increases in velocity only up to a certain point. After that "peak power" value, however, any increase in velocity causes a decrease in the maximum force the muscle can generate. The product of force × velocity increases on the left side of the curve leading up to a maximum value, and then the force ×

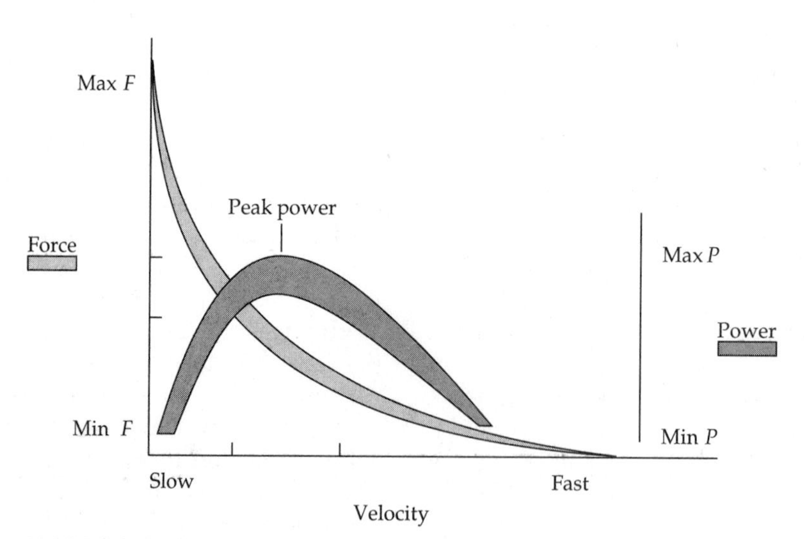

FIGURE 2.8

Interrelationship of force, velocity, and power, showing that peak power is produced by optimum force and optimum velocity, which vary from one muscle group to another.

velocity product decreases on the right side of the curve. The production of *maximum* power clearly depends on the production of *optimum* force and *optimum* velocity.

Muscular power is required in activities in which an object or the body must be forced to a high speed quickly. For example, bursts of muscular power are needed in the shot put, sprint starts, high-jump takeoff, segmental movements through water in competitive swimming, and dance leaps. An increase in power production can result from increasing strength or the speed of force application or both. More is said about muscular power in Chapters 7 and 16. For now, note that the power of muscular contraction varies with different combinations of muscle force and speeds of contraction.

Angle of Muscle Pull

A final consideration in determining the effectiveness of a muscle's tension is the angle at which the muscle pulls on the bone. The angle of pull influences the magnitude of muscle force needed to cause segmental movement. When a muscle's angle of attachment is 90 degrees to the bone on which it pulls, 100% of its force contributes to the bone's movement. Because most muscles cannot achieve a 90-degree attachment angle, the joint position for which the muscle's attachment angle is closest to 90 degrees provides the greatest rotary force for movement of the segment. A complete discussion of the reasons for force changes with changes in the muscle–bone angle is in Chapter 3.

Understanding the Neuromuscular System and Factors Affecting Muscular Tension

1. List the components of a motor unit.
2. List the ways in which a muscle may increase the total tension it can produce.
3. Identify the functional differences among SO, FOG, and FG muscle fibers.
4. Speculate about the relative proportions of fiber types in the muscles of championship performers: (a) the legs of a marathon runner, (b) the arms of a shot-putter, (c) the arms of a gymnast, (d) the arms of a baseball pitcher, (e) the legs of a downhill skier, (f) the legs of a cross-country skier, and (g) the legs of a mountain climber.
5. Explain how the elastic component in skeletal muscle is useful in movement.
6. List and describe three activities in which the force–velocity relationship of muscular contraction may be influential.
7. Describe three activities in which the length–tension relationship could influence performance.

2.5 Functional Aspects of the Sensorimotor System[3]

The Sensory Unit

The coordination of the muscular system, that is, the initiation of changes in individual muscles (e.g., contractions, relaxations, force adjustments), depends on a continuous supply of information about the immediate state of those muscles. Information on the position, length, tension, and speed of contraction of individual muscles is supplied through an extensive network of **sensory units** that are grossly similar in structure to the motor units. Each sensory unit consists of the following: (a) sensory **receptors**, which are located within various muscles and connective tissues (intrinsic) or cutaneous tissue (extrinsic), (b) an **axon**, and (c) a **nerve cell**. Some of the sensory systems provide the sensations necessary to produce appropriate muscular responses of a reflex nature. Others provide sensations to the brain to be integrated and acted on consciously and voluntarily by the performer depending on the purpose of the movement and the background experience of the performer. The mechanism by which the sensory system orders responses is similar to that of the motor system, but it works in a reverse manner. The variation of the strength or importance of the stimuli or both on the tissues determines the response of the sensory receptor as it transmits different strengths and frequencies of electrical impulses to the central nervous system (the spinal cord). There they either synapse with a motor nerve to produce an immediate movement or continue to travel to the brain, where they are integrated and sorted before being sent down the spinal cord to synapse with motor neurons to produce the appropriate movements. The former (immediate) motor responses are reflex in nature; the latter (consciously controlled) movements are voluntary in nature.

[3] The complexity of neuromuscular processes and sensorimotor control mechanisms preclude a detailed treatment of these aspects of movement within this text. Excellent coverage is provided by Schmidt (1988).

Proprioceptors

Proprioceptors are internal receptors that are located in and around the joints and muscles, in the skin, and in the inner ear. They respond to changes in position and acceleration of body segments. According to Schmidt (1988), the term *proprioception* traditionally has been used to mean the perception of the body's movement and its position in space whether it is still or moving. A term often used synonymously with proprioception is **kinesthesis**. As more is learned about the sensorimotor mechanisms, it appears that all sensory receptors can contribute to kinesthesis and that sense of position and movement is not restricted to only those classified as proprioceptors. Four of the more predominant types of proprioceptors used for movement perception are cutaneous receptors, joint receptors, musculotendinous receptors, and labyrinthine receptors.

CUTANEOUS RECEPTORS. In addition to specialized cutaneous receptors used for outside information (exteroceptors), those that sense deep pressure or touch and very light touch, including movement of hairs on the skin, serve as important proprioceptors for kinesthetic sense and an awareness of how the body is interacting with its environment.

JOINT RECEPTORS. Receptors located in the joint capsules and the surrounding ligaments respond to stimuli that produce sensations relating to positions, velocities, and accelerations occurring at the joints. The receptors are stimulated by pressure changes from within and surrounding the articulation and are sensitive to pressure changes resulting from as little as 2 degrees of change in joint position (O'Connell and Gardner, 1972). The receptors provide information about the position of the joint at any given time, velocity of the movement that may be occurring at that joint, and acceleration and deceleration of segmental movements.

Pressure receptors within joints, particularly in the lower extremity and especially in the foot, add to the sensations provided by the cutaneous receptors and are used for postural adjustments. Pressure changes within the intervertebral joints stimulate the muscles that maintain the upright position of the trunk and contribute to the balance of the body's weight. The distribution of the body's weight from one limb to the other has important implications for single-limb weight bearing. The pressure changes within the supporting hip joint when the other limb is lifted off the ground provide sensory information that stimulates the stabilizing muscles surrounding the hip joint. The body responds by pulling the body weight over the supporting limb so that the unbalanced body weight does not pull the body over to the unsupported side. The changes in the hip joint position then stimulate, through sensory pathways, the surrounding muscles to help lift the pelvis so that it does not tilt to the unsupported side.

Velocities and accelerations of body segments also serve to stimulate the joint receptors, and the sensations are frequently used to adjust the speed of these segmental movements. The most obvious example concerns an object that is perceived to be heavier than it actually is. When the object is lifted, too many motor units are activated in anticipation of the perceived weight, and the object is lifted much faster and farther than expected or desired. Sensing the inappropriate speed of the movement, the joint receptors send impulses to the brain to indicate the incorrect response. The brain then sends impulses down through the motor pathways to correct the error.

A similar feedback system is used in producing and correcting sport skill performance. Skilled performers can sense, by way of the joint receptors, movements that are too fast, too slow, or in the wrong direction, and consequently, correct their future responses. The skilled free-throw shooter can sense the success of a shot as the ball leaves the hand. The skilled golfer can sense the accuracy of a chip shot as the ball leaves the club face. On the other hand, the beginner has nothing with which to compare the sensations; thus performance becomes a trial-and-error procedure until adequate sensations of a successful performance are established as one passes through the intermediate and advanced stages of learning.

MUSCULOTENDINOUS RECEPTORS. A third type of proprioceptor used in muscular control and coordination is the musculotendinous receptor. There are two such types of receptors—the tendon receptors and the muscle receptors.

Specialized receptors located between the muscle and its tendon are called the **Golgi tendon organs**. These receptors are sensitive to the stretch of the muscle tendon produced by the *contraction* of its associated muscle (active stretch). Their response to this stretch is to inhibit the contraction of the associated muscle (autogenic inhibition) and thus relieve the tension produced at the musculotendinous junctions and to excite the antagonistic muscle group. This inhibition of the contraction of its muscle has important implications for physical activity and conditioning programs. The Golgi receptors in the tendon of a strongly contracted muscle respond by inhibiting the contraction of that muscle. Such a muscle can be stretched fully more easily than a muscle not under the influence of this response. Second, since the Golgi organs facilitate the contraction of the antagonists, the contraction of a muscle group to stimulate the Golgi organs enhances the contraction of the opposite muscle group. One use of these receptors may be when an individual attempts to stretch a muscle fully. A strong static (isometric) contraction of the muscle just before the stretch may stimulate the Golgi tendon organs and thus inhibit the muscle and enable it to be stretched to a greater degree. Flexibility training using proprioceptive neuromuscular facilitation (PNF) is discussed in Chapter 7.

If one wants a strong contraction of muscle group A, then one should contract the antagonistic group B just before the desired contraction of A to stimulate the Golgi tendon organs in group B; this Golgi stimulation will have an excitatory effect on muscle group A, enhancing its contraction.

Findings of research presented by Schmidt (1988) indicate that the Golgi tendon organs are sensitive detectors of tension in distinct, localized regions of their host muscle. As sensors of degrees of tension within parts of the muscle, they are probably important contributors to fine motor control, as well as protective agents.

The **muscle spindle** is a second type of musculotendinous receptor. Muscle spindles are located throughout the muscle and are situated between and parallel to individual muscle fibers. In this position they are stretched when their adjacent muscle fibers are stretched. When the whole muscle is stretched, the stretch of the muscle spindles causes a reflex contraction of their host muscle. This response is called the *myotatic* or *stretch reflex*. The knee-jerk reflex is an example.

The internal structures and functions of the muscle spindle are highly complex and are not yet fully understood. By no means should muscle spindles be remembered only for their elicitation of the stretch reflex. On the contrary, their functional substructure provides for constant monitoring and regulation of sensorimotor function that enables appropriate body movement, both reflexively and voluntarily. The role of muscle spindles in motor control is described in detail by Schmidt (1988). Although muscle spindles previously were not thought to contribute to the conscious perception of position or movement, recent work indicates that the spindles can be sources of information that contribute to kinesthesis when combined with other sensory input (Schmidt, 1988).

Muscle spindles respond to either active or passive stretch and are sensitive to the speed of the stretch; that is, the faster the stretch, the stronger the response of muscle contraction. The spindles provide information on the length of the muscle as well as the speed of the stretch. Evidence exists also that the acceleration (rate of speed increase or decrease) of spindle stretch and the speed of stretch are directly related to the stretch reflex response (Perot and Goubel, 1981). Because the muscle contracts more strongly immediately after it is put on stretch, the spindles also cause the inhibition of the antagonists, which is a process called **reciprocal inhibition**. Both the stretch reflex and the reciprocal inhibition of the antagonistic muscle group are used effectively in activities in which a strong contraction is desired.

A great contribution of the spindle's response to stretch in sport and dance activities is how it prepares the muscle for using its elastic energy. The work of Nichols and Houk (1976), cited by Schmidt (1988), provides evidence that the reflex muscle contraction caused by the stretch reflex (as in a backswing or in shock absorption) causes the contracted state necessary for a subsequent quick short stretch to store elastic energy in the muscle. A degree of stiffness is thereby provided the muscle to be stretched again so that it can recoil to help accelerate the movement.

Examples of use of the stretch reflex together with the recoil of stored elastic energy are found in throwing and striking activities in which a preparatory phase stretches the muscles that are going to do the striking or throwing just before their contraction. A slight movement of all segments to the right just before a swing of the bat in baseball, a long backswing in golf, or the backswing of a bowler not only gives the muscles more ROM during the force phase of the movement but also puts the motive muscles on stretch, thus enhancing their contractile force as well as facilitating the recoil of elastic tissue. For example, picture the stretch of the shoulder joint muscles of a baseball pitcher as he takes his windup. During the initial forward phase of the throw, the rotation of the trunk forward and the lagging back of the upper extremity put the anterior shoulder muscles on stretch just before the contraction of the muscles to bring the arm forward and consequently stimulate their contraction before the final quick stretch. In this manner, the forward speed of the arm is greater when the pitcher throws the ball.

Integrated Sensorimotor Responses

Posture and Equilibrium[4]

Body orientation and balance depend on a large combination of sensory units. The reflex nature of some movements initiated by stimulation of the cutaneous receptors include the righting reflexes used in balance activities and the postural reflexes used in balance and upright standing. Reflexes are involuntary sensory motor responses, although an individual is usually aware of such a movement as or after it happens.

[4] For a more detailed discussion of the postural and equilibrium responses, the reader is referred to Goodwin (1976), and Gowitzke and Milner (1988).

As a child begins to assume upright posture, the righting reflexes develop to help stimulate the appropriate muscles in the neck and trunk to assist in aligning the body segments correctly with gravity, and thus the child is able to sit and eventually stand. The joint receptors in the neck respond to positional orientations of the head and neck as well as to neck movements. Once the body can be held upright, other reflexes must predominate so that the body can maintain equilibrium. The labyrinthine receptors respond to movements of the head about all axes and in all directions. The position of the head is important in maintaining postural equilibrium, and the impulses generated provide the postural adjustments to body sway while standing.

Another reflex is the extensor thrust reflex, which elicits an extension reaction in the lower extremity when pressure is applied to the sole of the foot. This reflex may help in the extension of the lower extremity that is necessary in activities such as diving, rebound tumbling, running, and jumping. The extensor thrust reflex is one of the important reflexes used in postural alignment in standing, during which a continual sway of the body produces variations in pressure on the soles of the feet and thus results in motor adjustments for balance.

The equilibrium reflexes are aided by the stretch reflex and other regulatory information provided by the muscle spindles and Golgi tendon organs of the antigravity muscles, those extensor muscles of the trunk and lower extremity that counteract the force of gravity and prevent the body from collapsing or tipping over. As one sways while standing, the muscles antagonistic to the direction of sway are stretched, thus stimulating the muscle spindles, which cause a reflex contraction of those muscles to bring the body back into alignment. Further help is given by the cutaneous receptors in the soles of the feet, which respond to alterations in pressure from the ground, and by the joint receptors in the foot and ankle. External pressures and intrajoint pressures stimulate these receptors and allow a person to make appropriate adjustments to uneven terrain.

We can see that a complicated set of rules govern the equilibrium responses; however, in summary, these responses to movement stimulate the appropriate muscles so as to maintain the balance of the body and prevent it from tipping over.

Skilled Movement

Much of the awareness of one's own body position and movements is provided by visual feedback. For example, when one moves a body part, one can usually see the movement relative to the environment and thus judge its correctness and accuracy. The information provided by sensations from the cutaneous, joint, and muscle receptors is called kinesthetic sense, or kinesthesis. It is an important capability that is well developed in athletes and that increases the performance quality of daily activities of the nonathlete.

To develop a frame of reference for kinesthetic sensations, an individual must experience a variety of positions and movements in many environments. The perceptions associated with those situations are then stored in the brain for use as a reference or comparator for future movements. Precise and accurate motor responses are developed from the integration of information from the cutaneous, joint, tendon, muscle, and labyrinthine receptors. Visual cues may reinforce or be a means for correcting a kinesthetic perception; that is, if one stands in front of a mirror and views one's body positions, one can develop an awareness of the sensations produced when a given body position is assumed. The sensations are provided by the joint and musculotendinous receptors and are associated with the visual information provided by the mirrors. Correctness or incorrectness of a prescribed position (e.g., a fencing en garde position) may then be determined.

Teachers and coaches of movement skills teach their students to rely heavily on visual feedback, but it is important also to stress the kinesthetic feedback so that movement correctness is felt as well as seen.

Understanding the Sensorimotor System

1. Select a sports activity in which the extensor thrust reflex plays an important part, and describe how it does so.
2. Describe how the Golgi tendon organs facilitate and inhibit muscular contractions. Cite a sport-specific example.
3. Describe how the muscle spindles facilitate and inhibit muscular contractions. Cite a sport-specific example.
4. Stand in front of a large mirror (or facing a partner), close your eyes, and abduct your shoulder to bring your arm to a horizontal position. Open your eyes, and check your arm's position. Repeat with your other arm. What information did you depend upon to know where your arm was if you could not see it? Experiment with other segment positionings to qualitatively evaluate the accuracy of your kinesthetic information.
5. Describe two sport situations in which a teacher or coach may want to teach the performer to use information provided by kinesthesis.

References and Suggested Readings

Asmussen, E., & Bonde-Petersen, F. (1974). Storage of elastic energy in skeletal muscles in man. *Acta Physiologica Scandinavica, 91,* 385–392.

Astrand, P. & Rodahl, K. (1977). *Textbook of work physiology*. New York: McGraw-Hill.

Aura, O., & Komi, P. V. (1987). Coupling time in stretch-shortening cycle: Influence on mechanical efficiency and elastic characteristics of leg extensor muscles. In B. Jonsson (Ed.), *Biomechanics X-A* (pp. 507–512). Champaign, IL: Human Kinetics.

Basmajian, J. V., & MacConaill, M. A. (1977). *Muscles and movements—A basis for human kinesiology*. Huntington, NY: Robert E. Krieger Publishing Co.

Bosco, C. (1987). Mechanical delay and recoil of elastic energy in slow and fast types of human skeletal muscles. In B. Jonsson (Ed.), *Biomechanics X-B* (pp. 979–984). Champaign, IL: Human Kinetics.

Burke, R. E., & Edgerton, V. R. (1975). Motor unit properties. In J. H. Wilmore & J. F. Keogh, (Eds.), *Exercise and sports sciences reviews* (Vol. 3, pp. 31–81). Santa Barbara, CA: Journal Publishing Affiliates.

Caiozzo, V. J., Perrine, J. J., & Edgerton, V. R. (1980). Alterations in the in vivo force–velocity relationship. *Medicine and Science in Sports and Exercise, 12*(2), 134.

Campbell, C. J., Bonen, A., Kirby, R. L., & Belcastro, A. N. (1979). Muscle fiber composition and performance capacities of women. *Medicine and Science in Sports, 11*(3), 260–265.

Cavagna, G. A. (1977). Storage and utilization of elastic energy in skeletal muscle. In R. S. Hutton (Ed.), *Exercise and sports sciences reviews* (pp. 82–129). Philadelphia: Franklin Press.

Cavagna, G. A., Dusman, B., & Margaria, R. (1968). Positive work done by a previously stretched muscle. *Journal of Applied Physiology, 24,* 21–32.

Chapman, A. E. (1985). The mechanical properties of human muscle. In R. L. Terjung (Ed.), *Exercise and sports sciences reviews* (pp. 443–501). New York: Macmillan.

Chapman, A. E., & Caldwell, G. E. (1985). The use of muscle stretch in inertial loading. In D. A. Winter, R. W. Norman, R. P. Wells, K. C. Hayes, & A. E. Patla (Eds.), *Biomechanics IX-A* (pp. 44–49). Champaign, IL: Human Kinetics.

Dapena, J., & Chung, C. S. (1988). Vertical and radial motions of the body during the take-off phase of high jumping. *Medicine and Science in Sports and Exercise, 20*(3), 290–302.

Denoth, J. (1985). Storage and utilization of elastic energy in musculature. In D. A. Winter, R. W. Norman, R. P. Wells, K. C. Hayes, & A. E. Patla (Eds.), *Biomechanics IX-A* (pp. 65–70). Champaign, IL: Human Kinetics.

deVries, H. A. (1980). *Physiology of exercise*. Dubuque, IA: William C. Brown Co.

Edman, K. A. P., Elzinga, G., & Noble, M. I. M. (1982). Residual force enhancement after stretch of contracting frog single muscle fibers. *Journal of General Physiology, 80,* 769–784.

Eriksson, E. (1976). Sports injuries of the knee ligaments: Their diagnosis, treatment, rehabilitation, and prevention. *Medicine and Science in Sports, 8,* 133–144.

Funato, K., Ohmichi, H., & Miyashita, M. (1985). Electromyographic analysis on utilization of elastic energy in human leg muscles. In D. A. Winter, R. W. Norman, R. P. Wells, K. C. Hayes, & A. E. Patla (Eds.), *Biomechanics IX-A* (pp. 60–64). Champaign, IL: Human Kinetics.

Gans, C. (1982). Fiber architecture and muscle function. In R. L. Terjung (Ed.), *Exercise and sports sciences reviews* (Vol. 10, pp. 160–207). Philadelphia: Franklin Institute Press.

Garrett, W. E., Nikolaoi, P. K., Ribberck, B. M., Glisson, R. R., & Seaber, A. V. (1985). The effect of muscle architecture on the biomechanical failure properties of skeletal muscle under passive extension. *The American Journal of Sports Medicine, 16,* 7–12.

Gollnick, P. D., Armstrong, R. B., Saubert, C. W., Piehl, K., & Saltin, B. (1972). Enzyme activity and fiber composition in skeletal muscle of trained and untrained men. *Journal of Applied Physiology, 24,* 312–319.

Goodwin, G. M. (1976). The sense of limb position and movement. In J. F. Keogh & R. S. Hutton (Eds.), *Exercise and sports sciences reviews* (Vol. 4, pp. 87–124). Santa Barbara, CA: Journal Publishing Affiliates.

Gowitzke, B. A., & Milner, M. (1988). *Understanding the scientific bases of human movement* (3rd ed.). Baltimore: Williams & Wilkins.

Grabiner, M. D. (1985). Elastic responses of isotonically contracting skeletal muscle. In D. A. Winter, R. W. Norman, R. P. Wells, K. C. Hayes, & A. E. Patla (Eds.), *Biomechanics IX-A* (pp. 55–69). Champaign, IL: Human Kinetics.

Gray, H. (1966). *Anatomy of the human body* (28th ed.) (C. M. Goss, Ed.). Philadelphia: Lea & Febiger.

Harris, F. A. (1978). Facilitation techniques in therapeutic exercise. In J. V. Basmajian (Ed.), *Therapeutic exercise* (3d ed.), (pp. 93–137). Baltimore: Williams & Wilkins.

Hayes, K. C., Hayes, J. D. S., & Vandervoort, A. A. (1987). Time-dependent nonlinearities of the stretch reflex. In B. Jonsson (Ed.), *Biomechanics X-A* (pp. 485–490). Champaign, IL: Human Kinetics.

Heerkens, Y. F., Woittiez, R. D., Huijing, P. A., van Ingen Schenau, G. J., & Rozendal, R. H. (1987). Passive knee resistance as a function of actual muscle length. In B. Jonsson (Ed.), *Biomechanics X-B* (pp. 985–988). Champaign, IL: Human Kinetics.

Herzog, W. (1987). Considerations for predicting individual muscle forces in athletic movements. *International Journal of Sport Biomechanics, 3,* 128–141.

Hof, A. L., Geelen, B. A., & VanDen Berg, J. W. (1983). Calf muscle moment, work and efficiency in level walking; role of series elasticity. *Journal of Biomechanics, 16*(7), 523–537.

Homma, S. (1983). The stretch reflex as motor control. In H. Matsui & K. Kobayashi (Eds.), *Biomechanics VIII-A: Proceedings of the Eighth International Congress on Biomechanics* (pp. 189–205). Champaign, IL: Human Kinetics.

Hudson, J. L., & Owen, M. G. (1985). Performance of females with respect to males: the use of stored elastic energy. In D. A. Winter, R. W. Norman, R. P. Wells, K. C. Hayes, & A. E. Patla (Eds.), *Biomechanics IX-A* (pp. 50–54). Champaign, IL: Human Kinetics.

Huijing, P. A., Vossen, P. C., Rijnsburger, W. H., & Woittiez, R. D. (1987). Range of length for active force generation and *in situ* length range of human m. soleus and its fibers during maximal ankle excursion. In B. Jonsson (Ed.), *Biomechanics X-B* (pp. 973–978). Champaign, IL: Human Kinetics.

Hutton, R. S., & Nelson, D. L. (1986). Stretch sensitivity of Golgi tendon organs in fatigued gastrocnemius muscle. *Medicine and Science in Sports and Exercise, 18*(1), 69–74.

Ianuzzo, C. D. (1976). The cellular composition of human skeletal muscles. In H. G. Knuttgen (Ed.), *Neuromuscular mechanisms for therapeutic and conditioning exercise* (pp. 31–53). Baltimore: University Park Press.

Ikegami, Y., Miura, M., Matsui, H., & Hashimoto, I. (1981). Biomechanical analysis of the javelin throw. In A. Morecki, K. Fidelus, K. Kedzior, & A. Wit (Eds.), *Biomechanics VII-B: Proceedings of the Seventh International Congress of Biomechanics* (pp. 271–276). Baltimore: University Park Press.

Jaric, S., Ristanovic, D., Gavrilovic, P., & Ivancevic, V. (1985). A new method for determining the force–velocity relationship in human quadriceps muscle. In D. A. Winter, R. W. Norman, R. P. Wells, K. C. Hayes, & A. E. Patla (Eds.), *Biomechanics IX-A* (pp. 82–85). Champaign, IL: Human Kinetics.

Jones, N. L., McCartney, N., & McComas, A. J. (1986). *Human muscle power*. Champaign, IL: Human Kinetics.

Katsuta, S., & Takamatsu, K. (1987). Estimation of muscle fiber composition using performance tests. In B. Jonsson (Ed.), *Biomechanics X-B* (pp. 989–994). Champaign, IL: Human Kinetics.

Knuttgen, H. G. (1986). Quantifying exercise performance with SI units. *The Physician and Sportsmedicine 14,* 157–161.

Knuttgen, H. G., & Kraemer, W. J. (1987). Terminology and measurement in exercise performance. *Journal of Applied Sport Science Research, 1,* 1–10.

Komi, P. V. (1984). Biomechanics and neuromuscular performance. *Medicine and Science in Sports and Exercise, 16*(1), 26–28.

Komi, P. V. (1984b). Physiological and biomechanical correlates of muscle function: effects of muscle structure and stretch–shortening cycle on force and speed. In R. L. Terjung (Ed.), *Exercise and sport sciences reviews* (pp. 81–121). Lexington, MA: Collamore.

Komi, P. V. (1986). The stretch-shortening cycle & human power output. In N. L. Jones, N. McCartney, & A. J. McComas (Eds.), *Human muscle power* (pp. 27–39). Champaign, IL: Human Kinetics.

Komi, P. V., & Bosco, C. (1978). Utilization of elastic energy in jumping and its relation to skeletal muscle fiber composition in man. In E. Asmussen & K. Jorgensen (Eds.), *Biomechanics VI-A: Proceedings of the Sixth International Congress of Biomechanics* (pp. 79–85). Baltimore: University Park Press.

Kovanen, V., Suominen, H., & Heikkinen, E. (1984). Mechanical properties of fast and slow skeletal muscle with special reference to collagen and endurance training. *Journal of Biomechanics, 17,* 725–736.

Lamontagne, M., Dore, R., Drouin, G., & Meunier, A. (1987). Mechanical properties of tendons measured by strain gauges for different loading angles. In B. Jonsson (Ed.), *Biomechanics X-B* (pp. 955–960). Champaign, IL: Human Kinetics.

Muro, M., & Nagata, A. (1985). The effects of training on force–velocity characteristics. In D. A. Winter, R. W. Norman, R. P. Wells, K. C. Hayes, & A. E. Patla (Eds.), *Biomechanics IX-A* (pp. 86–90). Champaign, IL: Human Kinetics.

Nichols, T. R., & Houk, J. C. (1976). The improvement of linearity and the regulation of stiffness that results from the actions of the stretch reflex. *Journal of Neurophysiology, 39,* 119–142.

O'Connell, A., & Gardner, E. B. (1972). *Understanding the scientific bases of human movement.* Baltimore: Williams & Wilkins.

Osternig, L. R., et al. (1983). Function of limb speed on torque patterns of antagonist muscles. In H. Matsui & K. Kobayashi (Eds.), *Biomechanics VIII-A: Proceedings of the Eighth International Congress on Biomechanics* (pp. 251–257). Champaign, IL: Human Kinetics.

Otten, E. (1985). Morphometrics and force–length relations. In D. A. Winter, R. W. Norman, R. P. Wells, K. C. Hayes, & A. E. Patla (Eds.), *Biomechanics IX-A* (pp. 27–32). Champaign, IL: Human Kinetics.

Otten, E. (1988). Concepts and models of functional architecture in skeletal muscle. In K. B. Pandolf (Ed.), *Exercise and sport sciences reviews* (pp. 89–137). New York: Macmillan.

Perot, C., & Goubel, F. (1981). Control mechanism of bifunctional muscles: a reflex approach. In A. Morecki, K. Fidelus, K. Kedzior, & A. Wit (Eds.), *Biomechanics VII-A: Proceedings of the Seventh International Congress of Biomechanics* (pp. 84–88). Baltimore: University Park Press.

Perrine, J. L., & Edgerton, V. R. (1978). Muscle force–velocity and power–velocity relationships under isokinetic loading. *Medicine and Science in Sport, 10*(3), 159–166.

Pierrynowski, M. R., & Morrison, J. B. (1985). Length and velocity patterns of the human locomotor muscles. In D. A. Winter, R. W. Norman, R. P. Wells, K. C. Hayes, & A. E. Patla (Eds.), *Biomechanics IX-A* (pp. 23–38). Champaign, IL: Human Kinetics.

Powers, W. (1976). Nervous system control of muscular activity. In H. G. Knuttgen (Ed.), *Neuromuscular mechanisms for therapeutic and conditioning exercise* (pp. 1–30). Baltimore: University Park Press.

Schmidt, R. A. (1988). *Motor control and learning.* Champaign, IL: Human Kinetics.

Shi-Ping M. & Zahalak, G. I. (1985). The mechanical response of the active human triceps brachii muscle to very rapid stretch and shortening. *Journal of Biomechanics, 18,* 585–598.

Tsarouchas, E., & Klissouras, V. (1981). The force–velocity relation of a kinematic chain in man. In A. Morecki, K. Fidelus, K. Kedzior, & A. Wit (Eds.), *Biomechanics VII-A: Proceedings of the Seventh International Congress of Biomechanics* (pp. 145–151). Baltimore: University Park Press.

Van Leemputte, M., Spaepen, A. J., Willems, E. J., & Stijnen, V. V. (1983). Influence of pre-stretch on arm flexion. In H. Matsui & K. Kobayashi (Eds.), *Biomechanics VIII-A: Proceedings of the Eighth International Congress on Biomechanics* (pp. 264–270). Champaign, IL: Human Kinetics.

Van Leemputte, M., Spaepen, A. J., & Willems, E. J. (1987). Effect of the contraction history on the force–velocity relation of elbow flexors. In B. Jonsson (Ed.), *Biomechanics X-A* (pp. 271–276). Champaign, IL: Human Kinetics.

Viitasalo, J. T., & Komi, P. V. (1981). Rate of force development, muscle structure and fatigue. In A. Morecki, K. Fidelus, K. Kedzior, & A. Wit (Eds.), *Biomechanics VII-A: Proceedings of the Seventh International Congress on Biomechanics* (pp. 136–141). Baltimore: University Park Press.

Wells, R. P. (1988). Mechanical energy costs of human movement: an approach to evaluating the transfer possibilities of two-joint muscles. *Journal of Biomechanics, 21,* 955–964.

Woittiez, R. D., Rozendal, R. H., & Huijing, P. A. (1985). The functional significance of architecture of the human triceps surae muscle. In D. A.

Winter, R. W. Norman, R. P. Wells, K. C. Hayes, & A. E. Patla (Eds.), *Biomechanics IX-A* (pp. 21–26). Champaign, IL: Human Kinetics.

Woittiez, R. D., Huijing, P. A., Bobbert, M. F., Rozendal, R. H., & Rijnsburger, W. H. (1987). The power and force of slow and fast muscles during concentric and eccentric contractions. In B. Jonsson (Ed.), *Biomechanics X-A* (pp. 529–534). Champaign, IL: Human Kinetics.

Wolf, S. (1978). The morphological and functional basis of therapeutic exercise. In J. V. Basmajian (Ed.), *Therapeutic exercise* (3rd ed.), (pp. 43–85). Baltimore: Williams & Wilkins.

Wood, G. A., Lockwood, R. J., & Mastaglia, F. L. (1985). Reflex activity changes with human muscular strength development. In D. A. Winter, R. W. Norman, R. P. Wells, K. C. Hayes, & A. E. Patla (Eds.)., *Biomechanics IX-A* (pp. 368–371). Champaign, IL: Human Kinetics.

Wood, G. A., Pyke, F. S., LeRossignol, P. F., & Munro, A. R. (1983). Neuromuscular adaptation to training. In H. Matsui & K. Kobayashi (Eds.), *Biomechanics VIII-A: Proceedings of the Eighth International Congress on Biomechanics* (pp. 306–311). Champaign, IL: Human Kinetics.

Yamazaki, Y., Mano, T., Mitarai, G., & Kito N. (1985). Contribution of stretch reflex to knee bouncing movement. In D. A. Winter, R. W. Norman, R. P. Wells, K. C. Hayes, & A. E. Patla (Eds.), *Biomechanics IX-A* (pp. 346–351). Champaign, IL: Human Kinetics.

CONCEPT MODULE C

Looking at Movement: Some Mechanical Concepts

CONCEPT MODULE C

PREREQUISITES

Chapters 1, 2
Concept Module B

CONCEPT MODULE CONTENTS

C.1 The Movement of a System Within a Frame of Reference

The concepts related to force and the motion of bodies form the foundation for examining movement. Whether the movement is that of the human body or of some object or group of objects in the environment, its analysis should be based on an understanding of the ways in which things can move and how their movements are affected by different forces. A first step in examining movement is to identify the system. A **system** is a body or group of bodies or objects whose motion is to be examined; it might be an arm, a leg, a whole body, a pair of football players, a tennis ball, or a racket. To analyze the motion of the hand, for example, we would picture the hand as a segment subjected to the forces of muscle pull, the bone contact forces at the wrist joint, gravity, and other environmental forces acting on it. Figure C.1 shows the hand isolated as the system to be examined.

We also need to identify some frame of reference within which a system's

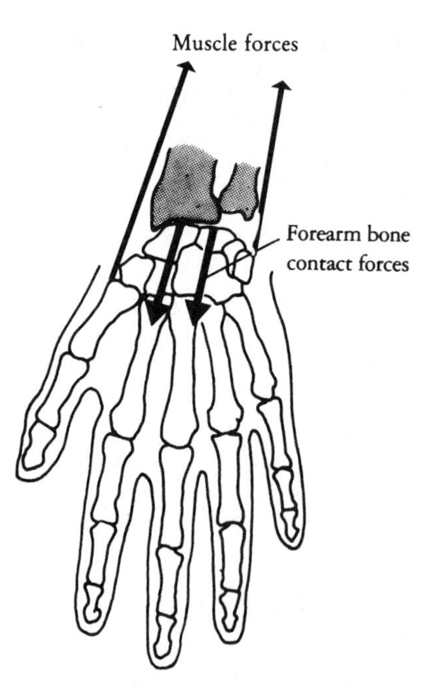

FIGURE C.1

The hand–wrist segment visualized as a system. Muscle forces acting across the wrist joint are external forces. The forearm bones contacting the wrist bones are also external forces acting on the system.

movement takes place. The frame of reference within which we view movement can be the stationary environment or something that is also moving. Using something stationary as a frame of reference is the least complicated. For example, we can describe the motion of a runner in terms of the body's change in location relative to the ground or relative to space. If a body segment is the system being studied, the frame of reference might be an adjacent segment, the midline of the body, a point on the head, or some other reference point on the body or in the environment. In the chin-up shown in Figure C.2a, the hand moves relative to the shoulder, but the hand does not move relative to the stationary environment (the bar).

The same arm movement occurs when a chin-up is simulated without a bar. The hand moves relative to the shoulder, but it also moves relative to the stationary environment (Figure C.2b). The body does not move relative to the bar as it did in the actual chin-up, however. Recognizing such relative motion of different parts of the body is important in analyzing movements of the body, its segments, and objects in the environment.

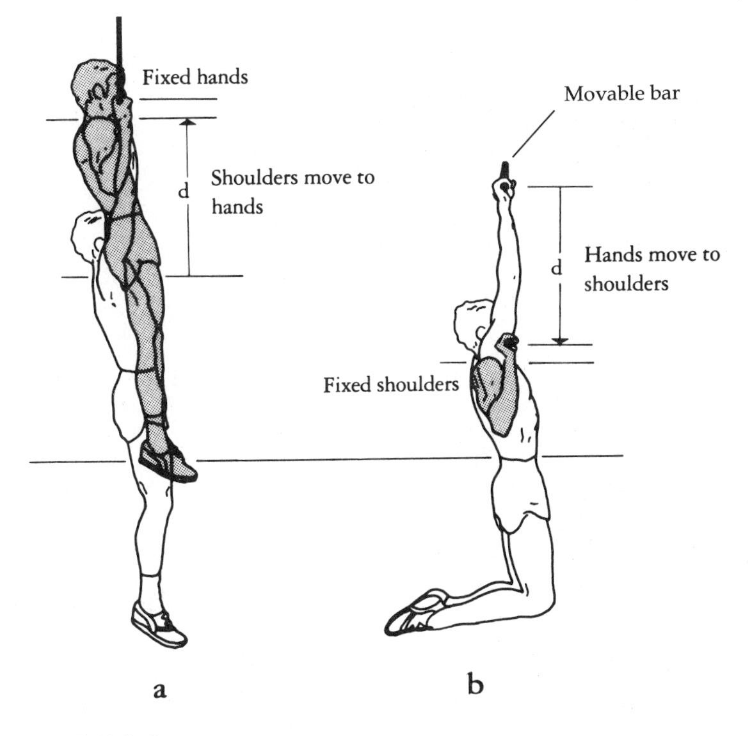

FIGURE C.2

Motion occurs relative to a frame of reference. (a) The shoulders are pulled up to the stationary hands. (b) The hands and movable bar are pulled down to the stationary shoulders with the same elbow and shoulder joint movements.

Orientation of the Moving Body in Space

Directions of movement may also be identified according to their relationship with the ground. For example, the person in Figure C.2a may be described in terms of the body's change in location relative to the ground. The axes and planes of movement are then identified as **spatial**. Any plane or axis that is parallel to the ground is identified as a **horizontal plane** or **axis**. Any plane or axis that is perpendicular to the ground is identified as a **vertical plane** or **axis**.

Because there are three dimensions in space, there are three planes and axes so named. The two axes that are perpendicular to each other and parallel to the ground are designated as the *x* and *y* axes. The *x* axis is parallel to the ground and directed forward and backward relative to the performer. Forward is designated as positive, and backward is designated as negative. The *y* axis is parallel to the ground and directed right and left relative to the performer. The left direction is positive, and the right direction is negative. The third axis, which is perpendicular to the other two, is vertical and is designated as the *z* axis. The upward direction is positive, and the downward direction is negative.

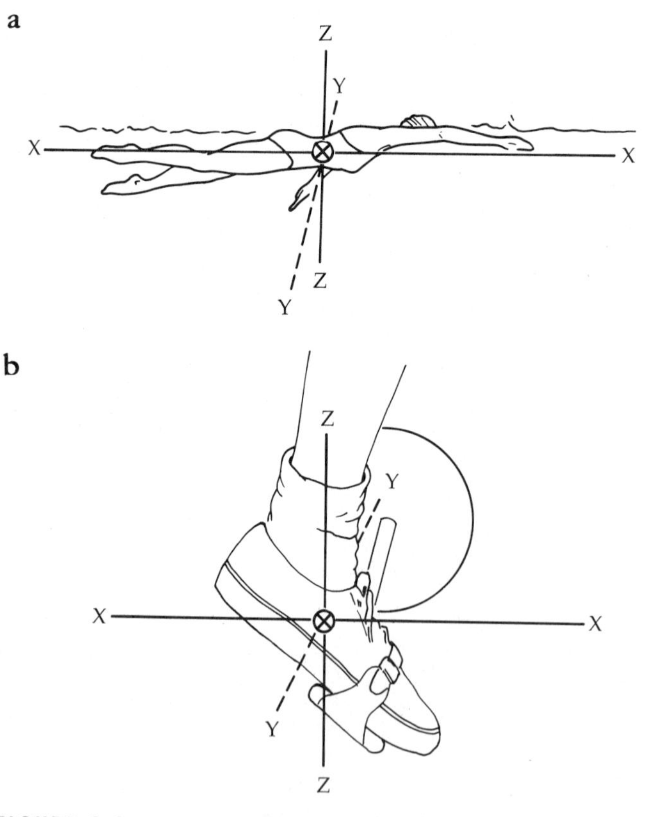

FIGURE C.3

Orientation of (a) a swimmer's body and (b) a cyclist's foot relative to spatial *x*, *y*, and *z* axes.

The x, y, and z axes may be used to decribe the direction of movement of the body or an object. For instance, the path of a golf ball may be described in this manner: The ball moves along the fairway in the positive x direction but may drop in the negative z direction or slice in the negative y direction.

When the body is standing, the longitudinal body axis corresponds to the spatial z axis, the ML body axis corresponds to the spatial y axis, and the AP body axis corresponds to the spatial x axis. When the body is lying down, as in swimming, the spatial x axis corresponds to the body's longitudinal axis. These spatial axes are shown in Figure C.3 relative to a cyclist's foot and a swimmer's body.

The spatial axes do not change with the body or object changing orientations in space. Rather, they are fixed, relative to the ground and the observer. In cases in which the body is tumbling or rotating, designating both the spatial axis and the body's axis is important for analyzing both segmental and total body movement in space. An understanding of the spatial, body, and segmental axes of movement facilitates description of any movement of the body or its segments occurring during human movement and sport activities.

Some basic definitions for quantities related to human movement activities are presented in the sections that follow. Measurements of quantities are made in British or metric units; Appendix I contains British and metric equivalents. The **Glossary** is a handy source for quickly reviewing definitions of common biomechanical terms.

C.2 Types of Motion

Linear Motion

When a system is forced to move in a path that is a line, it exhibits linear motion, that is, a change in location from one place to another within a spatial frame of reference. If the path is a straight line, it is called *rectilinear*, and if the path is curved, it is called *curvilinear*. For example, a dropped ball travels a rectilinear path toward the ground, or a jumper travels a curvilinear path over a high bar. The path of a pencil along a ruler is rectilinear; the path of a thrown discus moving through the air is curvilinear. Using only the word *linear* to mean rectilinear is customary.

The distance (**d**) that a system moves in a straight line is measured in linear *measurement units* such as meters, feet, centimeters, inches, kilometers, or miles, and is called the linear **displacement** of a system.

Rotary Motion

If some point within a system is restricted or secured so that the system rotates around this point when it receives a turning force, that point serves as an **axis**

of rotation, and the motion is called rotary or angular motion. The turning force responsible for rotation is called a **torque** and will be discussed in detail in Concept Module E. In the human body, each segment is connected to one or more adjacent segments to form joints. The joints serve as the locations of the axes of rotation for the body segments. For instance, if you hold your elbow in 80 degrees of flexion and receive a weight in your hand, the weight does not cause the entire forearm segment to move linearly downward because it is attached to the rest of the body or restricted from moving at the elbow joint, which provides resistance to downward motion of the elbow end of the forearm; the hand end of the forearm is free to move, however, and consequently rotates about the ML axis of the elbow joint (Figure C.4a).

The change in location of a rotating body is called its **angular displacement** and is designated by the Greek letter theta, θ. The path of a rotating body is measured in angular measurement units such as revolutions, degrees, and radians. (A radian is a proportion of a circle and is equivalent to approximately 57.3 degrees.)[1]

EQUATION C.1

$$\text{radians} = \frac{\theta \text{ in degrees}}{57.3 \text{ degrees}}$$

where θ is angular displacement (in degrees).

[1] For more information about the radian, see Appendix V.

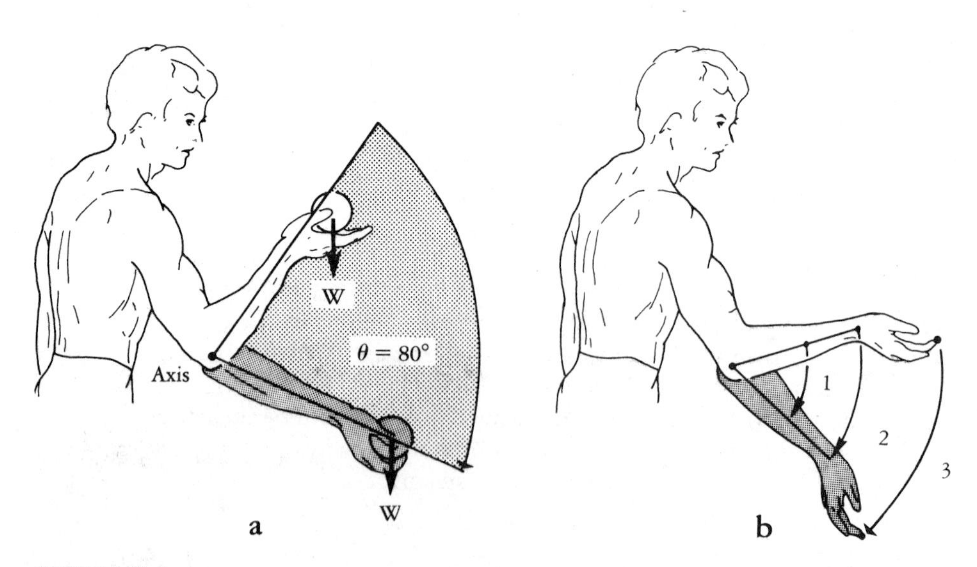

FIGURE C.4

Rotary motion of body segments occurs around joint axes. (a) The angular displacement of the forearm/hand is 80 degrees. (b) The greater the radius of rotation of a point, the longer the curvilinear or circular path it forms around the axis.

Examples of angular displacement of a system exhibiting rotary motion include the sweeping of the second hand on a clock through 360 degrees with each revolution, shaking the head "no" as it rotates about a longitudinal axis through the neck, and rotating the forearm through 80 degrees as a weight is lowered (Figure C.4a).

Due to the structure of the body's skeletal system and the articulation of the segments, nearly all segmental movements are rotary. As a segment rotates, each single point on the segment decribes its own circular path around the joint axis, as shown in Figure C.4b.

The distance from any point on the rotating segment to the axis of rotation is called that point's **radius of rotation** and is the radius of the circle formed by that point as the segment rotates. The radius of the point and the length of the circular arc that the point forms are measured in linear measurement units (e.g., feet, meters, inches, centimeters) because they are linear distances. The entire rotating segment sweeps along an imaginary surface or plane, and the segment's movement (displacement) is measured in angular measurement units (degrees or radians).

Linear and Angular Motion

In most human movements the whole body and/or its segments or both move linearly and rotate at the same time. For example, a diver falls linearly downward while simultaneously rotating in a somersault.

Other examples in sport include the legs of a runner moving forward as they rotate around the knee joints, the body of a pole vaulter as he rotates around his grip on the pole while being thrust upward in a curvilinear path, and a gymnast performing a somersault dismount from a balance beam.

C.3 Force

A force can be thought of as a push or pull; a rub (friction); a blow exerted by actual contact; or by gravity, the pull of the earth's gravitational attraction on a body within its field. In other words, a **force** is something that causes or tends to cause a change in the motion or shape of an object or body. It is measured in pounds (British units) or in Newtons (metric units). For example, you can push on a door with a force of 22 N (~5 lb) to open it.

C.4 Pressure

Although the terms *force* and *pressure* are used casually to mean the same thing, they are different. The same amount of force can cause significantly different amounts of pressure. The definition of **pressure** is the amount of force acting over a given area:

EQUATION C.2

$$\text{pressure} = \frac{\text{Force}}{\text{Area}}, \text{ or } \quad p = \frac{F}{A}$$

where p is pressure, F is force, and A is the area over which that force is distributed.

Pressure is expressed in Newtons per square centimeter or per square meter (N/m^2), or pounds per square inch ($lb/in.^2$ or psi). For a given force, if the area is halved, the pressure is doubled. For example, a person weighing 400 N standing on both feet (area = 200 cm^2) exerts a pressure of 400 N/ 200 cm^2 = 2N/ cm^2 on the floor. The same person standing on only one foot exerts a pressure of 400 N/ 100 cm^2 = 4 N/cm^2.

Air and water pressure are expressed also in pounds per square inch or Newtons per square centimeter. Pressure changes in fluids will be discussed in more detail later.

C.5 Mass, Gravity, and Weight

Any body is composed of a certain amount of matter, or **mass**. Mass is common to all material things, whether solid, liquid, or gaseous, and is a measure of a body's resistance to having its state of motion changed. The term given to such a resistance is *inertia*. Inertia is the property of a body or object that resists changes in the body's motion *in any direction*. The inertia of a body is related to how much mass the body possesses; the more mass, the more inertia. The inertia of the arm, for example, is less than the inertia of the leg because of the difference in their masses. The inertia, or sluggishness of a body, is measured by how many slugs (British) or kilograms (metric) of mass the body possesses.

Not to be confused with the mass of a body is the **weight** of the body. A body's weight is a measure of the force with which the earth pulls on the body's mass. Gravity (weight force) always acts in a downward direction, toward the earth's center. A body's mass and weight are directly proportional; that is, the more mass a body has, the greater the earth's force of attraction on it, and therefore, the more it weighs. Mass and weight, however, should not be thought of as the same quantity. Weight is a force; mass is not. A person with a mass of 58 kg would weigh ~569 N. (In British, a mass of 4 slugs would weigh ~128 lb.)

Mass does not have a direction associated with it, nor does the mass of a body change with changes in the gravitational force. Since the weight force is always directed downward toward the center of the earth, the magnitude of the body's weight will vary slightly according to differences in gravitational attraction. Gravitational attraction between two bodies is greater when the distance between their centers is less. Therefore, a body will weigh slightly more at the earth's poles than at the equator because the magnitude of gravitational force is slightly greater at the poles, where the earth is slightly flattened.

C.6 Center of Gravity

Recall that the earth exerts a force on each segment of a body in direct proportion to each segment's mass. The total effect of the force of gravity on a whole body, or system, is as if the force of gravity were concentrated at a single point called the **center of gravity** (CG). The center of a system's mass distribution also is called its **center of mass** and may be thought of as that point at which all the body's mass seems to be concentrated. The CG also can be thought of as the balance point of a system, since the system's mass balances out on all sides of this point. In a rigid object made of the same material throughout,

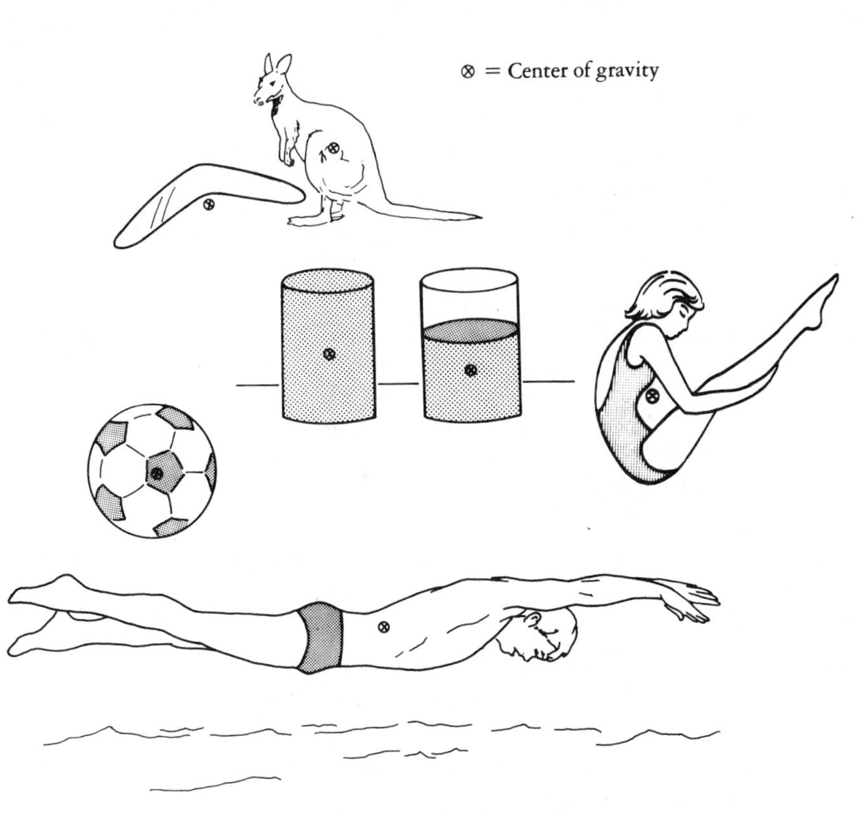

FIGURE C.5

Center of gravity (CG) locations of some common systems.

the CG is at the center of the object. In a segmental body whose parts can be rearranged, however, the location of the CG depends on the arrangement of the segments and their relative masses. Figure C.5 shows the locations of the CG for some familiar masses. Center of gravity is discussed in greater detail in Concept Module E.

Recall that when we analyze movement, we must identify the system whose motion we are interested in.

Regardless of what the system is—the human body, a body segment, two body segments, or some object in the environment—we need to use some point in the system that represents the total system and whose motion we can follow.

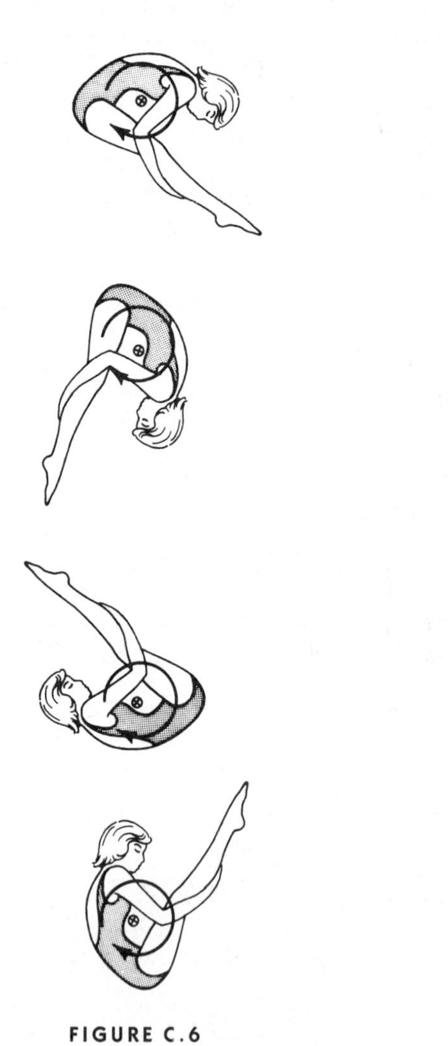

FIGURE C.6

A somersaulting diver falling linearly downward exhibits rotary motion around her center of gravity (CG).

That point is the CG of the system. For example, to describe the path of a jogger's body, we would track his CG from moment to moment, and it would form a wavy horizontal line as he bounced along his way. By examining the way a body's CG motion changes, we can identify the nature of the forces causing the changes.

In all activities in which a system is airborne and rotating (exhibiting linear and rotary motion), the axis for rotation is always the system's CG (Figure C.6).

Understanding a System and How It Moves

1. If the total human body is the system that you are examining, what is the single point you should follow to describe the body's motion?
2. If you run a distance of 50 m and a buddy runs 60 m alongside you from the same starting point, how far will your buddy have run relative to you?
3. Using an arm and a hand and a foot and a leg, demonstrate linear and rotary motion. Estimate the linear and angular displacements of the parts used. What measurement units are used for each type of motion?
4. Describe five examples in sport or dance in which something (a) moves linearly (rectilinearly or curvilinearly), (b) rotates about some axis, and (c) moves with both linear and rotary motion.
5. Rank the following objects in order of most massive to least massive: (a) a basketball, (b) a shot, (c) a baseball, (d) a volleyball, (e) a Frisbee, (f) a medicine ball, and (g) a golf ball.
6. Rank the objects listed in item 5 in order of most weight to least weight. Compare the two rankings of mass and weight. Are they the same, or do differences occur? Explain.
7. How is force defined?
8. How is weight defined?
9. How is pressure defined?
10. Give an example in sport in which the same force is applied to a body but the pressure is different.
11. By balancing each of the following items on the edge of your hand, find the CG for each: (a) a baseball bat, (b) a tennis racket, (c) a badminton racket, (d) a racquetball racket, (e) a fencing foil, (f) a ruler or meterstick.
12. From a piece of cardboard, cut out a shape similar to that shown in Figure C.7. Using a pencil, nail, pin, or paper clip, punch a hole anywhere in the cardboard. Suspend a string from this pencil, and attach a weight to the end. Hold the pencil so that the cardboard shape and the string are free to swing around it. Allow the cardboard and string to come to rest. With a felt-tip pen, draw a line that follows the line of the string across the cardboard. Repeat the previous procedure, using another hole anywhere

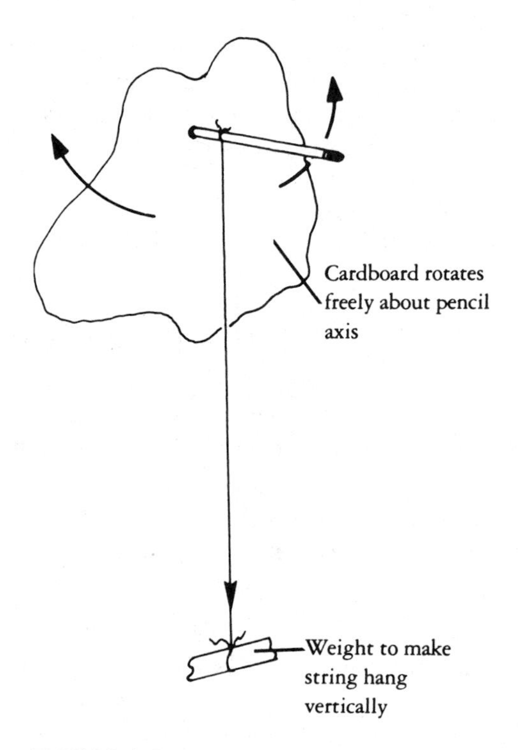

FIGURE C.7

If a flat body is suspended from an axis around which it can freely rotate, the center of gravity (CG) will come to rest directly under the support axis (somewhere along the string). Suspension from any other point will identify the CG at the intersection of the two string lines.

on the cardboard. Make a distinct mark at the intersection of the two lines. Repeat the procedure again. The third line should intersect at the same point. What is this point called? What is the definition of this point? Explain why this experiment works.

C.7 Work

Physical **work** is the product of force times the distance through which that force moves a load:

EQUATION C.3

$$\text{work} = \text{Force} \times \text{distance} \quad \text{or} \quad w = F \times d$$

where w is work, F is force, and d is distance force is applied.

If a force of 10 N moved a body 5 m across a floor, the work done on that body would be 10 N × 5 m = 50 Nm. If the object did not move when the 10 N of force were applied, zero work would have been done, even though the effort was felt. Work is expressed in units called joules (1 Nm = 1 joule).

A weight lifter who lifts a 100-N weight a distance of 1 m performs 100 Nm, or 100 joules, of work. The speed with which it is lifted does not affect the amount of work done, it affects the "power" output. If a force is applied to slow down or resist an object's motion through a distance, the work performed is called **negative work**.

C.8 Power

In sport, the term **power** is often misused to describe feats of strength, or force application. Correctly, power is the product of force times the speed with which that force is applied. A strong, fast muscle contraction is more powerful than that same muscle force applied slowly. From another perspective, power is the rate at which physical work is performed. Both expressions use the same quantities for determining power:

EQUATION C.4

$$\text{Power} = \frac{\text{work}}{\text{time}} = \frac{\text{Force} \times \text{distance moved}}{\text{time}} \quad \text{or} \quad P = \frac{F \times d}{t}$$

Which is the same as

$$\text{Power} = \text{Force} \times \frac{\text{distance}}{\text{time}} \quad \text{or} \quad P = F \times \frac{d}{t}$$

Since distance divided by time is speed,

EQUATION C.5

$$\text{Power} = \text{Force} \times \text{speed of force application} \quad \text{or} \quad P = F \times v$$

where P is power, F is force applied, d is the distance the force moves the load, t is the time the force is applied, and v is the speed the force is applied.

In biomechanics, looking at power in terms of force times velocity (speed) is meaningful. For example, a force of 100 N applied at a speed of 2 m/ sec is a power output of 200 joules/ sec, or 200 watts.[2] Recall the power–velocity curve in Chapter 2. The greatest power achieved for a particular effort is always a compromise between great force and great speed rather than a maximization of either one at the expense of the other. This is discussed further in the section on power in Chapter 7.

[2] 1 Nm/ sec = 1 joule/ sec = 1 watt

C.9 Energy

Energy is defined as the ability to do work. Many kinds of energy exist, such as heat, chemical, electrical, and mechanical. Mechanical energy is of primary significance in biomechanics. The more energy a body has, the greater the force with which it can move something (or change its shape) and/or the farther it can move it.

Mechanical energy has three forms that are useful in understanding how the body interacts with the environment: kinetic energy, gravitational potential energy, and elastic potential energy.

Kinetic Energy

Kinetic energy is the energy a body or object has because of its motion. It is expressed by:

EQUATION C.6

$$\text{Kinetic Energy} = \frac{1}{2}\,\text{mass} \times \text{velocity}^2 \quad \text{or} \quad KE = \frac{1}{2}mv^2$$

where KE is kinetic energy, m is mass, and v is speed.

The kinetic energy of a 10-kg medicine ball traveling at 3 m/sec is $1/2(10 \times 3^2) = 45$ Nm $= 45$ joules, which performs work on the catcher, who feels the force of impact during the time the ball's motion is brought to zero. The kinetic energy of the ball is used to move the catcher's arms and body, and some is transformed to heat energy as the ball is compressed in the catching process.

Note from Equation C.6 for kinetic energy, that speed of motion is squared and that a body's ability to do work on something is quadrupled if the speed is doubled, whereas the energy is simply doubled if the mass is doubled.

Gravitational Potential Energy

Whenever a body or object is in a position from which it can fall or be lowered by gravity, it possesses **potential energy** due to its height above the surface on which it will land. Quantitatively, gravitational potential energy is expressed by:

EQUATION C.7

$$\text{gravitational Potential Energy} = \text{Weight} \times \text{height} \quad \text{or} \quad PE = W \times h$$

where PE is gravitational potential energy, W is weight, and h is height above the surface.

A 400-N diver on a 10-m platform has 400 × 10 = 4,000 Nm (4,000 joules) of potential energy before the dive. As the diver descends, the body progressively loses its potential energy because it is being converted into energy of motion, or kinetic energy. By the time the diver hits the water surface, all the body's energy is kinetic, and this kinetic energy performs the work of moving a large mass of water during entry as well as creating heat.

Elastic Potential Energy

Elastic energy or *strain* energy, is the ability of a body or object to do work while it recoils (or reforms) after being stretched or compressed or twisted. For example, when an archery bow is deformed by the string pulling at each end, it has energy to move an arrow when the string is released, and the bow returns to its original shape.

Kinetic energy is the energy used to deform an object to create elastic potential energy and heat. When the deformed object reforms, it applies a force to anything it contacts and is capable of causing movement. The use of elastic force is discussed further in Concept Module D.

Understanding Work, Power, and Energy

1. What is mechanical work?
2. Describe an example of a performance of work.
3. Define power in terms of work performance.
4. Define power in terms of how force is used.
5. Energy is the ability to do work. Name the three forms of mechanical energy used in human movement activities.
6. Differentiate between a body with kinetic energy and one with gravitational potential energy.
7. What aspect is similar between a body with gravitational potential energy and one with elastic potential energy?
8. Which ball possesses more kinetic energy, a or b?
   a. A 10-N ball moving at 2 m/sec
   b. A 5-N ball moving at 4 m/sec

References and Suggested Readings

Barham, J. (1978). *Mechanical kinesiology*. St. Louis: C. V. Mosby.

Brancazio, P. J. (1984). *Sportscience*. New York: Simon & Schuster.

Feld, M. S., McNair, R. E., & Wilk, S. R. (1979). The physics of karate. *Scientific American, 240*(4), 150–158.

Hopper, B. J. (1967). Units and measurements in the mechanics of track. *Track Technique, 29*, 908–912.

Knuttgen, H. G., & Kraemer, W. J. (1987). Terminology and measurement in exercise performance. *Journal of Applied Sport Science Research, 1*, 1–10.

Resnick, R., & Halliday, D. (1977). *Physics: Part I*. New York: John Wiley & Sons.

Wheeler, G. F., & Kirkpatrick, L. D. (1983). *Physics: Building a world view*. Englewood Cliffs, NJ: Prentice-Hall.

CONCEPT MODULE D

Forces and Movement

PREREQUISITES

Chapter 1
Concept Module C

CONCEPT MODULE CONTENTS

D.1 Forces Acting on a System

Forces account for the motion and changes of motion of all things in the environment, including the whole body and the body segments. With this general understanding of the meaning of force, we can examine the nature of forces and the different kinds of forces.

Whenever a body or object moves, it changes its position or location relative to some frame of reference; it does so because some external force has been applied to it. Forces can be identified as **internal** or **external** forces.

Whenever we speak of internal or external forces, we mean internal or external relative to the system we are examining. For example, if we are interested in the motion of a baseball struck by a bat, our system is the ball, and the bat's force on it is external relative to the system (the ball). If we are interested in the motion of both the ball and the bat, the system is the ball and the bat, and the force of the bat on the ball is internal relative to the system (the ball and bat). If we consider the entire human body as a system, the external forces are those that originate from some source outside the body itself. Internal forces are those originating from some source within the body, or the system. Muscle forces are internal forces when the whole body is the system. If we choose a single body segment to be the system, however, a muscle's force acting on the bone is an external force relative to the system. The effects of internal and external forces are discussed in section D.6.

The Four Properties of a Force

All forces have four unique properties (Figure D.1): (1) **magnitude**, how much force is applied (e.g., 44 N, or ~ 10 lb), (2) **direction**, the way the force is applied (e.g., forward, south, vertically upward, perpendicular to a surface, at an angle of 60 degrees above the horizontal), (3) **point of application**, where the force is applied on the body or system receiving it (e.g., against the heel, on the fingertips, 1 m from the front edge of a table), and (4) **line of action**, the straight line extending through the point of application and extending indefinitely along the direction of the force. The line of action of a force is also called the line of force. A force may be represented on paper by an arrow.

D.2 Reaction Forces

We now focus on an important source of external forces, which can act to cause movement of the total body. We know that we can move our bodies from place to place by walking, running, jumping, diving, swimming, or rolling without the application of any *apparent* external force to cause this motion. If the body is the system, the muscles are internal forces that in themselves cannot cause movement of the system. Although internally initiated movements of the segments are necessary to move the body, the force responsible for displacement of the whole system (body) must be external to the body, and therefore it must be generated by the environment with which the body interacts. The body segments move to exert a force on something in the environment, and the environment delivers a **reaction force**, or counterforce, to the body. For example,

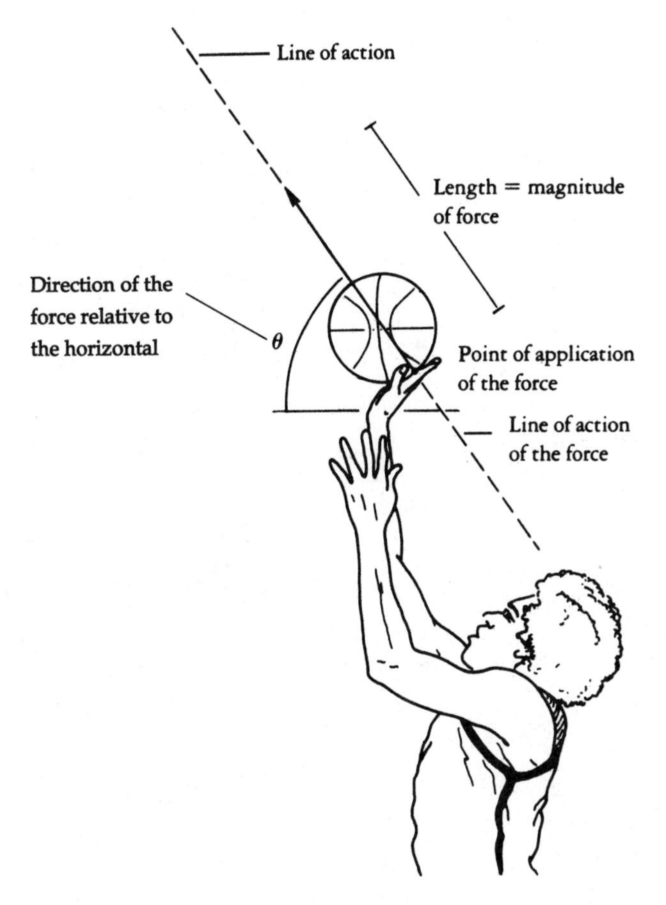

FIGURE D.1

The four properties of an external force applied to a basketball by the player's hand.

to move the total body upward in a push-up, the arm segments are caused to move by the internal forces created by the arm muscle contractions. The system is shown in Figure D.2a.

The hands placed against the ground are felt to exert a push downward as the joint actions in the arm occur. In response to the downward force of the hands on the ground, the earth exerts an equal and opposite force upward against the hands. This upward reaction force is the external force causing upward motion of the body. Another way to picture the situation is to think of the earth as preventing the hands from moving downward away from the body as the arms straighten so that the body must move upward away from the hands. Note that the earth, a system, experiences an external force applied to it by the hands, but the inertia (mass) of the earth is so great compared with that of the body that the earth's movement is negligible (too small to be measured). If this

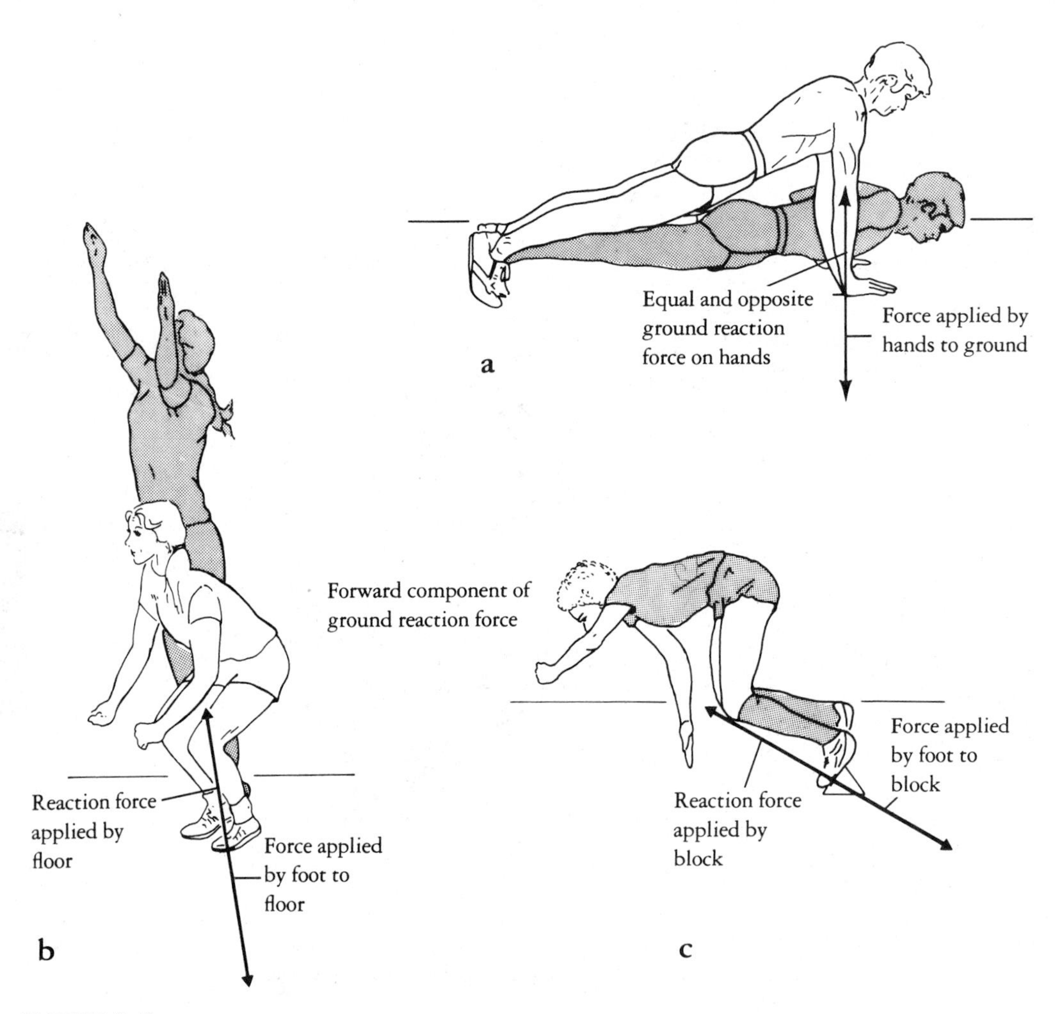

FIGURE D.2

External forces responsible for the motion of the body often are the reaction forces generated by the forces applied by the body against another less movable mass.

same push-up movement were performed in space where no physical contact with another mass exists, the arms would move away from the body and the body would move away from the arms so that the center of the system's mass would experience no change in position in space. In Figure D.2b, the jumper pushes against the ground and generates the resulting ground reaction force that pushes her upward. Similarly, the block reaction force generated by a sprinter's push-off propels her forward (Figure D.2c).

The phenomenon of an equal and opposite reaction force occurring in response to every force applied was described by Sir Isaac Newton about 300 years ago and is the subject of what is known as Newton's third law of motion:

For every force applied by one body on a second, the second body applies an equal and oppositely directed force on the first. Figure D.3 presents examples of oppositely directed applied forces and reaction forces. Visualization of the counterforces, or reaction forces, exerted by the ground in response to forces applied by the body is necessary for understanding movement of the whole body (of its center of gravity [CG]). Also keep in mind that the forces we exert on the ground have their effect, as illustrated by a ballerina who, following a demi-plié, "jumps two feet into the air. The Earth, balancing her momentum, responds with its own sauté and changes orbit by one ten-trillionth of an atom's width" (Lightman, 1983).

D.3 Friction Force

Friction is the force created between two contacting surfaces that tend to rub or slide past each other; it is the force that prevents the foot from slipping on the ground during walking. Friction force does not exist unless some sliding

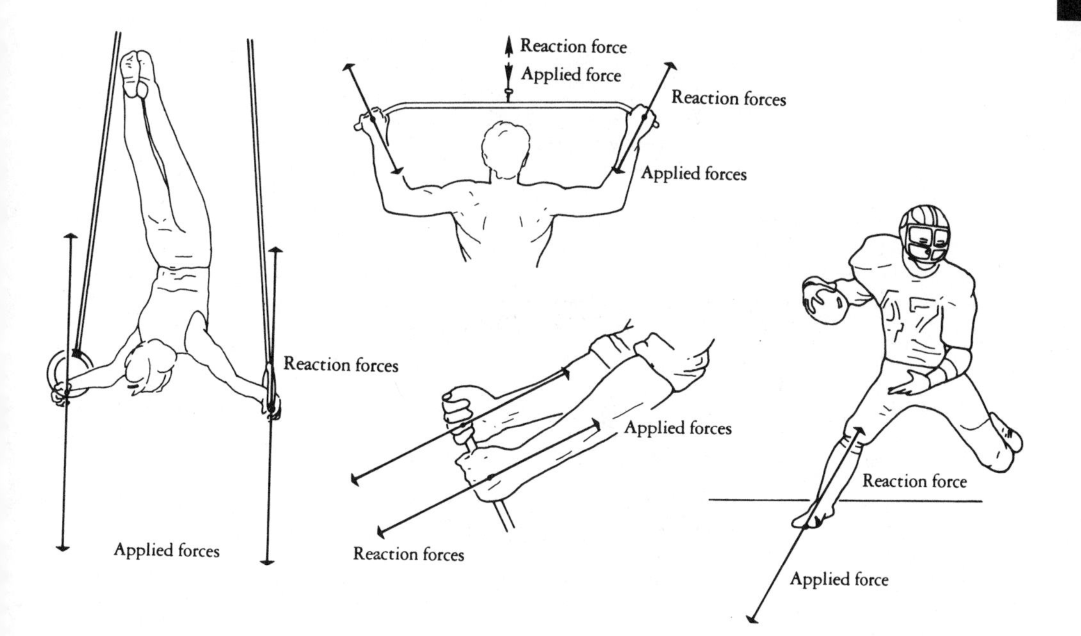

FIGURE D.3

Every reaction force by one mass on another is equal in magnitude and opposite in direction to the applied force (Newton's third law).

tendency exists or unless actual sliding occurs between two surfaces. When one surface exerts a friction force on a second surface, an equal and opposite friction force is applied by the second on the first. The amount of friction force that resists such a slipping or sliding tendency depends on the textures of both surfaces and on how much force is pressing the surfaces together. Wet or lubricated surfaces offer considerably less frictional resistance to each other than do dry ones. More friction force exists between rough-textured surfaces than between objects with smooth surfaces. When examined microscopically, even an apparently flawless surface has tiny bumps and grooves that "catch" the bumps and grooves of another contacting surface when there is cause for the surfaces to slide past each other.

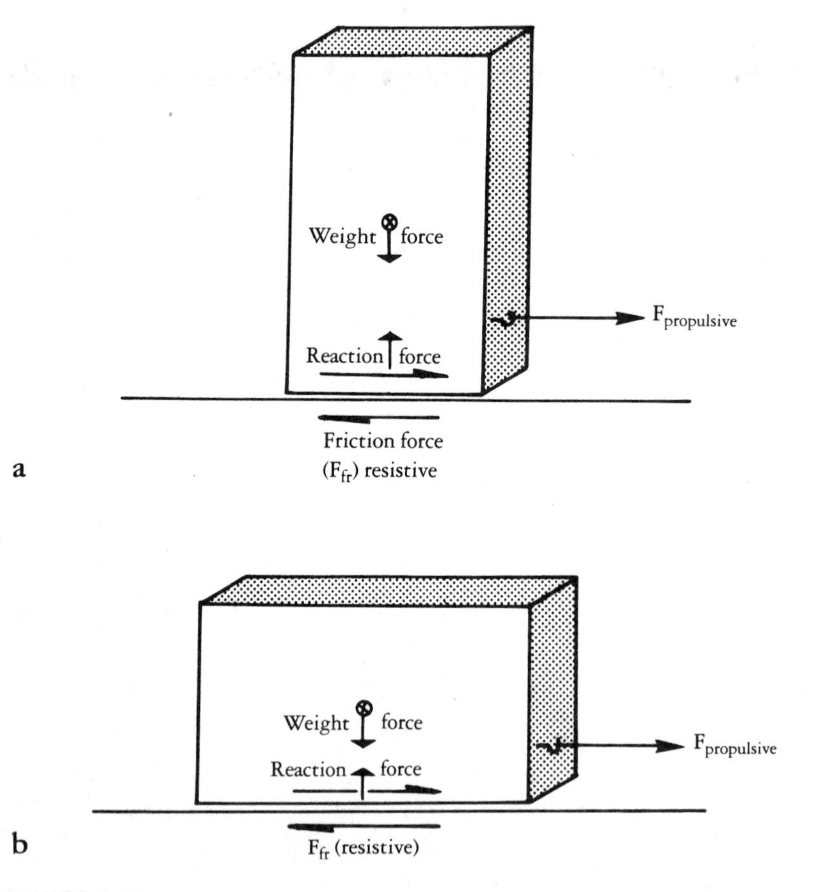

FIGURE D.4

Static friction force is increased if the force pressing the two surfaces together is increased (normal force) and/or the textures of the surfaces are rough. If the normal force and surface textures do not change, changing the apparent area of contact from that shown in (a) to that shown in (b) does not change the force of friction. In this example, weight is the normal force.

The greater the force pressing the surfaces together—called the *normal* or *perpendicular force*—the greater the resistance to sliding. Thus the force applied to move the box in Figure D.4 must be greater than the opposing frictional force applied against the box by the floor surface. Note also in Figure D.4 that the **apparent contact area** does not alter the frictional force.

The explanation for the independence of the surface area to the frictional force is based on the fact that the **actual contact area** is the area of contact of the bumps on each surface. The smaller the apparent contact area of a body on a surface, the greater the force per unit area (*pressure*). Greater pressure will cause greater flattening of the bumps, which creates a larger actual surface contact area, even though the apparent contact area is smaller. The decrease in apparent surface area when the box rests on its small end is compensated for by an increase in actual surface area, and thus the frictional force will remain the same. This occurs even with very hard surfaces (Resnick and Halliday, 1977).

The implication of this fact is that changing the apparent area of contact does not change the force of friction if the normal force and the surface qualities are the same. For example, a man wearing size 12 shoes has no more friction force on his feet than if he were wearing size 6 shoes. Nor does a car with wide tires have any greater friction force than that same car with narrow tires. (Large contact-area tires are used not for greater traction but for longer wear, owing to less friction per unit area of tread.)

A way in which the friction force can be modified by increasing the normal force is explained by this example: When you have a tight grip on a jar lid, the friction between your hand and the lid becomes greater than the friction between the lid and the jar, and the lid turns with your hand.

To quantify the force of friction between two surfaces, two facts must be known: (1) the normal force pressing the surfaces together and (2) a quantity known as the **coefficient of friction** for the two surfaces, which depends on their hardness and surface texture. The coefficient of friction for two surfaces that tend to slide under the influence of some motive force but have not yet begun to move is called the *coefficient of static friction*.

The coefficient of friction between two surfaces is the ratio of friction force to normal force:

EQUATION D.1

$$\text{Coefficient of friction} = \frac{\text{Friction Force}}{\text{Normal Force}} \quad \text{or} \quad C_{fr} = \frac{F_{fr}}{F_{\perp}}$$

where C_{fr} is the coefficient of static friction, F_{fr} is the maximum static friction force, and $F_{\perp}$ is the perpendicular ($\perp$) force pressing the surfaces together.

The coefficient for surfaces that are already sliding on each other is called the *coefficient of kinetic friction* (sliding friction), which is smaller than the coefficient of static friction. The static friction force between two surfaces that have not started sliding is greater than the kinetic friction force between the same two surfaces during sliding. As a force is applied to cause sliding of one body on another, the magnitude of the static friction force increases until it reaches a maximum value just before actual sliding. Any further increase in the force, therefore, would cause the body to slide past the other surface. Once sliding begins, the moving body experiences less frictional resistance to the applied effort because of the lower coefficient of friction. Such a drop in resistance of a moving load that has started to slide partially accounts for the claim that to start something moving is more difficult than to keep something moving.

The friction force acting on a rolling object is much less than the kinetic friction encountered in sliding. The reduced friction on a rolling object is a result of the relative ease with which the rolling surface is peeled away from the contacting surface in contrast to two sliding surfaces that have to resist the "catching" of bumps and grooves. For example, when one brakes a car that is rolling on an icy street, the braking force is applied to the rolling tires, and this force causes a sliding tendency. Thus, the tire–road interface assumes a static friction condition and has a higher frictional limit before actual sliding. Once sliding occurs, the situation involves kinetic friction, and thus the friction force decreases. The brakes should be released to resume the tires rolling and the brakes applied again.

In cycling, slipping usually occurs in cornering, especially on wet or gravelly surfaces. In wet weather, a thin film of water (and often road oil) exists between the tires and the road surface. This fluid acts as a lubricant to significantly reduce the friction available for turning, and the tires may slip out from under a leaning cyclist. Cyclists learn to slow down and make wide turns under such conditions. In dry weather, a cyclist taking a sharp corner may encounter a patch of gravel or oil that instantly decreases the friction needed for turning, and a mishap may occur.

Maximum static friction is also desirable in sports played on grassy surfaces, such as football, baseball, and golf. Players of these sports wear shoes with cleats because smooth-soled shoes provide little friction on grass and thereby cause slipping. In activities in which hand-implement friction must be increased, as in baseball, golf, or racquetball, the hand or implement or both are provided with a friction-increasing covering such as a glove, wrappings, or tacky resin. In some situations, friction is a disadvantage to performance, as in the case of the skater who wishes to slide across the ice with minimal resistance or the dancer who turns on the ball of the foot in a pirouette.

D.4 Centripetal and Centrifugal Forces

Whenever a body is restricted to traveling in a circular path around a central axis, such as a whistle on a string swinging in a circle, it experiences a force that is directed along the radius (string) toward the center of the circle (toward the hand holding the string). If the string is released, the center-seeking force disappears and the whistle is cast off in a linear path tangent to the circle. Such a center-directed force is called a **centripetal force** and is responsible for continually forcing the rotating object to follow the circular path. The equal and opposite reaction force is called the **centrifugal force**, and is the force that the rotating body exerts along the radius on the central axis. The hand swinging the string and whistle experiences this outward force as the whistle circles.

The greater the mass of the object or the faster it travels, the greater the centripetal and centrifugal forces. The larger the circle, the smaller these forces. Equation D.2 shows the factors that determine how much centripetal force is needed to constrain a turning mass in a circular path:

EQUATION D.2

$$\text{centripetal force needed} = \frac{\text{mass} \times \text{speed}^2}{\text{radius of circle}} \quad \text{or} \quad F_c = \frac{mv^2}{r}$$

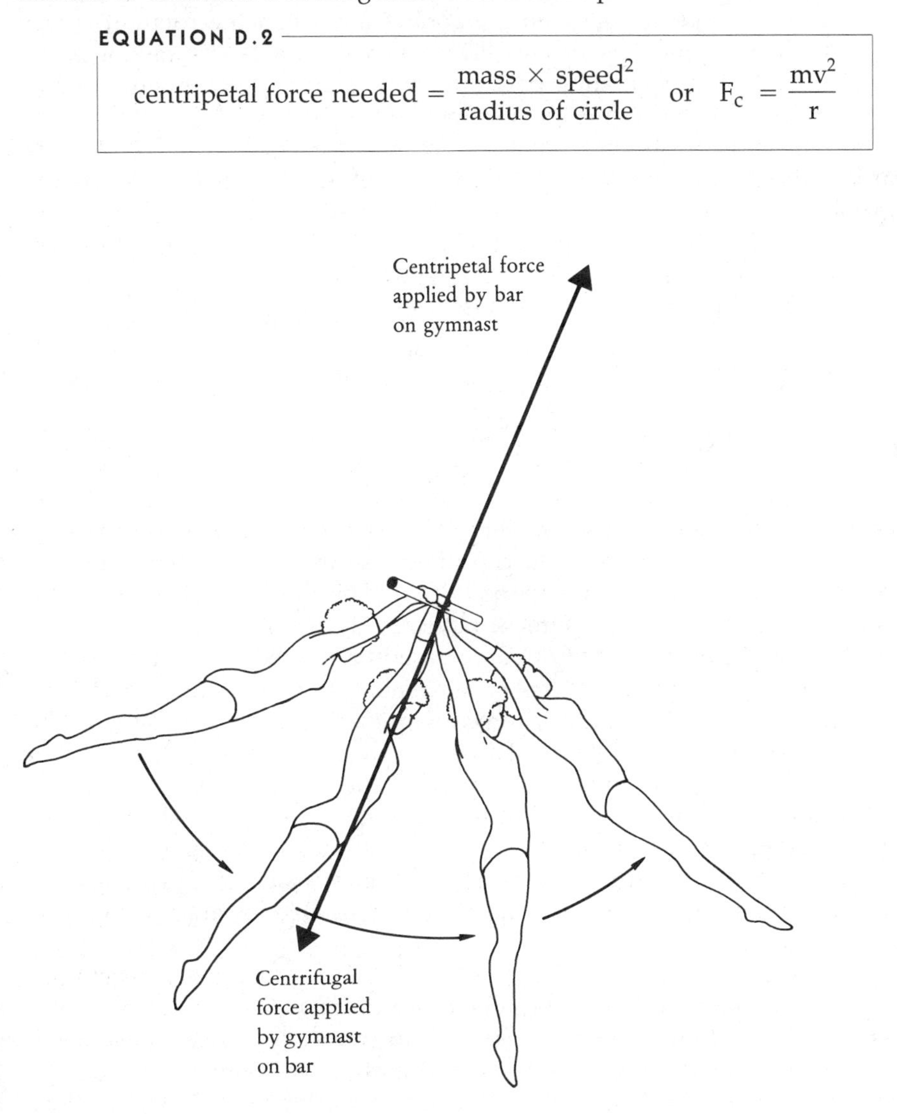

FIGURE D.5

The grip strength must be sufficient to withstand the centripetal and centrifugal force tending to extend the fingers.

It is evident from the squared value of speed that the centripetal force needs to be very great if that speed of travel is increased.

Many activities involve rotary motion in circles or parts of circles: a swing on a rope or rings, a giant swing on a high bar, a hammer throw, a discus throw, a swing of a skating or dancing partner, a swing of a baseball bat, a swing of a bowling ball or softball, and a swing of a lacrosse stick and ball. All these activities require a centripetal force to keep the body or object traveling circularly. A handgrip often provides the connection to the central axis, and it must be strong enough to maintain the connection when the rotation is fast or the body or implement is heavy. Figure D.5 illustrates the pulling-away tendency of a gymnast's rotating body and how she must maintain her grip on the bar so that the bar can exert centripetal force on her.

In going around curves on a track, a cyclist must lean inward and press outward on the ground (centrifugal force by cyclist on ground) to generate the centripetal ground reaction friction force (by ground on cyclist) to be able to travel in a curved path. If she travels too fast, however, the centripetal friction force is not great enough, and she slips. This friction problem is alleviated by banked curves that permit cyclists to press into a "wall" when they turn.

D.5 Elastic Force

The tension and subsequent "recoilability" of a stretched muscle was discussed in Chapter 2 as being an important contributor to the total force the muscle is able to apply to a bone to move a body part. This feature, called elasticity, of a body or material is a measure of how readily it will *reform* after it has been *deformed* by being stretched or compressed or twisted. It is an important source of force used in many aspects of movement, because when a deformed body or object reforms, it can apply force to another body or something in the environment. For example, when a falling ball hits the ground, the reaction force of the ground acting upward compresses the ball until its CG stops its downward motion. Then, the recoil of the ball back to its round shape causes its bottom surface to push against the ground, thus generating the ground-reaction force that moves the ball upward. The ball is said to possess **elastic energy** when it is in the deformed state because it is able to cause movement, or perform work.

All the implements and balls used in sport have some degree of elasticity, or recoil ability, that provide force. Springboards and trampolines and shoe soles have recoil ability. Human bones, muscles, and connective tissues have some elasticity, fortunately, that contributes significantly to movement efficiency in activities using wind-ups, or sudden stretching during muscle contraction. Kinetic energy is the energy used to deform the body or object, and it is converted into elastic energy and heat.

The degree to which the reformation, or restitution, of a deformed body will occur can be described by its **coefficient of elasticity (e)**. The greater the **e**, the greater its tendency to reform. The coefficients of elasticity can be determined for different balls in a bouncing experiment in which the height from which each ball is dropped to the floor is measured, and the height of rebound is measured. The **e** is then calculated as follows:

EQUATION D.3

$$\mathbf{e} = \sqrt{\frac{\text{height of rebound}}{\text{height of drop}}}$$

Visualize what the relative values of **e** would be for a superball compared to a ball of clay if they were dropped to the same floor from the same height. As you might imagine, the clay ball would hit and deform to a flattened shape and stay there. The superball would bounce very high and is said to be highly elastic. The clay is highly inelastic; that is, it stays deformed and resists reforming.

The elasticity of any body is determined by the materials that body is made of. It also can be influenced by temperature, which accounts for the greater liveliness of a warmer racquetball. The rebound or recoil behavior of an elastic body or implement, however, is influenced by the nature (elastic or inelastic) of the contacting surface or object and whether that object is also moving. The relative speed of both bodies coming together determines the force of impact and, therefore, the amount of deformation and reformation.

The rebound height used in Equation D.3 is directly related to the speed the ball hits and the speed it leaves the ground. It will always leave the ground with less speed than it hits because some of its mechanical energy is used to form heat during impact. If the impact is with another moving object the rebound speed will be determined by the elastic characteristics of the two bodies and their relative approach speed.

CONCEPT MODULE D

Understanding Different Kinds of Force

1. Name the four properties of any force.
2. Describe several examples of Newton's action–reaction law.
3. Describe an example of desirable friction in sport.
4. Place your hand flat on a table, press lightly, and slide it forward. Repeat, but press down hard as you slide. Explain why the resistance to sliding was different.
5. Arrange six of these textbooks side by side on a tabletop. Slowly push horizontally on the end book, and note the amount of resistance encountered just before motion begins. Stack the six books vertically on top of one

another, push on the bottom book, and note the amount of resistance as before. Compare the resistance encountered in the latter situation with that encountered when the books were aligned horizontally with more of their surface area touching the tabletop. In each case, relate the amount of resistive friction to the total weight of the load and to the surface area of the supportive base.

6. Describe the centripetal forces acting on (a) a rotating bat, (b) a gymnast doing a giant swing on a high bar, and (c) a person swinging on a playground swing. In each case, identify the centrifugal force that is acting.
7. What does it mean for one ball to have a large coefficient of elasticity compared to another ball?
8. Describe several examples of a force being applied to an elastic object to deform it so it gains elastic potential energy. For each example, tell how the object's reformation can do work on another body.

D.6 Internal and External Forces

The Effect of Internal Forces

The movements of a system's CG relative to the ground or to some other frame of reference in the environment must be accounted for by forces that are external to the system. For example, the CG location of a golf ball does not change relative to the ground until an external force is applied to the ball. The only internal forces occurring within the ball (the system) are molecular collisions, and the forces from these collisions cancel each other so that the overall effect is equivalent to no force influencing the motion of the ball's CG.

To further clarify this concept, picture a coiled spring bound tightly with string that is tossed into the air. The spring is isolated from physical contact and is subject to no external force except that of gravity (Figure D.6a). If the string breaks from the tension of the spring (an internal force), the spring expands to its normal, uncompressed length. The coils of wire comprising the spring (the system) change their positions relative to one another, but the location of the CG of the whole spring is not altered by the action of the internal force. Therefore, the only force influencing the motion of the spring's CG is the force of gravity acting downward. The human body behaves in the same manner when free of external forces (other than gravity).

A body free in space and in a "weightless" condition has no external force to act on it; therefore, it does not exhibit any change in the movement of its CG when its body segments move. (Note that gravity is an external force that causes the system to move toward the ground.) When the whole body is the system, its CG location is not affected by any forces generated within the body. For example, muscle forces cannot change the location of the CG of the unsupported body. When the whole body is the system, therefore, muscle

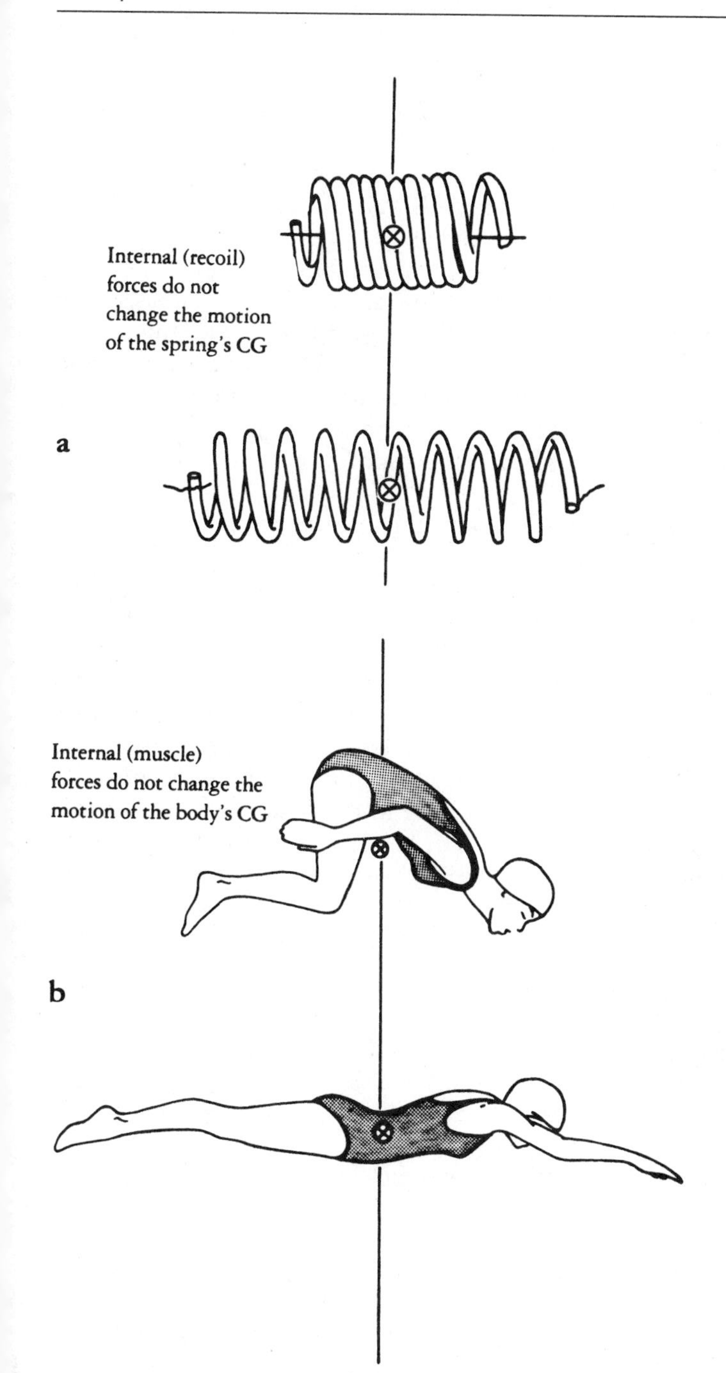
Internal (recoil) forces do not change the motion of the spring's CG
a
Internal (muscle) forces do not change the motion of the body's CG
b

FIGURE D.6

Forces that are internal to a system do not change the location or state of motion of the center of gravity of that system.

forces are internal forces that cannot cause movement or work against the existing movement of the whole body's CG. If a movement is performed while the body is free of support, as in the case of the person in a tuck position (Figure D.6b), her CG location in space does not change when she assumes a layout position. Internal muscle forces serve only to change the shape of her body; they cannot move her CG position relative to the ground.

Such is the plight of an isolated astronaut in a weightless condition, free of any contact with another object or surface. Any segmental movements only change the shape of the body around its CG; no segmental movements can cause the CG to move. The astronaut is stuck until help arrives. Such help must be in the form of an external force. The astronaut can move the body parts, arms and legs, in any way imaginable, but cannot change the body's CG location relative to the environment because the only forces acting upon the system are internal forces.

Returning to earth, recall that the external force of gravity acts on the CG of a system. If we consider the movement of a platform diver changing body positions during a descent, that is, from layout to pike to tuck to layout, we realize that gravity causes the body's CG to move downward no differently than if the diver were not changing shape. The body parts are merely changing their positions around the CG, and the CG is that which is responding to the external force of gravity. This same phenomenon occurs during any activity that includes a flight phase, or period of nonsupport, such as in long jumping, high jumping, rebound tumbling, running, aerial gymnastic stunts, and falling. The effect of the external force of air flow on the human body in many nonsupported aerial activities is so slight as to be insignificant. The effect of air resistance in some endeavors cannot be ignored, however. A sky diver, for example, encounters upward air resistance, which limits the rate of falling.

The Effect of External Forces

When a system is in contact with the ground or another support, it is capable of moving its CG. By changing its shape *against* the support, the system applies a force to the support that generates a reaction force against the system. This reaction force is the external force responsible for moving the system's CG. Consider the coiled spring bound with string (Figure D.7a) that is placed against a wall. When the string breaks, the spring uncoils, and because the wall is in the way, the spring pushes against it. The wall responds with an equal and opposite reaction force that is applied to the spring, causing the spring's CG to move in space. Similarly, the tucked body against the block in Figure D.7b uses internal muscle forces to change her shape to push against the block. The reaction force then serves as the external force propelling the person away from the block.

The difference in the movement of the CG while the body is supported and while it is not supported is an important concept that is used frequently in the analysis of the performance of various sport skills.

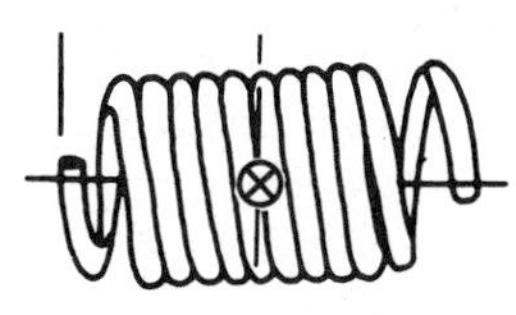

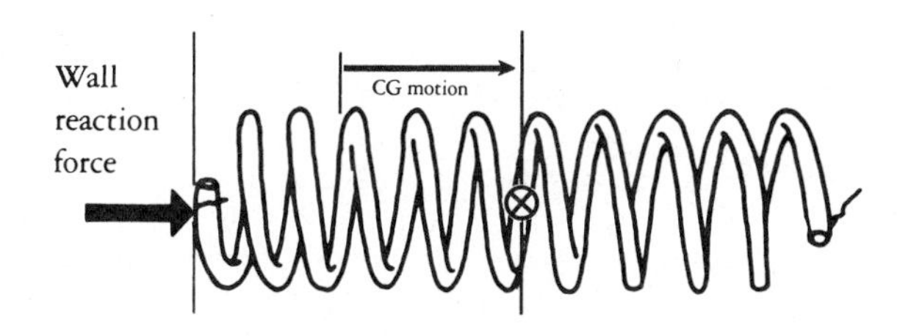
Reaction force of wall
on expanding spring
moves its center of
gravity
Wall
reaction
force
CG motion

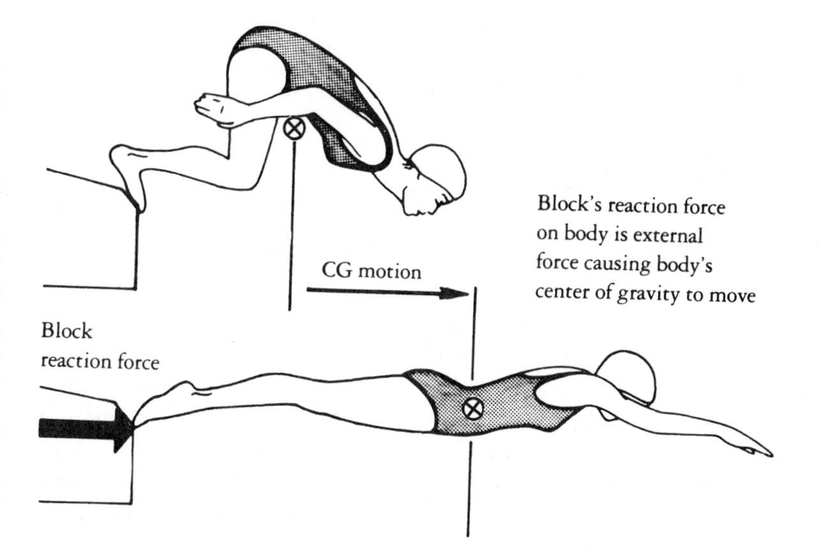
Block's reaction force
on body is external
force causing body's
center of gravity to move
CG motion
Block
reaction force

CONCEPT MODULE D

FIGURE D.7

Forces that are external to a system do change the location or state of motion of the center of gravity of that system. In this example, internal muscle forces change the shape of the system to generate the external reaction force from the wall.

Consider now the effects of other external forces on the movement of the whole body as a system. For example, a collision with an opposing player in basketball moves the body's CG in the direction of the force applied. The movement of the body's CG is readily apparent as the struck player moves from one location on the floor to another. Such changes in the location of the CG are not always so easily observed. For example, the force applied by a basketball to the hands and arms of a player catching it causes the arms to move backward, closer to the rest of the body mass, as the elbows flex. When this happens, the player does not change his or her location on the floor; however, the segmental masses of the body are forced into a new configuration, and the CG of the body shifts backward slightly. Thus, the external force moves the body's CG but not the location of the player's feet on the floor. Such a situation also serves as an example of how an external force can change the shape of a system.

D.7 Motive and Resistive Forces

A movement of some system (body, body part, or group of parts) can be examined in terms of the forces that act on that system. To affect the motion of the system, these forces must be external forces relative to the system. An external force that causes an increase in speed or change in direction is called a **motive**, or **propulsive**, force. An external force that acts to resist the motion caused by some other external force or to decrease the speed of a moving system is called a **resistive** force.

Resistive Forces

Resistive forces oppose or resist the movement of a system within a frame of reference, and they must be external relative to the system. If we select a body segment as our system, we can identify several sources of resistive forces that are external to the segment but are found within the human body. These sources are: (1) the tension in muscles acting to resist movement of a segment as it is acted upon by some force that succeeds in moving it or that merely attempts to move it, (2) the resistance of connective tissues to being stretched by some force, (3) the actual contact forces, such as the touching of two body segments as they approach one another (e.g., the contact of large biceps with the forearm muscles, bone on bone within a joint, or slack ligaments across a joint become taut), and (4) the degree of viscosity (thickness) of the fluids within muscle tissues and joint capsules.

These restrictive forces help provide joint stability because they restrict excessive joint movement that could cause injury to the joint surfaces. Occasionally, however, the tissues are torn when the motive force causing the segmental movement is greater than the resistive force that the internal structures can provide. This happens especially in contact sports.

Normally, before the movement of a segment is restricted or stopped by the structures of an *articulation* (joint), the muscles crossing over the articulation play an important role in gradually slowing the segment's movement. In arm wrestling, the descent of the loser's arm is slowed by the tension of the muscles of the shoulder and elbow. If these muscle groups were to relax instead of maintaining tension while they lengthened, the arm would be taken down much more rapidly. The role of muscle tension as a resistive force during lengthening should not be underestimated.

Resistive tension is as important—and as frequently employed—as the concentric muscle contractions that *cause* segmental movements. Resistive contractions (eccentric, or lengthening, muscular contractions) constantly control the extent and speed of segmental movements. Resistive muscular tensions control the speed of segmental movements that we normally take for granted. They are the braking forces applied near the end of the range of motion of body segments so that the movements slow gradually rather than halting abruptly as a result of tissue impact or the sudden stretching of joint structures.

The connective tissues in muscles provide resistance to stretch as a segment is moved toward the end of a joint's range of motion (ROM). In bouncing stretches done in a straight-leg sitting position on the floor, the forward-reaching movement of the arms and trunk is brought to a halt as their posterior muscles and associated tendinous tissues become stretched taut. The recoil of this *elastic* tissue subsequently provides the motive force to bounce the upper body segment back again. Stretches of a violent bouncing type are usually discouraged because of the possibility of overstretching by "throwing" the mass of the body parts with too much speed.

Another source of resistance to a system's movement is contact with any mass or force other than those within the system being considered. If the body is the system, such contact may be with the earth, another person or object, air and water resistance, or friction force. Resistive forces are specific as to direction. Recall that the weight of a body is a force that acts only in a downward direction and occurs only because of the earth's gravity. Weight, then, is a force that resists a body's upward movement. An outfielder running to catch a fly ball may encounter the horizontal resistive force of the center-field fence. A runner stealing second base slides to a halt as a result of the resistance provided by the ground friction on his body. Similarly, a rolling ball eventually stops as a result of the friction force of the ground against it. A swimmer moving through the water experiences the resistive force of the water flowing backward, and a skier experiences the resistive force of the air flow during a downhill run.

Tight-fitting clothes, often overlooked as a restrictive force, can be a significant source of resistance for segmental movements of the body. A vivid example of how restrictive such clothing can be is provided by a student wearing tight jeans attempting to climb stairs two at a time.

Motive Forces

Forces that cause a change in the way a body segment is moving can originate from within the human body or from outside the body, as long as the forces are external with respect to the segment (system). Motive forces for the segmental movements include those provided by muscle contractions and the recoil of elastic connective tissues. When a muscle is stretched and then released, the elastic recoil of the stretched connective tissues can cause the initial motion of the body segment. For example, when you sit on the floor with legs extended and vigorously reach forward toward your toes, the posterior spine and hip muscles and their connective tissues are stretched; consequently, you experience a rebound movement of your upper body.

Of the forces generated within the human body, contraction of the skeletal muscles yields the major motive force causing segmental movements. If a group of muscles contracts and shortens (concentric contraction), the attached body segment is caused to move by that motive force. For example, with your forearms and hands identified as the system, picture yourself opening a window by pulling upward with your hands. Your elbow flexor muscles would shorten to raise your forearms and hands. The external resistive force acting against the motive force in this situation is the weight (the force of gravity) of the segments and window and the friction force exerted on the window by the frame.

The most significant motive force causing motion of the total body is the pull of gravity. Gravitational force (for activities here on earth) is always present, always acts in a downward direction toward the center of the earth, and exerts its force on the CG of all masses. Gravity is the obvious motive force for the descent of a diver or the sinking of a lean swimmer attempting to float. Gravity is also the motive force for the downward movement of the forearms and barbell in a forearm curl.

Other motive forces occur in conjunction with some external object or another person, with which contact causes or tends to cause movement of the system. The recoil of a depressed diving board provides a motive force for projecting a diver upward (the whole body as the system). The loser in an arm-wrestling contest has his arm put down by the motive force applied by the opponent's hand. Note that from the point of view of the loser, his arm (the loser's system) has been subjected to the motive force of the opponent's push; whereas, from the winner's point of view, the pushing arm (the winner's) was caused to move by the motive force of muscle contraction and was resisted by the loser's applied force.

A strong wind propelling a skateboarder is another example of a motive force acting on a body. Water's force of buoyancy is normally thought of in terms of support; however, it acts as a motive force for the upward movement of a buoyant, submerged swimmer holding his breath.

In addition to being a resistive force, friction can be a motive force. A performer walking or running on an exercise treadmill finds that her leg is moved backward relative to her body by the frictional force applied to her shoe sole by the treadmill belt as it moves backward. If a person were running across the ground instead, the forward friction ground reaction force to the foot's backward friction force applied to the ground would be the external force causing the forward motion of the runner's CG.

Understanding Motive and Resistive Forces

1. Which type of force can be a motive force relative to the system—internal or external? Which can be a resistive force?
2. The total body is the system to be examined. Describe an example of (a) a motive force applied to the system, (b) a resistive force applied to the system. In each case, are muscle forces internal or external forces?
3. Consider the following forces that act on a system (be sure to define it) and cite examples in which each serves as a motive force and in which each serves as a resistive force: (a) muscular tension, (b) friction, (c) gravitational forces, (d) air flow, (e) water flow, and (f) contact with an external object.
4. Differentiate among these three "pairs" of forces by giving an example that clarifies each pair: internal and external, action and reaction, motive and resistive.

D.8 Force Diagrams and Vectors

When a force is shown in diagrams, each of its four properties may be represented by an arrow whose length is drawn to a scale selected to represent the magnitude of the force—for example, 1 cm = 5 N or 1 in. = 10 lb. The arrow is drawn to point in the direction in which the force is applied, and the tail of the arrow is placed at the force's point of application. The line of action of the force may be imagined as continuing indefinitely in both directions (head and tail end), although the actual arrow, if drawn to scale, must remain a given length. A graphic representation of a force of 100 N acting at a 30-degree angle to a table being pushed is shown in Figure D.8.

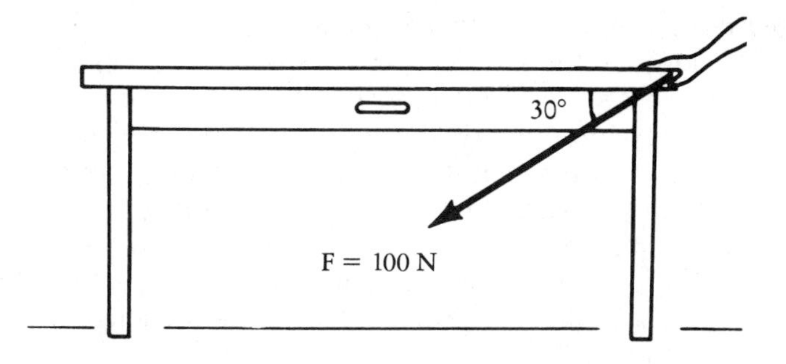

FIGURE D.8

A force vector drawn to show the application of a 100-N force directed 30 degrees below horizontal. The tail of the force vector is drawn at the point of application.

Because force has the two properties of magnitude and direction, it is known as a **vector** quantity. Vector quantities are different from **scalar** quantities because vectors always have direction specified and direction is an important characteristic of the quantity. Scalar quantities have only magnitude—a size or amount of a certain quantity—and no direction is associated with them. Mass, volume, and area are examples of scalar quantities.

The Effect of Two or More Forces on a System: Vector Composition

When more than one external force is applied to a system simultaneously, all the external forces must be considered, for it is their **net** effect that determines the motion of the body. The net effect, or **resultant**, of the applied forces, is determined by combining the force vectors. The combining of forces is called **vector addition**, or **vector composition**, and it differs from scalar addition (regular numerical addition) in that the direction of the force is just as important as the magnitude of the force. Vector addition can be accomplished graphically by drawing diagrams to scale and measuring or by using trigonometry.[1]

To demonstrate the concept of a single resultant force, or net force, formed by two different forces, we will consider the motion of a ball falling in a sidewind, as shown in Figure D.9. Gravity pulls vertically downward on the ball with a force equal to the ball's weight of 3 N (~ 10 oz). A wind blowing horizontally toward the west exerts a force in that direction equal to 9 N (~ 30 oz.). The motion of the ball is determined by the resultant of these two forces, which act simultaneously (Figure D.9a).

The force vectors are drawn to a convenient scale that is similar to a road map scale; for example, in Figure D.9, 1 cm represents 3 N of force. The force vector that represents the object's weight, therefore, is drawn 1 cm long in the vertical direction, and the wind-force vector is drawn 3 cm long in the horizontal direction to represent its 9 N of force.

Figure D.9b shows how the two forces are visualized as two sides of a rectangle and how the opposite sides are then drawn to form the whole rectangle. The resultant force, R, is represented by the diagonal that is drawn from the corner of the rectangle formed by the tails of the two force vectors (Figure D.9c). The direction of the resultant force is downward and westward, and the exact angle of the force can be measured with a protractor. The magnitude of the resultant force may be determined by measuring the length of R with a ruler and converting the length to Newtons of force, using the same scale to which the two forces were drawn originally (1 cm = 3 N). Since the length of R measures 3.16 cm, the magnitude of R equals:

$$3.16 \text{ cm} \times \frac{3 \text{ N}}{1 \text{ cm}} = 9.48 \text{ N}$$

[1] Trigonometric methods of vector composition are in Appendix V.

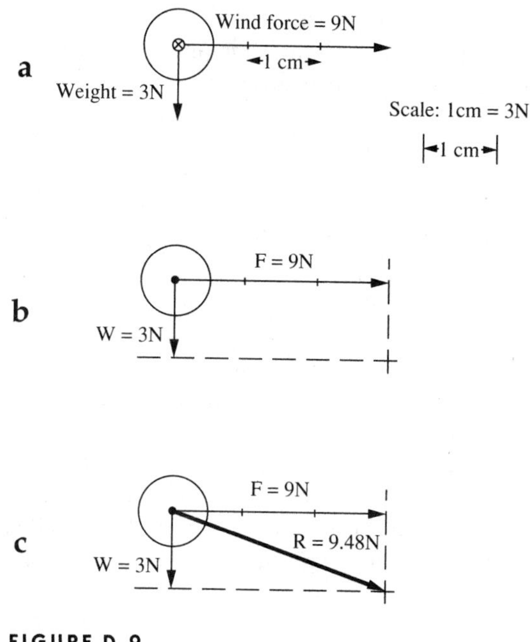

FIGURE D.9

The net (resultant) effect of two forces—the ball's weight and a sidewind—is found by vector composition. The resultant force, R, will determine the ball's motion.

CONCEPT MODULE D

The ball's motion, therefore, is in the direction of the resultant force, R, applied to it as shown. This example enables us to see the difference between scalar addition of quantities and vector addition. Vector addition is influenced greatly by the directions of the component vectors.

If the component force vectors (the two independent forces acting together) are perpendicular components (i.e., if they form a 90-degree angle with one another), the magnitude of the resultant can be calculated by the *Pythagorean theorem*. (See Appendix V.) The Pythagorean theorem specifies that the length of the hypotenuse of a right triangle is equal to the square root of the sum of the squares of the two sides:

EQUATION D.4

$$c = \sqrt{a^2 + b^2}$$

therefore

$$R = \sqrt{3^2 + 9^2} = \sqrt{9 + 81} = \sqrt{90} = 9.48\text{ N}$$

We can calculate in this manner only when the component force vectors are perpendicular to each other.

Let us look at another example in which a resultant force determines the motion of a body. Suppose a soccer ball is kicked simultaneously by two players. The force diagram of this situation is shown from an overhead view in Figure D.10a. We use the same process of drawing each applied force to a convenient scale. In this case, we will visualize the two force vectors as two sides of a nonrectangular parallelogram (i.e., a four-sided figure whose opposite sides are parallel) (Figure D.10b). A third vector is drawn from the tails of the two force vectors to the opposite corner of the parallelogram. This resultant vector is the net force applied to the soccer ball, and its magnitude is found by measuring the length with the same scale used for drawing the original forces. The ball's motion will be changed in the direction indicated by the resultant force and will be proportional to the magnitude of that force.

Another situation in which the motion of a body is determined by a **net force** acting on it is one in which a weight is lifted. Figure D.11 shows the vector diagram of the forces acting to cause the upward motion of the load.

For this situation, the resultant of two opposing forces must be found. Forces that lie along the same line as those in the example are called *colinear forces*. A parallelogram cannot be drawn as before, because no corner is formed at the tails of the two force vectors. Therefore, the algebraic sum of the two forces is found by means of vector addition, with the upward-directed force being termed a positive force and the downward-directed force, a negative one. The algebraic sum of the two is determined as follows:

$$(+25\ \mathrm{N}) + (-20\ \mathrm{N}) = +5\ \mathrm{N}$$

The positive 5 N is the net force experienced by the load, and it is moved upward (in the positive direction). If one were to exert only 20 N of upward force, the positive 20 N would match the negative 20 N of weight; that is, the sum of the forces in the vertical direction would equal zero, and with zero resultant force, no movement would occur.

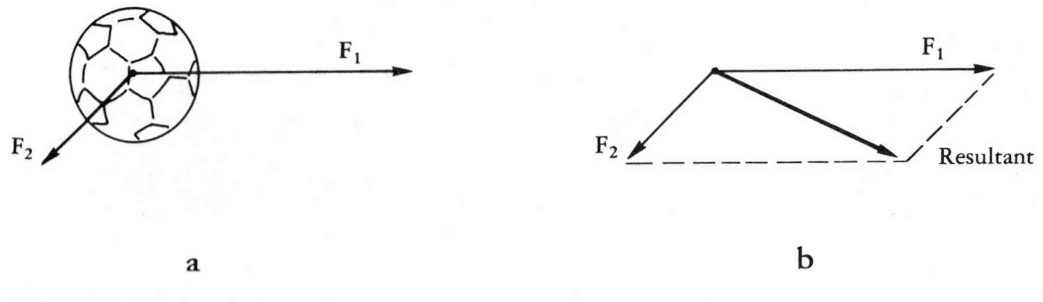

FIGURE D.10

The net effect of two nonperpendicular forces applied to a soccer ball, as seen from a top view.

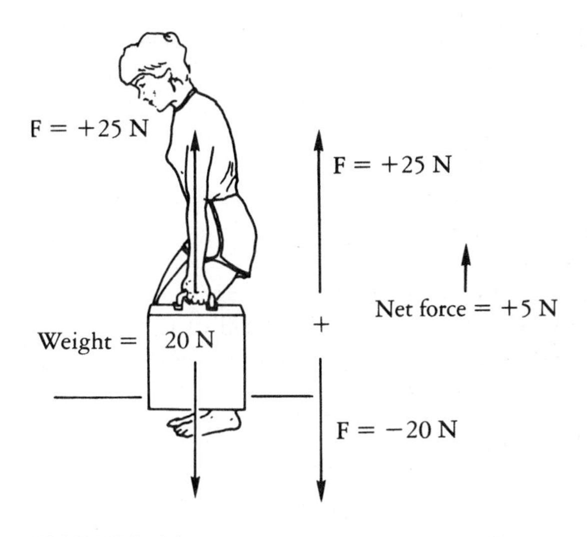

FIGURE D.11

When two opposite forces act on a system (the upward lift and the load's weight), the net force is the algebraic sum of the two forces. The system's motion is determined by the net (resultant) force.

The Directional Effects of a Force: Vector Resolution

To identify the resultant of two or more external forces on a system, we use a process called *vector addition* or *vector composition*. The multiple forces are then represented by a single net force acting on a system in some resultant direction. Recall that a vector is any quantity that has magnitude and direction associated with it. Three of the most common vector quantities with which we deal are forces, displacements, and velocities. Any single vector quantity may be added or composed with another of like kind.

Often an occasion arises in which the observed movement of a system or a single force acting on a system is to be analyzed in terms of identifying its component directions. In such cases, the single vector quantity given is divided into two components (e.g., a horizontal component and a vertical component). These two components, if composed, represent the single resultant vector. The directions of these components are relative to some reference, such as the ground or a body segment, and are usually perpendicular to each other. Such a procedure may be thought of as the reverse of the process of vector composition. The operation is called **vector resolution**, and is the method for determining two component vectors that form the one vector given initially.[2]

[2] Trigonometric methods of vector resolution are in Appendix V.

As an example of the general process of vector resolution, let us examine the linear path of a baseball projected horizontally. Figure D.12a illustrates the side view of the path of the ball downward and forward. The total displacement of the ball from the release to the ground is represented by the vector R and can be resolved into a forward displacement and a downward displacement. These are the horizontal and vertical components, respectively. The operation of resolving the given vector, R, into two mutually perpendicular components is illustrated in Figure D.12b–d.

Resolution consists of these steps: (1) draw the vector given initially to a selected scale (Figure D.12a), (2) from the tail of the vector, draw lines representing the desired directions of the two perpendicular components (Figure

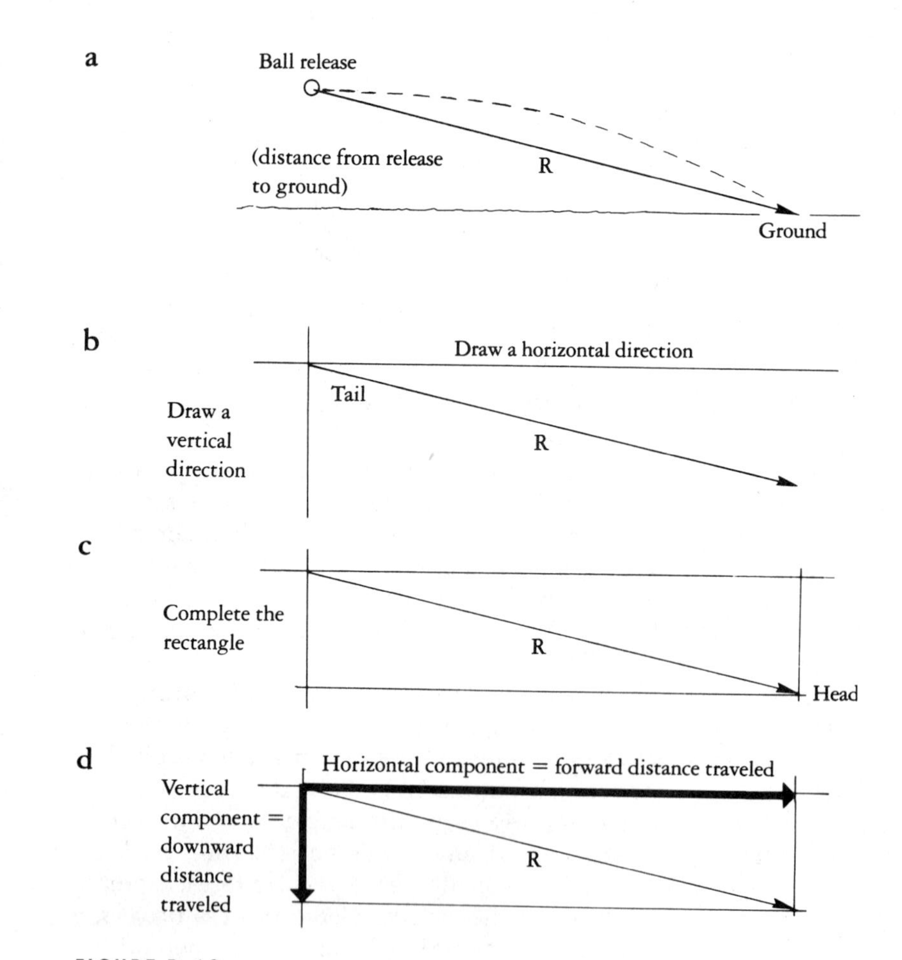

FIGURE D.12

The process of vector resolution. The distance a ball travels from release to the ground is resolved into its horizontal distance component and its vertical distance component.

D.12b), (3) from the head of the vector, draw lines parallel to each of the two directional lines so that a rectangle is formed (Figure D.12c), and (4) draw an arrow head on each directional line at the point formed by the intersection of each parallel line (Figure D.12d). Each head represents the head of one component vector.

The length of the component vector is measured from the tail to the head, according to the same scale used to draw the initial vector.

Basketball provides another example of the use of vector resolution and vector addition in analyzing motion occurring in sport. Figure D.13 shows a player dribbling the ball. The ball is the system subjected to two external forces, the downward force of gravity (the ball's weight) and a forward and downward push by the player. The push has two directional components: a forward horizontal force component, which makes the ball cover horizontal distance, and a downward vertical force component, which moves the ball to the floor faster than would gravity alone. The first step in the analysis is to draw the hand's forward and downward force components to scale (Figure D.13a). Next, the weight vector is added to the hand's downward component to obtain—to scale—the total downward force acting on the ball (Figure D.13c).

The significance of this kind of force-and-motion analysis also is apparent when the motion of a fastball pitch is explained. Such a pitch appears to rise near the plate; however, the upward force acting on the ball due to its backspin

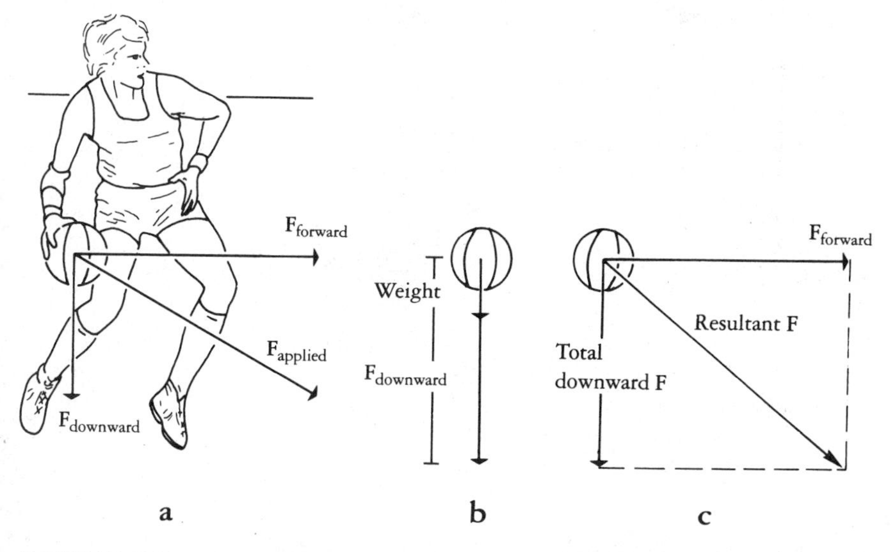

FIGURE D.13

(a) The resolution of the force applied by the hand to a basketball into its horizontal and vertical components. (b) The downward component of the hand's force is added to the downward force of the ball's weight. (c) The total downward force on the ball is combined with the hand's horizontal force to obtain the resultant (net) force acting on the ball at the time of contact.

is not as great as the downward force of the ball's weight. Therefore, the net force is downward, and it accelerates in that direction (Kreighbaum and Hunt, 1978). Because the ball does not drop as quickly as would be expected (that is, without backspin), the illusion of the ball rising is created.

In the human body, vector resolution is used extensively to visualize the effects of muscle pull on bone. Figure D.14a shows a force diagram for the biceps muscle pulling on the forearm. The muscle's force is represented by the force vector F, its length is drawn to scale (e.g., 1 cm = 40 N of force), and the angle it forms with the bone is the same as the muscle's line of pull in that position. The two directions of interest in muscle-bone activity are not horizontal and vertical as in the previous examples. A muscle's force is resolved into one component that is perpendicular to the bone and one component that

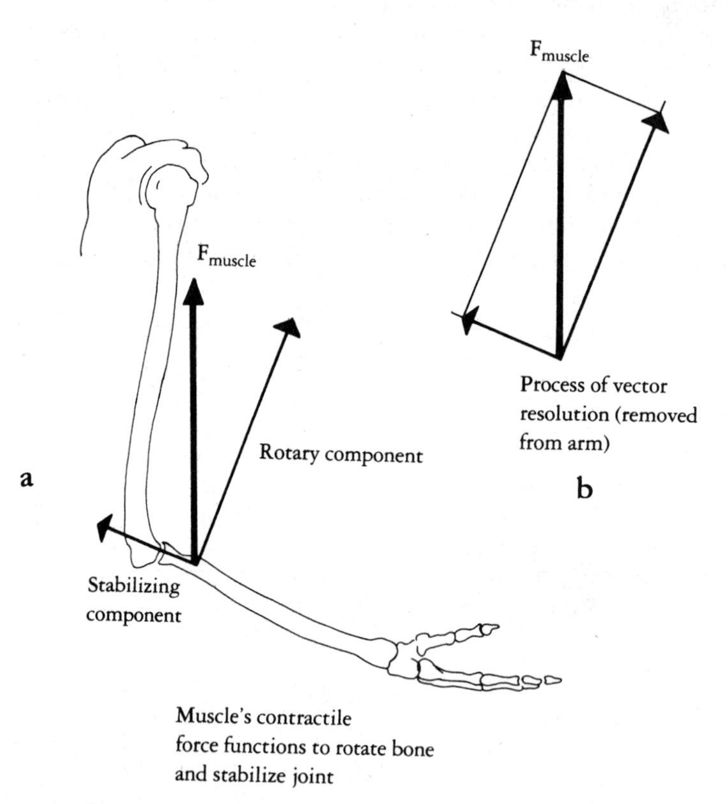

FIGURE D.14

A force vector diagram of the biceps muscle pulling on the forearm. The muscle force is represented by a vector whose length represents the amount of force applied in the direction shown. It is shown resolved into two components: one that is perpendicular to the forearm (it rotates the arm around the elbow axis) and one that lies along the bone and is directed through the elbow axis (it stabilizes the elbow joint).

is parallel to the bone (Figure D.14b). Using the same scale (1 cm = 40 N), each force component is then measured to determine how much is directed perpendicular to the bone, acting to turn the bone (the **rotary component**), and how much is directed along the bone to force it into the elbow joint (the **stabilizing component**). (Note that horizontal and vertical components in such muscle–bone diagrams would serve no purpose, hence we choose the directions of the components to be those of functional significance.)

Understanding Force Vectors

1. Identify or calculate the four properties of each of the forces shown in Figure D.15. Determine the magnitude, using a scale of 1 cm = 20 N of force. Identify the point of application with a P and the lines of action with dashes.
2. Draw a force diagram showing the force of a bat being applied to a baseball. If the baseball is the system, show the four properties of the bat's force.

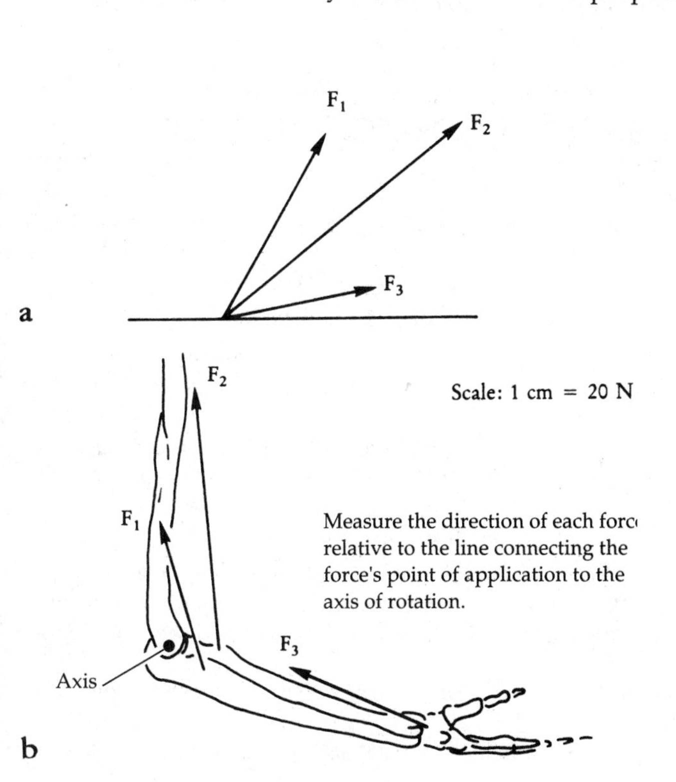
F1
F2
F3
a
F2
Scale: 1 cm = 20 N
F1
Measure the direction of each forc
relative to the line connecting the
force's point of application to the
axis of rotation.
F3
Axis
b

FIGURE D.15

3. A runner pushes downward with 300 N of force and backward with 400 N of force against the ground. Using a protractor and ruler, perform this vector analysis: Draw to scale the two force vectors, considering magnitude, direction, and point of application. Find the magnitude and direction of the resultant force applied to the ground by the foot.
4. According to Newton's third law of motion, what is the resultant force being received by the runner in question 3? What is the vertical reaction force received by the runner? What is the horizontal reaction force received?
5. Using a ruler and protractor, perform the following vector resolutions:
   a. The runner pushes downward and backward with a total force of 200 N at an angle of 40 degrees to the ground. Resolve this force to obtain the magnitude of the vertical and horizontal components. Draw the vertical and horizontal ground reaction forces acting on the runner.
   b. A desk is pulled with a force of 60 N in the following directions: (1) forward and upward at an angle relative to the horizontal of 45 degrees, (2) forward and upward at an angle of 30 degrees, (3) forward at an angle of 0 degrees, and (4) forward and downward at an angle of −10 degrees. Find the vertical and horizontal force components for each situation.

References and Suggested Readings

Alexander, R. M. (1975). *Biomechanics: Outline studies in biology*. New York: John Wiley & Sons.

Barham, J. (1978). *Mechanical kinesiology*. St. Louis: C. V. Mosby.

Brancazio, P. J. (1984). *Sportscience*. New York: Simon & Schuster.

Cooper, J. M., & Glassow, R. B. (1976). *Kinesiology* (4th ed.). St. Louis: C. V. Mosby.

Hopper, B. J. (1967). Units and measurements in the mechanics of track. *Track Technique 29*, 908–912.

Kreighbaum, E. F., & Hunt, W. (1978). Relative factors influencing pitched baseballs. In F. Landry & W. A. R. Orban (Eds.), *Biomechanics of sports and kinanthropometry* (pp. 227–236). Miami: Symposia Specialists.

Lightman, A. (1983). Pas de deux. *Science 83, 3*(3), 33.

Resnick, R., & Halliday, D. (1977). *Physics: Part I*. New York: John Wiley & Sons.

Stucke, H., Baudzus, W., & Baumann, W. (1984). On friction characteristics of playing surfaces. In E. C. Frederick (Ed.), *Sport shoes and playing surfaces* (pp. 87-97). Champaign, IL: Human Kinetics.

Tricker, R. A. R., & Tricker, B. J. K. (1967). *The science of human movement*. New York: American Elsevier Publishing Co.

Wheeler, G. F., & Kirkpatrick, L. D. (1983). *Physics: Building a world view*. Englewood Cliffs, NJ: Prentice-Hall.

CONCEPT MODULE E

Torque

PREREQUISITES

Concept Modules B, C.1–6, D

CONCEPT MODULE CONTENTS

E.1 Torque and Rotary Motion

When an external force acts on an unrestrained system, it causes that body's center of gravity (CG) to move in a linear path (e.g., a thrown softball, a pushed opponent, a projected jumper). If the system is restricted to moving around an axis, however, the body rotates when the force is applied, but only if the force does not act directly through the axis of rotation; that is, the point of application of the force must not be at the axis, and the line of action of the force must not pass through the axis. For example, to open (rotate) a door, the force is applied at a point far from the hinges, and the direction of the pull is not in toward, or away from the hinge axis. To turn a steering wheel, the hand applies a force at the rim, not the center. To turn a page in a book, it is lifted at the edge away from the binding.

Such an off-axis force is called an **eccentric** (off-center) **force**, and the turning effect it has on the body is called **torque**.[1] The off-axis distance to the force's line of action is called the **force arm** (or sometimes moment arm, lever arm, or torque arm). The greater this distance, the greater the torque produced by the force. The specifications of a force arm must be noted, however. *The force arm is the shortest distance from the axis of rotation to the line of action of the force.* The shortest distance is always the length of the line that is perpendicular (90 degrees) to the force's line of action ($d_{\perp}$). Recall that *perpendicular* is designated by the symbol $\perp$. This definition of a force arm is essential in determining how much torque is acting on a system.

Properties of Torques

All forces have unique properties that must be considered when dealing with rotating systems. The properties are *magnitude* (how much force), *direction* (the

[1]Another common term used to denote torque is *moment* or *moment of force*.

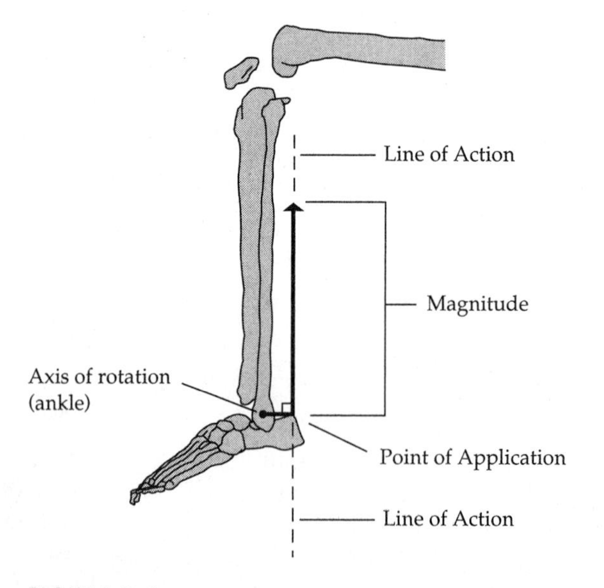

FIGURE E.1

A muscle force vector with its four properties identified: magnitude, direction, line of action, and point of application. The muscle force is torque-producing because its line of action does not pass through the axis of rotation.

way the force is applied), a *line of action* (the theoretical line represented by the dashes that go forevermore in either direction), and *a point of application* (the point at which the force is being applied to the system). These properties also are associated with torque: *magnitude*, $T = F(d_{\perp})$, and *direction*, clockwise (–) or counterclockwise (+). The *point of application* and *line of action* of a torque are the same as the point of application and line of action as the force producing the torque, and in addition, a torque must include a *force arm* ($d_{\perp}$).

Torque is produced by muscles when they pull on bones, and the result is rotary motion of the body segments. Figure E.1 illustrates a muscle force on a bone to produce torque. The stronger the contraction of a muscle, the greater the torque on the bone; the longer the force arm, the greater the torque. Although the point of application of a muscle's force on a bone (insertion point) cannot be changed, the line of action of a muscle's pull (in line with the direction of the muscle's shortening) changes relative to the bone (the angle of muscle insertion). As a result of the changing direction of muscle force on the bone as the bone rotates, the force arm changes length. Therefore, the torque that a muscle is able to exert on a bony segment changes as the segment position changes. Such changes are examined in detail in Chapter 3.

Calculating Torque

The amount of torque acting to rotate a system is found by multiplying the magnitude of the applied force by the force arm distance from the line of force to the axis of rotation[2]:

EQUATION E.1

$$\text{Torque} = \text{Force} \times \text{force arm} \quad \text{or} \quad T = F(d_{\perp})$$

where T is torque (Newton meters or pound-feet), F is force (Newtons or pounds), and $d_{\perp}$ is the length (meters or feet) of the shortest line drawn from the axis to where it meets the line of action of the applied force at 90 degrees.

Because force is measured in Newtons or pounds[3] and the distances are measured in meters or feet, the appropriate torque unit is Newton meters (Nm) or pound-feet (lb-ft). We could also refer to the units as torque units. For example, if a force of 50 N were applied with a force arm of 2 m, the torque produced by that force would be 100 Nm, or 100 torque units. The larger the force and/or the longer the force arm, the larger the torque. For this reason, with two equal forces applied, the force that has the longer force arm produces the larger torque.

[2]Trigonometric calculations of torque also can be used, and the appropriate concepts are included in Appendix V.
[3]Units in British and metric systems are listed in Appendix I.

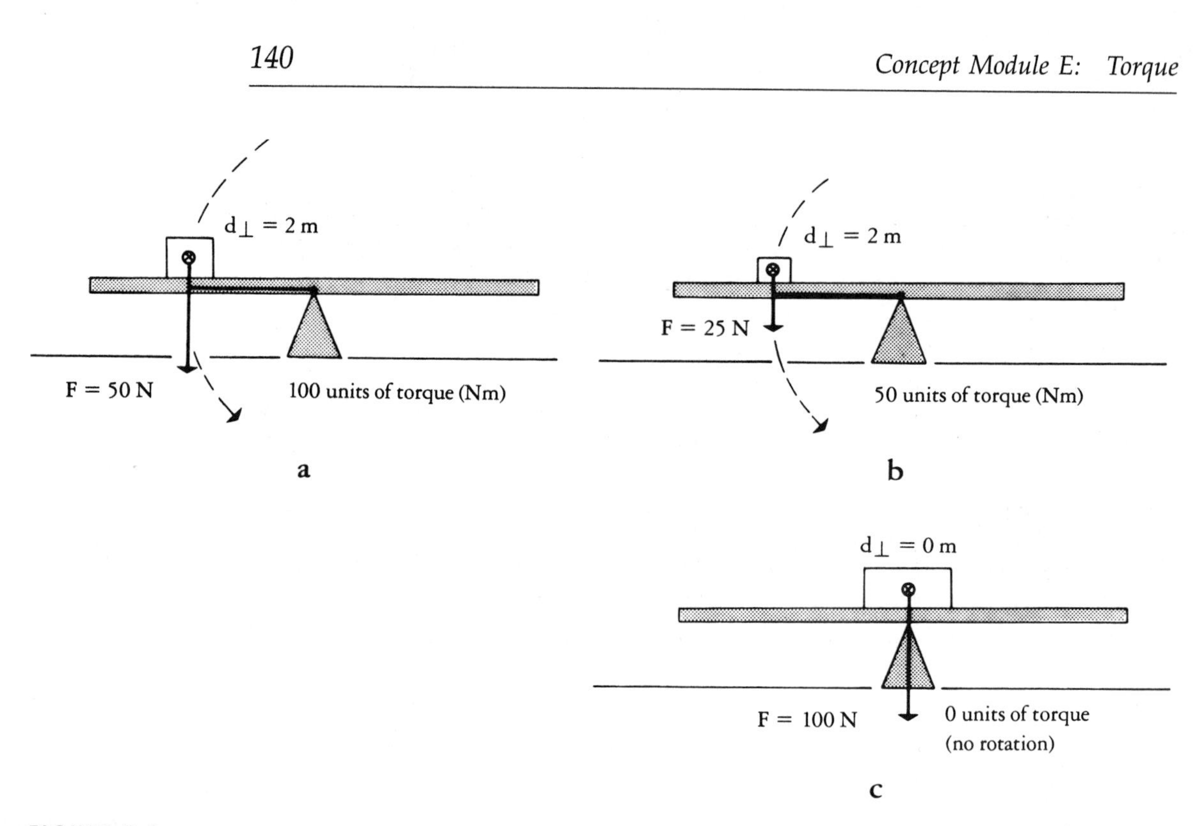

FIGURE E.2

(a and b) A plank balanced on a fulcrum is caused to rotate by a torque produced by a weight placed on the plank at some distance from the axis. (c) No torque is produced when the weight is placed over the axis, and the plank does not rotate.

Figure E.2 illustrates a teeter-totter as a system capable of rotation when a torque is applied to the plank. The torque of 100 torque units (Nm) is the product of the 50-N eccentric weight force times the perpendicular distance of 2 m from the line of force to the axis of rotation, and the plank rotates in the direction of the applied torque (Figure E.2a).

Note that if the weight is lighter, it creates less torque (Figure E.2b). Also, if the weight is placed closer to the axis, it creates less torque. In fact, if the line of force of the weight is directed through the axis itself, zero distance ($d_\perp$) results (therefore zero torque) and no rotation is produced (Figure E.2c).

Note that the distance between the axis and line of force is not necessarily the distance between the axis and the point of force application. An example illustrating this concept is shown in Figure E.3. The trapdoor is caused to rotate around its hinges (a horizontal axis) when a pulling torque ($F_1 d_\perp$) is applied at the door handle. The distance from the axis must be perpendicular to the line of force, which, as shown for F_1, is also perpendicular to the door itself. If the force is applied at some angle other than 90 degrees to the door, as in F_2, the $d_\perp$ is shorter and the torque is reduced.

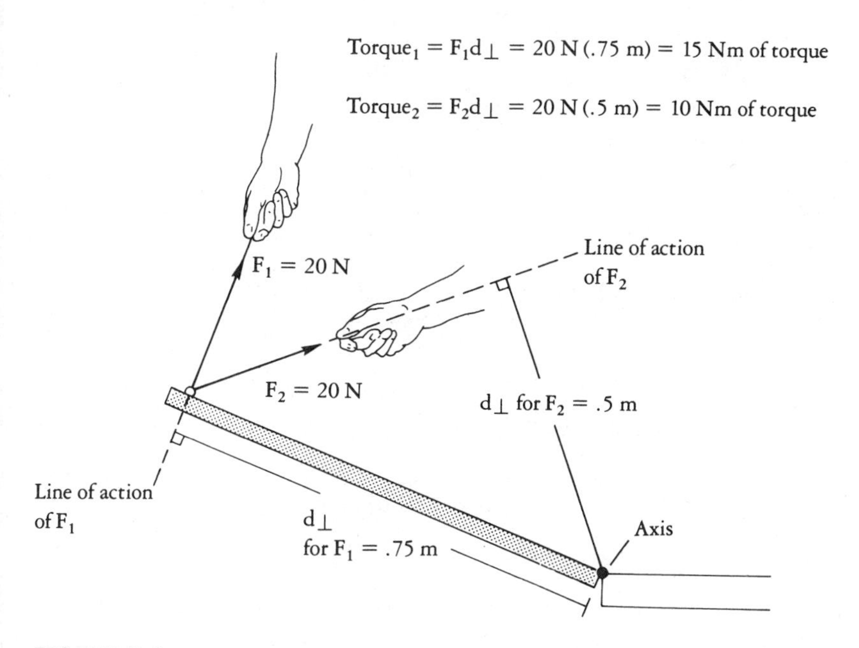

FIGURE E.3

The force arm ($d_\perp$) is the distance from the axis to the line of action of the applied force.

E.2 The Effect of Two or More Torques on a System: Vector Composition

When more than one external torque is applied to a system at the same time, all the torques must be considered, for their net effect determines the rotary motion of the body. Using the algebraic method to find the resultant torque on a system that has multiple torques applied to it is the same process as that discussed under net force for linear motion.

Figure E.4 shows a segment with two external forces applied to it and the torques they produce. Diagrams of such situations are drawn to a convenient scale, such as 2 cm on paper equals 40 N (~9 lb) for representing force, and 2 cm on paper equals 10 cm for representing distance.

The two applied torques in Figure E.4 tend to rotate the segment in opposite directions; they oppose one another. Conventionally, a counterclockwise (ccw) direction is called the positive direction of rotation, and a clockwise direction (cw), the negative direction. The rotary-motion response of the segment is determined by the **net torque** applied to it, just as the linear-motion response of a body is determined by the net force. Calculating the net torque is similar to finding the net force on a system and may be found by algebraically adding

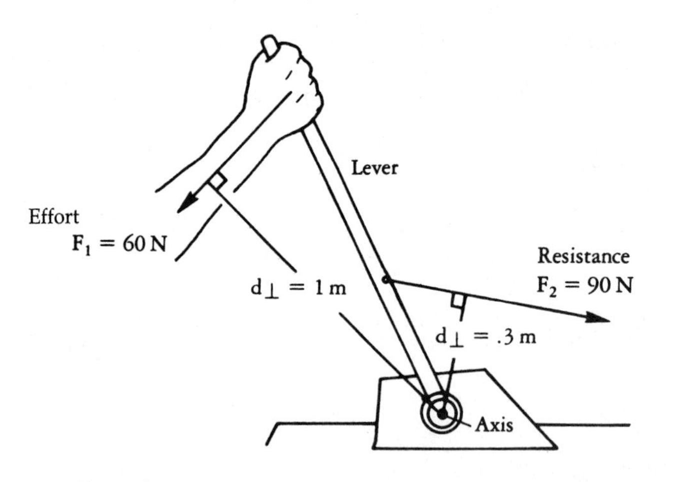

FIGURE E.4

Two forces are applied to a handle free to rotate around an axis. The torque produced by each force is $T = F(d_{\perp})$.

the two or more applied torques, with plus and minus signs used to indicate the directions of each:

$$(+60 \text{ torque units}) + (-27 \text{ torque units}) = +33 \text{ torque units}$$

The segment thus rotates ccw in response to a torque of +33 torque units (Nm).

The graphic method of vector composition of torques on a system is useful for analyzing a system in rotary motion, as is the vector composition of forces on a system in linear motion. Because torques are vector quantities, the direction of the torques on a system may be specified by what is known as the **right-hand thumb rule.** Figure E.5a illustrates this convention. If the fingers of the right hand are curved in the direction of the torque around an axis of rotation, the extended thumb points along the axis and represents the direction of the applied torque.

To determine the net torque on the system shown in Figure E.4, two arrows are drawn to scale to represent the magnitude and direction of each of the torques acting (Figure E.5b). In this case, the torque produced by Force 1 is +60 torque units ccw, and (applying the right-hand thumb rule) is directed out from the page in Figure E.4. Since the ccw direction is positive, this vector is drawn to scale toward the top of the page in Figure E.5b to represent +60 torque units. The torque produced by Force 2 is –27 torque units (cw), and (applying the right-hand thumb rule) is directed into the page in Figure E.4. Since the cw direction is negative, this vector is drawn to scale toward the

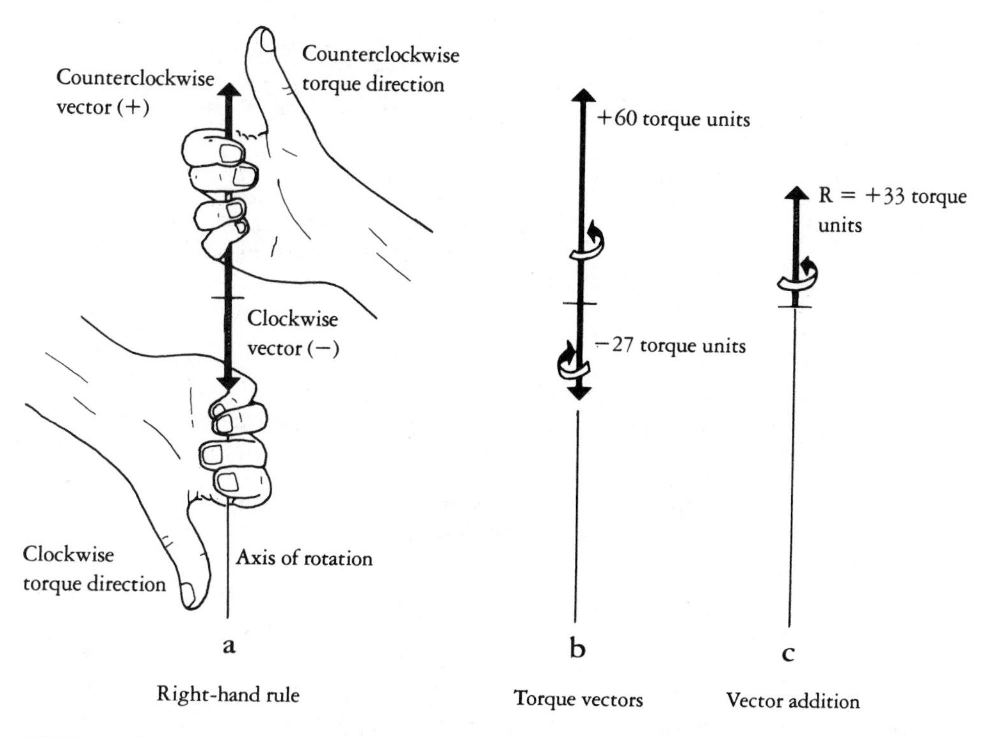

FIGURE E.5

(a) The right-hand thumb rule for finding the direction of a vector in angular quantities. When the fingers of the right hand are flexed about an axis of rotation in the same direction as the torque, the vector direction is indicated by the direction of the thumb. (b, c) The graphic method for determining the net torque on the lever in Figure E.4.

bottom of the page in Figure E.5b to represent –27 torque units. We now have illustrated the torque vectors in magnitude and direction and can now add, or compose, these vectors into a net torque acting on the system, as shown in Figure E.5c.

As can be seen, the resultant torque on the system as calculated by this method is the same as that calculated by the algebraic method, that is, a net torque of +33 torque units acting in the positive direction to cause ccw rotation with the vector directed out from the page.

E.3 Systems in Linear and Rotary Motion

Purely linear motion of a system's CG is caused by a net external force whose line of action passes through the body's CG. Purely rotary motion of a body occurs when a torque is applied on a body with a fixed axis. Both linear and

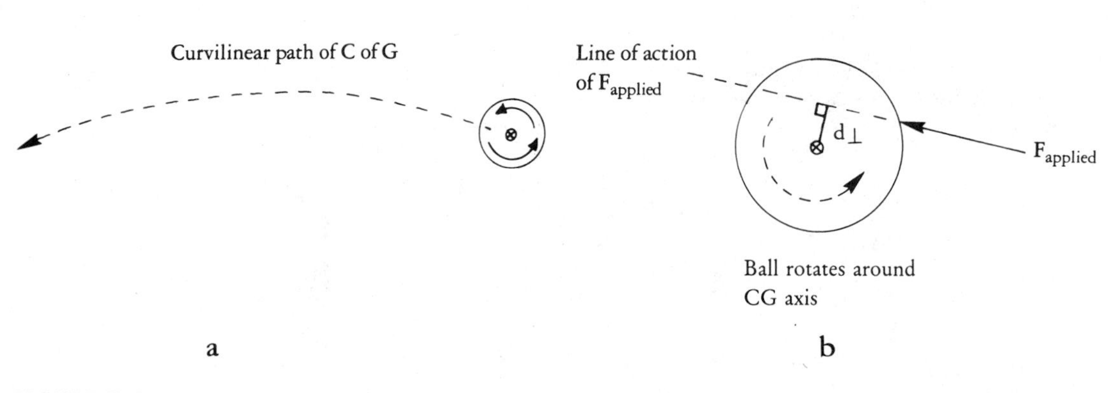

FIGURE E.6

(a) A volleyball displaying linear and rotary motion. (b) Application of torque to cause the ball to spin.

rotary motion occur for a body not restrained to a fixed axis, that is, when the system is free to travel. Such is the case in many sport events, particularly activities in which a system is projected into the air. Recall that any unsupported rotating system rotates around an axis that passes through its CG. As an example, a volleyball served with spin is shown in Figure E.6a. The ball's CG travels along a curvilinear path. At the same time, the ball rotates around an axis passing through the CG. A force must be applied eccentric to the axis of rotation to produce these linear and rotary motions (Figure E.6b).

During the time that the striking hand is in contact with the ball, the force of the hand acts forward and upward and eccentric to the axis of rotation at the CG. Thus, this force causes initial upward and forward linear motion of the ball's CG. The torque on the ball can be determined by multiplying the magnitude of the applied force produced by the hand and its perpendicular distance from the CG (axis of rotation). If the line of action of the hand's force passes through the axis of rotation, no torque is produced and therefore no spin.

The same principle applies to projecting the human body in aerial somersaults in gymnastics and diving and in twists and spins in dance leaps. In a push-off, the reaction force's line of action must not pass through the body's CG if rotation is desired (see Figure J.1b, p. 534).

Understanding Torque

1. Define torque, and draw a force diagram of torque being applied to an object to cause it to rotate. How could you increase the torque without increasing the magnitude of the applied force?
2. Draw a force diagram of two teachers on a teeter-totter in a balanced position. Insert magnitudes for each person's weight (force) and the dis-

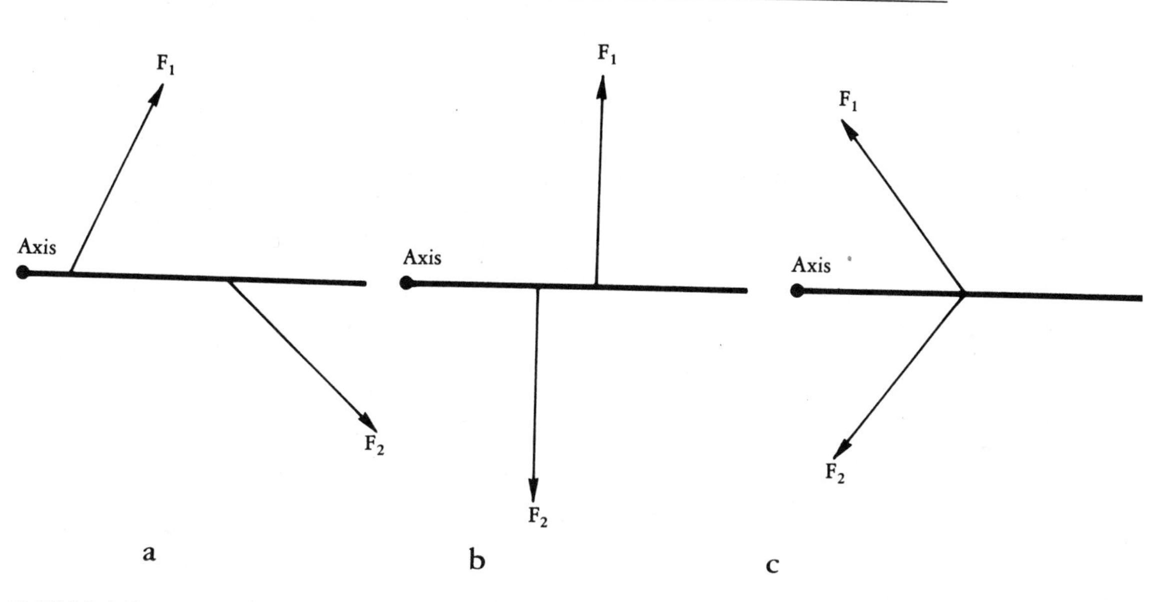

FIGURE E.7

The forces applied to each system in (a), (b), and (c) are all of equal magnitude.

tance that each force is from the axis (fulcrum). How much torque is being applied to the teeter-totter by each person? According to the numbers you selected, is the teeter-totter balanced? If not, what could be changed to make it balance?

CONCEPT MODULE E

3. For each of the illustrations shown in Figure E.7, calculate whether the torques shown are motive torques or resistive torques, and calculate the direction of motion of the system.

4. In each of the four systems shown in Figure E.8, use dashes to indicate the lines of action of each force. Draw the force arms for each of the forces in each system. Then answer these questions:

   a. Compare the force arms and directions of Force 1 (F_1) in each of the four systems. What generalization can be made about the relationship between the direction of the forces and the magnitudes of their force arms? (F_1 is equal in magnitude in all systems.)

   b. If all the forces are of equal magnitude in all the systems, compare the torque-producing capabilities of each, and then state a principle about the direction of pull of the force and the magnitudes of the torque that a force can produce.

   c. Using the convention of a plus sign for ccw and a minus sign for cw, indicate for each torque whether it is a ccw- or cw-producing torque by placing a plus or a minus sign by the vector.

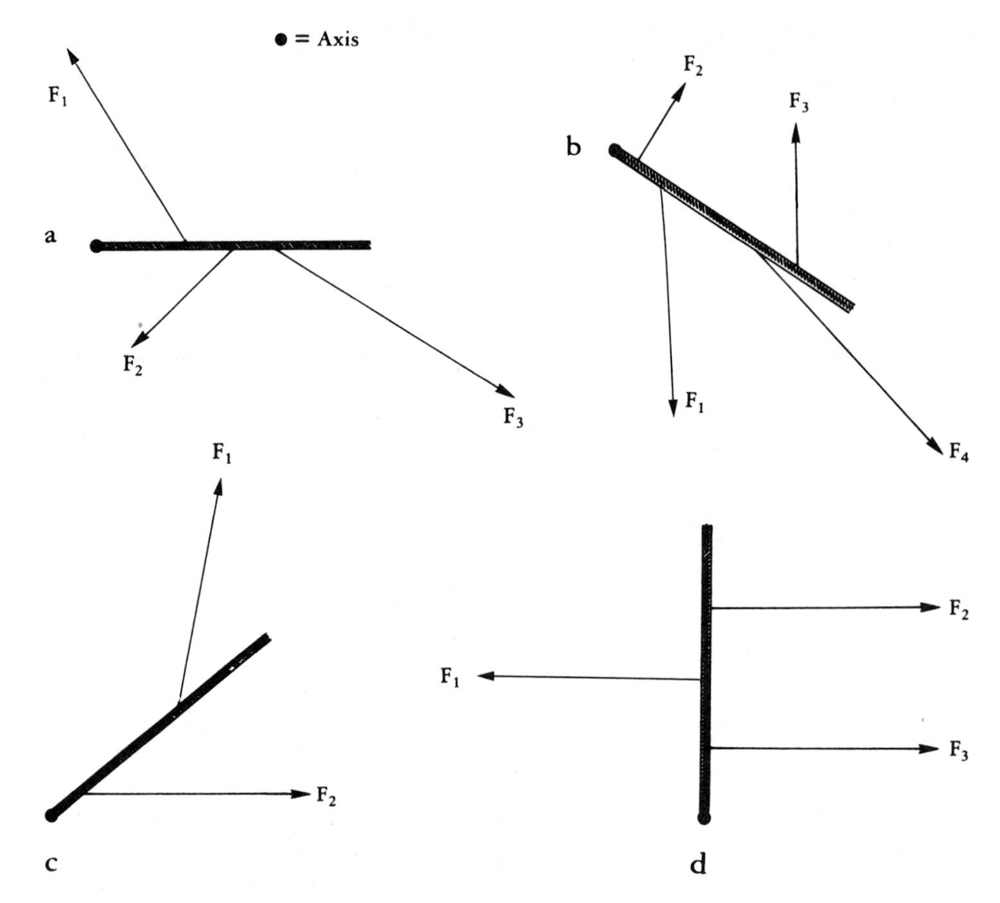

FIGURE E.8

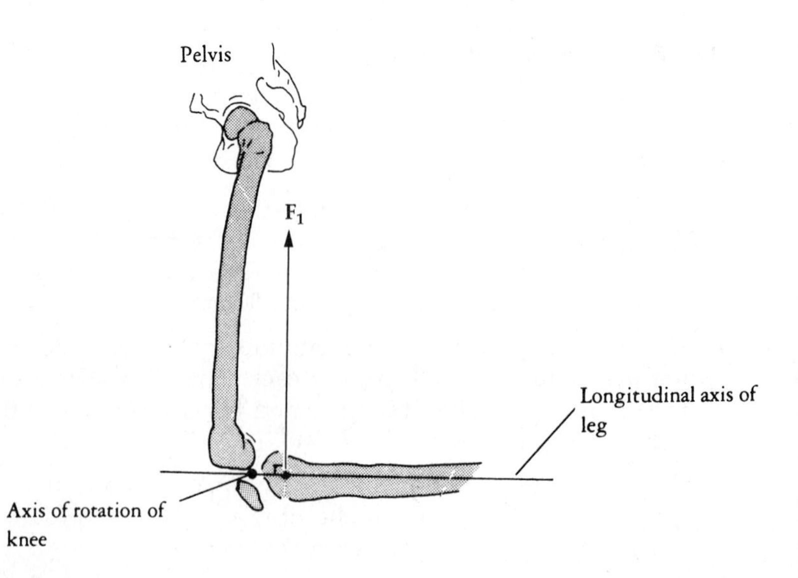

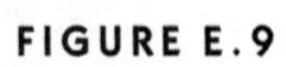

FIGURE E.9

d. Using a scale of 1 cm = 10 N of force, and 1 cm = 1 cm of force-arm distance, calculate each of the torque magnitudes for each of the systems. Place the appropriate sign in front of each torque, and sum the torques acting on each system.

5. In Figure E.9, identify the four properties of the muscle force and its force arm.

E.4 Torque and the Body's Center of Gravity Location

Because the CG is frequently the point in a system that is used for motion analysis, the factors that determine its location should be understood. The CG of a body is that point through which the force of gravity acts; all the body's weight seems to be concentrated at this point. The magnitude of the force of gravity acting on the body is called the weight of that body. The point of application of the force of gravity is at the body's CG; its line of action is vertical, toward the center of the earth.

The force of gravity actually pulls on every particle of mass in the body and thus gives each part weight. From a practical standpoint, it is useful to consider the individual movable body segments as the smallest divisions of the total body's mass. Each body segment, or mechanical link, has its own mass, weight, and CG, and gravity exerts a downward force through each CG. Figure E.10 illustrates the approximate locations of the CGs of the body segments, or links.

As would be expected, the CG of each segment is situated closer to the more massive end of the segment (usually toward the proximal end). The force of gravity pulling on each segment is equal to the weight of the segment, which may be estimated as being some proportion of the total body weight. Thus, the total body weight may be found by adding all the downward-directed forces (the segmental weights). The estimated proportion of the total body weight given to each segment (for males and females) is listed in Appendix III, Table III.1.

Each segment's weight influences not only the total body *weight* but also the total body's CG *location*. One definition of the CG of a system is that the CG is an imaginary point about which the body's weight is equally *distributed*. The key to this definition and to the relationship of the CG to torque is the word *distributed*. According to the definition, the sum of the segmental weight *torques* on all sides of the body's CG point must be equal to zero. The definition does *not* mean, however, that the sum of the *weights* on all sides are equal. For example, a 100-pound boy sitting 4 feet from the fulcrum of a teeter-totter would cancel the weight torque of a 50-pound girl sitting 8 feet from the fulcrum. Similarly, the body's CG serves as the point around which the body segments would balance. The body parts would balance on either side of the CG *without regard to the body's orientation in space*.

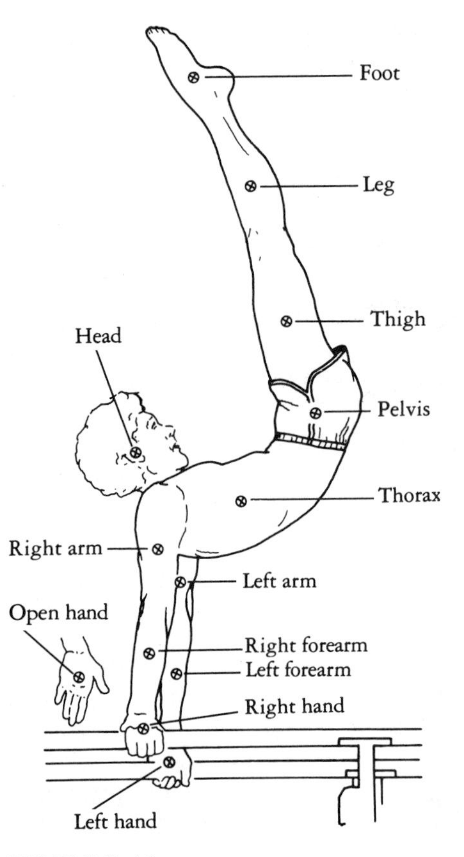

FIGURE E.10
Location of the CG of each body segment.

To understand the importance of segmental weight torques to CG location, we will examine first the illustration of blocks of various sizes and masses shown in Figure E.11. The weights of the various blocks are analogous to the various weights of the human body segments. For the system to be in balance, the opposing torques on each side of the fulcrum must be equal. That is, the blocks must be *equally distributed* about the axis of rotation or fulcrum, which means that the sum of the torques on one side is equal and opposite to the sum of the opposing torques on the other side. If the opposing torques are equal, the balance point, or fulcrum, is in line with the CG of the system. The condition illustrating that the CG is over the fulcrum is shown in Figure E.11a.

In Figure E.11b, the total weights on each side of the fulcrum are the same; however, the sum of the torques produced by these weights does not equal zero (they do not cancel out) because of the distribution of the weights relative to the axis of rotation. Each weight must be multiplied by its $d_{\perp}$ to determine its torque. For the arrangement in Figure E.11b, a new balance point for the system must be located. (Note that the system in Figure E.11a is divided into equal mass [weight] halves and that the opposing torques are also equal. In Figure E.11b, however, the opposing torques are *not* equal even though the system is divided into equal mass halves.)

Similarly, the balance point, or CG, of the human body for any given arrangement of body segments may be found by locating the point around which are equal opposing segmental torques on all sides. The CG point must be located in all three dimensions; that is, it is a representation of the point around which right and left, front and back, and top and bottom segmental torques would cancel out to zero.

If a transverse plane were inserted somewhere through the pelvic region, depending on the individual's body build, a fairly good estimate would be that the CG is located on that plane. This plane dividing the body into equally distributed masses would pass through the CG point. A body balanced on a

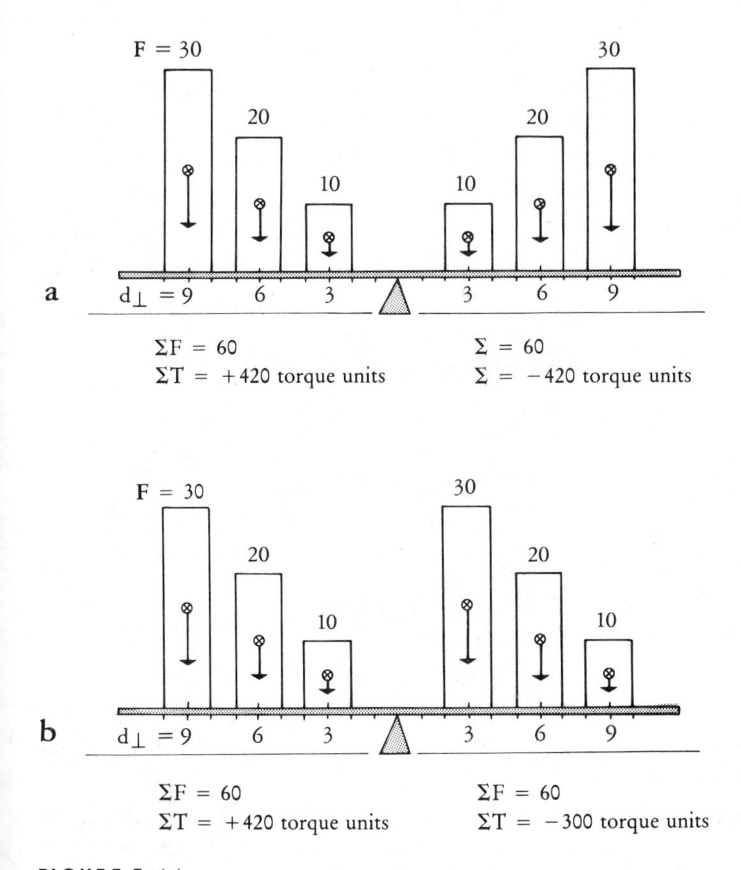

FIGURE E.11

The weight of each mass segment produces a torque about the axis of rotation. (Σ means "sum of.")

CONCEPT MODULE E

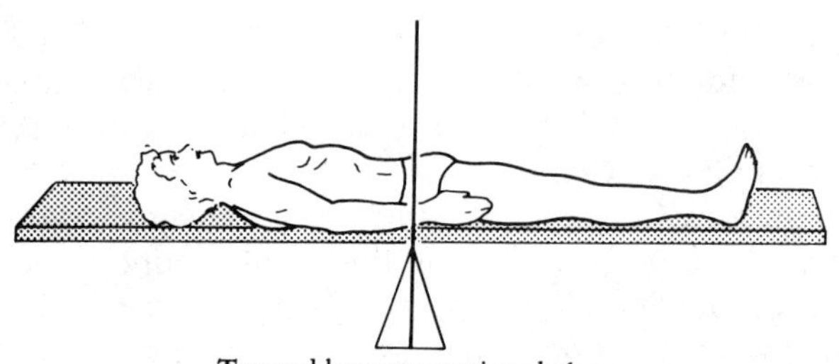
Top and bottom portions balance
a

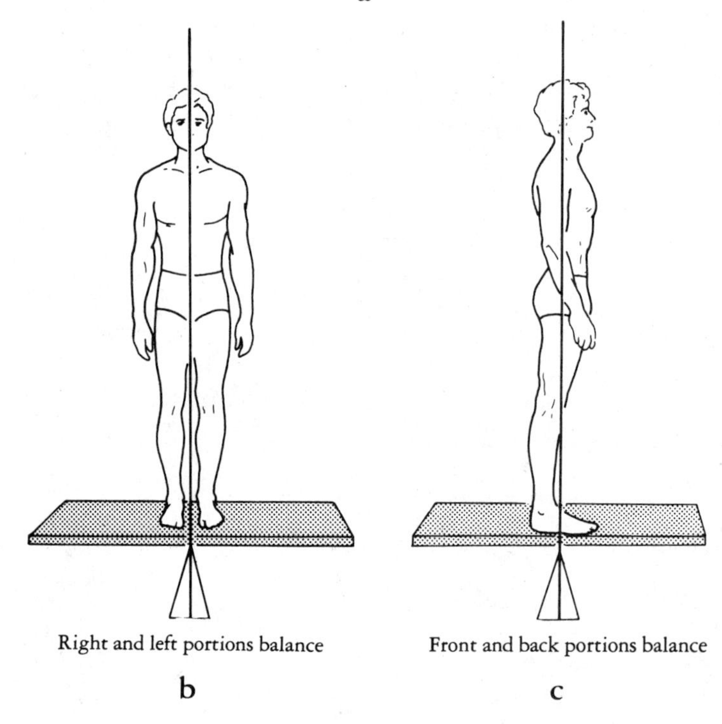
Right and left portions balance
b
Front and back portions balance
c

FIGURE E.12

A body will balance on the teeter-totter if the fulcrum is directly under the body's CG. (a) Top and bottom, (b) right and left, and (c) front and back portions balance.

board and fulcrum such that its CG is located directly above the fulcrum is shown in Figure E.12a.

Similarly, if a plane were located through the body that equally distributed the masses into right and left portions, these portions would balance on either side of the fulcrum, as shown in Figure E.12b. The plane dividing the body in this manner would pass through the CG point. Another plane that divided the body into equally distributed front and back parts would serve to balance the body in the position shown in Figure E.12c. This plane also would pass through the CG.

To visualize a body's CG location, the body does not need to be in a standing or lying position. The same visualization process is followed for determining the CG location in any body position. Picture three "cuts" through the body, each one being made at right angles to the others so that the segmental masses are equally distributed on either side of each plane. The point at which all three orthogonal (perpendicular) planes intersect is the location of the body's CG. Figure E.13 shows the approximate location of the body's CG for three bodies in different positions.

As with the boxes on the board shown in Figure E.11, the CG location of the human body depends on the arrangement of the body segments at any given time. When an arm or leg is moved, the trunk is flexed, the head is lifted or turned, or if some mass is added to the body, the location of the system's CG is changed according to the direction, distance, and amount of mass that is shifted or added. This is due to the fact that if one or more segments move relative to the others, the torque of that segment also changes; therefore, the location of the CG relative to the rest of the body segments changes accordingly as shown in Figure E.14. If the relative position of the body segments remains the same (the system is in the same posture), but the total body orientation is changed in space, the CG remains in the same location within the body.

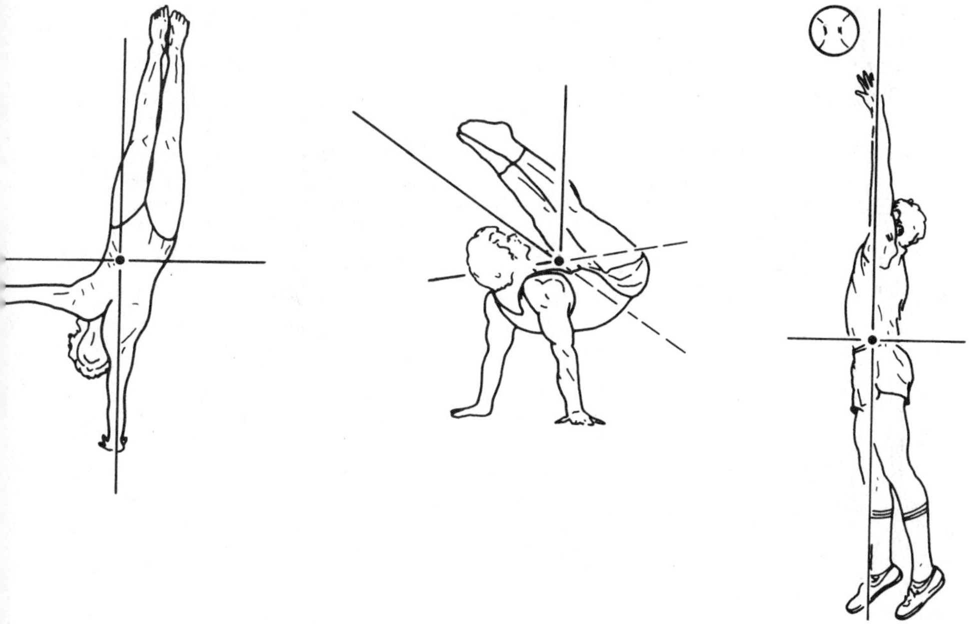

FIGURE E.13

Location of the CG of the body in three different positions.

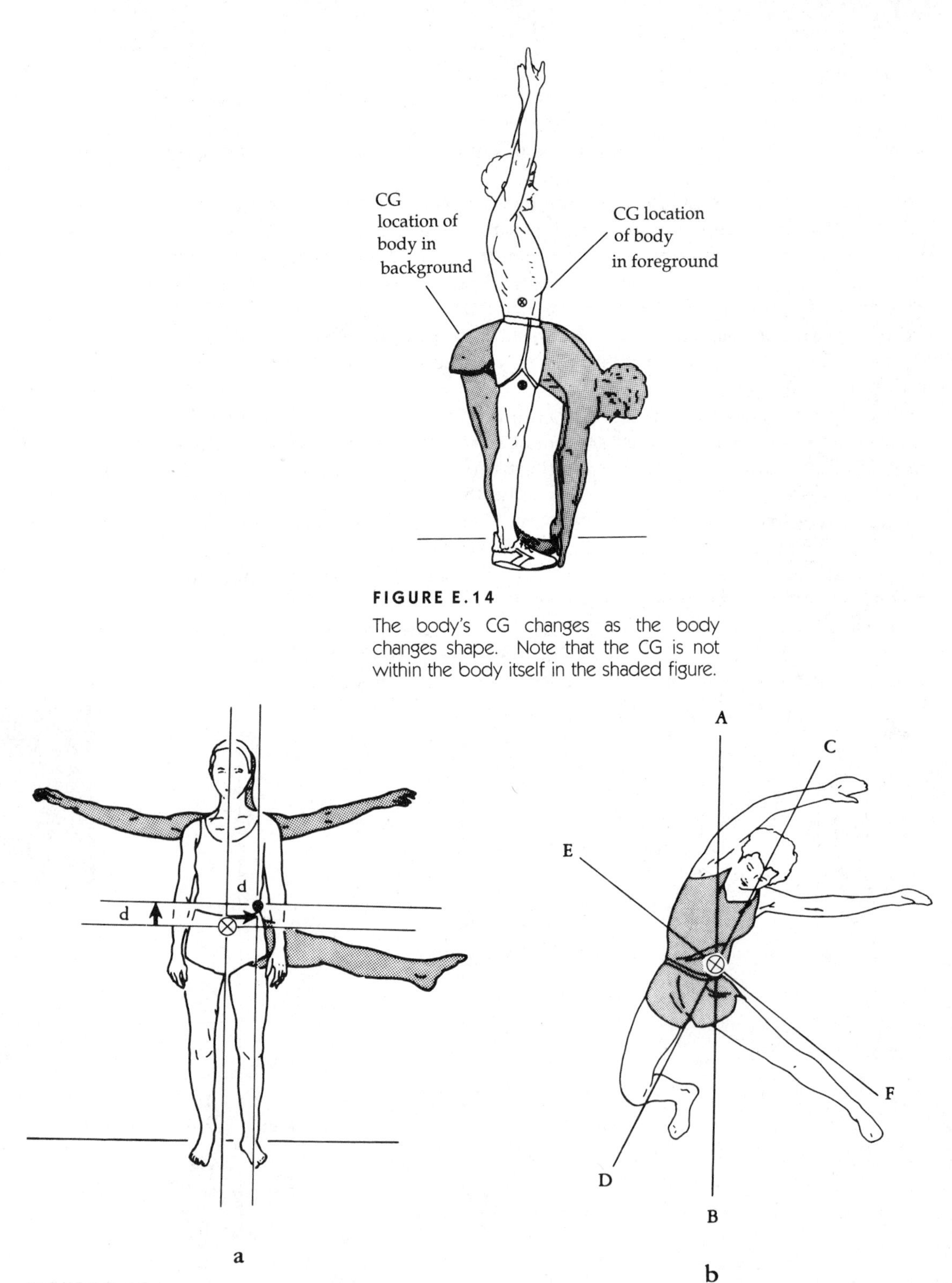

FIGURE E.14

The body's CG changes as the body changes shape. Note that the CG is not within the body itself in the shaded figure.

FIGURE E.15

(a) Estimating shifts in the body's CG as segments shift position. (b) The body would balance if its CG were placed on the edge of any plane (at the end of any line passing through the CG).

The CG location is an important factor in movement analysis, and therefore we should be able to locate the CG of the total body with the body segments in different positions. Since the segmental weights do not change during movement, the positions of the segments relative to one another is the important factor in producing a change in the segmental torques and thus the location of the CG of the body.

For teaching and coaching applications, we need not calculate the CG location but need only understand the concept of balancing the torques on each side of the CG. The two steps in this process are illustrated in Figure E.15a. First, one must imagine the location of the CG in anatomical position (usually in the pelvic region, and equidistant from front to back and from right to left sides). For each segment that is moved from this anatomical position, the CG location moves in the same direction. If the leg is moved up and to the right, the CG of the body moves up and to the right *relative to the rest of the body.*

One must also consider the relative weight of the segment being moved. For example, a leg being moved the same distance as an arm has a greater effect on the shift of the CG location.

Once an estimate of the CG location has been made, one may test the accuracy of the prediction by marking three lines through the estimated CG location and visualizing whether or not the body would balance on a fulcrum, or plane, along that line. For example, would the body in Figure E.15b balance if placed on plane A-B, C-D, or E-F? If the answer is yes, then the estimated CG location is approximately correct. If the answer is no, then the estimated CG location must be changed accordingly so that the body will balance.

The CG location of any system is based on distributed weights of the body's segments where distribution incorporates the torque produced by those segments relative to the CG location. The teacher or coach should be able to visualize and track the CG location of performers as they move in space.

If further calculations are desired, refer to Appendix IV, Methods of Calculating the Center of Gravity.

Understanding the Center of Gravity

1. What is the approximate location of your body's CG when you assume an upright (standing) position? How is your body's CG location changed when you assume the following positions: (a) raise both arms out to the side to a horizontal position, (b) raise both arms forward to a horizontal position, (c) raise only one arm as indicated in a and b. Select other positions and track the changing locations of your body's CG.
2. Figure E.16 shows several blocks of various weights placed on a board supported at each end. A scale is located under the knife edge on the left-hand side of the system. Determine the position of the CG for each of

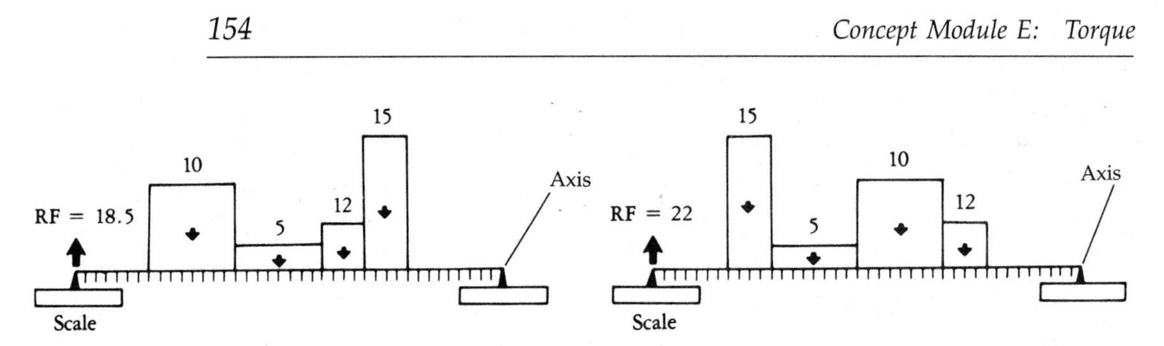

FIGURE E.16

Blocks of various weights placed at different distances from the axis. The scale reading reflects the amount of downward force produced at that end by the specific arrangement of the blocks.

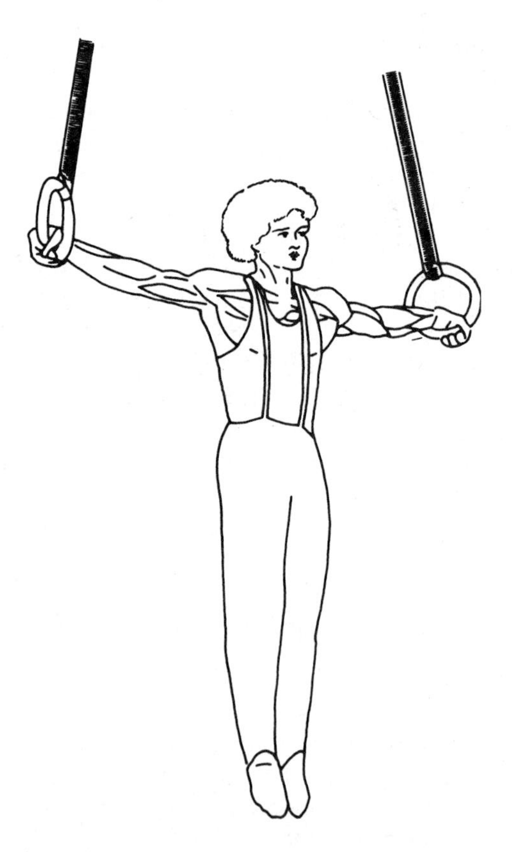

FIGURE E.17

A gymnast performing an iron cross on the rings.

the two block arrangements shown. Discuss why and how the CG changes when the arrangements are changed.

3. When a person moves a body segment or segments, the CG location changes. How far and in what direction the total body CG changes depends on two factors. What are these factors?
4. If a person measuring 64 cm and a person measuring 74 cm each has a CG height of 56% of standing height, what does that tell you about their relative body statures?
5. Three people, all the same height, have different CG locations—52%, 55%, and 58% of standing height. What can you say about their relative body builds?
6. Place a dot where the CG is located for the person shown suspended from the rings in Figure E.17. Discuss the change in CG location when the legs are raised to an *L* position.
7. On the following figures, place a dot where you would estimate the CG to be: E.18, E.19, 1.3.

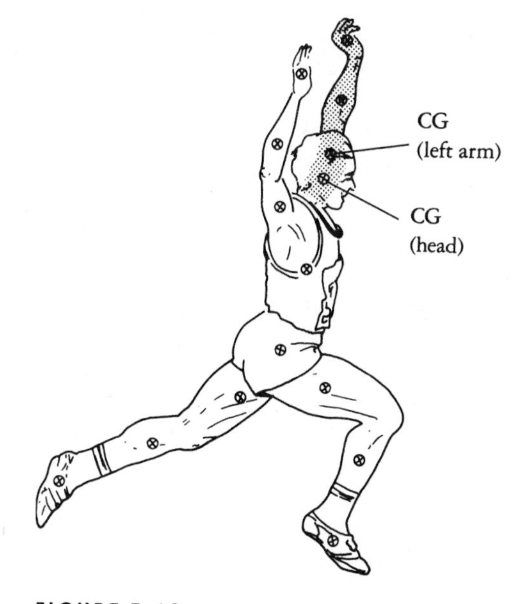

FIGURE E.18

A runner with his segmental CG locations identified.

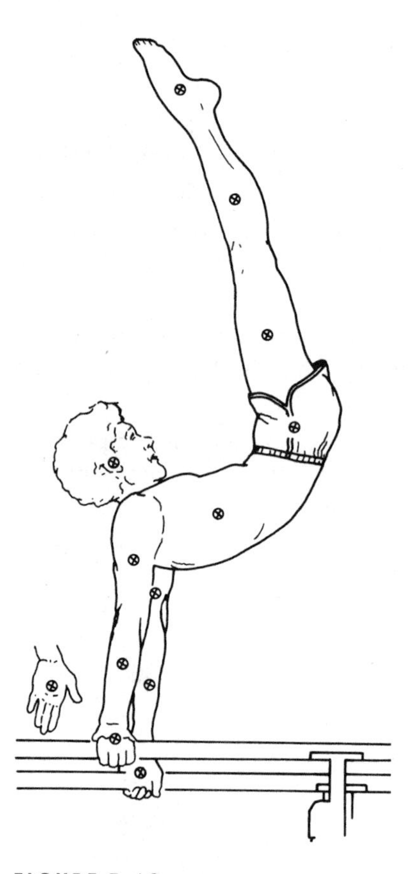

FIGURE E.19

A gymnast with segmental CG locations identified.

References and Suggested Readings

Brancazio, P. J. (1984). *Sportscience.* New York: Simon & Schuster.

Gowitzke, B. A., & Milner, M. (1988). *Understanding the scientific bases of human movement* (3rd ed.). Baltimore: Williams & Wilkins.

Hay, J. G. (1974). The center of gravity of the human body. In C. J. Widule (Ed.), *Kinesiology III* (pp. 20–44). Reston, VA: AAHPERD.

Tricker, R. A. R., & Tricker, B. J. K. (1967). *The science of human movement.* New York: American Elsevier.

Wheeler, G. F., & Kirkpatrick, L. D. (1983). *Physics: Building a world view.* Englewood Cliffs, NJ: Prentice-Hall.

CHAPTER 3

Biomechanics of the Musculoskeletal System

PREREQUISITES

Concept Modules A, B, D, E
Chapters 1, 2

CHAPTER CONTENTS

The mechanical aspects of bone–muscle arrangements should be used as the basis for developing exercises for musculoskeletal training and conditioning programs. Knowledge of the machinelike mechanisms used in producing segmental movements is necessary to understand the effect of various body positions during exercise, the effects of motive and resistive forces on the body segments, and the resulting movements during exercise and sport activities.

Any machine may be described as having one or more of these four functions: (1) to balance two or more forces, (2) to provide an advantage in force whereby less motive force is required to overcome a greater resistive force, (3) to provide an advantage in range of linear motion and speed of movement (i.e., to move the resistance or load a greater distance or faster than the motive force moves), or (4) to change the effective direction of the applied motive force (i.e., to move the resistance in one direction while the motive force is being applied in another direction). Three types of musculoskeletal arrangements serve as machines to provide the structure's ability to produce movement. These three machine-like structures are the lever, the wheel-axle, and the pulley.

3.1 Leverlike Arrangements

A lever system consists of an axis of rotation (sometimes called a fulcrum) around which a rigid lever moves and two categories of torques—motive torques, which act to move (rotate) the lever, and resistive torques, which act to resist the rotation of the lever.

The axes of rotation in the human body lever system are the same joint axes that were introduced in Concept Module B: anteroposterior, mediolateral, and longitudinal. Any movement of a segment around a given axis occurs along a segmental plane of motion: frontal, sagittal, and transverse, respectively. The plane of segmental movement is always perpendicular to the axis of rotation. For example, the forearm moving in flexion and extension rotates around a mediolateral axis through the elbow joint and moves along (in) an anteroposterior plane.

Forces and Torques Applied to Bones

Recall that all forces have four unique properties: (1) magnitude, (2) direction (relative to the bone), (3) a line of action, and (4) a point of application (the point of attachment to the bone). Figure 3.1 illustrates the four force characteristics of a muscle acting on a bone. Since the bone levers rotate around axes of rotation, the only forces that can affect the *rotation* of these segmental levers are those that are torque producing.[1] To be torque-producing, a force must have a force arm (sometimes called a moment arm, lever arm, or torque arm).

Because all lines of action for muscle forces are directed eccentric to the axis of rotation through the joint, all muscles can affect the rotational motions of the bones to which they are attached. It is important to remember that *each* muscle has its own unique force arm and thus its own *unique* torque.

In the case of a long bone, the direction of muscle pull is measured from the longitudinal axis of the bone. In the case of a flat or irregular bone as shown in Figure 3.2, the direction of muscle pull is measured from a line connecting the point of application of the

[1] A torque, in addition to affecting the rotary motion of a body, also affects the linear motion. For a body segment, linear effects of the torque are counteracted by an equal and opposite reaction force at the joint, thus the sum of the linear forces in all directions is zero.

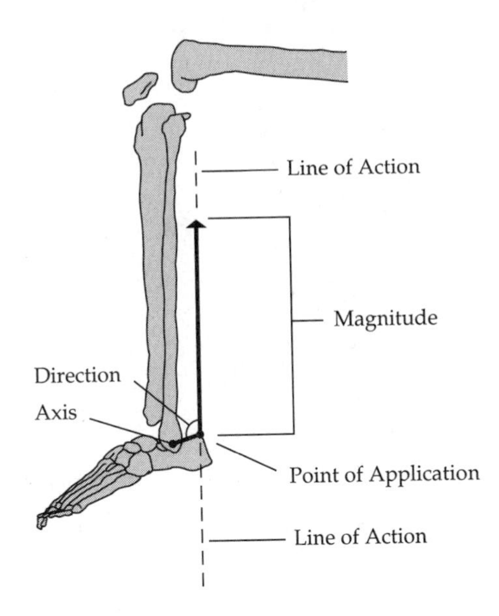

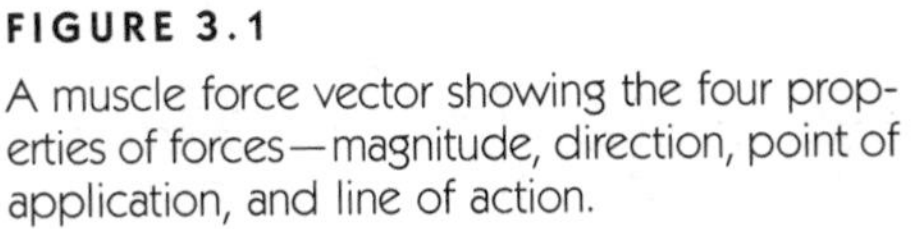
FIGURE 3.1

A muscle force vector showing the four properties of forces—magnitude, direction, point of application, and line of action.

muscle force to the axis of rotation. The flat bone shown, the scapula, has two forces acting on it at the same point of application, F_1 being at 90 degrees to the line connecting the point of application and the axis and F_2 being directed at a larger angle. The difference in the length of the force arms, $d_\perp$ for forces F_1 and F_2 should be noted.

Recall from Chapter 2 that muscle *tension* depends on the number of fibers (motor units) contracting, the tension developed in those fibers, the length of the muscle during its contraction (length–tension relationship), and the velocity of shortening. All of these factors influence the magnitude of force that a muscle applies to a bone. In addition, the magnitude of the torque produced by each force depends on the length of the force arm of that force.

Remember that a force's line of action extends forever. Thus, if a force arm drawn from the axis to the muscle force vector cannot be made perpendicular to that vector, then the force's line of action must be extended until such a perpendicular line may be drawn to the line of force. Figure 3.3 illustrates several examples of forces whose lines of action are directed at various angles to the longitudinal axis of the bone.

The effect of changing the direction of a force on the length of its force arm may be seen by comparing the force-arm lengths of the forces in Figure 3.3a–c. Note that as the force is directed closer to the axis, the force arm is reduced (Figure 3.3a); conversely, as the force is directed farther away from the axis—up to 90 degrees to the lever—the force arm becomes larger (Figure 3.3b). Beyond 90 degrees, the force arm again becomes smaller (Figure 3.3c). What would be the length of the force arm when the force is directed through the axis of rotation, as is the weight force of the forearm and hand in Figure 3.3d?

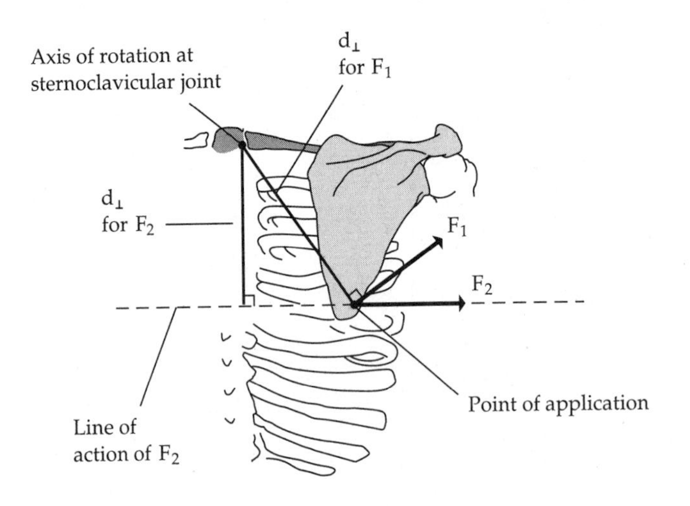

FIGURE 3.2

Forces acting at angles to a flat bone. F_1 acts at 90 degrees to a line connecting the point of application of the force to the axis of rotation in each bone. F_2 acts at an angle greater than 90 degrees.

This relationship between the length of the force arm and the direction of the force (angle from the lever) holds true for all forces acting on a system, whether they are motive forces or resistive forces, muscle forces or other external forces. Consider the force of gravity, which has a fixed downward direction. As the lever moves through various positions, the angle of the weight force changes with respect to the moving lever, but the force's direction is always toward the center of the earth. When the line of action of the weight force is directed closer to the axis of rotation (i.e., at a smaller angle, less than 90 degrees) or farther away from the axis of rotation (i.e., at an angle greater than 90 degrees), the length of its force arm is reduced; therefore, the torque production of that force is also reduced.

The lever system may be static (motionless) or it may be dynamic (moving). The forces influencing the lever system's movement must be external to that system, and thus any forces that serve to move a body lever, such as muscle forces, are considered external to the lever system itself. Most commonly, muscle forces are thought of as being the only forces that create segmental movement; however, other forces also act to move the body segments, such as gravity, external objects, and other people. A muscle torque acting on a lever not only may create or change movement but also may hold the bony lever in a particular position by resisting other external torques. If the muscles are functioning in this manner, they are in an isometric (static) tension state. If an imbalance of torques is operating on the bone, the lever assumes a dynamic state and begins to rotate in the direction of the greatest torque (net torque). Recall that if the muscle forces are acting on the lever in the direction opposite to its movement, the muscles are contracting eccentrically and are producing resistive

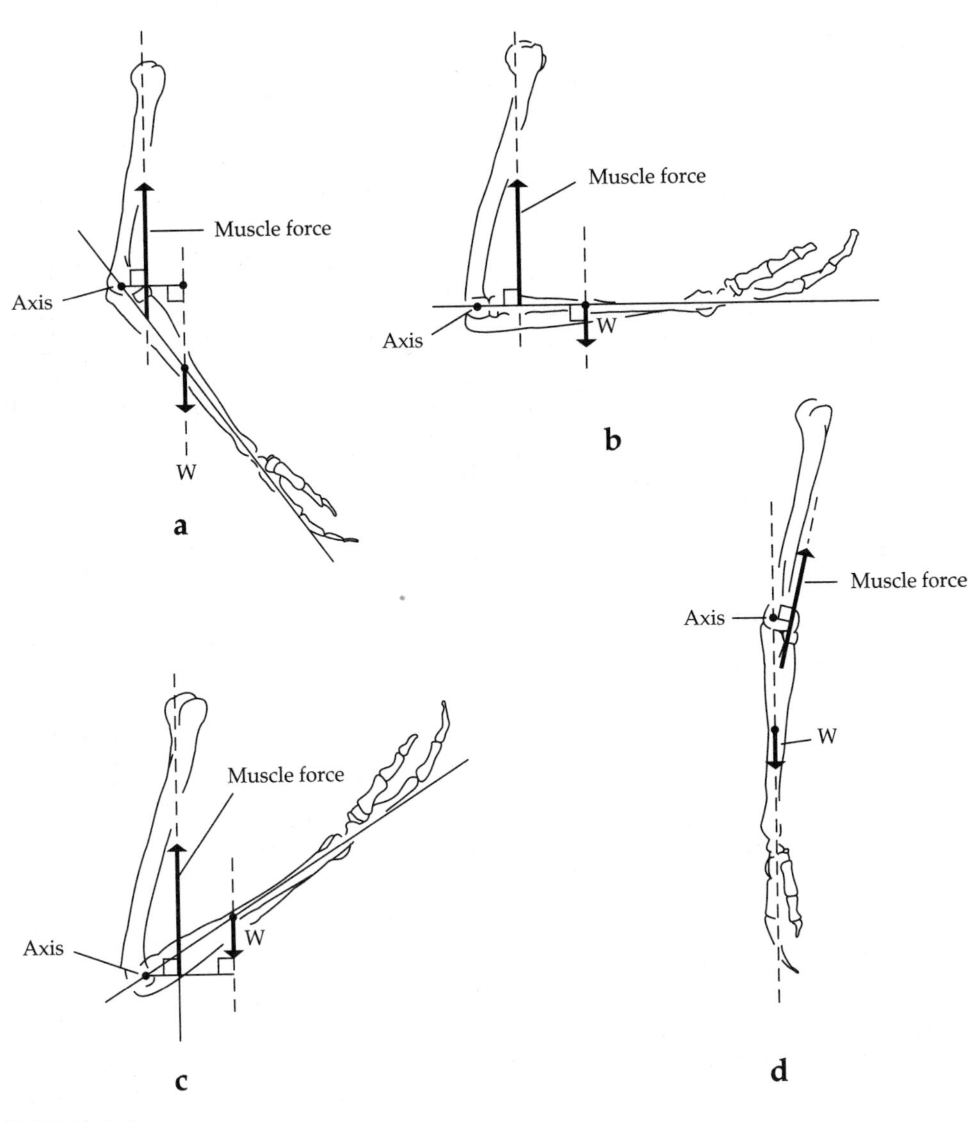

FIGURE 3.3

Four examples of muscle forces acting on the forearm.

torques. Figure 3.4 presents three examples of lever systems in static and dynamic tension states.

A static situation is shown in Figure 3.4a. The two external forces are represented by arrows. The external *torques* are equal in magnitude and opposite in rotary direction; therefore, the segment is not moving. Notice that the muscle-force arrow is directed along the belly of the muscle toward the nonmoving attachment and that the resistive torque produced by the weight of the segment is directed from the center of gravity (CG) of the segment toward the center of the earth.

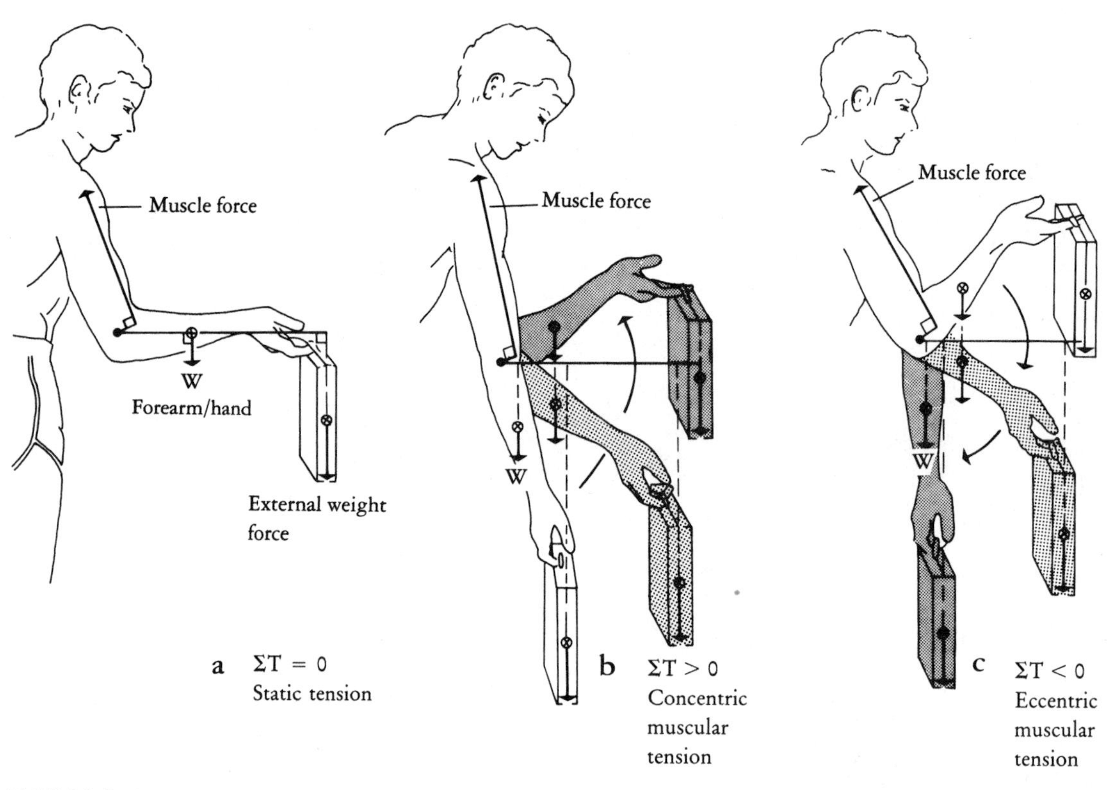

FIGURE 3.4

Lever systems representing static and dynamic states of motion: (a) static, (b) dynamic concentric, and (c) dynamic eccentric.

In Figure 3.4b, the muscle torque provides the motive torque; the muscle contracts concentrically and moves the segment in the direction of the muscle pull. The weight torque is a resistive torque. In Figure 3.4c, the segment moves in the direction of the weight force, and thus gravity serves as the motive torque. The muscle force provides the resistive torque, and the muscle contracts eccentrically.

Sometimes dynamic situations are analyzed as if they were a series of static positions, but such analyses are unrealistic. The inertial or resistive properties of a dynamic rotary system affect the way the body moves. These factors are examined in Concept Module I.

Effect of Net Torque on a Bone–Lever System

In the body's lever system, the torques are divided into motive torques (i.e., those that produce the motion) and resistive torques (i.e., those that resist the motion). If the segment is moving clockwise (cw), then all torques directed cw are motive torques and all those directed counterclockwise (ccw) are resistive torques. Likewise, if the segment is moving ccw, the ccw torques are motive torques, and the cw torques are resistive torques.

To determine whether a segment will move ccw, cw, or remain stationary, one must find which torques are greater by adding all the ccw torques and adding all the cw torques. (Note that the torques and not merely the forces are added.) If the sum of the ccw torques is equal to that of the cw torques, the segment will not accelerate in any direction; it will be in rotational equilibrium. If the opposing sums are not equal, the segment will accelerate in the direction of the larger sum. Figure 3.5 provides two examples. To maintain consistency in calculations, all cw torques are designated as negative (–), and all ccw torques are designated as positive (+).

To determine the net torque acting on the leg lever in Figure 3.5a, identify each force and multiply its magnitude by the length of its respective force arm. The weight torque of the leg equals the weight of the leg multiplied by its force arm.

leg weight torque = $W(d_\perp)$
−6 metric torque units = −40 N × .15 m

[−4.42 British torque units
= −8.98 lb × 0.49 ft]

The torque produced by the weight of the leg is designated as negative because the weight would make the leg rotate cw if no other torques were acting on it.

The torque produced by the ankle weight equals its weight multiplied by its force arm.

ankle weight torque = $W(d_\perp)$
−15 metric torque units = −30 N × .5 m

[−11.06 British torque units
= −6.74 lb × 1.64 ft]

The torque is negative because it would produce a cw movement of the leg.

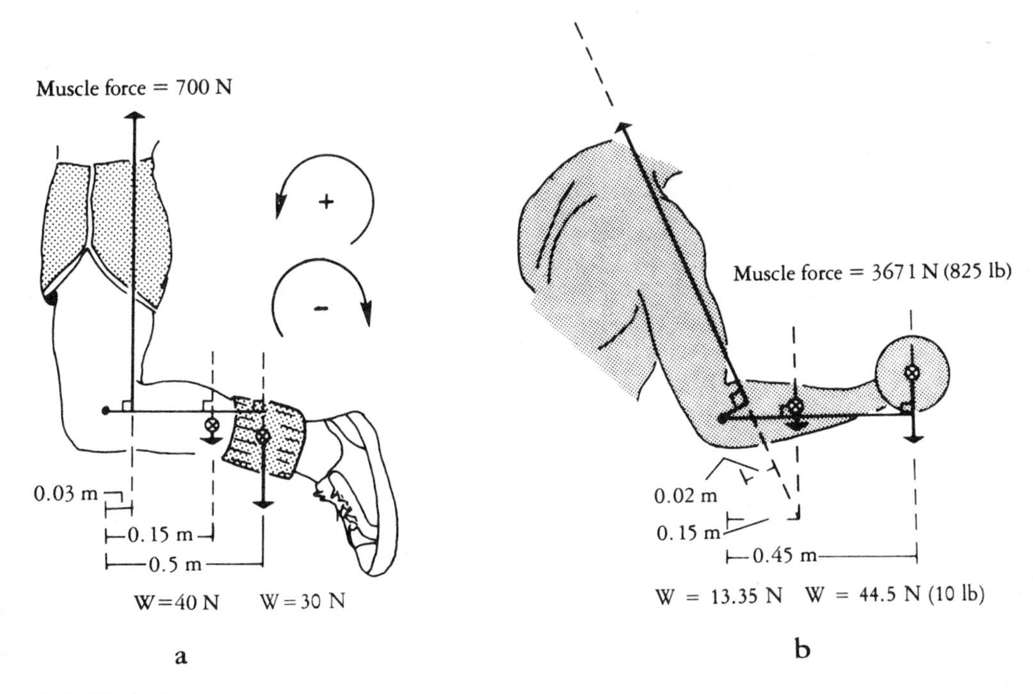

FIGURE 3.5
Forces acting on a lever system. (a) The leg holding an ankle weight. (b) The muscles of the forearm attempting to hold a barbell.

finally, the muscle torque equals the muscle force times its force arm if no other torques were acting on it.

$$\text{muscle torque} = F(d_{\perp}) + 21 \text{ metric torque units} = 700\ \text{N} \times 0.03\ \text{m}$$

$$[15.48 \text{ British torque units} = 157.30\ \text{lb} \times 0.10\ \text{ft}]$$

The muscle torque is positive, as illustrated in this example, because it would produce a ccw rotation of the leg if no other torques were acting on it.

Because the total cw torque—the sum of the leg weight torque plus the ankle weight torque—is equal to a negative 21 torque units and the total ccw torque—that of the muscle force—is equal to a positive 21 torque units, the leg will be held stationary. To make the comparison clearer, one can set up an equation and merely add the torques:

EQUATION 3.1

$$\Sigma T = 0$$

where Σ is "the sum of" and T is torque.

$$\text{leg torque} + \text{ankle weight torque} + \text{muscle torque} = 0$$

$$(-6 \text{ metric torque units}) + (-15 \text{ metric torque units}) + (+21 \text{ metric torque units}) = 0$$

$$[(-4.42 \text{ British torque units}) + (-11.06 \text{ British torque units}) + (15.48 \text{ British torque units}) = 0]$$

If the net torque (sum of all torques) is zero, no movement of the system will occur. If the sum is a negative number, the cw torques will be greater, and the segment will move cw. If the sum is a positive number, the ccw torques will be greater, and the segment will move ccw.

Figure 3.5b illustrates a similar case with the forearm. The amount of force necessary to hold the barbell in the position shown is determined by adding the torques acting on the forearm with the elbow serving as the axis of rotation. The magnitude or direction of the applied muscle torque is unknown, however. The solution to the equation will indicate a sign for the muscle torque, and thus a direction.

$$\Sigma T = 0$$

$$(F_m \times d_{\perp}) + (F_{bb} \times d_{\perp}) + (F_{fa}) \times d_{\perp}) = 0$$

where F_{bb} is weight of the barbell, F_{fa} is weight of the forearm and hand, and F_m is muscle force required.

$$F_{bb} \times d_{\perp} = -44.5\ \text{N} \times .45\ \text{m} = -20.03 \text{ metric torque units}$$

$$[-9.97\ \text{lb} \times 1.48\ \text{ft} = -14.76 \text{ British torque units}$$

$$F_{fa} \times d_{\perp} = -13.35\ \text{N} \times .15\ \text{m} = -2.00 \text{ metric torque units}$$

$$[2.93\ \text{lb} \times 0.49\ \text{ft} = -1.44 \text{ British torque units}]$$

$$F_m \times d_{\perp} = F_m\ \text{N} \times 0.02\ \text{m} = 0.02F_m \text{ metric torque units}$$

$$[F_m\ \text{lb} \times 0.07\ \text{ft} = 0.07F_m \text{ British torque units}]$$

The sum of the torques equal:

$$(-20.03 \text{ metric torque units}) + (-2.00 \text{ metric torque units}) + (0.02F_m \text{ metric torque units}) = 0$$

$$F_m = 1,101.5\ \text{N}$$

$$[(-14.76 \text{ British torque units}) + (-1.44 \text{ British torque units}) + 0.07F_m \text{ British torque units} = 0]$$

$$F_m = 810\ \text{lb}$$

Thus, 1,101.5 N (810 lb) of muscle force acting ccw (+) is necessary to hold the forearm, hand, and a barbell weighing 44.5 N (9.97 lb) in a static position. If the muscle applied more torque, the system would move ccw; if the muscle applied less torque, the system would move cw.

Understanding Torques on the Musculoskeletal System

1. What are the two requirements for producing a torque?
2. Figure 3.6 shows three boxes on a board with an axis (fulcrum) in the middle of it. The weights of the boxes are given as well as the distance that each is located from the axis: B is 1 unit from the axis, C is 2 units from the axis, and D is 3 units from the axis. How far does box A have to be from the axis to make the system balance if A weighs 100 units? If A weighs 40 units?
3. If a bone–muscle system is in static equilibrium, what does that tell you about the torques acting on the system?
4. If the shoulder joint is being abducted, what happens to the resistance torque of the arm's weight force as abduction takes place to the horizontal position? What happens to the resistance from a 90-degree abduction to a position over the head? Explain why.

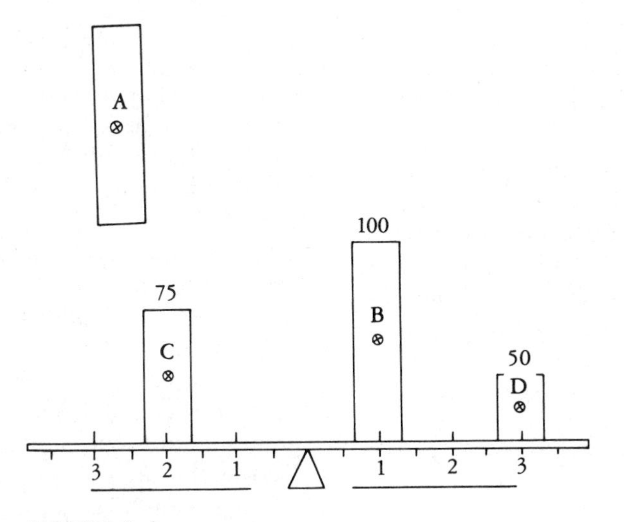

FIGURE 3.6
Balance the system.

5. If the person in Figure 3.17 (p. 179) were going to lift the object shown, what could the person do to make the lift easier, and why?

6. Consider the person doing leg levers in Figure 2.4f (p. 72) to answer the following questions:

   a. How much force does a hip flexor group have to exert to hold the lower extremity in its position? (The CG of the thigh is 10 units from the hip, the CG of the leg and foot is 28 units from the hip, the thigh weighs 20 units, the leg weighs 9 units, the foot weighs 3 units, and the line of action of the resultant muscle force passes 3 units from the hip joint.)

   b. If the person flexes the knee so that the CG of the leg and foot is 15 units from the hip, how much force does each hip flexor group have to expend?

   c. Would a short- or long-legged person have more difficulty in doing this type of activity? (Consider gymnastics and calisthenic activities in general.)

Analysis of the Musculoskeletal Lever System

The analysis of leverage in musculoskeletal systems has several applications that are useful for the physical education teacher or coach. The advantages derived from knowing muscle mechanics are emphasized in the sections that follow.

Rotary and Joint-Stabilizing Components of Muscle Forces

The effect of changing a muscle's angle of pull on the magnitude of muscular torques was briefly examined previously. The force of a muscle, or a muscle group, may be represented by an arrow. This arrow is called a resultant force vector and represents the magnitude (amount) and direction (angle of pull) of a muscle or muscle group on a bone. For example, if the arrow representing the magnitude of the muscle's force is drawn 4 cm long, based on the scale 1 cm = 20 N (89 lb), the muscle's force will be 80 N (356 lb). The direction will be the angle that the resultant force arrow makes with the longitudinal axis of the bony lever to which it is attached. (Recall this is represented by a line connecting the axis of rotation to the point of application of the muscle.)

Graphically, this force-vector arrow with its magnitude and direction can be resolved into two components that are perpendicular to each other and represent two important directions in which the muscle pulls in a two-dimensional muscle system. Figure 3.7 illustrates how one resolves a muscle force vector. The two components are labeled in this manner: **Ro** for the **rotary component** which is always perpendicular to the bone lever, and **Stabl** or **Disl** for the **stabilizing** or **dislocating component** which is parallel to the bone lever. The rotary component of a muscle's force is that part of the resultant force that acts to rotate the bony lever. The rotary component of a muscle's force is less than the resultant force, unless the resultant force acts at 90 degrees to the bone, in which case, the resultant and rotary forces are the same. The force arm for the rotary component is always directed parallel to the lever and is the distance between the line of force (also the point of application) of the Ro component and the axis of rotation.

The second component is called stabilizing if it is directed parallel to the long axis of the bone and *toward* the axis of rotation (i.e., if it tends to pull the two bones together at the joint). (See Figure 3.7a.) The second component is called dislocating if it is directed parallel to the bone and *away* from the axis of rotation (i.e., if it

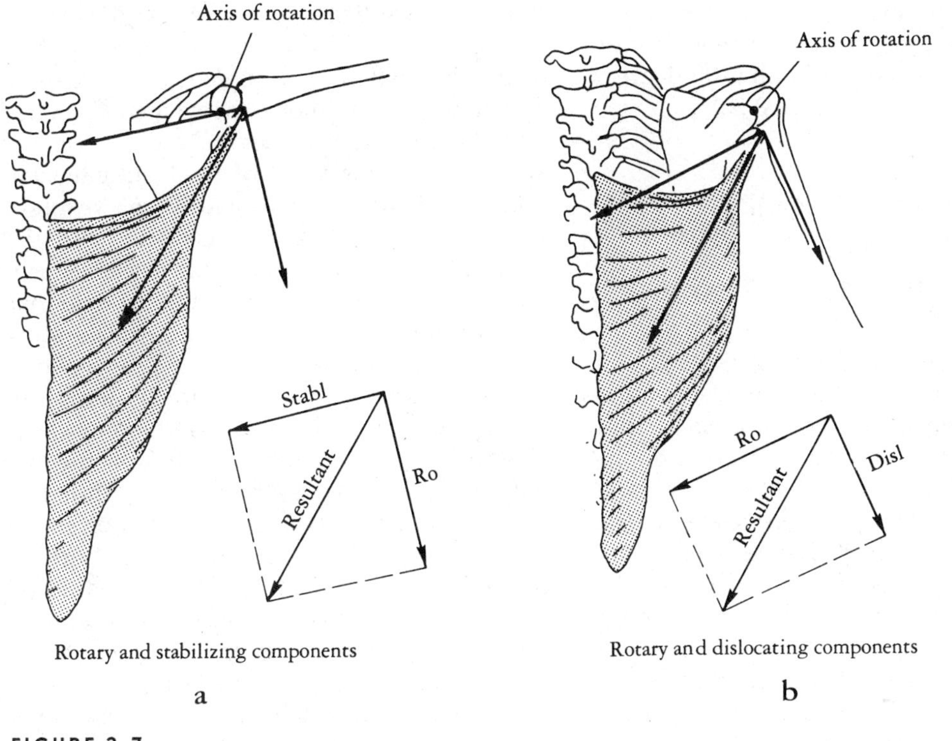

FIGURE 3.7
Resolving a muscle's resultant force into two perpendicular components in two positions of the shoulder joint.

tends to pull the bones apart at the joint). (See Figure 3.7b.) Because the stabilizing or dislocating components are parallel to the lever and their lines of action run through the axis of rotation, they never have a force arm; therefore, they never provide any torque. (There is no perpendicular distance between the line of action of these components and the axis.) Thus, stabilizing or dislocating components can never influence the rotation of the lever; they can only tend to stabilize or dislocate. Although one might think that a 100% rotary force would be ideal, the stabilizing functions of some muscles are important in maintaining the integrity of joints, particularly the more lax joints such as the shoulder and the knee.

Exact magnitudes of the components can be determined by rather sophisticated techniques, but what is important to understand here are the relative magnitudes of the two components of a muscle's force on a bone. These can be estimated by comparing the sizes of the two component vectors (the arrows). One should also understand how the muscle attachment angle to the bone affects the sizes of the two components. As the resultant muscular force (R) approaches 90 degrees to a line connecting the point of application of a muscle to the axis of rotation, its rotary component becomes larger, and the stabilizing or dislocating component becomes smaller, until, at 90 degrees, the resultant force is 100% rotary (Figure 3.3b). Thus, the greatest torque may be generated by a muscle whose direction of pull is 90 degrees to the shaft of its bone (Figure 3.5a).

Evaluation of Several Joint Positions in Relation to a Muscle's Effectiveness

As the joint angle changes in the course of a movement, the muscle-to-bone angle changes as well. The two angles are associated, but they should not be confused. Certain positions of the joints allow for greater rotary muscle torque; these muscle torques are related to the angle of the muscle attachment as well as the length of the muscle at any given instant.

To analyze the most effective joint angle for producing muscle torque, how the angle of attachment changes as the joint angle changes must be understood. For this purpose, the action of the hamstring muscle at the knee will be used as an example. Figure 3.8 illustrates three positions of knee flexion, each with its related hamstring angle of attachment to the leg. The largest force arm that can be achieved by the hamstring muscle at the knee is that produced when the muscle pulls at 90 degrees to the leg bones (position 2). In this position, the knee *angle* is slightly less than 90 degrees. Before the 90-degree muscle angle, the muscle has a rotary and stabilizing component of force (position 1). When the muscle pulls at 90 degrees to the bone, the muscle force is 100% rotary. When the muscle angle is greater than 90 degrees (continued flexion of the knee), the rotary component diminishes, and the dislocating component increases (position 3).

The length–tension relationship presented in Chapter 2 is significant here because the muscle is a two-joint muscle. If the hip joint is held in a position of extension, as in a standing position, the *tension* able to be produced in the muscle continues to decrease as the muscle shortens to flex the knee. The muscular tension is less when the muscle is at a 90-degree attachment to the bone than when the knee first started flexion. Assessing the net effect of what is lost as a result of the length–tension relationship with what is gained as a result of the changing attachment angle is difficult.

The increased tension at the lesser attachment angles occurs when the muscle has a stabilizing component, not a dislocating component. Because the muscle is a two-joint muscle, certain adjustments to the length of the muscle may be made as the muscle is shortening (i.e., as the knee flexes, the hip may be flexing). Thus, the length of the muscle may be maintained, because it shortens to flex the knee while it lengthens with flexion of the hip. Consequently, if the hip is flexed as the knee is flexed, the muscle does not lose the tension advantage of its resting length, as it would if it were merely flexing the knee. This ability to maintain the length of a two-joint muscle during its contraction may allow the muscle greater tension when it begins to generate a dislocating component (i.e., when its attachment is greater than 90 degrees to the bone).

Comparison of Several Muscles Performing a Similar Mover Function

Consideration of certain physical activity situations, physical therapy in particular, requires an understanding of the torque production of several different muscles that perform the same motive function. A prime example concerns the torque contributions of the three elbow flexors that act on the forearm segment: the biceps brachii, the brachioradialis, and the brachialis. These muscles are important in terms of their angles of pull, their force arms, and thus their rotary and stabilizing con-

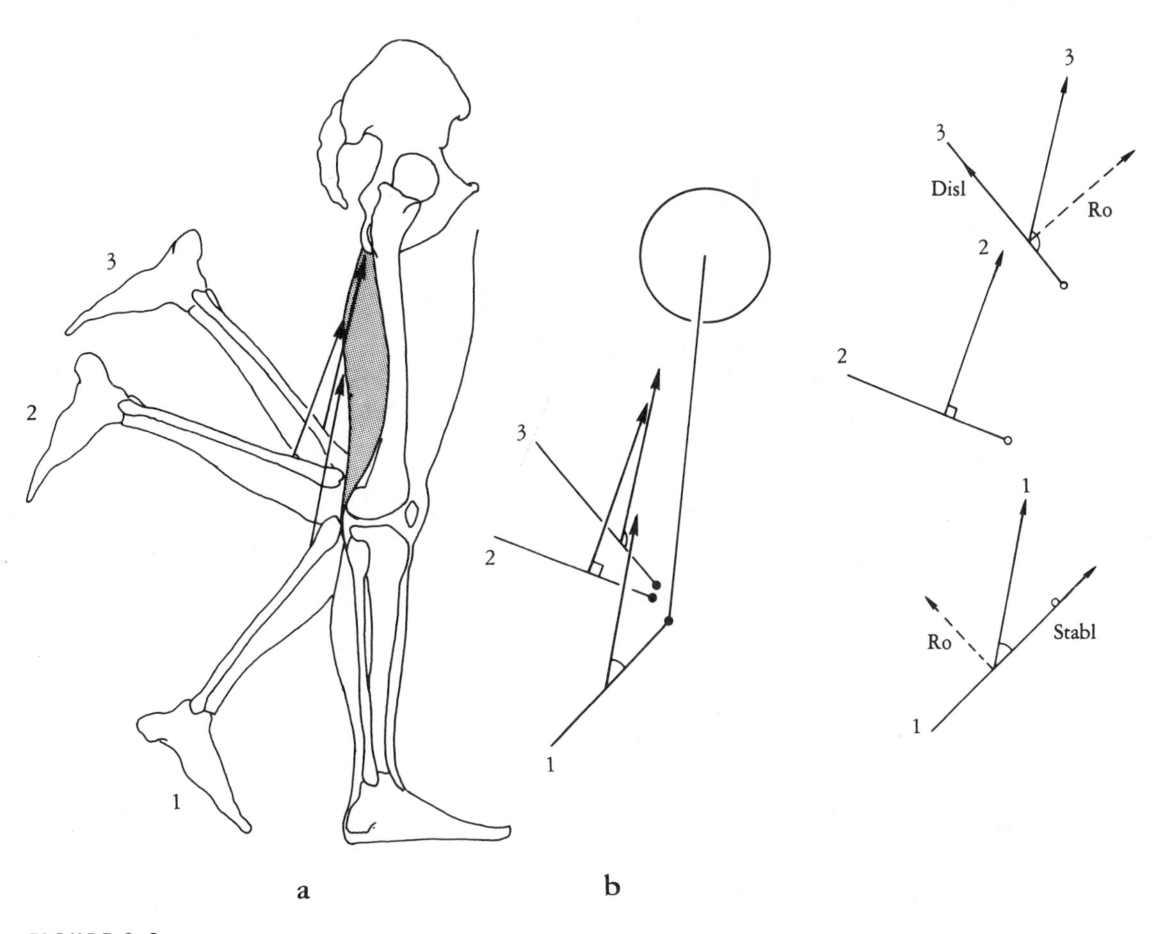

FIGURE 3.8

The angle of attachment of the hamstring muscle in three knee positions: (a) on a skeleton, (b) on a line drawing (note: axis moves as tibia moves along femoral condyles) and (c) with the muscle forces resolved into two components.

tributions to the working of the elbow joint. Figure 3.9 shows the positions of the three muscles.

Several previously introduced examples emphasized that the larger the force arm of the resistive force, the greater the motive force or force arm or both needed to hold a given resistance in equilibrium. The force arms of the muscles have certain restrictions as a result of the permanence of the muscles' attachments. The only practical adjustment is to place the resistance close to the joint axis to decrease the resistive torque and thus decrease the motive torque needed. As the angle of the muscle force approaches 90 degrees to a long bone lever, its force arm increases. The closer the joint angle is to the point at which the muscle angle of attachment is 90 degrees, the more effective the muscle force is in producing torque to hold the resistance. When comparing the relative effectiveness of several muscles performing the same function (e.g., elbow flexors), one may start by comparing the relative sizes of their respective force arms. Several points

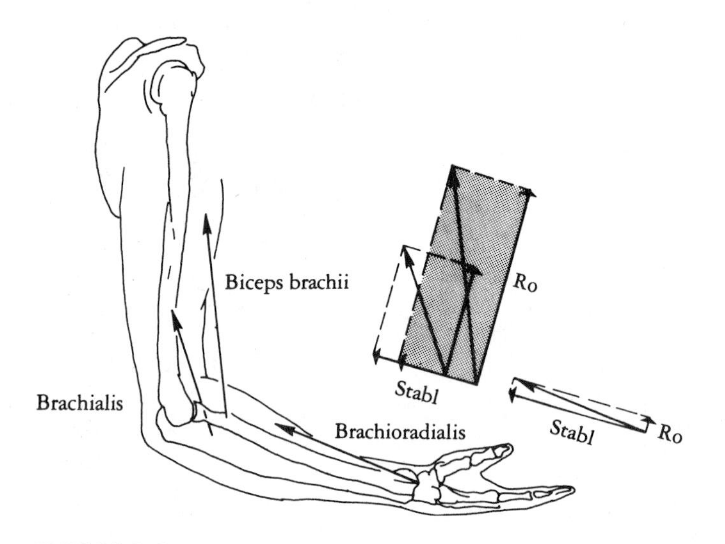

FIGURE 3.9
Vector resolution of the forces applied by three elbow flexors.

must be considered. First, the brachioradialis can never approach an angle of 90 degrees to the forearm, and, in fact, always has a small angle of pull. Thus, the brachioradialis always has a large stabilizing component in its resultant muscle pull when it is used for forearm motion. Furthermore, it can never achieve even a 45-degree angle of pull to the forearm; therefore, the stabilizing component is always greater than the rotary.

Second, when the attachment angles of the brachialis and the biceps brachii are compared, the attachment angle of the brachialis is always less than that of the biceps. During the range of motion of the forearm, however, both muscle attachments have the possibility of reaching a 90-degree attachment angle. Therefore, either can be put in a position where 100% of its force is rotary. Because the angle of attachment of the brachialis to the ulna is always slightly less than the angle of attachment of the biceps to the radius, the biceps achieves 100% rotary force at a larger elbow angle than does the brachialis. The larger elbow angle means that the biceps muscle is at a greater percentage of its length than the brachialis when its 100% rotary position is achieved.

On the basis of the discussion of the length–tension relationship in Chapter 2, the biceps is at a more favorable length than the brachialis when each achieves a 90-degree attachment to the bone. Furthermore, since the biceps is a biarticulate muscle, its proximal attachment can make the muscle more effective as a result of the length–tension relationship when it is stretched over the shoulder joint (in shoulder joint hyperextension) while attempting to contract. As a one-joint muscle, the brachialis does not have the ability to adjust its length. Thus, the concept of mechanical advantage of one muscle over another in producing torque is not the only relevant consideration. Other factors important to a muscle's force are the relative strength of the muscles involved,[2] the position of the other articulations of a biarticulate muscle (e.g., shoulder joint and radioulnar positions), and the muscle fiber arrangement.

[2]A muscle producing 100 units of force with 20% of it rotary is less effective than a muscle producing 50 units of force with 50% of it rotary.

Understanding Applications to Musculoskeletal Structures

1. Which of the following angles of muscle pull has the greatest rotary component: 40 degrees, 23 degrees, 35 degrees, 65 degrees, 100 degrees, or 140 degrees?
2. Which of the angles listed in question 1 has the greatest stabilizing component?
3. With the aid of a ruler and protractor, determine what the rotary and stabilizing (or dislocating) components are for a muscle exerting 50 N of force (222.5 lb) at a 30-degree angle to the bone.
4. Construct a diagram of each of the forces in Figure 3.10, using a scale of 1 cm = 20 N (4.50 lb) of force. Draw the rotary and stabilizing components to each force. Measure the force arms for each resultant force and its rotary component. Compare and discuss what you find.
5. The systems in Figure 3.11 are muscle forces on bones. For each system, draw the stabilizing or dislocating force and the rotary components of force. (The force parallel to the bone may be dislocating instead.)

   a. How does the direction of the resultant force change with changing angles of the joint?

   b. How does the direction of the resultant force change the stabilizing, dislocating, and rotary components of the muscle's force?

   c. In what position does one get 100% rotary force out of the muscle's tension?

   d. Complete this sentence: If one were going to lift (resist) a heavy load, one

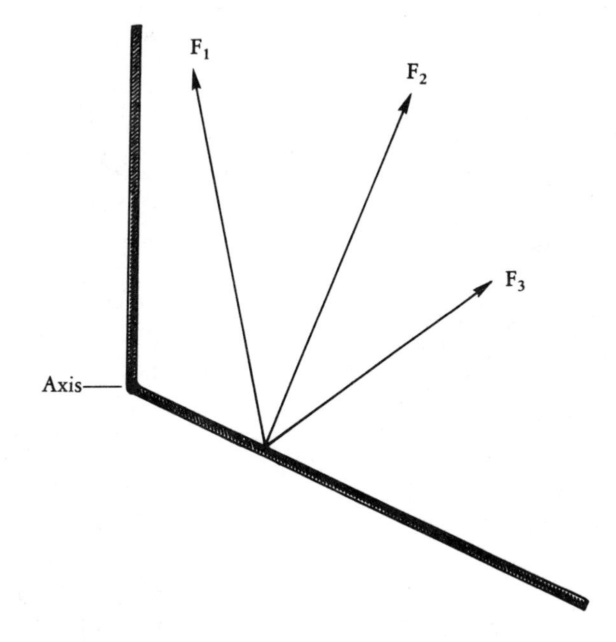

FIGURE 3.10

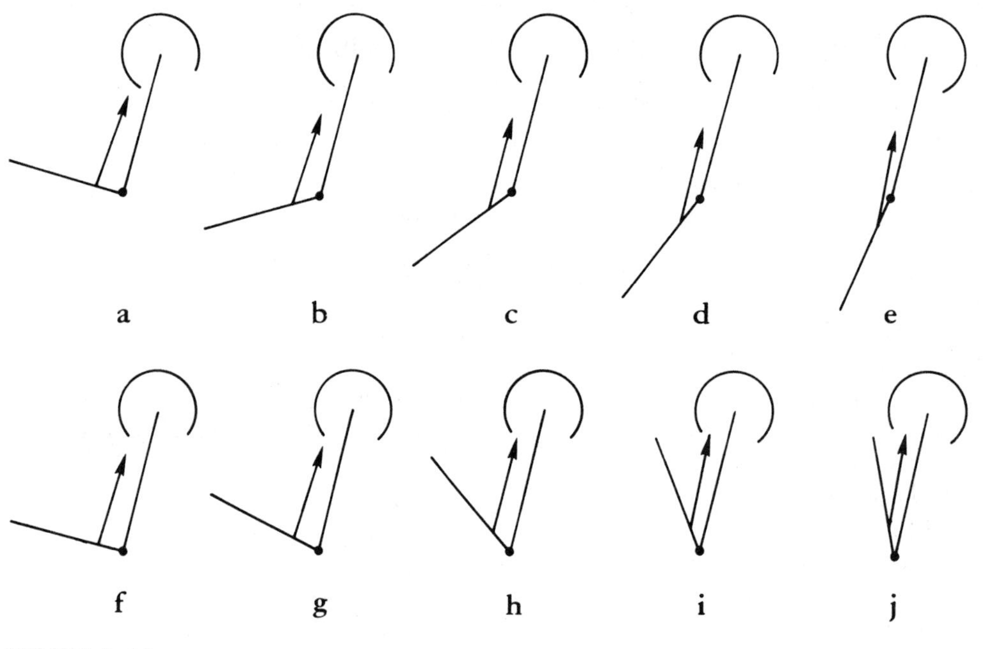

FIGURE 3.11

would want to position the body segments doing the lifting so that . . . .

e. After observing the situation of the stabilizing and rotary components of a muscle's pull, what can you say about the body segment's trade-off between the stability of the articulations (i.e., working against dislocation) and its mobility (i.e., the ability to produce motion)?

6. Figure 3.8 (p. 169) includes a representation of the hamstring muscles holding the leg in three different positions. Redraw the system, and perform the following operations for each position:

   a. Determine the magnitude of the force, using the scale 1 cm = 100 N of force (22.5 lb).

   b. Place a dot on the point of application of each force.

   c. Determine the direction of the pull of the forces by measuring the angle with a protractor.

   d. Draw the lines of action of the forces, using dashes.

   e. Draw the force arm for each force and measure it in centimeters.

   f. Answer the following questions:

      (i) Which position allows for the greatest magnitude of force?

      (ii) In which position does the greatest muscle torque act on the leg?

      (iii) In which position does the muscle force act at the greatest angle to the bone? the smallest angle to the bone?

      (iv) What is the relationship between the muscle's angle to the bone and the amount of torque that it produces?

(v) What angle is optimum for torque production of a muscle (or of any other force)?

(vi) Which factor is more important to the torque production of a muscle, the magnitude of the force or the length of its force arm?

7. In Figure 3.12 are musculoskeletal diagrams and stick diagrams. On the stick diagrams, draw and label the following:

   a. Resultant muscle force (R)

   b. Motive force arm (MFA)

   c. Rotary force component (Ro)

   d. Stabilizing or dislocating component (Stabl or Disl)

   e. Resistive force (RF)

   f. Resistive force arm (RFA)

Calculating Mechanical Advantages in Musculoskeletal Lever Systems

Typically, a considerable amount of muscle force is necessary to rotate or hold static a given resistive force. For the example illustrated in Figure 3.5b (p. 163), the muscle force necessary to hold static a 44.5-N (10-lb) barbell plus the weight of the forearm and hand was calculated as 700 N (157.30 lb), an astonishingly greater sum. The same is true for all muscles in the body. They are at a mechanical disadvantage in force; that is, they must exert a much greater force than the resistance they are manipulating. The reason for this lies in the fact that muscles' lines of action run very

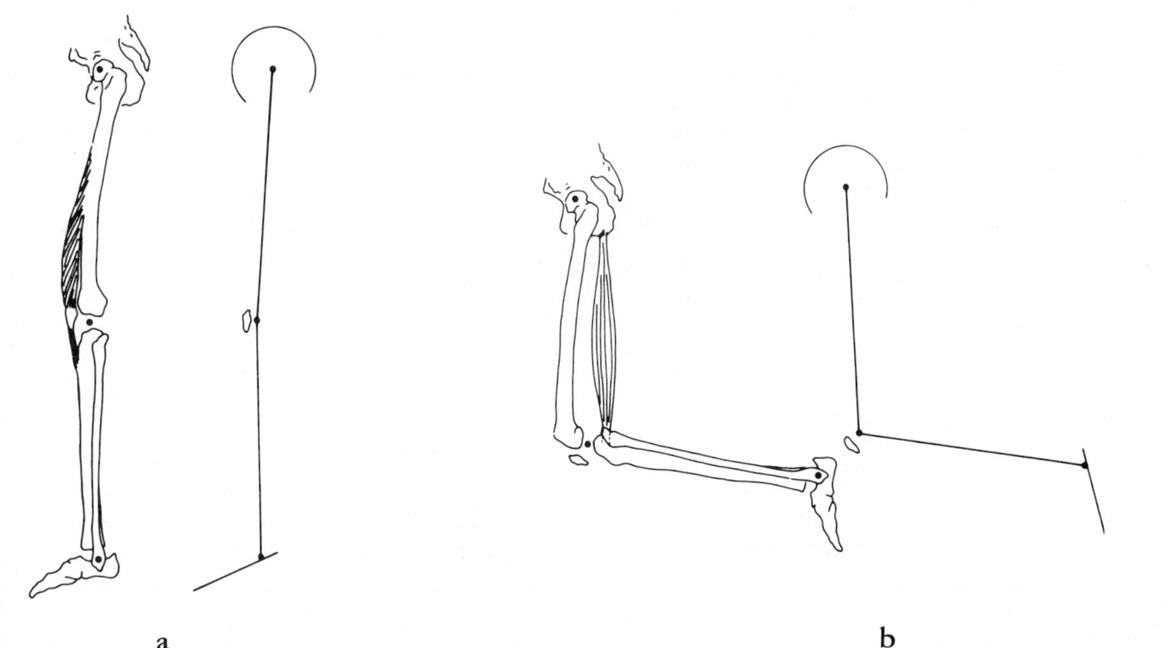

FIGURE 3.12

close to the axes of rotation for the joint movements. Consequently, muscles do not have very large force arms, and thus, they have a smaller torque-producing capability than the resistive forces. For the few muscles able to achieve the 90-degree angle of pull, which maximizes their force arms, the points of attachments are still relatively close to the joint axis of rotation, which reduces their force-arm distances.

The ratio that expresses the relationship of the resistive force to the motive force is called the *mechanical advantage* (MA), and is calculated by the following equation:

EQUATION 3.2

$$\text{Mechanical advantage (MA)} = \frac{\text{resistive force}}{\text{motive force}}$$

If the quotient is greater than one, the resistive force is greater than the motive force; if the two forces are equal, there is no advantage; and if the quotient is less than one, the motive force is greater than the resistive force. If a muscle is the motive force (concentric contraction), the muscle always has a mechanical disadvantage in force. The quotient is less than one because the muscle is exerting more force than the resistance it is moving.

Because of the one-to-one relationship between a force and its force arm and the production of torque, MA also may be defined by the ratio of the respective force arms:

EQUATION 3.3

$$\text{MA} = \frac{\text{motive force arm}}{\text{resistive force arm}}$$

Notice that the motive and the resistive terms are reversed in the two equations. This makes the MA ratio come out the same, since a motive force arm that is smaller than a resistive force arm produces a quotient less than one. In other words, either equation produces the same ratio.

Although the musculoskeletal lever system is designed for a mechanical disadvantage in force, it does have an MA in range of motion (ROM) and speed of movement. When two points on the rotating lever move the same *angular* distance, the point farthest from the axis of rotation moves the greater *linear* distance. This is shown in Figure 3.13. When the ankle dorsiflexes against a resistance, the dorsiflexors shorten a relatively small distance compared to the movement of the resistance itself, which is located toward the end of the lever. Thus, although muscles have a disadvantage in force, they have an advantage in ROM and speed of movement.

To calculate the MA in ROM and speed of movement, the reciprocal of the previous equations is used. If a muscle has a *disadvantage in force* (e.g., needing 5 motive force units to move 1 resistive force unit), it also has the *advantage in linear distance* of moving 1 unit of linear movement to move the resistance 5 units of linear movement (i.e., it moves the resistance 5 times farther than it moves itself).

Since the lines of action of the muscle forces in the human body pass closer to the axis of rotation than are the lines of action of the resistive forces, the muscles always have a disadvantage in force and an advantage in ROM and speed of movement. The speed of movement advantage serves well in activities in which linear speed at the end of a lever is important (e.g., throwing, kicking, and striking activities).

Classes of Levers

Recall that the basic components of a lever system are a rigid lever, an axis, a motive force, and a resistive force and that the musculoskeletal system often functions as a lever system. Three types of levers are classified according to the relative positions of the axis, the motive force, and the resistive force. The

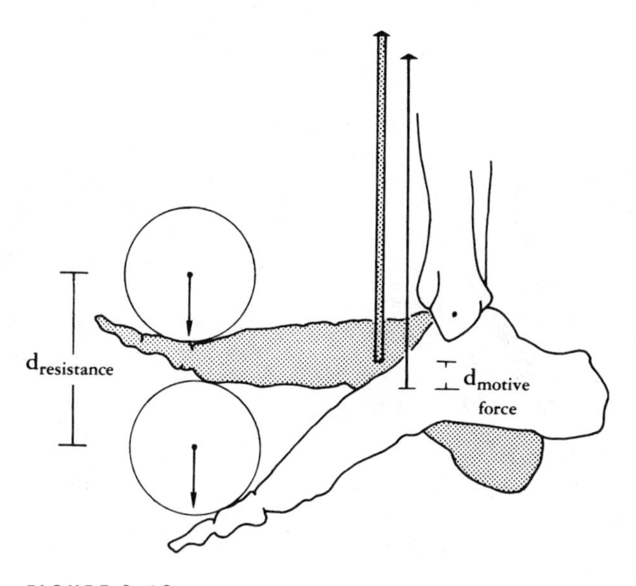

FIGURE 3.13

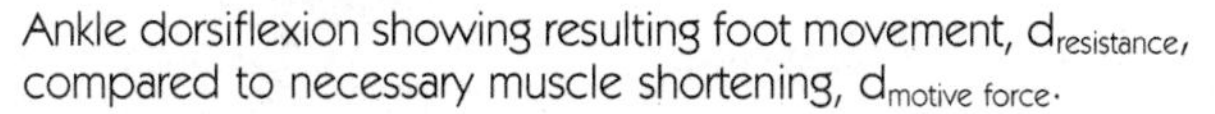
Ankle dorsiflexion showing resulting foot movement, $d_{resistance}$, compared to necessary muscle shortening, $d_{motive\ force}$.

function of each class of lever (see p. 158) depends on these relative positions and on the relative sizes of the force arms of all the forces being applied.

First-Class Lever Systems

Whenever the axis is located between the motive force and the resistive force, the lever is of the first class. Some of these relationships are shown in Figure 3.14.

Common examples include a teeter-totter, a crowbar, a pair of pliers, a pair of scissors, and a hammer removing a nail. The first-class lever is the most versatile of the levers because it can serve any one or more of the four functions of a machine, depending on the arrangement of its parts. In the teeter-totter, the lever is basically used for balancing two equal or unequal forces. When the two forces are unequal in magnitude, however, the torque production of the forces—and thus balance—may be obtained by a slight adjustment in the point of application of the force to give the smaller force a larger force arm (i.e., to position it farther away from the axis). The smaller force then has an advantage in torque (i.e., a smaller force can now exceed or equal the torque of a larger force), and the larger force has an advantage in ROM (i.e., the larger force moves the point of application of the smaller force a greater linear distance with equal angular displacements). Furthermore, in the first-class lever systems of the body, the motive and resistive forces act in opposite circular directions.

Not all first-class levers perform all four machine functions. The versatility of the first-class lever provides the user with the option of employing it in a variety of ways, which cannot be done with levers in the other two classes. The hammer removing a nail is an example of a first-class lever that is designed to give a force advantage by putting the motive

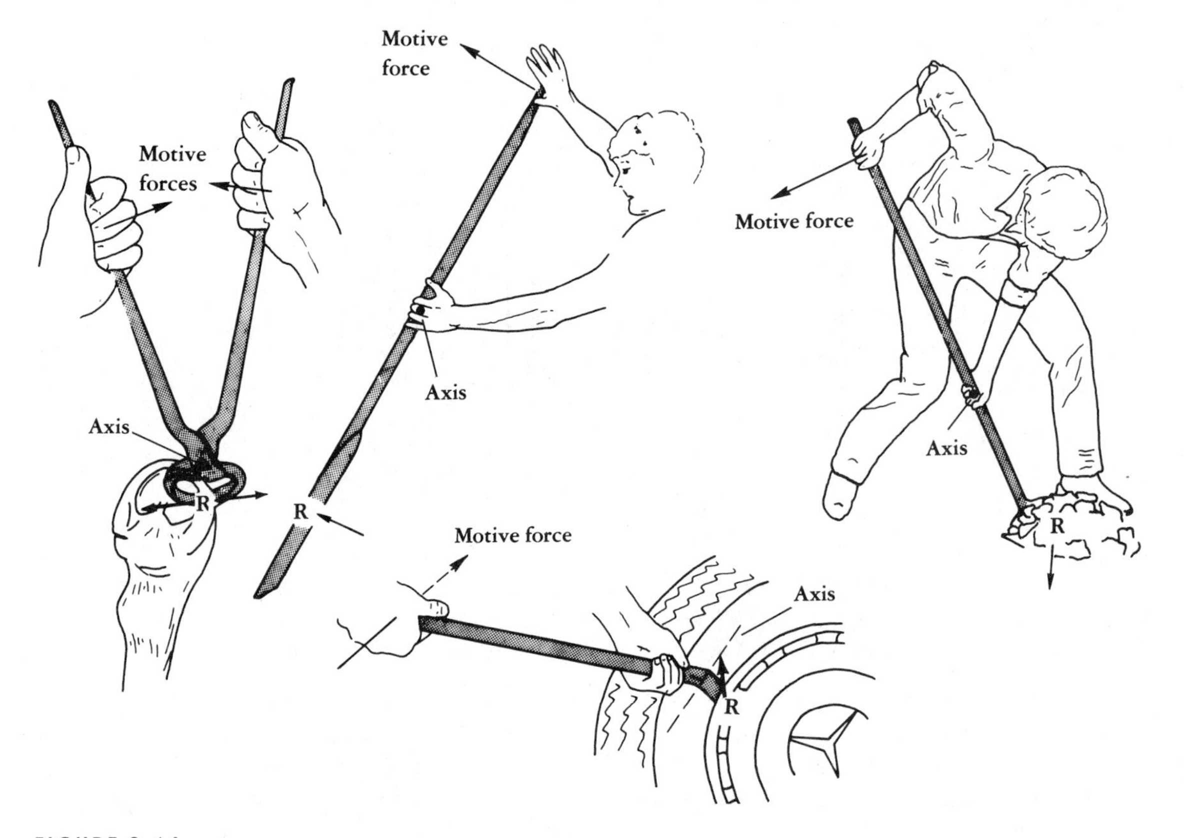

FIGURE 3.14
First-class lever systems.

force (pressure point on the handle) farther away from the axis than the resistive force (the head of the nail acting on the claw). Thus, one can remove the nail with less applied force than the resistance force of a nail, but the applied motive force must move a greater linear distance to do so. This disadvantage in distance is evident to anyone who has ever tried to pull out a long nail with a small hammer. Although the motive-force advantage is enough to pull the nail, the distance that the resistive force is moved during the pull is not enough to pry out the nail. This disadvantage may be overcome by putting a block of wood under the axis and pulling again.

First-class levers are unusual in the musculoskeletal system because the body levers are not generally arranged with the axis of rotation between the muscle attachment and the resistance. Two examples are the triceps acting to extend the elbow overhead and the posterior calf muscles working to plantar flex the ankle against resistance at the ball of the foot, as shown in Figure 3.15. In neither of these examples does the first-class lever arrangement give an advantage in force, because the motive force arms are relatively short as compared to the resistive force arms. Rather, these first-class lever arrangements provide an advantage in ROM and speed of

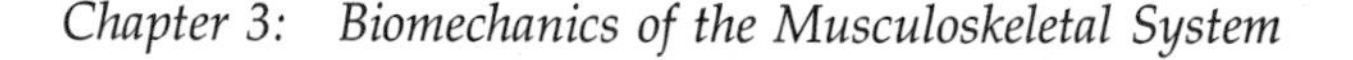

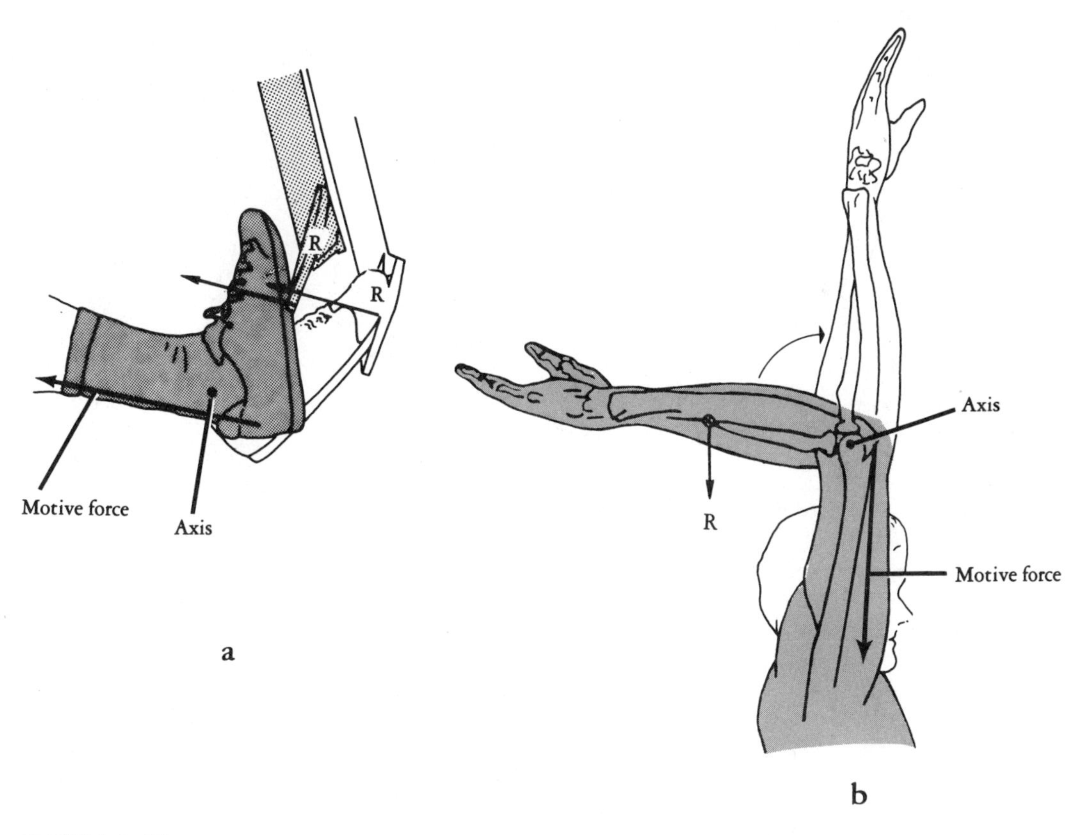

FIGURE 3.15

Examples of first-class levers in the musculoskeletal system. (a) Posterior calf muscles working to plantar flex the ankle against resistance at the ball of the foot. (b) The triceps acting to extend the elbow overhead.

movement and change the direction of the movement of the resistance compared to the applied force.

Second-Class Lever Systems

Whenever the rotary component of the resistive force is located between the rotary component of the motive force and the axis, the lever is of the second class. A diagram of this relationship is shown in Figure 3.16.

Common examples of second-class machines include a wheelbarrow, a rowboat, and a nutcracker. The relationship between the rotary components of the motive and resistive forces in the second-class lever system dictates that the force arm for the rotary motive force be larger than the force arm for the rotary resistive force. Therefore, in second-class lever arrangements, a force advantage usually exists for the motive force; that is, less motive force need be applied to move a greater resistance. With a wheelbarrow, one may lift a load of rocks that could not be budged otherwise. If one needs to lift a load a great distance, however, such as hay bales to a loft, the second-class lever system is not advantageous, because the lifting force must move a greater linear distance than the hay bales must be moved.

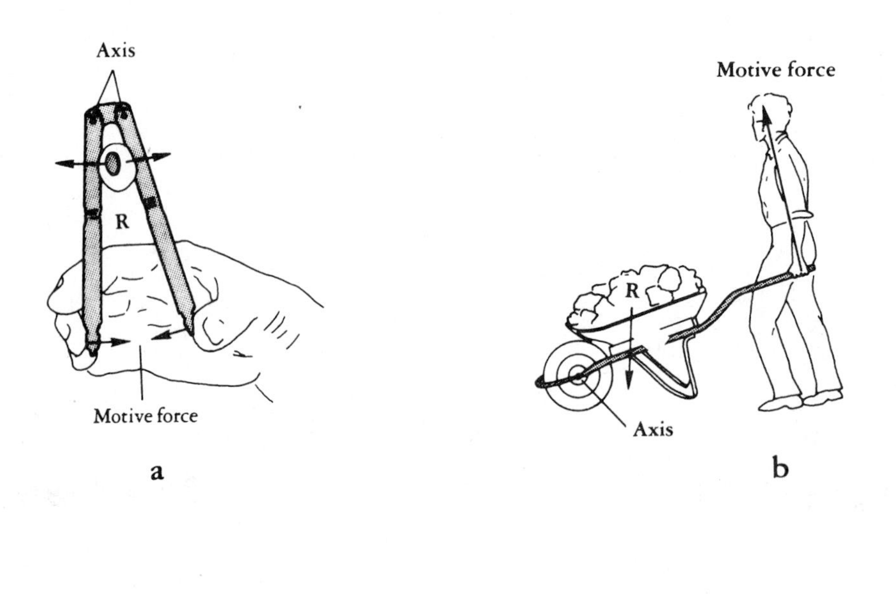

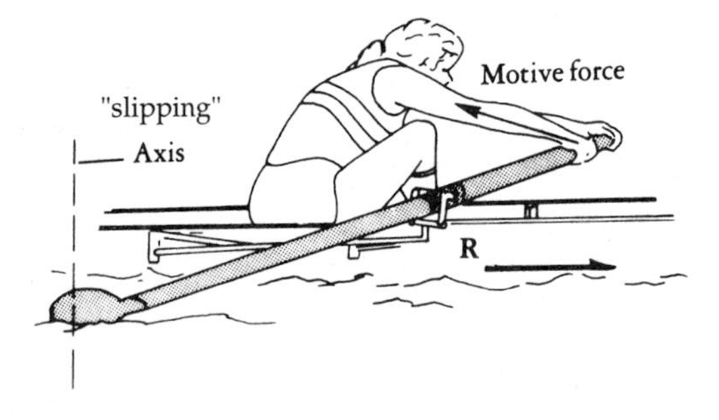

FIGURE 3.16
Second-class lever systems.

The second-class lever system is not as versatile as the first-class system because it cannot be used for gaining advantages in ROM or for changing the direction of movement of the resistance. It is designed to provide the user with an advantage in force. Second-class lever systems in the body with a muscle as the motive force are unusual. Consequently, the muscles rarely have an advantage in force when moving a body segment. The push-up, however, is an example of the *total body* acting as a second-class lever: the foot is the axis of rotation, the weight of the body at the CG is the external resistive force, and the reaction force of the ground pushing against the hands is the external motive force.

Sometimes, force other than muscle force is the motive force, and the muscles act to resist or control the movement (eccen-

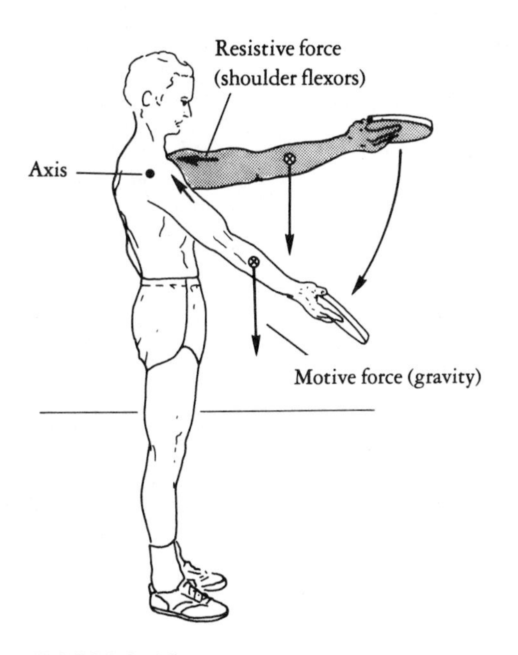

FIGURE 3.17

Eccentric tension of muscles as an example of a second-class lever system (muscles as resistive forces).

tric contraction). In these cases, the motive force has the force advantage, and the resistive force (muscle force) has the ROM advantage. For example, eccentric tensions in muscles are considered second-class levers, as shown in the lowering of the weight in Figure 3.17.

Third-Class Lever Systems

Whenever the rotary component of the motive force is located between the axis and the rotary component of the resistive force, the lever is of the third class. This relationship is illustrated in Figure 3.18.

Common examples of third-class levers are shovels, deep-sea fishing apparatus, and most implements used in sports events (e.g., rackets, fencing foils, bats). Most musculoskeletal lever arrangements are of the third class when the muscle is the motive force, because the muscle attachments are closer to the joint axis of rotation than are the resistances that they are moving.

The relationship of the location of the motive and resistive forces to the axis of rotation in a third-class lever system gives the system an advantage in ROM and speed of movement for relatively light resistances. The magnitude of the motive force (when the muscle force is the motive force) that must be applied to a given resistance is greater than the magnitude of the resistive force, but the resistive load is moved a greater linear distance than the point of application of the motive force. This particular advantage in ROM is important not only in moving resistances greater distances but also when one is concerned about the speed of movement.

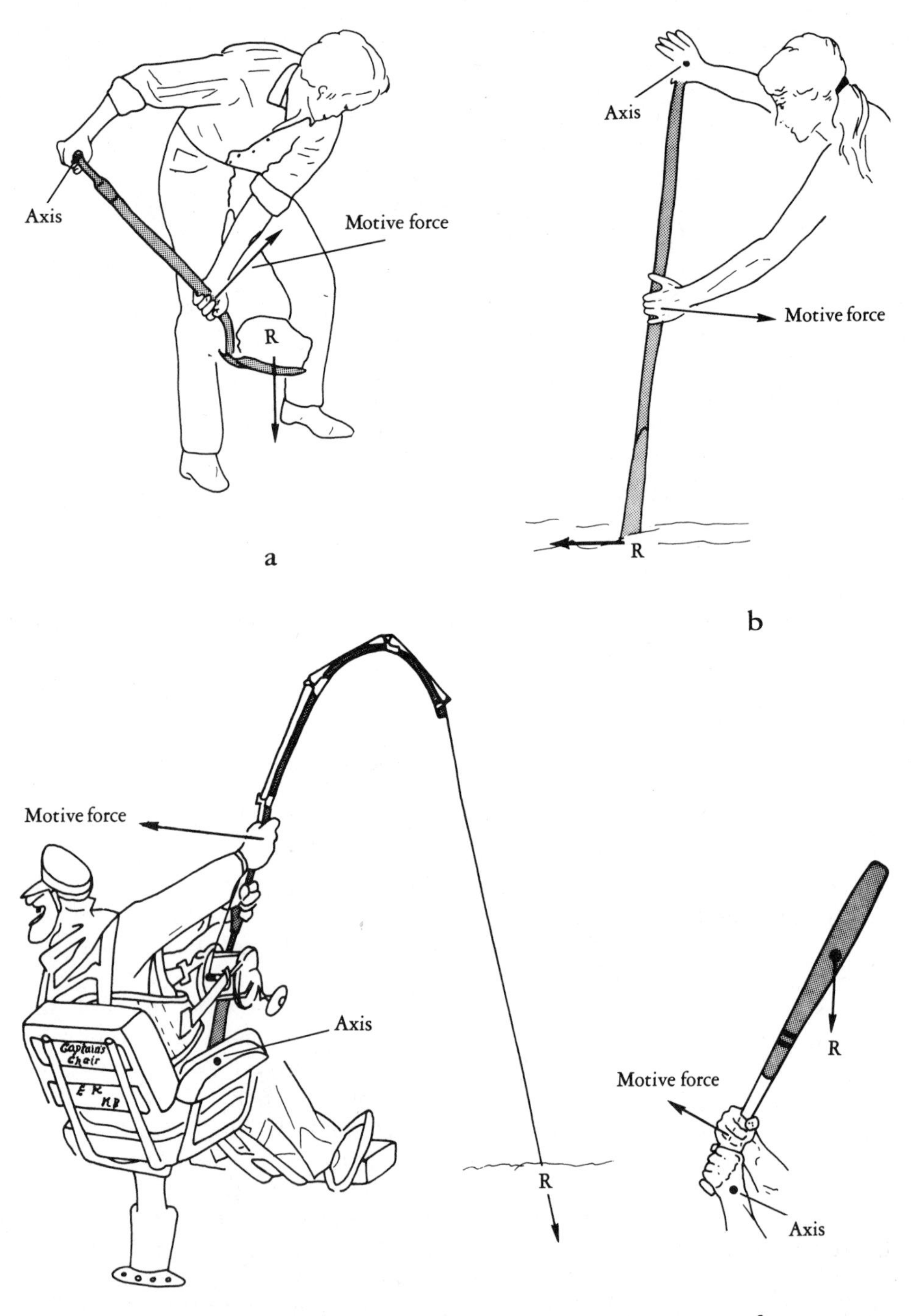

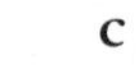

FIGURE 3.18
Third-class lever systems.

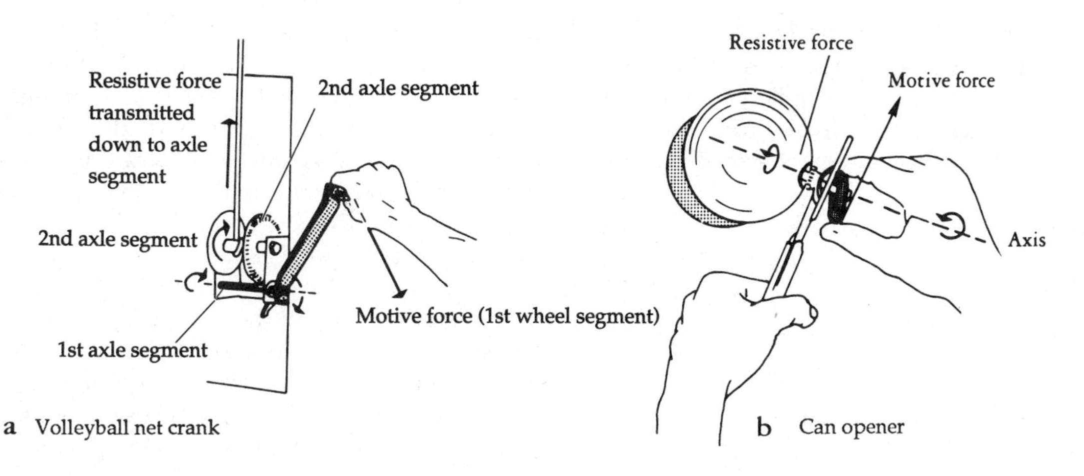

FIGURE 3.19
Wheel-and-axle mechanisms.

The speed advantage of the third-class lever arrangement has important implications for sport activities in which implements are involved, such as striking and throwing activities. The longer the lever, the greater the linear-speed advantage. When one needs an implement to move fairly light objects a great distance or when the amount of effort expended is not significant, one usually selects a tool of the first or third class. The longer the implement chosen, the greater the ROM and speed advantage of the body; however, the longer the implement, the greater the force the body needs to accelerate the implement.

3.2 Wheel and Axlelike Arrangements

A second type of machine found in the musculoskeletal system is the wheel and axle. This type of machine may function to give a mechanical force advantage (MA) in ROM/ speed of movement. Wheel-and-axle mechanisms are illustrated in Figure 3.19.

In both of the cases pictured in Figure 3.19, the motive force is applied to the wheel or rim segment (e.g., the handle of the volleyball net crank). The force arm in the case of wheel-and-axle machines is the radius of the wheel or the radius of the axle. The resistance is provided by the axle segment, which receives the resistive force of the gear wheel. As is seen in this example, the wheel (handle) has a larger radius than the axle. Thus, if the forces on the wheel and axle are equal, the torque on the larger radius wheel will be greater than the resistance torque on the smaller radius axle. The MA of a wheel–axle arrangement may be calculated by this equation:

EQUATION 3.4

$$\text{MA} = \frac{\text{radius of the wheel}}{\text{radius of the axle}}$$

Because the radius (force arm) for the force applied to the wheel is greater than the radius (force arm) of the force applied to the axle, the force advantage is always with the force applied to the wheel segment and is equal to the ratio of the two radii.

Another way to calculate the MA of a wheel–axle system is to compare the forces applied:

EQUATION 3.5

$$MA = \frac{\text{force on the axle}}{\text{force on the wheel}}$$

If the motive force is applied to the wheel that has a larger radius and the resistive force is applied to the axle with the smaller radius, then the force advantage is with the motive force; that is, a smaller force can move a greater resistive force. If the motive force is applied to the axle with the smaller radius, then the motive force is at a *disadvantage* in force but is at an *advantage* in the ROM/speed of movement given to the resistance.

A few wheel–axle arrangements exist in the body, whereby motive force is applied to a wheellike structure, but more frequent are examples in which motive force is applied to

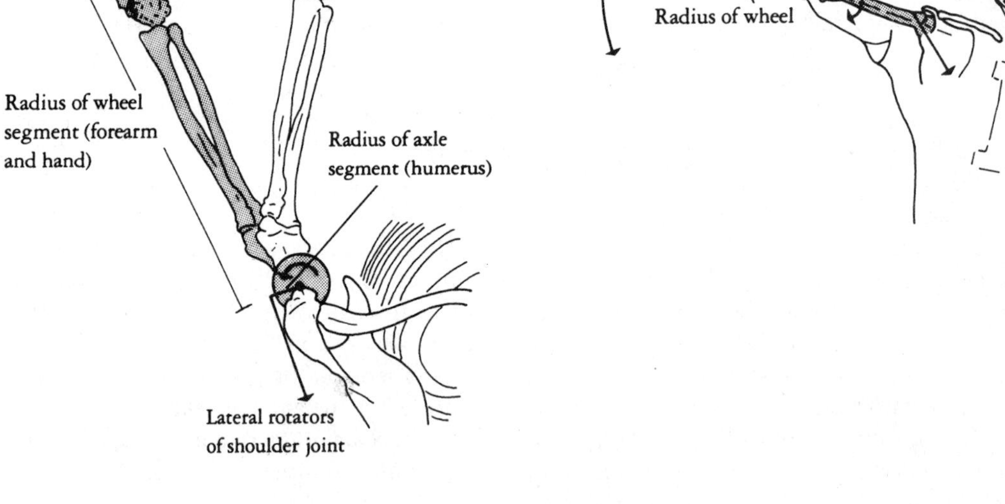

FIGURE 3.20

(a) Cross section of the humerus, showing the wheel-and-axle arrangement of the lateral rotators of the shoulder joint. (b) The humerus and forearm acting like a wheel-and-axle as a player throws a football pass.

an axlelike structure. Examples of the wheel–axle machine are found in rotations of body segments about their longitudinal axes. The muscles that cause these rotations act on the outside surface of the bone (axle) to cause distal segments that serve as the wheel to rotate. This is illustrated in a cross section of a humerus with the elbow flexed, as shown in Figure 3.20a.

The bone serves as the axle on which the muscle force is applied, and a second segment that articulates with the axle at some angle serves as the wheel. For instance, in medial rotation of the shoulder with the elbow flexed, the axle is the humerus, and the wheel segment is the forearm and hand. Because the motive force (muscle) is applied to the axle, the mechanical advantage of the mechanism is in the ROM/ speed gained by the hand as it sweeps through its motion, as seen in Figure 3.20b.

The wheel–axle arrangement of the body structures, when using rotations around the longitudinal axes of segments, is a great advantage in producing speedy segmental movement for throwing and striking activities. Additional sport-related examples are forearm pronation in badminton and medial rotation of the shoulder joint in the tennis serve. In these examples, the implement serves to increase the radius of the distal "wheel" segment and to provide an even greater MA in ROM/ speed of movement. Additional wheel–axle movements are discussed in Concept Module K.

3.3 Pulleylike Arrangements

The third type of machine that we find in the body is the single pulley. There are also arrangements of multiple pulleys, which create advantages in force and ROM/ speed of movement, but these are not found in the human body. The main function of pulley systems within the body is to change the effective direction of the applied motive force. Figure 3.21a illustrates a mechanical pulley that changes the direction of the resistive force. This application is used in the design of some weight-training machines. The most common example of a musculoskeletal pulley system is the muscular action around the malleolus on the lateral side and the medial side of the ankle. As the muscle contraction shortens the muscle toward the knee, the foot is moved downward into plantar flexion. Figure 3.21b illustrates this action.

Another example is the patella being pulled between the condyles of the femur during extension of the knee (Figure 3.21c). The quadriceps muscles pull proximally toward their bellies, and yet the knee action from flexion to extension causes the leg to move in the opposite direction. This is due to the single pulley function of the patellar tendon–patella–quadriceps arrangement.

3.4 General Considerations of Musculoskeletal Machines

As previously mentioned, most of the musculoskeletal levers in the body are of the third class, as a result of the close proximity of the muscle attachments and the result-

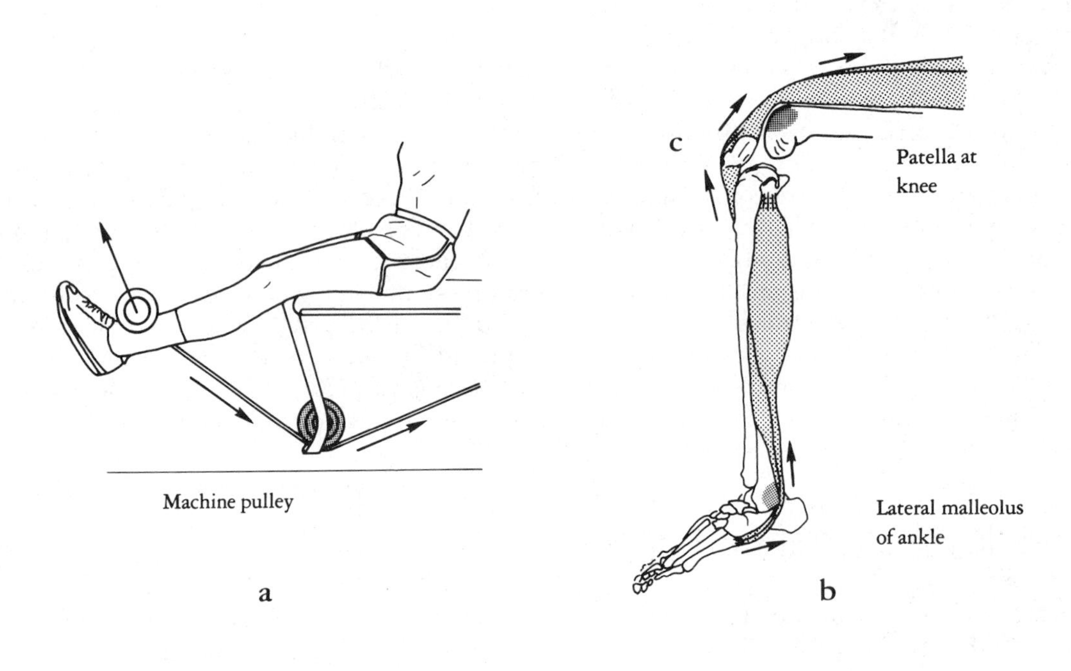

FIGURE 3.21

(a) A weight machine as a pulley system. (b) The malleolus of the fibula functioning as a pulley. (c) The patella at knee functioning as a pulley.

ing shorter force arms for the muscle forces than for the resistive forces. Consequently, the human body is built for ROM/ speed of movement advantages and not for force advantages. In most analyses, one finds that the muscle exerts more force to move the resistance than the force of the resistance itself. The resistance moves a greater linear distance, however, and hence has a greater lineal speed associated with its motion that does the motive force. This is true when the muscle functions as a motive force. Whenever the muscle is not the motive force (e.g., in eccentric contractions in which gravity moves the body and the muscle functions as the resistance), then the musculoskeletal lever mechanism is of the second class. In this case, the resistance (the muscle force) is located between the axis (the joint) and the motive force (gravity). The motive force (gravity) has a greater force arm than the resistive force (muscle), and therefore the muscle must exert more force than the resistive force.

Those who have undertaken the identification of the types of machines in the musculoskeletal system traditionally have assumed that the muscle is the motive force acting on the lever. This is not necessarily the case.

If one understands the relationships between the external forces involved and the advantages that those relationships provide in a system, then one can analyze the muscular contributions to athletic performance and conditioning exercises.

Understanding Musculoskeletal Machines

1. Draw a diagram of a first-class lever system that allows very fast movement of a light resistance. Show the axis of rotation, the magnitude of the force, the resistance, the force arm, and the resistance arm.
2. Repeat the task in question 1, showing a machine that is able to move a heavy resistance with a light force.
3. Repeat the task in question 1, showing a machine that provides for the balancing of the two forces.
4. Assuming the force is applied perpendicular to the bar, examine Figure 3.22 to answer the following questions:
   - **a.** What class lever system is it?
   - **b.** In which position, 1, 2, or 3, will the task be the easiest?
   - **c.** If it takes 100 force units to move the system at position 2, what will it take to lift it at positions 1 and 3?

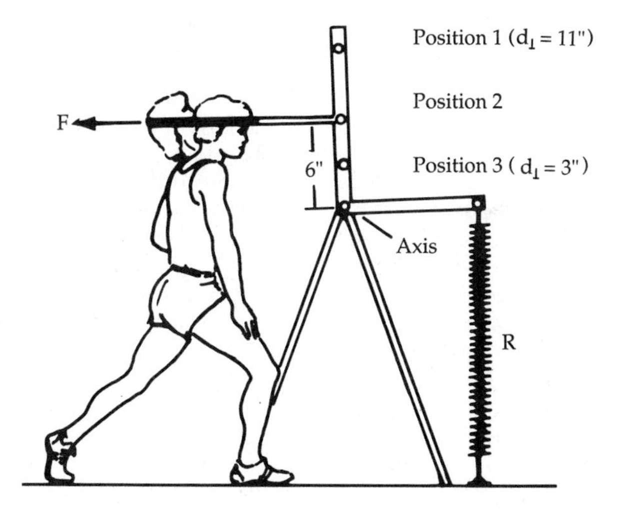

FIGURE 3.22

This person is exercising his neck muscles by hyperextending his neck against the resistance provided by the machine. The headstrap can be attached to positions 1, 2, or 3 on the vertical bar.

5. Complete the following sentences:
   a. In concentric muscular contractions against the force of gravity, _______ is the motive force.
   b. In concentric muscular contractions against the force of gravity, _______ is the resistive force.
   c. In eccentric tension of the muscle, _______ is the motive force.
   d. In eccentric tension of the muscle, _______ is the resistive force.
6. Musculoskeletal levers most generally have an advantage in what?
7. When a pulley-type mechanism is found in the human body, what is its main function?
8. It takes 50 units of force to lift 10 units of resistance.
   a. What is the advantage of this type of arrangement?
   b. Calculate the MA.
   c. What class of lever probably was used?
9. The motive force is 100 units. The balanced resistance is 50 units. The motive force arm is 0.6m.
   a. What is the motive torque?
   b. What is the MA?
   c. What is the advantage to this type of arrangement?
   d. What is the resistive force arm?

References and Suggested Readings

An, K. N., Hui, F. C., Morrey, B. F., Linscheid, R. L. and Chao, E. Y. (1981). Muscles across the elbow joint: A biomechanical analysis. *Journal of Biomechanics, 14*, 659–669.

Ariel, G. B. (1974). *Computerized biomechanical analysis of human performance: An application for exercise equipment design and athletic performance.* (Technical Report Uni-1). Fresno, CA: Universal Fitness Research Department, Universal Athletic Sales.

Basmajian, J. V., and MacConaill, M. A. (1977). *Muscles and movements—A basis for human kinesiology.* Huntington, NY: Robert E. Kreiger.

Coleman, J. E. (1977). The function of anatomical pulleys. In, C. Dillman and R. Sears (Eds.), *Kinesiology: A National Conference on Teaching* (pp. 63–76). Champaign-Urbana, IL: University of Illinois.

Degutis, E. W. (1971). A problem solving approach to the study of muscle action. In C. J. Widule (Ed.), *Kinesiology Review* (pp. 20–31). Reston, VA: AAHPERD.

Kume, S., and Ishii, K. (1981). Biomechanical analysis of isokinetic exercises. In A. Morecki, K. Fidelus, K. Kedzior, & A. Wit, (Eds.), *Biomechanics VII-B: Proceedings of the Seventh International Congress of Biomechanics* (pp. 404–10). Baltimore: University Park Press.

Oshimo, T. A. Greene, T. A. , Jensen, G. M., and Lopopolo, R. B. (1983). The effect of varied hip angles on the generation of internal tibial rotary torque. *Medicine and Science in Sports and Exercise, 15* 529–534.

Piscopo, J. (1974). Assessment of forearm positions upon upper arm and shoulder girdle strength performance. In J. G. Hay (Ed.), *Kinesiology IV* (pp. 53–57). Reston, VA: AAHPERD.

Van Eijden, T. M. G. T., deBoer, W., and Weijs, W. A. (1985). The orientation of the distal part of the quadriceps femoris muscle as a function of the knee flexion–extension angle. *Journal of Biomechanics, 18*, 803–809.

Van Zuylen, E. J., van Velzen, A., and van der Gon, J. J. D. (1988). A biomechanical model for flexion torques of human arm muscles as a function of elbow angle. *Journal of Biomechanics, 21*, 183–190.

CHAPTER 4

Biomechanical Relationships in the Upper Extremity

PREREQUISITES

Concept Modules A, B, D, E
Chapters 1, 2, 3

CHAPTER CONTENTS

It is assumed that the reader has studied basic musculoskeletal and ligamentous structures in a previous course; therefore, emphasis is on the mechanical aspects of these structural units. The upper extremity may be divided into five functional units: the shoulder girdle, the shoulder joint, the elbow, the radioulnar joint, and the wrist and hand.

The following structures are discussed in terms of stability and mobility of an articulation: bony structures, ligamentous structures, and muscular structures. The stability provided by a muscle depends not only on the strength of the muscle but also on the muscle's angle of pull on the bone. The angles of pull determine the line of action of the resultant muscle force as well as the rotary and stabilizing or dislocating components of that resultant force. The contribution of the muscular structure to the stability of the articulation depends on the relative magnitudes of these components.

4.1 The Shoulder Girdle Complex

Structure

The shoulder girdle is made up of the clavicle and the scapula. The articulations within the shoulder girdle are the sternoclavicular, the acromioclavicular, and the coracoclavicular joints. The shoulder girdle articulates with the trunk at the sternum (sternoclavicular), with the thorax (scapulothoracic), and with the upper extremity at the humerus.

The following bony landmarks of the scapula should be reviewed: the anterior and superior surfaces; the superior, lateral (axillary), and medial (vertebral) borders; the superior, inferior, and lateral angles; the supraspinous, infraspinous, subscapular, and glenoid fossae; the coracoid and acromion processes; and the spine. Figure 4.1 illustrates the articulations and surrounding structures of the shoulder girdle.

The sternoclavicular joint articulates the shoulder girdle complex with the trunk at the manubrium of the sternum. An examination of the configuration of the two bones at their articulation shows that no bony prominences help to hold the two bones together, and therefore the bony stability is weak. The sternoclavicular joint is the site of most of the movements of the scapula: elevation, depression, upward rotation, downward rotation, protraction (abduction), and retraction (adduction).[1]

In addition to being the location of the axis of rotation for the movements of the shoulder girdle, the sternoclavicular joint also absorbs the shock of trauma to the shoulder from sideward impact. Four main ligaments are associated with the sternoclavicular joint: the anterior sternoclavicular ligament, the posterior sternoclavicular ligament, the costoclavicular ligament, and the interclavicular ligament. These four ligaments in combination provide the ligamentous stability for the articulation of the two bones.

The acromioclavicular joint is located between the acromion process of the scapula and the distal end of the clavicle. It functions to absorb the stress of impact on the shoulder. There are no bony protrusions on any side of the articulation, and thus the bony arrangement is weak. The ligamentous arrangement comprises two main ligaments: the superior and the inferior acromioclavicular ligaments. Although these two ligaments provide stability to counter a pulling apart (tension) of the two bones, they cannot prevent the clavicle from riding up and over the top of the acromion process when the shoulder encounters a blow from the side. A dislocation in this manner is a common injury in football and wrestling. Two very strong ligaments of the coracoclavicular joint, the trapezoid and the conoid, help to counteract this dislocation. (See Figure 4.1.)

The third articulation of the shoulder girdle is the coracoclavicular joint. The joint is

[1] The terms *abduction* and *adduction*, meaning moving away from and moving toward the midline of the body, respectively, refer to movements of the scapula and are appropriate for physical therapy applications. The terms *protraction* and *retraction*, meaning to draw toward and draw back, respectively, refer to the movements of the shoulder and are more appropriate for sport activity situations. Both sets of terms refer to the same movements of the shoulder girdle complex as a whole, but the choice of which to use depends on the perspective of the user.

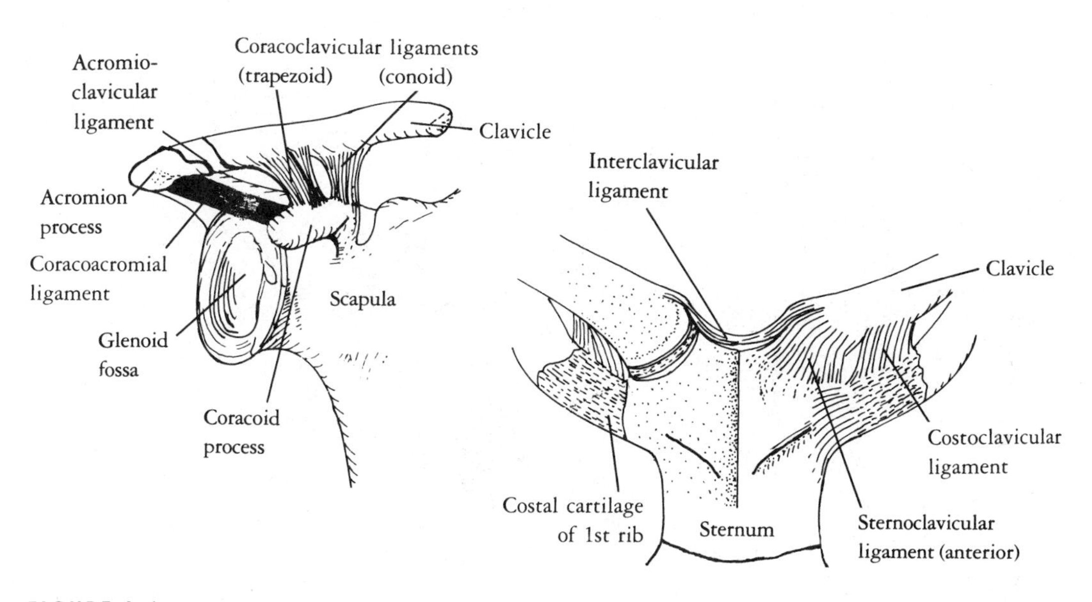

FIGURE 4.1
Articulations and ligaments of the shoulder girdle.

articulated by means of two main ligaments, the conoid and the trapezoid. The ligaments, which run downward and somewhat medially from the clavicle to the coracoid process of the scapula, help to reinforce the weak acromioclavicular articulation by transferring the force applied to the acromion process of the scapula to the clavicle. In this manner, the clavicle is "pulled along" when a force is encountered by the acromion.

Although there is no formal bone-to-bone articulation between the acromion process and the coracoid process, there is an important ligament, the coracoacromial (CA) ligament, that binds the two bony prominences. The CA ligament provides strong additional support for the acromioclavicular (AC) joint; however, it is probably more important in serving as a buffer for the rotator cuff muscle tendons that run anterior, superior, and posterior to the humeral head and can be subjected to pressure and impact from the acromion process during the performance of overarm throwing and striking skills. Salter, Nasca, and Shelley (1987) report finding irregular areas of bone on the inferior surface of the acromion process as well as rotator cuff tendon tears in those cadavers with markedly worn and degenerated CA fibers.

Muscles of the Shoulder Girdle as Stabilizers

Because the shoulder girdle is designed for mobility (i.e., for reaching and grasping types of activities), its stability is reduced. The muscular arrangements of the shoulder girdle and the shoulder joint are such that they can provide the stability lacking as a result of the weak arrangements of the bones and ligaments. The muscles must be strong enough to provide the necessary stability, however. The lack of upper body strength accounts for a great deal of injury to the shoulder region.

Figure 4.2 illustrates the muscles of the

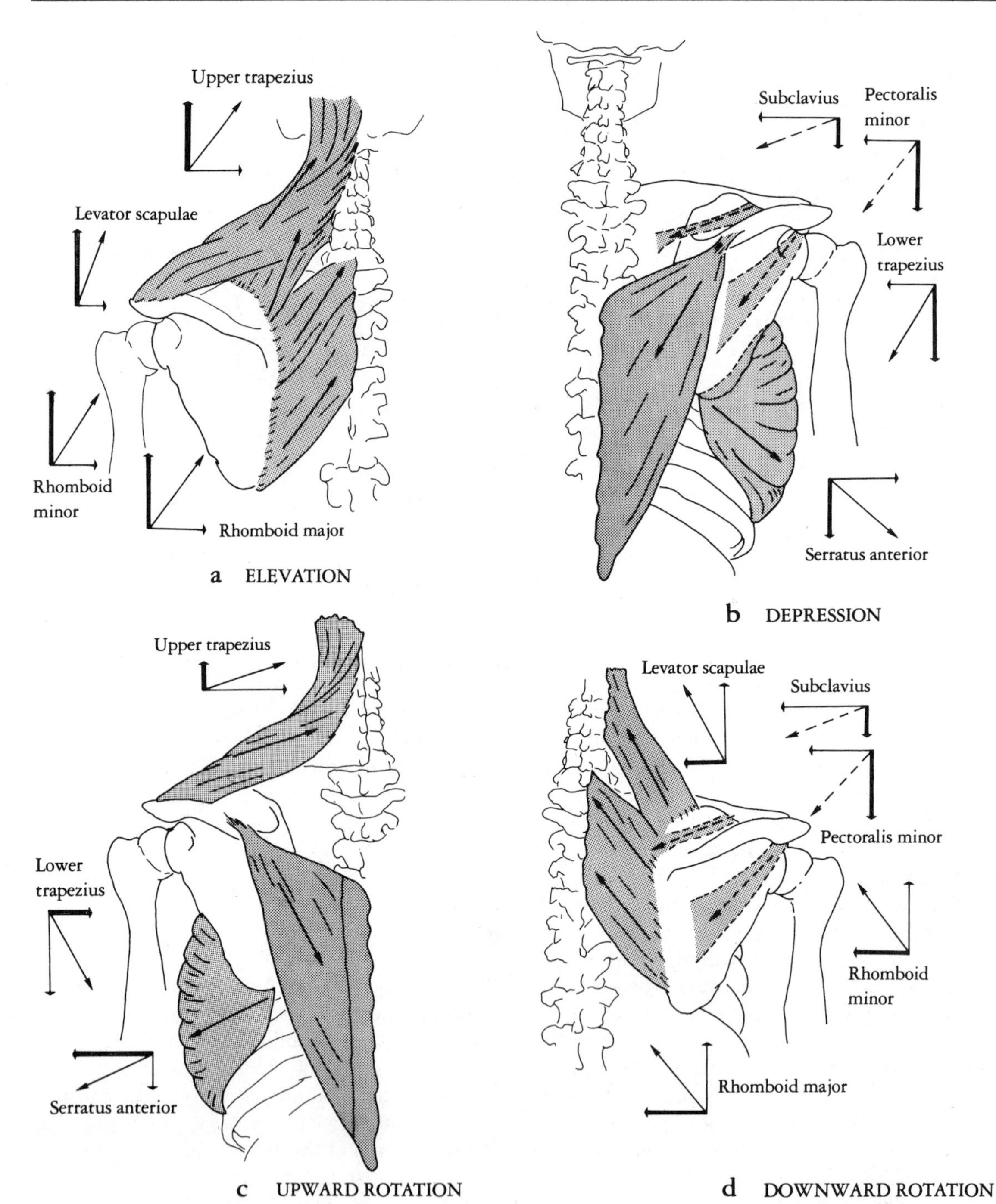

FIGURE 4.2

Shoulder girdle movements and musculature, with resultant and component muscular functions indicated. Dashed arrows represent muscle forces that are located beneath the structure.

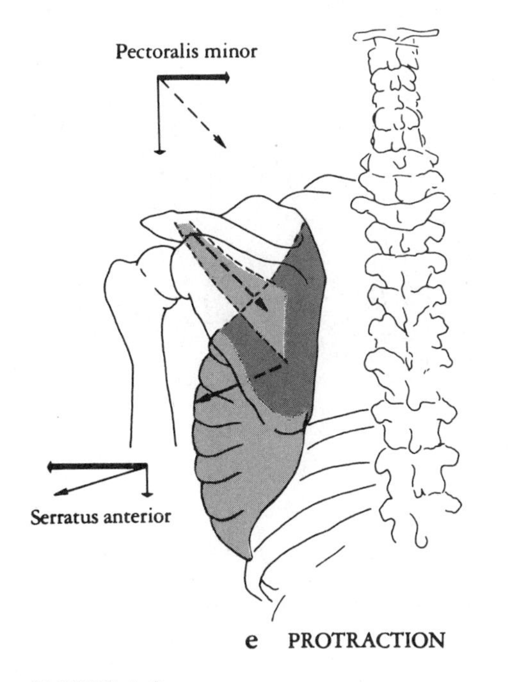

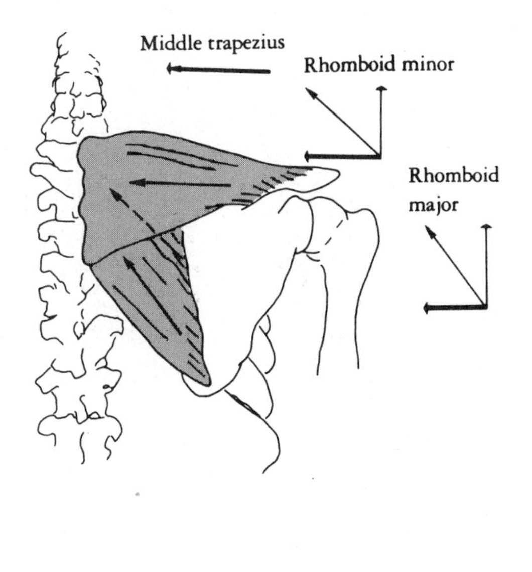

FIGURE 4.2
(continued)

shoulder girdle that produce the six movements of the scapula. In addition to the resultant directions of muscular force indicated on each muscle, the two perpendicular components of muscular force are indicated to the side. Through the use of these diagrams, the rotary and stabilizing components for each muscle may be determined.

Because the scapula is attached by muscles to the thorax and the vertebral column, a dislocating force would be one that would tend to pull the scapula away from these attachments. A muscle that has a component directed toward the midline of the body would have a stabilizing effect. The angles with which these muscles pull on the scapula determine the relative amounts of moving and stabilizing that they can do. One can make this determination by observing the relative lengths of the component force vectors. (The longer arrow indicates the greater component.) All the muscles can hold the scapula in place against the antagonistic motion given to it by the muscles that attach from it to the humerus. This is particularly true when one attempts to move the trunk relative to a fixed hand, as in performing a pull-up (Figure 4.3a). If one does a straight-arm support on the parallel bars, one wishes to stabilize the scapula in depression, as shown in Figure 4.3b.

Although a throwing injury is usually thought of as occurring at the elbow or shoulder joints, injurious stress on the ligaments or the musculature of the shoulder girdle is also possible if a muscle's stabilizing components are not strong enough to hold the articulations together. In addition, since the shoulder girdle is the beginning of the upper extremity and is fairly mobile relative to the trunk, it is called upon in many instances to be a stable base on which muscles of the shoulder joint may pull.

Numerous studies of overarm throwing and striking movements and selected swimming strokes reinforce the importance of the

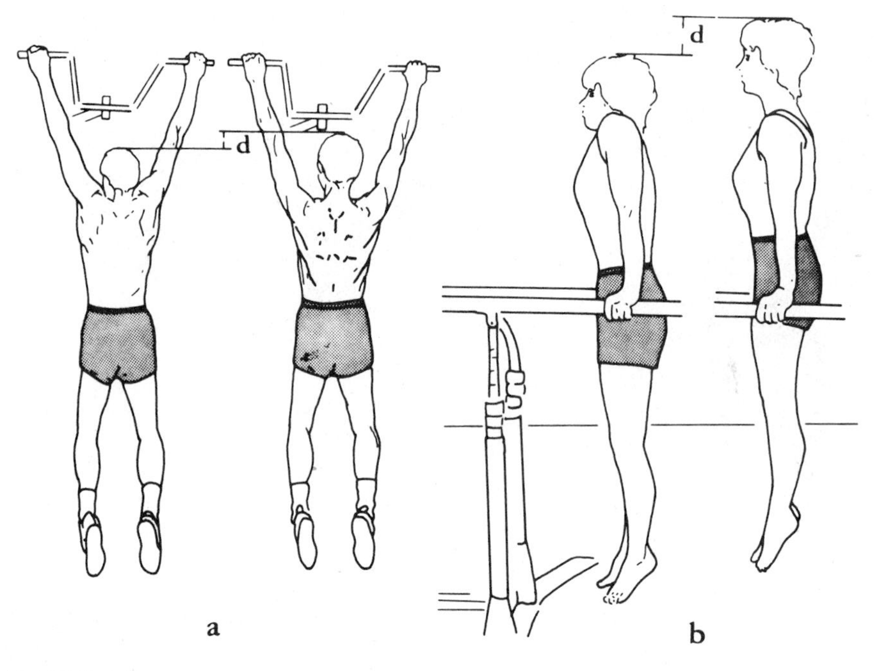

FIGURE 4.3

(a) The initial stage of a pull-up, showing the depression of the shoulder girdle. (b) Stabilization of the shoulder girdle in depression while a straight arm support is performed on the parallel bars.

shoulder girdle muscles acting with an isometric, stabilizing effect on the shoulder girdle or with an eccentric, decelerating effect on the shoulder girdle during these movements. Clearly, during forceful overarm motions, the strength of the agonistic and antagonistic muscles surrounding the shoulder girdle is necessary to prevent overuse strains on the surrounding tissues.

Muscles of the Shoulder Girdle as Movers

Because the stability of the shoulder girdle is important in providing a base from which other upper extremity functions can occur, the movement function of the shoulder girdle joints is often overlooked. In most activities involving the upper extremity, the shoulder girdle provides the initiation of the movements.

Observing the movements of the shoulder girdle enables understanding of the use made of the shoulder girdle motions in performing various tasks. The elevation movement is the initiation of lifting; the depression movement, of pushing downward; the protraction movement, of reaching, throwing, or pushing forward; the retraction movement, of pulling backward; the upward rotation movement, of increasing the range of overhead reaching; and the downward rotation movement, of forceful arm adduction at the shoulder joint.

The most common example is overarm throwing and striking, such as baseball pitching and tennis serves and smashes. During the latter stages of forward trunk rotation, shoulder girdle protraction is initiated. The protraction initiates the forward motion of the upper extremity in the throwing pattern. The serratus anterior is highly active during the deceleration phase at the end of the back swing and during the forward acceleration phase of these activities.

A second example is performing a pull-up. Many of us have watched a person, attempting to perform a pull-up, who cannot get beyond the initial stage, which is the depression of the shoulder girdle (Figure 4.3a). To determine why the depression part of the exercise is always easier than the rest of the arm movements, consider that the directions of pull of the muscles on the scapula for depression have large downward components. The muscles being used on the rest of the upper extremity pull at very small angles and thus have small rotary components when the shoulder joint is flexed and the elbow is extended.

Appendix VI, Tables VI.1 and VI.2, list the muscles and articulations of the upper extremity.

4.2 The Shoulder Joint

Structure

The shoulder joint is made up of the articulation of the glenoid fossa of the scapula and the head of the humerus. The reader should review the following bony landmarks of the humerus: the head; the neck; the greater and lesser tuberosities; the bicipital groove; the shaft; the deltoid tubercle; the radial groove; the medial and lateral epicondyles; the trochlea; the capitulum; and the coronoid, olecranon, and radial fossae. Figure 4.4 illustrates the bony and ligamentous arrangement of the shoulder joint.

The bony arrangement of the shoulder joint consists of a shallow socket (glenoid fossa) to which is joined the one-half-spherical head of the humerus. Less than half of the humerus is in the socket at any one time, and the bony arrangement is therefore weak. To assist in the task of stabilizing, a labrum or lip of cartilage circles around the outside of the fossa to increase the depth of the concavity and assist in holding the head in place. Because it is a ball-and-socket type of arrangement, the shoulder joint is a multiaxial joint that can move through the following movements: flexion, extension and hyperextension, transverse adduction and abduction, abduction and adduction, medial (inward) and lateral (outward) rotation. Two main ligaments stabilize the shoulder joint: the coracohumeral and the superior, middle, and inferior glenohumeral ligaments, named for their attachments and locations (see Figure 4.4). The coracohumeral ligament serves to stabilize the humerus against upward and lateral displacement, and the glenohumeral serves to stabilize the humerus against forward displacement. The glenoid fossa is directed slightly anteriorly. Thus, the back edge tends to prevent backward displacement of the humerus, but this leaves the forward position vulnerable to dislocations. As with the shoulder girdle, the shoulder joint is designed for mobility and therefore sacrifices bony and ligamentous stability.

Muscles of the Shoulder Joint as Stabilizers

The musculature surrounding the shoulder joint is arranged to produce large stability

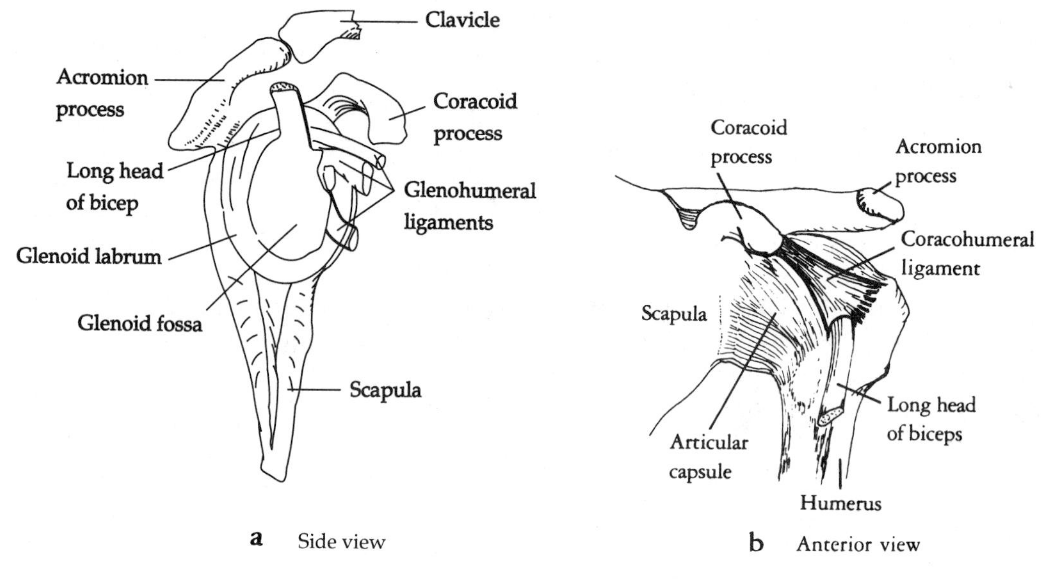

FIGURE 4.4
The body and ligamentous arrangements of the shoulder joint.

components of the muscle's forces. Figure 4.5 shows muscular arrangements that provide muscular stability to the shoulder joint. These muscles are called the **rotator cuff** muscles. There are four rotator cuff muscles, and all have large stabilizing components, as can be seen by the inserted component-force vectors. Three of these four muscles originate on the posterior surface of the scapula: the supraspinatus (above the spine), the infraspinatus (below the spine), and the teres minor. These three are sometimes remembered by their acronym as the SIT muscles. The fourth rotator cuff muscle is the subscapularis (below the scapula); it originates on the anterior surface of the scapula.

All four of the rotator cuff muscles have large stabilizing components as evidenced by the large vector components directed medially. Regardless of the position of the humerus, the deltoid muscle—anterior, posterior, and middle—has a large stabilizing component because of its small angle of pull. When the humerus is in extension, however, the deltoid's stabilizing component is directed to pull the humerus up into the acromion process. Thus, the components of depression of the rotator cuff muscles must counteract this tendency. As with the deltoid muscle, the upward components of these muscles must be counteracted by the depressive components of the rotator cuff muscles.

Further stability is provided by the long heads of the biceps brachii and the triceps, on the anterior and posterior sides, respectively. Because their tendons pass so close to the joint, each muscle's stabilizing component is large when the humerus is in the position antagonistic to the muscle's function (e.g., the stabilizing component of the biceps at the shoulder joint is great when the shoulder joint is extended). As with the deltoid muscle, the upward components of these muscles must be counteracted by the depressive components of the rotator cuff muscles.

Most of the other muscles surrounding the shoulder joint have force components that stabilize; however, their main function

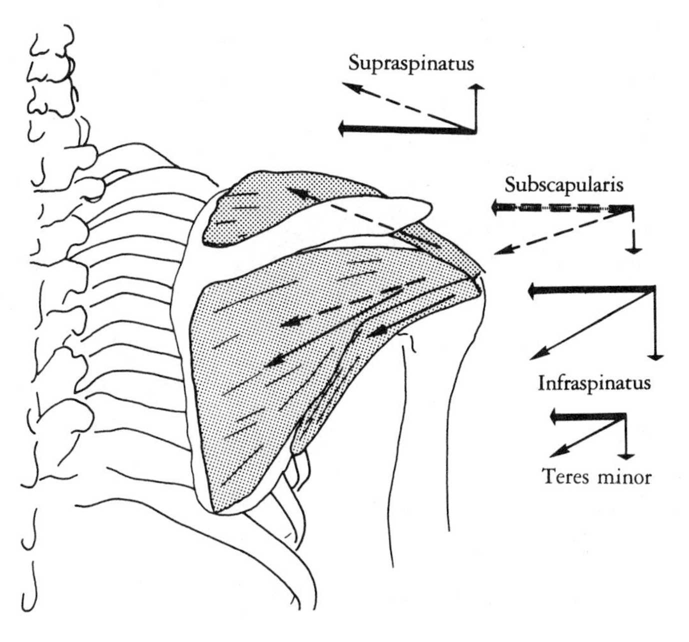

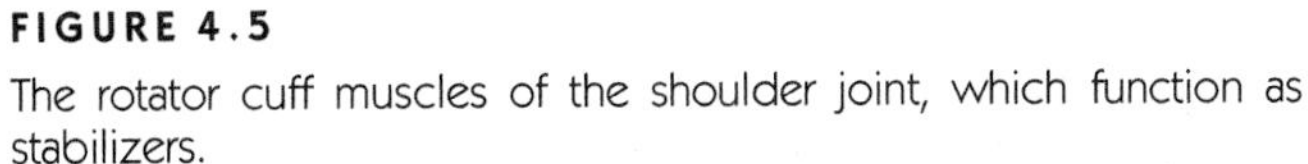

FIGURE 4.5
The rotator cuff muscles of the shoulder joint, which function as stabilizers.

is to move the humerus. In addition, as the humerus moves into various degrees of abduction, flexion, and extension, these mover muscles change their angles of pull considerably; thus, they are not generally classified as stabilizers of the shoulder joint. These mover muscles are discussed in the next section.

Muscles of the Shoulder Joint as Movers

The muscles that serve as the primary movers of the humerus at the shoulder joint are shown in Figure 4.6. They are the anterior deltoid, coracobrachialis, pectoralis major, latissimus dorsi, and teres major; the long and short heads of the biceps on the anterior side; the posterior deltoid and the long head of the triceps on the posterior side; and the middle deltoid on the lateral side.

Those muscles that are directed medially from their distal attachments on the anterior side function to forward flex and transverse adduct the shoulder joint. Those on the posterior side and directed medially cause extension and transverse abduction at the shoulder joint. Figure 4.7 illustrates a cross section of the humerus showing the muscles' resultant forces that cause medial and lateral rotation at the shoulder joint.

Two of the rotator cuff muscles insert on the posterior side of the humeral head, the infraspinatus and teres minor. Each functions in a wheel-axlelike mechanism to rotate the posterior side of the humerus medially, causing lateral rotation of the humerus around the longitudinal axis of the shoulder joint. The subscapularis is inserted on the anterior side of the humeral head and functions in a wheel–axlelike mechanism to rotate the anterior side medially, causing medial rotation of the humerus. The supraspinatus is attached distally to the top of the humeral head and

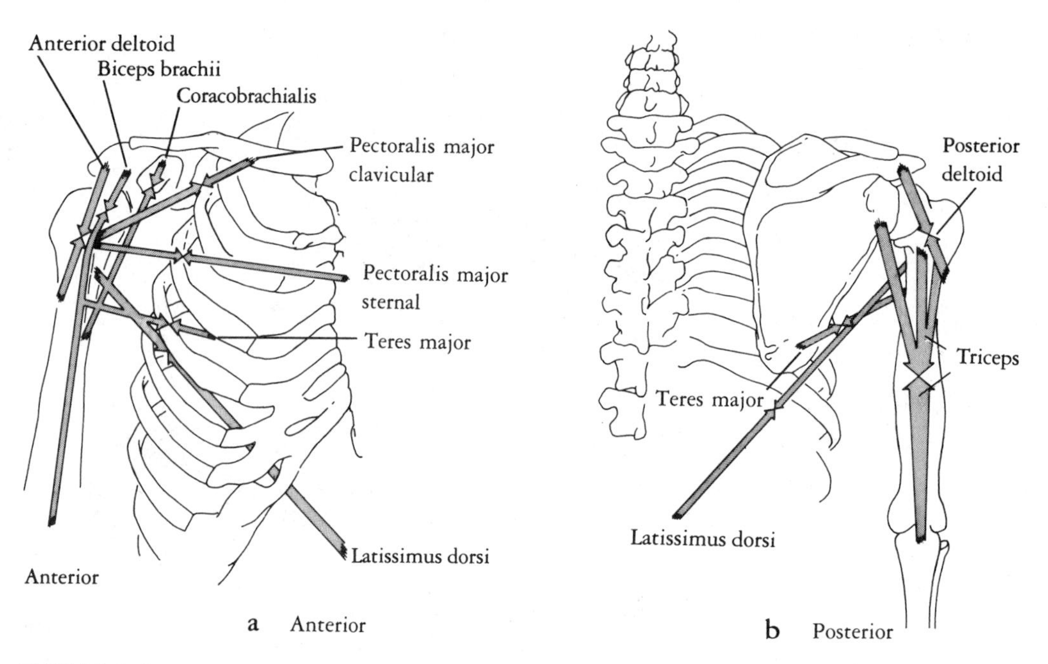

FIGURE 4.6
(a) Anterior and (b) posterior muscles of the shoulder joint that are used predominantly as movers.

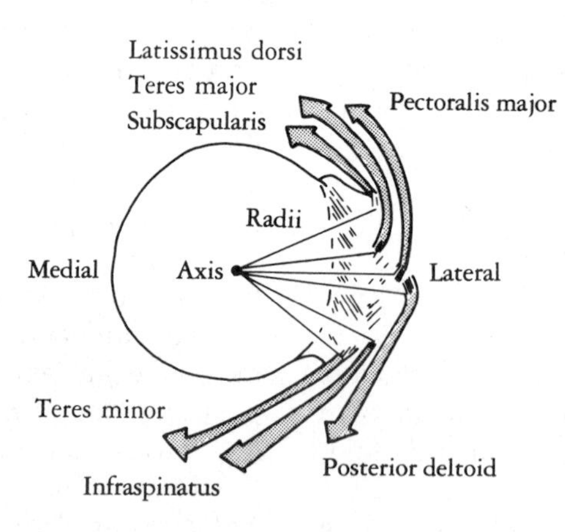

FIGURE 4.7
Cross section of the right humerus showing the force vectors for the medial and lateral rotators of the shoulder joint.

above the anteroposterior axis of rotation of the shoulder joint. Thus, it functions as a first-class lever mechanism, pulling the top of the humeral head medially to move the shaft of the bone laterally into abduction.

The latissimus dorsi and teres major originate posteriorly on the trunk and insert on the anterior side of the humerus. When they shorten, they pull the anterior humerus medially in a wheel–axle arrangement to cause medial rotation and extension of the humerus.

The humerus functions as the axle in a wheel–axle arrangement when it is medially and laterally rotated. The wheel is represented by the forearm and hand if the elbow is flexed, or by the hand if the elbow is extended and the wrist is flexed. The wheel–axle arrangement of the upper extremity is discussed in Chapter 3. The size of the radius of the axle (i.e., the radius of the bone) is fixed. The radius of the wheel depends on the degree to which the elbow or wrist is flexed; the greater the flexion—up to 90 degrees—the greater the radius, the more the resistance torque and also the greater the advantage in providing range of linear motion or speed of movement of the hand (wheel) segment. The range of linear motion of the hand when the elbow is flexed to various degrees during medial rotation of the shoulder joint is an important consideration in throwing and striking activities. A greater wheel radius may be had in activities using hand-held implements such as rackets, clubs, bats, or sticks.

As the arm is moved from anatomical position to a flexed or extended, abducted or adducted position, the resultant force vectors of the muscles surrounding the shoulder joint change as well. Because of the mobility of the shoulder joint, numerous positions of the upper extremity are possible, with each having its own resultant muscle directions.

For example, one must be aware of the muscles' resultant line of action and the two components of the muscle force vector when performing a bench-press movement. If the shoulder joint begins the press from a position in which the shoulder girdle is retracted and the shoulder joint is in extreme transverse extension (or hyperextension), then the initial tension of the muscles used to transverse flex the shoulder joint will produce a large component of force directed forward, such that it tends to displace the head of the humerus anteriorly.

This component must be counteracted by the posterior rotator cuff muscles or the humeral head will be dislocated anteriorly relative to the glenohumeral joint. Problems related to the anterior glenohumeral capsule, the glenoid labrum, and the attachment of the long head of the biceps will ensue if these stabilizing forces are not effective in counteracting the anterior dislocating forces of the muscles involved in the movement.

Mechanics of Arm Abduction

The nature of arm abduction is interesting and complex. The coordination of the shoulder complex (girdle and joint) not only allows the mobility necessary for the reaching, grasping, and throwing functions of the human upper extremity but also provides the structures necessary to stabilize an inherently weak, bony, and ligamentous arrangement. Figure 4.8 illustrates the important muscles that function to abduct the arm at the shoulder joint and to stabilize the humerus against the dislocating components of the mover muscles.

The supraspinatus initiates the first few degrees of shoulder joint abduction. Apparently, the first-class lever arrangement of the supraspinatus, even though it acts with a disadvantage in motive force, gives a better angle of pull than does the deltoid. The two components of force for the deltoid are always such that the stabilizing component is greater than the small rotary component throughout the entire range of abduction. At initial stages, the stabilizing component is actually a dislocating one, because it is directed in such

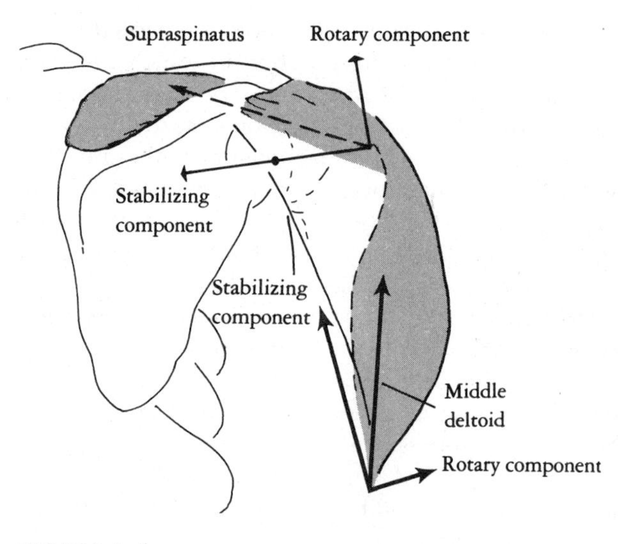

FIGURE 4.8
Muscles important in shoulder abduction.

a way that it pulls the head of the humerus upward into the inferior side of the acromion process. To overcome some of these problems, two things happen. First, as the humerus abducts, the scapula upwardly rotates at a ratio of about 3:2; that is, the glenoid fossa rotates upward 2 degrees for every 3 degrees of humeral abduction. This action moves the acromion process out of the way as the greater tuberosity of the humerus approaches it. Second, the rotator cuff muscles tense to stabilize the humeral head against the glenoid fossa so that the component of the deltoid that would tend to displace the humerus upward is neutralized. Three of the rotator cuff muscles—the infraspinatus, the subscapularis, and the teres minor—have downward components to perform this stabilization (Figure 4.5).

The activities that require the shoulder joint to be abducted or forward flexed and medially rotated cause shoulder pain in a significant number of athletes. This shoulder impingement syndrome, as it is labeled, is found in baseball pitchers, tennis players, and freestyle, butterfly, and backstroke swimmers. Since these athletes perform repetitive and forceful movements that involve shoulder abduction, flexion, and medial rotation, they are the most susceptible to the impingement syndrome. Some call this syndrome the rotator cuff impingement syndrome, although only one rotator cuff muscle is directly involved. Moreover, the long head of the biceps, which is not part of the rotator cuff group, also becomes impinged. Most sources agree that the inflammation occurs from the squeezing of the supraspinatus tendon, which passes over the head of the humerus and under the anterior part of the acromion process and the coracoacromial ligament. The latter two structures make up what is known as the coracoacromial arch. For a thorough discussion of pathologies of the shoulders of swimmers the reader is referred to McMaster (1986a).

The compression of the head of the humerus upward and inward to the glenoid fossa and the coracoacromial arch is motion that should be prevented by strengthening those muscles that stabilize the humerus against these motions. The muscles are those that have a downward force component when the arm is abducted or flexed.

Musculoskeletal Analysis of Overarm Throwing and Striking Skills

Most overarm throwing and striking skills use similar movements and muscles. When these skills are repeated forcefully a number of times, such as in baseball pitching or tennis serving, the stresses that these muscles must undergo frequently result in overuse-type injuries. The movement pattern consists of an abducted dominant arm, which is "cocked" into extreme lateral rotation, followed by a "force or acceleration" phase, which consists of forceful protraction of the shoulder girdle at the sternoclavicular joint and medial rotation of the humerus. These shoulder movements are followed by elbow extension and various radioulnar and wrist movements, depending on the skill being performed.

The high deceleration of the "cocking" motions followed by large acceleration of the force-phase motions places undue stress on the muscles involved in those actions. The shoulder girdle protractors and the humeral medial rotators serve to decelerate the retraction, lateral rotation motions by contracting eccentrically (a stretch-shorten movement). The protectors and medial rotators then contract forcefully to accelerate the protraction and medial rotation motions. The serratus anterior acts to protect the scapula and the pectoralis major, latissimus dorsi and and the pectoralis major, latissimus dorsi and teres major are active in medially rotating the humerus. The stabilization of the shoulder girdle through eccentric contractions of the trapezius and rhomboids while it is protracting is an important part of an effective throw or strike.

During the transition from "cocking" to "acceleration" phases many shoulder injuries occur. The anterior capsule may be ruptured or torn, the humeral head may be anteriorly dislocated, epiphyseal integrity can be threatened in the immature performer, and the anterior glenoid labrum may be torn, which may be associated with the stabilizing and decelerating function of the long head of the biceps at the shoulder and elbow (Andrews, Carson, and McLeod, 1985). In addition, the elongation of the medial rotators at the end of lateral rotation (estimated to be 160 degrees) followed by forceful concentric tension can lead to tendinitis in the medial rotators or tendon tears or ruptures. Forceful deceleration occurring in the follow-through phase may result in inflammation of those muscles responsible that are working in eccentric tension states.

It is obvious from the biomechanical analysis of the structures involved that strengthening only the muscles involved in the concentric portions of an overarm throw or strike will result in overuse-type injuries to the other surrounding structures. These considerations should be applied not only to baseball pitching but to tennis, volleyball, badminton, weight throwing, swimming, and other sports involving overarm movements.

Understanding the Shoulder Girdle–Shoulder Joint Complex

1. Stand in back of a partner. Place your hand on your partner's scapula. The partner slowly abducts the arm over his or her head.
   - a. What movement of the scapula occurs?
   - b. When do you start to notice it (in degrees)?

c. Why is this movement necessary?

d. What does it do to the glenoid fossa?

2. Have your partner shrug his or her shoulders.

a. From what articulation does this movement originate?

b. What bones are moving because of the direct muscle pull on them?

c. What bones move because they are attached to the moving bones?

d. Observe the lines of force for the muscles that perform this function in the illustrations of the scapula and clavicle in Figure 4.2. Discuss the effectiveness of each muscle group in performing the motion (in terms of relative magnitudes of the components).

3. Move the shoulder girdle in the following ways: elevation, depression, upward rotation, downward rotation, protraction, retraction. Place your hands over your partner's scapulae during the protraction and retraction movement. What is the motion of the scapulae relative to the vertebral column?

4. What muscles should be strengthened to aid in the deceleration of the shoulder girdle and humerus during a follow through in an overarm throw or strike?

5. Stand facing a partner who is in anatomical position. The partner flexes the shoulder joint as far as it will go without moving the arm or forearm around any other axis besides the frontal.

a. How far will it flex?

b. What stops the movement?

c. What movement allows the partner to get the arm overhead? Why?

6. Starting from the anatomical position, measure the range of motion (ROM) for medial rotation of the shoulder joint. Starting again from the anatomical position, measure the ROM for lateral rotation. Compare the ROMs.

7. Identify the movement of the shoulder girdle and the shoulder joints in the following activities (during the execution, or force, phases): (a) a butterfly stroke in swimming, (b) a back-crawl stroke in swimming, (c) a dip on the parallel bars, (d) a pull-up, (e) a volleyball spike, and (f) drawing the arm back in archery.

8. What is an example of a wheel-and-axle arrangement in the shoulder joint? Draw a diagram that represents this arrangement. Discuss the mechanical advantage of this arrangement in terms of force requirement and ROM advantages. What is the mechanical advantage in ROM, speed of movement, or force?

4.3 The Elbow Joint

Structure

Technically, the elbow joint is made up of the distal end of the humerus and the proximal end of the ulna. The proximal end of the radius also articulates with the distal end of the humerus and is therefore considered part of the elbow. The reader should review the bony landmarks of these particular areas: on the humerus, the medial and lateral epicondyles, the trochlea, the capitulum, and the coronoid, radial, and olecranon fossae; on the ulna, the olecranon and coronoid processes; and on the radius, the head and the radial tuberosity. Figure 4.9 illustrates the bony and ligamentous arrangement of the elbow joint and the radioulnar joint.

The only movements possible at the elbow joint are flexion and extension. These are accomplished by the movement of the olecranon process and coronoid of the ulna, which act as a partial cuff around the trochlea of the humerus. The bony structure makes the elbow stable and acts to prevent dislocations. The head of the radius is attached to the ulna by means of the annular ligament; thus, the movement of the ulna is transferred to the radius, which moves along with it. The articular surface of the radial head slides along the capitulum of the humerus. The humeral–radial bony stability is weak.

The annular ligament is a three-quarter, ring-shaped tissue. It is cup shaped, and it attaches to the proximal end of the ulna and surrounds the head of the radius. It functions to hold the head of the radius to the ulna, thereby transferring movement of the ulna to the radius and preventing distal or lateral displacement of the radius from the ulna. Because of its ring-shaped structure, the annular ligament allows the radius to rotate around the longitudinal axis of the forearm to provide for pronation and supination of the radioulnar joint.

Two ligaments help prevent sideward displacement of the elbow articulation: the lateral ligament (radial collateral) and the medial ligament (ulnar collateral), which prevent adduction and abduction, respectively. Because of the strong bony structure holding the elbow together, dislocation of the elbow is not as common as dislocation of the shoulder complex. However, because of excessive and continual stress at the humeral–radioulnar area as a result of repeated throwing or striking activities, the ligaments, tendons and even bony parts may become stressed beyond their capacity, and conditions such as tennis elbow or pitcher's elbow then result.

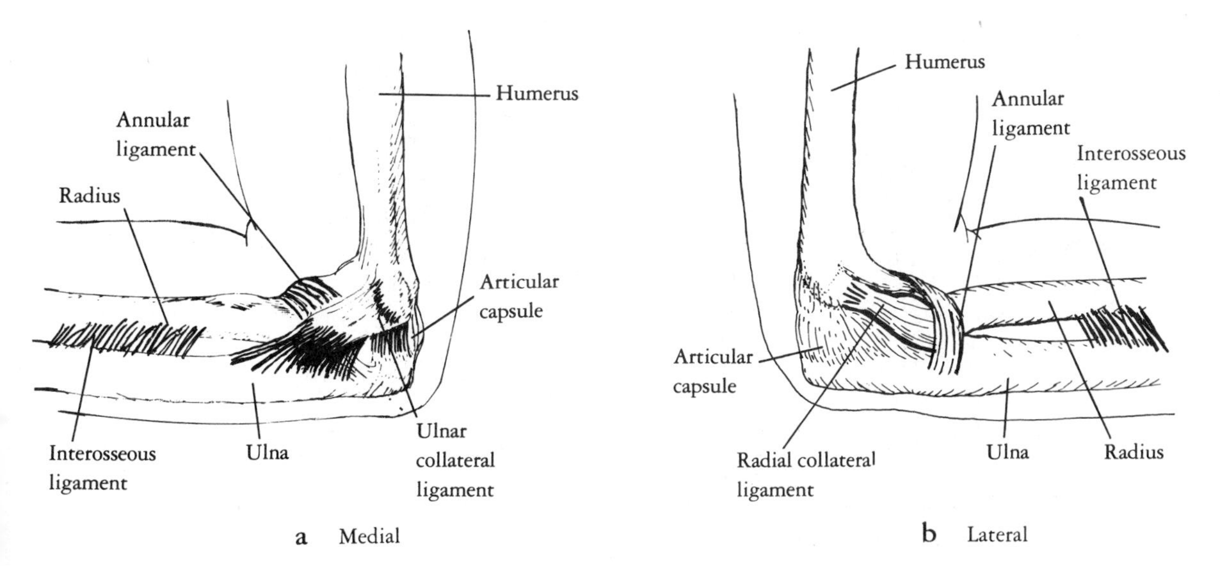

FIGURE 4.9

The bony and ligamentous arrangements of the elbow joint and the radioulnar joint.

Muscles of the Elbow as Stabilizers

The anterior and posterior sides of the elbow joint are stabilized by the stabilizing components of the muscle groups that cross the joint. The anterior muscles are shown in Figure 4.10a. We have already seen how the three main elbow flexors (biceps, brachialis, brachioradialis) are arranged mechanically around the elbow joint in Chapter 3. Other anterior muscles that pass over the elbow have their proximal attachments close to the elbow joint and their distal attachments, for the most part, below the wrist. Because the lines of force of these muscles pass so close to the elbow joint, their torques for rotary motion at the elbow are small; thus, their function at the elbow is mainly stabilizing. Notice that the muscles

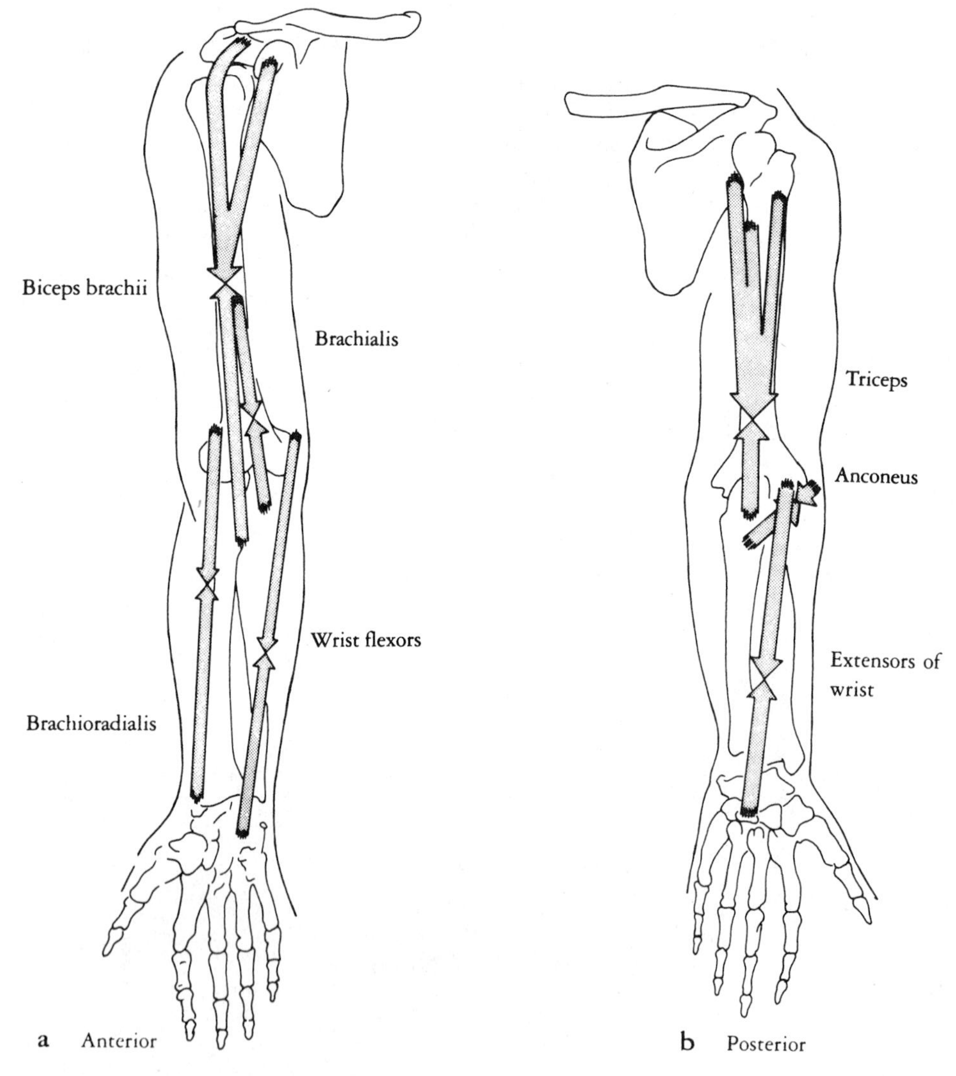

FIGURE 4.10

(a) Anterior muscles of the elbow joint. (b) Posterior muscles of the elbow joint.

that function mainly as stabilizers of the elbow and flexors of the wrist have their proximal attachments to the medial epicondyle of the humerus. Because of the number of muscles that act as stabilizers on the anterior side, the muscular stability of the elbow is considered strong. In positions in which the anterior muscles have dislocating components (i.e., when they achieve an angle greater than 90 degrees), the muscles are in such a shortened position that the tension is minimal owing to the length-tension relationship.

The posterior muscles of the elbow are shown in Figure 4.10b. The main muscle on the posterior side of the elbow is the triceps. From its attachment on the olecranon process of the ulna, it proceeds up the humerus, attaching two of its heads to the posterior humerus and one long head to the scapula. Although all the heads join together to form a common belly and attach to a single distal tendon, the two shorter heads only extend the elbow, while the longer head is a two-joint muscle that extends the shoulder joint as well as the elbow. Because of the particular structure of the posterior elbow (the hooklike configuration of the olecranon and coronoid process), the triceps stabilizes the elbow whether its pull is greater or less than 90 degrees to the long axis of the ulna.

The muscles that extend the wrist also function mainly as stabilizers of the elbow and are attached to the lateral epicondyle of the humerus (Figure 4.10b). The bony arrangement of the posterior elbow is strong, and so is its muscular stability.

Muscles of the Elbow as Movers

The main elbow flexors are discussed as an example of the stabilizing and rotary functions of muscles in Chapter 3. Refer to Figure 3.9 (p. 170) for a review if necessary. The biceps is a three-joint muscle (shoulder, elbow, radioulnar), and therefore should be strengthened over all of its articulations. The shoulder-joint flexion exercise with the elbow extended (Figure 2.4j, p.72) should be performed as well as the traditional forearm-curl. The elbow flexion against resistance should be performed with the shoulder joint held in extension to take advantage of the full ROM of the biceps.

The triceps as a mover of the elbow is an interesting and rare example of a first-class lever system in the body. It is shown in Figure 3.15b (p. 177).

The elbow may be positioned so that the triceps pulls at a 90-degree angle to the long axis of the ulna, thus contributing 100% of its effort to the rotary function. In addition, because it is a two-joint muscle, the long head of the triceps can lengthen at its proximal end when the shoulder flexes, while it shortens at its distal end to extend the ulna. Consequently, the length–tension relationship can be maintained.

To fully strengthen all three heads of the triceps one should perform resistance exercises by extending the shoulder joint with the elbow extended as shown in Figure 2.4d (p. 72) and by extending the elbow with the shoulder joint flexed as shown in Figure 2.4e. Appendix VI.1, Tables VI.1 and VI.2 show the muscles and their articulations of the upper extremity.

4.4 The Radioulnar Joint

Structure

The radioulnar joint is made up of three articulations: the proximal, middle, and distal radioulnar joints. Although their bony arrangements give virtually no stability, their ligamentous arrangements provide the necessary cohesion. The proximal radioulnar joint is supported by the annular ligament, which is shown in Figure 4.9, p. 201. It prevents downward and lateral displacement of the radius. The middle radioulnar articulation is supported by an interosseous membrane (Figure 4.9), which is located between the shafts of the radius and ulna along their entire length. Through the membrane pass blood vessels and nerves. The fibers of the membrane run obliquely from the radius distally to the ulna. This oblique direction of the fibers strengthens the articulation, helping to prevent a strong pull of the biceps and brachioradialis muscles on the radius from causing proximal displacement of the radius. Because muscles also are attached to the interosseous membrane, it functions generally to transfer stress between the radius and the ulna. The distal radioulnar joint is supported by a fibrodisc joining the two bones at the distal end and by the anterior and posterior radioulnar ligaments.

The movements of the radioulnar joints are pronation and supination around the longitudinal axis. Figure 4.11 illustrates the relative radioulnar positions in each of these movements. Notice that the proximal

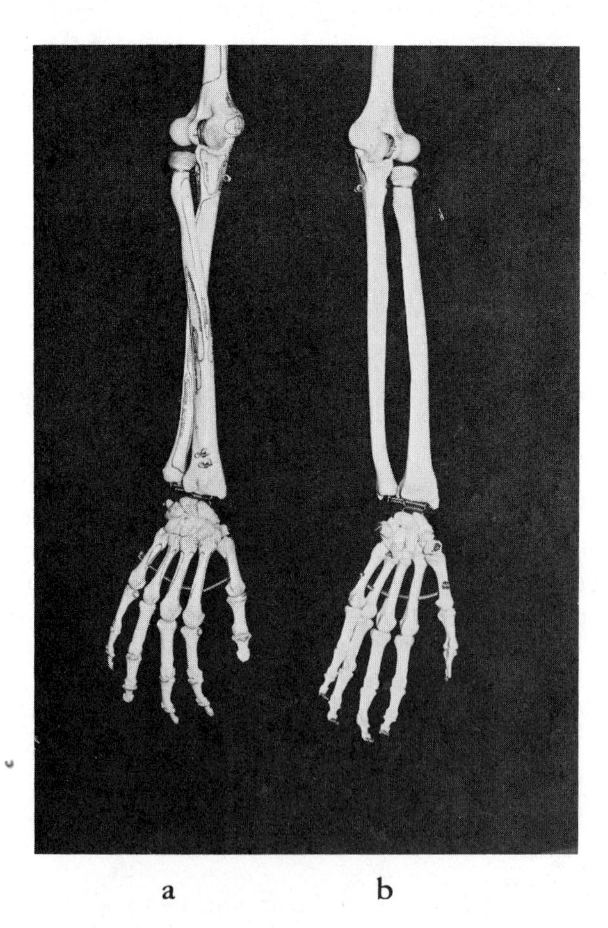

a b

FIGURE 4.11

Radioulnar positions in (a) right-arm pronation and (b) left-arm supination of the radioulnar joint.

head of the radius rotates around the annular ligament (represented by a wire) but does not shift positions relative to the ulna. When the forearm is in pronation, the radial shaft crosses over the ulna. The distal end of the radius is then on the medial side of the ulna. The fibrodisc moves with the radius and back under the ulna when the radius supinates.

Muscular Function at the Radioulnar Articulation

The muscles of the radioulnar joint are those that stabilize the two bones and produce either pronation or supination. The four muscles that are influential to radioulnar stabilization and movement are the biceps, the supinator, the pronator teres, and the pronator quadratus. These muscles are illustrated in Figure 4.12. (The muscles that are located along the forearm and produce movements at the wrist do not play a major role in elbow or radioulnar movements, even though they cross both joints; thus they are not discussed here.)

As can be seen from the force arrows of the radioulnar muscles, three of the four muscles contribute to the stability of the radioulnar articulation. The biceps, because it is attached only to the radius and not to the ulna, does not have this stabilizing function. The pronator quadratus and the supinator are located so as to permit the ulna to be used as a stable base from which the muscles can pull on the radius to produce pronation and supination. The pronator teres has a stabilizing component but also pulls across the elbow to the medial epicondyle of the humerus. All three have stabilizing components because all pull on the radius toward the ulna.

The mechanics of the biceps pull as it relates to supination of the radioulnar joint deserves

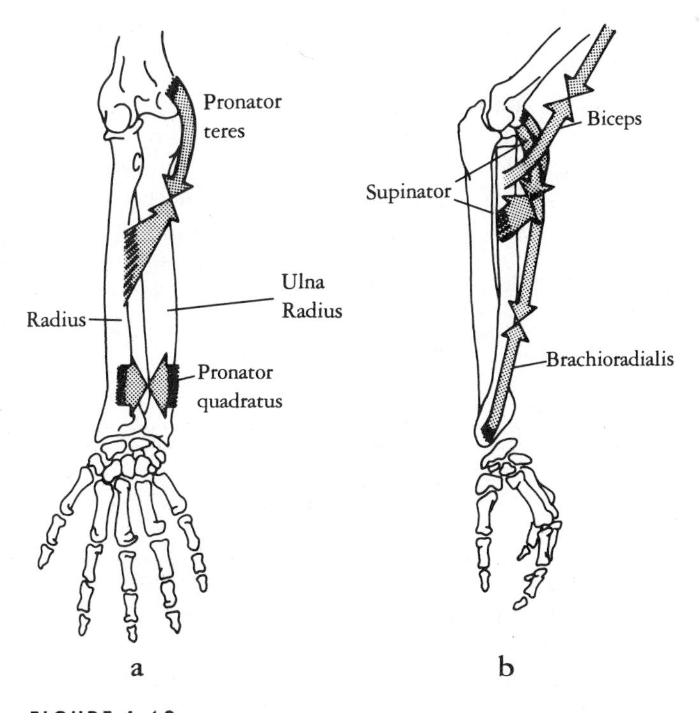

FIGURE 4.12
Muscles of the radioulnar joint: (a) pronators, (b) supinators.

special attention. The attachment of the biceps on the tuberosity on the medial side of the radius allows the biceps to act in a wheel–axle configuration to produce supination when the forearm is in pronation.

When the forearm is pronated, the tendon of the biceps is wrapped around the radius by following the radial tuberosity into medial rotation. This wrapping around the radius is presented as a reason for the decrease in the effectiveness of the biceps as an elbow flexor when the radioulnar joint is pronated. Thus, a flexed arm hang or a pull-up is easier to perform when the forearm is supinated. The wrapping, however, does allow the biceps to function as a strong supinator of the forearm when the forearm is pronated. Thus, elbow flexion with supination is an effective movement for using the biceps muscle at the elbow. A similar situation occurs with the brachioradialis, in that its attachments to the anterior humerus and the radius on the lateral side make it a pronator to midposition when the forearm is supinated and a supinator to midposition when the forearm is pronated. Thus, the biceps and the brachioradialis work together effectively in producing forearm flexion with supination from the extended, pronated position. The brachalis plays no role in the radioulnar movements, because it is attached to the ulna. This particular arrangement of the elbow flexors must be taken into account when analyzing elbow-flexion exercises such as pull-ups (forearm in pronation), chin-ups (forearm in supination), and many of the gymnastics skills. (See Appendix VI, Tables VI.1 and VI.2 for muscles and movements of the radioulnar joint.)

4.5 The Wrist Joint

The wrist consists of the articulation of the distal ends of the radius and ulna to the carpal bones. The movements of the wrist joint are those of flexion and extension and radial and ulnar flexion[2] or deviation. Although the bony stability of the wrist is weak, it has fairly strong ligamentous stability, and the numerous muscle tendons that cross the wrist on all sides provide the necessary muscular stability. The wrist motions of flexion and ulnar flexion are combined with forearm pronation and medial rotation of the humerus, as in throwing and overhead striking movements.

[2] Sometimes called adduction and abduction when moved from anatomical position.

Understanding the Elbow, Radioulnar, and Wrist Joints

1. Have a partner sit with his or her elbow supported, in a slightly flexed position, and relaxed. Supinate the forearm against resistance without increasing or decreasing flexion at the elbow. Palpate the triceps. Explain why there is tension in the triceps. Which one of the three functions of muscular tension is this muscle displaying?

2. Flex the forearm vigorously and stop the movement suddenly in the middle of

its ROM. Palpate the triceps during this action. Why is muscular tension felt in the triceps? What function of muscular contraction is it displaying?

3. Find a person who has a large ROM for elbow flexion and extension and a person who has a more limited ROM. Compare the ranges of elbow flexion and extension. Determine the reasons why this difference exists.

4. Starting from anatomical position for each measurement, measure the ROM for pronation and then for supination of the radioulnar joint. Have a partner combine the pronation motion of the radioulnar joint with the medial rotation of the shoulder joint and then combine the supination with the lateral rotation, keeping the elbow extended. Discuss the combined movements (shoulder and radioulnar) on the ROM of the hand in each direction.

5. Draw a diagram of the directions of force for the three elbow flexors. Discuss the change in the magnitude of force available for the flexion of the elbow (rotary component) with the elbow extended, flexed 100 degrees, and flexed 150 degrees.

6. Identify and draw an example of a wheel–axle arrangement in the radioulnar joint. How could one adjust the radius of the wheel segment? Discuss in terms of baseball pitching and badminton overhead shots.

7. Identify the movements of the elbow, radioulnar, and wrist joints during the force phase in these activities: (a) spiking a volleyball, (b) passing a football, (c) bowling, (d) batting a baseball (dominant arm), (e) swinging a golf club (dominant arm), and (f) the sculling motion of the arm used in treading water.

References and Suggested Readings

An, K. N., Morrey, B. F., & Chao, E. Y. (1985). Kinematics of the elbow. In D. A. Winter, R. W. Norman, R. P. Wells, K. C. Hayes, & A. E. Patla, (Eds.), *Biomechanics IX-A* (pp. 154–159). Champaign, IL: Human Kinetics.

Andrews, J. R., Carson, W. G., & McLeod, W. D. (1985). Glenoid labrum tears related to the long head of the biceps. *The American Journal of Sports Medicine, 13*, 337–341.

Basmajian, J. V., & Latif, A. (1957). Integrated actions and functions of the chief flexors of the elbow: A detailed electromyographic analysis. *Journal of Bone and Joint Surgery, 39*(A), 1106–1118.

Bowers, K. D. (1983). Treatment of acromioclavicular sprains in athletes. *The Physician and Sportsmedicine, 11*(11), 79–89.

Brunet, M. E., Hoddad, R. J., & Porche, E. B. (1982). Rotator cuff impingement syndrome in sports. *The Physician and Sportsmedicine, 10*(12), 86–94.

Cain, P. R., Murschler, T. A., Fu, F. H., & Lee, S. K. (1987). Anterior stability of the glenohumeral joint. *The American Journal of Sports Medicine, 15*, 144–148.

Duda, M. (1985). Prevention and treatment of throwing-arm injuries. *The Physician and Sports Medicine, 13*, 181–186.

Dvir, Z., & Berme, N. (1978). The shoulder complex in elevation of the arm: A mechanist approach. *Journal of Biomechanics, 11*(1/2), 219–225.

Fimrite, R. (1978, August 14). Stress, strain, & pain. *Sports Illustrated, 14*, 30–43.

Fowler, P. (1979). Shoulder problems in overhead-overuse sports: Swimmer problems. *American Journal of Sports Medicine, 7* (2), 141–142.

Gabbard, C., Gibbons, E, & Elledge, J. (1983). Effects of grip and forearm position on flexed-arm hang performance. *Research Quarterly, 54*(2), 198–199.

Gainor, B. J., Piotrowski, G., Puhl, J., Allen, W. C., & Hogen, R. (1980). The throw: Biomechanics and acute injury. *American Journal of Sports Medicine, 8*(2), 114–118.

Garth, W. P., Allman, F. L., & Armstrong, W. S. (1987). Occult anterior subluxations of the shoulder in noncontact sports. *The American Journal of Sports Medicine, 15*, 579–585.

Gowan, I. D., Jobe, F. W., Tibone, J. E., Perry, J., & Moynes, D. R. (1987). A comparative electromyographic analysis of the shoulder during pitching. *The American Journal of Sports Medicine, 15*, 586–590.

Grabiner, M. D., & Jaque, V. (1987). Activation patterns of the triceps brachii muscle during submaximal elbow extension. *Medicine and Science in Sports and Exercise, 19*, 616–620.

Hang, Y. (1983). Little league elbow: A clinical and biomechanical study. In H. Matsui & K. Kobayashi (Eds.), *Biomechanics VIII-A: Proceedings of the Eighth International Congress of Biomechanics* (pp. 70–86), Champaign, IL: Human Kinetics.

Hawkins, J. R., & Kennedy, J. C. (1980). Impingement syndrome in athletes. *American Journal of Sports Medicine, 8*(3), 151–158.

Hinton, R. Y. (1988). Isokinetic evaluation of shoulder rotational strength in high school baseball pitchers. *The American Journal of Sports Medicine, 16*, 274–279.

Hogfors, C., Sigholm, G., & Herberts, P. (1987). Biomechanical model of the human shoulder—I. Elements. *Journal of Biomechanics, 20*, 157–166.

Jobe, F. W., Moynes, D. R., & Antonelli, D. J. (1986). Rotator cuff function during a golf swing. *The American Journal of Sports Medicine, 14*, 388–391.

Jobe, F. W., Moynes, D. R., Tibone, J. E., & Perry, J. (1984). An EMG analysis of the shoulder in pitching. *The American Journal of Sports Medicine, 12*, 218.

Kennedy, J. C. (1978). Orthopaedic manifestations. In B. Eriksson & B. Furberg, (Eds.), *Swimming Medicine IV: Proceedings of the Fourth International Congress on Swimming Medicine* (pp. 93–100). Baltimore: University Park Press.

McMaster, W. C. (1986a). Anterior glenoid labrum damage: A painful lesion in swimmers. *The American Journal of Sports Medicine, 14*, 383.

McMaster, W. C. (1986b). Painful shoulder in swimmers: A diagnostic challenge. *The Physician and Sports Medicine, 12*, 116–122.

Nash, H. L. (1988). Rotator cuff damage: Reexamining the causes and treatments. *The Physician and Sports Medicine, 16*, 129–132.

Nuber, G. W., Jobe, F. W., Perry, J., Moynes, D. R., & Antonelli, D. (1986). Fine wire electromyography analysis of muscles of the shoulder during swimming. *The American Journal of Sports Medicine, 14*, 7–11.

Pettrone, F. A. (Ed.) (1984). *American Academy of Orthopaedic Surgeons Symposium on Upper Extremity Injuries in Athletes.* St. Louis: Mosby.

Priest, J., & Nagel, D. (1976). Tennis shoulder. *American Journal of Sports Medicine, 4*(1), 28–42.

Richardson, A. B., Jobe, F. W., & Collins, H. R. (1980). The shoulder in competitive swimming. *American Journal of Sports Medicine, 8*(3), 159–163.

Ryu, R. K. N., McCormick, J., Jobe, F. W., Moynes, D. R., & Antonelli, D. J. (1988). An electromyographic analysis of shoulder function in tennis players. *The American Journal of Sports Medicine, 16*, 481–485.

Salter, E. G., Nasca, R. J., & Shelley, B. S. (1987). Anatomical observations on the acromioclavicular joint and supporting ligaments. *The American Journal of Sports Medicine, 15*, 199–206.

Snijders, C. J., Volkers, A. C. W., Mechelse, K., & Vleeming, A. (1987). Provocation of epicondylalgia lateralis (tennis elbow) by power grip or pinching. *Medicine and Science in Sports and Exercise, 19*, 518–523.

Suzuki, R. (1983). Loose shoulder and suspension mechanism of the gleno-humeral joint. In H. Matsui & K. Kobayashi (Eds.), *Biomechanics VIII-A: Proceedings of the Eighth International Congress of Biomechanics* (pp. 57–69). Champaign, IL: Human Kinetics.

van Gheluwe, B. V., Ruysscher, I. D., & Craenhals, J. (1987). Pronation and endorotation of the racket arm in a tennis serve. In B. Jonsson (Ed.), *Biomechanics X-B* (pp. 667–672). Champaign, IL: Human Kinetics.

van Zuylen, E. J., van Velzen, A., & Denier van der Gon, J. J. (1988). A biomechanical model for flexion torques of human arm muscles as a function of elbow angle. *Journal of Biomechanics, 21*, 183–190.

Whiting, W. C., Gregor, R. J., & Finerman, G. A. (1988). Kinematic analysis of human upper extremity movements in boxing. *The American Journal of Sports Medicine, 16*, 130–136.

Woods, G. W., Tullos, H., & King, J. (1973). The Throwing Arm: Elbow joint injuries. *Journal of Sports Medicine, 1*(4), 43–47.

Yoshizawa, M., Itani, T., & Jonsson, B. (1987). Muscular load in shoulder and forearm muscles in tennis players with different levels of skill. In B. Jonsson (Ed.), *Biomechanics X-B* (pp. 621–628). Champaign, IL: Human Kinetics.

Zarins, B., Andrews, J. R., & Carson, W. G. (Eds.). (1985). *Injuries to the throwing arm*. Philadelphia: W. B. Saunders.

CHAPTER 5

Biomechanical Relationships in the Lower Extremity

PREREQUISITES

Concept Modules A, B, D, E
Chapters 1, 2, 3

CHAPTER CONTENTS

The lower extremity may be divided into four structural units: the hip joint, the knee, the ankle, and the subtalar joint. The structure of the lower extremity is similar to that of the upper extremity; however, important differences occur because of differences in function. Whereas the upper extremity is designed for reaching, throwing, and grasping activities, the lower extremity is designed for locomotor functions and weight bearing. Thus, although the structures are similar, the lower extremity is generally heavier and more stable. The one exception to this stability is the knee joint.

5.1 The Hip Joint

Structure

The hip joint is the articulation of the pelvis and the femur. It should not be confused structurally or functionally with the articulation of the pelvis and the vertebral column. The following bony landmarks of the pelvis and femur should be reviewed: the acetabulum, the ischial tuberosity, the pubis, and the anterior superior iliac spine of the pelvis; and the head, neck, greater trochanter, and medial and lateral condyles of the femur.

The bony arrangement of the hip joint consists of a rather deep hemispherical socket, the **acetabulum**, which is articulated with the hemispherical head of the femur. In itself, the bony arrangement of the hip joint is fairly strong. It is situated on the pelvis so that it is facing slightly anterior; thus, the bony arrangement is stronger in the prevention of posterior displacement of the femur than it is in prevention of anterior displacement. To assist in the task, a labrum or lip of cartilage, similar to the one found in the shoulder joint, circles around the outside of the acetabulum and the head of the femur. In addition, some strong ligaments surround the hip joint. They are shown in Figure 5.1.

Protection from anterior displacement of the femur is provided by the **iliofemoral (Y) ligament** and the **pubofemoral ligament**. The iliofemoral ligament originates on the anterior and lateral side of the ilium just above the acetabulum. As it proceeds anteriorly across the head of the femur, it splits, with one section attaching to the greater trochanter of the femur and the other running medially to the neck of the femur and the lesser trochanter. Hence, the name "Y" ligament. The pubofemoral ligament originates on the pubic bone and attaches to the lesser trochanter. The iliofemoral ligament is the strongest ligament in the body, and, along with the pubofemoral ligament, gives the hip the stability it needs on the anterior side as a result of the angulation of the acetabulum.

Protection from posterior displacement of the femur is provided by the **ischiofemoral ligament**, which originates on the ischium and attaches to the greater trochanter on the top of the femur.

To further stabilize the joint, another ligament, the **ligamentum teres**, is located within the joint capsule itself. It attaches from the inside of the acetabulum to the head of the femur. This combination of ligamentous and bony arrangement provides the hip joint with exceptional stability.

Muscles of the Hip Joint as Stabilizers

The bony and ligamentous arrangements of the hip joint are strong in themselves, but the muscular arrangement provides additional

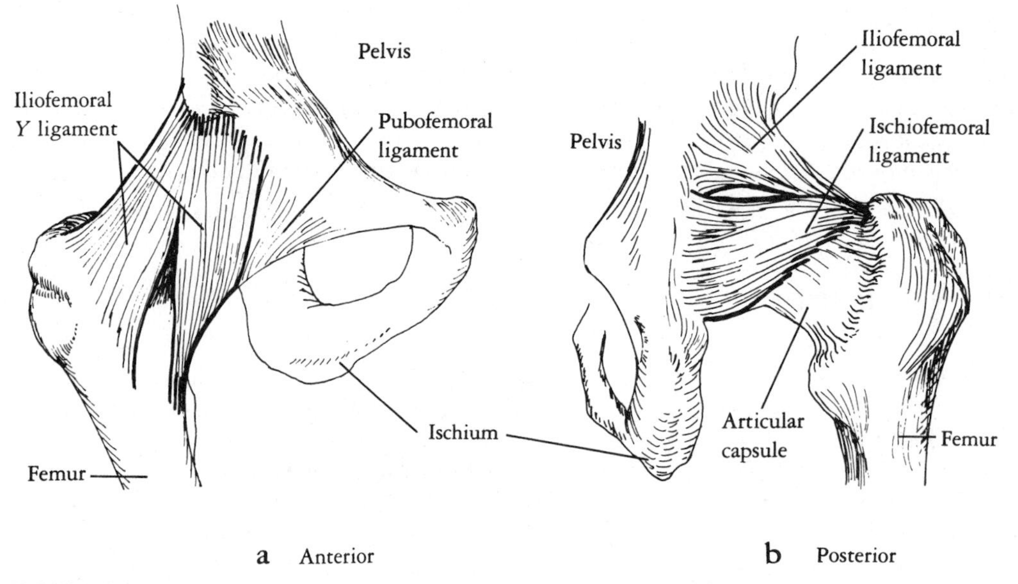

FIGURE 5.1
The bony and ligamentous arrangements of the hip joint.

stability. All the muscles that cross the hip have a stabilizing component at some point in the joint range of motion (ROM); however, the muscles emphasized here have large stabilizing components that are maintained throughout the ROMs.

The major muscles of the hip that provide stability are arranged in a similar manner to

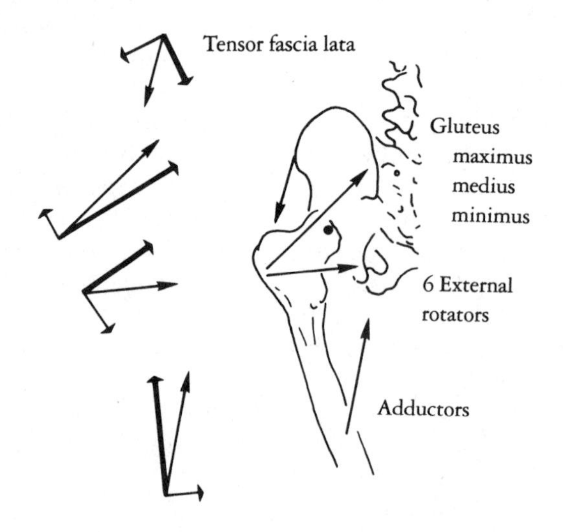

FIGURE 5.2
Posterior, medial, and lateral muscles of the hip that function as stabilizers.

those of the shoulder. They are the six external rotators and the three gluteals on the posterior side, the adductor muscles on the medial side, and the tensor fascia lata on the lateral side. These muscles are shown with their stabilizing and rotary components in Figure 5.2.

Of particular interest in considering stability of the pelvis is the function of the muscles during one-limb weight bearing. Classically, problems of one-limb weight bearing have been the concern of physical therapists and those working with amputees. In recent years, however, the lateral positioning of the pelvis during ambulation has become increasingly important for athletes, runners, and joggers as well. Figure 5.3 illustrates the situation in one-limb weight bearing.

During one-limb support, the pelvis acts as

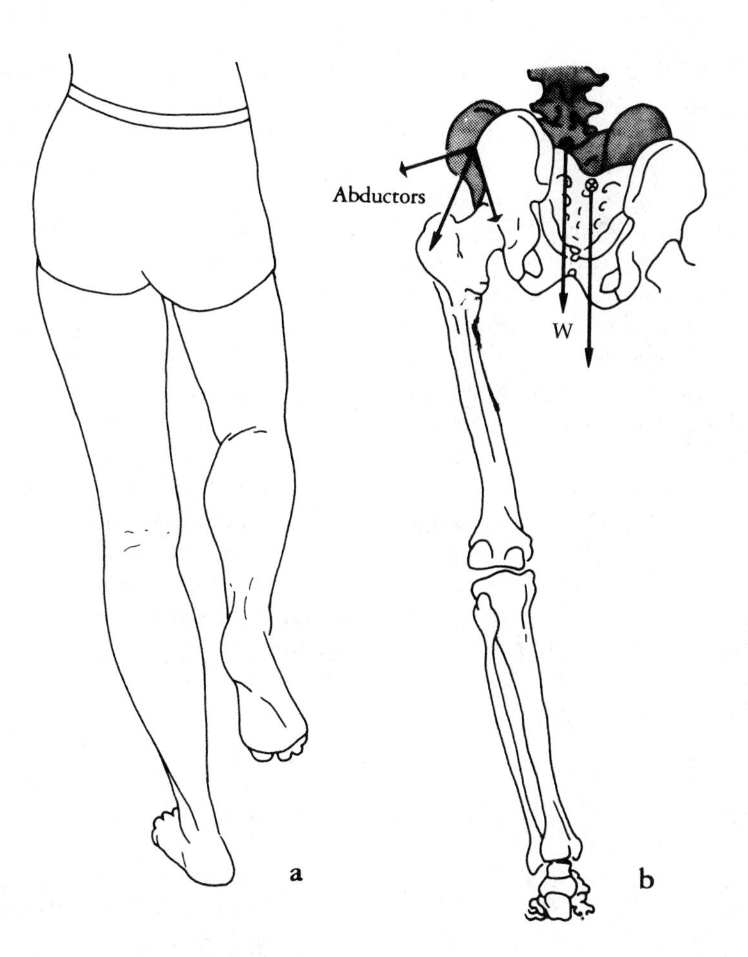

FIGURE 5.3

Muscle forces acting in one-limb weight bearing. (a) Lateral pelvic tilt. (b) Resultant force lines of the hip abductors, which act to prevent lateral pelvic tilt during one-limb weight bearing.

a first-class lever around an anteroposterior (AP) axis that passes through the hip joint. The movement of the pelvis is abduction and adduction of the pelvis relative to the thigh. The resistive force is considered to be the weight of the unsupported body acting through the center of gravity (CG) of the body, and the motive force is supplied by the muscles that work to hold the pelvis level (i.e., the hip abductors on the supported side). The necessity for stabilization of the pelvis by the hip abductors causes increased compression at the hip joint. Therefore, the ligaments surrounding the hip joint, and the hip joint itself, must assume an increased burden.

The weight of the body, minus the weight of the supporting leg, is represented by a force vector acting through the CG of the body. Because the line of gravity is located some distance from the axis at the hip, the resistive torque is quite great. Therefore, the body is drawn over the supporting limb by the action of the hip adductor muscles, and since the foot is fixed, the body shifts to the supporting side.

Prevention of the unsupported side of the pelvis from dropping is achieved by the hip abductors on the supported side. They exert force at their proximal attachments on the superior ilium of the pelvis, thus causing the pelvis to act as a first class lever around the hip joint and raising the unsupported side. The resultant force lines are shown for each of these muscle groups in Figure 5.3b.

The stabilizing and rotary components are indicated with arrows in Figure 5.2. The stabilizing components are relatively large, and thus the compression force on the hip joint is magnified beyond the actual weight of the unsupported body. It has been estimated to be 4.3 times the body weight (Crowninshield et al., 1978). In normal individuals and those who do not subject their bodies to continual trauma, this compression is handled easily.

In athletes engaged in activities that involve long-term running such as marathoners or cross-country runners, the compression at the hip joint may cause arthritis, fatigue fractures of the femoral neck, or develop into an overuse syndrome of the soft tissues. This is particularly common among runners who have one limb slightly shorter than the other.

One overuse syndrome that is prevalent is **lateral knee syndrome**, or tendinitis of the iliotibial band (sometimes called iliotibial band syndrome). This may be due to the excess tension required of the tensor fascia lata to help abduct the hip during one-limb weight bearing, particularly with the added impact from the foot plant during a downhill run.

Muscles of the Hip Joint as Movers

The movements occurring at the hip joint are numerous in spite of the bony and ligamentous stability. These movements are: flexion, extension, abduction, adduction, medial and lateral rotation, and transverse abduction (extension) and adduction (flexion).

Many muscles of the hip are longitudinal or unipennate in structure, which allows good ROM. Most are two-joint muscles, which ensures extra efficiency in movement of the lower extremity for locomotion. Further, several large bipennate muscles with large fleshy attachments provide the muscular strength needed to hold and move the body weight. Figure 5.4 illustrates the anterior and posterior lines of action of the muscles crossing the hip joint. The action of the muscles should be obvious from the resultant lines of action, and the components can be identified easily. As in the shoulder joint, the medial and lateral rotations of the thigh function as a wheel–axle mechanism, with the axle being the femur and the wheel being the leg (if the knee is flexed) or the foot (if the knee is straight). The medial and lateral rotators crossing the hip joint function as the motive forces. The radius of the wheel

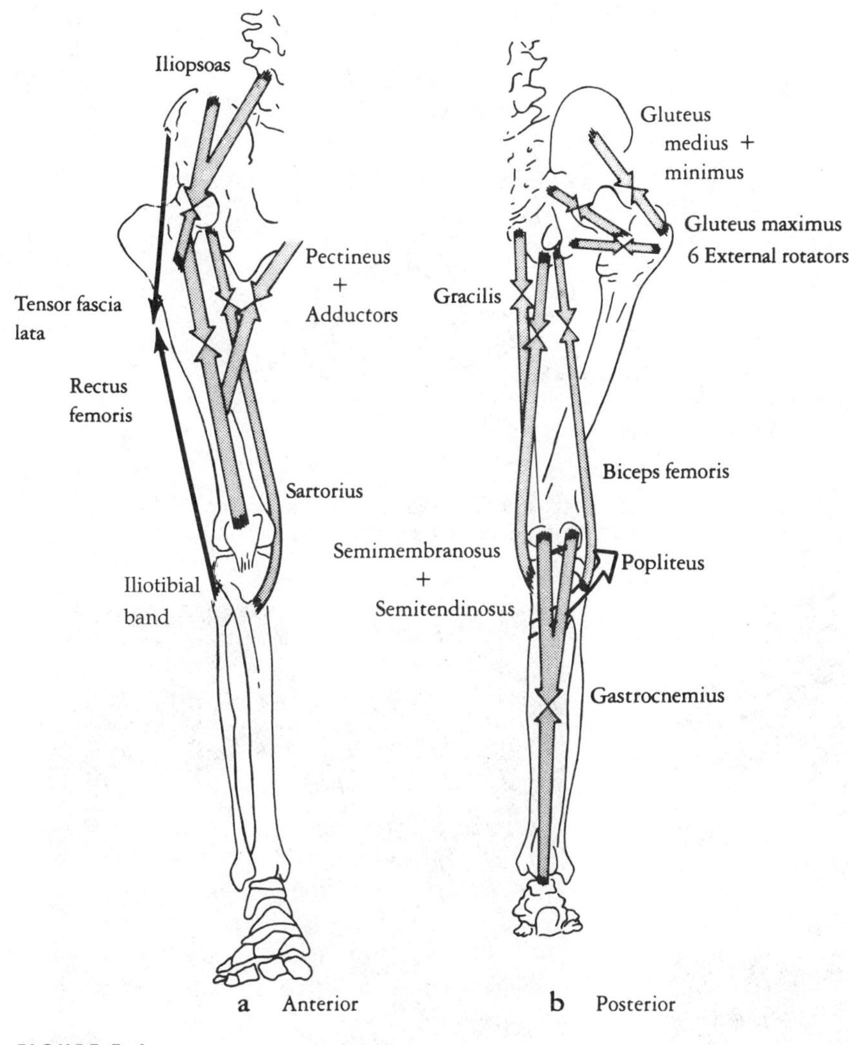

FIGURE 5.4

The lines of action of the (a) anterior and (b) posterior muscles that cross the hip joint.

portion varies in size depending on the degree of knee flexion or ankle flexion. Although not as prevalently used as the medial and lateral rotations of the humerus, the rotations of the hip joint can be noticed in many soccer kicks, tae kwon do kicks, or the breaststroke kick. A soccer pass is shown in Figure 5.5. The game of hacky-sack relies almost entirely on wheel–axlelike hip rotations to impart velocity to the footbag as seen in Figure 5.6.

Because the hip is a major weight-bearing joint, its function must be considered when the lower extremity is weight bearing and the joint is fixed, as well as when the thigh is free to move. The action of the hamstring group presents an interesting example of a first-class lever system when it acts as a resistance to the body's bending forward (hip flexion) or as a force to raise the body from a forward bending position. With various forms of

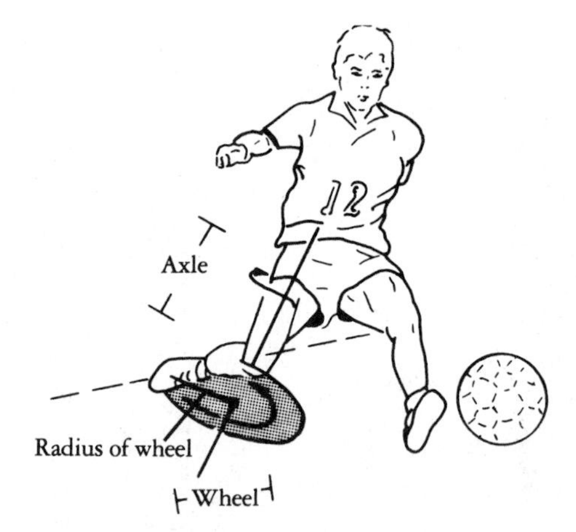

FIGURE 5.5

The wheel–axle arrangement of the lower extremity in performing a soccer pass.

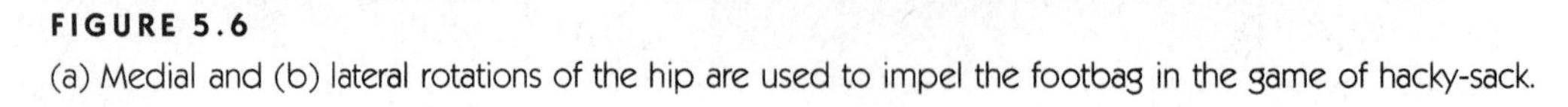

FIGURE 5.6

(a) Medial and (b) lateral rotations of the hip are used to impel the footbag in the game of hacky-sack.

bending and stooping over, the resistive force arm for the body's weight increases, and thus the resistive torque increases. With the increase in the resistive torque, a necessary increase in the motive torque arises not only for the hamstring muscles acting about the hip but also for the deep back muscles supporting the trunk.

During throwing and striking activities, the movement of the pelvis at the hip joint plays an important role as the pelvis medially rotates around the supporting extremity. For example, in throwing a javelin the initial plant of the forward foot positions the forward thigh in lateral rotation. Thus, as the thrower comes forward over the foot, the pelvis is medially rotated at the hip. The muscles that produce the rotation are the same ones that produce medial rotation of the thigh when the foot is not fixed. Different motor units and parts of the muscle may be used, however. Appendix VI, Tables VI.3 and VI.4 list the muscles and articulations of the lower extremity.

5.2 The Knee Joint

Structure

The knee joint is made up of the articulation of the distal end of the femur and the proximal end of the tibia. The following bony landmarks of the articulating bones should be reviewed: the lateral and medial condyles, the intracondular fossa, and the patellar surface of the femur; the medial and lateral articular surfaces, the intracondular eminence, and the tibia tubercle on the tibia; and the patella. Figure 5.7 illustrates the bony and ligamentous landmarks of the knee joint.

As can be seen, the bony articulation of the femur and the tibia consists of two shallow convex surfaces into which fit the semicircular-shaped femoral condyles. Because of these shapes, the bony stability of the knee is extremely weak. The ligamentous and cartilaginous structures are numerous. The two semicircular **menisci**, which surround the medial and lateral rims of the condular facets of the tibia, serve to increase the depth of the concavity and attempt to stabilize the movements of the femoral condyles on the tibia. The medial meniscus is the larger and more open of the two, and it allows for the slight rotary and locking mechanism of the knee joint.

The cruciate ligaments cross within the knee articulation. The **anterior cruciate** is attached to the anterior tibia and crosses to the posterior femur; the **posterior cruciate** is attached to the posterior tibia and crosses to the anterior femur. The role of the cruciates is predominantly to prevent forward displacement of the femur (posterior cruciate) on the tibia and backward displacement of the femur (anterior cruciate) on the tibia.

Stability on the medial and lateral sides is provided by the **medial** and **lateral collateral ligaments**. The **fibular collateral ligament** is located on the medial side and is attached to the femur and the fibula. The **tibiocollateral ligament** is located on the medial side and is attached to the femur and the tibia. Posteriorly, the knee joint is stabilized by the **popliteal ligament**, which is an oblique ligament running from the lateral posterior femur to the medial posterior tibia. Anteriorly, the knee is stabilized by the **patellar ligament**, which attaches the distal part of the patella to the tibia at the tibia tubercle.

The functional role of the knee joint in human motion is difficult to accommodate. The knee joint must be mobile enough to allow the movements and also stable enough to absorb the forces created by the weight of the body and the counterforce of the ground

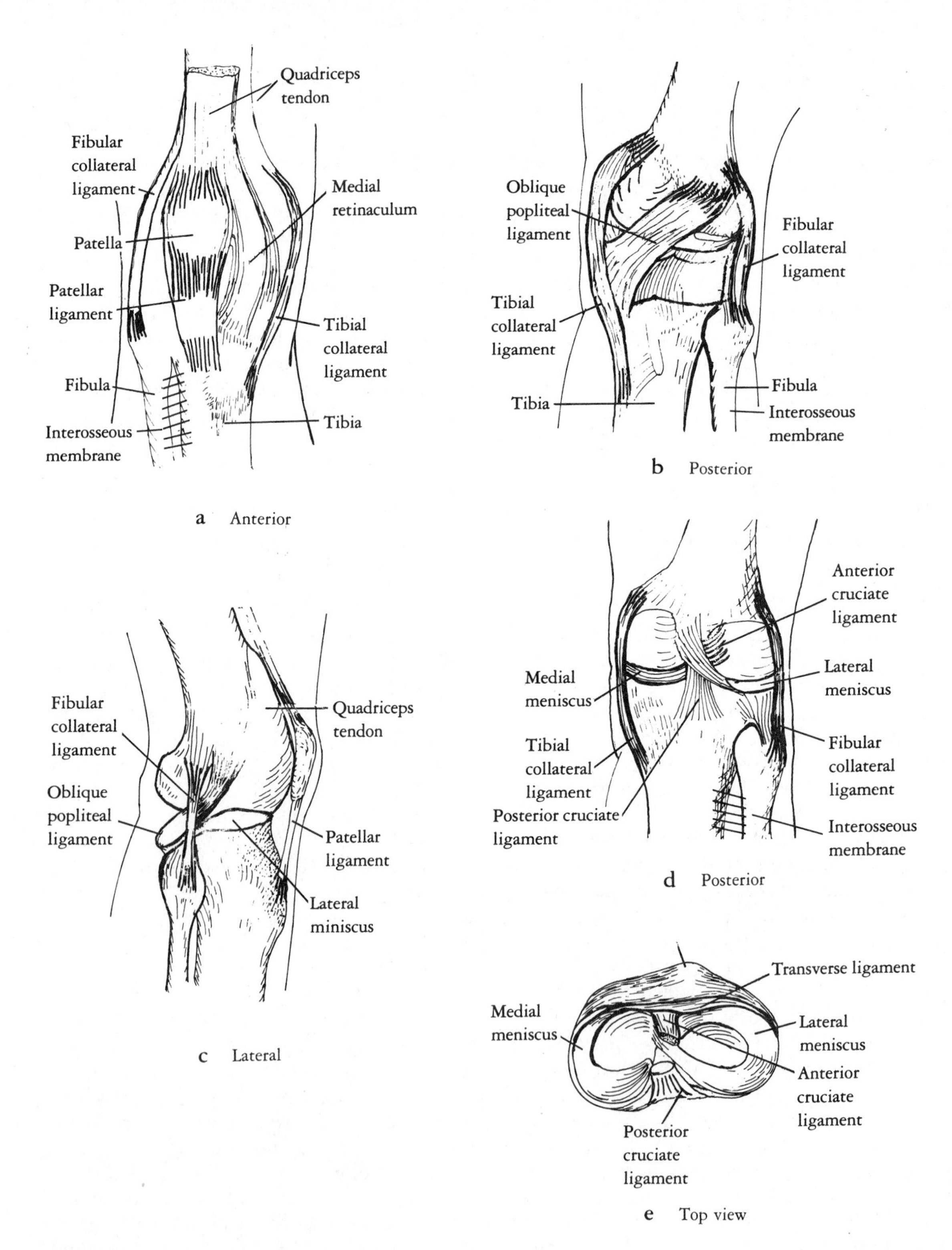

FIGURE 5.7

Bony landmarks and ligaments of the knee. (a) Anterior view. (b) Posterior view. (c) Lateral view. (d) Menisci and cruciates, from the posterior view. (e) Top view.

reaction acting upward during locomotion and weight-bearing activities. The knee is positioned between the forceful muscles acting over the hip joint and between the stable hip above it and the stable ankle joint below it. In addition, the foot usually acts as a propulsive agent against the counterforce of friction. The counterforce stabilizes the foot and thus provides for the transfer of forces from the foot up to the knee. Much of the rotation required for turning the body while the foot is fixed must come from the muscles that surround the knee. The knee must have the stability to withstand weight-bearing forces while remaining mobile enough to maintain the ROMs required for locomotor adaptability.

Because of the dual role that the knee must play, its ligamentous stability is not constant. When the knee is extended, its task is mainly to remain stable, and the ligaments are arranged so that the articulation is surrounded by fairly taut ligaments from all sides and from within. When the knee is flexed, however, it must be mobile enough to accommodate the need to change direction during locomotion; thus, some of the ligaments loosen slightly when the knee is in flexion to allow for greater ROM in rotation.

When the knee is flexed, the anterior cruciate ligament (ACL) and the collateral ligaments become less taut, allowing more mobility of the joint. Thus, the extended knee is fairly stable, but the flexed knee is less stable. A particularly violent rupture of the ACL occurs when skiers land from a jump or attempt to carve turns by sitting back on their skis. In this situation, the boot top pushes forward on the posterior side of the tibia (anterior drawer sign) and at the same time, the quadriceps act to prevent further flexion of the knee and to pull the femur forward. The rotary component of the quadriceps also acts to pull the tibia forward. Thus, these two strong forces acting on the tibia force it forward relative to the femur and frequently are injurious to the ACL (McConkey, 1986). When external trauma is applied, such as a blow to the thigh or leg from the side, the straight knee does not have an efficient way to absorb that force, and injury to the ligaments may result. The flexed knee is less stable and allows rotational movement. Because of its mobility, however, it may be pushed beyond its normal ROM; this, combined with the momentum of the moving segments, may result in injury.

Muscular Stability of the Knee

Because of the dual role for which the knee is designed, the muscular arrangement is of utmost importance in maintaining the stability necessary to prevent injury. The motions that are possible in the knee joint are flexion and extension, with medial and lateral rotation possible only when the knee is flexed and with maximum rotation possible at approximately 90 degrees of knee flexion. The flexion and extension motion is used in locomotor skills. The medial and lateral rotation seems to be a mechanism for allowing the nonweight-bearing foot to turn and for allowing the body to turn at the knee joint when the foot is fixed to the ground. The basketball player in Figure 5.8 illustrates the turning of the body with the foot fixed. The knee is stabilized on the anterior side by the quadriceps, on the medial side by the sartorius and the gracilis, on the lateral side by the tensor fascia lata, and on the posterior side by the hamstring group from above and the gastrocnemius from below. These muscles are shown in Figure 5.4 (p. 213).

Because of the permanently small angle of attachment of the quadriceps group to the tibia, a large stabilizing component is always acting at the knee joint. This is particularly important when the hamstrings are contracting strongly when the knee is flexed beyond 90 degrees, because the hamstrings then have

FIGURE 5.8

A basketball player changing direction by laterally rotating the leg at the flexed support knee.

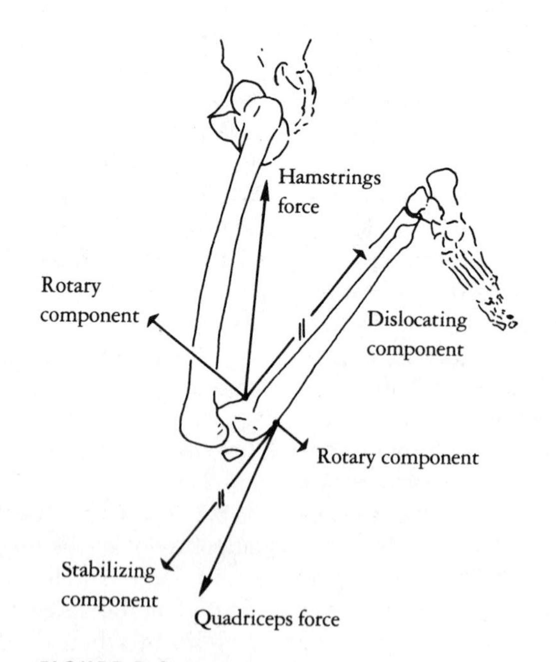

FIGURE 5.9

The stabilizing function of the quadriceps when the hamstrings are acting to flex the knee.

a backward dislocating component. The stabilizing function in this situation is shown in Figure 5.9.

In this example the stabilizing component of the quadriceps is set equal in magnitude to the dislocating component of the hamstrings. The hamstrings may also function as a stabilizer to counteract any forward displacement force on the tibia, in which case it assists the ACL in performing this task. Solomonow et al. (1987) demonstrated that in patients with deficient ACLs, the hamstrings assumed the role of knee joint stabilizer. In addition, Inoue et al. (1987) found that when the tibia is placed under a force causing abduction of the tibia, and the medial collateral ligament was deficient, the role was assumed by the ACL. The ACL stress may then be assisted by the hamstring muscles to prevent forward and lateral displacement of the tibia. Furthermore, Baratta et al. (1988) concluded that coactivation of these two muscle groups increases the area over which the force of compression acts and consequently reduces the intrajoint surface pressure and, in the long-term, the damage to the articular cartilage of the knee joint.

Although the hamstrings are shortened considerably when the knee is flexed beyond 90 degrees, a flexion of the hip serves to maintain their length somewhat so that tension may be maintained. Thus, the dislocating component of the hamstrings at the knee may be influential. The popliteal ligament also may provide some stability beyond this position.

The sartorius, gracilis, and gastrocnemius muscles can pull at an angle greater than 90 degrees to the leg; in this way, they create a slight dislocating component at the knee joint. During knee ROM from 180-degree to 90-degree flexion, most of the muscles crossing the knee provide a rotary and stabilizing effect. A greater-than-90-degree flexion of the knee, however, effects a dislocating component in some muscles, and these, along with the weak bony and ligamentous arrangements, put the knee in a weak position. Some lateral and medial stability is given by the hamstrings, and the tensor fascia lata provides additional stability on the lateral side.

Muscles of the Knee as Movers

Appendix VI, Tables VI.3 and VI.4 list the muscles of the knee. The muscles of the knee joint are predominantly two-joint muscles acting also at the hip: the hamstrings, the rectus femoris, the gracilis, the sartorius, and the tensor fascia lata. The two-joint muscle arrangement provides efficiency of movement for locomotion; however, the common insufficiency problem of two-joint muscle arrangements also is present—that is, **passive insufficiency**, attempting to stretch the muscle over both joints simultaneously, or **active insufficiency**, contracting a two-joint muscle to move both joints fully at the same time. The two-joint muscle cannot *stretch* enough to allow full ROM at both joints at the same time, nor can it *contract* enough to produce complete movement at both joints at the same time. A common example of this is seen when one tries to flex the hip and extend the knee fully at the same time or to extend the hip and flex the knee fully at the same time. The hamstrings cannot contract enough or stretch enough to allow either of these combinations to be performed in total.

Another characteristic of the two-joint muscles of the hip and knee is that they are a balanced combination of longitudinal muscles (gracilis, sartorius, tensor, and semitendinosus) and bipennate muscles (biceps femoris, semimembranosus, and rectus femoris). This combination provides for both strength and ROM needs of the main locomotor complex of the body.

In a study by Wells (1988), results indicated that using two-joint muscles in walking reduced the mechanical work cost over using

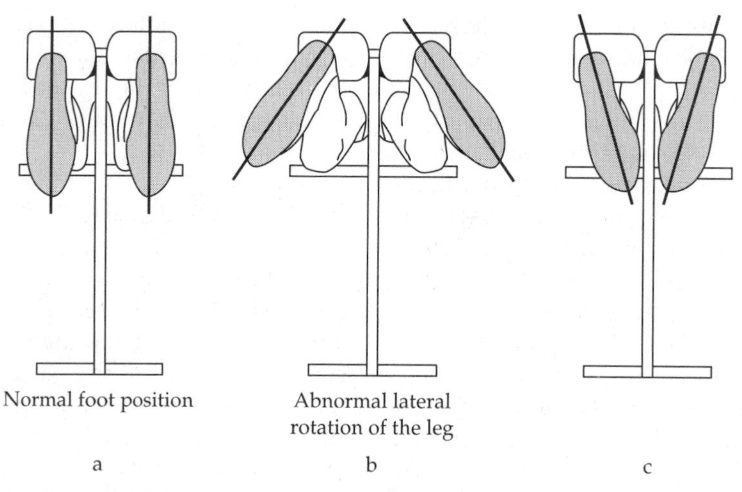

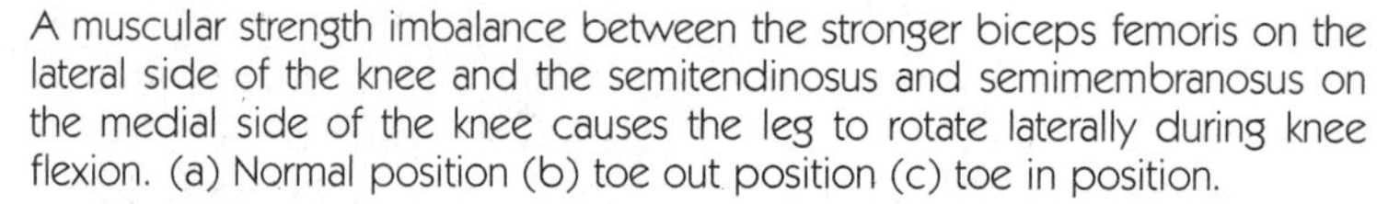

FIGURE 5.10

A muscular strength imbalance between the stronger biceps femoris on the lateral side of the knee and the semitendinosus and semimembranosus on the medial side of the knee causes the leg to rotate laterally during knee flexion. (a) Normal position (b) toe out position (c) toe in position.

one-joint muscles alone. Thus, although two-joint muscles are insufficient to cause or allow full ROM over both joints at the same time, they are cost-efficient while working within the middle ROMs.

Muscular Imbalances at the Knee

Although the hamstring muscles are usually considered as a group, important differences exist among them. The biceps femoris is attached on the lateral side of the fibula, while the semimembranosus and semitendinosus insert on the medial side of the tibia. Their locations allow them to produce lateral and medial rotation of the leg at the knee when the knee is flexed. If an imbalance in strength exists between the lateral and medial muscles (e.g., if the biceps is stronger than the medial muscles), a lateral rotation at the knee occurs when the knee is flexed. This is evident in the performance of the knee flexion against a resistance in a prone position on a weight-training machine, as shown in Figure 5.10b. As the group of hamstrings works to flex the knee, the biceps overrides the semis and causes the leg to be laterally rotated. To correct this condition, a performer should attempt to work the knee flexors while keeping the legs medially rotated. The performer will find that the resistive force probably must be decreased to flex in this medially rotated (toe-in) position (Figure 5.10c).

When the foot is free, the stability of the knee in abduction and adduction is reflected by the relative strengths of the medially and laterally attached hamstring muscles. If the medial muscles are weaker than the lateral muscle, the leg swings forward during gait in a laterally rotated position. The tension imbalance may be transmitted to the structures even when the knee is straight and may be evident in a laterally rotated leg during weight bearing.

Similar problems develop when an imbalance occurs in the strength of the quadriceps muscles, particularly when the vastus medialis is relatively weak. Because it must stabilize the position of the patella and keep it in the patellar groove during forceful contractions of the quadriceps in knee extension, it must be strong enough to fulfill this role. If it is not, the patella becomes displaced laterally with the pull of the vastus lateralis, and thus, the rubbing of the underside of the patella on the groove eventually wears on the inferior patellar surface, causing a parapatellar pain known as **chondromalacia**. The distal medial part of the vastus medialis serves to stabilize the patella during extension and during the last 20 degrees of knee extension and when the femur rotates medially during the locking mechanism of the knee at full extension. If chondromalacia is a problem, one should attempt to strengthen the vastus medialis against resistance during the last 20 degrees of extension. To further substantiate this concept, Percy and Strother (1985) report that the force exerted on the femur by the patella in walking is one third the body weight; force exerted at 90 degree flexion is more than three times the body weight; and the forces exerted in a deep squat are approximately seven times body weight. Activities that require forces on the patella in which the knee is at a greater than a 20- to 30-degree angle usually results in greater pain. The strength of all of the quadriceps muscles is crucial to maintaining the integrity of the knee in these situations.

Instability of the Knee

Because of its inherent mobility, the knee is prone to instability and injury when subjected to continual stress or when it goes beyond the limits of its ROM because of an impact. Some of the situations that commonly occur are described in the sections that follow.

Rotations with the Foot Fixed

During most sport situations, the foot is fixed on the ground during weight bearing. Exceptions to the norm are found in aquatic and aerial activities. Even during activities in which friction forces are assumed to be small, such as skiing and skating, if a performer "catches an edge," the continuing rotation of the thigh and trunk is transferred to the knee and must be counteracted by the ligamentous and muscular structures surrounding the knee. The knee is designed so that with flexion, rotation becomes possible. Its rotational capabilities make the knee a mobile adapter for turns and twists while the foot is fixed on the ground. Three common examples occur in basketball, soccer, and football, games in which turning to pass or to dodge is important. The knee has limits to its motion, and the ligamentous stability of the knee decreases with knee flexion. During normal turning and twisting motions, the knee fulfills its role quite well; however, if some external force is applied as the knee is reaching its limits of motion, causing it to move beyond its range, injury may be unpreventable, particularly when a shoe sole equipped with cleats enhances the friction force. With the hip and ankle stability as great as it is, the knee becomes the weakest link in the chain, so the structures of the knee must be strong enough to withstand the forces. For this reason, an athlete must maintain strong muscular stability and strength surrounding the knee to prevent any pushing of the knee beyond its safe limits.

Lateral Knee Pain

Two muscles that help stabilize the knee are implicated as the cause of lateral knee pain in runners. The tensor fascia lata muscle fibers are located lateral and slightly anterior to the ilium and attach distally by means of a long flat band of fascia on the lateral side of the tibia just below the knee joint. On its way to its distal attachment, the band passes over the lateral condyle of the femur. During knee flexion, while weight bearing, the tensor fascia lata muscle contracts to help stabilize the pelvis from lateral tilt. The added tension in the band enhances the friction force over the lateral epicondyle during repeated flexions and extensions, which results in what is called the **iliotibial band friction syndrome (ITBFS)**. The tenderness over the lateral knee is said to be diffuse, which differentiates this condition from a second source of lateral knee pain called **popliteal tendinitis (PT)**.

The popliteus muscle attaches proximally on the lateral condyle of the femur and passes medially behind the leg, where it attaches on the posteromedial side of the tibia. Its function is to medially rotate the tibia when the tibia is free and to help the posterior cruciate ligament stabilize the femur from anterior displacement on the tibia during weight bearing when the knee is flexed. It is particularly active during downhill walking or running. Because the popliteus is a medial rotator of the knee, it is subjected to tensile stress during lateral rotation of the knee. Runners whose weight-bearing posture includes a laterally rotated knee are particularly prone to tendinitis or inflammation of the popliteus tendon. This condition is enhanced especially with downhill running when the popliteus muscle actively stabilizes the femur from forward displacement and medial rotation on the tibia. In addition, those athletes who extend the knee while laterally rotating the knee against a resistance, such as in speed skating or skate-style cross-country skiing, are particularly susceptible. Tendinitis is distinguished from ITBFS by its soreness being focused at the popliteal attachment on the posterior lateral femoral condyle rather than the more diffuse pain in the iliotibial condition. People with tibial varus or a more inward-directed tibia and people who run on crowned roads, on hilly terrain, and long mileages are more prone to ITBFS (Noble et al., 1982; Messier and Piltala, 1988; Sutker et al., 1979).

Deep Squat

A second problem with regard to knee stability is in performing a deep knee bend or squat. Much controversy exists about whether the knee is designed to withstand the stresses put on it during the full squatting motion. Whether it can or cannot withstand those forces depends on the speed with which one descends, the size of the calves and thighs, and the strength of the controlling muscles. Figure 5.11 illustrates the position of the legs during the squat.

The main danger to the knee occurs when the center of rotation for knee flexion is altered because of the pressing together of the tissues of the calf and thigh. As one descends, the center of rotation is somewhere within the knee joint, but when the tissues touch, the center of rotation moves back to the contact area. This creates a dislocating effect at the knee, as shown in the illustration in Figure 5.11b. If the CG of the body system is kept forward of this center of rotation by altering the lifting position or if the muscles of the thigh are strong enough to prevent the body from resting on the calves or bouncing on the calves, then danger to the knees may be prevented. The danger is enhanced, however, when squats are done by individuals who cannot control their descent properly because of weak legs; when a large external weight is held on the shoulders, which brings the performer close to the maximum lifting force; when the squats are done as a timed event for maximum number of squats per time unit; or when the position brings the line of gravity of the body behind the center of rotation. In the first two cases, the possibility exists that the performer may lose control and descend too quickly beyond the safe point; in the third

FIGURE 5.11
(a) Safe position and (b) unsafe position in the deep squat.

case, especially near the end of the time frame when the muscles are fatigued, there is a tendency to bounce off the knees for quickening the change in direction from descent to ascent. With the gravity line located behind the center of rotation, a first-class lever situation with a wrenching, or separating, effect at the knee is produced. In any case, the deep squat is of little danger to the knee unless these restraints are not observed.

Breaststroker's Knee

Although in aquatic activities the foot is not fixed, the knee is still subjected to dislocation as a result of the muscular forces and the counterforces of the water against the leg as it contributes to the kicking motion (Figure 5.12). Particularly common is "breaststroker's knee," a syndrome characterized by tenderness on the medial side of the knee.

In the propulsive phase, the knee is extending and medially rotating with the foot in a pronated position (everted, abducted, and dorsiflexed). The medial rotation is performed by the two medial hamstring muscles. The hydrodynamic forces against the foot and the medial side of the leg, as it extends and medially rotates during the propulsive phase, are directed so that they tend to abduct the leg at the knee joint. The assumption is that the tender spot associated with breaststroker's knee is

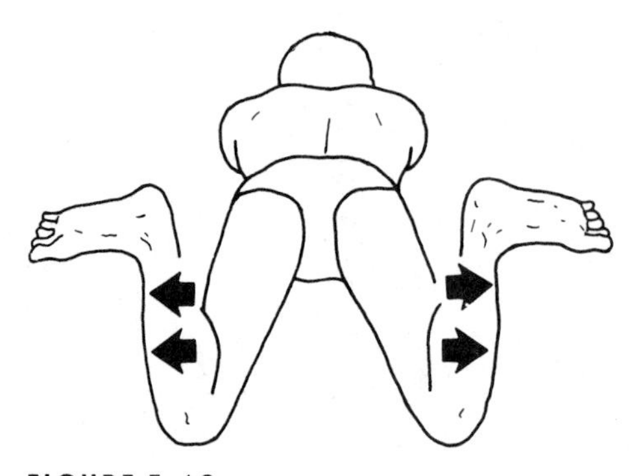

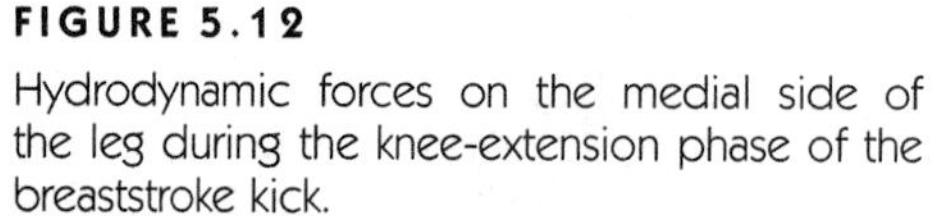

FIGURE 5.12
Hydrodynamic forces on the medial side of the leg during the knee-extension phase of the breaststroke kick.

located at the adductor tubercle and the associated medial collateral ligament (Kennedy and Hawkins, 1974).

Some evidence exists that breaststroker's knee syndrome is associated with the *technique* of performing the whip kick. Stulberg et al. (1980) found that painful knees occurred in breaststrokers that performed the whip kick with a greater hip joint abduction-adduction component to the kick. Vizsolyi et al. (1987) reported greater injury rates if hip abduction analysis were greater than 42 degrees or less than 37 degrees.

Because abduction or adduction of the knee is not thought of as a natural, useful movement, swimmers often forget that the position may be forced upon the leg and that some conditioning of the muscles that stabilize it is necessary. Strong contractions of the quadriceps and hamstring groups have been shown to significantly limit forced abduction and adduction of the leg (Goldfuss, Morehouse, and LeVeau, 1973).

Rovere and Nichols (1985) found that subjects with significantly reduced medial rotation of the hip most frequently had knee pain. Because the medial rotation of the hip is an important movement during the force phase of the kick, a restriction of the motion would place the tibia in a position to receive greater valgus (abduction) stress and thus, places a stress on the medial ligaments and tendons of the knee.

Knee Stability Related to Skiing

Medial and lateral rotations at the knee are rarely used as conditioning exercises, in spite of the fact that these motions are quite necessary to proper ski technique. Proper edging, balance, and control depend on the medial and lateral rotations not only of the hip but also of the knee, particularly when the knee is flexed, as it is in all skiing positions. The friction afforded by the edging of the skis stabilizes the foot somewhat; thus, the rotations take place generally by the superior segment rotating about the inferior segment.

The medial and lateral rotations of the knee are examples of a wheel–axle arrangement, with the foot or the ski being the wheel. If the ski tip is caught, a large torque is produced as a result of the length of the ski (radius of the wheel), and therefore stabilizing the knee against such wrenching is almost impossible. Such a situation highlights the importance of strong muscles surrounding the knee.

Understanding the Hip and Knee Joints

1. Measure the ROM of hip flexion with the knee extended and with the knee flexed. What is the difference? Why is there a difference?
2. Identify and draw a wheel–axle arrangement for the hip joint. What is the mechanical advantage of this arrangement? In what sport situations is this motion used?
3. Sit on a table with your knee flexed over the edge. Rotate the knee medially and then laterally. Be sure that the movement is from the knee and not from the hip or the foot. What is the ROM for each movement? Discuss this movement and its importance to changing directions in skiing and skating.
4. Flex your knee. Look down the longitudinal axis of your leg, and draw a diagram of the cross section of the leg at the approximate points of attachment of the hamstring group. Draw force arrows to indicate the points of application of the forces of each of these muscles on the bones. Discuss the wheel–axle mechanism of this muscle group in nonweight-bearing situations.
5. On a weight-resistance machine, perform leg curls (lie prone and flex the knee against resistance). Does the knee tend to rotate laterally? This reaction to strong knee flexion is quite common. What is the explanation?
6. Identify the movements of the hips and knees in hurdling during takeoff flight and landing phases. What biarticulate muscles are stretched during each phase? What muscles are concentrically contracted? Are there two-joint muscles that are stretched at one articulation and concentrically contracted at the other articulation? Discuss these movements in terms of injuries.
7. Identify the hip and knee movements during the execution (force) phase of the following activities: (a) a whip kick in the breaststroke, (b) mounting a horse, swinging the right leg over, (c) a football punt, (d) a headspring, (3) a soccer-style placekick. Are there two-joint muscles involved in the movements? Which joints do they cross? How could two-joint muscles be related to injuries?

5.3 The Tibiofibular Joint

Although the tibiofibular joint is similar to the radioulnar joint in the forearm, no particular movements are associated with it.

The tibiofibular joint is an important force transmitter and absorber of the force-and-counterforce exchange during weight-bearing activities. The following gross structures should be reviewed: the superior, middle, and inferior tibiofibular joints; the anterior and posterior tibiofibular ligaments on the superior and inferior joints; the interosseous membrane between the two bones and the transverse ligament; and the medial and lateral malleoli. The interosseous membrane, which is located between the tibia and fibula, allows blood vessels and nerves to pass through it, and generally absorbs and transmits force between the tibia and fibula. The fibers of the membrane run in just the opposite manner to the ones in the forearm; that is, they run from proximal on the tibia to distal on the fibula. One possible advantage for this arrangement may be that since the tibia is the main channel for weight-bearing forces, the counterforce of the ground is transmitted up through the tibia and tends to displace it superiorly. The membrane is structured so that this force is transmitted to the fibula. On the other hand, the fibula is the location of the origin of many of the weight-bearing, antigravity muscles that operate to plantar flex the ankle. Thus, the force of those muscles on the fibula tend to pull it downward, and the fibers of the membrane, which run in a position to counteract that downward displacement of the fibula, transmit the force to the tibia.

The malleoli, located on the distal ends of the tibia and fibula, function mainly as pulleys to alter the direction of the force of the muscles' pull on the foot.

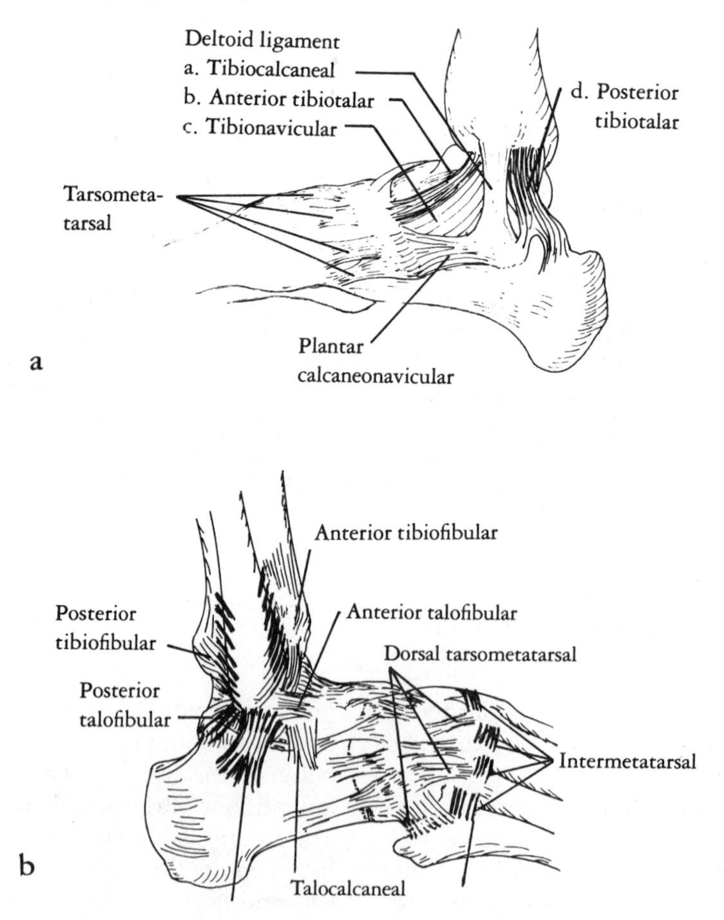

FIGURE 5.13

(a) Medial and (b) lateral ligamentous arrangement of the ankle.

5.4 The Ankle and Foot

Structure of the Ankle and Subtalar Joint Complex

Technically, the ankle joint is made up of the tibia and the talus, although the distal end of the fibula also articulates with the talus on the lateral side. Because of the bony arrangement of the ankle joint, with the distal end of the tibia being somewhat concave and the superior talus being convex, the bony stability is fairly strong. Because of the stress that the ankle must withstand, however, ligamentous stability is needed as well. The bony and ligamentous arrangements are shown in Figure 5.13. In addition to providing ankle stability, the articular capsule and the attached ligaments also extend down to the calcaneus.

The movements possible at the ankle are flexion and extension. The flexion movement is called dorsiflexion; the extension movement, plantar flexion. The axis of rotation for the ankle is not in a true frontal plane; it is oriented slightly backward from the frontal plane (13 degrees) on the lateral side and downward (7 degrees) from the medial to the lateral side (Figure 5.14a and c). The tilt creates a slight disorientation of the foot from true sagittal plane motion during plantar flexion and dorsiflexion.

The subtalar joint is located between the talus and the calcaneus. It is probably the most

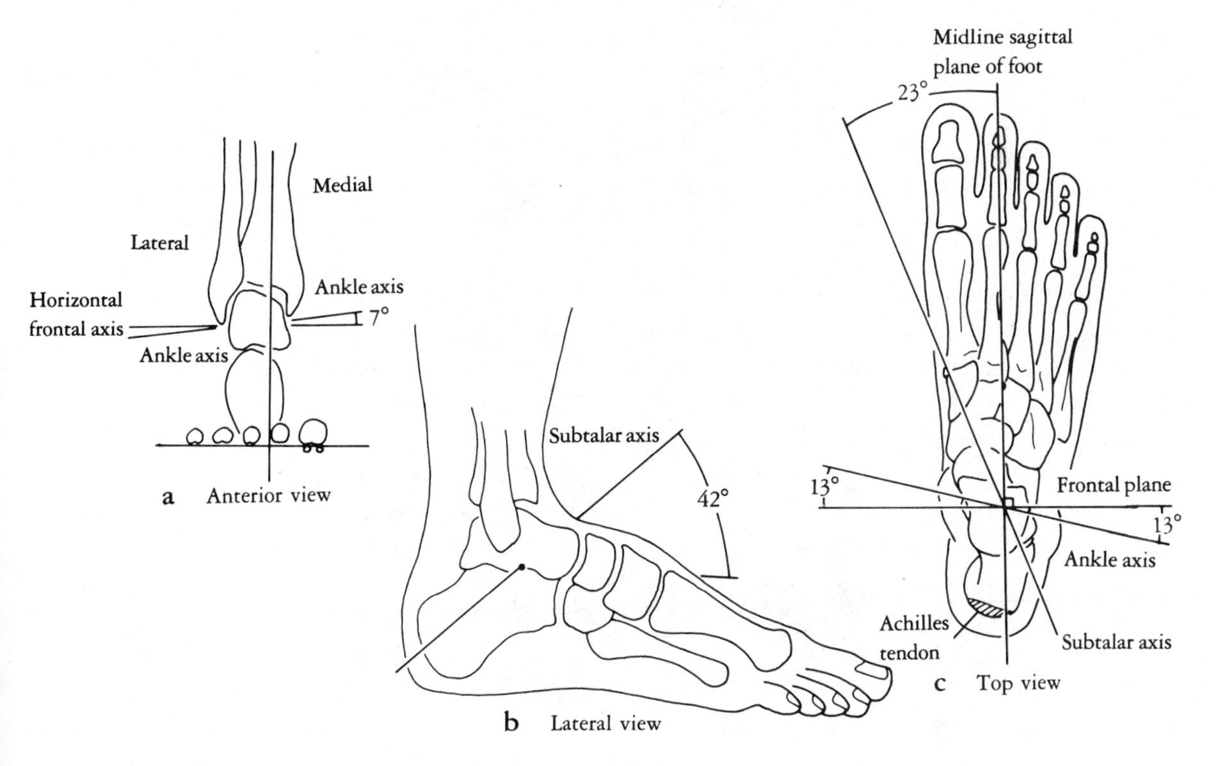

FIGURE 5.14
The obliquity of the ankle and the subtalar axes.

important articulation of the intertarsal group, because much of the inversion and eversion movements of the foot originate there. The subtalar joint is a uniaxial joint; however, its orientation is oblique, like that of the ankle joint. These two axes, ankle and subtalar, may be situated in many different combinations that vary from one individual's foot to another's. On the average, the subtalar axis is 42 degrees from the horizontal plane in the sagittal plane and 23 degrees from the midsagittal plane in the horizontal direction (Inman, 1976). These orientations are shown in Figure 5.14b and c.

Because of the obliquity of the ankle and subtalar axes, the foot displays movement in an oblique plane during the motions of inversion and eversion. The actual movements of the foot simultaneously around the subtalar and ankle axes are a combination of inversion at the subtalar joint with plantar flexion at the ankle joint and a combination of eversion at the subtalar joints with dorsiflexion at the ankle joint. The associated movements of the ankle and subtalar joints may be seen by observing the position of the foot and leg during rising on the toes, as shown in Figure 5.15.

The subtalar joint is the main interconnection between the mobility of the foot mechanism and the stability of the ankle and leg. Thus, it allows important adaptive positioning of the foot and leg during weight bearing, particularly in running and jogging on uneven or curved paths.

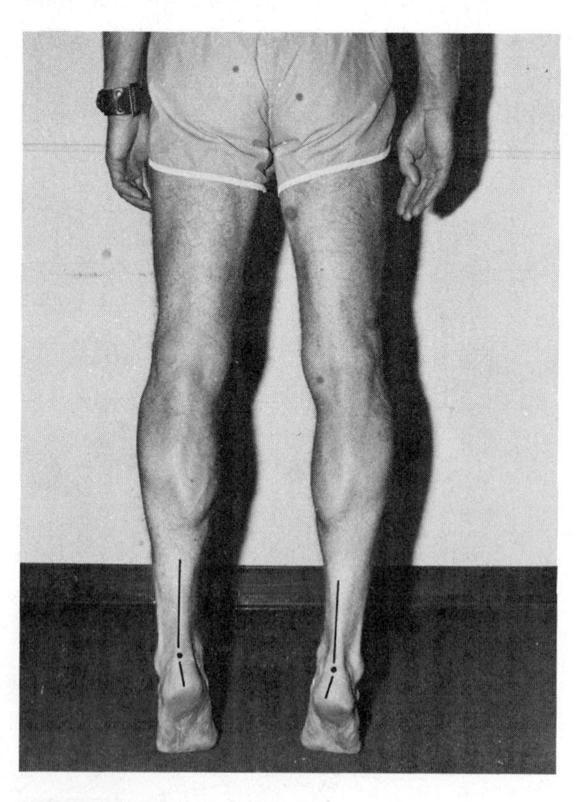

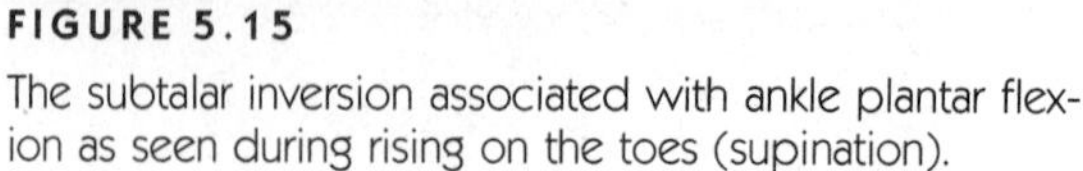

FIGURE 5.15

The subtalar inversion associated with ankle plantar flexion as seen during rising on the toes (supination).

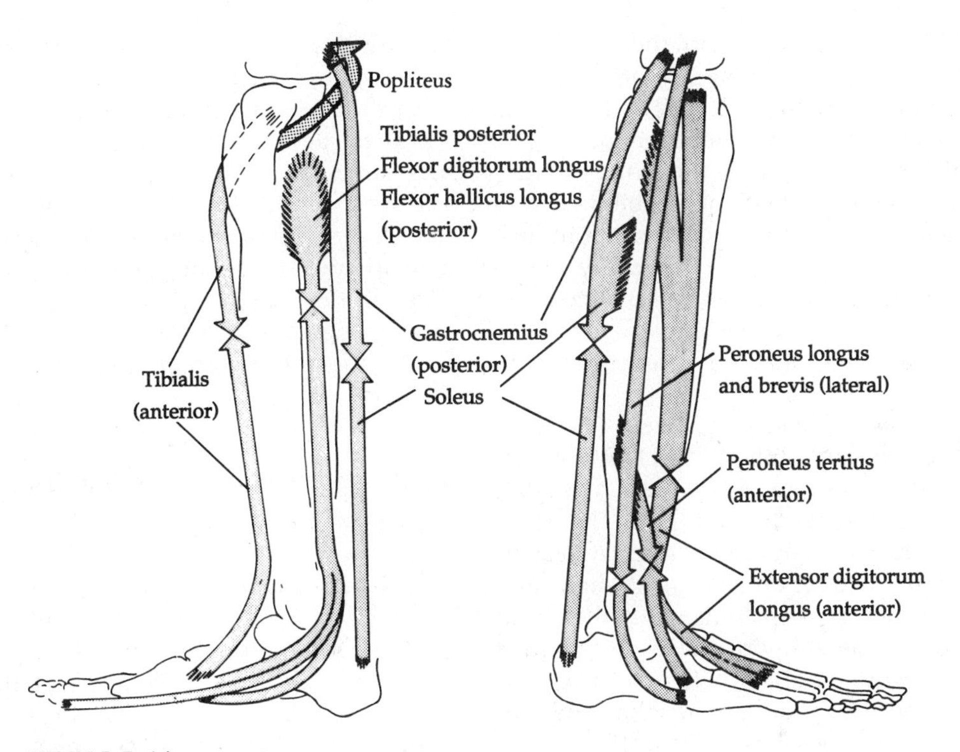

FIGURE 5.16
Medial and lateral muscular support for the ankle and foot complex. (Compartments indicated in parentheses.)

The Intertarsal and Tarsal–Metatarsal Joints

Since the tarsal–metatarsal articulations are of the nonaxial, gliding type, they have no particular motions. They glide on each other to produce a combination of movements that allows the foot to become a mobile adaptor for weight bearing, particularly on uneven surfaces and that continually adjusts the foot and leg alignment to maintain balance during weight bearing. The bones are shown in Figure 5.14.

Muscles of the Ankle and Foot as Stabilizers

The ankle's bony and ligamentous stability is good, and therefore the tendency for the ankle to be dislocated laterally or medially is remote. The muscles that cross the ankle and foot are needed for stabilizing the subtalar joint and the intertarsal joints and for supporting the arches. The musculotendinous insertions that support the medial and lateral aspects of the ankle and foot are shown in Figure 5.16.

Numerous intrinsic and extrinsic small muscles of the foot also support the intertarsal, metatarsal, and phalangeal articulations. Maintaining strength of the muscles on all sides of the ankle and foot is important in maintaining the integrity of the structure. Any imbalances in the strength or flexibility or both of the surrounding musculature result in misalignment, putting the line of gravity of the entire body eccentric to the articulations and thus creating a torque that must be counteracted by muscular effort or ligamentous tension.

For example, Gehlsen and Seger (1980) reported significant strength and ROM differences between subjects who had shin splints and those who did not. The shin-splint group had significantly greater plantar flexor strength than dorsiflexor strength and a greater angular displacement of the calcaneus during the support phase of gait than the group without shin splints. Michael and Holders (1985) concluded that when the foot is in a pronated position (calcaneal eversion), the medial aponeurosis of the soleus muscle is stretched, creating inflammation of the periosteum of the medial tibia and new bone formation in that area of the bone. Posteromedial "shin splints" may result.

Overdevelopment of the plantar flexors tends to cause a muscular imbalance between the strength of the plantar flexor-supinator muscles and the dorsiflexor-pronator muscles, which may result in lateral ankle sprains, particularly when landing after being airborne. Even during the swing phase of gait, the foot may tend to invert excessively and not return to its straightforward position before weight bearing.

Muscles of the Ankle and Foot as Movers

Tarsal and metatarsal movements occurring in the foot primarily position the foot according to the force of propulsion and the ground-reaction force. The movements of the ankle and tarsal joints that are relevant to the foot function are dorsiflexion and plantar flexion of the ankle and inversion and eversion of the subtalar joints. These movements, for the most part, are performed by multijoint mus-

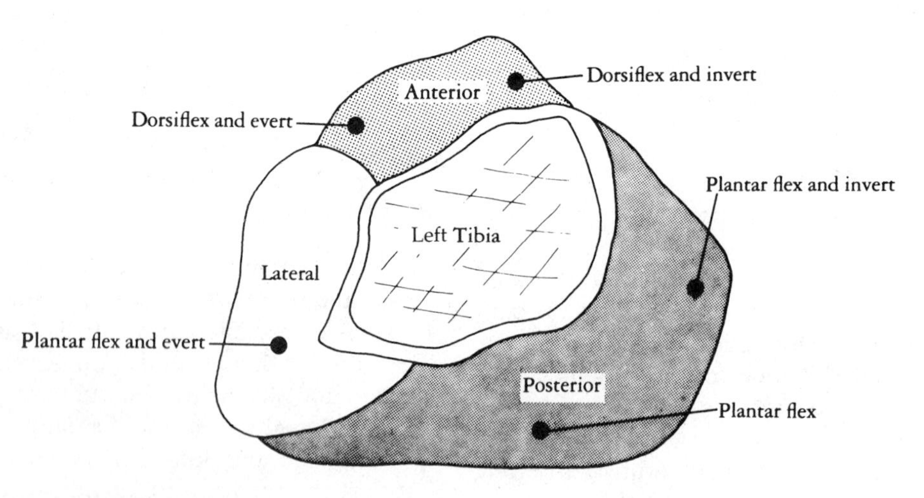

FIGURE 5.17

Cross section of the left tibia with the anterior, lateral, and posterior muscle compartments shown.

cles that cross the ankle and insert on the foot. Many of these small muscles relate to specific movements of the metatarsals and the phalanges, which are not covered in this text.

The malleoli function as pulleys to alter the direction of force of the muscles that pull on the foot. Those muscles that pass posterior to the malleoli will plantar flex, and those that pass anterior to the malleoli will dorsiflex. Similarly, those muscles that go around the medial malleolus will invert the foot, while those that go around the lateral malleolus will evert the foot.

Three compartments can be seen in a cross section of the leg: anterior, posterior, and lateral (Figure 5.17). In the **anterior compartment** are two groups of muscles; one group dorsiflexes and inverts the foot, while the other group dorsiflexes and everts the foot. In the **posterior compartment**, all of the muscles plantar flex; additionally, one group, which passes around the medial malleolus, inverts the foot, and the other (gastrocnemius and soleus) merely plantar flexes with no inversion or eversion function. The **lateral compartment** houses the muscles that plantar flex and evert the foot, because they pass around the lateral malleolus. The gastrocnemius and soleus, the muscles that make up the Achilles tendon, function to plantar flex the ankle, but because the gastrocnemius also crosses the knee, it secondarily serves to flex the knee joint. To stretch the Achilles tendon properly, both the gastrocnemius and the soleus must be stretched. Most of the stretching exercises advocate dorsiflexing the ankle while the knee is straight, which seems to be a logical way to stretch the gastrocnemius. This exercise leaves the soleus in a position of little stretch, however, since the tightness of the gastrocnemius occurs before any great dorsiflexion occurs. Proper stretching of the Achilles tendon, therefore, should include dorsiflexion with the knee straight and with the knee flexed.

Ligamentous and muscular tissues are affected by foot posture, and certain foot postures during weight bearing put excessive stress on muscle groups that are antagonistic to those positions. For instance, a foot that is weight bearing in a position of eversion and dorsiflexion stretches the invertor and plantar flexor groups (posterior compartment muscles). Similarly, an inverted foot stresses those muscles that evert the foot (anterior compartment). A more detailed discussion of leg and foot alignment is presented in a following section. Appendix VI, Tables VI.3, and VI.4 list the muscles and articulations of the ankle and foot.

Understanding the Ankle and Foot

1. Using several subjects, measure the ROMs of dorsiflexion and plantar flexion of the ankle. Do differences occur among the measurements of subjects? Why?

2. Sit on a table with the knee flexed over the edge. Gently plantar flex the ankle. What other movements occur in the foot during this plantar flexion? In what articulations do these movements occur? Why are they associated with plantar flexion?

3. Sit on the edge of a table and gently dorsiflex the ankle. Do other movements of the foot tend to occur? In what articulation do these occur? Why are they associated with dorsiflexion of the ankle?

4. In general, five groups of muscles are attached to the foot and leg. Group 1 attaches to the lateral foot and lateral leg; Group 2, to the lateral foot and anterior leg; Group 3, to the medial foot and pos-

terior leg; Group 4, to the medial foot and anterior leg; and Group 5, to the posterior foot and posterior leg. List and discuss the ankle and foot movements caused by each group's resultant muscle forces.

5.5 Biomechanics of the Lower Extremity During Locomotion

The walking pattern in the lower extremity primarily consists of a 35% swing phase and a 65% stance phase. During the swing phase, the leg approaches heel contact with a slight lateral rotation at the hip joint. The foot is in slight inversion. Through midstance and push-off, the weight-bearing point shifts from a position that is slightly lateral to the principal longitudinal axis of the foot to a position slightly medial to the longitudinal axis. This is shown in Figure 5.18. The darkest dots indicate the greatest weight-bearing pressure. The final propulsion is given by the distal heads of the first and second metatarsals. This weight transfer is accomplished by an eversion of the subtalar joint.

In running and jogging, the walking pattern is modified slightly by adding a phase of nonsupport between the stance and swing phases and by exaggerating the ROMs of the movements in the articulations to accommodate the increased force absorption necessary during weight bearing and to increase the force of propulsion. During running, the vertical force on the supporting foot is estimated to be approximately three times the body weight. In running, as in walking, the movements of the leg and foot generally follow a prescribed sequence. The rearfoot is positioned in a slight inversion before heel contact; this is followed by a slight plantar-flexion motion as the forefoot moves to the ground. The axes of rotation of the ankle and subtalar joints are positioned so that the inversion motion is associ-

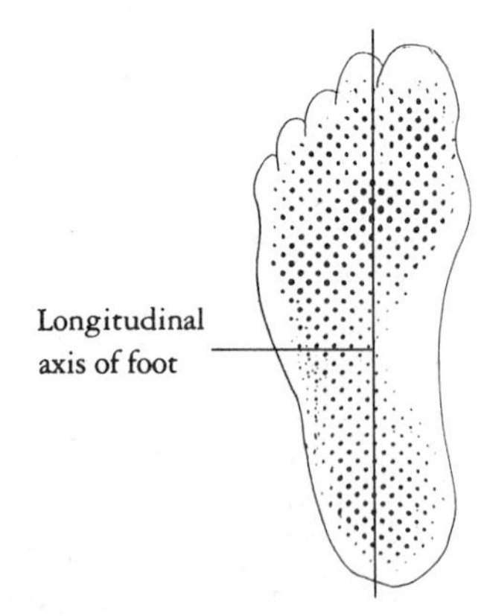

FIGURE 5.18

Dynamic weight bearing during the stance phase of a walking gait (from Ferrandino, 1978).

ated with plantar flexion of the ankle. These motions occurring with slight adduction of the forefoot are called **supination**.

As the body weight comes forward over the foot, the subtalar joint begins to evert, the ankle dorsiflexes, and the knee and hip joints also flex. These movements help absorb the shock of the body weight coming down on the lower extremity. Because of the inclination of the ankle and subtalar axes, the movements of eversion of the subtalar joint and dorsiflexion of the ankle are associated with an abduction movement of the foot when the foot is free. This combination is called **pronation**. When the foot is fixed, however, the abduction of the foot is prevented to some extent by the friction forces acting at the foot–ground interface. Consequently, the abduction motion manifests itself by a medial rotation of the tibia. The tibial rotation is possible because of the accompanying knee flexion. These motions continue as the body weight passes over the center of the foot.

As soon as the line of gravity passes forward of midstance, the ground-reaction force on the foot can provide an accelerating rather than a decelerating force to the body. The hip and knee joints begin to extend sequentially. The ankle continues to dorsiflex until the heel is lifted off the ground. With a taut Achilles tendon, the stretch reflex may be elicited, which will initiate the plantar flexion of the ankle joint.

In the normally extending knee, the medially rotated tibia must laterally rotate back to a neutral position. If excessive pronation is produced, the large medial rotation of the tibia, produced by extreme subtalar eversion, results in an abnormal torsion at the knee. The knee may not be able to accommodate this torsion, and various overuse injuries to the structures surrounding the knee may result. The plantar flexion of the ankle produces the associated inversion of the subtalar joint, and, consequently, the adducted foot position. A slightly supinated position is seen at push-off, with the weight-bearing point somewhere between the first and second metatarsal heads.

5.6 Lower Extremity Postures

The tightness or laxity of the ligaments in the lower extremity and the relative strengths and weaknesses of the antagonistic muscle groups surrounding the articulations produce lower extremity postures that are unique to each individual. For example, the relative position of the axes of the ankle and the subtalar joints varies so markedly between individuals that even an average measure cannot be used confidently in analyzing leg and foot postures and planes of foot motion. For practical purposes, however, a few simplified observations are useful in determining lower extremity postural conditions.

Ideal Alignments of Lower Extremity Segments

Normal alignment does not necessarily mean **ideal alignment**, because what is *normal* is a measure of what occurs on the average in the population, not necessarily an *ideal* measure. In fact, most people do not have ideal alignments for one reason or another.

The ideal alignment of the lower extremity segments may be likened to that of a column that supports a roof. Such a column should be as straight and as vertically aligned as possible

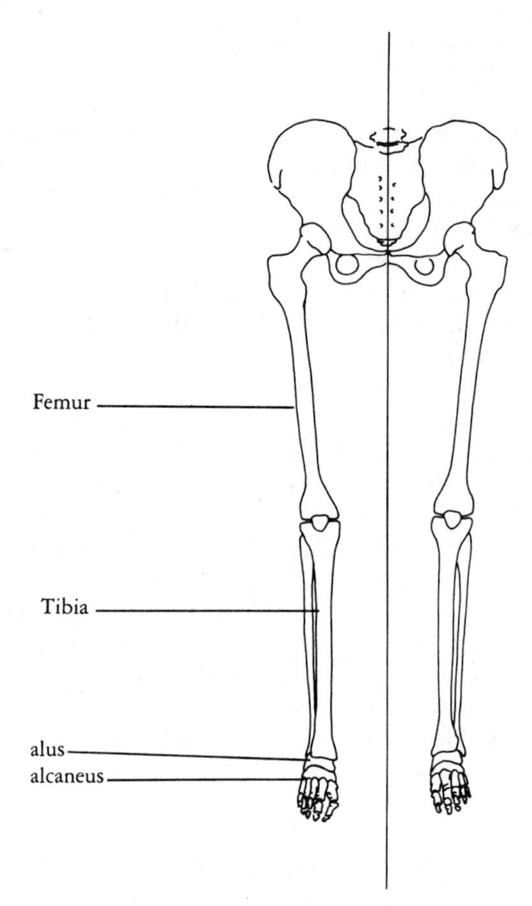

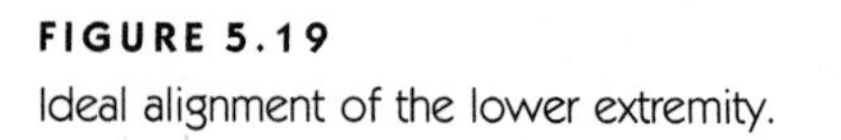

FIGURE 5.19
Ideal alignment of the lower extremity.

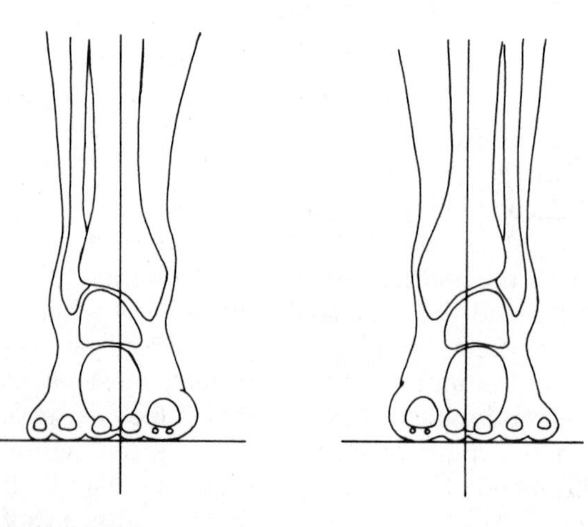

FIGURE 5.20
Ideal alignment of the forefoot.

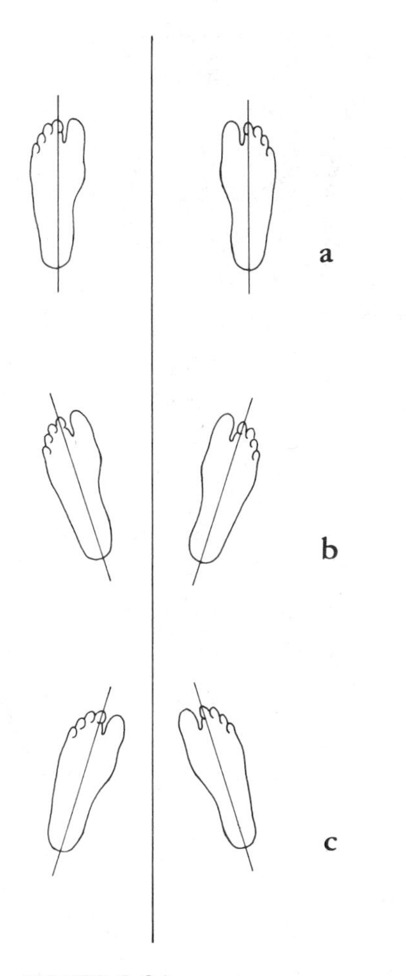

FIGURE 5.21
(a) Ideal, (b) toe-out, and (c) toe-in positions of the foot, relative to the body's midsagittal plane.

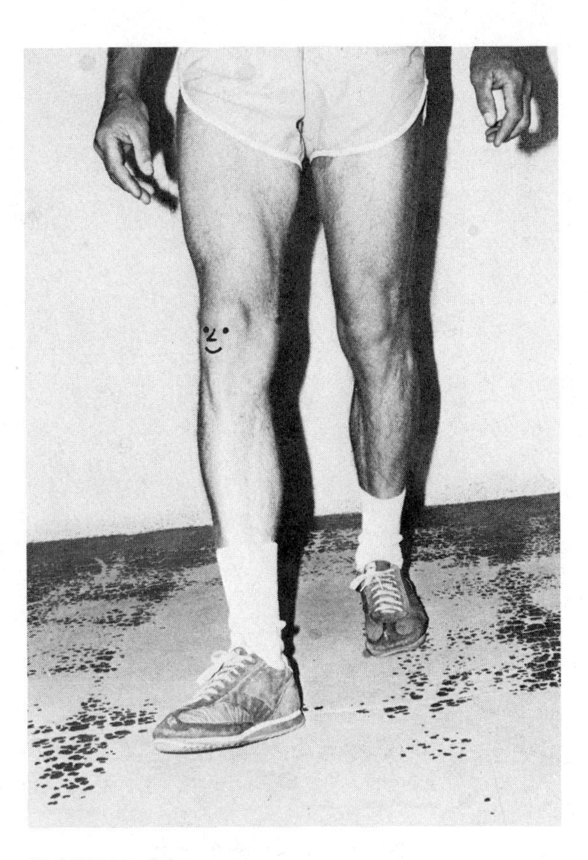

FIGURE 5.22
A view of the lower extremity during the swing phase of gait, showing lateral rotation at the hip joint.

to support the forces from above. Likewise, the midsagittal axes of the femur and tibia should be as parallel as possible with no twisting between adjacent segments. Figure 5.19 is a drawing of an ideal alignment.

In Figure 5.19, the pelvis is supported on either side by two columns made up of four structures. The structures making up the columns are the femur, the tibia, the talus, and the calcaneus. Normally, the femur is inclined slightly medial from its proximal to distal end. The inclination is somewhat dependent on the width of the pelvis.

Figure 5.20 shows an ideal arrangement of the forefoot. The circles represent the cross-sectional ends of the metatarsals.

Ideal alignment of the feet can be determined by observing how the feet align themselves with the midsagittal body plane. These possible positions are shown in Figure 5.21. In the ideal alignment in Figure 5.21a, the longitudinal axis of the foot segment is parallel with

FIGURE 5.23
An extremely flexible hip joint is beneficial in certain wrestling holds.

the midsagittal body plane. A **toe-out position** is shown in Figure 5.21b and a **toe-in position**, in Figure 5.21c. In each of these positions, the longitudinal axes of the feet are oblique to the midsagittal plane of the body.

Abnormal Alignments of the Lower Extremity

Lateral Rotation of the Hip Joint

If the four column segments are aligned with one another (i.e., no angulation or twisting of one segment occurs relative to an adjacent segment), a person may still display a toe-out gait. This situation is due to a laterally rotated hip joint. Figure 5.22 illustrates this position. Notice that the subject's right toe is directed outward and that his right leg and knee are aligned with his outwardly directed toe. This means that the rotation must be occurring at the hip joint. The lateral rotators of the hip are causing the lower extremity to rotate laterally during the swing phase of the gait. This condition may be corrected by strengthening the medial rotators of the hip joint and stretching the lateral rotators. Lateral rotation of the hip joint is commonly found in ballet dancers and wrestlers. Figure 5.23 illustrates how a laterally rotated hip is beneficial in a wrestling hold.

Torsional Conditions

Torsional conditions occur when one segment is rotated about its longitudinal axis relative to the next adjacent segment; that is, the midsagittal plane of the femur is not aligned with the midsagittal plane of the tibia.

FEMORAL TORSION. Figure 5.24a shows a femoral torsion condition; that is, the femur is rotated medially relative to the tibia. Identification of femoral torsion may be made by standing with the inside borders of the feet parallel and observing the position of the patellae. The patellae should be facing forward. If they are facing medially or inward slightly, then a femoral torsion condition exists. Often this condition is referred to as "cross-eyed" or "squinting" patellae. It is seen most frequently in females and is believed to indicate an imbalance of strength in the vasti muscles of the quadriceps group. Since the vastus medialis and lateralis muscles control the patella, an imbalance of strength between these muscles may cause the patella to be pulled to one side or the other. More likely, the cause is due to misalignment. If the femur is medially rotated relative to the tibia, the patellar tendon, and the femoral intracondular groove are not aligned. This misalignment manifests itself in parapatellar pain, a condition called **chondromalacia**, which is roughening of the undersurface of the patella that is due to its rubbing on the lateral condyle of the femur during extension of the knee. When this condition exists, strengthening the lateral rotators of the hip joint and stretching the medial rotators are advised.

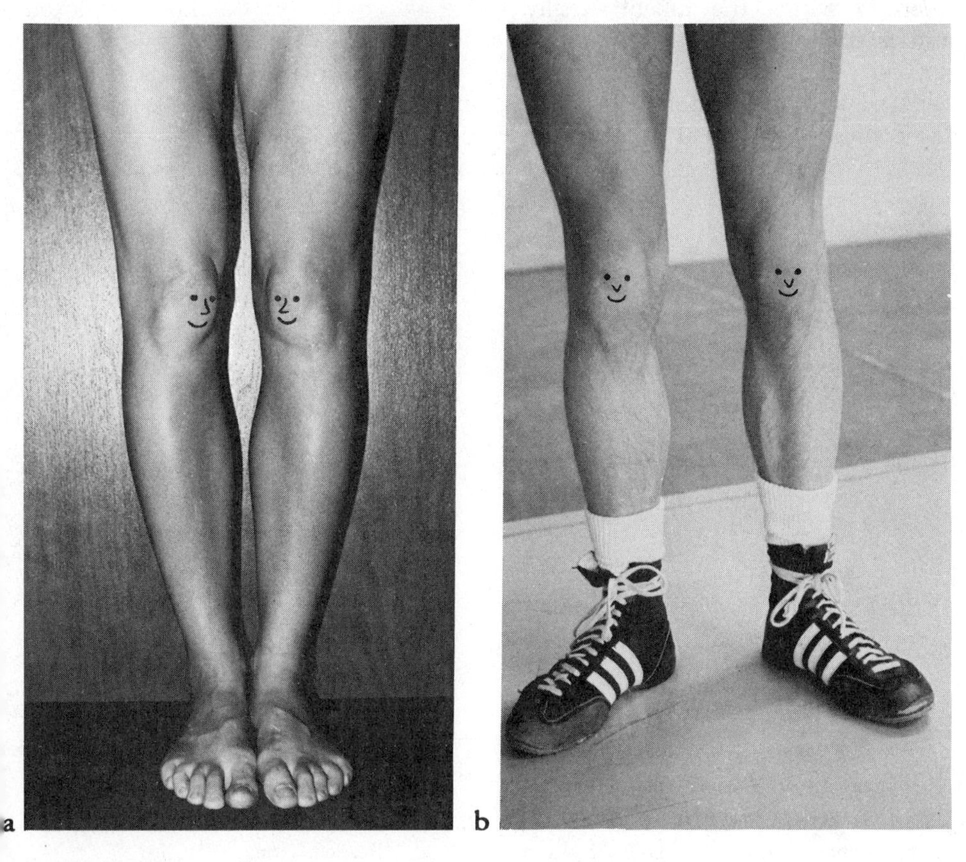

FIGURE 5.24
(a) Femoral torsion. (b) Tibial torsion.

Because of the force vectors on the patella during knee extension, quadriceps strengthening against a resistance should be done only through the last 30 degrees of knee extension. In addition, electromyographic studies have shown that the vastus medialis is more dominant during the last 30 degrees of extension.

Recall that the popliteus muscle is located to prevent lateral rotation of the tibia when the foot is free or to prevent medial rotation of the femur when the foot is fixed. Because those with femoral torsion have medially rotated femurs during the support phase, the popliteus muscle would be stressed (stretched) with each foot support. Another example of stressing the popliteus muscle is of a dancer taking off in a leap in which the takeoff foot is pushing and laterally rotating at the same time. Because femoral torsion is commonly associated with females, female dancers are particularly susceptible to popliteal syndrome.

TIBIAL TORSION. The second torsional condition is called **tibial torsion** and is seen more frequently in males. As in femoral torsion, the midsagittal axes of the femur and tibial are rotated relative to each other such that the midsagittal plane of the femur is directed forward but the midsagittal plane of the tibia is rotated outward. Figure 5.24b illustrates the condition.

An imbalance in the individual hamstring muscles has been implicated as a possible cause of this condition. Recall that the lateral hamstring muscle, the biceps femoris, has a second function to laterally rotate the tibia and that the medial hamstring muscles, the semimembranosus and tendinosus, medially rotate the tibia. As the knee is flexed against resistance, lateral rotation of the leg at the knee commonly occurs as a result of the stronger biceps femoris (Figure 5.10). If the knee is flexed against resistance without control of lateral functions, the tibia will be

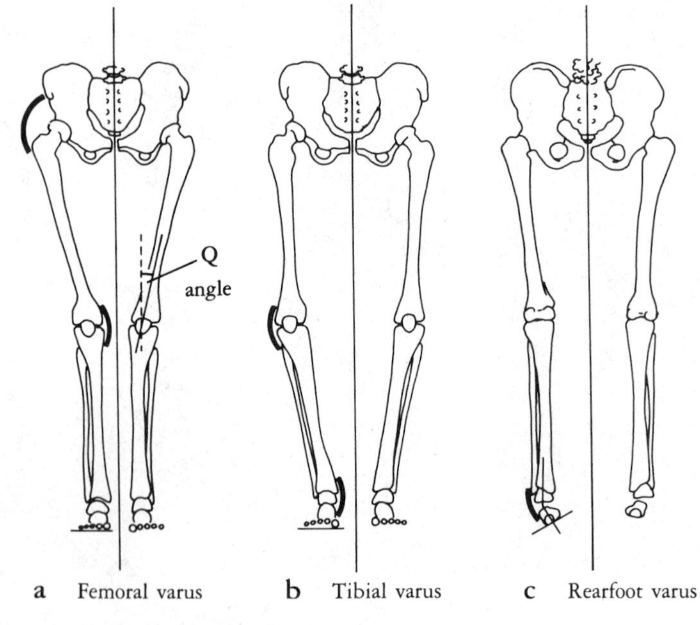

FIGURE 5.25
Common misalignment of the lower extremity: varus conditions.

pulled into lateral rotation as the knee is flexed and these movements will be reinforced. The tibial torsion condition causes stress on the medial structures of the knee joint. It also is associated with lateral knee pain (Flood, 1980) and causes a toeing-out position of the foot during weight bearing.

Varus and Valgus Conditions

In varus and valgus conditions of the segments of the lower extremity, the longitudinal axis of the segment angles or tends outward or inward from its proximal end to its distal end.

VARUS. **Varus conditions** are alignments in which the segment angles or tends inward from its proximal end to its distal end. Figure 5.25a shows femoral varus; Figure 5.25b, tibial varus; and Figure 5.25c, rearfoot varus. For clarity, all the other segmental alignments are placed in a neutral position. Curved lines about the lateral or medial articulations represent the location of tensile (pulling-apart) stress to the surrounding structures. With a varus condition, tensile stress is placed on the lateral side of the segment's proximal articulation and on the medial side of the segment's distal articulation. The muscles located on the stressed side should be strengthened, while the muscles located on the opposite side should be stretched.

Femoral varus (Figure 5.25a) contributes to a knock-knee-like posture. It is seen predominantly in females and frequently is associated with the wider pelvis in females although this has not been substantiated. Femoral varus affects what is called the **Q-angle**, or quadriceps angle, that is, the angle formed between the longitudinal axis of the femur that represents the pull of the quadriceps muscles and a line that represents the patellar tendon. A Q-angle greater than 15 degrees is reported as being a major factor in patellofemoral disorders due to a more lateral pull on the patella by the quadriceps (James et al., 1978). Tensile stress is placed on the lateral side of the hip joint and on the medial structures of the knee joint. To alleviate this condition, the hip abductor muscles, which are under tensile stress, should be strengthened, and the quadriceps should be strengthened to stabilize the knee against abduction of the leg. The hip adductors also should be stretched.

The femoral varus condition may be **compensated**; that is, a second misalignment condition may arise to help correct the first. With femoral varus, a compensated tibial valgus (outward tending) is prevalent. This puts additional tensile stress on the medial knee and lateral tissues of the ankle and subtalar joints when the foot is weight bearing.

Tibial varus is shown in Figure 5.25b. It causes a tensile stress to be applied to lateral knee joint, and it may produce lateral knee pain due to iliotibial-band friction syndrome. If the condition is not compensated, the feet or ankles may impact one another during the swing phase of gait. This is common with tibial varus runners who wear running shoes with flared outsoles. Tibial varus often induces rear foot valgus (tending outward), upon weight bearing, which places a tensile stress on the medial side of the ankle-subtalar joints. An inward tending tibia is particularly problematic in edging and control in downhill skiing (Matheson and Macintyre, 1987).

Rearfoot varus (Figure 5.25c) is a positional description of inversion of the subtalar joint. It is associated with tensile stress at the lateral side of the ankle and subtalar joint. The muscles that evert the subtalar joint should be strengthened so as to pull the rearfoot into a neutral position. A rearfoot varus position leads to susceptibility of ankle sprains upon landing from being airborne.

VALGUS. **Valgus** is the condition in which the segment tends outward from its proximal end to its distal end. Valgus conditions cause tensile stress to be placed on the medial side of the segment's proximal articulation and on

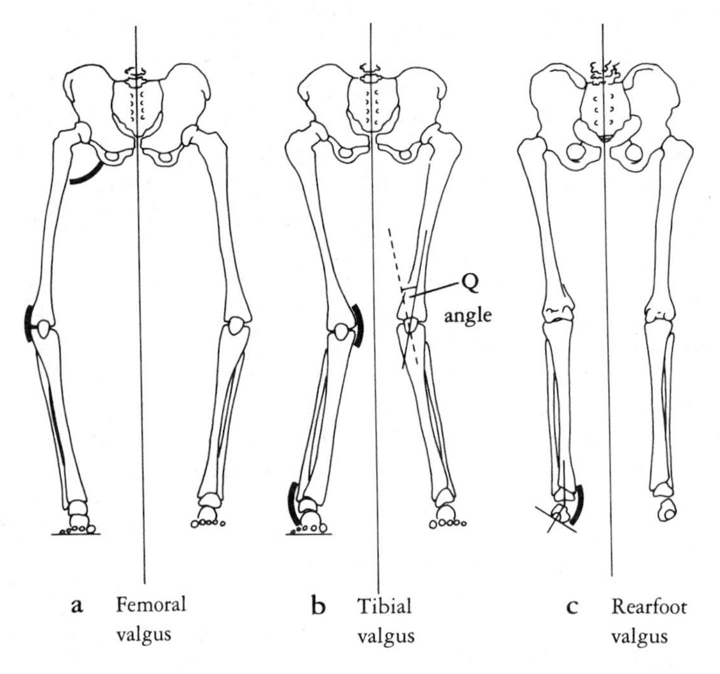

FIGURE 5.26
Common misalignment of the lower extremity: valgus condition.

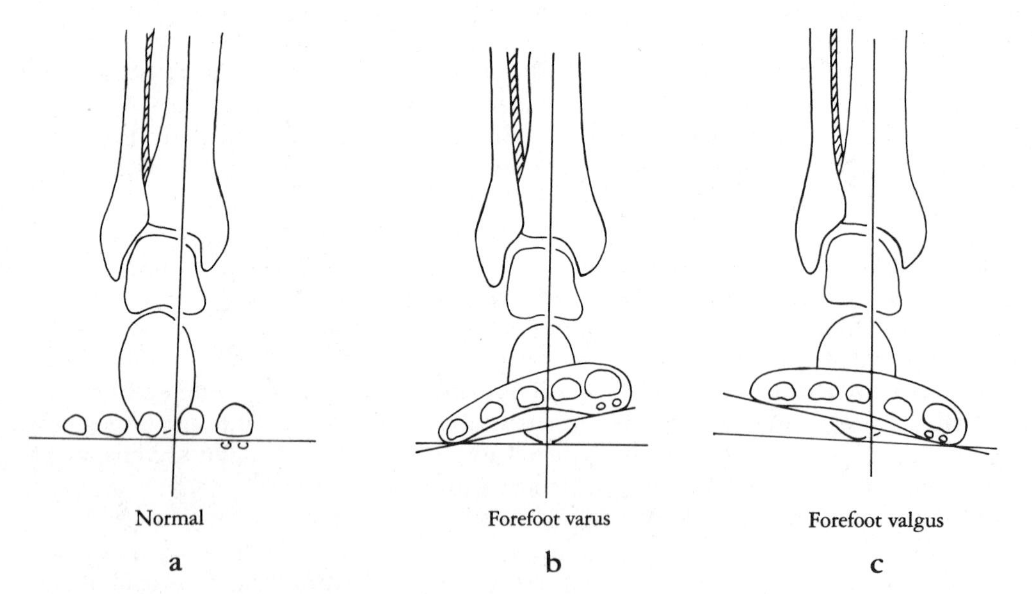

FIGURE 5.27
Forefoot positions. (a) Normal position. (b) Forefoot varus. (c) Forefoot valgus.

the lateral side of the segment's distal articulation.

Femoral valgus (Figure 5.26a) gives the impression of bowed legs. It is not a common condition, and it is usually associated with a skeletal deformity rather than a mechanically induced condition. Consequently, little can be done to correct it by strengthening or stretching surrounding tissues. Often it is seen in conjunction with tibial varus, which in this case, is also a skeletal deformity rather than mechanically induced. If the condition is compensated, rearfoot valgus is induced, and a tensile stress is placed on the medial side of the ankle–subtalar joints.

Tibial valgus (Figure 5.26b) appears as a knock-kneed alignment. Stress is placed on the medial side of the knee. To alleviate this condition, those muscles (quadriceps) that stabilize the knee from allowing abduction of the leg should be strengthened. Because stress is also placed on the lateral side of the ankle–subtalar joint complex if rearfoot varus is induced, the muscles that support the lateral side should be strengthened.

Rearfoot valgus (Figure 5.26c) is a positional description of eversion of the subtalar joint. Tensile stress is placed on the medial side of the ankle–subtalar joint complex. Strengthening exercises should be performed for the inversion muscles of the subtalar joint.

Forefoot varus and valgus conditions are best seen in a cross-sectional view of the metatarsal heads. Figure 5.27 illustrates forefoot varus and forefoot valgus. **Fore-foot varus** (Figure 5.27b) is a condition in which the first metatarsals are in a more dorsiflexed position than metatarsals 3 through 5. **Forefoot valgus** (Figure 5.27c) is a condition in which the first metatarsals are in a more plantarflexed position than metatarsals 3 through 5. Forefoot varus and valgus conditions cause important misalignments in the articulations above them during dynamic weight bearing, most noticeably in the subtalar joint. Forefoot varus results in the eversion of the subtalar joint upon weight bearing as there is little support on the medial side of the forefoot. Forefoot valgus results in the inversion of the subtalar joint as the person is weight bearing. Little can be done to correct these conditions by the use of muscle strengthening or stretching, but the use of an orthotic on the dorsiflexed side of the metatarsal heads may provide the support needed to prevent the inversion or eversion of the subtalar joint when walking or running.

Pronation and Supination of the Foot

When the ankle, the rearfoot posture, and the forefoot posture are considered, the foot may be in a neutral alignment, or it may be pronated or supinated.

PRONATION. **Pronation** of the foot is a particular combination of ankle dorsiflexion, subtalar eversion, and forefoot abduction. Forefoot abduction is a position in which the forefoot is in a toe-out position and is usually forced onto the metatarsals by misaligned weight bearing. Pronation is one of the most common misalignment conditions of the lower extremity, and causes runners and joggers many problems. Exaggerated pronation is often induced by other misalignment conditions, especially when a person runs in a toe-out position. The toe-out position may be the result of lateral rotation at the hip, tibial torsion (lateral rotation of the tibia), or forefoot abduction. The toe-out position causes the line of force of the body's weight to move from the initial contact point of the foot with the ground to the medial side of the foot as a person transfers his or her weight forward. This results in the collapse of the medial side of the foot and puts undue stress on the muscles that support the arch of the foot, the tibialis anterior and posterior. The stress on the tibialis muscles is associated with a condition known as shin splints, a stressful condition along the anterior medial or lateral side of the leg or both.

Some pronation of the foot occurs naturally because of the obliquity of the ankle and subtalar axes; however, excessive pronation is harmful. Excessive pronation not associated with a toeing-out position may be caused by weakness in the structures supporting the medial arch, by an abnormally flat foot with little natural arch, or by forefoot varus. Artificial support in the form of an arch support may be used; however, one should work gradually into the use of the support so that the supporting structures can adapt.

SUPINATION. **Supination** of the foot is the condition of plantar flexion of the ankle, inversion of the subtalar joint, and adduction of the forefoot. It is not as common as pronation, and it is not associated with causing any particular problem in runners. If the inversion aspect is severe enough, lateral ankle sprains may occur upon landing from a jump or when running on uneven surfaces. The muscles on the lateral side of the ankle–subtalar joint complex should be strengthened; those on the medial side should be stretched.

5.7 Lower Extremity Misalignments in Dance

In modern and ballet forms of dance, lateral rotation of the lower extremity is very important. From proximal to distal articulations, the hip, knee, ankle, and subtalar articulations can contribute to achieving the outwardly directed foot. Technically, however, complete turnout should be accomplished in the lateral rotation of the thigh at the hip joint so as to not stress tissues around the more distal articulations. It is unlikely that many can achieve the amount of turnout needed by using the hip joint alone.

Patellofemoral Problems Related to Dancers

Silver and Campbell (1985) report that the major cause of meniscal tears in dancers is related to poor technique. The dominant lower extremity positions of the lower extremity depend on excessive lateral rotation of the thigh at the hip joint. If the hip joint ligamentous and muscular arrangements do not allow for full lateral rotation, then the knee, ankle, subtalar, and forefoot joints must provide additional ranges of motion in the lateral direction so that the feet eventually can be 90, 180, or even 360 degrees from each other.

Bony, ligamentous, and musculotendinous constraints must be considered. The angle of the femoral neck to the shaft determines in part the amount of lateral rotation possible at the joint. The term *anteversion* is used to describe a femoral neck that tends forward of the frontal plane of the femur. *Retroversion* is used to describe a femoral neck that tends backward of the frontal plane of the femur. An anteverted femoral neck makes it very difficult for a person to laterally rotate the femur. While much of the angle is genetically determined, Clippinger-Robertson (1987) reports findings of Samarco, which indicate that increased retroversion may be achieved before the age of 11 years, after which increased turnout must be achieved with soft tissue allowances

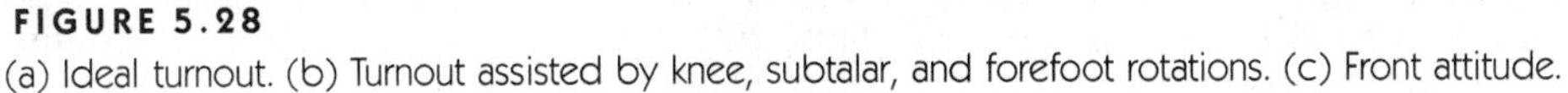

FIGURE 5.28
(a) Ideal turnout. (b) Turnout assisted by knee, subtalar, and forefoot rotations. (c) Front attitude.

alone, the articular capsule and surrounding ligaments.

If lateral rotation at the hip joint is insufficient to allow full 90 degree excursion of the foot, then it may be supplemented by lateral rotation of the leg at the knee joint (tibial torsion), eversion of the subtalar joint (rearfoot valgus), or abduction of the forefoot (pronation). Figures 5.28a and b illustrate the ideal turnout from the hipjoint and turnout created by additional contributions from the knee, subtalar, and forefoot articulations.

The supplemented turnout puts stress on the medial aspect of the knee in the plié posture, and when the dancer returns from the position by extending the knee, the structures do not allow for rotation, and thus are placed under extreme tension. Eventually, the tissues may stretch to accommodate the rotation, which contributes to the knee joint laxity. The knee is prone to stress from the distal segments' position as well. Recall that when the subtalar joint everts, a medially directed torque is placed on the tibia. The forced pronated position of the foot places stress on the muscles surrounding and supporting the instep. The tibialis anterior and posterior are stretched and may lead to shin splints, a frequent malady of dancers. The tibial torsion pulls the patellar tendon laterally and along with it, the patella. Because the femur does not rotate outward as far, the Q-angle is increased beyond the 15-degree range, which enhances patellar pathomechanics and frequently results in chondromalacia. Furthermore, in the plié position, the knee may be flexed to 90 degrees. In this position, the compression force at the patellofemoral joint can be seven to nine times body weight (Bishop, 1977).

In addition to the proper alignment, the use of different muscle groups may be important in avoiding problems. Clippinger-Robertson (1987) stresses the use of one-joint muscles or at least two-joint muscles that do not have a second antagonistic function to the movement being performed. She uses the example of performing a front attitude, in which the hip is flexed and laterally rotated (Figure 5.28). While the gluteus maximus is a strong lateral rotator of the hip, it also extends the hip, a movement that is antagonistic to the flexion required in performing this skill. One would be well advised to focus on using the six deep external rotators to perform this move so that the hip-flexion motion is not limited by the tension in the gluteus maximus. As a second example of controlling the muscles used, pliés in the second position may be performed using the hip adductors and hamstrings rather than emphasizing the quadriceps to stabilize the knee. The reduction in quadriceps force would reduce the compression force of the patella on the femoral condyles.

5.8 Analyzing Lower Extremity Misalignments

Often a physical education teacher or coach is asked by students, athletes, or fitness-minded people to provide analysis and prescription. The biomechanically oriented teacher or coach should be able to analyze and diagnose simple misalignment problems. In less severe cases, exercises to correct or prevent problems may be suggested. For complex cases, consultation may be advised with a podiatrist experienced in dealing with lower extremity alignment during exercise.

The easiest way to analyze a dynamic posture is with the help of videotape or films that are developed on the spot. If these tools are not available, conditions may be recognized during the performance, although this takes practice and experience.

An observer should begin by viewing static-stance front and rear views. The following can be observed from a front view of the static stance:

1. Femoral torsion if the patella is medially aligned
2. Tibial torsion if the patella is forward looking and the leg and foot face outward
3. Lateral rotation of the hip if the patella, the leg, and the foot are rotated outward slightly but all aligned with each other
4. Forefoot abduction if the femur, patella, and tibia are forward looking, and the foot points outward slightly
5. Valgus conditions if the femur or tibia tends outward from proximal to distal ends
6. Varus conditions if the femur or tibia tends inward from proximal to distal ends

The following can be observed from a rear view of the static stance:

1. Femoral and tibial varus and valgus

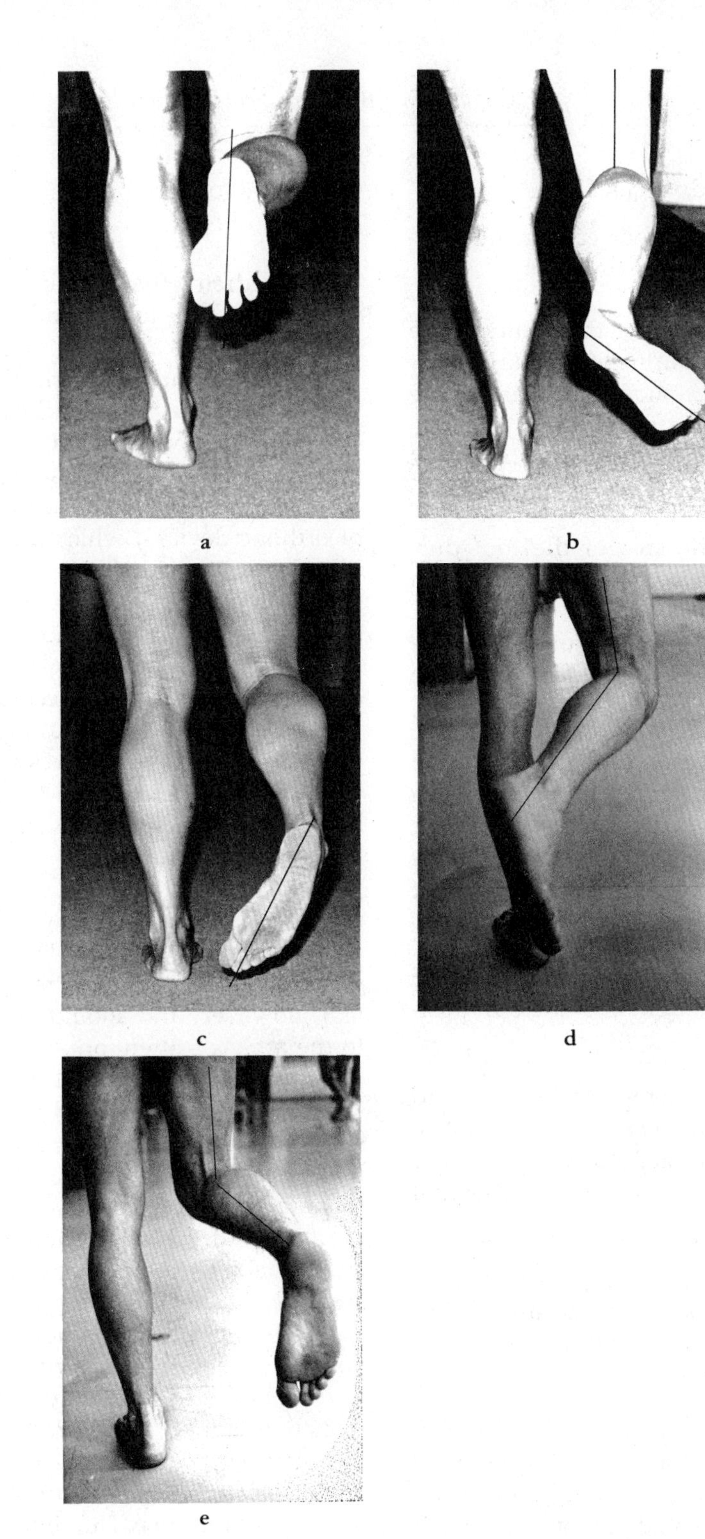

FIGURE 5.29

The nonweight-bearing leg and foot. (a) Neutral position. (b) Tibial torsion. (c) Supinated position. (d) Laterally rotated hip. (e) Femoral torsion (medially rotated hip).

2. Rearfoot varus and valgus
3. A combination of rearfoot valgus and forefoot abduction in a pronated foot

Whatever is viewed from the static stance may be exaggerated during dynamic weight bearing. The observer should be alert to the conditions noted in the static-stance position and to the following dynamic postures, as shown in Figure 5.29:

1. The position of the sole of the foot during the swing phase provides information on where the misalignment is occurring. A neutral position is seen in Figure 5.29a. The long axis of the foot is perpendicular to the ground.
2. Evidence of a tibial torsion condition (lateral rotation of the knee) is shown in Figure 5.29b. The midsagittal plane of the femur is aligned but the midsagittal plane of the tibia is rotated outward. The long axis of the foot is in a toe-out position.
3. A supinated foot is shown in Figure 5.29c. The long axis of the foot is in a toe-in position.
4. Lateral rotation of the hip causes the leg to look like that shown in Figure 5.29d. Notice that the knee is pointed outward and that the leg and the foot cross the midsagittal plane.
5. A medial rotation of the hip (femoral torsion) causes the knee to point inward and the leg and the foot to tend outward from the midsagittal plane (Figure 5.29e).
6. From the front, the path of the foot and leg during the swing forward should be observed. A paddling motion occurs from either a medially rotated hip joint, in which case the knee rotates inward and the leg segment moves outward, or from tibial torsion, in which case the knee remains straight forward but the long axis of the foot tends outward.

A final factor that may cause athletes some problems is the quality and type of footwear worn. Advances are being made by shoe manufacturers in understanding the mechanics of the foot and leg during heavy use and in developing shoes to suit a variety of foot and leg postures. Some shoes may be modified to the individual user through the use of orthotic devices, which are shaped specifically for an individual's foot and placed in the shoe to counteract any misalignment or imbalance problems. For instance, excessive pronation may be prevented by using an orthotic device that provides the necessary support for the arch. An extra heel support may prevent excess dorsiflexion and restrict the subtalar joint to a more normal ROM. Sometimes a heel lift will provide the necessary adjustment to relieve Achilles tendinitis.

Orthotic devices should be individually prescribed by a podiatrist who is knowledgeable about sport injuries. The trainer or coach should be able to recognize a potential problem, however, and should bring the problem to the athlete's attention.

A way to prevent problems is to select the shoe that is best designed to give the proper support. Several factors are important: a cushioned heel to absorb shock, a firm heel counter to provide rearfoot control, a firm arch support, a sole that is wide enough to provide the lateral stability needed for running on uneven surfaces but not too wide to cause collisions with the opposite foot, enough rise in the toe box to prevent excessive rubbing on the tops of the phalanges, flexibility in the metatarsal–phalangeal area, good traction on dry and wet surfaces, permeability to control temperature and humidity and durability over long-term use of the shoe. Durability includes shock absorption durability for impact and outsole wear.

Understanding Postural Alignments

1. Observe some joggers. From a front view, identify these positions or movements: (a) a toe-out position caused by lateral rotation of the hip joint, (b) a toe-out position caused by abduction of the foot.
2. From a rear view of the joggers, identify these positions or movements: (a) a pronated foot, (b) a supinated foot. What combination of movements produces pronation of the foot? supination of the foot?
3. Figure 5.30 shows two athletes about to land after being airborne. Describe for each: the misalignment, the cause of the misalignment, which muscles should be strengthened, which muscles should be stretched, and what injuries may develop from continually landing in this position.

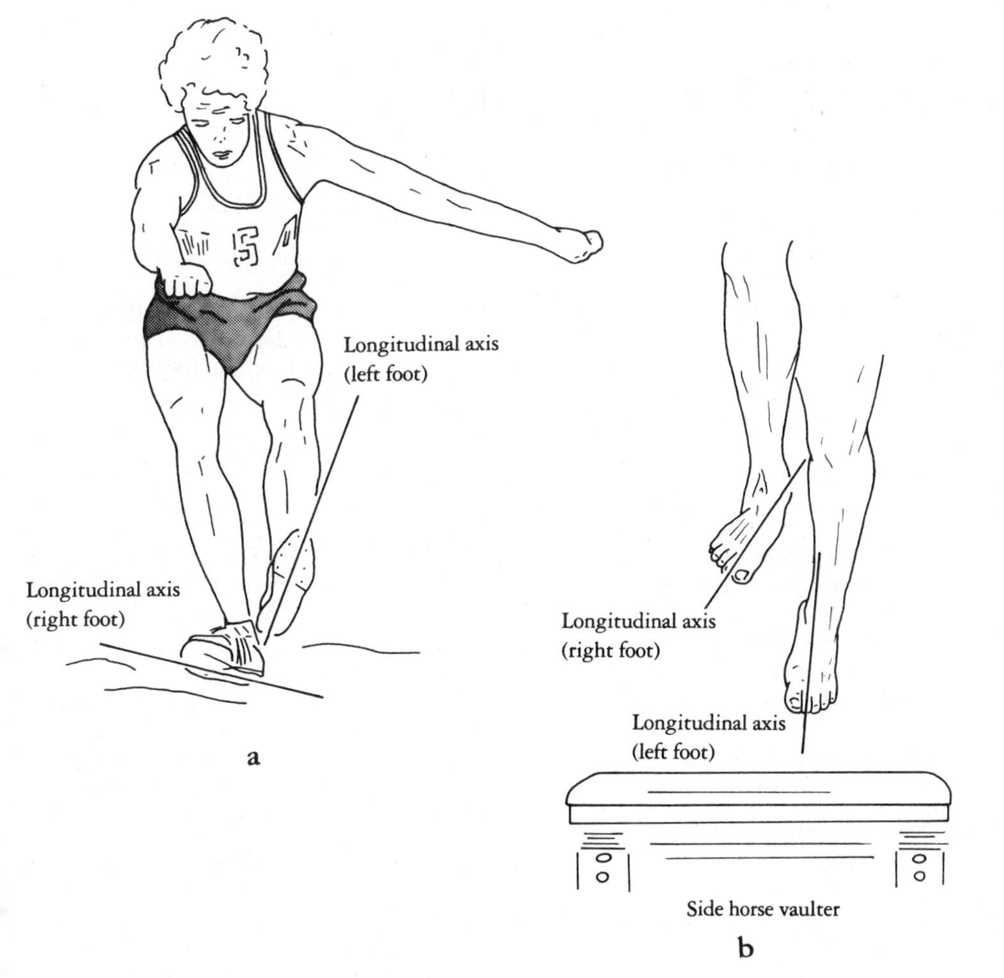

FIGURE 5.30

(a) A long jumper about to land and (b) a gymnast approaching the Reuther board, showing impending weight bearing in nonneutral positions.

4. Observe three people in each of the following activities: (a) static stance, (b) walking, (c) jogging, and (d) running. Watch the hip, knee, ankle, and subtalar positions and movements. For all the gaits, observe differences in positions during the contact phase, midstance phase, and the propulsive or toe-off stage. Answer these questions:
   a. Do you notice any deviations in the static stance that were also observed in the dynamic gaits?
   b. Do changes occur when the gait is accelerated? How?
   c. Are there differences within one subject between the right and left leg and foot? What are they?
   d. What muscles need to be stretched or strengthened or both to bring the leg into neutral alignment?

References and Suggested Readings

Abramowitz, J., Shiavi, R., Frazer, M., & Limbird, T. (1987). Kinematics of the anterior cruciate ligament deficient knee. In B. Jonsson (Ed.), *Biomechanics X-A* (pp. 93–94). Champaign, IL: Human Kinetics.

Adelaar, R. S. (1986). The practical biomechanics of running. *The American Journal of Sports Medicine, 14*, 497–500.

American Academy of Orthopaedic Surgeons, (1982). *Symposium on the Foot and Leg in Running Sports*. R. P. Mack (Ed.). St. Louis: C. V. Mosby.

Ariel, B. G. (1974). Biomechanical analysis of the knee joint during deep knee bends with a heavy load. In R. C. Nelson & C. A. Morehouse (Eds.), *Biomechanics IV: Proceedings of the Fourth International Seminar on Biomechanics* (pp. 44–52). Baltimore: University Park Press.

Baratta, R., et al. (1988). Muscular coactivation: the role of the antagonist musculature in maintaining knee stability. *The American Journal of Sports Medicine, 16*, 113–122.

Bates, B., et al. (1978). Lower extremity function during the support phase of running. In E. Asmussen & K. Jorgensen (Eds.), *Biomechanics VI-B: Proceedings of the Sixth International Symposium on Biomechanics* (pp. 30–39). Baltimore: University Park Press.

Bishop, R. (1977). On the mechanics of the human knee. *Engineering in Medicine, 6*, 46–52.

Bojsen-Moller, F. (1978). The human foot—a two speed construction. In E. Asmussen & K. Jorgensen (Eds.), *Biomechanics VI-A: Proceedings of the Sixth International Symposium on Biomechanics* (pp. 261–266). Baltimore: University Park Press.

Brody, D. M. (1980). Running injuries. *Clinical Symposia, 32*(4), 2–36.

Brown, T. R. M., Kelly, I. G., Paul, J. P., & Hamblen, D. L. (1985). The influence of pathological and reconstructed anatomy at the hip joint on loads in the lower limb joints. In D. A. Winter, R. W. Norman, R. P. Wells, K. C. Hayes, & A. E. Patla (Eds.), *Biomechanics IX-A* (pp. 539–543). Champaign, IL: Human Kinetics.

Cabri, J. M., Bollens, E. C., & Clarys, J. P. (1987). Muscle activity in the foot–calf system under different loads in female subjects. In B. Jonsson (Ed.), *Biomechanics X-B* (pp. 277–282). Champaign, IL: Human Kinetics.

Cavanagh, P. R. (1980). *The running shoe book*. Mountain View, CA: Anderson World.

Cavanagh, P. R., & Rodgers, M. M. (1987). The arch indes: a useful measure from footprints. *Journal of Biomechanics, 20*, 547–551.

Cavanagh, P. R., Williams, K. R., & Bednarski, K. N. (1985). Foot angles during walking and running. In D. A. Winter, R. W. Norman, R. P. Wells, K. C. Hayes, & A. E. Patla (Eds.), *Biomechanics IX-A* (pp. 451–456). Champaign, IL: Human Kinetics.

Clancy, W. G. (1983). Knee ligamentous injury in sports. *Medicine and Science in Sports and Exercise, 15*(1), 9–14.

Clippinger-Robertson, K. (1987, May–June). Biomechanical consideration in turnout. *JOHPERD*, (pp. 37–40).

Costigan, P. A., & Reid, J. G. (1985). Radial torque of the tibia during a deep knee bend. In D. A. Winter, R. W. Norman, R. P. Wells, K. C. Hayes, & A. E. Patla (Eds.), *Biomechanics IX-A* (pp. 420–23). Champaign, IL: Human Kinetics.

Cova, P., Pedotti, A., Rodano, R., & Vigano, R. (1985). Comparative analysis of some pathologies of the foot by the vector diagrams. In D. A. Winter, R. W. Norman, R. P. Wells, K. C. Hays, & A. E. Patla (Eds.), *Biomechanics IX-A* (pp. 534–538). Champaign, IL: Human Kinetics.

Crowninshield, R. D., et al. (1978). A biomechanical investigation of the human hip. *Journal of Biomechanics, 11*(1/2), 75–85.

Engsberg, J. R. (1985). A biomechanical analysis of the talocalcaneal joint—in vitro. *Journal of Biomechanics, 20*, 429–42.

Engsberg, J. R., & Andrews, J. G. (1987). Kinematic analysis of talocalcaneal/talocural during running support. *Medicine and Science in Sports and Exercise, 19*, 275–284.

Ferrandino, G. (1978). *Descriptive analysis of biomechanic and radiographic characteristics of the lower extremity of runners with Morton's foot syndrome.* Unpublished masters thesis, Montana State University, Bozeman, MT.

Flint, M. M. (1971). A differential study of the hip extensor muscles. In C. J. Widule (Ed.), *Kinesiology Review* (pp. 39–48). Reston, VA: AAHPERD.

Flood, J. E. (1980). *Selected lower extremity alignment and range of motion measurements and their relationships to lateral knee pain*. Unpublished master's thesis, Montana State University, Bozeman, MT.

Fuglevand, A. J. (1987). Resultant muscle torque, angular velocity, and joint angle relationships and activation patterns in maximal knee extension. In B. Jonsson (Ed.), *Biomechanics X-B* (pp. 559–566). Champaign, IL: Human Kinetics.

Fujita, M., Matsusaka, N., Chiba, T., Norimatsu, T., & Suzuki, R. (1985). The role of the ankle plantar flexors in level walking. In D. A. Winter, R. W. Norman, R. P. Wells, K. C. Hayes, & A. E. Patla (Eds.), *Biomechanics IX-A* (pp. 484–488). Champaign, IL: Human Kinetics.

Galea, V., & Norman, R. W. (1985). Bone-on-bone forces at the ankle joint during a rapid dynamic movement. In D. A. Winter, R. W. Norman, R. P. Wells, K. C. Hayes, & A. E. Patla (Eds.), *Biomechanics IX-A* (pp. 71–76). Champaign, IL: Human Kinetics.

Gehlsen, G. M., & Seger, A. (1980). Selected measures of angular displacement, strength, and flexibility in subjects with and without shin splints. *Research Quarterly, 51*(3), 478–485.

Gertsch, P., Borgeat, A., & Walli, T. (1987). New cross-country skiing technique and compartment syndrome. *The American Journal of Sports Medicine, 15*, 612–613.

Goldfuss, A., Morehouse, C., & LeVeau, B. (1973). Effect of muscular tension on knee stability. *Medicine and Science in Sports, 5*(4), 267–271.

Harrison, R. N., Lees, A., McCullagh, P. J. J., & Rowe, W. B. (1987). Bioengineering analysis of muscle and joint forces acting in the human leg during running. In B. Jonsson (Ed.), *Biomechanics X-B* (pp. 855–862). Champaign, IL: Human Kinetics.

Henning, C. E. (1988). Semilunar cartilage of the knee: function and pathology. In K. B. Pandolf (Ed.), *Exercise and sport sciences reviews* (pp. 205–213). New York: Macmillan.

Inman, V. T. (1947). Functional aspects of the abductor muscles of the hip. *Journal of Bone and Joint Surgery, 29*(A), 607–619.

Inman, V. T. (1976). *The joints of the ankle.* Baltimore: Williams & Wilkins.

Inoue, M., McGurk-Burleson, E., Hollis, J. M., & Woo, S. Y. L. (1987). Treatment of the medial collateral ligament injury. I: The importance of anterior cruciate ligament on the Varus-Valgus knee laxity. *The American Journal of Sports Medicine, 15*, 15.

Isakov, E., Mizrahi, J., Solzi, P., Susak, Z., & Lotem, M. (1986). Response of the peroneal muscles to sudden inversion of the ankle during standing. *International Journal of Sport Biomechanics, 2*, 100–109.

James, S., Bates, B., & Osternig, L. (1978). Injuries to runners. *American Journal of Sports Medicine, 6* (2), 40–50.

Karas, V., Valenta, J., Straus, J., Otahal, S., & Komarek, P. (1987). The magnitude and direction of reaction forces operating in the hip joint in walking. In B. Jonsson (Ed.), *Biomechanics X-B* (pp. 387–392). Champaign, IL: Human Kinetics.

Kennedy, J. C. (1978). Orthopaedic manifestations. In B. Eriksson & B. Furberg (Eds.), *Swimming Medicine IV: Proceedings of the Fourth International*

Congress on Swimming Medicine (pp. 93–100). Baltimore: University Park Press.

Kennedy, J. C., & Hawkins, R. (1974). Breaststroker's knee. *The Physician and Sports Medicine, 2*(1), 33–35.

Kennedy, J. C., Hawkins, R., & Krissoff, W. B. (1978). Orthopaedic manifestations of swimming. *American Journal of Sports Medicine, 6*(6), 309–322.

Keskinen, K., et al. (1980). Breaststroker swimmer's knee. *American Journal of Sports Medicine, 8*, 228–231.

Kinoshita, H., Bates, B. T., & DeVita, P. (1985). Inertial variability for selected running gait parameters. In D. A. Winter, R. W. Norman, R. P. Wells, K. C. Hayes, & A. E. Patla (Eds.), *Biomechanics IX-A* (pp. 499–502). Champaign, IL: Human Kinetics.

Kravitz, S. R. (1987, May–June). Relating the foot and ankle to overuse injuries. *JOHPERD*, (pp. 31–33).

Louie, J. K., & Mote, C. D., Jr. (1987). Contribution of the musculature to rotatory laxity and torsional stiffness at the knee. *Journal of Biomechanics, 20*, 281–300.

Leach, R. E., & Corbett, A. (1979). Anterior tibial compartment syndrome in soccer players. *American Journal of Sports Medicine, 7*(4), 258–259.

Lloyd-Smith, R., Clement, D. B., McKenzie, D. C., & Taunton, J. E. (1985). A survey of overuse and traumatic hip and pelvic injuries in athletes. *The Physician and Sports Medicine, 13*, 131–141.

Luethi, S. M., Frederick, E. C., Hawes, M. R., & Nigg, B. M. (1986). Influence of shoe construction on lower extremity kinematics and load during lateral movements in tennis. *International Journal of Sports Medicine, 2*, 166–174.

Mann, R. A., Moran, G. T., & Dougherty, S. E. (1986). Comparative electromyography of the lower extremity in jogging, running, and sprinting. *The American Journal of Sports Medicine, 14*, 501–509.

Mansour, J. M., & Pereira, J. M. (1987). Quantitative functional anatomy of the lower limb with application to human gait. *Journal of Biomechanics, 20*, 51–58.

Matheson, G. O., & Macintyre, J. G. (1987). Lower leg varum alignment in skiing: relationship to foot pain and suboptimal performance. *The Physician and Sports Medicine, 15*, 163–173.

McConkey, J. P. (1986). Anterior cruciate ligament rupture in skiing. A new mechanism of injury. *The American Journal of Sports Medicine, 14*, 160–164.

Messier, S. P., & Pittala, K. A. (1988). Etiologic factors associated with selected running injuries. *Medicine and Science in Sports and Exercise, 20*, 501–505.

Michael, R. H., & Holders, L. E. (1985). The soleus syndrome. A cause of medial tibial stress (shin splints). *The American Journal of Sports Medicine, 13*, 87–94.

Mubarak, S., et al. (1978). Acute exertional superficial posterior compartment syndrome. *American Journal of Sports Medicine, 6*(5), 287–290.

Naver, L., & Aalberg, J. R. (1987). Avulsion of the popliteus tendon. A rare cause of chondral fracture and hemarthrosis. *The American Journal of Sports Medicine, 13*, 423–424.

Nichols, J. (1978). Report of the Committee on Research and Education. *The American Journal of Sports Medicine, 6*(5), 295–304.

Nigg, B. M., Bahlsen, H. A., Luethi, S. M., & Stokes, S. (1987). The influence of running velocity and medsole hardness on external impact forces on heel–toe running. *Journal of Biomechanics, 20*, 951–960.

Nigg, B. M., & Morlock, M. (1987). The influence of lateral heel flare of running shoes on pronation and impact forces. *Medicine and Science in Sports and Exercise, 19*, 294–302.

Nobel, C. A. (1980). Iliotibial band friction syndrome in runners. *The American Journal of Sports Medicine, 8*(4), 232–35.

Noble, H. B., Hajek, M. R., & Porter, M. (1982). Diagnosis and treatment of iliotibial band tightness in runners. *The Physician and Sports Medicine, 10*, 67–74.

Olmstead, T. G., Wevers, H. W., Bryant, J. T., & Gouw, G. J. (1986). Effect of muscular activity on valgus/varus laxity and stiffness of the knee. *Journal of Biomechanics, 19*, 565–578.

Olney, S. J. (1985). Quantitative evaluation of cocontraction of knee and ankle muscles in normal walking. In D. A. Winter, R. W. Norman, R. P. Wells, K. C. Hayes, & A. E. Patla (Eds.), *Biomechanics IX-A*, 431-436. Champaign, IL: Human Kinetics.

Osternig, L. (1980). Patterns of tibial rotary torque in knees of healthy subjects. *Medicine and Science in Sports and Exercise, 12* (3), 195–199.

Percy, E. C. & Strother, R. T. Patellalgia. (1985). *The Physician and Sports Medicine, 13*, 43–56.

Pretorius, D. M., Noakes, T. D., Irving, G., & Allerton, K. (1986). Runners knee: what is it and how effective is conservative management? *The Physician and Sports Medicine, 12*, 71–75.

Pronk, C. N. A., van Nieuwenhuyzen, J. F., & Stam, H. J. (1985). A comparative study of isokinetic force measurements of the knee extensors. In D. A. Winter, R. W. Norman, R. P. Wells, K. C. Hayes, & A. E. Patla, (Eds.), *Biomechanics IX-B* (p. 22). Champaign, IL: Human Kinetics.

Reid, D. C., Burnham, R. S., Saboe, L. A., & Kushner, S. F. (1987). Lower extremity flexibility patterns in classical ballet dancers and their correlation to lateral hip and knee injuries. *The American Journal of Sports Medicine, 15*, 347–352.

Reinecke, S., Arms, S., Renstrom, P., Johnson, R. J., & Pope, M. H. (1987). In B. Jonsson (Ed.), *Biomechanics X-B* (pp. 111–118). Champaign, IL: Human Kinetics.

Renstrom, P., Arms, S. W., Stanwyck, T. S., Johnson, R. J., & Pope, M. H. (1986). Strain within the anterior cruciate ligament during hamstring and quadriceps activity. *The American Journal of Sports Medicine, 14*, 83–87.

Robertson, D. G. E. (1987). Functions of the leg muscles during the stance phase of running. In B. Jonsson (Ed.), *Biomechanics X-B* (pp. 1021–1028). Champaign, IL: Human Kinetics.

Robertson, L. (1976). The effect of two exercise routines on the movement of medial rotation of the leg. *Medicine and Science in Sports, 8*(4), 253–257.

Roels, J., et al. (1978). Patellar tendinitis (jumpers knee). *American Journal of Sports Medicine, 6*(6), 362-368.

Rovere, G. D., & Nichols, A. W. (1985). Frequency, associated factors, and treatment of breaststroker's knee in competitive swimmers. *The American Journal of Sports Medicine, 13*, 99–104.

Salathe, E. P., Jr., Arangio, G. A., & Salathe, E. P. (1987). A biomechanical model of the foot. *Journal of Biomechanics, 19*, 989–1002.

Shell, C. G. (Ed.) (1986). *The dancer as athlete.* Champaign, IL: Human Kinetics.

Silver, D. M., & Campbell, P. (1985). Arthroscopic assessment and treatment of dancers' knee injuries. *The Physician and Sports Medicine, 13*, 75-81.

Solomonow, M., et al. (1987). The synergistic action of the anterior cruciate ligament and thigh muscles in maintaining joint stability. *The American Journal of Sports Medicine, 15*, 207–213.

Stafford, M. G., & Grana, W. A. (1984). Hamstring/quadriceps ratios in college football players: A high velocity evaluation. *The American Journal of Sports Medicine, 12*, 209.

Steiner, M. E. (1987). Hypermobility and knee injuries. *The Physician and Sports Medicine, 15*, 159–165.

Stulberg, S. D., et al. (1980). Breaststrokers knee: pathology, etiology, and treatment. *The American Journal of Sports Medicine, 8*(3), 164–171.

Styf, J. (1988). Diagnosis of exercise-induced pain in the anterior aspect of the lower leg. *The American Journal of Sports Medicine, 16*, 165–169.

Sutker, A. N., Jackson, D. W., & Pagliano, J. W. (1979). Iliotibial band syndrome in distance runners. *The Physician and Sports Medicine, 9*, 69–73.

Teitz, C. C. (1987, May/June). Patellofemoral pain in dancers. *JOHPERD*, pp. 34–36.

Tsarouchas, E., Tokmakidis, S. P., & Giavroglou, A. (1985). The influence of the kinetic chain drive systems of the lower extremities on the acceleration work. In D. A. Winter, R. W. Norman, R. P. Wells, K. C. Hayes, & A. E. Patla (Eds.), *Biomechanics IX-B* (pp. 414–419). Champaign, IL: Human Kinetics.

Twardokens, G. (1975). *Rotations of lower limbs in skiing*. Unpublished paper.

van Eijden, T. M. G. J., de Boer, W., & Weijs, W. A. (1985). The orientation of the distal part of the quadriceps femoris muscle as a function of the knee flexion-extension angle. *Journal of Biomechanics, 18*, 803–810.

Vizsolyi, P., et al., (1987). Breaststroker's knee: An analysis of epidemiological and biomechanical factors. *The American Journal of Sports Medicine, 15*, 63–71.

Voloshin, A. S., Burger, C. P., Wosk, J., & Arcan, M. (1985). An in vivo evaluation of the leg's shock-absorbing capacity. In D. A. Winter, R. W. Norman, R. P. Wells, K. C. Hayes, & A. E. Patla (Eds.), *Biomechanics IX-B* (pp. 112–116). Champaign, IL: Human Kinetics.

Wall, J. C., & McDermott, A. G. P. (1985). Intracompartmental pressure changes during the walking cycle. In D. A. Winter, R. W. Norman, R. P. Wells, K. C. Hayes, & A. E. Patla (Eds.), *Biomechanics IX-A* (pp. 441–445). Champaign, IL: Human Kinetics.

Wells, R. P. (1988). Mechanical energy costs of human movement: An approach to evaluating the transfer possibilities of two-joint muscles. *Journal of Biomechanics, 21*, 955-963.

Wiklander, J., Marketa, L., & Lysholm, J. (1985). The correlation between running movements and muscle strength/joint mobility in the lower extremity. In B. Jonsson (Ed.), *Biomechanics X-B* (pp. 813–818). Champaign, IL: Human Kinetics.

Williams, K. R. (1985). Biomechanics of running. In R. L. Terjung (Ed.), *Exercise and Sport Sciences Reviews* (pp. 389–441). New York: Macmillan.

Wilson, J. M. J., Robertson, D. G. E., & Stothart, J. P. (1988). Analysis of lower limb muscle function in ergometer rowing. *International Journal of Sport Biomechanics, 4*, (pp. 315–325).

Yang, S. M., Kayamo, et al. (1985). Dynamic changes of the arches of the foot during walking. In D. A. Winter, R. W. Norman, R. P. Wells, K. C. Hayes, & A. E. Patla (Eds.), *Biomechanics IX-A* (pp. 417-422). Champaign, IL: Human Kinetics.

Young, P. (1986). The athlete's foot: Structure and function. *NSCA Journal, 8*(6), 52–55.

CHAPTER 6

Biomechanical Relationships in the Trunk

PREREQUISITES

Concept Modules A, B, D, E
Chapters 1, 2, 3

CHAPTER CONTENTS

The axial skeleton includes the skull, the thorax, the vertebral column, and the pelvis. Collectively, the thorax, the vertebral column, and the pelvis are called the **trunk**.

6.1 The Skull

The skull incorporates 29 bones into its structure. Although the specific bones of the skull are of little concern in the study of body movement, a knowledge of the skull's anatomy is valuable in understanding the effects of impact on the head. Impact injuries may occur when the head strikes a hard surface during a fall. In sports such as football, soccer, wrestling, skateboarding, and stock-car racing, the factor of fast deceleration of the head during an accelerated falling movement adds to the danger of impact.

6.2 The Thorax

The thorax includes 25 bones: the sternum, which consists of the manubrium, the body, and the xyphoid process, and 12 pairs of ribs. The biomechanical function of the thorax is to absorb the forces of impact and to serve as an attachment for many of the muscles that control various movements of the entire trunk. Of course, the thorax is of considerable importance in respiratory mechanisms and in protection of the vital organs. The latter two functions are not considered here.

6.3 The Vertebral Column

Structure

Probably the most significant functional unit of the body—from the biomechanist's point of view—is the vertebral column, for it is this progressive series of bones that provides the framework and foundation for most of the movements of the body and extremities. It consists of a total of 33 bones, which are categorized by function into five groups. The groups are shown in Figure 6.1.

All the vertebrae are similar in general structure, but two factors change when the vertebrae are viewed from the cervical through the lumbar: (1) the structures become progressively larger; (2) the direction of the articulating surfaces changes.

Parts of a typical vertebra are shown in Figure 6.1b. The cylindrical body of the vertebra is the main load-bearing section of the spine. Radiating posteriorly from the body are two pedicles, which form the neural arch. The posterior aspect of the body and the two pedicles surround the spinal cord. The articular facets are flattish surfaces. There are four facets on each vertebra, two superior and two inferior.

The superior facets on one vertebra articulate with the two inferior facets on the next adjacent superior vertebra. These joints are synovial joints, are lined with articular cartilage, and are surrounded by a joint capsule. The facets in the cervical region are horizontal, and the articular surfaces are directed headward and footward, while facets in the lumbar area are oriented vertically and face medially and laterally. These different orientations allow varied degrees of movements in the three sections of the column and also assume different roles in dealing with tension, compression, and shear forces. Radiating from the vertebra are two transverse and one spinous process.

The spinous and transverse processes serve as handles for the attachments of the deep and superficial muscles of the back. Depending on the direction of the lines of action of these muscles, the force of pull on the processes may cause forward, backward, and lateral bendings, or small amounts of rotation of one

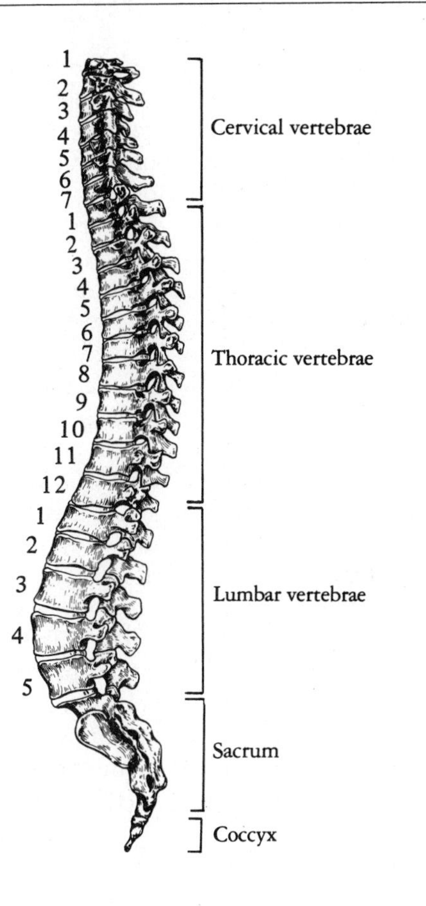

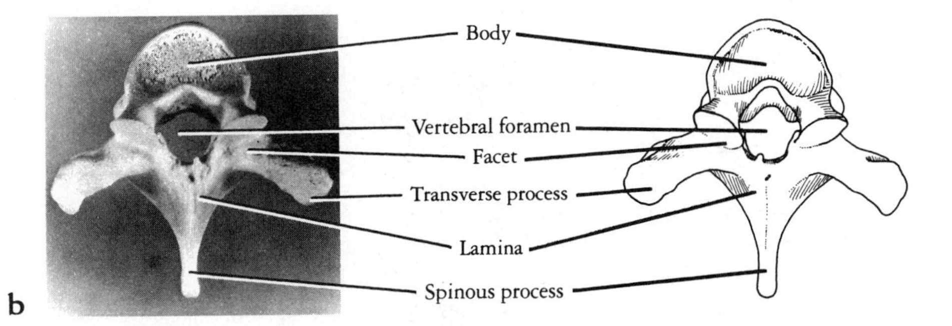

FIGURE 6.1
(a) Side view of the vertebral column. (b) Typical parts of a vertebra.

vertebra on the next, to make possible a fairly large amount of rotation of the trunk. The increasing size of the process down the vertebral column is an indication of the increased muscle size in the lower back and an indication of the line of action of the particular forces in bipedal movements. Comparatively, the relative sizes of the structures of the vertebrae in a dog, cat, elephant, or bison indicate the differing lines of force produced in quadripedal movements. The same relationship may be observed in the relative sizes of the bodies of the vertebrae, the lower vertebrae being the ones that sustain the most vertical compression during bipedal activities.

The articulations of the vertebral column are generally of two kinds: the articulations of one vertebral body on another, separated by the intervertebral discs; and the articulations between the articular facets of successive vertebrae. The invertebral discs allow slight movement but function mainly for shock absorption. The articular facets between the vertebrae are freely movable joints. Although the joint between any two adjacent vertebrae does not allow a great deal of motion, a series of vertebral joints has quite a large cumulative effect in producing forward flexion, extension, hyperextension, lateral flexion to both sides, and rotation. (Action is somewhat limited in the thoracic region as a result of the attachments of the ribs and the longer spinous processes on the thoracic vertebrae.)

Because it lacks great bony stability, the vertebral column relies to a large extent on ligamentous and muscular structures for support. The three types of soft tissue in and around the vertebrae are intervertebral discs between the bodies, joint capsules surrounding the articular facets, and ligaments.

The stability of the vertebral column is maintained by six kinds of ligaments that hold the bones together. These ligaments are shown in Figure 6.2. The anterior longitudinal ligament spreads along the anterior surfaces of the vertebral bodies and discs. It serves to pre-

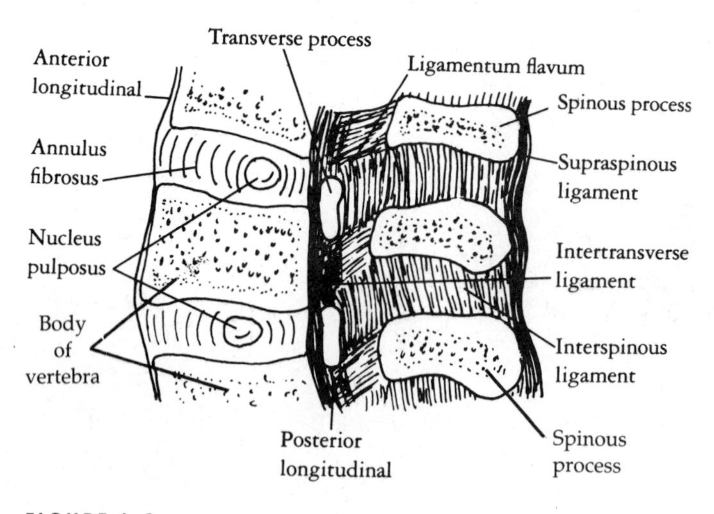

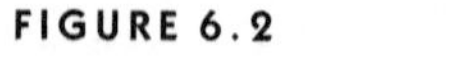
FIGURE 6.2

Sagittal cross section of the vertebral column and ligaments.

vent forward displacement of either structure. The posterior longitudinal ligament serves a similar function on the posterior sides of the vertebrae and discs. The intertransverse ligament holds the transverse process of one vertebra to the transverse process of the adjacent vertebrae on each side. The ligamentum flavum joins the laminae of successive vertebrae. The interspinous ligament joins the spinous processes of successive vertebrae, and the supraspinous ligament runs over the posterior ends of spines.

The ligaments can hold the vertebral column together, but continual reliance upon the ligaments, as a result of weak musculature or a strength imbalance between antagonistic groups of muscles, can result in a stretching of the ligaments and eventual damage to the discs, bodies, articular capsules, and spinous processes.

Muscular Stability of the Vertebral Column

The abdominal musculature acts to prevent the vertebral column from being continually hyperextended. The three abdominal muscles and their resulting directions of pull are shown in Figure 6.3a. The rectus abdominis acts to pull the anterior pelvis toward the sternum or to pull the thorax down toward the anterior pelvis, resulting in vertebral column flexion. In either case, the rectus abdominis has a large stabilizing component since its line of action is

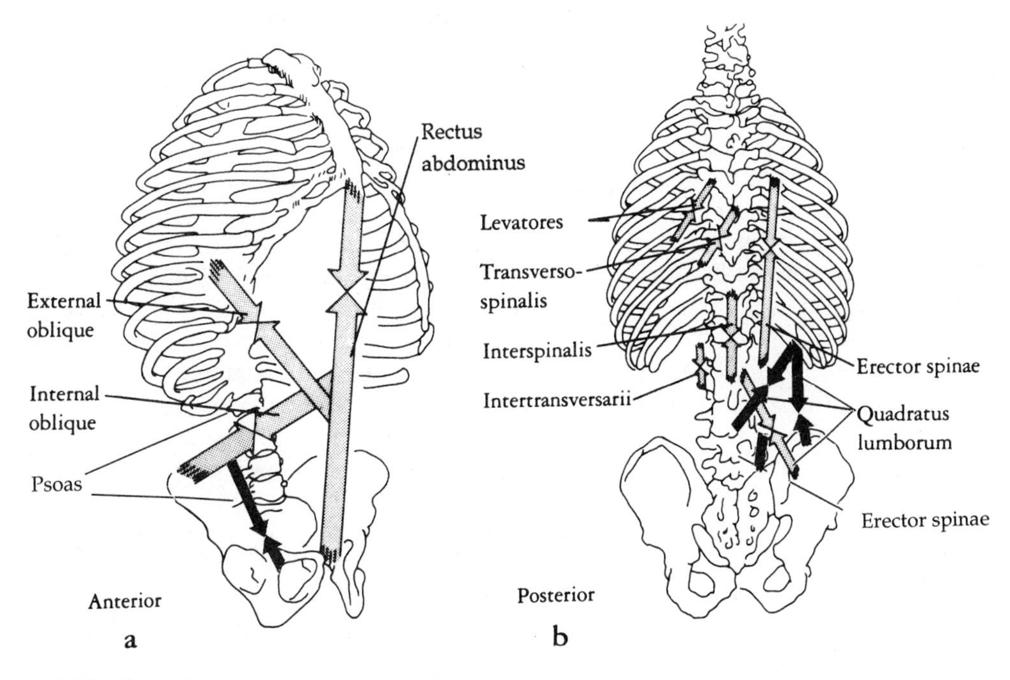

FIGURE 6.3
(a) Musculature of the anterior trunk. (b) Musculature of the posterior trunk.

parallel to the vertebral column itself. In addition to the abdominis, the psoas muscle originates on the transverse processes of the lumbar vertebrae, and joins the iliacus in its insertion on the lesser trocanter of the femur. While the psoas is a stabilizer of the lumbar vertebrae, it also pulls the lumbar vertebrae forward. If the lumbar vertebrae are pulled out from under those above this action will pull the lower back into hyperextension.

The posterior column is stabilized by several small muscles that run between the ribs, the vertebrae and, in some cases, the ilium of the pelvis. Some of the muscles are repeated for the cervical, thoracic, lumbar regions, and others are repeated between each of the adjacent vertebrae. These muscles are shown in Figure 6.3b.

The erector spinae group consists of several muscles that generally originate on the posterior ribs and insert on the lumbar vertebrae, sacrum, or ilium. When they contract, their stabilizing components will compress the vertebral column, thus helping to hold it together. The interspinalis group runs from one spinous process to the next; the intertransversarii muscles run from one transverse process to the next; the transversospinalis runs from the laminae of one vertebrae to the transverse processes of several of the underlying vertebrae on both sides; the levatores connect the transverse processes to the ribs; the rotatores connect the transverse processes inferiorly to the laminae superiorly; and the multifidi run on either side of the spinous processes along the length of the vertebral column.

Muscles of the Vertebral Column as Movers

The vertebral column is a multisegment system. The intersegmental movements between the vertebrae have small ranges of motion but combine to create larger ranges of motion of flexion, extension, lateral flexion, and transverse rotation of the entire trunk. The stabilizing muscles of the back all function as movers. The movements that they produce depend on the location of the origins and insertions. Although one can determine the exact movements caused by each muscle group, the posterior muscles will extend the vertebral column, the lateral muscles will laterally flex, and the muscles that originate either medially or laterally from their insertions will also cause rotation to either the contralateral (opposite) or ipsilateral (same) side depending on their exact lines of action.

The quadratus lumborum consists of three sections, one from the last rib to the iliac crest, one from the last rib and the transverse processes of the lumbar vertebrae, and one from the transverse processes of the first four lumbar vertebrae to the iliac crest. All three sections serve to flex the trunk laterally. The serratus posterior runs from the last four ribs to laminae of T11 to L2. It serves to laterally flex and rotate the trunk to the same side.

The psoas muscle on the anterior side originates from the transverse processes of the lumbar vertebrae on each side and the vertebral bodies of T12-L5 and runs laterally and inferiorly to meet the iliacus. The insertion of these two muscles is on the lesser trochanter on the medial aspect of the femur. Thus, when it contracts it serves to laterally flex, contralaterally rotate, and forward flex the trunk when the femur and pelvis are stabilized. Specifically, the psoas forward flexes the lumbar area. This action can ultimately result in hyperextension of the lumbar area by pulling the lumbar vertebrae forward, relative to the rest of the trunk. This action becomes important when analyzing abdominal strengthening exercises because the muscle along with the iliacus also flexes the hip. The three abdominal muscles serve to flex the trunk. In addition, the external oblique causes contralateral rotation by pulling the lateral rib cage downward and medially. The internal oblique causes ipsilateral rotation by pulling the medial abdominal wall laterally to the iliac crest.

In many sports, and especially during the performance of overarm throwing and striking skills, the vertebral column not only contributes force to the movement but serves to position the upper extremity for most effective use. The trunk does this through contralateral flexion and transverse rotation. Three common sport examples are given in Figure 6.4.

Of course, strengthening the contralateral flexors and transverse rotators is indicated; however, these muscles are often ignored in strength-conditioning programs. Moreover, the antagonist muscles to these trunk movers serve to decelerate the trunk during the later stages of the throw or strike and to stabilize the trunk so that it may serve as a stable base on which the shoulder girdle and upper extremity muscles pull. The strong eccentric muscle tensions used to decelerate and the isometric tensions used to stabilize are muscles to which attention should be given in any weight-training program.

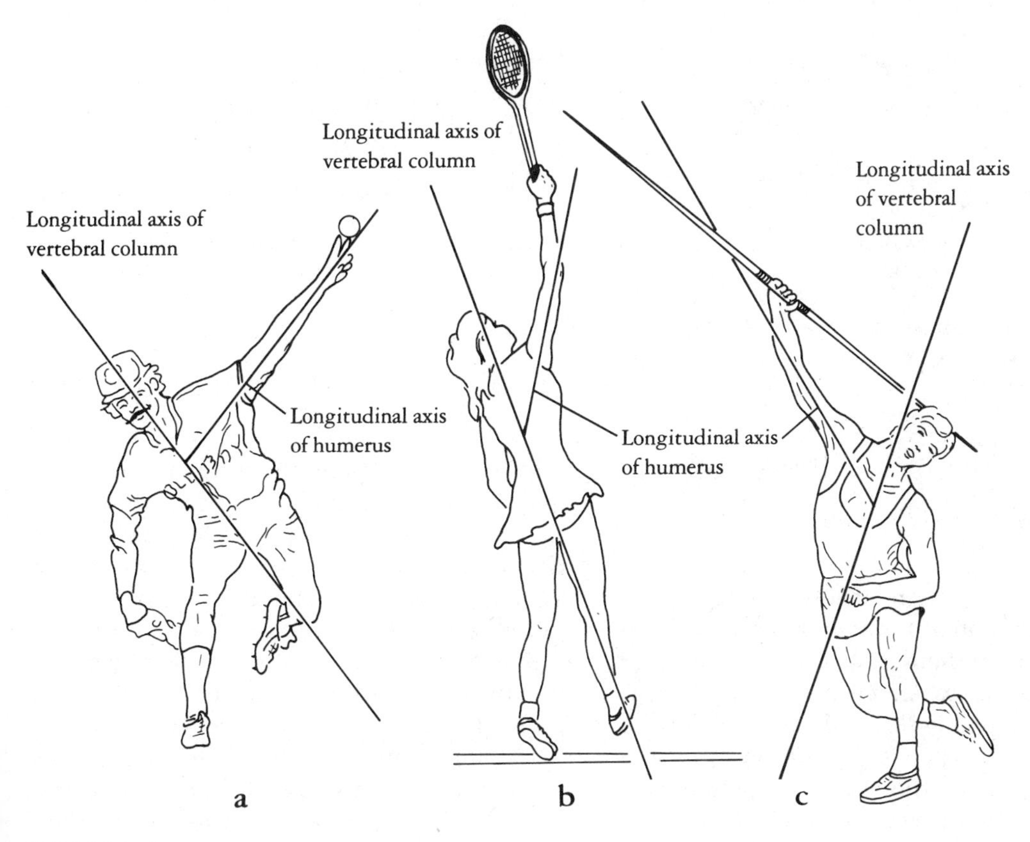

FIGURE 6.4
Sport activities in which the trunk repositions itself relative to the pelvis and lower extremity.

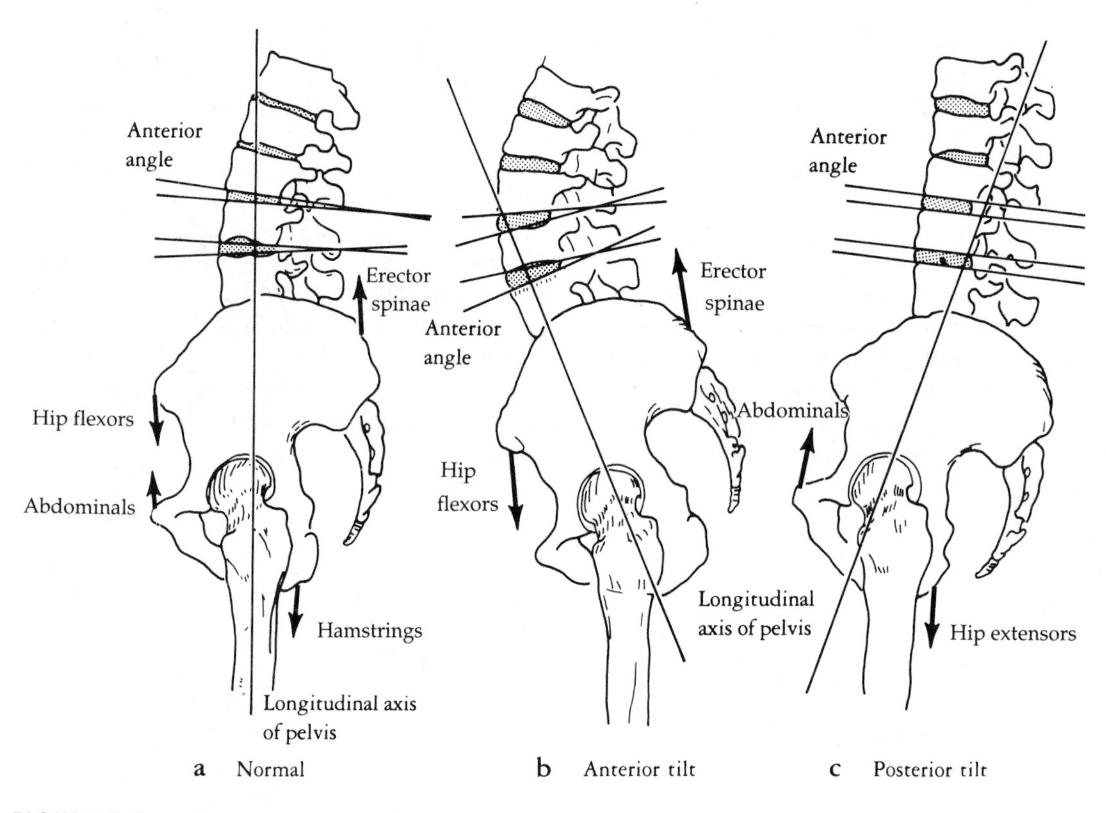

FIGURE 6.5

Pelvic inclinations. (a) Normal inclination. (b) Anterior tilt. (c) Posterior tilt.

Postural Alignment of the Vertebral Column and Pelvis

The pelvis is a segmental link bridging the vertebral column at the sacrum and the lower extremity at the femur. During weight bearing, the lower extremity is somewhat fixed. The pelvis is free to flex and extend relative to the femur. If the pelvis flexes or extends relative to the femur, the vertebral column must realign itself relative to the pelvis. Figure 6.5a illustrates the ideal vertebral–pelvis posture.

The major antagonistic muscle groups that play a role in aligning the trunk are the rectus abdominis, to prevent lumbar hyperextension; the erector spinae group, to prevent lumbar flexion; the iliacus and rectus femoris, to prevent hip extension; and the hamstrings and gluteals, to prevent hip flexion. The resistive functions of these muscles are listed because gravity will serve to move the vertebral column at the intralumbar articulations or the pelvis at the hip joint. Thus, the muscles serve as "antigravity" muscles to align these segments, and the muscles' relative strengths and flexibilities determine in part the segments's stability.

Three less than ideal relationships may also exist: lordosis, scoliosis, and kyphosis. The basis of these abnormalities in cases in which they are correctable is the balance or imbalance in the strength and flexibility of antagonistic muscles groups.

Lordosis

In the anterior-tilt position, the antero-superior iliac spines of the pelvis are located forward and downward from the ideal position, in which the antero-superior iliac spine and the symphysis pubis are in the same frontal plane. If part of a person's posture, this condition is called **lordosis**. The lordotic, or anterior-tilt position, is shown in Figure 6.5b. The hip joint is slightly flexed; therefore, the muscles that produce hip flexion are shortened in the anterior-tilt position. In addition, the muscles that flex the vertebral column, the abdominals, must be lengthened or at least relaxed enough to allow the pelvis to rotate forward and downward. The posterior muscles of the vertebral column and lower extremity are also affected by the anterior-tilt position. Since the hip is partially flexed, the hip extensors, the gluteals, and particularly the hamstring muscles are stretched, while the back muscles, those that extend the lumbar vertebrae, are shortened.

Generally, people who are predisposed to a lordotic posture are those with weak abdominals. Middle-aged people with large abdomens and pregnant women are particularly susceptible. If the lordotic position is maintained for a period of time, the muscles adapt to the posture, so that the back extensor and hip flexor muscles shorten and the hamstring and abdominal groups stretch, thus fixing the anterior tilt. The low back pain felt during pregnancy or in middle age is often a result of the anterior-tilt position, which causes compression of the posterior side of the intervertebral discs and a stretching of the anterior vertebral ligaments. The consequences of anterior tilt are intensified during lifting; during landing on the feet after being airborne, as in jumping; or during hitting another with your head and shoulders, as in the blocking action of football linemen; or pushing another, as in a standing wrestling posture.

The **posterior-tilt** position (Figure 6.5c) is brought about by or results in the opposite condition of the muscle groups involved. The hip flexor muscles and the low back muscles become stretched, while the abdominal and hamstring muscles are shortened. The posterior tilt is not as common as the anterior tilt and is rarely brought about by lack of muscular strength. The position is seen predominantly in females and has been likened to a typical model's posture. If the hip flexor muscles become stretched enough, the person begins to "ride on" the anterior hip joint ligaments. In addition, the posterior pelvic tilt puts the lumbar region in a precarious position. The backward leaning displaces the line of gravity of the trunk posteriorly, which increases the weight force arm and increases the resulting torque about the mediolateral (ML) axis. If the abdominal muscles are not strong enough to hold this position, the person tends to hang on the ligaments and spinous processes of the lumbar vertebrae and experiences pain and discomfort.

To correct an anterior-tilt position, one should strengthen the hip extensors and stretch the lower back and hip flexors. To correct a posterior-tilt position, one should strengthen the back extensors and hip flexors and stretch the hip extensors.

Scoliosis

A mediolateral curve of the vertebral column is called **scoliosis**. Functional scoliosis is caused by some mechanical problem (e.g., one short leg or muscle spasms on one side) and is gravity dependent. Functional scoliosis can be partially or entirely corrected by exercises or bracing. The most common and probably the most correctable scoliosis is caused by one short leg. The pelvis on the short side is dropped, and the pelvis on the long side is elevated relative to the sacrum. Thus, the lumbar vertebrae progress headward from the sacrum by being directed laterally to the short-leg side and curve back toward the midline around L1 and L2. A secondary curve then is forced to develop above the lumbar vertebrae as the column progresses headward. The secondary curve develops so that alignment can be maintained. An easy test for scoliosis is to drop a plumb line from the seventh cervical vertebra as the person is standing erect and note the deviations on either side down to the sacrum. Some leg length discrepancies may be corrected with a lift in the heel of one shoe.

If minor deviations are present, exercises that stretch the concave side and strengthen the convex side of the curve usually produce some relief. If rotation of the vertebral column is also present, the appropriate oblique abdominals must be strengthened.

Excessive scoliosis, particularly when vertebral rotation is also present, is not within the realm of treatment for the layperson or physical educator. If tests reveal major deviations, the person should be referred to a physician.

Kyphosis

An exaggerated anterior-posterior curvature of the vertebral column is termed **kyphosis**. Most frequently, it is an excessive forward bending of the thoracic area. Kyphosis occurs in older adults, usually women, and hence has been associated with osteoporosis and osteoarthritis. Some believe that kyphosis is possibly hereditary although this has not been substantiated. The term *flat back* is used frequently in conjunction with a kyphotic condition and is derived from the fact that with an exaggerated curvature of the thoracic vertebral column is an associated reduction of the natural lumbar curvature. To put the vertebral column and pelvis in alignment, a thoracic kyphosis that induces lumbar flattening creates a posterior-tilt condition of the pelvis so that the line of gravity passes through the middle of the base of support rather than toward the toes. As with posterior pelvic tilt, the condition creates a shortening of the hip extensors and a stretching of the hip flexors.

In gravity-induced cases of kyphosis, a strengthening of the vertebral column extensors and a stretching of the vertebral column flexors, particularly in the thoracic area, are advised. If a posterior-tilt condition coexists, a strengthening of the hip flexors and a stretching of the hip extensors are also appropriate.

Although the term *round shoulders* is sometimes associated with kyphosis of the thoracic vertebrae, it is not the same condition. A round-shoulder condition is technically abduction or protraction of the scapula. One may have abducted scapulae as a postural misalignment without having a kyphotic condition. With kyphosis, however, the forward bending of the thoracic vertebrae produces a greater gravitational force component on the shoulder girdle such that it works to abduct (protract) the shoulder girdle over time unless the retractor muscles continually counteract the movement.

To prevent a round-shoulder condition, one would strengthen the shoulder girdle adductors (retractors) and stretch the shoulder girdle abductors (protractors).

Abdominal Strengthening Exercises

Through knowledge of the relationship between the vertebral column, pelvis, and thigh and how the movement of one segment affects the other parts, one can comprehend the important factors to consider in selecting abdominal exercises. Physical educators and coaches prescribe abdominal exercises with unceasing regularity, and yet all exercises are not proper for all groups of individuals.

Recall that the pelvis is the segmental link between the upper body and the lower body. In a general way, it acts as the fulcrum, while the upper and lower bodies act as movable levers. Also keep in mind that a muscle group may produce movement at the proximal or distal segment. Consequently, when the body is supine, the abdominals may operate to move the thorax if the pelvis is fixed, or they may operate to move the pelvis toward the thorax if the thorax is fixed. The hip flexors may act to move the thigh if the pelvis is fixed, or they may move the pelvis if the thigh is fixed. Thus, abdominal exercises may create any number of segmental positions, depending on which segments are stabilized and which are movable. Body parts may be stabilized by using external forces, such as objects or other people; in many cases, stabilization may be obtained through the use of muscle force. By observing the series of illustrations in Figure 6.6, the torques and resultant movements for several common abdominal exercises may be analyzed.

In Figure 6.6, the forces acting on the upper and lower segments are illustrated. By multiplying the weights of these segments by their respective perpendicular distances from the fulcrum (in this case, the pelvis at the coccyx), the resistance torques can be determined. The muscles involved are the hip flexors and the abdominal muscles acting to flex the vertebral column. The fingers are placed at the sides of the head in all cases. Hyperflexion of the cervical area should be discouraged in all cases.

Figure 6.6a is the sit-up from the long-lying supine position. Ideally, as one sits up from this position, the abdominals work to flex the vertebral column, and the hip flexors work to prevent the pelvis from being pulled into a posterior-tilt position by the action of the abdominals. As the upper body rises, nearing the vertical position, the hip flexor muscles flex the hip joint by pulling the pelvis upward. If this sequence of events is accomplished as described, then this exercise for abdominal strength will be an effective one. Consider the requirements, however.

First, the abdominal muscles must be strong enough to flex the vertebrae to lift the trunk and strong enough to counteract the torques produced by the hip flexors and in particular, the psoas, in attempting to pull the pelvis into a flexed position to the thigh. If the abdominals are not strong enough to accomplish this, the pelvis initially flexes at the hip joint, which puts the lumbar vertebrae into a hyperextended position. Not only does this hyperextension compress the posterior vertebrae, but the performer probably exercises the hip flexors more than the abdominals in this sequence.

A similar situation occurs during leg lifts, as shown in Figure 6.6b. The hip flexors contract to lift the lower extremity. The abdominals must stabilize the pelvis so that it is not tilted forward by the proximal attachments of the hip flexor group, and the abdominals must, therefore, maintain a strong static (isometric) contraction. If the abdominals are not strong enough to stabilize the pelvis on the vertebral column, the pelvis flexes and puts the low back into a hyperextended position. Thus, the performer must have quite strong abdominals in the first place to accomplish leg lifts without injury to the back.

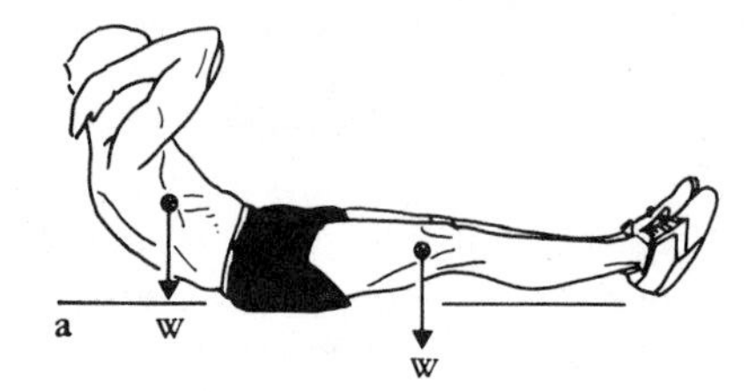

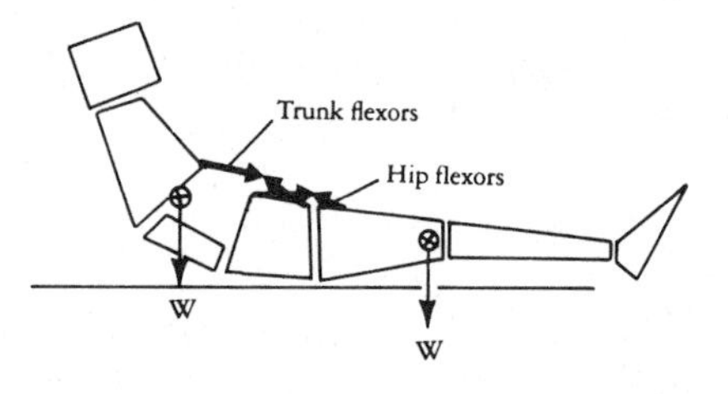

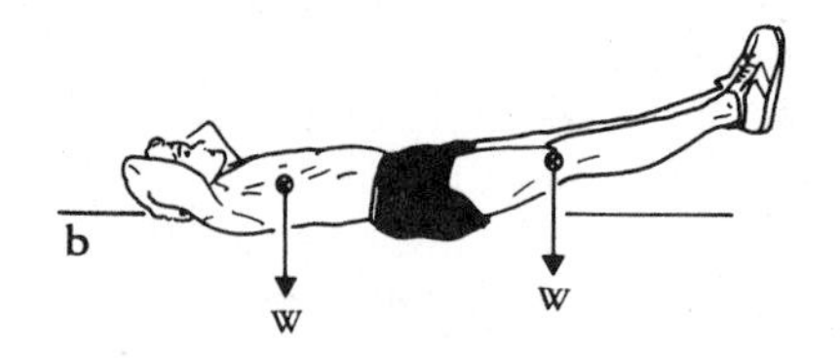

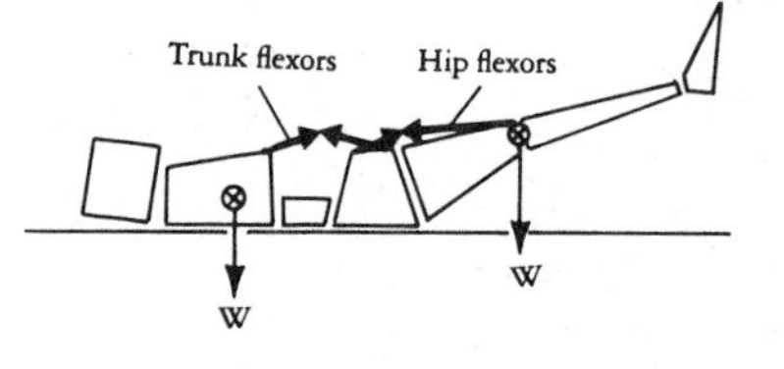

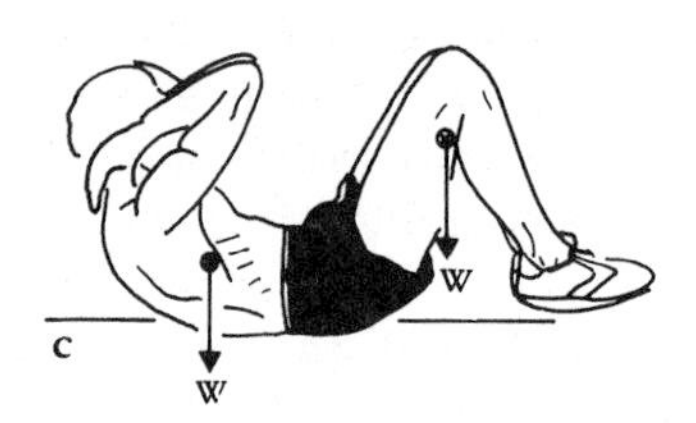

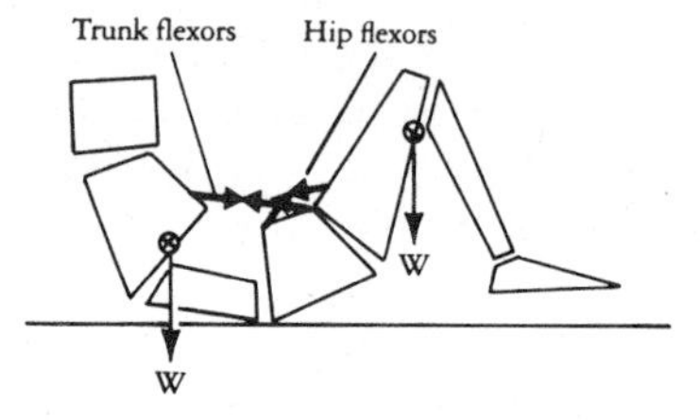

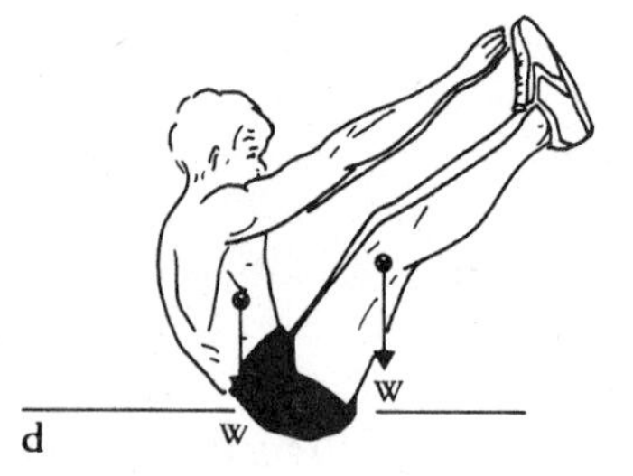

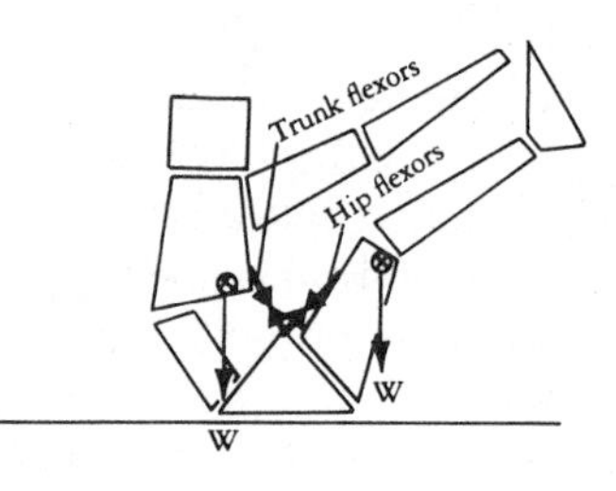

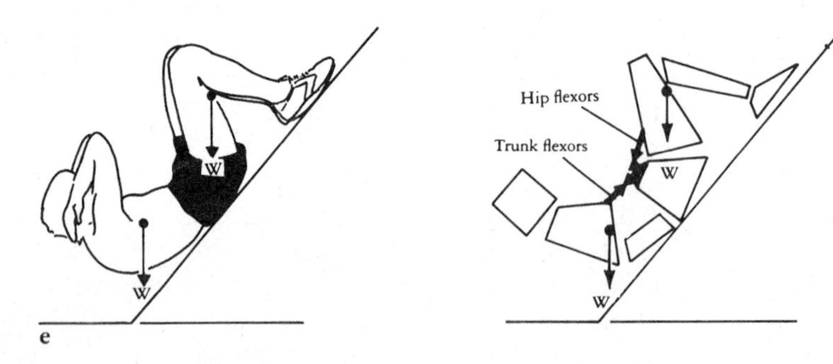

FIGURE 6.6

Abdominal exercises. (a) Sit-up. (b) Leg lifts. (c) Curl-up. (d) Snap-up. (e) Curl-up on an inclined board.

Of course, the heavier or longer the legs or both, the greater their resistive torque and the greater the motive torque necessary to lift them; therefore more stabilization is needed to hold the pelvis in line with the trunk. By putting the hands or a pillow under the pelvis, a posterior-tilt position is achieved, giving a better angle of attachment of the abdominals to the pelvis and helping them stabilize the pelvis.

In a third exercise called a curl-up (Figure 6.6c), the lower extremity is placed in a flexed position, thereby shortening the hip flexor muscles and reducing their contractile capabilities as a result of the length–tension relationship. In this manner, the abdominals contract to lift the trunk, the hip flexors stabilize the pelvis somewhat, and the body curls up without the danger of hyperextension of the lumbar vertebrae occurring because of early pelvic flexion at the hip joint.

The physical condition of the performer dictates how safe and effective these exercises will be in strengthening the abdominals. No exercise should be continued if the performer experiences pain when attempting it. For a performer with weak abdominals, the hip-flexed position is the best. In addition, the performer may bring the arms down to the sides, thereby reducing the resistance torque of the upper body, and curl into the upright position by lifting only the head and neck before attempting to lift the thorax region, the lumbar region, and, finally, the pelvis.

A word of caution: in timed events when the goal is the greatest number of repetitions, even the moderately strong may be forced into the hyperextended position toward the end of the time period, because the abdominals may become weakened by fatigue.

Two advanced methods that may be used to increase the necessary torque of the abdominals are snap-ups (Figure 6.6d) and sit-ups on an inclined board (Figure 6.6e). The inclined board places the body in such a position that the resistance torque is acting over a greater range of motion (ROM); that is, when the performer's upper body is perpendicular to the lower body in a floor sit-up, the resistance torque approaches zero because the line of action of the gravity force is through the fulcrum (coccyx). On an inclined board, however, the vertebral column must achieve a greater flexion before the resistance torque approaches zero, and thus the abdominals must be contracted over a greater ROM.

The snap-up, an exercise combining hip and knee flexion with simultaneous vertebral flexion, increases the torque that must be produced by the abdominals. The hip flexors must act with a large torque to produce the greater rotational acceleration of the lower extremity as it flexes at the hip. Hence, the abdominals, in addition to flexing the vertebrae, must also counteract that torque produced by the hip flexors to stabilize the pelvis and prevent it from rotating upward around the hip.

Electromyographic studies of the abdominal and hip flexor muscles during various abdominal exercises show that lumbar hyperextension occurs before the execution of all sit-ups: long-lying, hook-lying, legs-elevated, and curl-up positions (Ricci, Marchetti, and Figura, 1981).

One method of reducing the possibility of lumbar hyperextension even further is to perform curl-ups with the hips and knees flexed to 90 degrees with the legs resting on a bed, chair, or a stack of mats. Thus, the pelvis is prevented from flexing to a greater extent than when performing hook-lying curl-ups. The common term for this exercise is the "crunch." Although it appears to alleviate the lumbar problem, if performed with the hands behind the head, it can encourage hyperflexion of the cervical and thoracic areas.

Changes in the Vertebrae With Maturation and Advancing Age

The S-shaped curves in the vertebral column of an adult are not present in the infant; they develop as the infant progresses through the crawling and walking stages of maturation. Gravitational lines of force acting on the bipedal structure change the vertebral alignment until eventually a mature structure has been created. A lordosis (convex) curve is developed at the cervical and lumbar regions, and a kyphosis curve (concave) is located at the thoracic and sacral areas.

In the bipedal position, the vertebral column supports the weight through the bodies and discs of the vertebrae and through the articular facets between the vertebrae. King and Yang (1986, p. 213) describe the structure:

> *The load was transmitted by the bottoming out of the tip of the inferior facet on the pars interarticularis of the vertebra below. The spine is thus composed of a stack of three-legged stools, one leg of which (the disc) bears more load than the other two (the facets).*

Because the articular capsules surrounding the facet joints contain an abundant supply of nerves, it has been thought that low back pain may center around these areas (Ashton-Miller and Schultz, 1988).

Through the years, the vertebral discs dehydrate. In fact, during the course of the day, the discs dehydrate under the weight of the body, and consequently, one is shorter at the end of the day than at the beginning. Disc degeneration begins and continues from the 20s throughout life. Although this "natural" degeneration does not cause back pain itself, as one ages, one loses strength and flexibility as well. Protrusions of a vertebral disc, commonly known as a slipped disc, can cause pressure pain on the nerves, usually on the posterior side of the vertebral disc. Whether this protrusion is caused by disc degeneration, fatigue failure, or eccentric loading from lifting is not clear; however, all have been implicated.

Because of the secondary curves of the column, an anterior angle is created between each of the cervical and lumbar vertebrae. If the muscular stability is not maintained to support the column anteriorly, the cervical or lumbar region may increase its curvature, thereby increasing the anterior angles. If the angles are greatly increased, the discs may be forced anteriorly, causing pressure on the surrounding anterior longitudinal ligament.

During the course of a lifetime, continual bending, twisting, and lifting can erode the vertebrae's ability to accommodate (Ashton-Miller and Schultz, 1988). The ability of a body to withstand repetitive loads before failing is a mechanical property called **fatigue**. With continual lifting activities, the probability of fatigue failure of the vertebral end-plates increases. Lifing subjects the vertebrae to loads far in excess of the weight of the body or that of the load lifted as a result of compressive components of force of the muscles involved.

Three factors that increase the compressive forces on the vertebra are the amount of weight that is lifted, the distance that the weight is held from the vertebral column, and the amount of trunk flexion used when lifting. Without flexing forward, one may expect the compression force to be at least three times the weight of the object and superior body weight.

Low back pain (LBP) plagues a high number of people at times throughout their lives. Although it is a common ailment, it is idiopathic because it is hard to determine or locate the exact cause. The upright posture, the tendency to gain weight in the abdominal region, and the tendency to neglect the strength of the abdominal muscles combine to make anterior tilt of the pelvis and hyperextension of the lumbar vertebrae a common pathology. Problems may occur at the lumbosacral articulation or at the sacroiliac; however, the lumbosacral articulation is far more susceptible to pathologies. Although the term *sacroiliac* has come into common usage to describe LBP, it is usually only with some type of impact that the integrity of this joint is lost.

Back Pain in Athletes

High-impact sports such as football, wrestling, and gymnastics and sports that involve high-force repetitive lifting, pushing, and pulling have been implicated in the high incidence of back pain. Although athletes have well-conditioned bodies, so that typical back problems resulting from low strength and flexibility surrounding the vertebral column are not a general problem, high impacts such as landing from heights may stress the integrity of the supporting tissues.

In a study of retired wrestlers and heavyweight lifters, Granhed and Morelli (1988) found that the lifetime prevalence of low back pain was higher among wrestlers than among the control group. Also, the wrestlers had a higher incidence of old vertebral factures. The heavyweight lifters had a greater decrease in disc height than controls. Both groups displayed signs and symptoms of impaired back mobility.

Ashton-Miller and Schultz (1988) report that Brinckmann et al. estimated that 2 weeks of athletic conditioning could impose 5,000 load cycles on lumbar vertebrae through lifting. The probability of fractures of the endplates of the vertebral bodies resulting from fatigue failure is estimated to be 36% if lifting is done at a 30%–40% maximum load and 92% at a 60%–70% maximum load. Furthermore, they report that 8% probability of failure was evident at the higher percentage of maximum loads after only 10 repetitions. These levels and repetitions are not uncommon for most athletes in a conditioning program.

Impact from landing has been implicated in those athletes who are subjected to such. Hall (1986) studied lumbar hyperextension and impact forces in female gymnasts performing front and back walkovers and handsprings and the handspring vault. Although the handspring vault produced the highest vertical impact forces, the back handspring and walkover required the greatest amounts of back hyperextension. Maximum lumbar hyperextension occurred very close to the time that the impact force was sustained by the hands or the feet. Clearly, vertebral endplate fracture and pathologies associated with the articular facets and the surrounding bone are related to fatigue as well as to the magnitude of those forces encountered. Female gymnasts are at greater risk as long as lumbar hyperextension plays such a large role in the performance of the skills. Interestingly, no reports have been found regarding back problems of male gymnasts.

Two specific low back problems associated with athletes are spondylolysis, a defect in the pars interarticularis in the neural arch of the vertebra, and spondylolisthesis, a condition in which one vertebra slips forward relative to the vertebra below it. L5 is usually the affected vertebra but others have been affected as well. Jackson et al. (1976) found spondylolysis in 11% of female gymnasts, whereas the control group reported 2.3%. In a variety of male athletes, spondylolysis occurred 3.4 times more frequently as nonathletes. Of course, young athletes whose neural arches are not yet ossified are more susceptible.

Because of the flexed position of the lumbar area in the ready position of football linemen and because the drive forward and upward results in hyperextension of the lumbar area during the collision of blocking and tackling, an extreme shearing force is placed on the intervertebral structures. McCarroll, Miller, and Ritter (1986) found that football players in general had a greater incidence (15.2%) of spondylolysis and spondylolisthesis than a control group (6.0%). Other studies report a 50% incidence for interior linemen, 21% for football players in general. Of the collegiate players showing spondylolysis, only 2.4% of them developed it during their college years. A majority of these players developed the problem during adolescence, thus emphasizing the importance of care before ossification.

Hyperflexion was related to LBP in a study by Howell (1984) on women rowers. Seventy-five percent of these rowers demonstrated hyperflexion of the lumbar area, and 86% suffered LBP. It is estimated that 25%–30% of females in the normal population of this age group experience LBP. The strong back extension pulling motion while the trunk is hyperflexed, as rowers experience, can produce excessive intervertebral disc pressures. Furthermore, the reduction of the anterior angle may force a posterior protrusion of the disc, which places pressure on the nerves and consequently, produces LBP. For rowers in particular, special consideration should be given to muscle strength and ligament flexibility imbalances because in rowing the single oar forces inboard and outboard extremities and consequently, right to left strength and flexibility differences. Holden and Jackson (1985) report stress fractures in the lower ribs of female scullers, which they attribute to the continual protraction and retraction of the shoulder girdle. However, the serratus posterior inferior attached to the last four ribs and the spinous processes of T12-L3 is used to aid forced expiration. It is conceivable that right to left muscle imbalances in the serratus posterior inferior could pull the attached vertebrae to one side or the other.

Understanding the Trunk

1. On a partner, identify the cervical, thoracic, lumbar, and sacral sections of the vertebral column. Demonstrate isolated flexion of the cervical vertebrae, of the thoracic vertebrae, and of the lumbar vertebrae.
2. On a skeleton, observe the changes in the sizes and shapes of the vertebrae from the cervical to the lumbar sections. Discuss the functional significance of the changes.
3. Identify the lumbosacral and the sacroiliac articulations. What movements occur in each?
4. Discuss the back trauma impending on the weight lifter shown in Figure 6.7.
5. Figure 6.8 illustrates two people performing backward stunts. Discuss the difference in the vertebral column positions. Which tends to be more injurious? Why?
6. Discuss the movements of the vertebral column and hip in a dolphin kick in swimming in terms of the movements that occur and the sequence in which they occur (see Figure 12.24, p. 500).
7. Observe the S-shaped curves of the vertebral column in the sagittal plane. Between which two lumbar vertebrae is the greatest anterior angle located? Which muscles act to prevent this angle from increasing? Relate the lumbar angle to back injury.

FIGURE 6.7
A weight lifter.

8. For the following activities, describe the contribution of the vertebral column (i.e., what movement is used): (a) a flop-style high jump, (b) a back handspring, (c) a tennis backhand stroke, (d) a golf swing.

9. Discuss the differences among the long-leg sit-up, the hook-lying sit-up, the curl-up, and leg lifts in terms of the following:

   a. The resistive torques

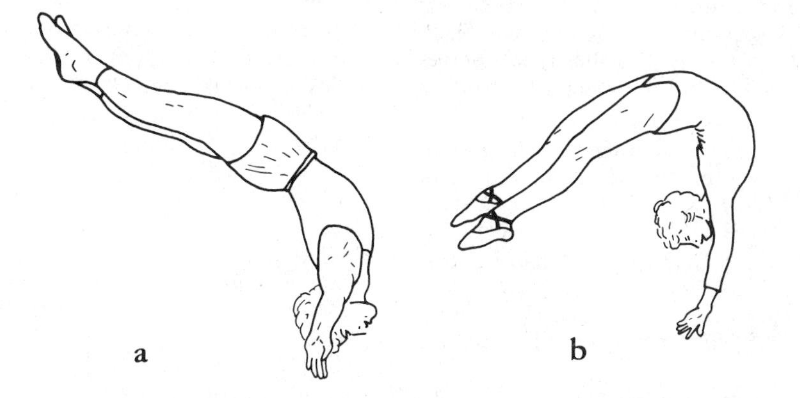

FIGURE 6.8
Body positions during (a) a back layout somersault and (b) a back handspring.

b. The muscle forces and their lines of action for those crossing the hip and for those crossing the vertebral column

c. The stabilization function of the abdominal muscles

d. The relative effectiveness and appropriateness of each exercise for (1) people with weak abdominals, (2) people with strong abdominals

e. The effect of introducing the time factor in sit-up tests, in relation to potential injury of the back

References and Suggested Readings

Anderson, C. K., Chaffin, D. B., Herrin, G. D., & Matthews, L. S. (1985). A biomechanical model of the lumbosacral joint during lifting activities. *Journal of Biomechanics, 18*, 571–584.

Anderson, G. B. L., Ortengren, R., Nachemson, A. L., & Schultz, A. B. (1983). Biomechanical analysis of loads on the lumbar spine in sitting and standing postures. In H. Matsui & K. Kobayashi (Eds.), *Biomechanics VIII-A: Proceedings of the Eighth International Congress on Biomechanics* (pp. 543–552). Champaign, IL: Human Kinetics.

Andersson, E., Sword, L., & Thorstensson, A. (1980). Trunk muscle strength in athletes. *Medicine and Science in Sports and Exercise, 20*, 587–593.

Andersson, G. B. J. (1985). Loads on the lumbar spine: In-vivo measurements and biomechanical analyses. In D. A. Winter, R. W. Norman, R. P. Wells, K. C. Hayes, & A. E. Patla (Eds.), *Biomechanics IX-B* (pp. 32–37). Champaign, IL: Human Kinetics.

Ashton-Miller, J. A., & Schultz, A. B. (1988). Biomechanics of the human spine and trunk. In K. B. Pandolf (Ed.), *Exercise and Sport Sciences Reviews* (pp. 169–204). New York: Macmillan.

Bachrach, R. M. (1986). Diagnosis and management of dance injuries to the lower back: An osteopathic approach. In C. G. Shell (Ed.), *The Dancer as Athlete* (pp. 83–90). Champaign, IL: Human Kinetics.

Barlow, D. A., & Neeves, R. E. (1978). Biomechanical assessment of partial iliopsoas isolation in women and its implications for athletic training. In F. Landry & W. A. R. Orban (Eds.), *Biomechanics of Sports and Kinanthropometry: Proceedings of the International Congress of Physical Activity Sciences* (pp. 53–60). Miami: Symposia Specialists, Inc.

Bejjani, F. J., Gross, C. M., & Pugh, J. W. (1984). Model for static lifting: Relationship of loads on the spine and the knee. *Journal of Biomechanics, 17*, 287–298.

Cappozzo, A., & Berme, N. (1985). Loads on the lumbar spine during running. In D. A. Winter, R. W. Norman, R. P.. Wells, K. C. Hayes, & A. E. Patla (Eds.), *Biomechancis IX-B* (pp. 97–100). Champaign, IL: Human Kinetics.

Cappozzo, A., Felici, F., Figura, F., & Gazzani, F. (1985). Lumbar spine loading during half squat exercises. *Medicine and Science in Sports and Exercise, 17*, 613–620.

Flint, M. M. (1965). An electromyographic comparison of the function of the iliacus and the rectus abdominis muscles. *Journal of the American Physical Therapy Association, 45*, 248–253.

Granhed, H., & Morelli, B. (1988). Low back pain among retired wrestlers and heavyweight lifters. *The American Journal of Sports Medicine, 16*, 530–533.

Hall, S. J. (1985). Effect of attempted lifting speed on forces and torque exerted on the lumbar spine. *Medicine and Science in Sports and Exercise, 17*, 440–444.

Hall, S. J. (1986). Mechanical contribution to lumbar stress injuries in female gymnasts. *Medicine and Science in Sports and Exercise, 18*, 599–602.

Harms-Ringdahl, K., & Ekholm, J. (1987). Influence of arm position on neck muscular activity levels during flexion–extension movements of the cervical spine. In B. Jonsson (Ed.), *Biomechanics X-B* (pp. 249–254). Champaign, IL: Human Kinetics.

Holden, D. L., & Jackson, D. W. (1985). Stress fracture of the ribs in female rowers. *The American Journal of Sports Medicine, 13*, 342–347.

Howell, D. W. (1984). Musculoskeletal profile and incidence of musculoskeletal injuries in lightweight women rowers. *The American Journal of Sports Medicine, 12*, 278–282.

Jackson, E. W., Wiltse, L. L., & Civincione, R. J. (1976). Spondylolysis in female gymnasts. *Clinical Orthopaedics, 117*, 68–73.

Kennedy, J. C. (1978). Orthopaedic manifestations. In B. Eriksson & B. Furberg (Eds.), *Swimming Medicine IV: Proceedings of the Fourth International Congress on Swimming Medicine* (pp. 93–100). Baltimore: University Park Press.

King, A. I., & Yang, K.-H. (1986). Biomechanics of the lumbar spine. In S. L. Y. Woo & Y. C. Fung (Eds.), *Frontiers in Biomechanics* (pp. 211–224). New York: Springer-Verlag.

Klausen, K., Nielson, B., & Modsen, L. (1981). Form and function of the spine in young males with and without back trouble. In A. Morecki, K. Fidelus, K. Kedzior, & A. Wit (Eds.), *Biomechanics VII-A: Proceedings of the Seventh International Congress on Biomechanics* (pp. 174–180). Baltimore: University Park Press.

Kumar, S., & Turner, A. A. (1985). EMG of erectors spinae in structured lifting tasks. In D. A. Winter, R. W. Norman, R. P. Wells, K. C. Hayes, & A. E. Patla (Eds.), *Biomechanics IX-B* (pp. 9–14). Champaign, IL: Human Kinetics.

Leskinen, T. P. J., Takla, E. P., Stalhammar, H. R. Kuorinka, I. A. A., & Troup, J. D. G. (1985). Hip torque, lumbosacral compression, and intraabdominal pressure in lifting and lowering tasks. In D. A. Winder, R. W. Norman, R. P. Wells, K. C. Hayes, & A. E. Patla (Eds.), *Biomechanics IX-B* (pp. 55–59). Champaign, IL: Human Kinetics.

Leskinen, T. P. J., Takala, E-P., & Stalhammar, H. R. (1987). Lumbar and pelvic movements when stooping and lifting. In B. Jonsson (Ed.), *Biomechanics X-B* (pp. 195–200). Champaign, IL: Human Kinetics.

LeVeau, B. F. (1974). Axes of joint rotation of the lumbar vertebrae during abdominal strengthening exercises. In R. C. Nelson & C. A. Morehouse (Eds.), *Biomechanics IV: Proceedings of the Fourth International Seminar on Biomechanics* (pp. 361–364). Baltimore: University Park Press.

Lloyd-Smith, R., Clement, D. B., & McKinzie, D. C. (1985). A survey of overuse and traumatic hip and pelvis injuries in athletes. *The Physician and Sports Medicine, 13*, 131–141.

Marymont, J. V., Lynch, M. A. & Hennings, C. E. (1986). Exercise-related stress reaction of the sacroiliac joint. *The American Journal of Sports Medicine, 14*, 320–323.

McCarroll, J. R., Miller, J. M., & Ritter, M. A. (1986). Lumbar spondylolysis and spondylolisthesis in college football players. *American College of Sports Medicine, 14*, 404–405.

McGill, S. M. & Norman, R. W. (1985). Dynamically and statically determined low back moments during lifting. *Journal of Biomechanics, 18*, 877–886.

Metzmaker, J. N., & Pappas, A. M. (1985). Avulsion fractures of the pelvis. *The American Journal of Sports Medicine, 13*, 349–356.

Miller, J. A. A., Schultz, A. B., Warwick, D. N., & Spencer, D. L. (1986). Mechanical properties of lumbar spine motion segments under large loads. *Journal of Biomechanics, 19*, 79–83.

Mutoh, Y. (1978). Low back pain in butterfliers. In B. Eriksson & B. Furberg (Eds.), *Swimming Medicine IV: Proceedings of the Fourth International Congress on Swimming Medicine* (pp. 115–123). Baltimore: University Park Press.

Mutoh, Y., Mari, T., Nakamura, Y., & Mujashita, M. (1983). The relation between sit-up exercises and the occurrence of low back pain. In M. Matsui & K. Kobayashi (Eds.), *Biomechanics VIII-A: Proceedings of the Eighth International Congress on Biomechanics* (pp. 180–185). Champaign, IL: Human Kinetics.

Neeves, R. E., & Barlow, D. A. (1975). Torque, work and power differences in bent-knee straight-knee sit-ups in women. *Medicine and Science in Sports, 7*(1), 77.

Panjabi, M. M. (1985). The human spine; story of its biomechanical functions. In D. A. Winter, R. W. Norman, R. P. Wells, K. C. Hayes, & A. E. Patla (Eds.), *Biomechanics IX-A* (pp. 219–223). Champaign, IL: Human Kinetics.

Pope, M. H., Andersson, G. B. J., Broman, H., Svensson, M., & Zetterberg, C. (1987). Electromyography of the lumbar trunk musculature with axial torque development. In B.

Jonsson (Ed.), *Biomechanics X-B* (pp. 177–182). Champaign, IL: Human Kinetics.

Ricci, B., Marchetti, M., & Figura, F. (1981). Biomechanics of sit-up exercises. *Medicine and Science in Sports and Exercise, 13*(1), 54–59.

Roozbazar, A. (1974). Biomechanics of lifting. In R. C. Nelson & C. A. Morehouse (Eds.), *Biomechanics IV: Proceedings of the Fourth International Seminar on Biomechanics* (pp. 37–43). Baltimore: University Park Press.

Rovere, G. D. (1987). Low back pain in athletes. *The Physician and Sports Medicine, 14*, 105–117.

Schlotten, P. J. M., Veldhuizen, A. G., & Sterk, J. C. (1987). The bending stiffness of the human trunk in vivo. In B. Jonsson (Ed.), *Biomechanics X-B* (pp. 213–218). Champaign, IL: Human Kinetics.

Takala, E.-P., Leskinen, T. P. J., & Stalhammer, H. R. (1987). Electroymographic activity of hip extensor and trunk muscles during stooping. In B. Jonsson (Ed.), *Biomechanics X-B* (pp. 255–258). Champaign, IL: Human Kinetics.

Thorstensson, A., Oddsson, L., Andersson, E., & Arvidson, A. (1985). Balance in muscle strength between agonist and antagonist muscles of the trunk. In D. A. Winter, R. W. Norman, R. P. Wells, K. C. Hayes, & A. E. Patla (Eds.), *Biomechanics IX-B* (pp. 15–20). Champaign, IL: Human Kinetics.

Vincent, W. J., & Britten, S. D. (1980). Evaluation of the curl-up—a substitute for the bent knee sit-up. *JOPERD, 51*, 74–75.

Volski, R. V., Bourguignon, G. J., & Rodriguez, H. M. (1986). Lower spine screening in the shooting sports. *The Physician and Sports Medicine, 14*, 101–106.

Wielki, C. (1987). Anatomical functional deviation of the spinal curves of athletes. In B. Jonsson (Ed.), *Biomechanics X-B* (pp. 567–574). Champaign, IL: Human Kinetics.

CHAPTER 7

Application of Biomechanics to Neuromuscular Fitness Activities

PREREQUISITES

Concept Modules A, B, C, D, E
Chapters 1, 2, 3, 4, 5, 6

CHAPTER CONTENTS

7.1 Aspects of Fitness

Biomechanical and physiological understandings form a base for the selection and evaluation of activities used for improving or maintaining personal fitness. People engage in fitness activities for psychophysical well-being or to modify the level of caloric expenditure for weight control. Individuals who need rehabilitative movement therapy following injury or inactivity and performers who wish to increase specific aspects of fitness or to raise stress tolerances for improving skill performance also engage in various forms of fitness activities.

The nature of the movements selected for accomplishing the special purposes of each individual varies widely. Most common are exercises designed to improve flexibility of tissues for increased range of motion (ROM) of the joints, muscular strength, muscular power, muscular endurance, and cardiovascular-respiratory function. The specific movement activity chosen for a given aspect of fitness should be performed according to the best available guidelines reported by those who have investigated that fitness variable. Such guidelines are found in professional journals and books or are presented at professional meetings and workshops. Other guidelines are in the form of personal word-of-mouth testimonials, claims by equipment manufacturers and salespersons, and other commercially oriented agents of products and services. The consumer of fitness activities is exposed primarily to media advertisements for fitness businesses, whose practices frequently are inconsistent with scientifically based programs for improvement. The average consumer usually cannot distinguish the good programs or information from those that are unsound.

The content of this chapter focuses on methods and devices for improving selected aspects of fitness: muscular strength and power, muscular endurance, and flexibility. Cardiovascular fitness training, usually in the form of running, swimming, and cycling, is evaluated more in terms of physiological measures rather than biomechanical measures. Such training methods and their concomitant biochemical adaptations are more appropriately studied in exercise physiology, and the reader is referred to up-to-date sources in that area for cardiovascular fitness information.

The Specificity of Movement Fitness and Training

Resistance training usually focuses on improving some kind of muscular quality—strength, endurance, or power. The changes observed as a result of training, however, are not limited to the muscular system alone. A substantial amount of evidence indicates that changes within the nervous system (neural adaptations) are induced by the training. The neural adaptation is a major factor contributing to the change in performance (Sale, 1988). Because neural control is essential for motor unit activation in voluntary movement, it is the *neuromuscular* changes that are the goal of training.

Training specifically for the movement pattern, joint position, speed, and type of contraction (concentric, eccentric, isometric) has been shown to produce improvement specifically in those movement parameters. In addition, the manner in which training exercises are performed determines the nature of the improvements. For example, heavy resistance training improves the amount of maximum force that can be developed by a muscle group, and explosive drop-jump training improves the speed with which maximum force can be generated. An individual needing to develop a large force very quickly at the initiation of a throw or strike would need to develop the ability to generate large forces fast (i.e., generate a high-frequency motor-unit firing rate for a brief amount of time); training should be designed to produce that ability.

In advanced strength training, one often sees the athlete focusing on developing individual muscle groups for hypertrophy, making the assumption that no further neural adaptation will occur. However, hypertrophy of nonspecific motor units could be counterproductive when the athlete's skill is performed (Sale, 1988). The selective recruitment of motor units demanded for the quantity and quality of muscular tension throughout a performance is dictated by the overall coordinated neuromuscular pattern.

Exercises performed for developing strength or some other quality in specific muscle groups also should be done with appreciation of the "weakest link" concept. For example, the knee extensor muscles are used in vertical jumping, but training this muscle group only in isolation, by raising and lowering a resistance while seated on a bench, does not exercise the knee extensor muscles *as they are used* while the hip is also extending, as in the vertical jump. If the knee extensors are unusually weak, however, it would be appropriate to train to increase their strength so that they would not be the "weak link" in the total coordinated pattern. As expressed by Vorobiev (1976) (cited by Jesse, 1979), "The greatest load falls on the weakest link during execution of the entire sports movement, and an overload occurs."

Therefore, training for improving a fitness variable should replicate as closely as possible the actual total body movement for which that variable is needed. Such specificity includes not only range of joint motion but also speed. In activities that require static strength, one should train for static strength in the position demanded in performance. In activities that require moving a heavy load slowly, one should train for strength to handle the load with approximately the same speed of movement. Many activities, such as throwing, kicking, and other striking activities, require maximum speed of distal limb segments. The kinematics of the segmental movements must be understood and incorporated into the strength, power, and flexibility training program to plan the most effective exercises. For example, to train the muscles for an overarm throw, one would want to duplicate as closely as possible the speeds and the sequence of segmental movements. To train the muscles to operate effectively for such a purpose, strength training must be performed at an accelerating rate; that is, as the resistance (the object and the body segments) starts to move, it gains speed, and the muscles must be able to shorten at a faster rate than the segment is moving at any given time; otherwise, the muscles will go "slack," or will not be pulling against any resistance. Therefore, the muscles should be used in a situation requiring a force application that increases speed from the initial joint position throughout the application range.

We also know that toward the end of the joint ROM each rotating segment slows down as a result of the "braking" force provided by eccentric contraction and the elastic components of antagonist muscle groups. Concentric contraction of the initial mover muscle groups is no longer needed; eccentric contraction of the antagonists is needed, however. Strength training using eccentric contraction of the antagonistic muscle groups may serve to protect the muscles against possible strain as they must abruptly decelerate a rapidly moving segment. An exercise plan for such a situation includes identification of the segments being decelerated during a throw, kick, or striking pattern and the antagonistic muscle groups used to perform that deceleration. Strengthening those muscle groups during fast eccentric contraction is need-specific training.

7.2 Resistance Devices Used in Training

When a person uses some device for developing muscular strength, power, or endurance, the exact amounts of resistance acting or muscle torques applied are difficult to determine. Throughout the range of a segment's motion, several things are changing. The muscles' angles of pull on the bones are changing, and, as a result of a smaller force arm, less torque can be applied at the beginning and end of a segment's rotation than toward the middle. The magnitude of contractile force is also changing as a result of changing muscle length. The muscle force–velocity relationship is evidenced in variable muscle torques produced during variable speeds of movement. The degree of fatigue influences the muscle torques that can be produced and also the recruitment pattern of muscle fiber-type motor units (Viitasalo and Komi, 1981). Another variable is whether the movement incorporates any ballistic characteristics or motion resulting from "follow through" of movements established at the beginning of the action—that is, whether a part continues to move because of its momentum. Postural control and balance also frequently affect the way a movement is performed, often to the point of compromising the effort.

To try to account for or control all of these variables during resistance training would be impractical, if not impossible. However, those variables that are significant to the *purpose of training* should be focused on when using any training device.

Devices used for muscle training come in many forms. Some merely provide more or less resistance to muscular efforts; others provide adjustable resistance that can be quantified and even recorded or stored in computer memory. The selection of a particular device should be guided by the capability of the device to allow for specific training of the variable needing improvement.

Simple and Handy Devices

Casual overloading of muscles can be achieved by making use of common items that may exist for purposes other than weight training.

Any item of sufficient weight and manageable size can be used as a free weight, or barbell. Plastic bottles filled with sand or water or large containers of food or drink are excellent for high-repetition weight work. A 5- or 10-lb bag of sugar in each hand is convenient and may provide more weight than books. Shock cord or rubber tubing are versatile resistance devices arranged in single or multiple bands for easy to difficult pulls. Tied to a doorknob or between the door and frame, elastic cord or tubing can provide resistance in almost any direction, and the initial tension can be increased by starting the pull with the cord at some prestretched length. Similarly, inexpensive resistance devices can be made from door or gate springs. These items can be helpful in school or club activities when funds for expensive equipment are limited.

Small equipment is also available commercially. Springs, shock absorbers, surgical tubing, and friction devices of many shapes and forms have been designed specifically for exercise purposes. Some have no advantage over that which can be made by the user. Others, however, have extra features (e.g., handles, hooks, pulleys) that make them more convenient to use and versatile. Some also have means for setting the resistance level, so that a quantified account of progress may be kept. No matter what the device, however, the user is the one who determines the progress made through the effort given.

Free Weights

The use of free weights or "constant" resistance such as barbells for training muscle strength and endurance is common at all levels of fitness work from rehabilitation and therapy settings to competitive weight-lifting training centers. Lifting free weights is convenient: the resistance is known and is readily modified, and the weights are portable. Usually, when one part of the body is being worked, other parts must be stabilized muscularly to balance and guide the weights. When extremely heavy weights are used to work one part of the body, however, other parts of the body could be injured if control is lost and the weights fall or if sudden great force demands are placed on weak muscles. Under these circumstances spotters should assist the weight lifter, or barbell racks should be close enough for the lifter to release the weight onto that support (Figure 7.1a).

Compared to exercising on specially designed weight machines, which come in various configurations, lifting free weights limits to some extent the segmental movements that can be used against resistance. This is because weight is gravity dependent (i.e., the only direction in which the resistive force acts is downward). Therefore, the user must often assume different body positions during the workout so that the motive muscle force moves the weight in an upward, or partially upward, direction (Figure 7.1b). The force exerted to start moving any weight upward must be *greater* than the downward weight force. The faster the acceleration upward, the greater must be the force of the concentric contraction of muscles. At the end of the upward movement, however, the muscle force decreases to less than the weight so that the load decelerates at the top of the lift. Static contraction holds the weight before lowering, and then eccentric contraction of the *same* muscle group controls the weight's descent (Figure 7.1c). This pattern of contraction for the same muscle group is used for up-and-down movements with weights; that is, the motive force for lifting is the same muscle group that is the resistive force during lowering. The pattern of contraction for the muscle group used is concentric, static (isometric), eccentric, static, concentric, and so on. Note that the antagonistic muscle group does not come into action when gravity (weight) does its work for it (see Chapter 2).

Moving weights horizontally overloads the motive muscle group only to the extent that the muscles must start moving the mass. The deceleration of the mass is then accomplished by the resistive force of the *antagonist* muscle group. For example, the shoulder transverse adductors (transverse flexors) concentrically contract to accelerate the arms and the mass forward from a sideward abducted position. The shoulder transverse abductors (transverse extensors) then eccentrically contract to stop the moving arms and mass. The most effort, however, is exerted by the static contraction of the shoulder abductors as they hold up the weight of the arms and mass in the horizontal position.

The body itself and its parts can be used as free weights for muscle work also. Different orientations of the body are necessary, however, so that movements are done against the force of gravity, as shown in Figure 7.2a and b. Because the weight of the body and its segments remains constant, the magnitude of the weight lifted cannot be increased, as when using external weights. Instead, the intensity of the exercise is increased by moving the weight faster during the initial lift. The eccentric (braking) contraction of the muscles at the end of a fast descent is also of greater intensity than for slow movements. Total work load also could be increased by doing more repetitions.

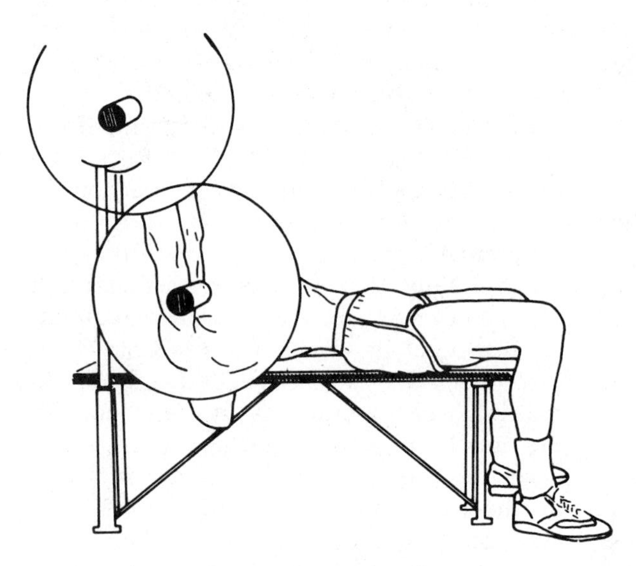

a Barbell rack provides safety for lifter of heavy weights

b Body orientation must be adjusted to control direction of weight's resistance

Concentric muscular tension used when raising weight (shoulder flexors)

Eccentric muscular tension used when lowering weight (shoulder flexors)

c Same muscle group is used during raising and lowering weight

FIGURE 7.1

The use of free weights for resistance during muscular effort.

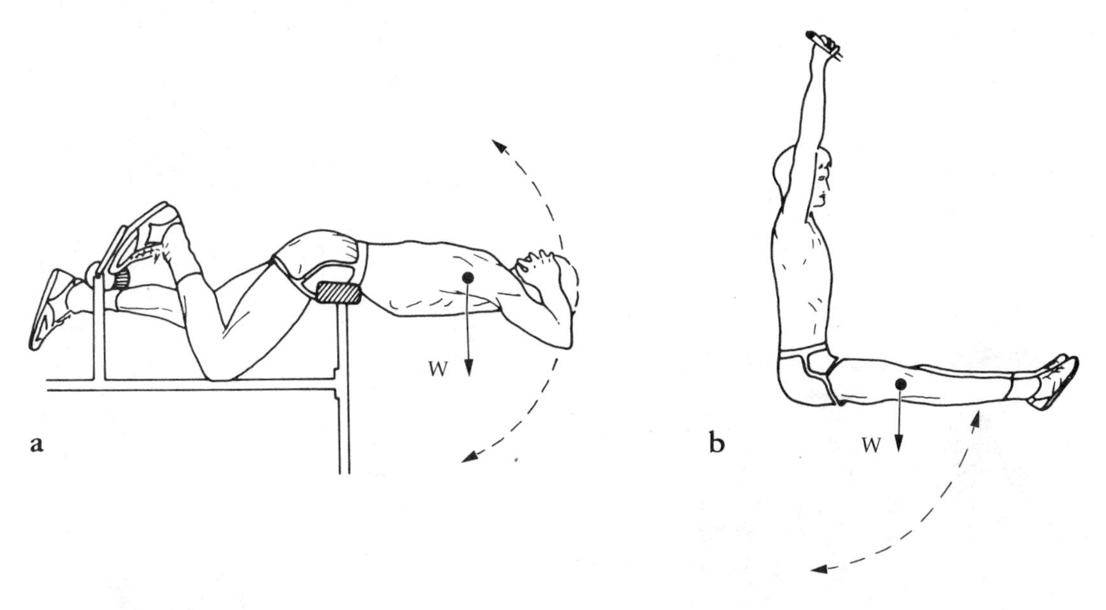

FIGURE 7.2
Body segments can be used as free weights. The location of the segment's CG determines the resistance torque provided by the segment (W, weight of segment).

Moving body segments horizontally, as mentioned earlier, places less demand on the motive and resistive muscles involved than moving them up and down. Faster reciprocating rotations (accelerations and decelerations) of the parts increase the intensity (power output) of the concentric and eccentric contractions of the active agonist–antagonist groups.

Gravity-Dependent Resistance Machines

Special devices have been designed to guide the path of weights along channels or rods so that the user need not be so concerned with the balance or stabilization of other parts. The weights to be lifted are attached to a cable that passes around one or more pulley wheels to place the handle or cuff in a convenient location for the user. The purpose of the pulleys is to change the direction the user's force may be applied to raise the weights (Figure 7.3a).

Elaborate weight-training stations such as the Universal Gym also allow the user to assume positions that are convenient for working on designated muscle groups (Figure 7.3b, c, and d). The weights are stacked on a rod and can be pinned together for selection of resistance (Figure 7.3e). The user pulls or pushes on a given lever or bar that is attached to a cable running through a system of pulleys to the weights. The pulley systems make such machines convenient and versatile for working various isolated muscle groups. Also, the ROM through which the user can move the body segments may be limited or restricted to that which a particular station permits. A disadvantage of using such stations for training for skill improvement is that the balance and stabilization requirements are removed from the efforts, and in real-life muscle activity, these abilities are necessary for total movement effectiveness.

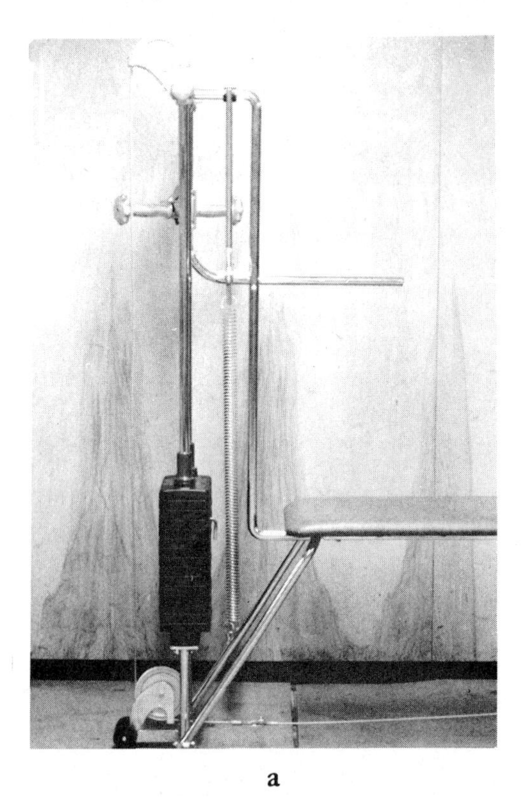

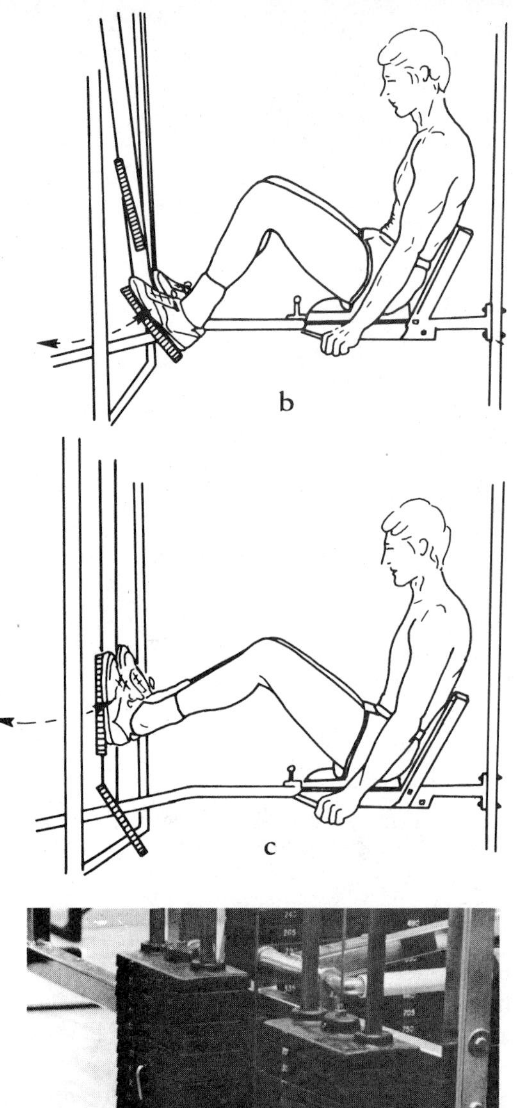

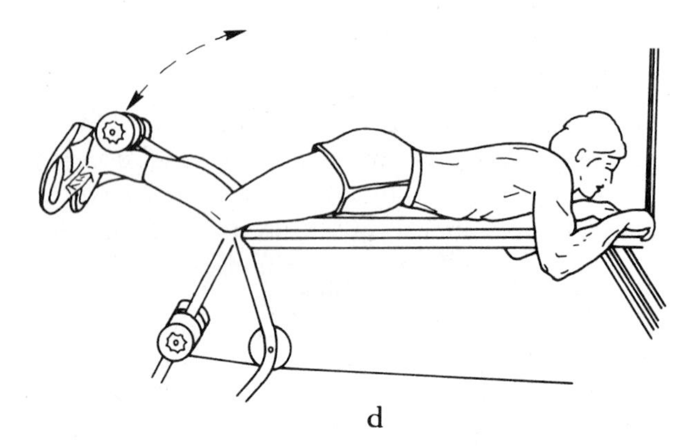

FIGURE 7.3

Gravity-dependent resistance machines provide for adjustable weights and the use of different body positions for exercising.

As with free weights, the acceleration of the weights initially influences the muscle overload. Also, as with free weights, the return movements are controlled with eccentric contraction of the same muscle group that concentrically contracted to lift the weights.

Variable Resistance Training Devices

The torque produced on a segment by a group of muscles depends on the angle of muscle attachments to the bone as well as the length–tension properties of the muscle and the speed of shortening. The muscle torque required depends on the magnitude of the resistance *torque*. In lifting free weights, the weight torque to be overcome varies throughout the joint's ROM, being greatest when the weight's line of action is farthest from the joint axis (see Chapter 3). Consequently, the tension of the muscles also varies with the resistance demands. It has been proposed that the strengthening stimulus would therefore vary throughout the ROM, it being the greatest at the points where the greatest tension is required to move the load. In an attempt to overcome this apparent deficiency in free-weight, or constant-resistance, training, **variable resistance** devices have been designed to alter the amount of resistance torque within the ROM of a single repetition. Instead of simply redirecting the user's applied force via cables and pulleys to free weights, chains or cables are directed around circular gear wheels or pulleys that are kidney shaped or egg shaped (cams). In this way, the resistance force arm changes as the cable turns the cam and thus provides more or less resistance during different parts of a single movement (Figure 7.4). The proposed advantage is that the resistance can be arranged to increase at the joint position where the muscles can exert greater torque. The torque capabilities of muscles throughout an ROM vary from one segmental movement to another, however. Consequently, the cams must be shaped specifically for the segmental movement to be resisted. Although this method has not been perfected, the variable resistance device is one approach to increasing the opportunity for the muscles to develop greater tension than in constant-resistance work.

Isokinetic Devices

The term **isokinetics** was originally coined to mean a constant speed of muscle shortening as a segment worked against a device set to move at a constant speed. It has been shown, however, that a constant speed of segment rotation is not associated with a constant speed of muscle shortening (Hinson, Smith, and Funk, 1979). The current use of the term applies to the muscle contraction that accompanies constant angular velocity of a limb. Isokinetic, or **accommodating resistance**, machines govern the maximum *rate* of joint movement because they can be set at a predetermined speed. A large range of speeds is available for selection in most isokinetic devices, such as the Cybex or Orthotron (Figure 7.5).

The advantage proposed for such machines is that the user can generate as much force as desired throughout the whole ROM and the resistance will not increase speed or gain momentum as in isotonics. Therefore, the maximum muscle tension available at that *approximately* constant muscle-shortening speed can be exerted throughout the repetition. (Recall that muscle force generated decreases with speed of shortening.) The user *need not* apply maximum effort throughout the ROM to move at that constant speed; half the effort would still move it the same way. For the motivated user, however, the *opportunity* for maximal exertion without

FIGURE 7.4

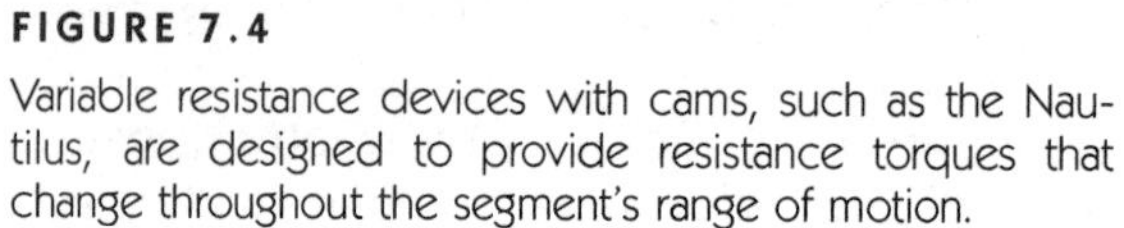

Variable resistance devices with cams, such as the Nautilus, are designed to provide resistance torques that change throughout the segment's range of motion.

much acceleration of the resistance is available with such devices. The resistance developed is designed to equal the force that the mover applies to it; that is, after movement reaches the preset speed, no matter how much muscle force is used against the machine, the machine matches it in the opposite direction (equal reaction force) but does not move faster. If maximal tension is developed in regular isotonic exercise, the resistance load is accelerated, and the muscle is then contracting against a load that has momentum and therefore does not develop maximum tension because it must shorten at a faster rate. Isokinetic resistance devices overcome this deficiency in normal isotonic resistance exercise by allowing maximal muscle tension to be used throughout an ROM. If such a device is not available, one can work with another person who accommodates to the changing maximal muscle torques applied.

Any acceleration and deceleration phases occur only during the initiation of a movement and at the termination of the movement. If the speed on the device is set to move slowly (20 degrees/sec), the acceleration and

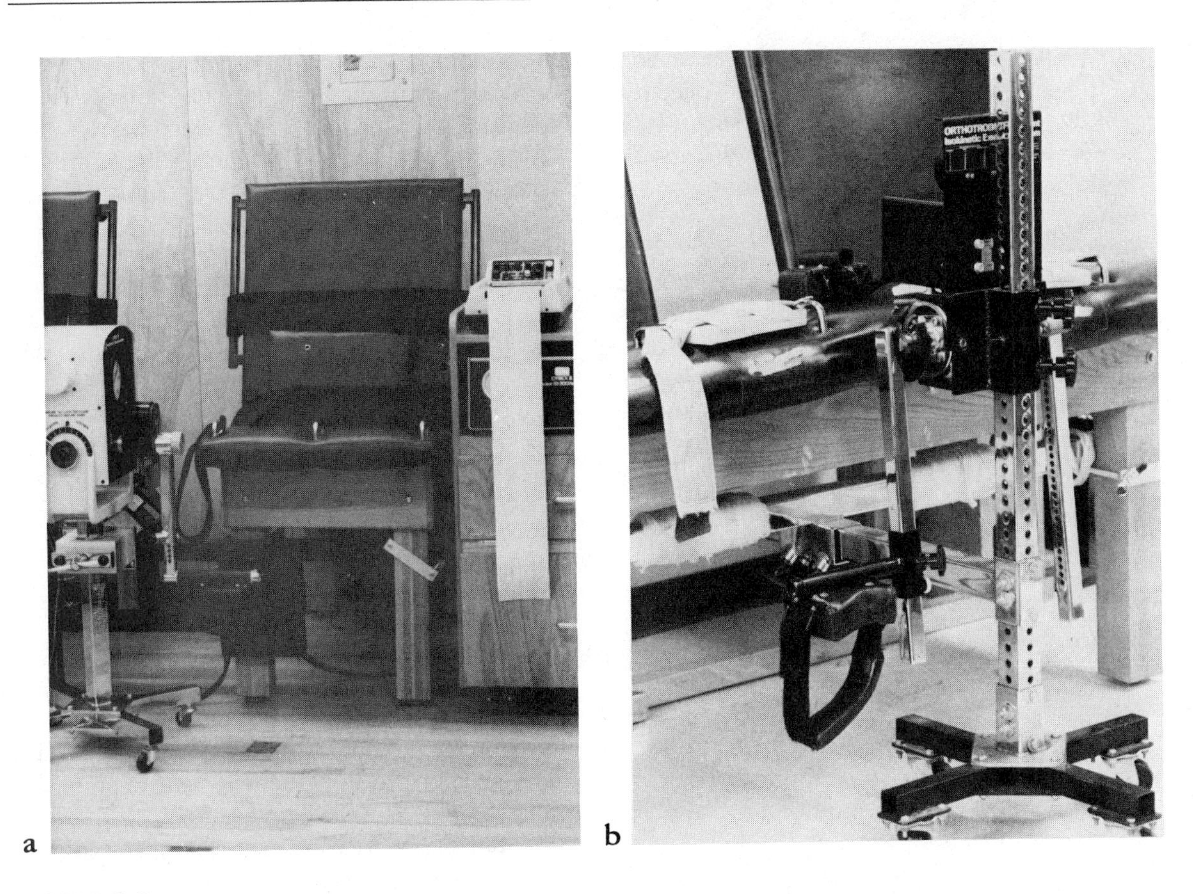

FIGURE 7.5
Isokinetic devices provide for relatively constant speed of the resistance moved. The speed of movement of the crank arm is independent of the force applied by the user. (a) A recording Cybex unit. (b) An Orthotron.

deceleration periods are extremely brief. At the faster speeds (300 degrees/sec), however, most of the ROM contains accelerated movement because the body segment must first be moving as fast as the device (lever) to apply force to it. This means that for faster set speeds, the muscles are not exerting their maximal tension against the load until a large part, if not most, of the ROM has been used. This factor influences the accomplishment of the purpose for which the machine might be used. For example, if the leg cannot be extended (accelerated) fast enough to exert force on the lever set to rotate at 300 degrees/sec until halfway through the ROM, no resistance occurs until halfway through.

Isokinetic devices are useful not only in resistance training but also in diagnosis of muscle weakness and in evaluating progress of rehabilitation. The Cybex, for example, incorporates a recorder for a continuous printout of the range of joint angles as well as the continuously changing torque values. The peak-torque angle changes, depending on the speed set for segment movement. For example, from a sitting position on an Orthotron isokinetic dynamometer, hamstring peak torque occurred at 32 degrees into knee flexion when the machine speed was set at

50 degrees/sec; the peak torque occurred at 61 degrees into flexion when the speed was 400 degrees/sec (Osternig et al., 1983). Similar results also were found for the quadriceps during knee extension: the peak torque occurrence changed from 87 degrees at a speed of 50 degrees/sec to 63 degrees at 400 degrees/sec. These results indicate that when working on isokinetic machines whose speeds can be set, the angle in the ROM where maximum torque is achieved will depend on the speed. Because we know from the muscle force–velocity relationship that the faster a muscle shortens, the less force it produces, less torque would be produced at higher speeds.

Isokinetic torque can be recorded only after the segment reaches the machine's set speed. Therefore, at fast speeds, the ROM through which the muscles can be worked truly isokinetically (at constant speed) becomes limited. Also, at high speeds, as the segment rapidly accelerates to catch the set speed and then impacts with the roller or pad, the segment is rapidly decelerated. High torque values occur at this point, and the relative safety of repetitive impacts on the joints has yet to be studied (Osternig et al., 1983). Recent work has shown, however, that this impact of the segment on the lever causes an initial artifact in the peak torque printout. Such spikes should be recognized as "overshoot" artifacts rather than as representing true muscular tension development (Sapega et al., 1982).

Understanding Resistance Devices

1. Inspect the items in the room around you, and demonstrate how three things could be used as resistance devices with little or no modification.
2. What is the feature of variable-resistance devices that makes them different from free-weight resistance?
3. Explain the force and speed features of isokinetic devices.
4. Diagram on a stick figure the resistance force and force arm involved in holding a weight in your hand with the shoulder abducted a) 90 degrees, b) 45 degrees, c) 180 degrees. Which muscle group exerts tension to hold these positions?
5. Repeat question 4 for holding a weight with the shoulder flexed 90 degrees. Which muscle group holds this shoulder position? What is the specific responsibility of the vertebral extensors?

7.3 Strength

Results of strength-training studies indicate that a number of neural adaptations occur in the process of getting stronger. Among these are the ability to recruit more motor units and/or the ability to increase the frequency of motor unit discharges and the ability to increase the synchronization of motor units. Further, the possibility exists that the excitability level of the motor neurons involved in specific muscle activities is increased by strength training (Sale et al., 1983).

Heavy resistance training also leads to muscle *hypertrophy*, or enlargement of the muscle mass. The cross-sectional area of both fast-twitch and slow-twitch fibers has been shown to be increased significantly. Such increases in active tissue mass are associated with the quality we call *strength*.

Muscular strength can be defined as "the ability to develop force against an unyielding resistance in a single contraction of unrestricted duration" (Atha, 1981, p. 7). This definition is suitable for describing strength in terms of muscle forces on bones or in terms of body segments applying force on devices that measure isometric (static) strength, such as tensiometers and dynamometers. The greater the force with which a muscle can pull on its connecting bones, the greater its strength. Operationally, however, the actual force of contraction of a muscle does not determine how strong an individual is; rather, the contractile force available is used to produce torque on bones, the magnitude of which depends on the muscle's angle of attachment at the time of force application, its length, and its shortening velocity. Muscle strength could be defined as the property that is measured by the device or particular test administered to measure it.

Routines for strength training have been refined from the original *progressive resistance exercise* (PRE) programs developed by DeLorme and Watkins (1951) in rehabilitation work. Since then, a multitude of weight-training routines has emerged using many possible combinations of repetitions, resistance, frequency, and weeks of training, all focused on what combination would produce the greatest strength gains.

Out of all this, three main categories of strengthening exercises have been neatly packaged for use, although, within each category, many possibilities for variation exist. These popularized categories are **isometric**, **isotonic** (dynamic), and **isokinetic**. The biomechanical aspects of each of these types are described in the following section.

Isometric (Static) Strength

Isometric exercise is muscle contraction that exerts its force against an immovable resistance; that is, the length of the muscle does not change (*iso* + *metric* = *same length*). Such static tension development in muscles has been most successful for strength gain and maintenance in rehabilitation cases when movement is not possible. Gains in strength through static contractions have been most notable for relatively weak individuals, and the gains usually taper off after the first 5 weeks of daily isometrics.

The main variable responsible for strengthening muscle isometrically has been identified as the magnitude of the tension developed. In general, for more effectiveness of isometric strengthening the tension developed should be near maximum effort. The duration of each contraction should be long enough to fully recruit all fibers in a muscle group (about 3 to 6 sec perhaps), and the contractions should be repeated several times daily (Atha, 1981). The variability allowed in the duration of each contraction for strengthening to occur stems from the differences among muscle groups in the time required to reach maximum tension. For example Atha (1981) reported the time to reach peak isometric tension for the knee extensors was 4.42 sec, while for the elbow flexors, it was only 1.61 sec. Such variance in the speed of maximum tension development justifies the recommended longer duration of each contraction so that this "general prescription" would take care of the slower muscle groups. Atha observed that muscle damage may occur from *instantly* generating peak force against an unyielding resistance and that some kind of stiff shock absorption device used in such relatively violent *instant-maximum* efforts might provide a margin of safety.

The optimum rest interval between isometric efforts has not been firmly established, partly due to researchers' focus on the other variables in isometric training. Rest intervals ranging from less than 1 sec to 2 min and over have been used, but their relative effects are unknown. One of the most popularized features of isometric training is that it produces gains in strength at specific joint angle held during contractions and that those strength gains are not detected at other joint angles (Clarke, 1973). Although such joint-angle specificity has been demonstrated, evidence also indicates some transfer of strength gains throughout the ROM (Whitley, 1967; Raitsin, 1974, cited by Atha, 1981). Explanations for the joint-specific strength gains so widely observed include not only the possibility of "neuromuscular learning" mechanisms but also a biomechanical reason: that synergist muscles are called into action at certain joint angles as they are needed and therefore become strengthened as members of that joint angle's "team" of muscles. At other joint angles, these same synergists may not be recruited to assist in the effort, thereby leaving the team shorthanded.

Isometric training, if undertaken for skill improvement, should be done with the recognition of its apparent joint angle specificity as well as its decreasing effectiveness for individuals whose state of training is more advanced than that of beginning or intermediate performers. Examples of sport situations using static strength include the iron cross on the rings or leg levers on the parallel bars. Such performances in which specific joint angles must be maintained, rather than passed through, may well be aided by isometric training in addition to other types of strengthening. The strength produced by a muscle group during a single *maximal voluntary contraction* (**MVC**) against an immovable resistance can be quantified by some type of tension-measuring device, such as a cable tensiometer or strain gauge dynamometer, which is stressed by the force exerted by a body segment in contact with it (Figure 7.6).

The *actual* strength, or force of contraction, of the muscle group can only be derived by using estimates of angles of muscle attachments to the bone, force-arm distances from the lines of action of the participating muscles to the joint axis, and the resistive force measured on the device multiplied by the distance to the joint axis from its line of force. These variables demand careful measurement during a test. The joints must be positioned at the same angles and the cuff is placed on the segment the same distance from the joint so that they can be replicated for any subsequent tests on the same performer. Careful attention also must be given to the other body parts that apparently are not involved in the task. Recall the concept of a chain being as strong as its *weakest link*. If an MVC of the elbow flexors is to be determined for the performer shown in Figure 7.6c and the vertebral extensor muscles are relatively weak, the effort to flex the elbow may be less than maximum as a result of the inability of the vertebral extensors to maintain (stabilize) the upright trunk position against the vertebral-flexion tendency caused by elbow flexion. (If the hand cannot be raised to the shoulder, the shoulder is pulled down toward the hand.) As the trunk flexes forward, the elbow moves into flexion, the elbow flexors shorten, and the force generated decreases with decreases in muscle length as they concentrically contract. Moreover, the angles of attachment of the flexors are changing, which produces changes in the torques produced on the bone.

Other body segments can be stabilized by strappings or by being held by an assistant.

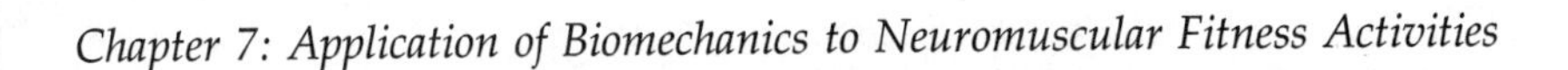

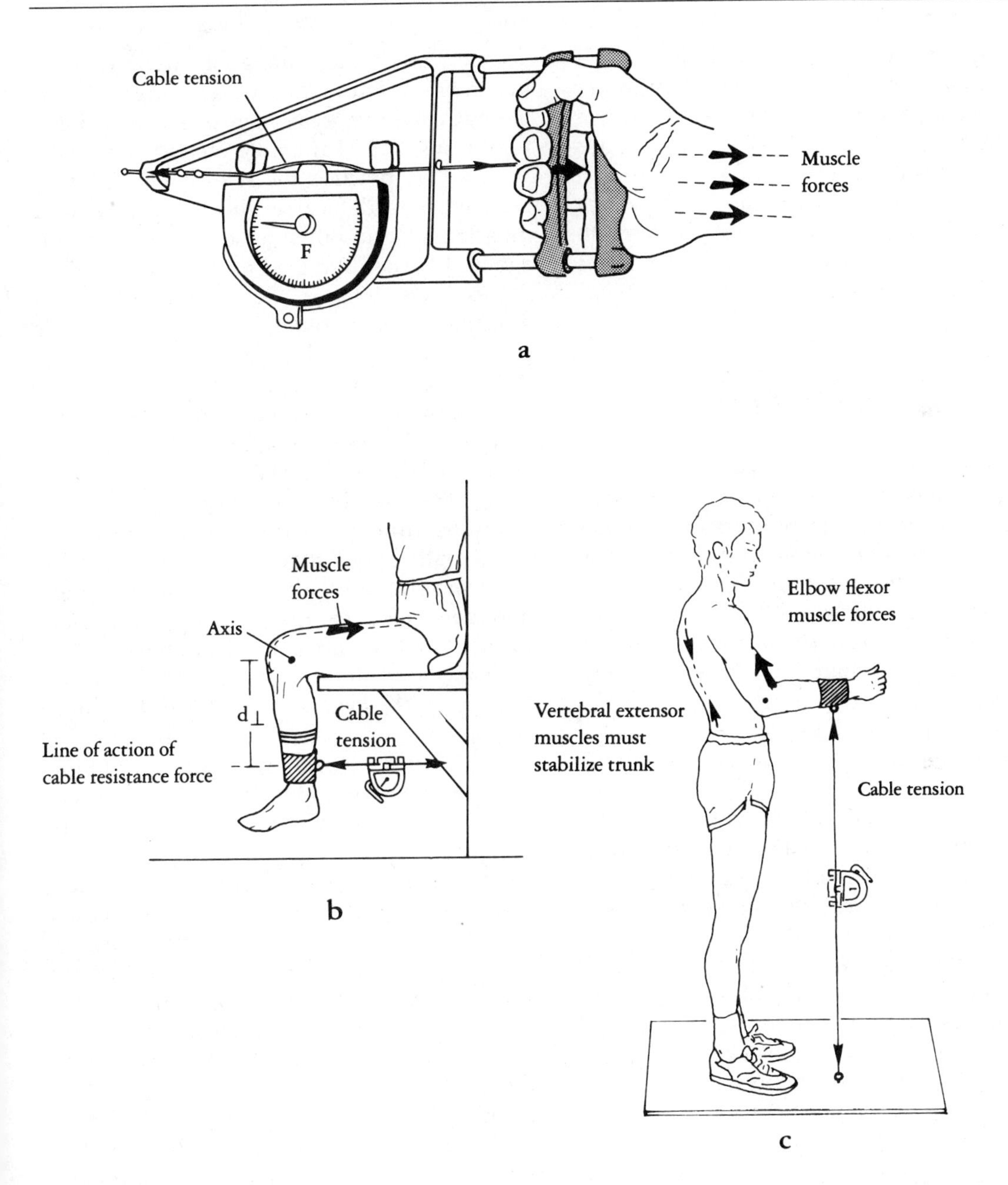

FIGURE 7.6

The use of a static-strength testing device, such as a cable tensiometer, provides an indirect measure of the contractile force of a muscle group. (a) Grip strength is measured by the tension in the cable. (b) Knee extensor strength is measured by the tension in the cable. (c) Elbow flexor strength is measured only to the extent that the back extensor muscles are able to prevent vertebral flexion.

Elbow flexor strength measured in such a manner, however, has limited application for activities that demand not only elbow flexor strength but also strength of muscle groups that serve to stabilize other body parts so that the arm may have a solid base from which to move. Attempts to measure the strength of isolated muscle groups should be made with full understanding of the *weakest link concept*.

Dynamic Strength Training

The term *isotonic* has been used to describe the presumably constant tension in muscles undergoing concentric or eccentric contraction (*iso* + *tonic* = *same tension*). Although this is not strictly true, the term remains in use, but many now prefer to use the term *dynamic contraction* or *dynamic tension*. Variations in dynamic exercise include the following: heavy resistance, progressive resistance, variable resistance, speed loading, eccentric and plyometric training, and hybrid (combination) methods (Atha, 1981).

The variables that can be manipulated in dynamic training are (1) load intensity, (2) the rate or speed of repetitions, (3) number of repetitions, and (4) the rest interval between repetitions.

During the last 40 or 50 years, many combinations of these variables have been used in research in attempts to identify "recipes" appropriate for specific strength needs. Probably the most important variable is the load, or the amount of resistance moved. To standardize the load to individual capacities, the term *repetition maximum* (**RM**) has gained acceptance. It refers to the weight, or magnitude, of the load that can be moved a given number of times and no more. For instance, a 10-RM load is a weight that can be lifted only 10 times and is too heavy to be lifted more than 10 times. The resistance selected at the beginning of strength training must be found by trial and error. As the strength of the muscles increases, the load that represents 10 RM must be increased (usually weekly). A practice for selecting weights to be used is to determine the 1-RM load for a muscle group and then to use a given percentage of that standard load for more repetitions. For example, 80% of 1 RM could be the load selected for eight repetitions. Selecting loads in this way provides an initial frame of reference within which progress can be evaluated. Using a maximum isometric load as a reference, researchers have found a 10-RM load to be about 50% of the MVC (Clarke and Herman, 1955, cited by Atha, 1981).

Results from the large number of studies of strength show that the major factor for achieving strength gains is the amount of tension developed by the muscle group used. Studies on load intensity also show that 2-RM loads and 10-RM loads are not as effective as 5- to 6-RM loads for producing the most muscle tension and, therefore, the greatest strength gains. Identification of an optimum rest interval, however, could alter the load intensity and number of repetitions recommended.

Increasing the speed of lifting a given load is a way to increase the load intensity, although the results of studies do not support increasing speed as being better than using greater weights slowly for improving strength (Atha, 1981). The greatest load intensity and greatest tension developed by the muscles in speed lifting occurs during the initial *acceleration* of the load at the beginning of the ROM. This acceleration period does not last long because of the deceleration that must occur at the end of that ROM. Consequently, the high tension produced by the muscles (the strengthening stimulus) may not be long enough to produce the strengthening effect. If the high tension of isotonic speed loading could be prolonged through the entire ROM, the method could be more effective for building strength. Hay, Andrews, and Vaughn (1980) *tentatively* proposed on the basis of their studies that "for a given load, a fast rate of lifting is likely to yield a better rate of strength development than are slower rates of lifting" (p. 9). Note that this proposal states "for a given load" and means that the load should be lifted fast rather than slow for better strength gains.

Variable Resistance Training

Strength gains from using free weights and variable resistance devices have been reported by Atha (1981). From his review of others' experimental findings, he notes that variable resistance devices do not produce greater strength improvements than does standard resistance training. There is some indication, however, that greater strengthening may occur at the *specific joint angle* at which *peak muscle tension* is produced by the variable resistance machine.

Eccentric Training

Muscle's force–velocity relationship shows that as the shortening (concentric contraction) speed becomes slower, the tension produced in the muscle increases, and the tension at zero speed is greater than during shortening. Further, greater tension can be elicited during muscle lengthening (eccentric contraction) than in the concentric or static states. Arguing that if greater tension acts to provide the strengthening stimulus, the proposal has been made that overloading (e.g., 120% 1 RM) an eccentrically contracting muscle would provide greater stimulus for strengthening and greater strength gains. A number of researchers have shown that eccentric training does cause strength gains. Eccentric training, however, has been shown to be *no more* effective in producing strength gains than concentric or isometric training. Whether eccentric training is superior to isokinetic training for producing greater strength has not been clearly established, although some investigators have reported that it is.

One aspect of eccentric training that has been noted is greater muscle soreness than that which normally results from concentric and isometric training. Such soreness, although nonpersistent, may serve to decrease effort put forth in the early sessions of training.

Prestretch Training

Drop-jumping, depth-jumping, bounce-loading, and rebound jump training are other names given to "plyometric," or prestretch, training. It uses the stretch-shortening cycle to store and use elastic energy and to elicit the stretch reflex.

In jumping, the method of loading is by using the kinetic energy of the body landing from a jump to put the muscles on stretch and elicit an eccentric contraction followed immediately by a concentric contraction. For other muscle groups, the kinetic energy of dropped weights, caught by the lifter, serves to stretch the targeted muscle groups. The force exerted by the muscles following this storage of elastic energy has been shown to be greater than without such storage (Komi and Bosco, 1978). Such observations have led to using prestretch type of loading in strength-training studies. Results so far have shown little indication that greater gains in strength occur, but the criterion measure has usually been vertical jump height, not strength (Blattner and Noble, 1979; Clutch et al., 1983).

According to the specificity of training principle, such results should be expected. Sale (1988) reported that explosive jump training increased the *rate* of force development, rather than the *maximum* force development. Also, the specific technique used for drop-jumping influences the concentric contraction force exerted following the stretch. A quick bounce-jump executed as soon as possible after landing from a drop-jump elicited greater contractile force than a landing followed by a deeper knee bend jump (Bobbert et al., 1987a). In a subsequent report (1987b), these investigators concluded that quick-bounce drop-jumping should be done from a height of no more than 20 cm (8 inches). From this height, the joint surfaces receive relatively small forces, and therefore the risk of injury is also small. Further, Atha (1981) questions the advisability of subjecting muscles to potentially injurious forces in attempts to determine whether the elastic recoil properties of the tissue can be improved.

Isokinetic Strength Training

Isokinetic strength training is a method in which the speed of segment rotation is limited to a constant value during ROM.

The speed factor influences the tension a muscle can develop according to the force–velocity relationship (Chapter 2, pp. 80–82). Greater strength gains, however, are reported for moderately slow isokinetic speeds than for ultra-slow isokinetic speeds that approach isometric contractions. Faster speeds are found to be less effective for building strength than slower speeds. This is not to say that fast-speed isokinetic training is not valuable for uses other than maximal strength development. Little evidence shows, however, that fast-speed isokinetic training is *more* effective than slow training for improving performance in explosive-type skills such as vertical jumps or sprint starts. Its potential continues to be studied.

Since the speed of the isokinetic device is set, there is no adjustment for the resistance, or load. The user exerts, or should try to exert, maximal tension throughout each repetition. As fatigue occurs, the force exerted decreases, but the device continues to move at the same speed, giving an equal and opposite reaction force to the user. The "load" is adjusted by the decreasing muscle tension that accompanies increasing speed. The load also may be adjusted by the user who chooses not to exert maximal effort. In such a case, the strengthening stimulus, high muscle tension, may not be produced, and gains in strength would not be realized.

The interrelated effects of contraction duration, tension level produced, speed of movement, frequency of repetitions, and rest intervals have yet to be fully explored to arrive at the best combination for maximum strength gains.

Comparison of Strength-Training Methods

For increasing strength, the consensus is that a well-designed dynamic training program is better than the isometric programs reported. No optimum combination of variables has been established, however, for either type of training. Conflicting reports exist on the superiority of isotonics and isokinetics for strength improvement (Atha, 1981).

Two problems inherent in isometric programs are that motivation limitations exist during maximal attempts and that the strength gains made are joint-angle dependent. Isokinetic programs also have the motivation limitation with regard to generation of maximum effort.

The reader should bear in mind that these methods and their variations have been examined only with respect to their value in *strength* training. Variations in load, repetitions, and speed that have been shown to be less effective for strength gains may be *desirable* variations when training for other purposes.

Recalling the specificity of training principle, the improvement of neuromuscular features other than strength may be the goal of training; and the mode or method of training can, indeed, determine the outcome.

7.4 Muscular Power

Muscular *power*, the ability of a muscle group to contract forcefully with speed, is used in power-lifting events and also in less strength-demanding activities that require segmental speed. Power equals force applied times the speed that force is applied. An increase in strength (force) and/or an increase in the speed of muscle shortening produces greater power output. An attitude that seems to prevail is that great amounts of strength give the power required for any power event. Most often, however, the kind of power needed is that produced by the speed of muscle shortening.

We can think of power, then, as being *strength*-dominated or *speed*-dominated. Strength-dominated power is that exhibited by an Olympic weight lifter raising 400 lb overhead in a two-hand snatch (Jesse, 1979). Speed-dominated power is that exhibited by a shot-putter, thrower, jumper, runner, or swimmer. Recall that prestretch and use of elastic energy help produce speed-dominated power. Strength increases are developed at very slow speeds, and the strength, or force of contraction, is not the same at higher speeds. (Recall the force–velocity curve in Chapter 2.)

The specificity of training can be seen in the results of many research studies. Strength gains made by isokinetic training at slow speeds were not demonstrated when the speed of contraction was greater than that used in training (Lesmes et al., 1978; Moffroid and Whipple, 1970). Ikai (1973, cited by deVries, 1980) demonstrated that force, not velocity, improved from training at 100% MVC. Training with no load and maximum velocity improved velocity but not force, whereas training at 30%–60% MVC improved both force and velocity. Such findings are consistent with reports of Daish (1972), who noted that optimum power is produced when the load is about one-third maximum and the speed is about one-third maximum. He also noted, however, that large muscle masses tend to move toward the slow-speed ranges using more force. Large muscle-mass individuals, then, tend to be more powerful in the strength-dominated power events (e.g., weight lifting, football blocking). Smaller muscle-mass performers tend to be more powerful in speed-dominated events (e.g., jumping, running, swimming, throwing).

For speed-dominated power events in which the performer is required to maximally *accelerate* a given mass (to start the body or implement moving very quickly), the performer must train at both the force and velocity specific to the event (Lesmes et al., 1978).

Power training for improving vertical jump performance takes specific and nonspecific forms. The results of two different modes of training (both fairly specific to jumping) were reported by Blattner and Noble (1979). Training on the isokinetic "Leaper" (for leg extensors) and drop-jump training both produced equal improvement in jumping ability. Such results are probably indicative of the improvement to be expected from training in any mode that provides for specificity of movement demands in the skill.

Greater improvement might be expected from individuals whose fiber types predispose them to either strength-dominated power events or speed-dominated power events. For example, the predisposition of individuals with a greater percentage of fast-twitch muscle fibers for fast power performance has been shown (Coyle, Costill, and Lesmes, 1979). Moreover, these investigators proposed that as the velocity demand of movement increases, the fast-twitch muscle fiber composition becomes increasingly important for fast-power production. Biopsies showed that as muscle shortening velocity increased, the decrease in muscle torque (as per force-velocity relationship) was significantly less for those with more fast-twitch fibers.

The athlete, however, should not make any assumptions regarding his or her fiber-type percentage; rather, the athlete should train specifically for the type of power required for a particular event or skill.

Understanding Muscular Strength and Power

1. Select three different movement activities that depend on static muscular strength of a particular muscle group for success. Devise a skill-specific, strength-improving exercise for each.
2. Define muscular power.
3. Select three different activities that depend on strength or power of concentric muscular tension of specific muscle groups. Devise an exercise that would improve such muscular strength or power specifically for each skill in question. Do the exercises contain movements that are actually performed in the skill or are similar in terms of (a) the ranges of segmental motion, (b) the positions of other body segments, (c) the spatial orientation of the body, (d) the speed of repetitions, and (e) the phases of increasing and decreasing speed of segmental rotation?
4. Repeat the steps in question 3 for three activities demanding strength during eccentric muscular tension.
5. For each of the exercises shown in Figures 7.1, 7.2, and 7.3, state an activity or skill for which it would be useful. Explain why.
6. Considering the same exercises as in question 5, how could you modify each exercise (spatially or temporally) to better match the exercise with the activity or skill?
7. Devise a series of exercises for strengthening the triceps. Recall its functions at both joints, its three proximal attachments, its possibility for length–tension adjustments, and its potential for active and passive insufficiency.
8. Compare the difficulty of doing straight-leg sit-ups with (a) the hands beside the neck, (b) the arms crossed over the shoulders, and (c) the arms at the sides. Explain the differences in difficulty in terms of motive and resistive torques and muscle groups being used (see Chapter 6).
9. Perform a muscular analysis of the straight-leg sit-up, the flexed-leg sit-up, and the straight-leg snap-up. What are the differences in terms of motive and resistive torques, the muscular requirements, and the stabilization function of muscle groups? (See Chapter 6.)
10. Identify three calisthenic-type exercises that can be done with less muscular effort by using stored elastic energy.
11. Identify four specific skills that require strength-dominated power for success.
12. Identify four specific skills that require speed-dominated power for success.

7.5 Muscular Endurance

The term **endurance** usually connotes the ability to continue for a long period of time. In human movement activities, two principal, but not altogether inseparable, types of endurance are recognized: cardiorespiratory endurance (also referred to as circulorespiratory or cardiovascular endurance) and local muscular endurance.

Cardiorespiratory endurance is the degree to which the heart, vascular system, and respiratory system can provide the oxygen necessary to enable the mover to continue. Movements that tax the cardiovascular system are those using large muscle groups or total body movement or both at a submaximal intensity for prolonged periods (e.g., running, jogging, cycling, distance swimming). For these aerobic, or oxygen-utilizing, activities, the limiting factor is the failure of the oxygen-transport system to keep up with the demand of the muscle tissues. Training for cardiovascular endurance is one focus within exercise physiology, and the principles for improving such fitness are studied within that discipline.

Local muscular endurance is the degree to which specific muscle groups can continue contracting against a given load for a certain period of time or a certain number of repetitions. Such endurance is specific to the type of muscular tension exerted, the speed of contraction, the cadence of repetitions, and the magnitude of the resistance. The limiting factor is not the body's cardiorespiratory functioning, but the physiological limitations within the muscle groups being exercised and the neural control of the muscular contractions. The exact causes of muscular fatigue are not known, although several different factors have been incriminated.

Muscular fatigue can be defined as "a reduction in the capacity of the neuromuscular system to generate force or to perform work" (Bigland-Ritchie, 1981, p. 85). Changes in muscle excitation, metabolism, and contractile properties all contribute to fatigue. Circulatory patterns within the muscles do not seem to be the most important determinant of muscular endurance. A number of intramuscular physiological variables play a role, including energy stores, enzyme levels, and muscle fiber type composition (Heyward, 1975). In addition, fatigue during a held MVC is associated with a decline in electrical activity reaching the muscle fibers from the motor neuron, indicating the source of fatigue may lie in inadequate transmission of electrical activity within the muscle cells (Bigland-Ritchie, 1981).

Absolute and Relative Muscular Endurance

The degree of endurance can be expressed in *absolute* terms, such as: How long can you hold 50 N at a 90-degree position of elbow flexion? The endurance also can be expressed in *relative* terms, such as: How long can you hold a weight that equals 50% of your maximal static strength? Such relative expressions of endurance are used frequently to make meaningful comparisons among individuals' endurance levels. Absolute static muscular endurance is related to the maximum strength of the individual (Tuttle, Janney, and Thompson, 1950). For example, holding a 50-N weight for 1 min is easier for a person who can exert an MVC of 500 N than for a person who can exert only 250 N MVC. The first person uses only 10% of maximum strength, whereas the second uses 20%. Static muscle endurance may be important for maintaining a particular position demanding stabilization of body parts during performance. The static strength used need not be of great magnitude, as, for example, when a certain posture is to be maintained. The important variable in a performance may be the length of time a static tension can be maintained without fatigue, rather than how much static strength is manifested. Such endurance is seen in dance performances and in activities requiring continual trunk stabilization to allow extremity movements, such as weight lifting.

Measurements of absolute and relative endurance are used also in evaluating dynamic muscular endurance. Many tests set forth procedures for evaluating dynamic muscular endurance and for comparing individuals. Concentric-tension endurance tests as well as eccentric-tension endurance tests are based on number of repetitions, which are measured in different ways. One method is to count the number of repetitions done in a specific time interval; such a procedure may be termed a "power endurance" test based on the definition of power as force times velocity of force application. Another method is to count the total number of repetitions performed at a prescribed cadence. Either of these or other approaches can be carried out with an absolute resistance, such as 40 N, or with a relative resistance, such as with a load equal to 50% of maximum static strength at midrange of the joint movement. Note that either relative or absolute endurance measures are appropriate measures of improvement for a given person pretraining and posttraining, whereas relative endurance measures are more appropriate for comparing different individuals.

Training for Local Muscular Endurance

The concept of overload used for strength training applies also to improving muscular endurance. When training for strength, the strengthening stimulus is the intensity of muscle contraction, and the number of repetitions is kept small (5 or 6 RM). For muscular endurance, the repetitions are increased, but the load is less than that used for strength work. Studies show that muscular strength and endurance improve simultaneously with overload repetition work. Keeping the total number of repetitions constant, training with a greater load produces greater muscle endurance than training with light loads. Also, with the total work output held constant, training with fewer repetitions using greater loads (25 RM) produces greater muscle endurance than more repetitions at lighter loads (Hellebrandt and Houtz, 1956). Moffroid and Whipple (1970) lend support to observations that high-power exercise increases muscular endurance at high speeds.

For improving muscular endurance, the specificity principle should serve as a guide for designing muscle endurance training. The performer should ensure that the joint movements, resistances used, body positions, speed or cadence of repetitions, ROM used, and total number of repetitions are *specific to* and *appropriate for* the activity for which the training is being done.

Understanding Muscular Endurance

1. With a heavy book in your hand, stand and abduct the shoulder to position the upper extremity horizontally. Time how long you can maintain that horizontal position, and compare your time with the results of others.

2. Repeat the steps in question 1 with the other arm but with the elbow fully flexed. Compare the muscular endurance with your previous results. Explain the difference in terms of the magnitude of the resistance (consider mass, weight, and weight torque).
3. Using a specific muscle group, measure the maximum static force you can exert against a scale, tensiometer, or some other device for estimating force. How long can you continue to exert that maximum force?
4. Using the result obtained in question 3, determine 80% of your maximum static force. How long can you continue to exert 80% of that maximum force? Allow several minutes' rest and repeat for 50% of your original maximum force.

7.6 Flexibility

The term **flexibility** is defined as the ROM available at an articulation. Flexibility is specific to each joint; that is, a person may have a large ROM at the wrist joint in all directions, but a limited ROM in one movement within the vertebral column (e. g., lumbar vertebral flexion). A lack of ROM at certain joints is a common cause for inefficient movement or skill performance. For example, swimmers who pursue more than adequate shoulder flexibility can injure the anterior capsule of the shoulder joint by stretching the muscles (and capsule) to a point at which the elbows cross behind the back. Joint laxity produced by such extreme stretching can lead to anterior dislocation, especially in the backstroke turn when the hand contacts the wall (Marino, 1983). In developing flexibility at joints for a particular activity, the ROM *necessary* for performance should be the target condition, keeping in mind that the dynamic ROM may be greater than the static ROM during the actual performance.

Passive and Active Flexibility

Passive flexibility refers to the ROM available when an outside force (e.g., gravity, momentum, another part of the body, or another person) is the motive force (Figure 7.7a). **Active flexibility** is the ROM available when internal muscle forces cause the movement range (Figure 7.7b). If a segment must be moved through a ROM by the motive force of muscles during an activity, weakness in those muscles may cause the ROM to be less than what it would be passively. Holt and Smith (1982) report differences averaging 18 degrees in hip flexion ROM in favor of passive flexibility.

Both passive and active flexibility can be important variables in the performance of a skill. Passive dynamic stretching of a muscle group in the course of a movement often results in joint positions that cannot be reproduced with slow voluntary muscular effort. For example, stretching the shoulder medial rotators in the initial execution phase of a baseball pitch allows the shoulder to be moved into extreme lateral rotation with abduction. Figure 7.7c shows the ROM that should safely be available to the thrower. In this case, the passive ROM is caused by the lag of the forearm and hand as the more proximal segments rotate forward.

In wrestling, passive flexibility is extremely important for injury avoidance. The body is subjected to extreme outside forces applied by the opponent. Flexibility enables the wrestler

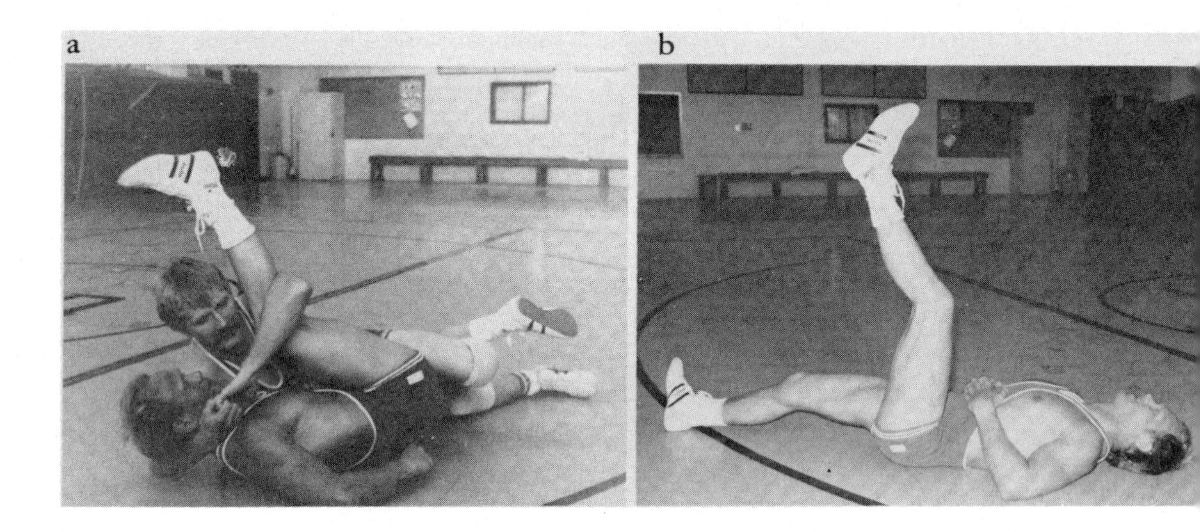

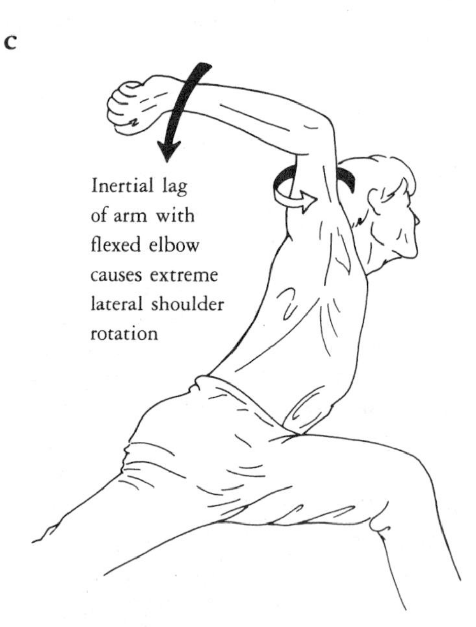

FIGURE 7.7

(a) Passive flexibility of the wrestler's hip joint. (b) Active flexibility of this wrestler's hip joint. (c) Passive flexibility fo lateral rotation of the shoulder joint of this pitcher.

to "deform" enough to avoid tissue tearing as well as to slip out of positions that would otherwise be disabling (Figure 7.8). Adequate strength in extreme joint positions also is necessary to prevent joint structure damage by the outside force. The greater the wrestler's active flexibility, which incorporates enough muscular strength to cause the ROM, the better able he is to wrap his body, arms, and legs around the opponent.

FIGURE 7.8
Passive flexibility of the shoulder joint subjected to extreme forces provides a margin of safety.

Factors Limiting Range of Motion

ROM may be limited by four factors: (1) connective tissue restrictions, (2) bone configurations at the articulations, (3) contact of tissue masses of adjacent segments, and (4) clothing restrictions. In addition to these factors, active flexibility is also limited by the strength of motive force muscle groups. Physical contact of the tissue masses of adjacent segments as well as the physical contact of the bony shapes forming the joint itself are conditions that must be accepted and cannot be changed unless the tissue masses that interfere with normal movement happen to be unnecessary adipose tissue. For such a restriction, a program of weight control would be in order.

A common cause of inflexibility is inactivity. When the total ROM at a particular joint is not used over a period of time, the connective tissues crossing that joint become shortened; that is, the tissues adapt to permit the ROM that is routinely performed. If a certain activity demands a large ROM at one or more articulations and that activity is performed frequently, the tissues crossing the articulation become accustomed to being lengthened and tend to remain flexible enough to impose no restriction on joint ROM. Inactivity, however, need not be total, whole-body inactivity for tissue shortening to occur. The areas of the body in which large movements of the segments are neglected because normal daily activity does not require them will exhibit the greatest movement restriction or inflexibility. Once a range is established at a particular joint and in a particular direction, that ROM will be maintained merely by frequent use.

If flexibility training is to be undertaken for greater ROM for skill improvement, the joint movements that need to be increased should be determined exactly; that is, the exercise should be specific to the need. For example,if a swimmer cannot swing both arms above the surface of the water in the butterfly stroke, the restriction of movement may be caused by the tissues crossing the anterior side of the shoulder joints. A stretching exercise that lengthens the transverse adductors of the shoulder joint would decrease the restriction of shoulder transverse abduction. In addition, strength may be lacking in the muscles that adduct the scapula (retract the shoulder girdle) and cause transverse extension of the shoulder joints. In this case, strengthening exercises would be in order. The total flexibility available to perform a given skill is as great as the flexibility of all the joints involved. This weakest link concept is applicable, therefore, in skills demanding flexibility as well as strength or endurance of various muscle groups.

Connective Tissue Properties and Implications for Stretching Methods

In most joints ROM is limited by connective tissue structures such as tendons, ligaments, muscle fascial sheaths, aponeuroses, and ligamentous joint capsules. The connective tissue network within and around muscle tissue provides most, if not all, of the resistance to stretching of the muscle (Sapega et al., 1981). Following injury, the scar tissues forming in connective tissues restrict ROM unless therapeutic stretching is done.

Because connective tissue is the target for flexibility programs, understanding the response of connective tissue to stretching, or elongation, is important. When connective tissue is elongated, its lengthening has two components: **elastic** stretching (i.e., it will recoil) and **plastic** stretching (i.e., it will not recoil). The permanence of the stretch depends on how much plastic stretch occurs; the elastic-type lengthening merely causes recoil of the tissue to its original length. The plastic elongation is directed toward the viscous (thick fluid) property of the *viscoelastic* nature of connective tissue. The proportion of the contribution of each kind to the lengthening depends on *how* the stretching is performed. The three variables concerned are (1) the magnitude of the stretching force, (2) the duration of the stretching force, and (3) the tissue temperature when the stretching is taking place (Sapega et al., 1981).

Summarizing the information provided by Sapega et al., permanent lengthening of connective tissue is produced best by lower force of longer duration applied at elevated tissue temperatures (40°C or 104°F and above in therapeutic settings). Cooling the tissue (by ice packs about 15 min) before releasing the tension seems to increase the permanence of the plastic elongation. In addition, under these conditions, any structural weakening of the tissues is minimized.

Elastic lengthening (elongation that will recoil) is produced by high-force, short-duration stretching at normal or colder tissue temperatures.

Methods of Stretching

Stretching methods are categorized as static or dynamic. Static stretching is also known as passive or slow, sustained stretching. Dynamic stretching is called active, ballistic, bouncing, or fast stretching. Research studies on flexibility improvement indicate that both methods are effective but that static methods are safer and result in less muscle soreness (Beaulieu, 1981; Corbin and Noble, 1980; deVries, 1980; Holt and Smith, 1982).

Dynamic Stretching

Dynamic stretch of a muscle group brings into play neuromuscular mechanisms involving the muscle spindle (myotatic, or stretch reflex) and the Golgi tendon organ. In fast stretching, the momentum of the moving body segment, rather than an external force, is used to push the articulation beyond its present ROM. Common examples of this are the bouncing method of toe touching, the lateral flexion of the trunk with the arms raised over the head, and the pectoral stretch done by flinging the arms in transverse abduction.

The speed of stretch elicited by the dynamic method stimulates the muscle spindles, and the stretch reflex causes a contraction of the same muscles that are being stretched. The greater the speed of muscle stretching, the greater the stretch reflex. Therefore, even though the momentum brings the articulation beyond its normal range, the ballistic method causes a countercontraction immediately following or during the stretch and seems to reduce stretching benefits. Reports associate muscle soreness with this type of stretching; that is, before motion stops, the sudden tension put on the tissues can result in small tears in the connective tissue or muscle fibers, which lead to swelling or pain.

Slow or Static Stretching

To reduce the velocity of stretch that the muscles undergo and thus reduce the intensity of the stretch reflex, an external object, such as a wall or a partner, or an antagonistic muscle group is used to apply tension slowly to the muscles to be stretched and thus causes these muscles to stretch beyond the normal range. Some muscle-spindle activity results from the stretched position, but it is minimal because the spindles react more to the speed of stretch than to the stretched position. Maintaining tension in a lengthened muscle group for 10 to 30 seconds has been found practical.

In light of the properties of connective tissue, the static, or long duration low-force, methods seem most favorable. In addition, an easy muscle warm-up activity before stretching increases the plastic elongation of the tissues as a result of the increase in temperature. Sapega et al. (1981) propose that stretching exercises *not* be done at the beginning of a warm-up routine. Stretching before running or swimming, for example, should be done *following* a minimum of 5 min of light but progressive muscular activity such as walking or jogging. Stretching should be performed immediately after the main part of a workout because it is then that the tissue temperatures are the highest. Sapega et al. also point out that the clinical use of "cryostretching," or stretching combined with cold application, should be used only for limited therapeutic goals. These include cases in which adhesions are to be torn, when muscle spasticity interferes with lengthening, and when stretching is so painful that cold must be used for analgesia. Otherwise, the tissues should be warm for the best stretching.

Proprioceptive Neuromuscular Facilitation Stretching

Proprioceptive neuromuscular facilitation (PNF) techniques developed for physical therapy use by Kabat (1950) subsequently have been modified by a number of researchers and practitioners. The PNF methods have been shown to be more effective than the traditional methods (Corbin and Noble, 1980; W. L. Cornelius and Hayes, 1987; Holt and Smith, 1982; Moore and Hutton, 1980; and Salem, 1980). Reports indicate also that less post-stretching soreness results with these methods.

The two most popular forms of PNF stretching are contract–relax (CR) and contract-relax, agonist contract (CRAC). Although the neuromuscular bases for the effectiveness of these techniques are still unclear (Moore and Hutton, 1980), the following rationale has been proposed. In the CR method, the tight muscle group is lengthened and then isometrically contracted maximally to elicit the Golgi tendon organ's autogenic inhibition. This is followed immediately by relaxation and passive stretching (usually by a partner) to further lengthen the muscle group. The slow prolonged stretch is used to prevent muscle spindle stimulation, which would increase the tension in the tight muscles through the stretch reflex.

In the CRAC method, the CR procedure is followed for the first phase, then, while the partner is applying gentle stretch, the agonist muscle group is concentrically contracted to increase the stretch. In addition to the autogenic inhibition and suppression of the stretch reflex, reciprocal inhibition may contribute to the effectiveness of the CRAC method. Recall from Chapter 2 that when an agonist contracts, its antagonist is inhibited.

The PNF method is discussed and applied to sport activities by Holt (n.d.). The procedure is in three parts:

1. The muscle group that needs to be stretched is identified, and the performer places the articulation at the limited end of its ROM (i.e., the muscle group is placed in its longest position).
2. The performer forcefully contracts (isometrically) the muscle group to be stretched against an immovable force such as a wall, the ground, or a partner.
3. Immediately thereafter, a gentle external source is applied to the segment as the performer contracts the muscle group opposite to the group being stretched.

The external source exerts a force against the segment so that the previously contracted muscle group is slowly stretched. The advantages to the slow stretch are the same as those given for the static stretch method. The slowness of the stretch minimizes the suddenness of stretch on the muscle spindles so that little or no reflex contraction is elicited in the stretched muscle. The example presented in Figure 7.9 demonstrates the PNF procedure for stretching the anterior shoulder muscles to allow greater ROM for full flexion and transverse abduction.

According to Holt (n.d.), the procedure would be the following:

1. Position the shoulder joint and arms so that the "tight" muscle group is at its longest length (Figure 7.9a).
2. Gradually contract the tight muscle group against the immovable resistance provided by the partner, building to maximum isometric tension in 2 sec and then holding at maximum for 4 sec.
3. Immediately contract the agonist muscle group to cause increased ROM in the desired direction and to stretch the tight muscle. At the same time, the partner assists with a gentle pull for 4 sec (Figure 7.9b).

Holt and Smith (1982) compared this procedure for hamstring stretch with a slow active stretch and an isometric contraction control group. Five 10-sec stretches with 10-sec rests between stretches produced greater ROM with the PNF method. Other investigators report successes with other protocols, as well. Etnyre and Lee (1987) present an informative overview of PNF investigations and their results.

Planning a Flexibility Program

Flexibility exercises should be planned for a specific purpose—to increase a given joint's ROM. An analysis plan should include these steps:

1. Identify the articulation that has limited ROM within a movement sequence.

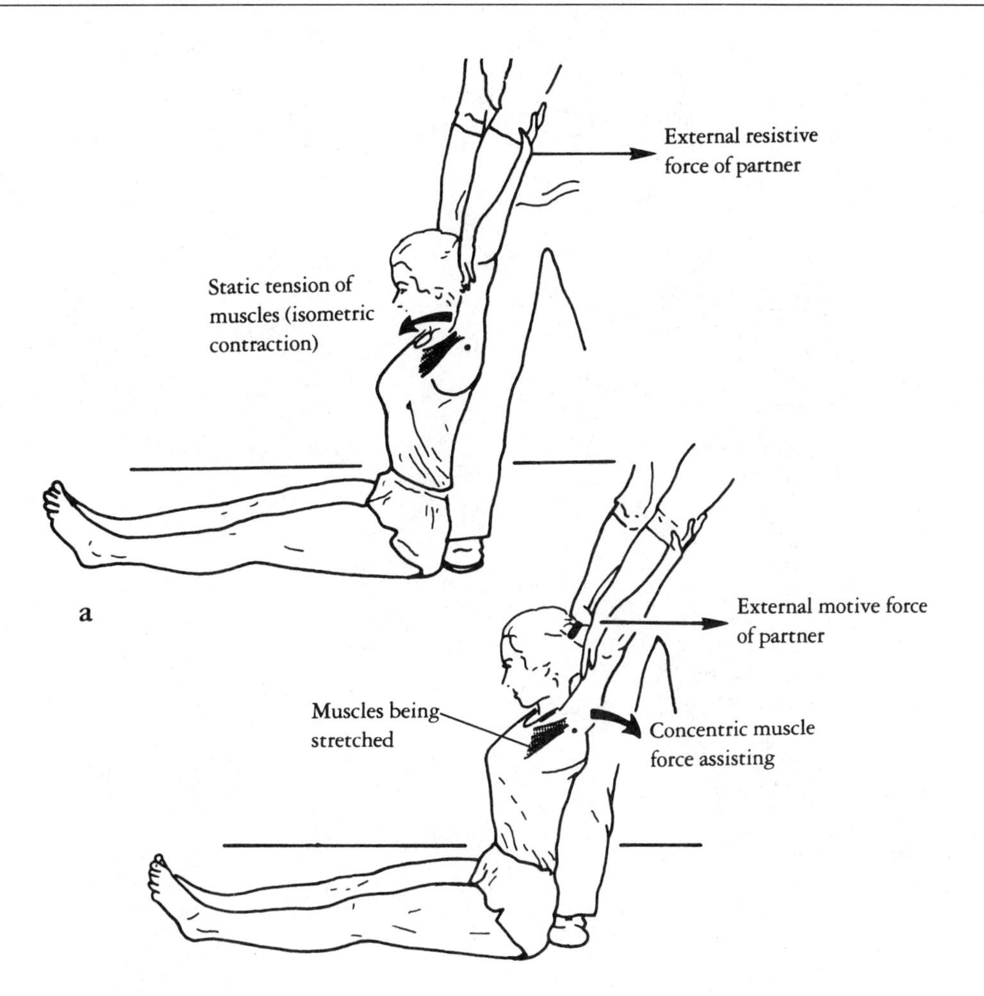

FIGURE 7.9
The procedure for a proprioceptive neuromuscular facilitation (PNF) method of stretching using a partner to provide resistance then assistance.

2. Identify the position of the joint at the point of restriction (e.g., hip flexed, shoulder abducted).
3. Note the positions of surrounding articulations that may influence the restriction of the given joint. (Look for those joints that share common muscles.)
4. Identify the general muscle group needing stretch (e.g., hip extensors, shoulder adductors).
5. Select a method of stretching, preferably static or PNF or another based on the best, most up-to-date information available.
6. Devise a plan of stretching so that the movements simulate the positions identified in steps 2 and 3; that is, stretching should be specific to the need within the activity.

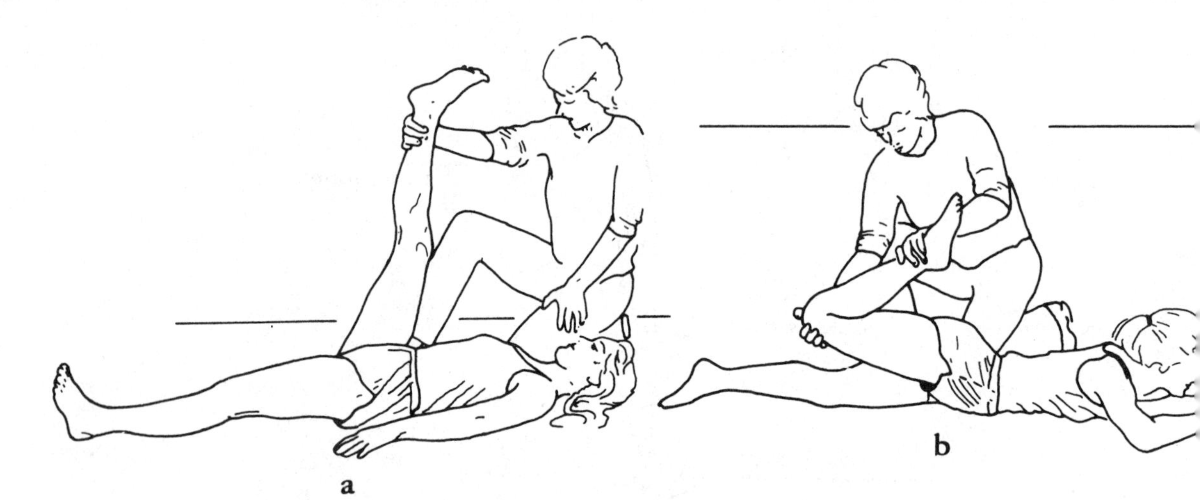

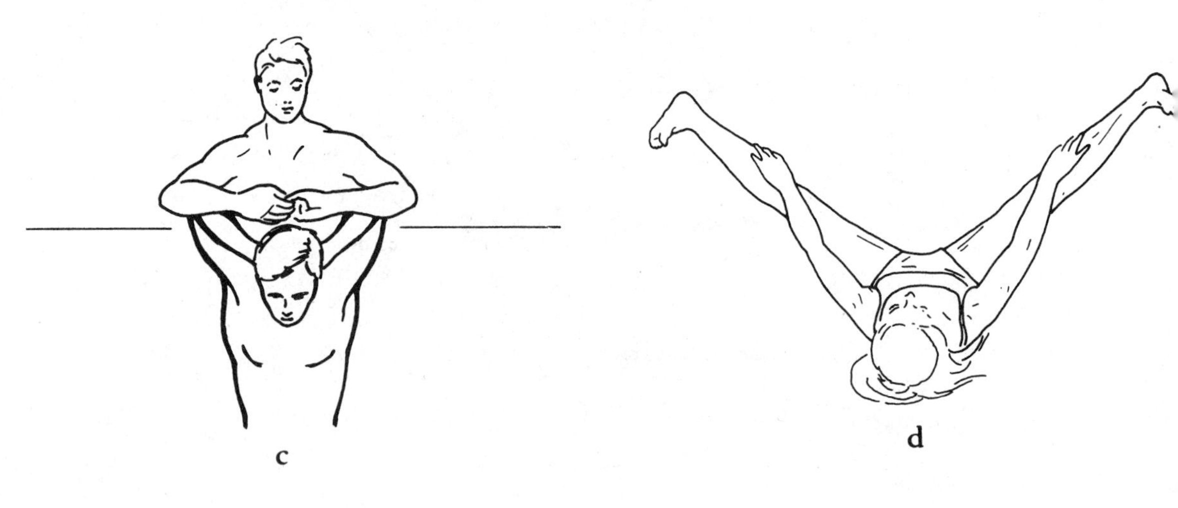

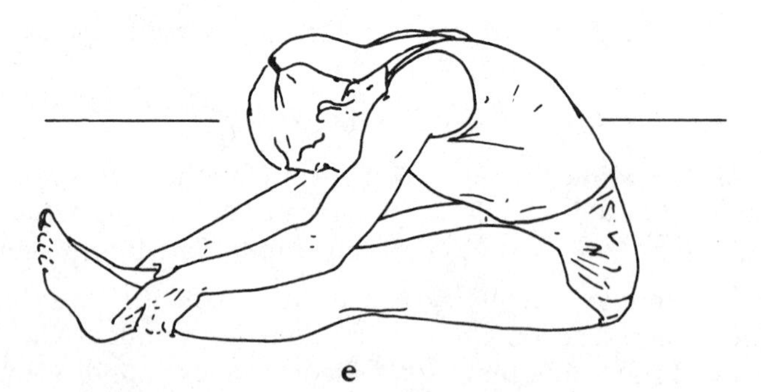

FIGURE 7.10

Exercises for stretching selected muscle groups.

Understanding the Biomechanical Aspects of Flexibility

1. Differentiate between active and passive flexibility.
2. Explain the difference between the elastic and plastic elongation of connective tissue.
3. Identify three stretching exercises, each commonly used in preparing for a different sport. Evaluate the effectiveness of the exercise in relation to the movements that require a given ROM in that sport.
4. Using the same three sports as in question 3, devise a different flexibility exercise for each, using the six steps in the analysis plan on pages 302–303.
5. Sit on the floor with your knees extended, ankles plantar flexed. Reach toward your toes, and hold the position of maximum reach. While holding this reach position, dorsiflex the ankles and note the result. Flex the knees slightly and note the result. Discuss these results in terms of specificity of joint positions in an activity for which stretching exercises are given.
6. For each of the flexibility exercises shown in Figure 7.10, indicate the following: (a) the articulation being stretched, (b) the position of the articulation, (c) the muscle group being stretched, (d) the force used to push or pull the joint beyond its range, (e) the muscle group antagonistic to the group being stretched.
7. Devise an exercise using the PNF procedure for extending the ROM of the following movements: (a) shoulder girdle retraction and shoulder joint transverse extension, (b) plantar flexion of the ankle, (c) dorsiflexion of the ankle, (d) lumbar vertebral flexion, (e) medial rotation of the hip, (f) lateral rotation of the hip. For each exercise, state the following: (a) the muscle group to be stretched, (b) the passive force used to stretch, (c) the antagonist muscle group used to facilitate the stretching, (d) the procedure for the stretching.

References and Suggested Readings

Andrews, J. G., & Hay, J. G. (1987). Strength curves for multiple-joint, multiple degree-of-freedom exercises. In B. Jonsson (Ed.), *Biomechanics X-A* (pp. 535–540). Champaign, IL: Human Kinetics.

Ariel, G. B. (1984). Resistive exercise machines. In J. Terauds, K. Barthels, E. Kreighbaum, R. Mann, & J. Crakes (Eds.), *Sports Biomechanics: Proceedings of the International Symposium of Biomechanics in Sports* (pp. 295–306).

Asmussen, E. (1979). Muscle fatigue. *Medicine and Science in Sports, 11*(4), 313–321.

Astrand, P., & Rodahl, K. (1977). *Textbook of work physiology*. New York: McGraw-Hill.

Atha, J. (1981). Strengthening muscle. In D. I. Miller (Ed.), *Exercise and sports science reviews*, (Vol. 9, 1–74). Philadelphia: The Franklin Press.

Barnes, W. S. (1980). The relationship between maximum isokinetic strength and isokinetic endurance. *Research Quarterly, 51*(4), 714–717.

Bartels, R., Coyle, E., Huesner, W., & McLaughlin, T. (1980). Scientists talk about strength training. *Swimming Technique, 17*(2), 14–25.

Beaulieu, E. (1981). Developing a stretching program. *The Physician and Sportsmedicine, 9*(11), 59–69.

Berger, R. A. (1970). Relationship between dynamic strength and dynamic endurance. *Research Quarterly, 41*(1), 115–116.

Bigland-Ritchie, B. (1981). EMG/force relations and fatigue of human voluntary contractions. In D. I. Miller (Ed.), *Exercise and sports science reviews* (Vol. 9, pp. 75–117). Philadelphia: The Franklin Press.

Blattner, S. E., & Noble, L. (1979). Relative effects of isokinetic and plyometric training on vertical jumping performance. *Research Quarterly, 50*(4), 583–588.

Bobbert, M. F., Huijing, P. A., & Van Ingen Schenau, G. J. (1987a). Drop jumping. I. The influence of jumping technique on the biomechanics of jumping. *Medicine and Science in Sports and Exercise, 19*, 332–338.

Bobbert, M. F., Huijing, P. A., & Van Ingen Schenau, G. J. (1987). Drop jumping. II. The influence of dropping height on the biomechanics of drop jumping. *Medicine and Science in Sports and Exercise, 19*, 339–346.

Booth, F. W., & Gould, E. W. (1975). Effects on training and disuse on connective tissue. In J. H. Wilmore & J. F. Keogh (Eds.), *Exercise and Sports Science Reviews* (Vol. 3, pp. 95–103). Santa Barbara, CA: Journal Publishing Affiliates.

Carlton, L. G., & Newell, K. M. (1985). Force variability in isometric tasks. In D. A. Winter, R. W. Norman, R. P. Wells, K. C. Hayes, & A. E. Patla (Eds.), *Biomechanics IX-A* (pp. 128–132). Champaign, IL: Human Kinetics.

Cavagna, G. A. , Citterio, G., & Jacini, P. (1975). The additional mechanical energy delivered by the contractile component of the previously stretched muscle. *Journal of Physiology, 251*, 658–668.

Chang, D. E. (1988). Limited joint mobility in power lifter. *The American Journal of Sports Medicine, 16*, 280–284.

Clarke, D. H. (1973). Adaptations in strength and muscular endurance resulting from exercise. In J. H. Wilmore (Ed.), *Exercise and sports science reviews* (Vol. 1, pp. 74–102). New York: Academic Press.

Clutch, D., Wilton, M., McGown, C., & Bryce, G. R. (1983). The effect of depth jumps and weight training on leg strength and vertical jump. *Research Quarterly for Exercise and Sport, 54*(1), 5–10.

Corbin, C. B., & Noble, L. (1980). Flexibility, a major component of physical fitness. *JOHPER, 51*(1), 23–24, 57–60.

Cornelius, C. J., & Seireg, A. A. (1986). Optimum human power. *SOMA, 1*, 21–29.

Cornelius, W. L., & Hayes, K. K. (1987). A comparison of single vs repeated MVIC maneuvers used in PNF flexibility techniques for improvement in ROM. *Journal of Applied Sport Science Research, 1*, 71–73.

Coyle, E. F., Costill, D. L., & Lesmes, G. R. (1979). Leg extension power and muscle fiber composition. *Medicine and Science in Sports, 11*(1), 12–15.

Daish, C. B. (1972). *Learn science through ball games.* New York: Sterling.

de Haan, A., van Doorn, H. E., Huijing, P. A., Woitiez, R. D., & Westra, H. G. (1987). The effect of muscle length on the economy of repetitive isometric contractions. In B. Jonsson (Ed.), *Biomechanics X-B* (pp. 967–972). Champaign, IL: Human Kinetics.

DeLateur, B. J. (1978). Exercise for strength and endurance. In J. V. Basmajian (Ed.), *Therapeutic exercise* (3rd ed.) (pp. 86–92). Baltimore: Williams & Wilkins.

DeLorme, T. L., & Watkins, A. L. (1951). *Progressive resistance exercise.* New York: Appleton-Century-Crofts.

deVries, H. A. (1980). *Physiology of exercise* (3rd ed.). Dubuque, IA: William C. Brown.

Duchateau, J., & Hainaut, K. (1988). Training effects of submaximal electrostimulation in a human muscle. *Medicine and Science in Sports and Exercise, 20*, 99–104.

Ellenbecker, T. S., Davies, G. J., and Rowinski, M. J. (1988). Concentric versus eccentric isokinetic strengthening of the rotator cuff. *The American Journal of Sports Medicine, 16*, 64–69.

Etnyre, B. R., and Abraham, L. D. (1988). Antagonist muscle activity during stretching: a paradox re-assessed. *Medicine and Science in Sports and Exercise, 20*, 285–289.

Etnyre, B. R. and Lee, E. J. (1987). Comments on proprioceptive neuromuscular facilitation stretching techniques. *Research Quarterly for Exercise and Sport, 58*, 184–188.

Gans, C. (1982). Fiber architecture and muscle function. In R. L. Terjung (Ed.), *Exercise and sports science reviews* American College of Sports Medicine Series, (pp. 160–207). Philadephia: Franklin Institute Press.

Garhammer, J. (1980). Power production by Olympic weightlifters. *Medicine and Science in Sports and Exercise, 12*(1), 54–60.

Gregoire L., Veegher, H. E., Huijung, P. A., & van Ingen Schenau, G. J. (1984). Role of mono-

and biarticular muscles in explosive movements. *International Journal of Sports Medicine, 5,* 301–305.

Harman, E. (1983). Resistive torque analysis of 5 Nautilus exercise machines. *Medicine and Science in Sports and Exercise, 15,* 113 (abstract).

Hay, J. G., Andrews, J. G., & Vaughn, C. L. (1980). The influence of external load on the joint torques exerted in a squat exercise. In J. M. Cooper & B. Haven (Eds.), *Biomechanics: Proceedings of the Biomechanics Symposium* (pp. 286–293). Indiana State Board of Health.

Hay, J. G., Andrews, J. G., & Vaughn, C. L. (1983). Effects of lifting rate on elbow torques exerted during arm curl exercises. *Medicine and Science in Sports and Exercise, 15*(1), 63–71.

Hay, J. G., Andrews, J. G., Vaughn, C. L., & Ueya, K. (1983). Load, speed, and equipment effects in strength-training exercises. In H. Matsui & K. Kobayashi (Eds.), *Biomechanics VIII-B: Proceedings of the Eighth International Congress of Biomechanics* (pp. 940–949). Champaign, IL: Human Kinetics.

Hellebrandt, F. A., & Houtz, S. J. (1956). Mechanisms of muscle training in man. *Physical Therapy Review, 36,* 371–383.

Heusner, W. (1980). *The theory of strength development*. Paper presented at the 1980 American Swimming Coaches Association Clinic, New Orleans.

Heyward, V. H. (1975). Influence of static strength and intramuscular occlusion on submaximal static muscle endurance. *Research Quarterly, 46*(4), 393–402.

Hinson, M. N., Smith, W. C., & Funk, S. (1979). Isokinetics: A clarification. *Research Quarterly, 50*(1), 30–35.

Holt L. E. *Scientific stretching for sport (3-S)*. P. O. Box 7045, North Halifax, Nova Scotia: Sport Research Limited, B3K 5J4, n.d.

Holt, L. E., & Smith, R. K. (1982). The effect of selected stretching programs on active and passive flexibility. In J. Terauds (Ed.), *Biomechanics in Sports: Proceedings of the International Symposium of Biomechanics in Sports* (pp. 54–67). Del Mar, CA: Research Center for Sports.

Holt, L. E., Travis, T. M., & Okita, T. (1970). A comparative study of three stretching techniques. *Perceptual and Motor Skills, 31,* 611–616.

Ikegawa, S., Tsunoda, N., Yata, H., Matsuo, A., & Fukunaga, T. (1987). The absolute muscle strength of various muscle groups. In B. Jonsson (Ed.), *Biomechanics X-A* (pp. 519–522). Champaign, IL: Human Kinetics.

Ikegawa, S., Tsunoda, N., Yata, H., Matsuo, T., Fukunaga, T., & Asami, T. (1985). The effect of joint angle on cross-sectional area and muscle strength of human elbow flexors. In D. A. Winter, R. W. Norman, R. P. Wells, K. C. Hayes, & A. E. Patla (Eds.), *Biomechanics IX-A* (pp. 39–43). Champaign, IL: Human Kinetics.

Jesse, J. P. (1979). Misuse of strength development programs in athletic training. *The Physician and Sportsmedicine, 7*(10), 46–51.

Johnson, B. L., Adamczyk, J. W., Tennoe, K. D., & Stromme, S. B. (1976). A comparison of concentric and eccentric muscle training. *Medicine and Science in Sports, 8*(1), 35–38.

Kabat, H. (1950). Studies on neuromuscular dysfunction: XIII. New concepts and techniques of neuromuscular re-education for paralysis. *Permenente Foundation Medical Bulletin, 8,* 121–143.

Kedzior, K., Kotwicki, E., & Niwinski, W. (1987). Testing module for static and dynamic measurements of muscle group characteristics. In B. Jonsson (Ed.), *Biomechanics X-B* (pp. 1127–1130). Champaign, IL: Human Kinetics.

Knuttgen, H. G. (1976). Development of muscular strength and endurance. In H. G. Knuttgen (Ed.), *Neuromuscular mechanisms for therapeutic and conditioning exercise* (pp. 97–118). Baltimore: University Park Press.

Knuttgen, H. G. (1986). Quantifying exercise performance with SI units. *The Physician and Sportsmedicine, 14,* 157–161.

Knuttgen, H. G., & Kraemer, W. J. (1987). Terminology and measurement in exercise performance. *Journal of Applied Sport Science Research, 1,* 1–10.

Komi, P. V. (1984). Physiological and biomechanical correlates of muscle function: Effects of muscle structure and stretch-shortening cycle on force and speed. In R. L. Terjung (Ed.), *Exercise and sport sciences reviews: Vol. 12* (American College of Sports Medicine Series) (pp. 81–121). Lexington, MA: The Collamore Press.

Komi, P. V. (1986). Training of muscle strength and power: Interaction of neuromotoric, hypertrophic and mechanical factors. *International Journal of Sports Medicine, 7*(Supplement), 10–15.

Komi, P. V., & Bosco, C. (1978). Utilization of stored elastic energy in leg extensor muscles by men and women. *Medicine and Science in Sports, 10*(4), 261–265.

Kornecki, S., Kulig, K., & Zawadzki, J. (1985). Biomechanical implication of the artificial and natural strengthening of wrist joint structures. In D. A. Winter, R. W. Norman, R. P. Wells, K. C. Hayes, and A. E. Patla (Eds.), *Biomechanics IX-A* (pp. 118–122). Champaign, IL: Human Kinetics.

Kulig, K., Andrews, J. G., & Hay, J. G. (1984). Human strength curves. In R. L. Terjung (Ed.)., *Exercise and sport sciences reviews: Vol. 12* (American College of Sports Medicine Series) (pp. 417–466). Lexington, MA: The Collamore Press.

Kumar, S., & Chaffin, D. B. (1987). Static and dynamic lifting strength of young males. In B. Jonsson (Ed.), *Biomechanics X-B* (pp. 25–28). Champaign, IL: Human Kinetics.

Kume, S., & Ishii, K. (1981). Biomechanical analysis of isokinetic exercises. In A. Morecki, K. Fidelus, K. Kedzior, & A. Wit (Eds.), *Biomechanics VII-B: Proceedings of the Seventh International Congress on Biomechanics* (pp. 404–410). Baltimore: University Park Press.

Lesmes, G. R., Costill, D. L., Coyle, E. F., & Fink, W. J. (1978). Muscle strength and power changes during maximal isokinetic training. *Medicine and Science in Sports, 10*(4), 266–269.

A look at equipment. (1980, August). *Swimming Technique*, pp. 36–38, 41–43.

Marino, M. (1983). Sports medicine in action: A look at swimming [Monograph]. *Muscle and Bone, 3*(3). Published by Current Concepts, Inc., 850 Third Ave., New York.

Marshall, J. L., et al. (1980). Joint looseness: A function of the person and the joint. *Medicine and Science in Sports and Exercise, 12*(3), 189–194.

Moffroid, M., & Kusiak, E. (1975). The power struggle: Definition and evaluation of power of muscular performance. *Physical Therapy, 55*(10), 1098–1104.

Moffroid, M., & Whipple, R. H. (1970). *Specificity of speed of exercise*. Paper adapted from master's thesis, New York University.

Moore, M. A., & Hutton, R. S. (1980). Electromyographic investigation of muscle stretching techniques. *Medicine and Science in Sports and Exercise, 12*(5), 322–329.

Mueller, K., & Buehrle, M. (1987). Comparison of static and dynamic strength of the arm extensor muscles. In B. Jonsson (Ed.), *Biomechanics X-A* (pp. 501–506). Champaign, IL: Human Kinetics.

Osternig, L. R., Sawhill, J. A., Bates, B. T., & Hamill, J. (1983). Function of limb speed on torque patterns of antagonist muscles. In H. Matsui & K. Kobayashi (Eds.), *Biomechanics VIII-A: Proceedings of the Eighth International Congress on Biomechanics* (pp. 189–205). Champaign, IL: Human Kinetics.

Perrine, J. J., & Edgerton, V. R. (1978). Muscle force-velocity relationships under isokinetic loading. *Medicine and Science in Sports, 10*(3), 159–66.

President's Council on Physical Fitness and Sports (1975). Joint and body range of motion. *Physical Fitness and Research Digest, 5*, 17.

Raitsin, L. M. (1974). The effectiveness of isometric and electro-stimulated training on muscle strength at different joint angles. *Theory and Practice of Physical Culture, 12*, 33–35.

Rutherford, O. M., & Jones, D. A. (1986). The role of learning and co-ordination in strength training. *European Journal of Applied Physiology, 55*, 100–105.

Sale, D. G. (1986). Neural adaptation in strength and power training. In N. L. Jones, N. McCartney, & A. J. McComas (Eds.), *Human muscle power* (pp. 281–305). Champaign, IL: Human Kinetics.

Sale, D. G. (1987). Influence of exercise and training on motor unit activation. In B. Pandolf (Ed.), *Exercise and sport sciences reviews* (Vol. 15, pp. 95–151). New York: Macmillan.

Sale, D. G. (1988). Neural adaptation to resistance training. *Medicine and Science in Sports and Exercise, 20*(Supplement), S135-S143.

Sale, D. G., & MacDougall, J. D. (1981). Specificity in strength training: A review for the coach and athlete. *Canadian Journal of Applied Sports Sciences, 6*(2), 87–92.

Sale, D. G., MacDougall, J. D., Upton, A. R. M., & McComas, A. J. (1983). Effect of strength training upon motorneuron excitability in man. *Medicine and Science in Sports and Exercise, 15*(1), 57–62.

Salem, N. (1980). Selected stretching techniques based on mechanical and neurophysiological principles and muscle tightness. In J. M. Cooper & B. Haven (Eds.), *Biomechanics: Proceedings of the Biomechanics Symposium* (pp. 352–353). Indiana State Board of Health.

Sapega, A. A., Nicholas, J. A., Sokolow, D., & Saranti, A. (1982). The nature of torque "overshoot" in cybex isokinetic dynamometry. *Medicine and Science in Sports and Exercise, 14*(5), 368–375.

Sapega, A. A., Quedenfeld, T. C., Moyer, R. A., & Butler, R. A. (1981). Biophysical factors in range of motion exercise. *The Physician and Sportsmedicine, 9*(12), 57–65.

Schmidtbleicher, D., & Buehrle, M. (1987). Neuronal adaptation and increase of cross-sectional area studying different strength training methods. In B. Jonsson (Ed.), *Biomechanics X-B* (pp. 615–620). Champaign, IL: Human Kinetics.

Shimoshikiryo, K., Nagata, A., Muro, M., & Sunamoto, H. (1985). Relationship between fatigability and stretching levels of lower leg muscles during maximal voluntary and nerve stimulated contractions in humans. In D. A. Winter, R. W. Norman, R. P. Wells, K. C. Hayes, & A. E. Patla (Eds.), *Biomechanics IX-A* (pp. 112–117). Champaign, IL: Human Kinetics.

Sjøgaard, G. (1978). Force–velocity curve for bicycle work. In E. Asmussen & K. Jorgensen (Eds.), *Biomechanics VI-A: Proceedings of the Sixth International Congress of Biomechanics* (pp. 93–99). Baltimore: University Park Press.

Skyne, K. (1982). Richard H. Dominguez, MD: To stretch or not to stretch? *The Physician and Sportsmedicine, 10*(9), 137–40.

Smith, T. K. (1984). Preadolescent strength training: some considerations. *JOPERD, 55*(1), 43–44, 80.

Stevens, R. (1980). Isokinetic versus isotonic training in the development of lower body strength and power. *Scholastic Coach, 49*(6), 74–76.

Strass, D. (1987). Effects of fatigue on isometric force-time characteristics in human muscle. In B. Jonsson (Ed.), *Biomechanics X-A* (pp. 265–270). Champaign, IL: Human Kinetics.

Thigpen, L. K., Moritani, T., Thiebaud, T., & Hargis, J. L. (1985). The acute effects of static stretching on alpha motor neuron excitability. In D. A. Winter, R. W. Norman, R. P. Wells, K. C. Hayes, & A. E. Patla (Eds.), *Biomechanics IX-A* (pp. 352–356). Champaign, IL: Human Kinetics.

Tihanyi, J., Apor, P., & Petrekanits, M. (1987). Force-velocity-power characteristics for extensors of lower extremities. In B. Jonsson (Ed.), *Biomechanics X-B* (pp. 707–712). Champaign, IL: Human Kinetics.

Thrash, K., & Kelly, B. (1987). Research notes: Flexibility and strength training. *Journal of Applied Sport Science Research, 1*, 74–75.

Tuttle, W. W., Janney, C. D., & Thompson, C. W. (1950). Relation of maximum grip strength to grip strength endurance. *Journal of Applied Physiology, 2*(12), 663–670.

Urbanik, C., & Ubukata, O. (1985) The influence of training with concentric and eccentric work on the force and velocity characteristics of muscle. In D. A. Winter, R. W. Norman, R. P. Wells, K. C. Hayes, & A. E. Patla (Eds.), *Biomechanics IX-A* (pp. 77–81). Champaign, IL: Human Kinetics.

van Ingen Schenau, G. J. (1984). An alternative view to the concept of utilization of elastic energy in human movement. *Human Movement Science, 3*, 301–36.

Viitasalo, J. T. (1985). Effects of training on force-velocity characteristics. In D. A. Winter, R. W. Norman, R. P. Wells, K. C. Hayes, & A. E. Patla (Eds.), *Biomechanics IX-A* (pp. 91–95). Champaign, IL: Human Kinetics.

Viitasalo, J. T. (1985). Measurement of force-velocity characteristics for sportsmen in field conditions. In D. A. Winter, R. W. Norman, R. P. Wells, K. C. Hayes, & A. E. Patla (Eds.), *Biomechanics IX-A* (pp. 96–101). Champaign, IL: Human Kinetics.

Viitasalo, J. T., & Komi, P. V. (1981). Rate of force development, muscle structure, and fatigue. In A. Morecki, K. Fidelus, K. Kedzior, & A. Wit (Eds.), *Biomechanics VII-A: Proceedings of the Seventh International Congress of Biomechanics* (pp. 136–141). Baltimore: University Park Press.

Vorobiev, G. (1976). *Track and Field* (Moscow, USSR), *7*, 31–32.

Wachowski, E. (1981). Speed and useful power in strength exercises. In A. Morecki, K. Fidelus, K. Kedzior, & A. Wit (Eds.), *Biomechanics VII-B: Proceedings of the Seventh International Congress of Biomechanics* (pp. 379–385). Baltimore: University Park Press.

Wallin, D., Ekblom, B., Grahn, R., & Nordenborg, T. (1985). Improvement of muscle flexibility: A comparison between two techniques. *The American Journal of Sports Medicine, 13*, 263–267.

Warren, C. G., Lehmann, J. F., & Koblanski, J. N. (1976). Heat and stretch procedures: An evaluation using rat tail tendon. *Archives of Physical Medicine and Rehabilitation, 57*(3), 122–126.

Weltman, A., & Stamford, B. (1982). Strength training: Free weights versus machines. *The Physician and Sportsmedicine, 10*(11), 197.

Whitley, J. D. (1967). The influences of static and dynamic training on angular strength performances. *Ergonomics, 10*(3), 305–310.

Wolf, S. (1978). The morphological and functional basis of therapeutic exercise. In J. V. Basmajian (Ed.), *Therapeutic exercise* (3d ed.) (pp. 43–85). Baltimore: Williams & Wilkins.

Wolf, S. L., Ariel, G. B., Saar, D., Penny, M. A., & Railey, P. (1986). The effect of muscle stimulation during resistive training on performance parameters. *The American Journal of Sports Medicine, 14*, 18.

CHAPTER 8

Body Balance and Stability Control

PREREQUISITES

Concept Modules A, B, C, D, E
Chapters 1, 2

CHAPTER CONTENTS

8.1 Balance, Equilibrium, and Stability

Balance and *Equilibrium* are often used synonymously to describe a system in a state of rest (motionless). Strictly speaking, **equilibrium** describes the state of a system that is not changing its speed or direction. Equilibrium can exist for a body at rest, in which case it is called **static equilibrium**; or for one that is moving with unchanging speed and direction, in which case it is called **dynamic equilibrium**. In human activities, however, the body is always experiencing some kind of movement change, whether movements of individual segments or the expansion of the thorax for breathing.

For the human body, **balance** implies some kind of movement control, either for short or long periods of time. When we discuss the concept of balance, therefore, we will consider it to be a *process whereby the body's state of equilibrium is controlled for a given purpose.*

The concept of **stability,** or the difficulty with which equilibrium can be disturbed, is encountered frequently in activities demanding static postures as well as in many skills and tasks involving the moving system. Neuromuscular mechanisms play an important role in body balance when we are apparently motionless and during movement. This chapter focuses on the mechanical determinants of equilibrium. Attention is drawn to the fact that the human being is not a rigid unit; it is capable of changing shape, thereby complicating some of the simple principles of balance normally applied to inanimate objects.

Linear and Rotary Equilibrium

A body can move in linear or rotary paths, and it can experience a state of equilibrium in either of these types of motion. For equilibrium to exist, no net external force can be causing a change in the system's state of linear motion and no net external torque can be causing a change in its state of rotary motion. In other words, all the forces in all directions must cancel out to zero, and all the torques in all directions must cancel out to zero.

An example of a body at rest in which no net force or torque acts upon it is provided by a student asleep in a chair in a noon class. The body-weight force of 712 N (160 lb) acts downward and is opposed by the equal and opposite ground-reaction force of 712 N that is transmitted by the chair. No other forces or torques are acting to upset the body's equilibrium, so no motion change occurs. The student has static equilibrium.

A moving system could be in a state of rotary equilibrium but not linear equilibrium, and vice versa. A diver performing a pike somersault dive spins with constant speed of rotation, while her center of gravity (CG) is increasing speed downward during the time she is falling.

An example in which linear equilibrium exists while rotary equilibrium does not is provided by a dancer who spins about his body's longitudinal axis on one foot. His body's CG is not moving linearly, but the floor friction provides a net resistive torque on the foot to slow his rate of turning. The most common situations encountered in human movement are those in which the body is in neither linear nor rotary equilibrium; that is, the body or its parts are changing linear and rotary speed from one moment to the next. Therefore, some net force and/or net torque are acting on the system. The body increases or decreases linear speed according to the direction of the net force and increases or decreases rotary speed according to the direction of the net torque.

Linear and Rotary Stability

Although the conditions for true equilibrium are practically nonexistent in human situations, they are often the target conditions of the mover who is attempting to control static balance or the static balance of another person or object. The goal may be to assume a precarious position from which equilibrium may be easily disturbed, as in a track or swimming start, in which falling forward is coupled with forceful leg extension to accelerate the body.

Stability is the term that describes a body's resistance to losing its static or dynamic equilibrium, or its resistance to changing its state of motion. The body can have a certain amount of stability in terms of moving in a straight line (**linear stability)** as well as stability against tipping over (**rotary stability)**. Each of these types of stability is encountered in most movement activities, and often one type influences the other.

Linear Stability

Linear stability of a stationary body is a condition in which the body is resistant to being moved in a given direction by an external force; such would be the target condition of a tug-of-war team member. Linear stability of a moving body is a condition in which the body is resistant to being stopped or having its direction changed, the target condition of an offensive lineman in football.

Whether a body should have maximum, minimum, or optimum linear stability depends on the purpose of the activity. If a straight line course is the target condition (sailing, rowing, canoeing, swimming, cycling) we would want maximum directional linear stability; that is, we would want to eliminate forces that would alter the straight line path of the body. If direction change is the goal (surfing, water skiing, ice skating, skateboarding, skiing), then minimum linear stability is wanted; that is, the body should not be resistant to direction change.

A general principle concerning linear stability may be expressed as follows: *The greater the force needed to upset the linear equilibrium of a body, the more stable the body.* It follows, then, that the greater the body's inertia, or mass, the greater the linear stability, for a greater force is necessary to cause it to change speed or direction. It also follows that the greater the resistive friction acting on the body, the more resistant it is to being moved linearly. Linear stability in sliding situations depends on the frictional forces generated between the contacting surfaces of the body and the environment. An ice skater is easily moved linearly by a small push, which is sufficient to overcome the small resistive frictional force of the ice against the skate blades. The motive force of wind on sails moves a boat through water but not across the sand because of the difference in resistive friction force on the hull. In each of these cases, the resistive friction force is an important determinant of the system's linear stability.

In situations in which the weight of the system is the force with which the two frictional surfaces are pressed together, heavier (more massive) bodies will have more friction force to enhance their linear stability. Thus, an increase in the body's mass increases its linear stability in two ways. Such stability is manifested by a football lineman whose mass requires a greater force to make him move and whose weight gives better friction between the shoes and the ground.

Whether or not linear stability is desirable depends on the objective of the mover. Perhaps only minimal linear stability will be advantageous to performance, as in surfing or water skiing. For situations involving linear stability or instability, it is mainly the linear mobility that should be considered as desirable or undesirable in the performance. Rotary stability is different and may or may not be desirable.

Rotary Stability

The *rotary stability* of a body is the condition in which a body is resistant to being tipped over or falling. When we speak of stability in human movement, we are almost always referring to some rotary balancing situation, such as walking a balance beam, holding a headstand, or stepping into a floating canoe. Therefore, much more attention will be given to the determinants of rotary equilibrium and stability than was given to linear stability situations.

Rotary stability refers to a body's resistance to losing its equilibrium by being rotated about some fixed point by a net motive torque. A general principle regarding rotary stability may be expressed as follows: *the more stable a body, the greater the torque needed to upset its rotary equilibrium.* A body may be upset by applying a large torque for a brief amount of time or by applying a smaller torque for a longer period of time. For example, if one were to push on a block standing on edge, it would tip over or rotate about the axis formed by the bottom edge touching the supporting surface (Figure 8.1a). This push would be enough to rotate the block into a position where its weight torque would cause it to rotate farther away from its original balanced position. If a lesser push were applied, its weight torque would cause the block to return to its original position.

If the block were pushed only enough to raise one edge and then held there (Figure 8.1b), the opposing torques from the push force and the weight force would cancel out and the block would remain in that position. Once the line of gravity (the line of action of the weight force) falls to one side or the other of a body's axis of rotation, it becomes torque producing. If there is no opposing torque, it causes a rotation of the body about this axis. If you were to stand rigid and someone were to push you from behind, your body would rotate about an axis formed by your toes on the ground, and you would fall forward and downward. Such a fall seldom occurs, however, because you would not remain as a single rigid unit; you would step forward automatically as you started to lose your balance, and you would be able to stop your motion as you planted your foot in front of you (Figure 8.1c). Comparing these two situations, we can see the fundamental difference in potential stability between inanimate objects and human beings. We can move or rearrange our segments to maintain rotary stability either by changing the location of the line of gravity relative to the base of support or by changing the base of support itself.

When the body is stationary, *the body will be balanced if its line of gravity passes through some part of the body's base of support.* The importance of the body's source of support warrants a careful definition. The body's **base** (area) **of support** is that region bounded by body parts in contact with some resistive surface that exerts a counterforce against the body's applied force. For example, the outline of the foot defines your base of support when you stand on one foot. If you are hanging by both hands from a horizontal bar, the area enclosed between the outer limits of your fingers forms your base of support. If some object or surface exerts a counterforce against some part of your body in some direction other than the vertical, it also forms part of your base of support in that it is a force that contributes to your state of equilibrium. Figure 8.2a shows such a position of balance created by vertical and horizontal forces, similar to the block in Figure 8.1b. Other examples of support bases are illustrated in the figure as well. Note that the area of each base of support is always defined by the outermost limits of the contacting surfaces.

STABLE, UNSTABLE, AND NEUTRAL EQUILIBRIUM. The state of rotary equilibrium of a body is often classified into three categories: *stable, unstable, and neutral.* The physical characteristics and the nature of the base of support determine a body's vulnerability to upset or loss of balance. Figure 8.3a–c represents each of the three categories. A body in stable equilibrium requires a greater torque to upset its balance than does a body in unstable equilibrium. A torque applied to a spherical shape does not cause a rotary balance upset, because there is no tendency of the ball to "fall" one way or another; its CG is said to be in neutral equilibrium and remains directly over the base of support on a level surface. If the surface were inclined, however, the ball would "fall" downhill, because the gravity line would pass beyond the base of support on the downhill side, and the net torque on the ball would cause the ball to roll (Figure 8.3d).

Various static body positions in movement

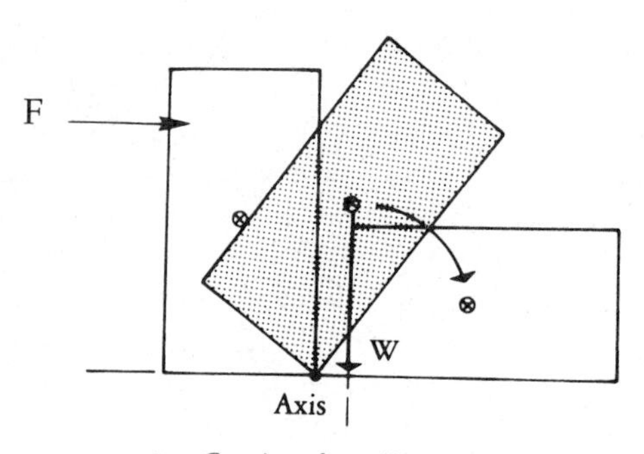

a Continued application of torque causes line of gravity to fall beyond base edge

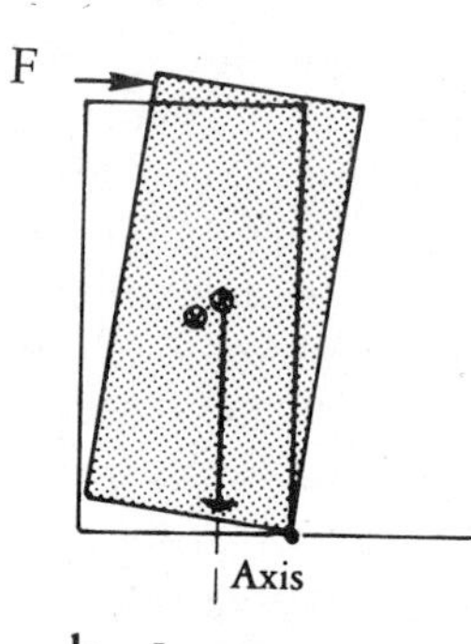

b Lesser torque can balance weight torque to hold block in position

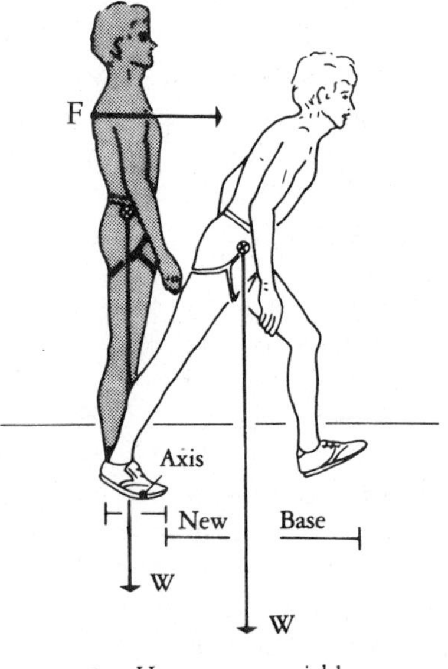

c Humans can quickly adapt to upsetting forces by forming new base of support

FIGURE 8.1

How much torque or how long it is applied determines whether an object or body will rotate and fall.

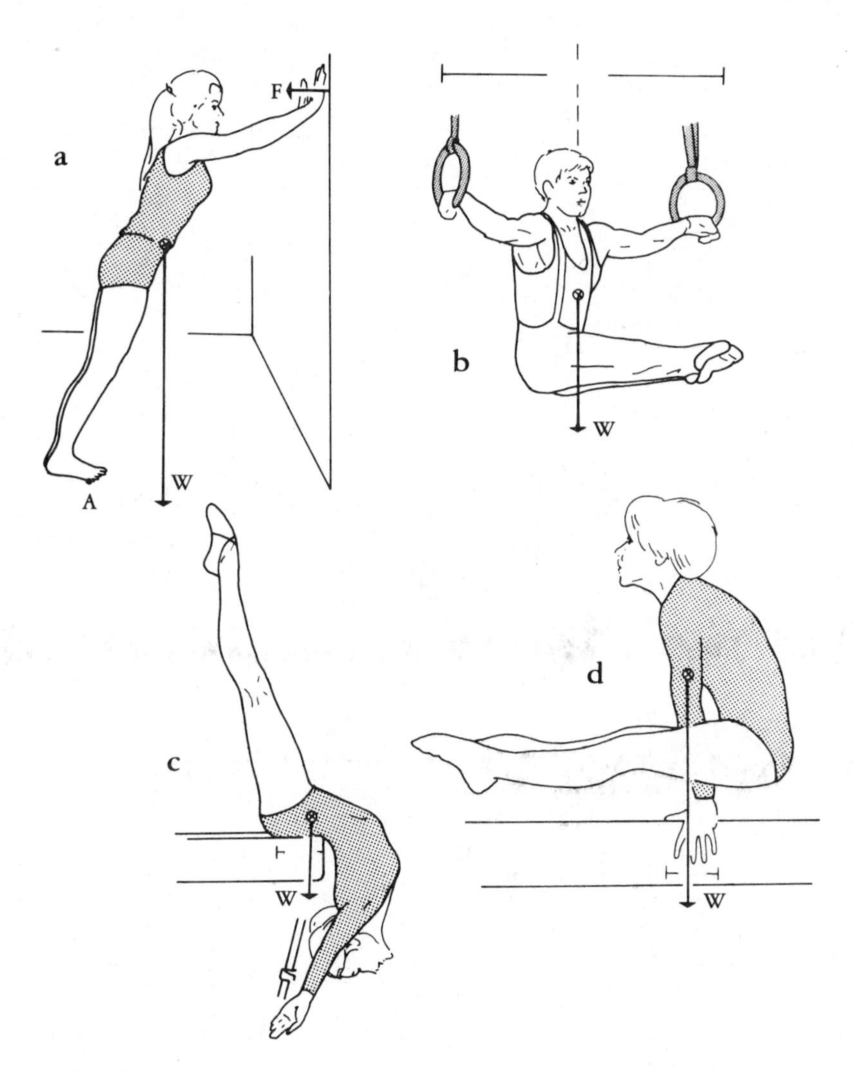

FIGURE 8.2
The base or area of support of a body in equilibrium is formed by the boundaries of contact with sources of support.

activities simulate these three basic types of equilibrium. The degree of rotary stability of a body is *specific to direction.* "Specific to direction" means that a body may be stable, or resistant to upset, in one direction but not necessarily in another. For example, while balancing on a narrow beam, you are relatively stable in a forward–backward direction, but not in a sideward direction because of your narrow lateral base.

For situations in which gravity and the upward ground-reaction force are the only external forces acting on a stationary body, balance exists if the gravity line passes vertically within the area of support. For situations in which there is some external torque applied

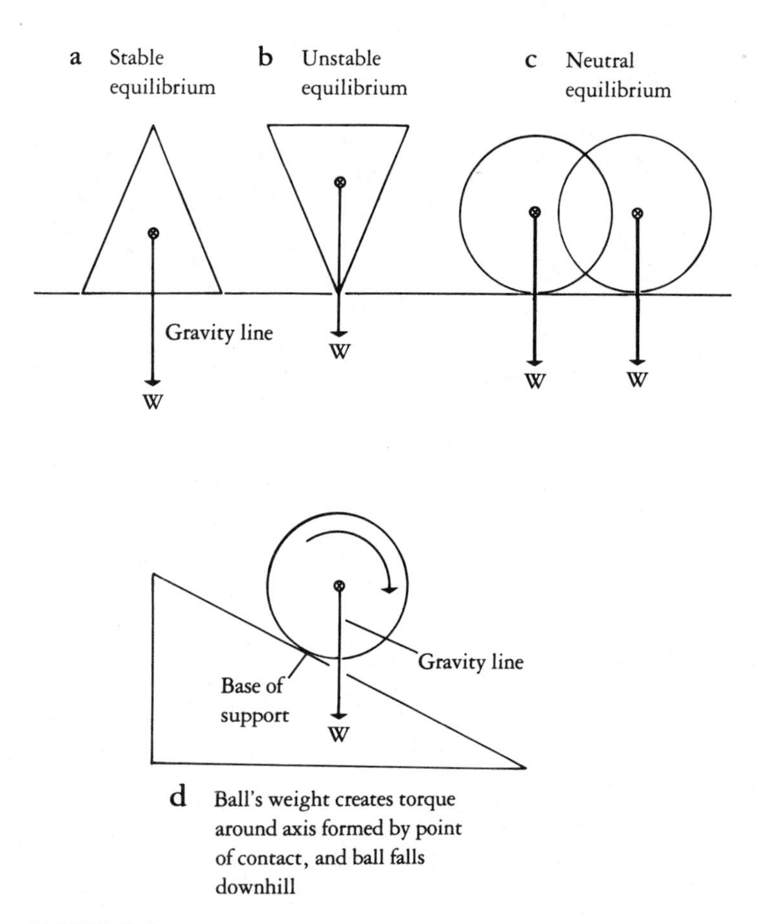

FIGURE 8.3

(a, b, and c) States of rotary equilibrium and (d) the loss of balance of a ball on a slope causes it to "fall" downhill.

in addition to gravity and the ground-reaction force, balance may be upset temporarily or permanently, depending on the magnitude of this torque and/or how long it is applied.

Rotary instability of a wide-base body can occur depending on where the gravity line passes within the base. If the torque is large enough to impart sufficient rotary motion to the box in Figure 8.4a, this momentum will carry the CG up and beyond the base edge. If the gravity line passes close to the base edge that will be the axis, a relatively small torque will cause it to tip and fall (Figure 8.4b).

Figure 8.4c illustrates a sprinter in his set position in the starting blocks. His weight produces a clockwise (cw) torque about an axis at the fingers. The ground-reaction force acting on each foot forms a counterclockwise (ccw) torque about the axis. Both of these ccw torques from the blocks serve to cancel the cw-weight torque, so that the net torque acting on the body is equal to zero; the body is in equilibrium. He is in *unstable* equilibrium because he has positioned his CG forward so that his gravity line falls vertically near the forward edge of the base formed by the fingers.

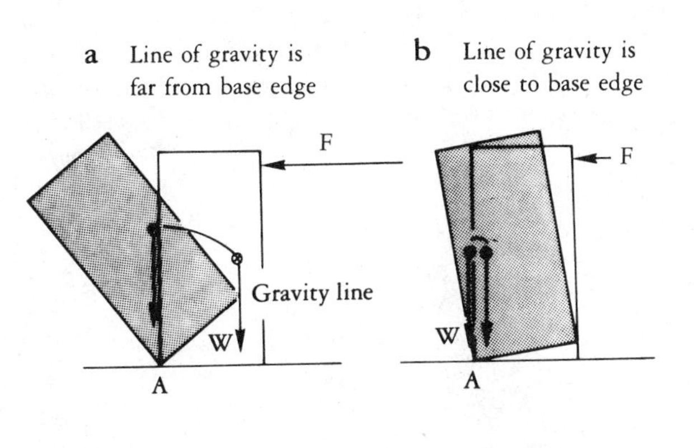

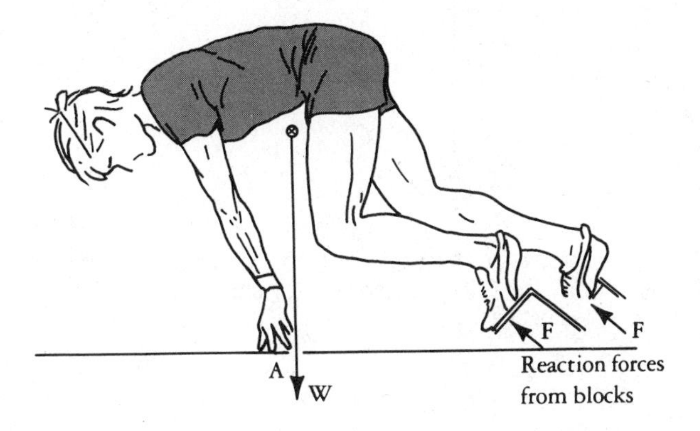

FIGURE 8.4

(a) When the gravity line is initially farther from the potential axis of rotation, a greater push is required to make it rotate and fall to the left. (b) If the gravity line is initially near the axis of rotation, a small push will upset it. (c) A sprinter's gravity line falls vertically through the base of support, close to the leading edge, and represents the unstable equilibrium condition.

In such a position, he is still balanced, but only a small push is necessary to upset his equilibrium and cause him to rotate (fall) in the forward direction, a move advantageous to quick starting. At the gun, the hands are withdrawn as a source of support, and the CG moves forward and upward from the reaction forces generated by the legs against the blocks. Of course, the mass of the sprinter would determine how much force would be required to start moving quickly. *The greater the mass of the body, the greater the torque necessary to cause a loss of equilibrium.* This is true because the CG of the body must be elevated to tip it over (Figure 8.4a).

Another principle for increasing stability is that *the lower the body's CG, the greater its stability.* As for other balance principles, however, this is true only if other changes that may cancel this benefit do not occur (Figure 8.5). It is

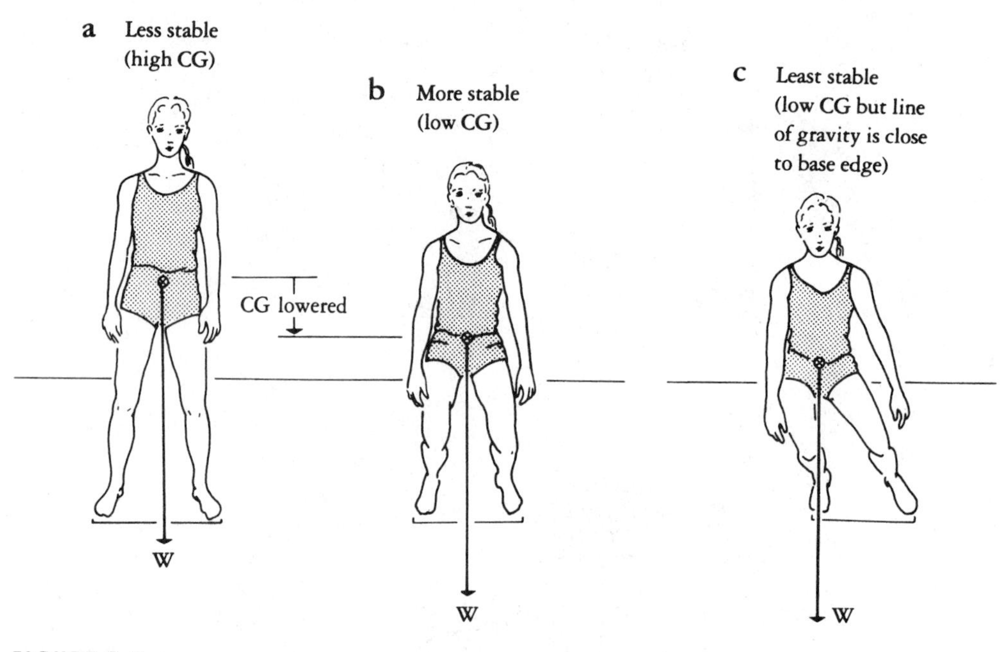

FIGURE 8.5

(a and b) The lower its CG, the more stable the body, but only if all other variables do not change to offset this stability advantage. (c) If the CG is also moved toward the edge of the base, the body actually becomes less stable in that direction.

not true if the CG is lowered and is also moved toward one edge of the base of support (Figure 8.5c); for if that base edge is the potential axis for body rotation caused by an external blow or push, the body will become less stable. Also, if the right foot of the person in Figure 8.5b were to be lifted as the CG is lowered, balance would be lost immediately because of the withdrawal of that source of support. We see, therefore, that for human movement situations in which a number of balance variables can change simultaneously, an understanding of all the mechanical factors governing stability, not just a few, best equips the teacher or coach for analyzing performance.

ROTARY STABILITY DURING SUSPENSION. We can evaluate the stability of a body suspended from a "base" above the body using the concepts of torque and position of the gravity line relative to the base of support. *A suspended body free to rotate in a vertical plane will seek a position in which the CG is at the lowest point possible within the constraints present.* If the base of support is below the body, the gravity line must pass somewhere through the area of the base for balance to exist if no other source of support is present. The gymnast in Figure 8.6a illustrates the fact that the body will rotate until its CG is at the lowest point before static equilibrium

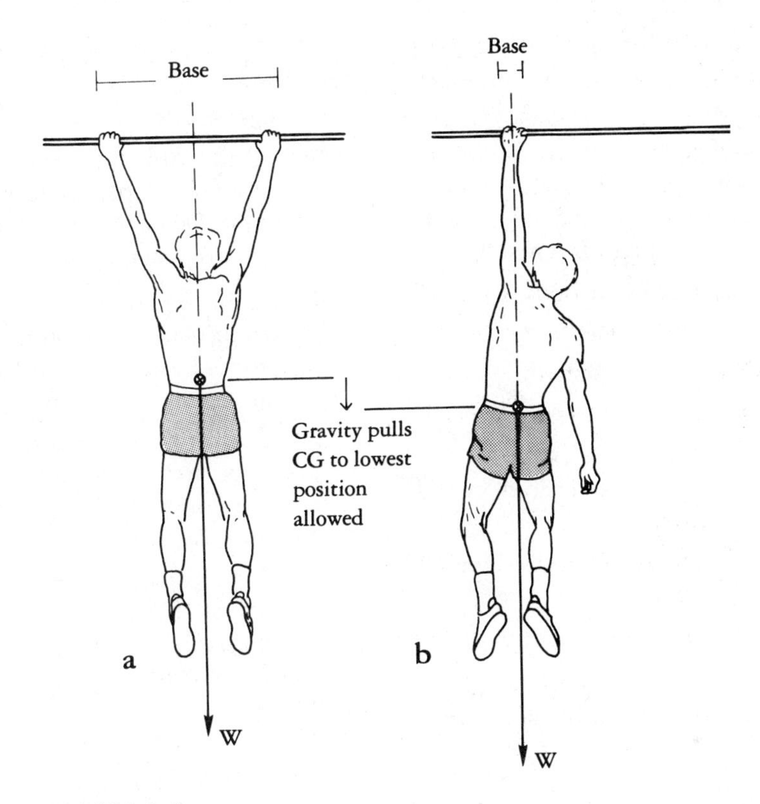

FIGURE 8.6

A body free to rotate about a point of suspension comes to rest in an equilibrium position where the (CG) is at its lowest possible point.

is reached. If one hand is released from the bar, the body will again rotate in response to the weight torque about the axis of the single hand, and the CG will find its new lowest position (Figure 8.6b). The center of gravity behaves like a weight on a string.

Understanding Balance, Equilibrium, and Stability

1. Stand with your heels and back against a wall. Keeping your legs straight, reach down toward your toes without moving your base, and explain the results in terms of the CG, the line of gravity, and the base of support.
2. Repeat the exercise in question 1, standing away from the wall. Explain the dif-

ference in results. What is the movement that occurs at the ankle joints as you reach toward your toes?

3. Stand with your side against the wall so that your legs are parallel to your body's longitudinal axis. Slowly lift your outside foot. Describe and explain the results.
4. State a principle relating a body's line of gravity to its base of support . . .
   a. if balance is to be maintained.
   b. if minimum stability is desired in the forward direction.
   c. if maximum stability is desired in the backward direction.
   d. if optimum stability is desired in all horizontal directions.
5. Perform the following movements:
   a. Stand with your feet, heel to toe, along a line directed from front to back. Close your eyes, rise to your toes, and determine the direction in which you are least stable.
   b. Stand with your feet spread laterally about 1 m apart, and repeat the steps in (a). Determine the direction in which you are least stable.

   Using the experience gained in (a) and (b), state a principle relating the body's gravity line to its base of support for maximizing stability.
6. Assume a push-up position. What adjustment is necessary for balance if you remove one hand from the base of support? Explain why.
7. Using the concept of CG location and noting how it changes with segmental movements, explain how "pumping" in a suspended swing can initiate the swinging motion.

8.2 Controlling Balance in Static Positions

If the body is in a state of static balance, the situation is relatively simple to analyze as compared to one in which the body is in motion, and the variables may be continually changing. We must realize, however, that all our static positions are achieved by movement and that the momentum of the body or its parts may carry it beyond the position to be held (e.g., pushing into a handstand). Once the body is in a static position, it is subject to upset by its own segmental movements and by external forces that may act on it. Keep in mind the condition necessary for maintaining a position in equilibrium: *The net external torque acting on the body must be equal to zero.* This condition, one in which all opposing torques cancel, should be used as the target condition to be sought by the performer wishing to maintain a static position.

The apparently simple act of standing motionless is actually a continuing process of minute adjustments of body position to keep the CG over the base of support. The smaller the base (e.g., standing on one foot), the more accurate such adjustments must be to maintain rotary stability. Pressure of the supporting surface on the soles of the feet serves to maintain tension in the extensor muscles via the extensor thrust reflex. Because this tension prevents the lower extremities from flexing under the force of body weight, these extensor muscles are called the antigravity muscles. When relaxation occurs in these postural muscles, the joints of the vertebral column and lower extremities are forced into flexion by the body weight. When the body leans forward slightly, the gravity line moves toward the front edge of the base, increasing pressure on the balls of the feet and at the same time lengthening the ankle extensors (plantar flexors). When these muscles are so stretched, the myotatic, or stretch, reflex causes them to contract concentrically to rotate the body backward at the ankle. The gravity line is thereby shifted backward, removing the threat of a forward fall. Although the gravity line rarely shifts behind the ankle joint in normal standing, if the backward shift is too great, the ankle flexors are stretched, and subsequent reflex contraction causes the body to be rotated forward at the ankle. Alternating contraction and relaxation of the ankle extensors results in what is called postural sway. The amplitude of sway is less precisely regulated if visual or auditory cues or both are not present. With increased sway, other body segments are brought into action to prevent loss of balance. The labyrinthine organs of the inner ear, which are sensitive to changes in head position and changes in speed of head movement, also reflexly initiate such compensatory movements in the upright position. The head-righting reflexes must be voluntarily suppressed in inversion skills, such as headstands and handstands.

In most land activities, the body weight and ground-reaction force press the body and the ground together and thus provide the normal force, which is one factor in how much friction force would be present if the body were forced to slide. This is true when the ground is horizontal, or level. If one were standing on a slope, however, the normal force would be only that component of body weight directed perpendicular to the slope. Figure 8.7a and b illustrates the vector resolution of the force of body weight, showing the normal force component directed into a hillside (perpendicular) and the component directed parallel to the hillside surface.

In each case, as the slope becomes steeper, two changes occur: (1) the friction force is decreased, owing to a smaller normal force pressing the body into the hill, and (2) the component of the body weight that is directed parallel to the hillside becomes greater, thereby becoming a greater and greater motive force for sliding. When the force of friction of the ground acting on the body is surpassed by the weight component parallel to the surface of the hill, the feet and body slide downhill.

If the body's CG remains vertically above the base of support, the body slides downhill without rotating (falling over). The tendency to lean backward (toward the uphill side) on a slope often shifts the gravity line outside the base, thus producing a torque and causing rotation or falling into the uphill side of the slope. As the body mass falls downward into the slope, there is an unweighting of the feet, which means that the normal force pressing them to the hill decreases. Thus, the friction force is decreased, and the feet slip out from under the rotating body.

8.3 Controlling Balance During Movement

Even though human beings are influenced by the mechanical laws that govern the motion of inanimate objects, we are not rigid or single-shape bodies. We can move our segments independently, thereby redistributing the body mass and changing the location of the body's CG. Moreover, we can, and usually do, move our segments to change our base of

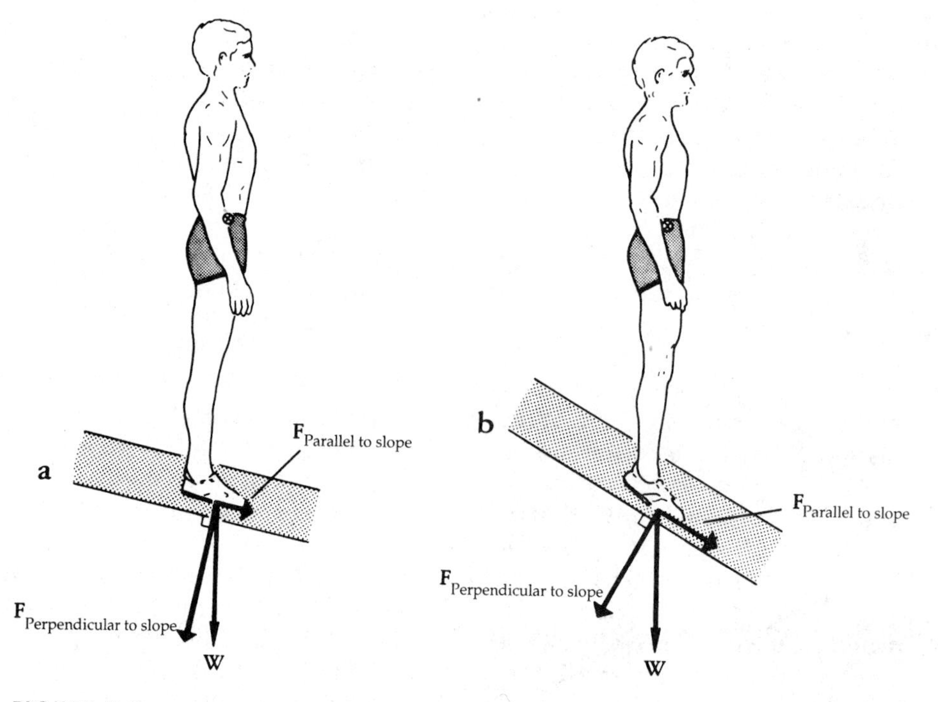

FIGURE 8.7

The normal force pressing the body to the hillside in (a) progressively decreases as the hill becomes steeper in (b). When the normal force decreases, so does the force of friction. When the component of body weight directed parallel to the slope becomes greater than the slope's friction on the feet, the feet slide downhill.

support from one moment to the next to cope with impending loss of balance. Fortunately, any sudden movement of body parts caused by external influences can cause reflex movements that prevent loss of balance. Several general categories of balance-threatening situations are presented in the following sections.

Applying and Receiving Forces

When a force is applied to an object, the reaction force of the object on the body must be anticipated. If this reaction force serves as part of the body's support and is abruptly removed, loss of balance results. In one familiar case of this kind, you push against a heavy door to open it, and someone on the other side suddenly opens it, pulling it away from you; this causes you to fall forward unexpectedly with no forward base prepared. A general rule for maintaining balance in horizontal force-giving and force-receiving situations is: *The body's base of support should be enlarged horizontally in the direction of the horizontal force to be given or received.* This rule applies whenever the force given or received has a horizontal component, even if the force is not entirely horizontal. Consider how you might position your feet to prepare to push a heavy object forward. With a forward–backward stride position, you can step forward with the load's ensuing motion as well as apply the push in the desired direction. Similarly, if you are expecting a push from your right side, a wide lateral stride allows your moving gravity line to remain within your base for a longer period of time, thus allowing you more opportunity to make an adjustment to keep your balance.

When throwing a ball or other object, your leading foot must be moved in the direction of your body motion to provide a new base for the forward-moving CG. Striking objects that yield to your force often leads to the same type of weight shift, and a new base must be ready to support this weight. In catching, most objects that are caught do not have sufficient mass or speed (kinetic energy) to cause body imbalance on impact. Even so, by forming a large base in the direction of the approaching object, the body position can be shifted more freely over the feet to facilitate accurate contact with the hands. If the object (e.g., a baseball) is to be caught and then immediately thrown again in a different direction, anticipation of the need to form an appropriate base for the next move contributes not only to the speed of the performance but also to better control of the segmental movements.

Pushing and Pulling

If a heavy object must be moved across some surface, the resistive friction is greatly reduced when the object can be rolled on wheels. When the object must be slid, however, the direction of the force applied can increase or decrease the friction force and thus influence the degree of difficulty of the task. Recall that any applied force can be resolved into its horizontal and vertical components. If the force is entirely horizontal, the vertical component is zero. Because the friction force is directly proportional to the force pressing the two surfaces together, the friction force resisting the hori-

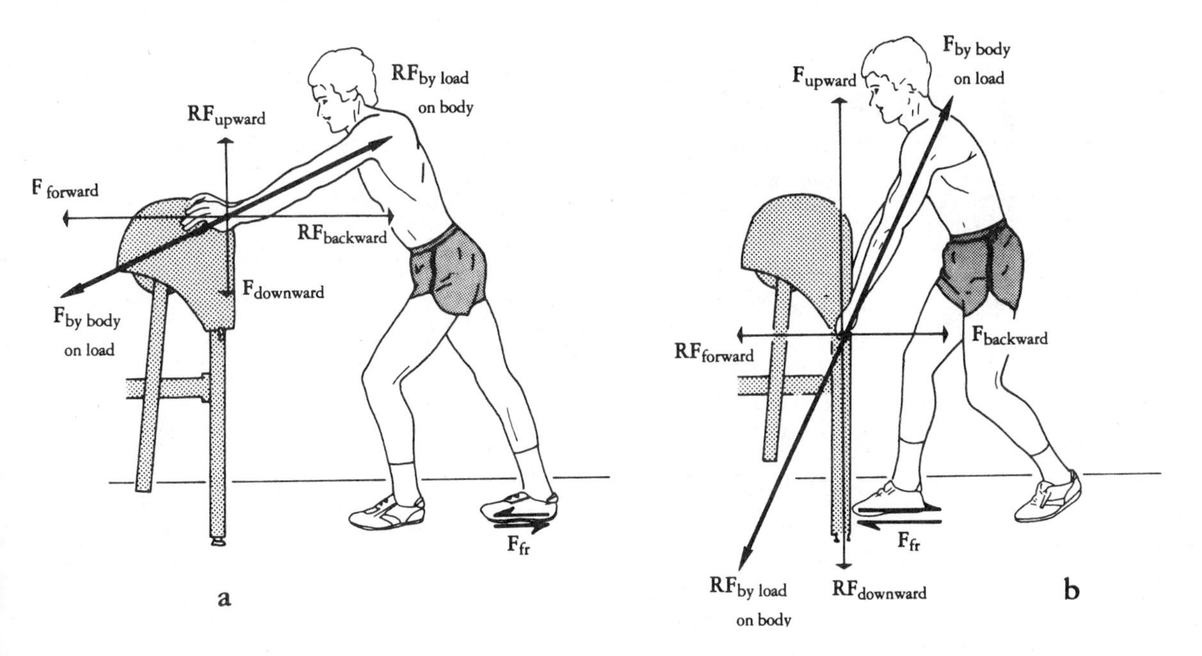

FIGURE 8.8

(a) The upward reaction force to the downward component of a push decreases the foot–ground friction. (b) The downward reaction force to the upward component of a pull increase foot–ground friction. A double advantage is gained by pulling in situations where load–ground friction is great. (F, force; RF, reaction force).

zontal motion of the load is neither increased nor decreased as it would be by a vertical component from the external force.

Note in Figure 8.8a that when a load is pushed forward and downward, the upward reaction force on the body causes a loss of friction between the feet and the floor. Such a situation compounds the difficulty of moving the load because the feet have less friction force applied to them and thus tend to slip; the load has more friction force applied to it and thus tends to stick. If the load is pulled, however, the downward reaction force applied by the load to the upward component of pulling force serves to increase the foot friction as the load's friction decreases (Figure 8.8b).

For certain pulling tasks, the use of a rope can change the line of force application and magnitudes of the components of the pull. Figure 8.9 shows the use of a long rope for a low-friction load and a short rope for a high-friction load. Once the load (or the foot) has started to slide, the friction force decreases, having been at its maximum just before sliding began. Therefore, decreased friction on the load during motion requires less applied force to maintain the speed of sliding than is required to start the load moving against greater static friction. In addition, recall that the force necessary to maintain the speed of an object already sliding is less than that required to start it mov-

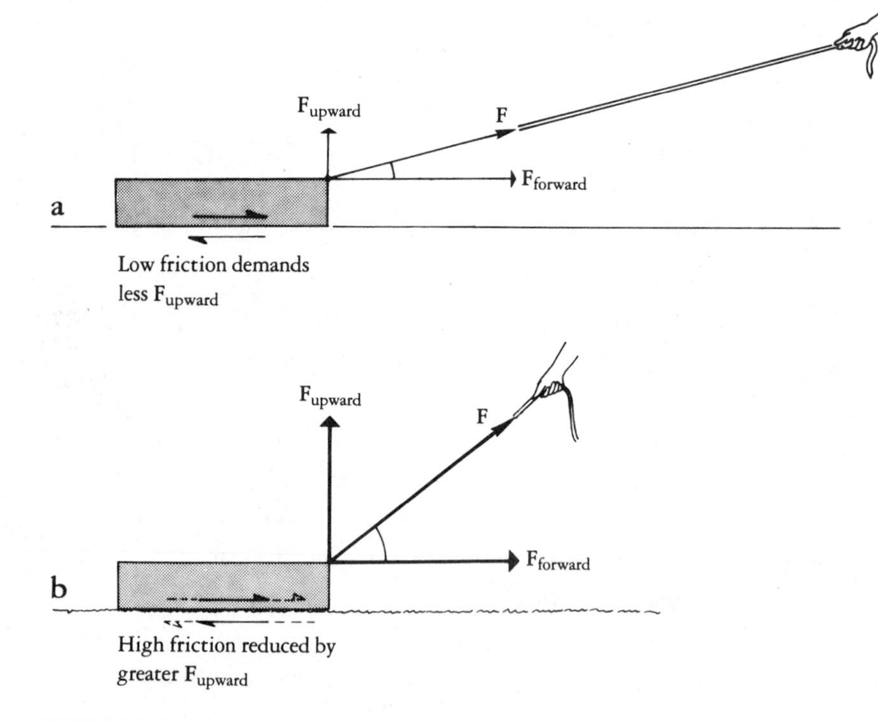

FIGURE 8.9

(a) Using a long rope to pull low-friction loads enables the mover to exert more force in the desired horizontal direction. (b) If the load friction is great, pulling it with a short rope produces a friction-reducing upward component, so that the horizontal component has less friction force to overcome.

ing (to accelerate it). Advance preparation of the body's base of support in the direction of movement, therefore, is important for maintaining control of body balance, especially if the movement of the load might be sudden or unexpected.

The point of application and line of action of a pulling or pushing effort can greatly affect the resulting movement of a load (Figure 8.10). If no load-floor friction exists, no fixed axis forms at one edge, and the load slides easily across the surface. Such a case is rarely seen, unless the surface is extremely slippery. Often, the friction against the base of the load together with the applied push or pull serve to cause rotation instead of sliding. If the friction between the load and the supporting surface is great enough or if there is some obstruction in the path, the leading edge tends to stick and becomes an axis for the load's rotation (picture a heavy cart with very small wheels "catching" in a crack in the path).

If a horizontal force is applied at a low point, less torque is produced; therefore, there is less tendency for the load to rotate as it is pulled or pushed. Figure 8.10a and b shows how a horizontal force applied at a low level on a high-friction load can produce sliding and how a lesser force applied at a higher level can produce rotation.

When a force is applied at a higher level, its force arm is large enough to allow a small force to cause rotation if this *applied* torque exceeds the load's *weight* torque. Such a small force often is not great enough to overcome the load-floor friction and therefore causes no sliding, only rotation. A greater force, enough to overcome the force of load-floor friction,

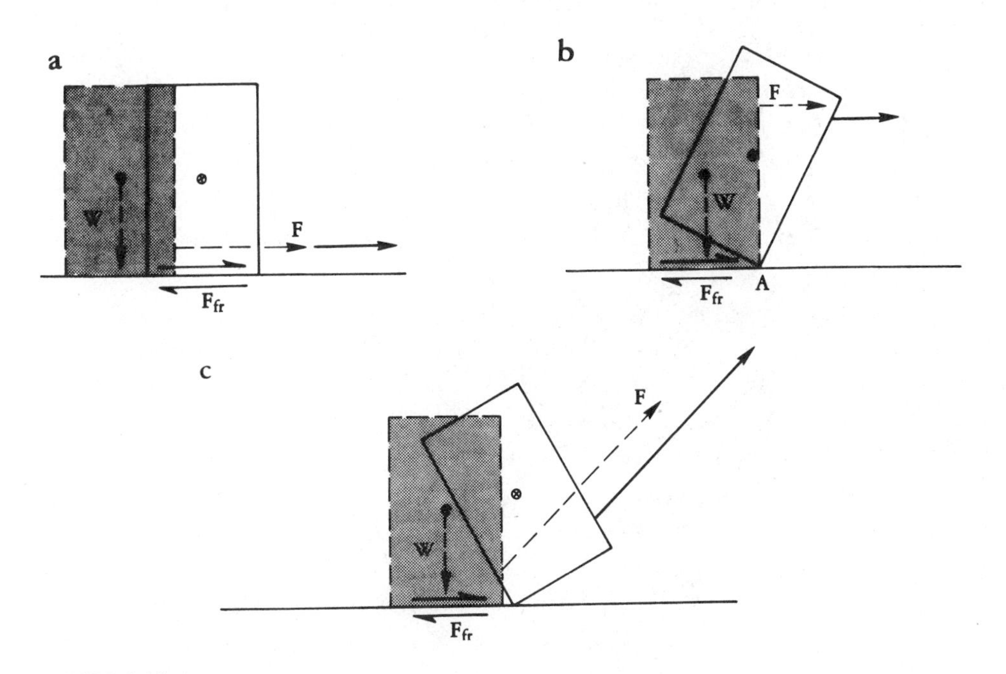

FIGURE 8.10

(a) A force applied at a low level on a high-friction load produces sliding but no rotation. (b) A force applied at a high level on the load produces rotation but no sliding. (c) A force applied to produce both sliding and rotation, in accordance with the principles of net force and net torque.

can be applied at a lower level to cause only sliding; no rotation is produced from such a force because its force arm, and therefore its torque, is so small. Both sliding and rotation are caused by the pull shown in Figure 8.10c, provided the horizontal component of the pull is greater than the load-floor friction and the pull torque is greater than the weight torque.

If the CG of the load is high and close to the edge of the base of support, a force applied at a low level, great enough to overcome the load-floor friction, can cause the base to slide out from under the CG, thus allowing the load's weight torque to cause its rotation. This situation can be demonstrated by standing a book on end and striking the side sharply at a low level. Similarly, when a football player is tackled near his support base, he is upended.

In wrestling, as well as in the martial arts, the points of application and the lines of application of the forces relative to the opponent's base and CG location are primary factors for success in upsetting his balance (Figure 8.11).

Often, a load can be moved more easily by intentionally rotating it than by sliding it, especially if the support surface creates large friction forces. Figure 8.12a shows the application of torque about one leg of a heavy file to "walk" the load to its new location. A hand truck should be available for moving large or multiple loads to prevent muscle strain that could result from attempting to move one large load or from attempting to move a moderate load when the muscles are fatigued from repeated light efforts. Figure 8.12b shows the torques applied to a load on a hand truck to carry out the move.

For controlling the path of a load being pushed or pulled, the effort should be applied at two points of contact, with adequate

FIGURE 8.11

A wrestler applies force on the opponent to cause the opponent's body to rotate under the influence of his own body weight.

a Rotating a heavy load to "walk" it instead of carrying or sliding it.

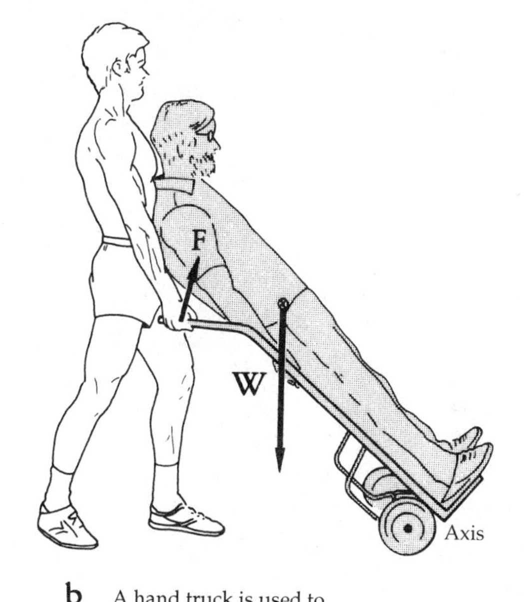

b A hand truck is used to balance and move heavy or awkward loads (second-class lever).

FIGURE 8.12
Two methods used for moving heavy loads.

horizontal distance between these points. Compensation for any irregular friction spots that cause a change in sliding resistance can be made by adjusting the force on each point.

Lifting and Carrying

When a load is lifted or carried, the mass of the load becomes part of the body mass and therefore makes a contribution to the CG location of the system (body and load). For balance to be maintained, the body's base of support must be adjusted for the new location of the gravity line if the load is sufficiently large or held far enough away from the body to cause imbalance. Alternatively, body segments can be moved in a direction away from the load to balance its effect. Figure 8.13a illustrates changing the base of support to receive the new gravity line of the body and load. In Figure 8.13b, the body must lean to the side to produce a balancing countertorque for the load. Note that leaning from near the base (the subtalar joints) eliminates the need for lateral flexion of the vertebral column and therefore results in less chance of muscle strain in the lumbar region of the vertebral column.

As Figure 8.13c illustrates, carrying *two* loads (even if the total weight is greater) can produce *less* chance of muscle strain on one side, and very few—if any—compensatory segment movements are necessary to keep the gravity line over the base (LeVeau, 1977).

Improper body mechanics employed in lifting tasks often lead to back injury. To perform any lifting task safely, observe these steps:

1. Test the load to estimate its weight and to determine whether you are capable of lifting it without the help of another person or of some device.

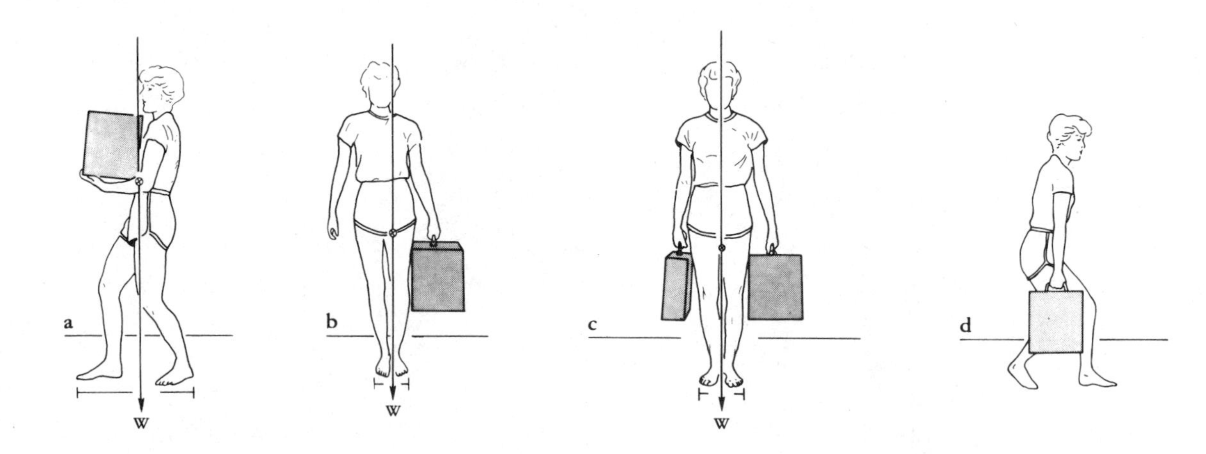

FIGURE 8.13

(a and b) A base change and compensatory body movements allow one to maintain balance while lifting a load. (c) The load is balanced on either side. (d) Stooping to raise a load with the strong lower extremity muscles.

2. Establish your support base as close as possible to the load, stabilizing the vertebral column in an extended position and lowering the body into position by flexing the hips, knees, and ankles (Figure 8.13d).
3. Raise the load, using the extensor muscles of the lower extremities concentrically without letting the extensors of the vertebral column eccentrically contract because the load is too heavy for them to hold static contractions.

Changing Speed or Direction

Skills involving starting, stopping, changing or reversing direction, running around curves, and twirling on a narrow base are balance-important activities. An analysis of such skills must consider the control of linear and rotary stability during each phase of the activity.

Starting and Stopping

If the purpose of a start is to move the body mass linearly by extending the lower extremities, the body's CG should be positioned along the line of action of the support's reaction force against the feet. If rapid horizontal movement is desired, the CG should be positioned as low and as far forward as possible for the initial push-off (see Figure 8.4c). Balanced starting positions used in track and swimming are designed to put the body in a position of unstable equilibrium. In such a position, only a small push is required for rotary imbalance. For fast starting, therefore, the body's gravity line is generally positioned as close as possible to the base edge in the direction of intended motion (Figure 8.14a). The converse holds true for the backward body lean, causing the body to slow down (decelerate); the gravity line is taken as far away as support friction permits from the leading base edge (Figure 8.14b).

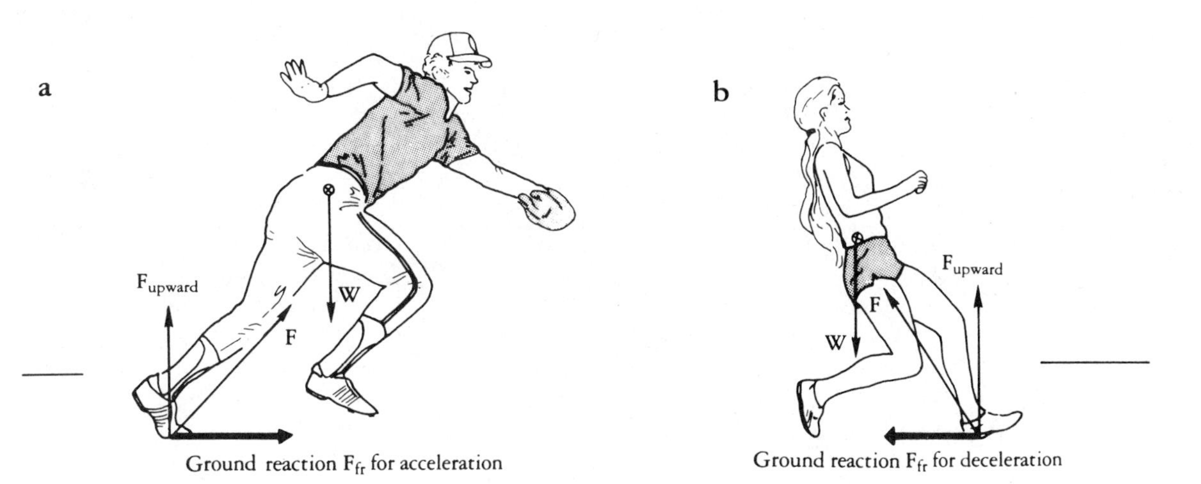

FIGURE 8.14

The body's gravity line is shifted (a) toward the leading base edge for forward acceleration and (b) away from the leading edge for deceleration.

Once the body is moving in a given direction, its momentum must be considered to maintain balance if it is to change speed or direction. This is true in such activities as walking, running, surfing, skiing, skating, biking, dancing, rotating, falling, and landing. Whenever one is moving forward and wants to slow to a stop, the backward component of the ground reaction force causes that deceleration.

To decelerate, the leading foot is planted ahead of the line of gravity, causing resistive ground friction and a weight torque that tends to rotate the CG backward and downward about the foot base. The friction provides some resistive force to slow the forward body momentum, which tends to carry the body forward over the foot. Several steps may be required before the friction reaction force and the backward-rotation component that is due to gravity stop the body's forward motion.

A similar balance situation is seen in landing from a broad jump or long jump. Because the body has forward momentum, it continues moving forward when the heels strike the ground, and the jumper's body tends to fall forward over the feet. A longer jump distance can be achieved by using this momentum—by planting the heels farther forward, flexing the hips and knees, and allowing the body's momentum to carry it forward over the base without danger of falling backward (Figure 8.15).

If a runner wishes to reverse directions with a pivot and a 180-degree body turn, the backward lean and foot plant used to decelerate the forward-moving body becomes the forward lean and foot plant necessary for acceleration in the opposite direction. If too much lean occurs at the foot plant, the feet become unweighted to the extent that the ground friction is insufficient to secure the foot, the base of support slides out from under the body, and the body slides linearly in the direction of existing momentum. Such a maneuver is the slide executed by a runner in baseball to drop below the opponent's tag while moving linearly forward to touch the base.

Loss of balance (e.g., tripping and falling) can occur unexpectedly when the preplanned base of support cannot be formed in time to accommodate the weight of the forward-

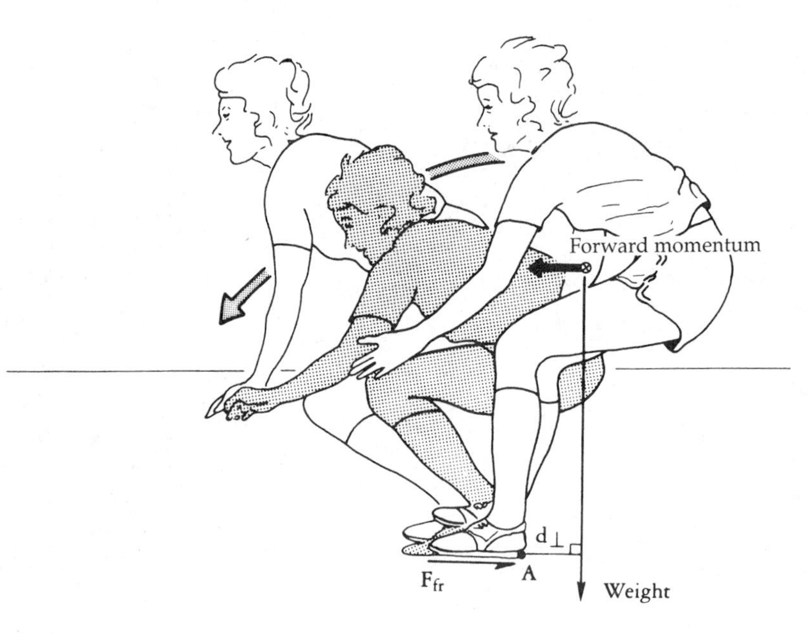

FIGURE 8.15

A jumper can delay the heel plant and gain distance with no danger of losing balance backward at the landing if the forward momentum of the body is sufficient and is used to rotate it forward over the base formed by the feet.

moving body. Such a situation exists during running when the body's CG is moving forward at a rather steady speed while each foot alternately forms a new base of support for the weight that is passing above it. If the recovering lead foot catches a resistance, the expected base is not formed, the momentum of the body continues, the weight produces a torque around an axis formed by the lead foot, and rotary instability results. If the forward speed of the CG is not too great, the reflex-initiated movement of the trailing leg may be swift enough to form a new base below the weight. A series of recovery steps may be necessary to regain balance. Frequently, if the support surface is not hazardous, the forward momentum of the body mass can be used to transform an otherwise injurious or awkward landing into one less destructive, and the body is rolled to an upright position. The shoulder roll is an example of such a maneuver. It is used often in activities that demand immediate restoration of a position of readiness in the event of a fall.

The side roll, observed in volleyball, is a similar technique. The off-balance body rolls to the side and over onto the feet, rather than falling to a position that would require precious time to recover to a position of effectiveness. This technique is illustrated in Figure 8.16.

Changing Direction

A change in body velocity that is more common than merely starting and stopping is one in which the body changes its direction of motion, as in running a curve, dodging other players, and maneuvering in games. To execute an abrupt change in direction, the body maintains some motion in the existing direction while introducing some lateral component of motion.

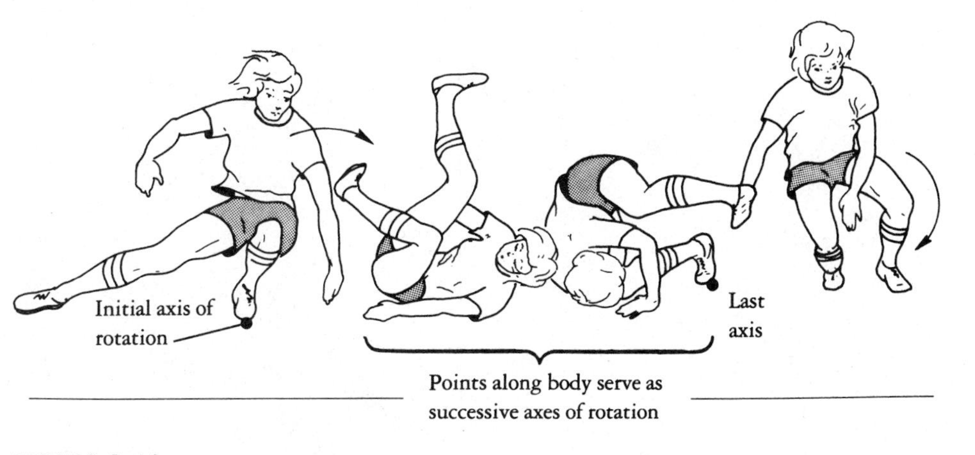

FIGURE 8.16
The side roll employs the sideward momentum of the body as it rotates about the foot and then the body as axes of rotation.

Any change in linear motion is caused by an external force in the direction of that change. To change direction without stopping requires a ground-reaction force in a direction that gives the body the desired resultant motion. If a runner is moving forward and pushes to the right, the ground-reaction (friction) force pushes him to the left, but does not stop his forward motion. His new path is diagonally forward and to the left. Maneuvering abruptly while in motion requires the foot-ground friction to be great enough to prevent slipping when the foot push is made. The greater the direction change desired, the greater must be the push and ground reaction force; therefore, little maneuvering can be done on slippery surfaces or with wornout shoes. The heavier the player, the greater the friction, but the greater the ground-reaction force must be to change the motion of a more massive player. Because of the mass factor, larger players are not known as well for their maneuverability as are smaller, "quicker" players. The "dancing" of a tennis serve receiver in the ready position may provide more friction for forceful fast starts in any horizontal direction owing to the "overweighting" during foot push-off (up and horizontally). The same may be said for a basketball player who jumps and lands to a stop—overweighting increases friction.

Running around curves is a series of direction changes executed by making small changes from a straight line path. When rounding the end of a track or running bases in softball, the runner must push outward on the ground to generate an inward ground-reaction force to make the body turn (Figure 8.17a). The tighter the turn (the smaller the radius of the curve), the greater the reaction friction force must be; also, the faster the run, the greater the friction must be to make a given turn.

Experience tells us that if a turn is too sharp or speed is too great on horizontal ground, the support friction will be insufficient to provide the necessary force for each intended turn, and the feet will slip out from underneath the body as the body leans inward. Before the feet slip away, however, the runner usually slows or enlarges the turning radius or does

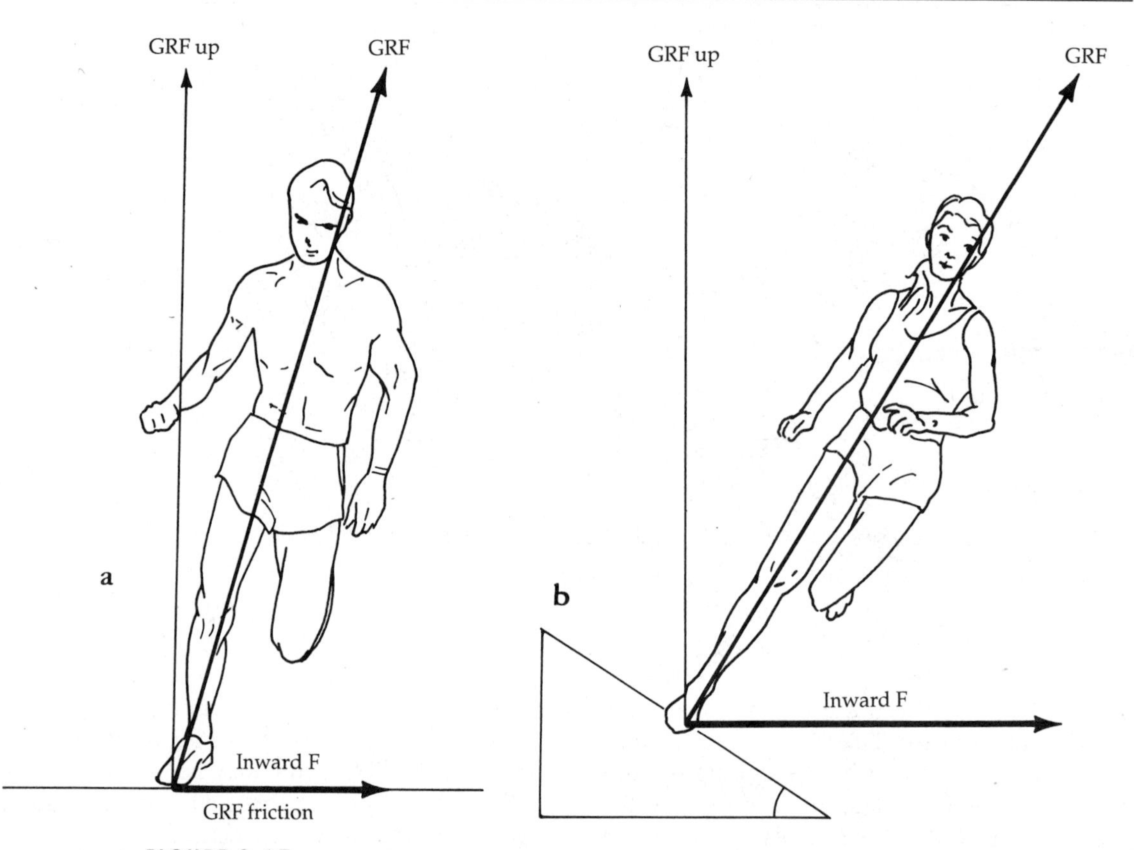

FIGURE 8.17

(a) The inward-directed component of the ground-reaction friction force on the foot provides the inward force necessary to change the runner's direction from a straight line to a curved path (the inward force is totally friction-dependent). (b) Running on a banked (inclined) surface enables the runner to continue moving fast as she curves without concern for inadequate friction and slipping.

both, thereby reducing the need for such a large friction force and body lean. Such leaning postures are assumed by track cyclists, roller skaters, and other performers who must negotiate curves, and the high speeds at which they travel require a large inward force. Usually the surface friction is not sufficient on a horizontal track, and therefore the curves are banked, or inclined, for circular paths (Figure 8.17b). The slope of the banked curves allows the force application and reaction force to be directed more perpendicular to the surface, thus eliminating most—if not all— of the friction dependency for the inward force.

It should be noted that a runner's body does not follow a smooth curved path, as a cyclist's would. During each air (nonsupport) phase, the body travels a straight path; then, with each foot plant, a sideward thrust is made to change direction, a few degrees off the previous linear path.

Understanding Balance Control

1. By throwing and catching a basketball with a partner eight strides away, experiment with various foot positions and determine which position enables you to throw the ball the fastest and to catch the ball most easily without losing your balance.
2. Stand facing a partner one stride away. Placing your palms against your partner's, experiment with different foot positions and directions of force. What determines maximum stability when you attempt to push each other off balance? Explain the differences.
3. Run at a moderate speed in a circle approximately 6 m in diameter. Have an observer note the changes in inward body lean as you increase the running speed. Explain.
4. Repeat the steps in question 3, and increase the diameter of the circle without increasing the speed. Have an observer note the changes in body lean. Explain.
5. Run as fast as you can to a line 20 m away and stop on it while maintaining balance. Have an observer note the changes in body lean and foot placement as you approach and stop at the line. Explain your success or failure to stop on the line without losing your balance.
6. Explain the advantage in the baseball rule that allows a batter to overrun first base. Explain why such a wide turn is made if the hitter continues on to second base.

References and Suggested Readings

Anderson, C. K. (1985). A biomechanical model of the lumbosacral joint for lifting activities. In D. A. Winter, R. W. Norman, R. P. Wells, K. C. Hayes, & A. E. Patla (Eds.), *Biomechanics IX-B* (pp. 50–54). Champaign, IL: Human Kinetics.

Broer, M. R., & Zernicke, R. F. (1979). *Efficiency of human movement* (4th ed.) Philadelphia: W. B. Saunders.

Chaffin, D. B., & Andersson, G. B. J. (1984). *Occupational biomechanics.* New York: Wiley.

Freivalds, A., Chaffin, D. B., Garg, A., & Lee, K. S. (1984). A dynamic biomechanical evaluation of lifting maximum acceptable loads. *Journal of Biomechanics, 17*, 251–262.

Gagnon, M., Sicard, C., & Drouin, G. (1985). Evaluation of loads on the lumbar spine with motion analysis techniques and a static planar model. In D. A. Winter, R. W. Norman, R. P. Wells, K. C. Hayes, & A. E. Patla (Eds.), *Biomechanics IX-B* (pp. 44–49). Champaign, IL: Human Kinetics.

Kobayashi, K., et al. (1974). A study of stability of standing posture. In R. C. Nelson & C. A. Morehouse (Eds.), *Biomechanics IV: Proceedings of the Fourth International Seminar on Biomechanics* (pp. 53–59). Baltimore: University Park Press.

LeVeau, B. (1977). *Williams and Lissner: Biomechanics of human motion* (2d. ed). Philadelphia: W. B. Saunders.

Luttgens, K., & Wells, K. F. (1982). *Kinesiology: Scientific basis of human motion* (7th ed). Philadelphia: Saunders College Publishing.

Ohtsuki, T., Yanese, M., & Aoki, K. (1987). Quick change of the forward running direction in response to unexpected changes of situation with reference to ball games. In B. Jonsson (Ed.), *Biomechanics X-B* (pp. 629–35). Champaign, IL: Human Kinetics.

Sodeyama, H., et al. (1976). Study of the displacement of a skier's center of gravity during a ski turn. In P. V. Komi (Ed.), *Biomechanics V-B: Proceedings of the Fifth International Symposium on Biomechanics* (pp. 271–76). Baltimore: University Park Press.

CONCEPT MODULE F

Linear Movement Responses to Applied Forces

PREREQUISITES

Concept Modules C, D

CONCEPT MODULE CONTENTS

The movement response of a system to forces and torques applied to it is determined by the magnitudes of the net force and net torque applied, the inertial characteristics of the system, and the pathway available for the system's movement.

F.1 Linear Speed and Velocity

An identifiable change in a body's state of motion, or how it is moving, is the *movement response* of a system to the forces acting on it. The state of motion of an object can be described in terms of its being at rest (i.e., not moving relative to the ground or some other reference frame) or in terms of its being in a state of motion (i.e., moving with some given speed). **Speed** is defined as how fast a body is moving, that is, the distance covered divided by the time it takes to cover that distance.

Another factor that describes a moving object's state of motion is the direction in which it is moving at the time we are looking at it. For example, a ball is released by a pitcher at an angle of 2 degrees below the horizontal with a speed of 30 m/sec. When both speed and direction are specified for a system, it is the **velocity,** not just the speed, which is defined. Velocity is a vector quantity because it has magnitude and direction; speed is a scalar quantity that designates only how fast something is moving. The speed of a cross-country runner may be 5 m/sec on the average, whereas the velocity of a sprinter in a 100-m run might be 8 m/ sec in a straight line south. The equation used to find the magnitude of a body's speed or velocity is:

EQUATION F.1

$$\text{velocity} = \frac{\text{distance}}{\text{time}} \quad \text{or} \quad v = \frac{d}{t}$$

where v is linear speed (or velocity in a given direction) of the object, d is distance (or displacement in a given direction) that the object moved, and t is time during which the distance was covered.

So far, we have two tools necessary to describe a system's state of linear motion: d, how far, and v, how fast and in what direction.

The most useful information about the state of linear motion of a system is obtained by describing the linear motion of the CG of the system. Such a reference point is especially valuable when the motion of the human body, with all of its segmental motion, is examined as a single point.

F.2 Linear Acceleration

In addition to how far and how fast a system is moving, we can further describe its motion by stating whether it is speeding up (**accelerating**) or slowing down (**decelerating**) or changing direction (also a form of accelerating). This happens whenever a body changes its state of motion (as in starting, stopping, changing its speed or direction).

To change the state of linear motion of a body, a net force must be applied to it. The greater the net force, the greater the motion change. How fast it is changing its speed or direction is the magnitude of its acceleration. More precisely, linear acceleration is the time rate of change in a body's velocity.

Quantitatively, linear acceleration is expressed as follows:

EQUATION F.2

$$\text{acceleration} = \frac{\text{change in velocity}}{\text{time}} \quad \text{or} \quad a = \frac{v_2 - v_1}{t} \quad \text{or} \quad a = \frac{\Delta v}{t}$$

where a is linear acceleration of the system, v_2 is final velocity of the system, v_1 is initial velocity of the system, t is time during which the velocity was changed, and Δ is "change in."

The magnitude of a body's acceleration is large if a large change in velocity occurs in a small amount of time (e.g., going from 0 to 50 km/hr in 5 sec in a sports car). The acceleration is small if the same velocity change occurs over a longer period of time (e.g., 20 sec in an old truck). Inserting these numbers in Equation F.2, the linear acceleration for the sports car is expressed as follows:

$$a = \frac{v_2 - v_1}{t} = \frac{50 \text{ km/hr} - 0 \text{ km/hr}}{5 \text{ sec}} = \frac{10 \text{ km/hr}}{\text{sec}}$$

The car is increasing its speed at the rate of 10 km/hr during each second that the driver applies the gas pedal force. For the truck, the change in speed would be at the rate of 2.5 km/hr per second (sluggish). Note that time is mentioned twice in acceleration values; this is essential for the correct understanding of the meaning of acceleration.

A body experiences acceleration only for the time during which force is being applied. When the force ceases, the body has reached its new speed and moves with that speed until that speed or direction is changed by another force. Also, the direction of the acceleration is in the same direction as the net force applied. As would be expected, a large force is necessary to produce a large acceleration, and a small force produces a small acceleration. For example, a large force applied by the club is necessary to drive a golf ball off a tee so that its velocity is changed greatly in an instant. Since a smaller change in velocity is required for putting the ball on the green, less force is applied by the putter.

Qualitatively, when a body is gaining speed under the influence of an external force, it is accelerating; if it is losing speed, it is decelerating. The convention also is to designate speeding up as positive (+) acceleration and slowing down as negative (−) acceleration. Positive and negative signs must be used carefully, however, because the direction of movement also can be designated as positive or negative. For example, a body could be described as negatively accelerating when it is speeding up in the negative direction. Acceleration is discussed more qualitatively than quantitatively here, however, so that the concept may be readily applied to things we are watching. Therefore, the words *acceleration* and *deceleration* usually are used, and the direction is specified (e.g., forward, backward, to the right).

g: The Acceleration Caused by the Force of Gravity

An external force that is constantly acting on our bodies and everything else on earth is the force of gravity, or weight force. Newton's law of universal gravitation declares that the force of attraction between two masses is directly proportional to the masses of the two bodies. This means that the greater the masses of the bodies, the greater the force of attraction between them. Because the mass of the earth is so large compared to any of the masses on or near its surface, the attraction between any body and the earth appears to be manifested only by a constant force of the earth exerted on that body. This force holds us down to the ground, which exerts an upward reaction force on us so that in standing, for example, zero net force is acting on us and no acceleration occurs.

If a body is free from any contact support, however, it will be forced downward by the gravitational attraction force. As it falls, the body will continue to gain downward speed (accelerate) at the rate of 9.8 m/sec (32 ft/sec) *during each second it is falling*. Similarly, an object projected upward will decelerate at the same rate as a result of the downward force of gravity. For example, if a ball is tossed upward, it will decelerate at the rate of 9.8 m/sec (32 ft/sec) during each second it is rising until its vertical velocity reaches zero. The ball will then start to fall and will accelerate downward at the rate of 9.8 m/sec (32 ft/sec) during each second it is falling until it lands. The symbol **g** represents this acceleration of 9.8 m/sec/sec (m/sec^2) (32 ft/sec/sec), which is due to the force of gravity.

The gravitational force of attraction of two bodies is also inversely proportional to the squared distance between the two bodies, so that the farther apart the centers of mass, the smaller the force of attraction between them. This is why one hears that a body's weight (force of attraction on the body's mass) is less at high elevations (Denver) than at low elevations (Los Angeles). However, this difference would be practically negligible.

Zero Acceleration: Constant Velocity

If a net external force acts on a system, that system will accelerate in the direction of that net force. Therefore, if no net force is acting (if all opposing forces cancel out), the linear acceleration of the body is zero. If acceleration is zero, no change in velocity occurs. Therefore, if a body is at rest, it does not change its velocity if all the external forces acting on it cancel out to zero. Similarly, when a body is moving at some given velocity and experiences opposing forces that are equal, the result is the same: there is no change in the system's velocity. In such a situation the velocity is said to be constant, because there is no net force to cause acceleration or deceleration. This is true for a body at rest (v = 0) and for a body moving with a constant speed and not changing its direction of motion (v ≠ 0). For the body to change its direction, a net force must be applied to that body.

A sky diver presents a suitable example for clarifying the above concept. Figure F.1 shows the forces acting on the diver at the time he jumps from the aircraft and then during the descent. The earth exerts a downward force equal to the diver's weight of 712 N (160 lb), and his body accelerates downward,

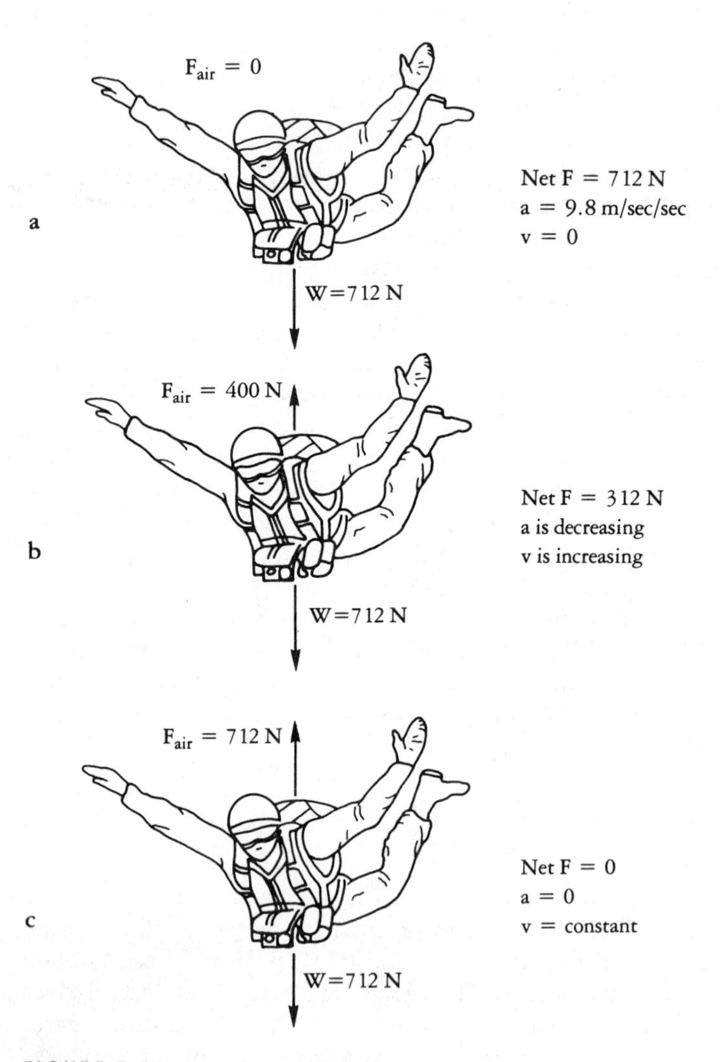

FIGURE F.1

The net force acting on a falling body changes as a result of increasing upward air resistance while downward velocity increases. Constant velocity is reached when the air resistance is as great as body weight.

gaining speed at the rate of 9.8 m/sec (32 ft/sec) during each second of descent. As speed is gained, however, the air through which he is falling flows faster and faster past his body and exerts a greater and greater upward force on his body (Figure F.1b). Air resistance increases dramatically as speed increases and as the area of the body encountering the flow increases. The upward air resistance force increases progressively with downward speed, which is also increasing, until the diver is falling so fast that the air resistance force becomes equal to his body's weight, including his uniform and equipment (Figure F.1c). When this happens, the sum of the forces in the vertical direction is equal to zero; therefore, no acceleration or deceleration occurs, and the diver falls with a constant velocity (called **terminal velocity**). If the diver changes body position by moving the limbs, he will expose a different body surface area to the air flow and thereby alter the air resistance against it. If the air resistance increases, there will be a momentary net force upward, and the body's falling velocity will be slowed until, once again, the upward and downward forces are equal. Similarly, if he changes position so that the air resistance becomes less (assumes a streamlined position), he will accelerate downward because of the net force in that direction until the air resistance increases to a value equal to the body's weight.

The movement response of a body subjected to such force conditions was summarized by Newton and put forth in what is known as Newton's first law, the **law of inertia.** This law predicts that a body at rest will stay at rest until a net external force acts to move (accelerate) it and that a body in motion will continue to move at a constant velocity and in the same direction until a net external force acts to change that velocity. This concept may be summarized in the phrase "resistance then persistence."

Variable Velocity: Average and Instantaneous Values

Only in rare circumstances do we perform activities in a state of constant velocity. Even the sky diver in Figure F.1 experiences variable velocity, resulting from the changing body positions and air forces acting on his body. We usually experience fluctuating net forces acting on us, and our movements are always accelerating or decelerating in response to these changing forces. Thus, an important concept concerning a system's motion is that of how its velocity changes over some period of time.

The following example will serve to illustrate velocity changes. Consider a sprinter, motionless in the starting blocks, ready to start a 100-m run. On the gun, she accelerates out of the blocks and changes her speed from zero to her top running speed. After her initial acceleration period, she runs at a steady speed, going neither faster nor slower, for a short time interval. During the time when she is running at a steady speed and does not change her direction, she is traveling at a constant velocity; that is, her acceleration (change in velocity) is equal to zero. Although the sprinter tries to maintain her top speed, some deceleration usually occurs near the final section of the distance.

From the start to the finish line of the 100-m distance, her time was 13 sec. Therefore, her velocity was 100 m/13 sec or 7.7 m/sec. The figure of 7.7 m/sec, however, represents only her **average velocity,** because she did not run that fast for the entire 100-m distance. At the start, her velocity was zero, and she had to accelerate to her top running speed. She then decelerated through the finish line and subsequently slowed to a stop. If we look at a graph of her velocity for the time she was active (Figure F.2), the variable velocity is evident. We can see that at the instant 2 sec after starting, her instantaneous velocity was approximately 5 m/sec. At the instant 6 sec after starting, her instantaneous velocity was 9 m/sec. If we take all the instantaneous values and average them, we will obtain the average velocity. Note that the average velocity of 7.7 m/sec for the whole run is below the maximum velocity and above the starting velocities. Also note that the smaller the time interval, or instant, the more precise the velocity value.

Very small time intervals in sport are often the focal point in analyzing skills. Examples of these time intervals are the instant of foot plant in a run, the instant of contact of a racket and ball, and the instant of release of a thrown discus or ball. One of the most valuable pieces of information used in the analysis of projecting objects (e.g., throwing, kicking, striking, jumping) is the **instantaneous velocity** of projection (also called release velocity, rebound velocity, takeoff velocity).

We can use an example of ball throwing for examining the instantaneous velocities of the ball as it leaves the hand and travels along its path. Note in Figure F.3a that the flight path is curvilinear. If we measure the velocity of the ball at any given instant, we can draw a vector to represent the ball's instantaneous velocity. The immediate direction of the ball at any instant is tangent to that point on the arc formed by the ball's path.[1] The ball's direction at

[1]Tangential direction is one that is perpendicular to the radius of the circular arc at any given point on the arc.

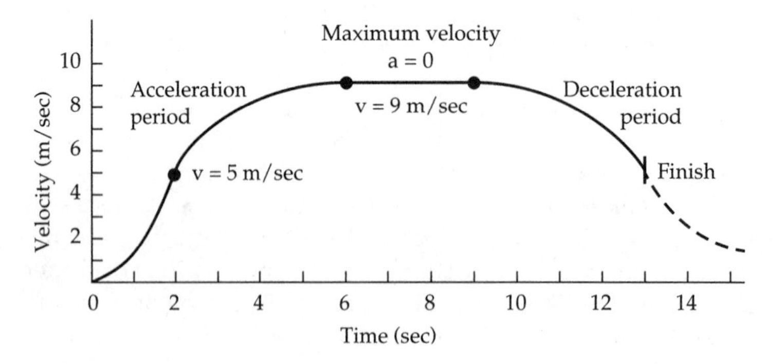

FIGURE F.2
The velocity of a sprinter changes from the start to the finish. The average velocity for the total distance is less than the fastest running velocity.

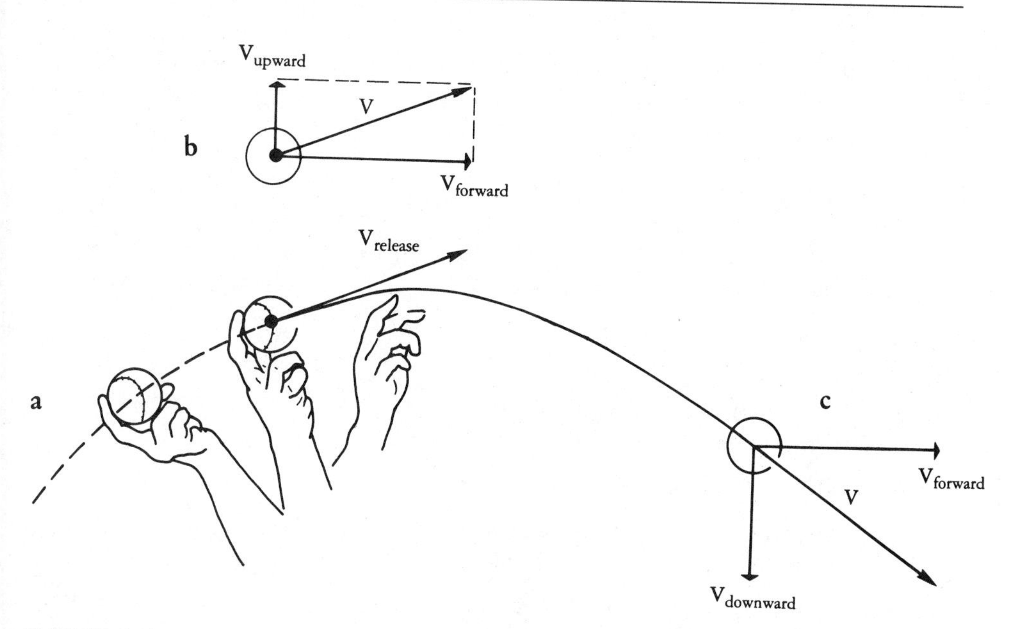

FIGURE F.3

The resolution of an instantaneous velocity vector into its horizontal and vertical components. (a) The direction of instantaneous velocity of the ball is tangent to the path of the ball at that instant. (b) The instantaneous values of the vertical and horizontal components are determined by constructing a rectangle around the original vector. (c) The velocity vector and components at a point during the ball's descent.

the instant of release is shown in Figure F.3a and b. The vertical and horizontal components of the ball's velocity at this instant will be shown to be important for evaluating the effectiveness of the throw relative to its purpose. These components are graphically determined by vector resolution.

To review this process, construct to scale a rectangle, the diagonal of which is the original velocity vector (Figure F.3b). Draw in the direction lines (horizontal and vertical) from the point on the path where the tail of the original velocity vector lies. Next, starting from the head of the original vector, draw lines parallel to the horizontal and vertical lines. The points at which the vertical and horizontal lines intersect determine the lengths of the vertical and horizontal velocity components, and the arrow heads are drawn at these locations. The magnitudes of these components are then found by measuring their lengths using the same scale. Figure F.3 also shows the ball's velocity vector and its components at an instant in the descending path.

Note that as soon as the ball is released at a given speed and direction, its velocity starts to change as a result of two external forces: gravity (the force of the ball's weight) and air resistance. Gravity acts downward to reduce the upward speed of the ball until it reaches its peak, and then gravity acts to accelerate its descent. The horizontal and vertical speeds of the ball are decreased by air resistance during the ball's flight. If the forces were not acting on the ball, it would travel forever in a straight line (release velocity direction) with a constant speed (release velocity speed). This does not happen, because gravity is always acting downward.

Understanding Linear Motion Changes

1. What factor distinguishes a body's velocity from its speed?
2. Clarify the definition of linear acceleration of a body by giving examples in human movement.
3. If a body is moving with a constant velocity, what is its acceleration?
4. Which runner has the greater acceleration, the one who gains a velocity of 5 m/sec in 4 sec or the one who gains a velocity of 4 m/sec in 3 sec?
5. What is the direction of the instantaneous velocity of release of a bowling ball if the straight arm is vertical when the fingers release the ball? What is the release direction when the arm is 45 degrees beyond the vertical when the ball is released?
6. What does the symbol **g** represent?
7. What is the vertical velocity component at release of a horizontally projected ball?

F.3 The Relationship of Force, Mass, and Linear Acceleration

According to Newton's law of action–reaction, whenever two bodies or objects come into contact with each other, they exert the same amount of force on each other, but in opposite directions. If the movement of one of these bodies is to be the object of analysis, that body is the one you define as your system, or your central focus. The other object then becomes merely the source of the external force being applied to your system. The effect of this external force on the movement of the body being analyzed may easily be predicted if we are familiar with another law set forth by Newton. He observed that the movement response (acceleration) of a system depends not only on how much net external force is applied to it but also on the resistance to movement change (inertia) possessed by the system. Recall that the inertia of a body is the property that accounts for the body's tendency to remain still or continue to move at a constant speed without changing magnitude or direction. This "sluggishness" is directly related to the mass of the body for linear motion. Consequently, for linear motion, the greater the mass of a body, the greater its resistance to motion change caused by some force.

The interrelationship of the variables of external force, mass, and acceleration of the mass was set forth by Newton in his second law, the law of acceleration:

EQUATION F.3

$$\text{acceleration} = \frac{\text{Force}}{\text{mass}} \quad \text{or} \quad F = ma \quad \text{or} \quad a = \frac{F}{m}$$

where a is acceleration of the body (system), F is net external force applied to the body, and m is mass of the body receiving the external force.

This relationship of factors means that the acceleration of a system is directly proportional to the net external force applied to the system and is in the same direction as that applied force. If the acceleration of the body is directly proportional to the applied force, then the greater the force applied to the body, the greater its acceleration. The law also states that the acceleration of the body is inversely proportional to the mass of the body, which means that the greater the body's mass, or inertia, the less its acceleration in response to a given applied force. A **Newton** (N) is the unit of force of such size that under its influence a body whose mass is 1 kg would experience an acceleration of 1 m/sec/sec (4.45 N = 1 lb). Also, 9.8 N would accelerate a 1-kg mass 9.8 m/sec/sec (the acceleration that is due to gravity).

In performing a motion analysis of a system, we must consider simultaneously the interdependency of all three variables. To fully appreciate the relationship between any two of the quantities, we will look at how those two vary while we assume the third quantity stays at a constant value. The process of kicking a ball will serve as an example. The ball is the body, or system, being analyzed, and the impact of the foot is the external force applied to the ball. In each of the three cases that follow, one of the three variables (m, F, or a) will be held constant:

Case I: Varying the magnitude of the kicking force applied to the same ball and observing the magnitude of the resulting acceleration is shown in Figure F.4.

Case II: Varying the magnitude of the mass of the ball (using different types of balls), applying the same amount of kicking force to each, and observing the resulting acceleration is shown in Figure F.5.

Case III: Varying the mass of the ball (using different types of balls), specifying that each ball be accelerated the same amount, and observing the magnitude of the applied force required to produce that acceleration is shown in Figure F.6.

The following example will serve to relate the concept of external reaction force and motion change of the body exerting the action force. A sprinter accelerates out of the blocks and by doing so increases velocity from zero to

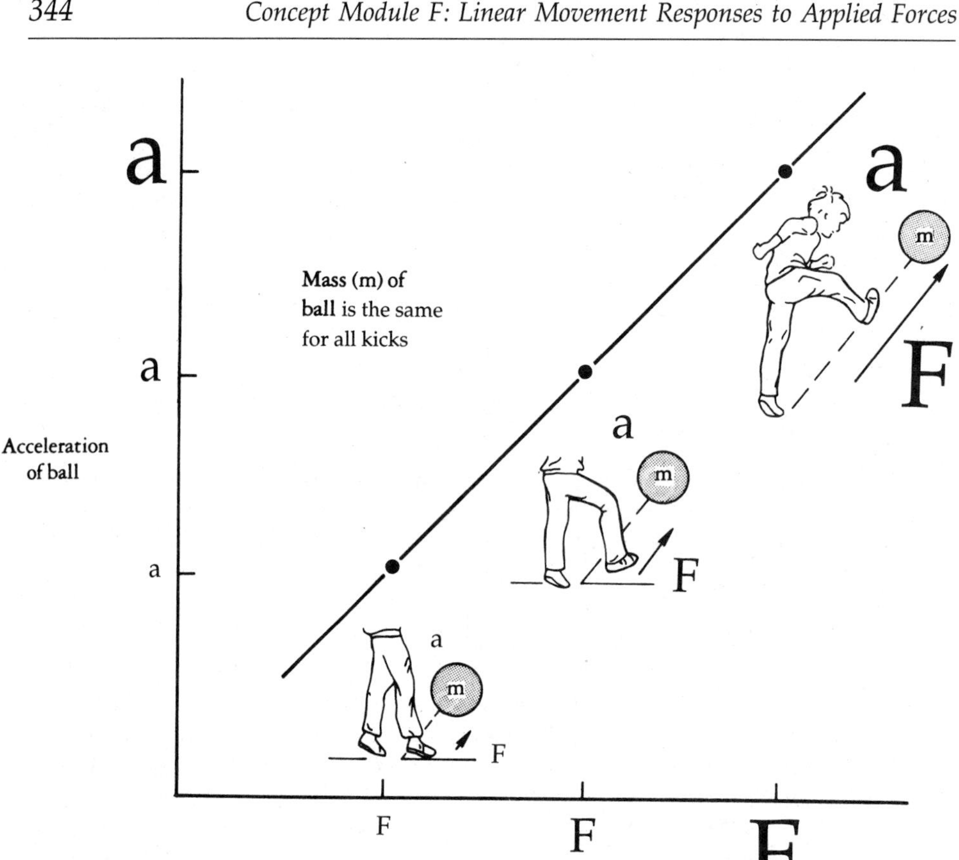

FIGURE F.4

The greater the force applied to a constant mass, the greater the acceleration of that mass.

maximum running velocity. Such acceleration of the body forward is caused by an equal and opposite reaction force from the stable blocks when the legs extend to push backward against them. Recall that the leg muscles are internal forces to the system that change the shape of the system so that it exerts an action force against another mass that will exert an equal and opposite reaction force against the system. Figure F.7a illustrates the action force vector and the ground-reaction force vector that is the external force acting to accelerate the runner during the start. The amount of acceleration that the runner will display is inversely proportional to the amount of the runner's mass.

After the initial start, the run is continued near top speed until the end of the race. After the finish, the sprinter comes to a stop. Such a decrease in speed is the system's deceleration, and this change in velocity must be caused also by an externally applied force. The external force that decelerates the body is the backward-directed ground-reaction force responding to the forward-directed component of the force of the feet on the ground during the final steps (Figure F.7b). Recall that the magnitude of the runner's deceleration that is due to a

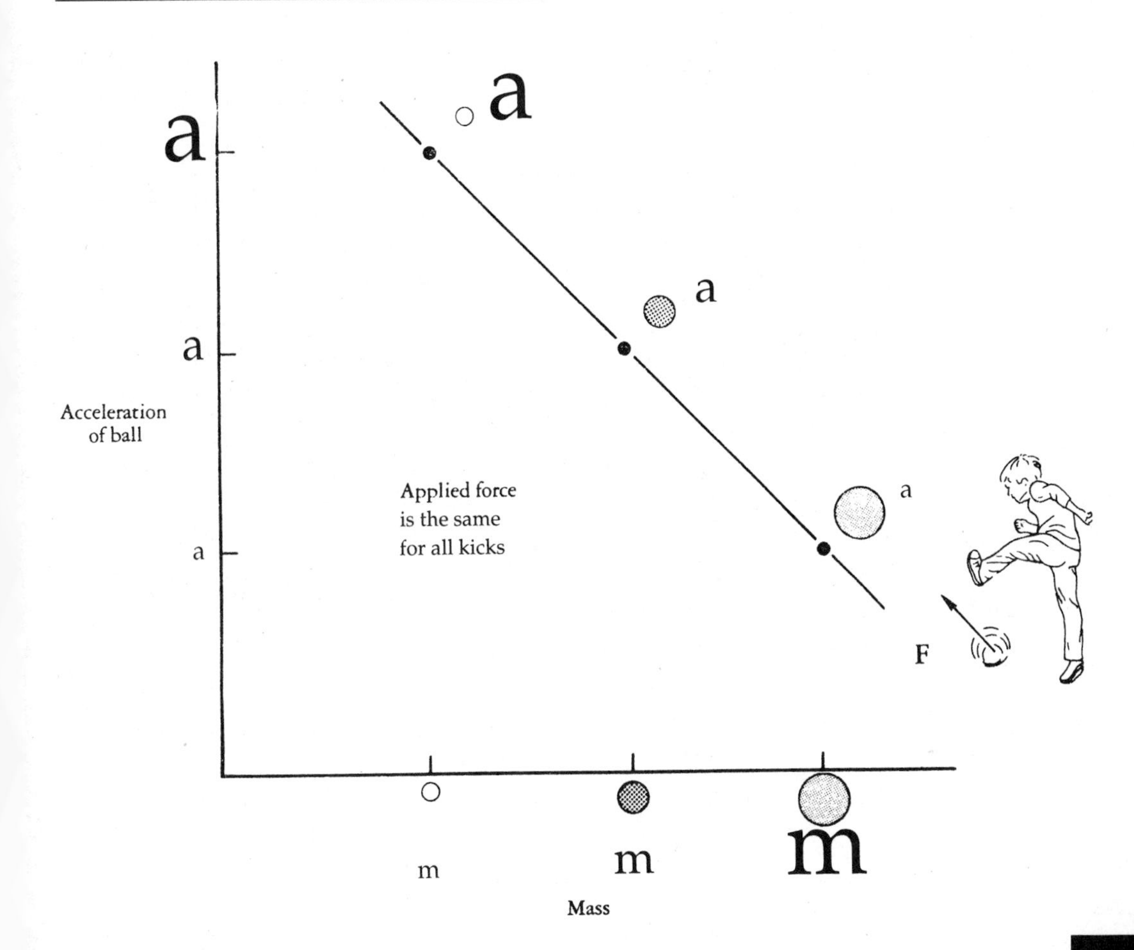

FIGURE F.5
As the mass of an object or body is increased, it experiences less acceleration from a given amount of force applied to it.

given force depends on his mass. Such forward force and backward reaction force occurring on level ground depend on the forces of friction. The friction force depends on the roughness of the ground and the shoe soles, together with the force of the body against the ground.

To integrate the concepts of someone's applied force, weight force, ground-reaction force, net external force, and acceleration of a system, we will use a situation in which a load is lifted from a spring scale, as shown in Figure F.8. The dial reading represents the equal and opposite reaction force of the earth upward on the 80-N (18-lb) load. (Consider that if the scale platform were not there to apply an upward reaction force, the weight would move downward.) In this initial condition (Figure F.8a), the sum of the external forces is equal to zero (i.e., a negative 80-N weight force downward and a positive 80-N force upward). If a person were to apply an upward force of 60 N on the load (Figure

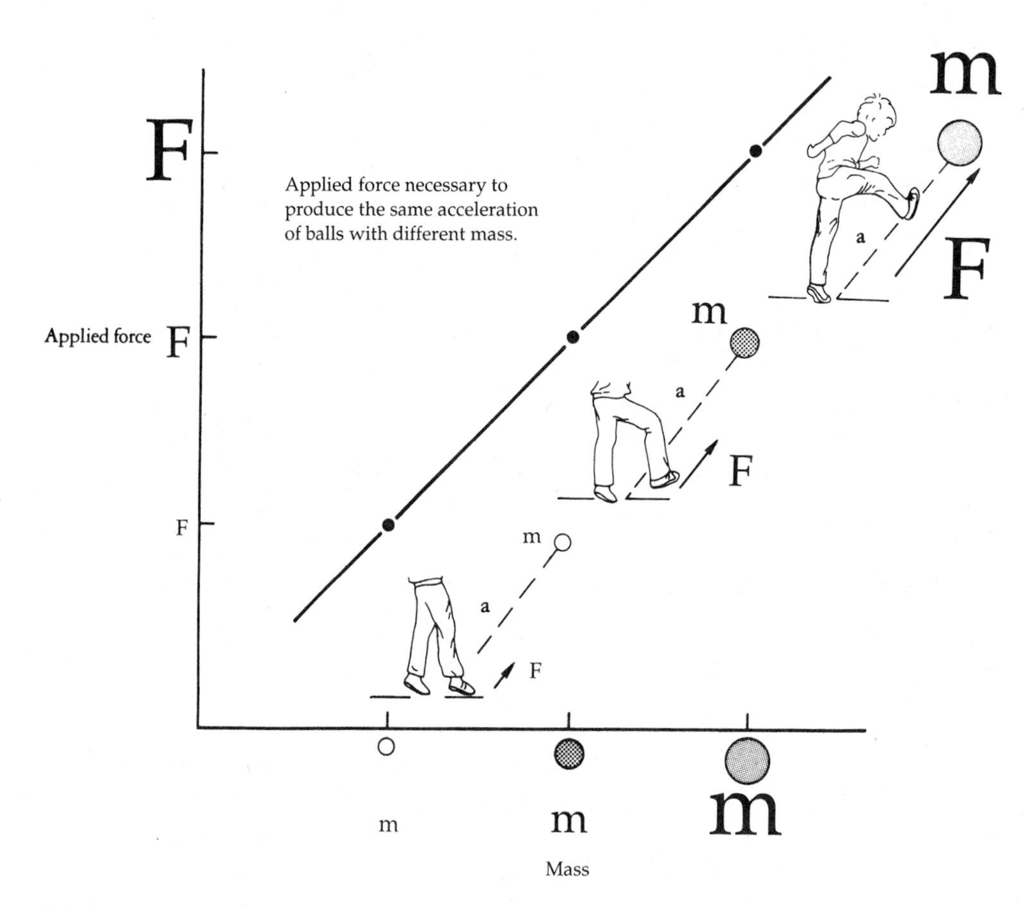

FIGURE F.6

As the mass of an object or body is increased, the force applied to it must be increased to produce a given acceleration.

F.8b), the scale would read 20 N. Again, no motion of the load would occur because the net force acting on the load is still zero: $(+20) + (+60) + (-80) = 0$. When the person applies an upward force of 80 N (Figure F.8c), the scale reads zero. As soon as the force applied by the lifter becomes greater than the weight force of the object, the object begins to accelerate upward in proportion to the amount of net force (lifting force minus weight force).

In this lifting example, a distinction must be made between the mass of the object and the weight of the object. The weight of the object is downward force and thus figures in calculating the net force on the system. The magnitude of the weight force downward is always subtracted from the force upward to arrive at the net force. Therefore, the greater the weight of the object, the less the net force available to accelerate the object upward with a given amount of lifting force. Once the net force is determined, the amount of acceleration of the object depends on the subject's mass, following the relationship $a = F/m$.

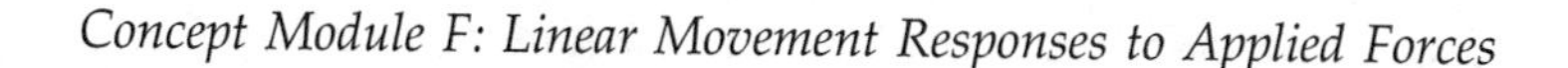

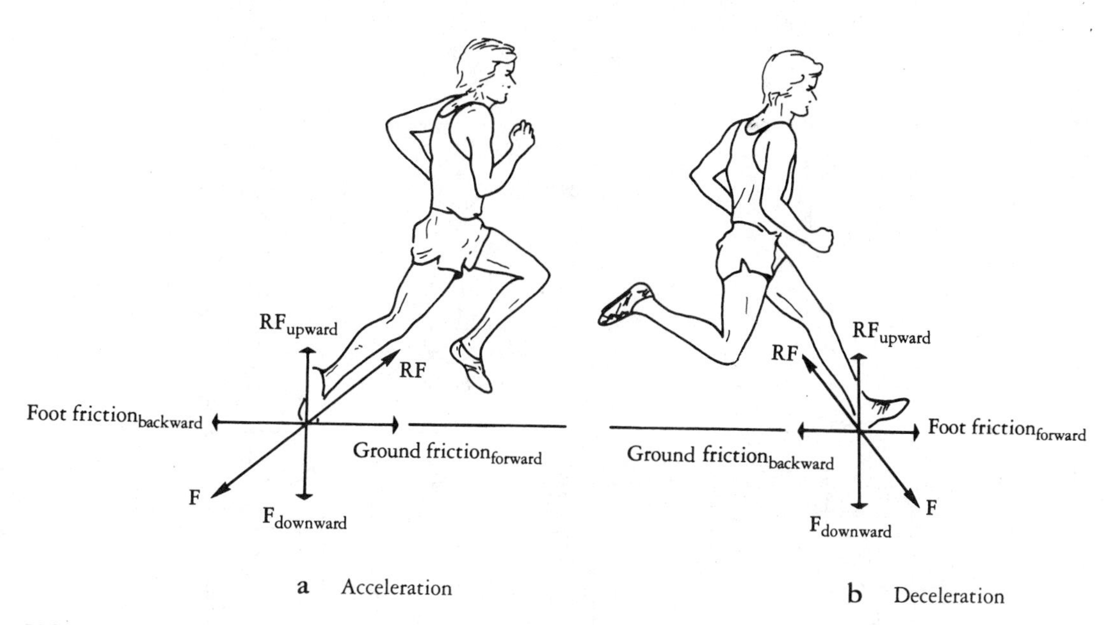

FIGURE F.7

(a) The friction component of the ground-reaction force against the foot of the runner serves to accelerate the runner forward. (b) The friction component of the ground-reaction force against the foot serves to decelerate the runner after the race.

Toward the end of the range of motion (ROM) of the lifter, the upward force applied on the load by the lifter decreases. Thus, the upward net force decreases until it reaches a value equal to the weight of the object, and its upward motion stops.

So we see that whenever one force is greater than the force opposing it, the body will change its velocity according to the direction of the larger force. This means that (1) a body initially at rest will gain speed in the direction of the net (larger) force, (2) a body already moving in the direction of the net force will gain more speed (accelerate) in that direction, and (3) a body already moving in a direction opposite to that of the net force will lose speed (decelerate), and depending on how much force is applied or for how long it is applied, the body may decelerate to a stop and then begin to accelerate in the direction of the net force. An object projected upward into the air responds in this manner under the net force of gravity.

Newton's law of acceleration, $F = ma$, also can be used to relate the quantities of a body's weight, its mass, and its acceleration due to gravity. Because weight is a force, and g is the acceleration of 9.8 m/sec^2 (32 ft/sec^2), a body's weight can be expressed in the form of $W = mg$ ($F = ma$). Similarly, a body's mass is expressed by $m = W/g$. For example, the weight of the sky diver in Figure F.1 is 712 N, so his mass would be $712/9.8 = 72.6$ kg. In the British system, his mass would be 5 slugs $= 160/32$.

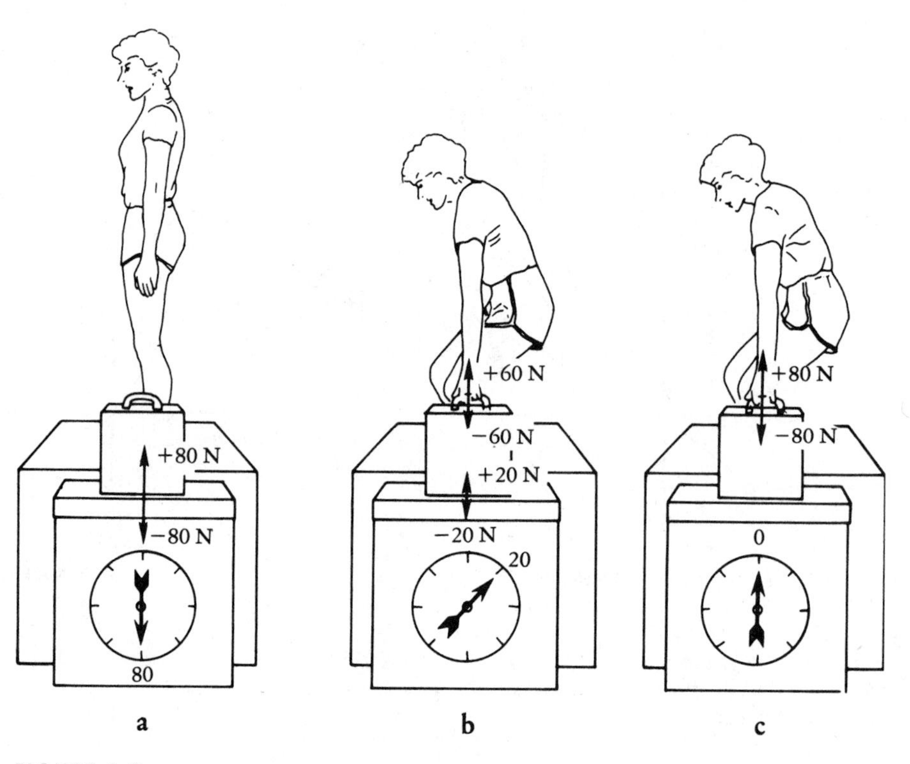

FIGURE F.8
A load being lifted from a spring scale, showing the changing ground (scale) reaction force to the load's changing downward force.

F.4 Centripetal Force and Radial Acceleration

We have seen how a body can be accelerated by having its speed changed by an external force. Now we can apply the concept of acceleration to explain the direction change of a moving body.

Centrifugal force and centripetal force were presented in Concept Module D. Now the relationship between force and acceleration can be presented to further clarify the concept. Picture a cyclist rounding a corner. She leans inward as she curves, causing the tires to contact the road at an angle, which exerts a centrifugal force (friction) outward on the ground. The ground friction reaction force exerted on the tires is directed toward the center of the circular arc she is traversing. This centripetal force is directed along the imaginary line that is the radius of the circular path at any given instant and is the force responsible for the cyclist's direction change. The centripetal force, therefore, causes a **radial acceleration** (direction change) of the cyclist. How fast the rider can travel the arc without slipping depends on the mass of the person and bike, the friction available, and the "tightness" (radius) of the turn.[2] The greater the tire–ground friction available, the faster the rider can travel without slipping; however, the speed is limited by the maximum friction available. The greater the mass of the rider and bike, the greater the centripetal friction force needs to be (using F = ma, centripetal force = mass × radial acceleration). Also, the tighter the turn (smaller radius of curvature), the greater the direction change (radial acceleration) and therefore, the greater the friction needs to be.

The same combination of variables—mass, radius of curvature, speed of travel, and the centripetal force causing the radial acceleration of the body—can be used to understand the forces operating in other activities in which motion is constrained to circular arcs, such as when a gymnast swings from a bar. The grip taken to swing around a bar must be strong enough to prevent the centrifugal force (exerted by the body on the bar) from "uncurling" the fingers. Because the centrifugal force is the equal and opposite reaction force to centripetal, it is influenced by the same factors. Therefore, the greater the mass and the greater the speed of swinging, the stronger the grip must be. Likewise, when a ball player swings a heavy baseball bat, the grip strength is usually adequate to hold onto the bat so that it doesn't fly off in a straight line.

In the field event of the hammer throw, the centripetal force exerted by the athlete on the hammer must be great enough to *radially* accelerate the faster and faster moving hammer before release. The leaning position of the thrower allows the feet to push diagonally downward and outward on the ground, thereby generating a ground friction reaction force (centripetal force) acting on the thrower and hammer system. The importance of adequate foot–ground friction in this event is evident if you think of trying such a performance on ice.

Understanding Force and Motion Relationships

1. If your hand applies a force of 45 N (10 lb) on a cart, how much force is the cart applying on your hand? According to Newton's third law of action–reaction, if you are receiving that force from the cart, why are you not forced to move backward?

[2]The quantitative relationship of these variables is presented in Concept Module D and Appendix II, Table II.2.

2. Under the same circumstances given in question 1, how would your motion response differ if you were on wheels and the cart were not? What would be the motion response of your body and the cart if you both were on wheels?
3. How is a system's acceleration related to the net force applied to it? to the amount of mass that the system has?
4. If you were to collide with another person who weighed twice as much as you, what would be the relative magnitude of acceleration of the two bodies?
5. If the force of a tennis racket causes the ball to accelerate forward, what effect does the ball's reaction force have on the racket? Which object's motion is more easily changed? Why?
6. Answer the following questions:
   a. If you apply an *upward* force of 80 N on a shot that weighs 50 N, what is the net external motive force and direction of the shot's acceleration?
   b. If you apply a *horizontal* force of 80 N on a shot that weighs 50 N, what is the net external motive force and direction of the shot's acceleration?
   c. In which situation, a or b, is the net motive force greater? Explain why.
   d. In which situation, a or b, is the shot's acceleration greater? Explain why.
7. What might be an injurious effect of your swinging an infant around you in faster and faster circles while holding its hands? Explain in terms of radial acceleration and the associated forces.
8. Using F = ma, calculate your body weight in Newtons, using kg for body mass.
9. If your mass is 70 kg and you are accelerating at a rate of 2 m/sec/sec, what is the magnitude of the net external force acting on your body in the direction of your acceleration?

References and Suggested Readings

Barham, J. (1978). *Mechanical kinesiology.* St. Louis: C. V. Mosby.

Brancazio, P. J. (1984). *Sportscience.* New York: Simon & Schuster.

Gustafson, D. R. (1980). *Physics: Health and the human body.* Belmont, CA: Wadsworth.

Hopper, B. J. (1973). *The mechanics of human movement.* New York: American Elsevier.

Tricker, R. A. R., & Tricker, B. J. K. (1967). *The science of human movement.* New York: American Elsevier.

Wheeler, G. F., & Kirkpatrick, L. D. (1983). *Physics: Building a world view.* Englewood Cliffs, NJ: Prentice-Hall.

CONCEPT MODULE G

Linear Momentum and Kinetic Energy

PREREQUISITES

Concept Modules C, D, E, F

CONCEPT MODULE CONTENTS

G.1 Linear Momentum

The way in which external forces affect the motion of a system has been described in terms of Newton's law of acceleration, a = F/m. The change in a system's state of motion can be examined from a different perspective, that of the momentum of the system. **Momentum** incorporates a system's resistance to change in its motion (its inertial properties) and its velocity (speed with a specified direction). The greater the mass or the greater the velocity of a system, the greater its momentum.

In the situations described for Newton's law of acceleration, some force was applied to a mass, and that force caused the mass to change its velocity. As Newton first explained it, an object's "quantity of motion" is the product of that object's mass and velocity and can be changed by the application of an outside force. This "quantity of motion" of a system is an expression of the system's momentum, a measure of its "persistence" in its state of motion.

EQUATION G.1

$$\text{Momentum} = \text{mass} \times \text{velocity} \quad \text{or} \quad M = mv$$

where M is momentum of the system (kg-m/sec), m is mass of the system (kg), and v is velocity of the system's center of gravity (m/sec).

Momentum is a vector quantity because it has direction as well as magnitude. Whenever a system moves, it has momentum, and at any given instant it has momentum in the direction it is moving at that instant. Therefore, when an external force acts on a system to change its velocity, that force is acting to change the system's momentum. The greater the force applied and/or the longer the time the force is applied, the greater the change of momentum. The change in momentum of a particular system can occur very slowly or very quickly. For instance, picture the difference between the starts used for a 100-m run and the Boston Marathon. Both starts require a change in the momentum of the system (in this case, the body), but the rate of change must be much more rapid for the sprint, thus the force necessary to produce the faster rate of momentum change must be greater.

G.2 Linear Impulse

The concept of the rate of change of a system's momentum and the force necessary to produce such a change leads us back to the first expression of Newton's law of acceleration, $a = F/m$, or $F = ma$. We can do some rearranging and substituting in the expression, $F = ma$, to show how the momentum of the system and the force on the system are related:

$$F = ma$$

where acceleration is the change in velocity divided by the time it took to change it,

$$a = \frac{(v_2 - v_1)}{t}$$

then if we substitute for a in $F = ma$:

$$\text{Force} = \frac{\text{mass} \times \text{its change in velocity}}{\text{time it took to change}} \quad \text{or} \quad F = \frac{m(v_2 - v_1)}{t}$$

then, by multiplying the mass of the system, m, by its change in velocity:

$$F = \frac{(mv_2 - mv_1)}{t} \quad \text{which is} \quad \frac{\text{the change in momentum}}{\text{time it took to change}}$$

Note that the change in the system's momentum, ($mv_2 - mv_1$), is the final momentum of the system minus the initial momentum it had before the force was applied. It can be seen from the rearrangement of these quantities that the momentum change is directly proportional to the force applied to the system. Also, it can be seen that the longer the time a given force is applied, the smaller that force need be for a given change of momentum; that is, a small force applied for a long period of time will produce the same momentum change as a large force applied for a short period of time. We can write Newton's law of acceleration, therefore, so that it expresses something about a system's momentum change:

EQUATION G.2

Force × the time that force is applied to a system

= the change in the momentum of the system

$$\text{or} \quad F(t) = M_2 - M_1 \quad \text{or} \quad \Delta M$$

where F is force, t is time during which the force is applied, M_2 is final momentum of the system, M_1 is initial momentum of the system, and Δ is "change in."

The preceding equation is known as the impluse-momentum equation, in which the quantity F(t) is the **impulse** that the system received to change its momentum. As mentioned above, the magnitude of the force can vary, and the length of time during which the force is applied can vary for any given momentum change. Frequently, and perhaps unknowingly, we manipulate these two variables as we manage equipment, implements, and our own body.

Impulse for Absorbing Shock

Two shock absorption examples in which we can apply the impulse-momentum concept are those of catching a ball (Figure G.1a) and landing from a jump (Figure G.1b).

Both of these activities require reducing the velocity of the moving mass to zero: in catching, by bringing the ball to a stop in the hands, and in jumping, by stopping the downward body motion as the feet strike the ground. From experience, we know that both skills involve flexion of the limbs to "absorb the shock." If no flexion occurred, we would feel extremely large forces of impact as the velocity of the mass was reduced to zero in a short period of time (large force times small time equals momentum change). "Absorbing the shock" is the common phrase for receiving the moving mass gradually so that the deceleration of the moving mass is not as large as it would have been if the object had been brought to a more abrupt halt. In this way we use smaller forces for longer periods of time for the reduction of momentum to zero. Whenever the momentum of a mass needs to be changed gradually, the

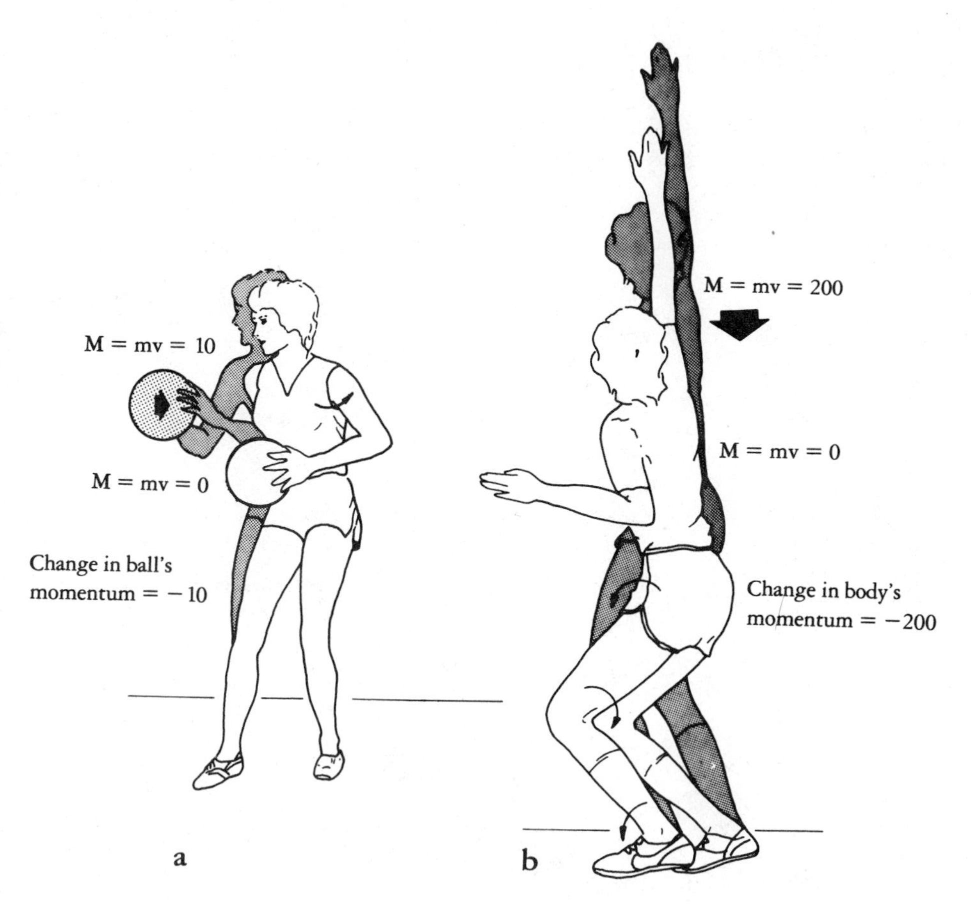

FIGURE G.1

Absorbing the shock means gradually decelerating a moving mass by applying smaller forces over a longer period of time. (a) The player receives less force of impact by "giving" with the ball to stop it. (b) By flexing the joints during landing, the body's momentum is stopped by a smaller ground-reaction force acting for a longer time.

time for deceleration can be increased and less force is required at any instant. Conceptually, we can picture the shock-absorption principle for a given change in momentum as follows:

The change in momentum can be caused by a smaller force applied over a longer time:

$$\Delta M = \text{force} \times \mathbf{TIME}$$

rather than causing that same change in momentum by applying a large force for a short time (a very sudden stop):

$$\Delta M = \mathbf{FORCE} \times \text{time}$$

Impulse for Changing Directions

Linear impulse, F(t), is also responsible for how we change direction while moving. Figure G.2 shows a runner in a maneuvering situation.

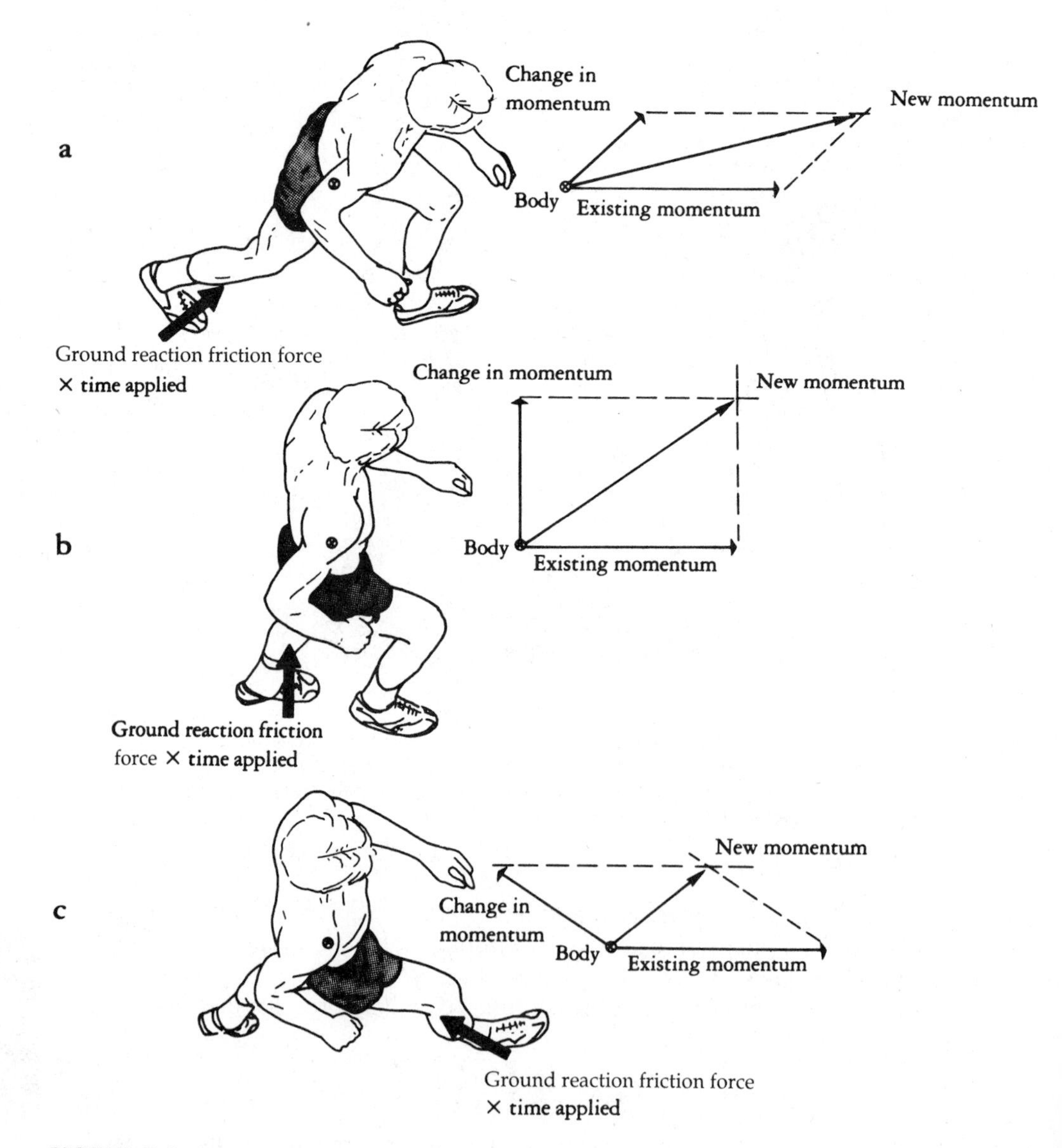

FIGURE G.2

The body's existing momentum (magnitude and direction) can be changed only by an external impulse. (a) The impulse given by the ground-reaction friction force adds momentum forward and to the left. (b) Friction acting leftward changes the runner's existing forward momentum to diagonally leftward. (c) Ground-reaction friction force acts to slow the runner and move him diagonally to the left.

When the runner has motion (momentum) in one direction and wishes to change the direction of motion, a force must be applied to do it. The ground reaction friction force generated by a sudden jabbing of his foot against the ground provides the motive impulse F(t) to give a new momentum component to his existing momentum. Such a foot-jab reaction force is applied for a short time and accounts for the change in momentum. The resulting momentum is found by the vector composition method of completing the parallelogram of momentum vectors. Note in Figure G.2a that the impulse (and change in momentum) has a forward and leftward direction to it. The forward part augments the existing forward momentum, and the leftward part causes the sideward momentum that accounts for the direction change.

In Figure G.2b, the impulse contributes nothing to the forward momentum but is responsible for adding sideward momentum and a greater direction change.

In Figure G.2c, the braking impulse causes a decrease in forward momentum, which is due to the backward component of the impulse, and it creates momentum to the left, which is due to the leftward component. Because the mass of the runner does not change, the impulses serve to change the momentum by changing the velocity.

If a heavier runner wanted to do this same kind of quick maneuvering, he would have to generate a greater ground-reaction impulse F(t) because he would have more mass to accelerate. If he did not, his velocity change would be smaller than that of a lighter person. Visualize the following relative magnitude of mass and velocity change (Δv) for the same F(t):

$$F(t) = \begin{matrix} m(\boldsymbol{\Delta v}) \\ \text{or} \\ \mathbf{m}(\Delta v) \end{matrix}$$

Impulse for Accelerating a Body

In Newton's second law, a = F/m, the acceleration of a body is caused by the applied force, and this applied force is the result of all the forces acting. It is important to realize that for a body to keep accelerating, the force giver must keep up with the moving body and continue to apply a *net* force. For example, to move a stalled car, the motive push force must be greater than the resistive friction force. To keep the car moving after its initial acceleration, the push force need only be equal to the friction. But to increase the speed, the push force must continue to be greater than the resistive force. The net impulse, then, is the product of the force × time only for the time during which the push force was acting as an accelerating force, not merely a motion-maintaining force. So, the longer the time the force acts as an *accelerating force*, the greater the change in momentum (velocity change).

In throwing and pushing activities for which the purpose is to cause an increase in velocity of a body or object, one should maximize the time during which the applied force can act as an *accelerating* force. For example, taking a backswing or windup or crouch before applying a force to accelerate an implement or the body serves to provide more time for that force to accelerate it to a faster release, impact, or take-off velocity (acceleration × time = velocity achieved).

In ball and implement collisions where the performer cannot control the time of force application because of the elastic characteristics of the objects, the velocity of the ball off the implement depends on the force of elastic recoil plus implement force. Contact times for impacts in golf, baseball, and tennis vary from approximately 1 to 5 milliseconds (0.001 to 0.005 sec) for deformation and reformation (Plagenhoef, 1971). The impulse causing the change in momentum (velocity) of a struck ball, therefore, must have the applied force maximized because the time applied is so brief.

The foregoing applications of the impulse momentum relationship show that in human movement activities (1) the best shock absorption technique is based on receiving a small force spread over a long period of time, (2) starting and abrupt changes of direction are based on the application of a large force applied for a short period of time, and (3) the larger the applied force and/or the longer the time a force is applied, the greater is the change in velocity of a body or object receiving that force.

G.3 Conservation of Linear Momentum

The first part of Newton's law of inertia, which states that a system at rest will remain at rest until forced to move, seems obvious from our everyday experience. The second part of the law, which declares that a system will continue to move in a straight line with constant speed if no external force acts on it, is not as self-evident, because the motion of our bodies and of the objects with which we deal is governed by the external force of gravity and much of the time by friction. Consequently, we rarely find things moving with constant speed. However, the persistence of motion of moving things is easily observed. Follow-through movements of body segments in throwing and kicking, for example, happen because the segments have momentum. Such persistence of motion is explained by Newton's law of inertia expressed in terms of momentum: A system will continue to move with constant momentum unless an external force acts to change that momentum. This statement is known as the principle of conservation of linear momentum. So, as we have seen in the preceding examples, the momentum of a system can be changed gradually or abruptly by the application of an external force; otherwise, the momentum would not be changed, it would be conserved. Likewise, a body at rest with zero momentum would keep that zero momentum unless it were caused to move by an external force.

We can use this conservation principle to help us identify the external forces that must be the cause of motion change. Or, if we want to cause a motion change (accelerate, decelerate, or change direction), we know an external force must be applied to do so. Also, if a motion change is undesirable, we can identify the forces that need to be eliminated or reduced, such as wind resistance on a downhill skier.

Momentum Conservation in Collisions

The conservation of linear momentum is used to predict the outcome in impact situations in which two colliding bodies are considered to be the isolated system. For example, a 90-kg football receiver in Figure G.3 collides with the ball as he jumps vertically in the air and catches a pass.

The horizontal momentum of the football itself is changed from its initial value with a high velocity to a smaller value when its velocity is reduced as it is caught. The horizontal momentum of the receiver is changed from its initial value of zero before the catch to some value after the catch because of the horizontal force applied by the ball. The total momentum of the system, composed of the ball and the receiver, is not different after the catch from the total momentum of the system before the catch. This is because the momentum

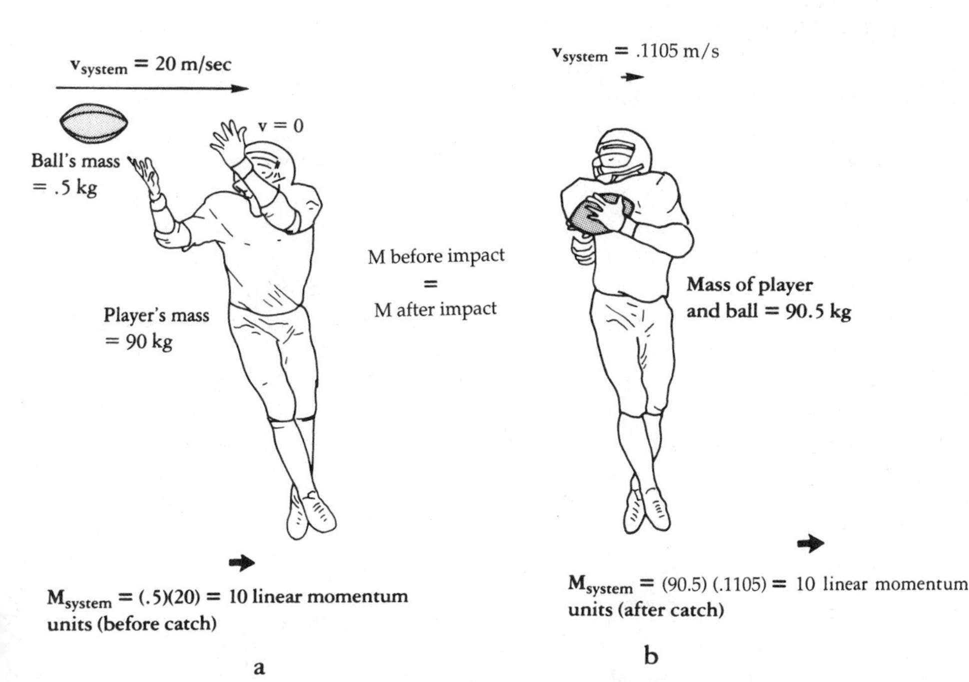

FIGURE G.3

The total momentum of the system (player and ball) (a) before the catch is equal to the total momentum of the system (player and ball) (b) after the catch.

lost by the ball (when it lost most of its velocity) is gained by the receiver at impact (he gained a small amount of velocity as he caught the ball).

The total momentum (a vector quantity) of the system after collision is equal to the total momentum of the system before collision. Quantitatively, this concept is expressed as follows:

$$\underset{\text{before collision}}{(m_1v_1 + m_2v_2)} = \underset{\text{after collision}}{(m_1v_1 + m_2v_2)}$$

For the football example, call the ball m_1 and the player m_2 with their velocities v_1 and v_2, respectively. The numbers would look like this:

$$[(.5 \times 20) + (90 \times 0)] = [(.5 \times .1105) + (90 \times .1105)]$$

$$\underset{\text{before impact}}{\text{momentum} = 10} = \underset{\text{after impact}}{\text{momentum} = 10}$$

In collisions involving rebound of two bodies that make up an isolated system, the same conservation rule applies. When a bat collides with a ball, for example, each receives an equal and opposite impulse from the other. Therefore, each would experience an equal and opposite momentum change. The ball's velocity change is much more dramatic than the bat's because it is stopped, compressed, then accelerated back in the other direction. The total change in velocity is the sum of the stopping and rebound velocities. The more massive bat, however, experiences a small velocity change as it is merely decelerated by the ball's impulse. The momentum lost by the bat is gained by the ball so that the system's momentum is the same before and after impact.

In such ball-rebound collisions, we usually are interested in the resulting motion of the ball rather than what happens to the bat, so we would consider only the ball as the system and then focus on its response to the impact force.

If the response of each of the two colliding bodies is important, as in soccer, basketball, baseball, football, and other activities with body contact (intentional or not), the conservation principle tells us that whatever momentum change one body causes for another, that body can expect an equal and opposite momentum change. The changes are in the form of velocity changes because the masses do not change. The greater the mass of one body, the less its velocity change will be (and the less the damage or injury potential). So, if a 200-pounder collides with a 150-pounder, each will receive the same impulse, but the 150-pounder will experience the greater velocity change and is more susceptible to injury.

G.4 Kinetic Energy

Note that a body's momentum (mv) involves the same two quantities as that body's kinetic energy (1/2 mv^2). Both momentum and kinetic energy are involved in all movement situations. Which quantity has more significance in any particular situation depends on whether the body's motion change is of interest or if it is important to know how much work it is capable of doing on another body or object (force × distance moved). If our interest is in a body's change in velocity (caused by an accelerating or decelerating force), then we would focus on its momentum, knowing that its mass will determine how much force must be applied to it to achieve a certain velocity. A body's kinetic energy would be our focus if we were interested in the body's ability to apply force to move or deform another body.

The importance of how much kinetic energy a body possesses, rather than how much momentum it has, is illustrated by the following example of a diver's body landing on a springboard following the last approach step (hurdle step). The body's kinetic energy serves to depress the board so it can recoil and apply its elastic force back up on the diver; the greater the diver's downward velocity at board contact, the greater the board deformation (the greater the work done on the board). Because the velocity factor is squared for kinetic energy, it is the more significant factor in the body's ability to cause the board deflection. The downward momentum of the body would not be an indicator of the amount of work (deflection) that could be done on the board because the body's mass is of equal importance as velocity in momentum. Table G.1 shows a simplified situation in which two divers are landing on a springboard with the same momentum but different kinetic energy.

The momentum of each diver in Table G.1 is 240, but the lighter diver, Kelly, has more kinetic energy to convert into elastic potential energy of the board. Thus, the lighter weight diver who achieves a higher hurdle step height can do more work on the springboard than the slower, heavier diver, Pat, who has the same momentum. An added advantage for the lightweight Kelly is that the net upward force (recoil force minus body weight) is greater than for heavyweight Pat. Also, Kelly's mass is smaller, and with the larger force, her upward acceleration before takeoff leads to a faster takeoff velocity for greater height. Although the diver's push-off timing during depression and recoil also influences takeoff velocity, it would be beneficial for any given diver to concentrate on achieving a higher hurdle step to obtain a faster touchdown velocity for board depression. Greg Louganis, 1984 and 1988 Olympic gold medalist, displayed just such an ability, typically achieving enough hurdle height to attain touchdown velocities ranging from 4 to 4.5 m/sec (Miller and Munro, 1985).

TABLE G.1

THE RELATIVE IMPORTANCE OF MOMENTUM AND KINETIC ENERGY

Diver	Mass	Velocity ↓	Momentum (mv)	Kinetic Energy ($\frac{1}{2}mv^2$)
Kelly	60	4	240	**480***
Pat	80	3	240	**360**

* Both divers have the same momentum, but the one with greater kinetic energy can depress the board a greater distance for a greater recoil force.

CONCEPT MODULE G

Understanding Momentum and Energy

1. What is the difference between inertia and linear momentum?
2. Determine how much momentum your body has when you are running with a velocity of 4 m/sec. (Convert your weight to mass.)
3. Define linear impulse.
4. What is the magnitude of the force that would have to be applied to you to stop your motion in 2 sec? In 4 sec? (Refer to question 2.)
5. Identify several sport examples in which an object continues to move because of its momentum after the accelerating force has been removed. For each, identify the impulse that is responsible for reducing the object's momentum to zero.
6. Using the concepts of impulse and momentum, cite several examples of how body movements are used to absorb shock when a body must stop a moving object or when the moving body meets a stationary object.
7. Use the conservation of linear momentum principle to tell the direction of motion of these two players (the system) after they collide: Player A's mass is 80 kg and runs north at 5 m/sec into player B. Player B's mass is 90 kg and runs south at 2 m/sec into player A.
8. Describe two movement activities in which the kinetic energy of a body or object is more significant than its momentum.
9. Explain how gravitational or elastic potential energy is given to a body or object by something possessing kinetic energy.
10. If you double your running speed, how is the magnitude of your momentum changed? How is the magnitude of your kinetic energy changed?

References and Suggested Readings

Barham, J. (1978). *Mechanical kinesiology.* St. Louis: C. V. Mosby.

Daish, C. B. (1972). *Learn science through ball games.* New York: Sterling.

Hopper, B. J. (1973). *The mechanics of human movement.* New York: American Elsevier.

Miller, D. I., & Munro, C. F. (1985). Greg Louganis' springboard takeoff: II. Linear and angular momentum considerations. *International Journal of Sport Biomechanics, 1,* 288–307.

Plagenhoef, S. (1971). *Patterns of human motion.* Englewood Cliffs, NJ: Prentice-Hall.

Resnick, R., & Halliday, D. (1977). *Physics: Part I.* New York: John Wiley & Sons.

Sanders, R. H., & Wilson, B. D. (1988). Factors contributing to maximum height of dives after takeoff from the 3M springboard. *International Journal of Sport Biomechanics, 4,* 231–259.

Tricker, R. A. R., & Tricker, B. J. K. (1967). *The science of human movement.* New York: American Elsevier.

Wheeler, G. F., & Kirkpatrick, L. D. (1983). *Physics: Building a world view.* Englewood Cliffs, NJ: Prentice-Hall.

CHAPTER 9

Observing and Analyzing Performance

PREREQUISITES

Concept Modules A, B, C, D, E, F, G
Chapters 1, 2

CHAPTER CONTENTS

9.1 The Nature of Skills

Constraints to Movement

A **movement pattern** is a general series of anatomical movements that have common elements of spatial configuration, such as segmental movements occurring in the same plane of motion. Examples of movement patterns are walking, running, jumping, throwing, striking, and kicking. These general movement patterns are not limited by any external influence, assuming that the performer has the ability to do them and is not impaired. When a general movement pattern is adapted to the constraints of a particular activity or sport, it is called a **skill**. For example, a high jump is a skill within the general movement pattern of jumping; a racquetball shot is a skill within the general movement pattern of striking. When a particular type of the same skill is performed, it is called a **technique**. The straddle and the flop are two techniques of the skill of high jumping. Different segmental movements are used to perform the two high-jumping techniques, and each technique may be recognized by the series of segmental movements used to perform it. Within each technique, a performer may use individual modifications such as unique timing or specialized movements. These individual adaptations of a technique are defined as the **style** of performance. The curved run-up and the straight run-up to the high bar are two styles of approach used in the back layout (flop) technique of high jumping.

Skills, techniques, and styles develop because of the **constraints** or limitations of an event or a performer. Event constraints are the three-dimensional physical boundaries of the playing area, the rules governing an event, costume requirements for participants in the event, the equipment specifications, and the environment in which the event is held (e.g., weather, noise, visibility).

Human constraints include the size, muscular strength, power, endurance, flexibility, and skill level of the performer and of the opponents; the position or movement of the performer immediately preceding execution of the skill; and the movements to be executed immediately after performance of the skill.

Both human and event constraints may require a performer to alter the movements of a skill from a mechanically ideal performance to a less than ideal performance. Some constraints may be incorporated into practice sessions. For example, basketball players *know* the height of the hoop, and divers *know* if they are diving off a 3-m or a 1-m board. Some constraints are difficult or impossible to introduce into practice sessions, such as the speed of an opposing basketball team's guards or the limited visibility on a ski hill. The experienced teacher or coach incorporates as many constraints as possible into practice situations.

Classification of Skills

Open And Closed Skills

All movement is performed in some environment, whether it is a racquetball court, an open field, or with or without other people or equipment. The easiest skill to learn, in terms of the environmental constraints imposed on the performer, is a closed skill. A **closed skill** is one that is done in a *predictable* environment, or one in which the performer is free to execute a skill without having to make quick decisions that would be required if unexpected changes occurred (Schmidt, 1982). Examples of closed skills include a tennis serve, a javelin throw, a softball pitch, a high jump, a basketball free throw, a 100-m backstroke, a synchronized swimming routine, springboard diving, and weight lifting. In these activities, the environment is predictable, and no quick decision must be made in response to some unexpected variable.

A mover must often perform in response to changes in the environment, as when executing a backhand rather than a forehand return of a tennis serve. When a skill is performed in response to an unpredictable, changing environment, it is called an **open skill**. Examples of open skills include those done in response to the unpredictable action of an opponent or situation as encountered in wrestling or in responding to a serve or a fastball pitch (Schmidt, 1982). In situations such as these, the perception and quick decision-making abilities of the performer may well be more important for success than the isolated biomechanical features of the skill performance. Therefore, evaluation of skill performance within an actual *game* situation should take into account the *perceptual* components of the performance. Variance from using the "ideal" biomechanics of an open skill should be expected in a game because the skill usually must be initiated within a severe time constraint. The time available for actually performing the movements of the skill is only the time that remains *after* the decision to do the skill has been made.

By the time a performer is expected to move in response to environmental demands, the *skills* that are used as tools in the game should already have been drilled for basic mechanical effectiveness. Drills or practice sessions can be constructed for learning the movements that will be used as open skills in games. Practice of these skills in an out-of-game situation frees the learner from the unexpected and allows attention to be given to the biomechanics of performance.

Discrete And Continuous Skills

From a kinematic point of view, skills can be labeled by how they appear. A movement performance can be classified according to whether it has a definite beginning and ending—a **discrete skill**—or whether it appears to have no specific beginning and ending—a **continuous skill**.

Examples of discrete skills include a tennis serve, a vertical jump, a handstand, a forward roll, a basketball free throw, throwing, kicking, batting, and catching. The high jump, the pole vault, the long jump, diving, gymnastic stunts, the soccer kick, the javelin throw, pass receptions, and the flip turn in swimming are also discrete skills, but they usually are initiated from a running or moving approach, which is an important part of the total performance.

Running is a skill that is of a most continuous nature; it is composed of repeated cycles of the same movement pattern. Swimming strokes, the eggbeater kick in water polo, walking, skipping, bicycling, speed skating, and rowing are also examples of continuous skills.

Observing Skill Performances

For ease in observation, a discrete skill may be divided into major phases that make up the total act. These phases are the **preparation phase, execution phase**, and **recovery, or follow-through, phase**. The movements in one phase influence the movements in the next phase; therefore, the purpose of each phase is evaluated in terms of its contribution to the effectiveness of the total skill. In the overarm throw in Figure 9.1a, for instance, the preparation phase includes the backswing movements of the body parts, or the windup. The backswing is a preparation for executing the actual movements that result in the projection of the ball (Figure 9.1b). The recovery or conclusion phase of a discrete skill is the continuation of segmental or body movements following the completion of the execution phase. After the ball has been released from the hand of the thrower, the rotating segments continue forward and around in the same direction they were moving before ball release (Figure 9.1c).

The recovery phase of one skill may blend into the preparatory phase of a subsequent skill. This occurs in activities in which a readiness position is assumed between skill executions, as in tennis, racquetball, volleyball, fencing, wrestling, gymnastics, and the martial arts. In games of strategy, the next skill often cannot be predetermined, and the performer must use the recovery movement of the previous skill to move into a general position of readiness that best permits the execution of the next skill. The use of shoulder rolls and backward rolls in volleyball allows the player rapid recovery to a position of readiness. A

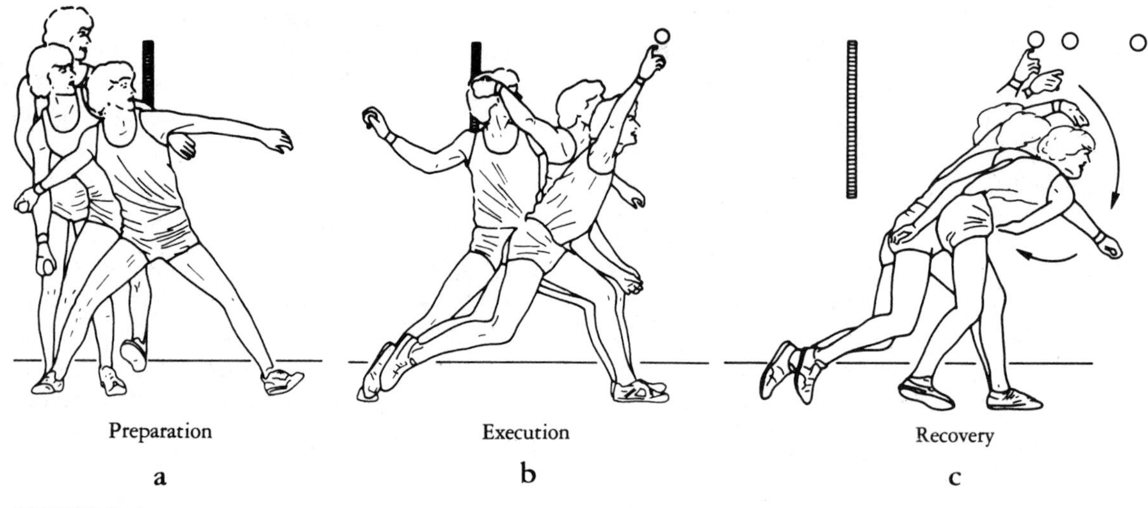

FIGURE 9.1
The three phases of an overarm throwing pattern.

choreographer for dance, synchronized swimming, skating, or gymnastic routines must consider that recovery movements from one skill serve to prepare the performer for a subsequent skill.

When one examines continuous skills such as running and swimming, one usually assumes that each *cycle* of the pattern is performed in the same way; that is, the one cycle selected for analysis is representative of all the cycles in the total performance. This is not an unreasonable assumption if the performer is highly skilled, because a skilled mover exhibits less variation among repetitions of a movement pattern than does a less-skilled mover. Usually, the observer can more easily analyze a continuous skill because the cycles are repeated continuously for observation. Discrete skills also must be repeated over and over again for observation, but the variability among the performances is usually greater than in continuous skills. The time elapsed between each observed performance is also greater, and the observer lacks the benefit of the "instant replay" provided by the performance of a continuous skill. Each cycle of a continuous skill may be divided into phases, as discrete skills are, but divisions are arbitrary and vary from skill to skill. For example, in cycling typical phases have been defined as the following:

1. Part 1: zero degrees top crank angle to 90 degrees counterclockwise
2. Part 2: from 90 degrees counterclockwise to 180 degrees
3. Part 3: from 180 degrees counterclockwise to 270 degrees
4. Part 4: from 270 degrees counterclockwise to 360 degrees

Each phase identified in a continuous skill may have its own particular purpose that contributes to the overall performance. Because the purposes differ between phases, it is important to specifically define each phase, in a type of master plan, before proceeding to the analysis of the total skill.

Understanding The Nature Of Skills

1. Consider a runner who is free from all but the minimum constraints; that is, she has the ability to run, an open space in which to run, and an even surface on which to run.
   a. List general categories of constraints that may be imposed on the runner to alter her running technique by placing her in different sport or recreational activities or both. (For example, a general category of constraint that would force her to run differently would be boundaries.)
   b. For each category of constraint listed in part a, provide three examples. (An example of a specific boundary constraint would be the lane lines in a 200-m run.)
2. Select a general movement pattern. List five different sport skills that use that movement pattern. For each skill, list as many different techniques of performance as you can. Describe two or three specific and unique stylizations of the techniques listed.
3. Identify five skills that should be classified as open skills. For each skill, describe how the mechanics of the skill could be practiced for skill improvement.
4. Select three **discrete** closed motor skills. For each, identify the following phases: preparation, execution, and recovery.

9.2 The Overall Performance Objective of a Skill

To facilitate a biomechanical analysis and to focus our attention on improvements to enhance the effectiveness of a movement performance, we must know the objective of the movement. Each sport event, sport skill, or movement may be classified according to its **overall performance objective (OPO)**. The overall performance objective (OPO) is usually expressed in mechanical terms; for instance, the objective of spiking a volleyball is to project the ball downward into the opponent's court (accuracy of projection). Such an objective is primary, because if the ball does not land in the opponent's court, it is out of play. Secondary objectives of the spiking movement are to project the ball with maximum speed and to send the ball to a particular location within the court (accuracy with speed enhancing the effectiveness).

With a multiobjective movement, an order, or priority, of importance should be established. Is the speed of the ball more important than its placement (direction)? If not, then the secondary objective (speed) may have to be compromised to allow for the necessary control of ball direction. Such speed–accuracy trade-offs, or compromises, are frequently encountered in sport situations in which the speed of the projectile means little if the target is not hit. Table 9.1 suggests a list of overall performance objectives that serve as the basis for the analysis of movement skills.

Some examples in Table 9.1 are discrete skills that are entire events in themselves, such as the javelin throw, pole vault, dart throw, and diving. Other examples are skills that are part of a larger event but that have overall performance objectives in themselves. The objec-

TABLE 9.1

OVERALL PERFORMANCE OBJECTIVES OF HUMAN MOVEMENT ACTIVITIES

Overall Performance Objective	Examples
1. To project an object or the body for maximum horizontal distance	Discus, javelin, long jump, triple jump
2. To project an object or the body for maximum vertical distance	High jump, pole vault, vertical jump, Eskimo high kick, weight throw (Highland Games)
3. To project an object for maximum accuracy	Horseshoes, darts, archery, slow-pitch softball pitch
4. To project an object for maximum accuracy when the speed of the projectile enhances its effectiveness	Baseball pitch, volleyball serve, tennis serve, badminton smash
5. To manipulate a resistance	Weight lifting, wrestling, judo, football block, catching
6. To move the body over a prescribed distance with or without time constraint	Cross-country running and skiing, kayaking, swimming, bobsledding, orienteering
7. To move or position the body or its segments in a prescribed pattern with the intent of achieving an ideal or model performance	Gymnastics, diving, trampoline tumbling, body building, dancing
8. To move the body with the intent of interacting with the natural environment	Scuba diving, hang gliding, surfing, mountain climbing

tive of a single skill may or may not be the same as the overall performance objective of the entire event, such as a basketball free throw, a single tumbling stunt in floor exercise, and a football block. One may argue that the objective in a game of basketball is to score more points than the opponent. One cannot analyze biomechanically how many points were scored, however; one can analyze the effectiveness of the shots taken. Thus, a basketball shot is a discrete skill with an overall performance objective of accuracy, #3, and is the overall performance objective of the entire game of basketball, as well.

On the other hand, many entire events have unique performance objectives, and each component of the event contributes to it. For instance, a timed freestyle event in swimming is composed of several skills with the overall performance objective of moving the body as quickly as possible over a prescribed distance, #6. One may argue that the primary purpose of the start is to begin to swim as soon as possible; however, one must keep in mind that the mechanical purpose of the start is to reduce the time necessary to move over the prescribed distance.

Generally, one attempts to improve by accomplishing the overall performance objective better than one accomplished it in the past or better than another performer. The approach to the process of biomechanical analysis in this text is based on mechanical principles applicable to groups of skills that share a common **OPO**. Skills that have the same **OPO** have the same mechanical principles governing their effectiveness. Knowledge of these principles and their appropriate application allows one to develop a system of analysis that may be used for many different skills without having to learn the specifics of each.

9.3 The Analysis Process

After categorizing the event into one of the **overall performance objectives (OPO),** one should divide the skill into discrete parts, if appropriate. For example, a dive has an OPO of "moving or positioning the body or its parts in a prescribed pattern with the intent of achieving an ideal or model performance." However, there are several parts or phases of the dive, each of which contributes to the successful achievement of the OPO. The forward $1\frac{1}{2}$-somersault dive has the following discrete parts: approach, hurdle step to touchdown, board depression to takeoff, air phase, and entry.

Recall that a discrete skill has a specific beginning and ending. It is very important for the analyst to specifically detail the exact beginning and ending of each discrete part of the skill before analysis. The following descriptions will be used to define the beginning and ending of the five discrete parts of the forward $1\frac{1}{2}$-somersault dive:

Approach begins with the first observable forward movement of the body toward the end of the board and ends just before the first observable contact with the board of the step before the hurdle step. (Figure 9.2a–b.)

Hurdle Step to Touchdown begins with the first observable contact with the board of the step before the hurdle step and ends just before the first observable contact following the hurdle. (Figure 9.2c-e.)

Board Depression to Takeoff begins with the first observable contact with the board following the hurdle step and ends with the last observable foot contact with the board. (Figure 9.2f–h.)

Air Phase begins with the first moment of noncontact with the board and ends just before the first observable hand contact with the water. (Figure 9.2i–n)

Entry begins with the first observable hand contact with the water and ends with the first moment of complete submersion. (Figure 9.2o–p.)

Each of the discrete parts of the skill has its own **mechanical purpose (MP),** which may or may not be the same as the OPO of the entire event but which enhances or contributes to the successful accomplishment of the OPO or the other discrete parts. Achieving the MP of one part must provide the ideal conditions necessary for achieving the MP of the next discrete part. The following lists the MP of each discrete part of the dive:

Approach—to produce a greater ground-reaction force into the hurdle step at the end of the approach.

Hurdle step to Touchdown—to achieve the greatest possible touchdown velocity onto the board for board depression.

Board Depression to Takeoff—to project the body vertically as high as possible (or to optimum height for the dive being performed).

Air Phase—to move the body or its parts in a prescribed or ideal manner.

Entry—to perform a safe and aesthetic deceleration of the body into the water.

Next, to successfully accomplish the MP of each discrete part, one must identify the **biomechanical factors (BF)** that determine or influence its achievement. Once these biomechanical factors are identified, **biomechanical principles (BP)** incorporating these BFs can be used as a guide to achieve the mechanical purpose. After identifying the biomechanical principles that guide performance, the **critical features (CF)** should be listed. The critical features are those specific body or

FIGURE 9.2
A forward $1\frac{1}{2}$ pike somersault dive.

k
j
l
m
Air phase
i
n
Entry
o
p

segmental actions that can be observed by the coach and that help the performer achieve the mechanical purpose. While more specific biomechanical factors, biomechanical principles, and critical features are introduced in subsequent chapters, a more generalized list is presented here to serve as an example of how the analysis system works. The performance is a forward $1\frac{1}{2}$-somersault dive in a pike position. It must be emphasized that much of the technical analysis requires information presented in subsequent chapters. Thus, this section has been worded to circumvent the use of unfamiliar terminology and concepts.

Table 9.2 shows the Biomechanical Factors, Biomechanical Principles, and Critical Features of the first discrete part of the forward $1\frac{1}{2}$-somersault dive, the Approach which was shown in Figure 9.2a–b.

Table 9.3 lists the Biomechanical Factors, Biomechanical Principles and Critical Features of the second discrete part of a forward $1\frac{1}{2}$-somersault dive, the Hurdle to Touchdown, which was shown in Figure 9.2c–e.

Table 9.4 presents the Biomechanical Factors, Biomechanical Principles, and Critical Features of the third discrete part of a forward $1\frac{1}{2}$-somersault dive, the Board Depression to Takeoff which was shown in Figure 9.2f–h.

Table 9.5 presents the Biomechanical Factors, the Biomechanical Principles, and the Critical Features of the fourth discrete part of the forward $1\frac{1}{2}$-somersault dive, the Air Phase. The air phase was shown in Figure 9.2i–n.

Table 9.6 presents the Biomechanical Factors, Biomechanical Principles, and Critical Features of the fifth discrete part of the forward $1\frac{1}{2}$-somersault dive, the Entry, which was shown in Figure 9.2o–p.

In summary, the analysis process consists of the following six steps:

1. Identify the Overall Performance Objective (OPO).
2. Divide the skill into discrete parts.
3. Identify the Mechanical Purpose (MP) of each discrete part.
4. List the Biomechanical Factors (BF) that determine the accomplishment of the mechanical purpose.
5. Identify the Biomechanical Principles (BP) that relate the biomechanical factors to the performance.
6. List the Critical Features (CF) of each part, that is, the movements that should be performed to satisfy the biomechanical principles, which then would contribute to the successful accomplishment of the mechanical purpose of that discrete part, and ultimately the overall performance objective of the event.

The constraints to the free performance of each skill may or may not be manipulated by the coach and performer. The rules that govern performance change infrequently. New techniques may grow out of an individual's stylization of the performance. Whether a unique stylization will continue to be legal, as was the back layout (flop) high jump, or to be ruled illegal, as was the somersault long jump, is not determined by the coach or performer, but unique performances may be attempted within the constraints of existing rules.

Events that are performed in routines such as those in gymnastics affect the analysis. For example, effective movements in performing a successful back handspring from a standing position differ from effective movements in performing a successful back handspring from a roundoff. As the approach was important for the dive, preceding movements are important to consider as part of the skill to be analyzed.

The physical environment in which the skill is performed must be considered in the analysis but usually cannot be manipulated or changed by the coach or performer. Features of the physical environment include the surface of play (grass or wood), wind direction and speed, and wet or dry surfaces.

TABLE 9.2

BIOMECHANICAL FACTORS, BIOMECHANICAL PRINCIPLES, AND CRITICAL FEATURES OF THE APPROACH

Mechanical Purpose: To Produce a Greater Ground Reaction Force for Initiating the Hurdle Phase

Biomechanical Factors	Biomechanical Principles	Critical Features
1. Accelerated approach with sudden deceleration 2. Activation of the stretch–shorten cycle of lower extremity muscles	1. Greater vertical force can be achieved by a sudden stretch–shorten cycle of the muscles involved 2. The greater the action (applied) force, the greater the ground-reaction force	1. Short, quick blocking action of the plant foot 2. Low amplitude of hip, knee, and ankle flexion 3. Quick extension of the hip, knee, and ankle 4. Opposite (free) knee and arms accelerate upward

TABLE 9.3

BIOMECHANICAL FACTORS, BIOMECHANICAL PRINCIPLES, AND CRITICAL FEATURES OF HURDLE TO TOUCHDOWN

Mechanical Purpose: To Achieve Greatest Possible Kinetic Energy at Touchdown in Order to Create Greatest Board Depression

Biomechanical Factors	Biomechanical Principles	Critical Features
1. Maximum vertical height of center of gravity at takeoff 2. Optimal angle of takeoff 3. Maximum vertical velocity of takeoff at the optimal angle	1. The higher the CG, the greater the velocity and the steeper the angle at takeoff, the greater the CG at its peak 2. The greater the peak height of the CG, the greater distance to accelerate downward 3. The greater distance to accelerate downward, the greater the velocity at contact 4. The greater the velocity at contact, the greater the KE 5. The greater the KE, the greater the board depression 6. The greater the board depression, the greater the stored elastic energy in the board.	1. A rigid body position into the hurdle at takeoff 2. Elevated arms and free leg at takeoff into the hurdle 3. Small horizontal excursion of the body during the hurdle 4. Arms down at contact with the board 5. Legs extended at touchdown

CG, Center of gravity; KE, Kinetic energy

TABLE 9.4

BIOMECHANICAL FACTORS, BIOMECHANICAL PRINCIPLES, AND CRITICAL FEATURES OF BOARD DEPRESSION TO TAKEOFF

Mechanical Purpose: To Project the Body Vertically as High as Possible at the Optimum Angle

Biomechanical Factors	Biomechanical Principles	Critical Features
1. Maximum board depression 2. Maximum board recoil force 3. Maximum height of the CG at takeoff 4. Maximum vertical velocity of CG at takeoff 5. Optimum angle of the CG for the dive being done	1. The greater the KE at touch-down, the greater the board depression 2. The greater the board depression, the greater the force of recoil 3. The greater the force of recoil, the greater the vertical velocity of takeoff 4. The greater the vertical velocity of takeoff, the greater the height of the CG at its peak	1. Accelerate the arms upward and extend the hip, knee, and ankle joints near the bottom of board depression to cause maximum board depression 2. Extended rigid body during initial board recoil 3. High arms at takeoff 4. Large vertical excursion of CG and optimal horizontal excursion for required maneuvers

CG, center of gravity; KE, kinetic energy

TABLE 9.5

BIOMECHANICAL FACTORS, BIOMECHANICAL PRINCIPLES, AND CRITICAL FEATURES OF THE AIR PHASE

Mechanical Purpose: Move or Position the Body or Its Parts in a Prescribed Pattern With the Intent of Achieving an Ideal or Model Performance

Biomechanical Factors	Biomechanical Principles	Critical Features
1. Flexibility (active and passive) 2. Kinesthetic perception 3. Coordination and timing of segmental movements 4. Stylized (individualized) movements	1. The greater the flexibility, the easier to change body configuration 2. The greater the kinesthetic awareness, the more likely the athelete will perceive the attainment of ideal performance 3. The timing and coordination of segmental movements will dictate the aesthetic quality of the performance 4. The body will somersault slowest in a layout, faster in a pike, and fastest in a tuck position 5. The tighter the body position, the faster the spin	1. Leave board with extended body as rigid as possible with arms fully flexed overhead 2. Body lean should be optimal (for a $1\frac{1}{2}$ forward, researchers report an optimal angle of 12°) 3. Body movements should display right–left symmetry 4. Body segments should be aligned near the vertical at initial entry

TABLE 9.6

BIOMECHANICAL FACTORS, BIOMECHANICAL PRINCIPLES, AND CRITICAL FEATURES OF ENTRY

Mechanical Purpose: To Safely and Aesthetically Decelerate the Body Into the Water

Biomechanical Factors	Biomechanical Principles	Critical Features
1. Safe deceleration 2. Aesthetic appearance 3. Body configuration 4. Kinesthetic perception	1. The more vertical the body segments as they feed into the water, the more gradual (safe) the deceleration 2. The more streamlined the body entering the water, the less the splash 3. The more rigid the articulations, the less chance of segmental movement on deceleration at entry 4. The greater the kinesthetic perception, the more able the diver is to correct misalignments during entry	1. Vertical orientation of the body segments as they feed into the water 2. Rigid straight body with arms flexed overhead 3. Streamlined body position 4. Small splash

Although the size and strength of the performer may be changed with training, at any given time the coach must analyze the performance in light of the size and strength of the performer at the time. For example, one coaches a tall javelin thrower or basketball player differently than a short performer, and one coaches a 12-year-old tennis player differently than a 22-year-old college player.

Because of the wide variations in structure and movement capabilities of human beings, no two individuals perform the same skill in exactly the same way. No two individuals possess the same structural proportions, joint ranges of motion, musculoskeletal qualities, neuromuscular pathways developed, or motivation. Each performer may exhibit some unique, personalized movements that may be of limited importance in the execution of the skill or that may limit improvement of the performance. Therefore, one should not assume that *all* the stylized movements of one skilled performer, even a performer of national caliber, contribute to a biomechanically ideal performance. The analyst should be familiar with the movement characteristics that are common among *several* successful performers of the skill and should incorporate these common components into an ideal or model performance. What may be of benefit to one may not be of benefit to all.

One should always question the mechanical effectiveness of any new style that emerges on the scene, for the new style may be merely a psychological boost for the user and may serve no mechanical advantage to anyone else. A model performance must be fairly general and should exhibit effective mechanical components.

On the other hand, an analyst should be

flexible enough in his or her thinking so as not to feel limited to only those techniques that are prevalent in national competition. One should always be alert to the possibility of making modifications based on sound biomechanical principles to improve the model performance or of creating a unique or novel approach for accomplishing the overall performance objective of the event.

Any overall analysis should incorporate biological abilities, constraints of the neuromuscular and musculoskeletal system, and the mechanical laws of physics. Thus, information about how the human body functions is merged with mechanical concepts to provide biomechanical principles that govern performance effectiveness. Knowledge in both areas is prerequisite to a proper analysis. The application of these biomechanical relationships are covered specifically and more thoroughly in subsequent chapters.

Understanding the Analysis Process

1. For the following list of activities, state the overall performance objective from Table 9.1 (p. 368).
   - **a.** Pole vault
   - **b.** Baseball batting
   - **c.** Fielding a line drive
   - **d.** 50-m sprint
   - **e.** Bobsledding
   - **f.** Mountain climbing
   - **g.** Body building
   - **h.** Shot put
   - **i.** Long jump
   - **j.** Basketball jump ball
   - **k.** Free throw
   - **l.** Softball pitch (fast pitch)
2. For the following discrete skills, divide the performance into discrete parts or phases: pole vault, long jump, shot put.
3. For the skills listed in question 2, write the overall performance objective and the mechanical purpose of each of the discrete parts.
4. Repeat questions 2 and 3 for the following continuous skills: 50-m dash, speed skating, sidestroke.
5. List three national caliber athletes in different sports whose performance includes some stylization. State the overall performance objective of the skill. How does the stylization influence the effectiveness of the performance?
6. List three examples of relatively new techniques for performing a skill. Are these new techniques mechanically more effective than the traditional techniques? How?
7. List three instances in which rules or rule changes prevented or outlawed a new technique for the performance of some skill. Were these new techniques more effective than the accepted techniques in achieving the overall performance objective of the skill?

References and Suggested Readings

Hay, J. G. & Reid, J. G. (1988). *The anatomical and mechanical bases of human motion*. Englewood Cliffs, NJ: Prentice-Hall.

Hensley, L. G. (1983). Biomechanical analysis. *JOPERD, 50*(5), 21–23.

Hochmuth, G. & Marhold, G. (1978). The further development of biomechanical principles. In E. Asmussen & K. Jorgensen (Eds.), *Biomechanics VI-B* (pp. 93–108). Baltimore: University Park Press.

Hoffman, S. (1977). Toward a pedagogical kinesiology. *Quest*, NAPECW-NAPEAM monograph 28.

Miller, D. I. (1978). Body segment contributions to sport skill performance: Two contrasting approaches. *Research Quarterly for Exercise and Sport, 49*, 219–233.

Sapega, A. (1978). Sport specific performance factor profiling, fencing as a prototype. *The American Journal of Sports Medicine, 6*, 232.

Schmidt, R. A. (1982). *Motor control and learning*. Champaign, IL: Human Kinetics.

Sprigings, E. J. (1988). Sport biomechanics: Data collection, modelling, and implementation stages of development. *The Canadian Journal of Sport Science, 13*(1), 3–7.

CHAPTER 10

Analysis of Projectile-Related Activities

PREREQUISITES

Concept Modules C, D, F, G
Chapters 1, 9

CHAPTER CONTENTS

10.1 Properties of Motion Related to Projectiles

To create the best possible set of circumstances in which to project objects, one must understand the factors governing the motion of projectiles and the relative importance of those factors. Four primary mechanical purposes are used in projecting: (1) to project an object or the body for maximum horizontal distance, (2) to project an object or the body for maximum vertical distance, (3) to project an object for maximum accuracy, and (4) to project an object for maximum accuracy when the speed of the projectile enhances the projectile's effectiveness.

Resolution of the Velocity Vector

Velocity is a vector quantity that possesses magnitude as well as direction. Any given velocity may be resolved into two perpendicular components. For a resultant muscle force, the two components are labeled the rotary component and stabilizing component. For a velocity vector, both components are spatial; they are named according to their direction relative to some stationary frame of reference, such as the ground. At any given instant, a projectile's velocity may be resolved into a vertical velocity component and a horizontal velocity component. A projectile's motion is represented by the motion of its center of gravity (CG). When the object is traveling parallel to the level ground, its velocity has only a horizontal component; if the object is traveling straight up or down, its velocity has only a vertical component.[1] Figure 10.1a and b illustrate how two objects' projection velocities may be resolved into vertical and horizontal components at the instant of projection.

The magnitude of each component of the object's velocity depends on the angle of projection of the object. Figure 10.1c illustrates various angles of projection and how the angle changes the relative sizes of horizontal and vertical velocity components. Note that the projection velocity has equal magnitude in all of these cases. The angle of projection at which the vertical and horizontal velocity components are equal is 45 degrees from the horizontal. Any angle greater than 45 degrees produces a vertical component that is greater in magnitude than the horizontal component; conversely, an angle less than 45 degrees produces a greater magnitude of horizontal velocity than of vertical velocity. The arithmetic sum of the magnitudes of the two components will always be greater than the magnitude of the resultant.[2]

Forces Influencing Projectiles

Once a projectile is released from the projecting surface, only two forces can influence its motion until it reaches its destination—gravity and air force.

Gravity

Gravity, the force of the earth's attraction for other masses, is always directed toward the center of the earth. The force of attraction of two masses depends on the magnitude of the masses as well as the distance of their CGs from each other.

If gravity were the only external force acting

[1]For trigonometric calculations of an object's location in space, the reader is referred to Appendix V, Section III, p. 730–731.

[2]Because the resultant and the two components describe a right triangle, the magnitudes of the sides may be determined by the Pythagorean theorem, $a^2 + b^2 = c^2$, or they may be determined using trigonometric relationships. (See Appendix V, Section II, p. 723)

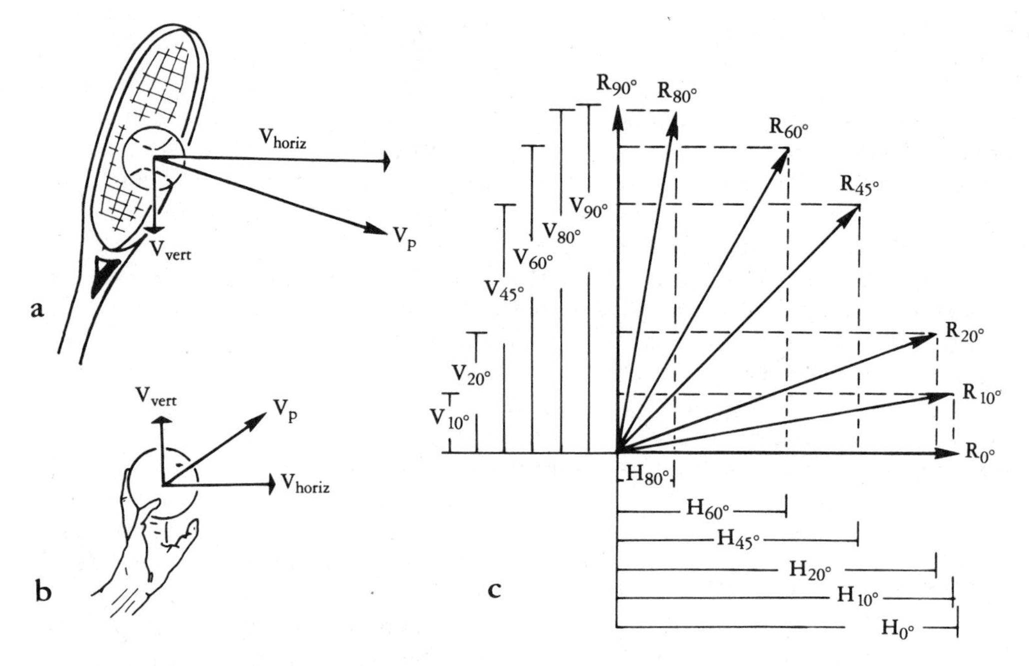

FIGURE 10.1

Resultant projection velocities (V_p) of (a) a tennis serve and (b) a shot put with resolution into respective vertical (V_{vert}) and horizontal (V_{horiz}) components. (c) Relative sizes of the vertical (V) and horizontal (H) velocity components for various angles of projection.

on a projectile, the path the object would take would be in the shape of a parabola, which is a type of curve resembling a hill, the steepness of which can vary (Figure 10.2a). The greater the angle of projection, the steeper the parabola.

Recall that the acceleration (g) of an object due to gravitational force is approximately 9.8 m/sec^2 (32 ft/sec^2). The velocity of the CG of an object (i.e., a projected object's velocity) will be altered in the vertical direction by 9.8 m (32 ft) per second for every second that it is nonsupported. This acceleration that is due to the force of gravity on any object is constant regardless of the object's weight. Figure 10.2b illustrates the changes that gravity causes in the magnitude and direction of the projection velocity (and its horizontal and vertical components) of a projectile as it travels along its path or trajectory.

Recall that the magnitude of velocity is indicated by the length of the vector and that the shape of the path reflects only the effect of gravitational force on its vertical motion after release. Note also that the vertical component decreases during ascent and then *increases* during *descent*, reflecting gravity's decelerating and accelerating effects. In this example, the horizontal component remains the same throughout the trajectory, because the force of gravity can only alter vertical motion, and we have assumed that no other forces are acting on the object.

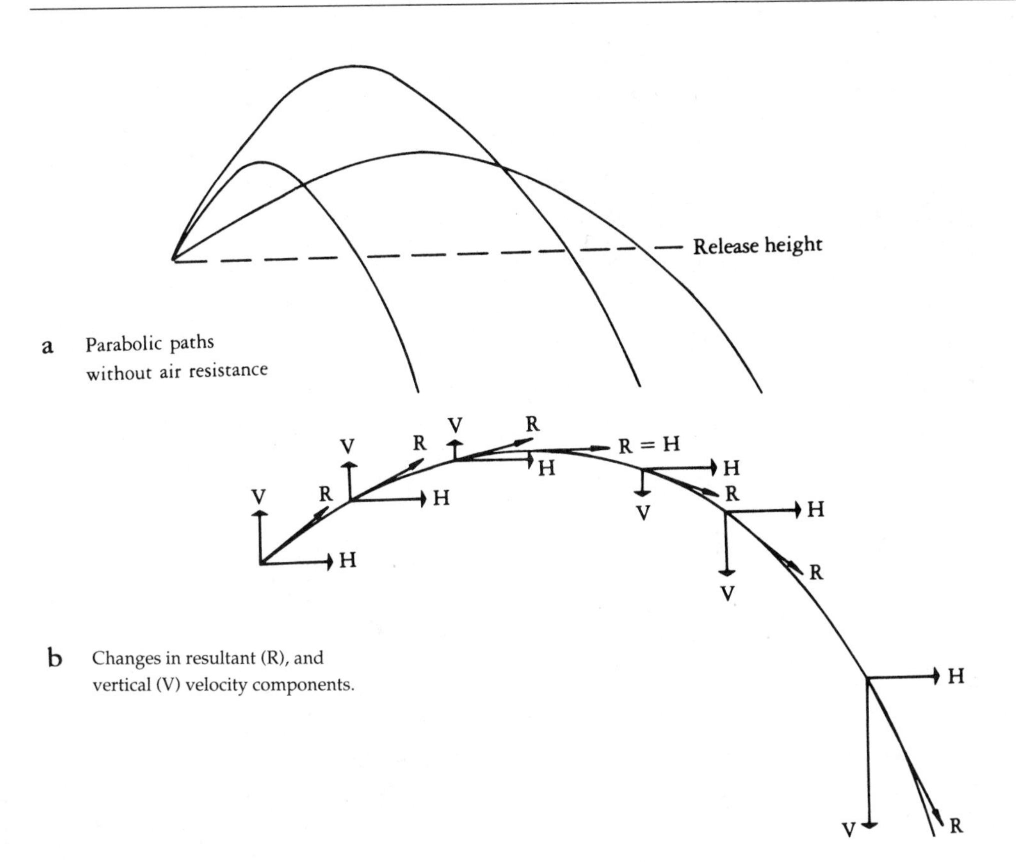

FIGURE 10.2

(a) The shape of a projectile's trajectory is a parabolic curve if gravity is the only influential external force. (b) Gravity's downward force causes a change in its vertical component velocity and thus its resultant velocity.

Aerodynamic Forces

Gravity is not the only external force acting on a projectile. Forces generated by the flow of air relative to the moving projectile can play a major role in determining how a projectile will travel in its aerial path, particularly if the object is not massive. Figure 10.3 illustrates the difference in trajectory of an object traveling with and without significant influence of air forces.

In some cases, air forces are so small, as compared with other mechanical factors, that they can be ignored; for example, one can overlook air forces when jumping over a puddle. Fluid forces (air being one) are important and must be given more than superficial mention if the motion of projectiles is to be fully understood. More in-depth coverage is found in Module H and Chapter 11.

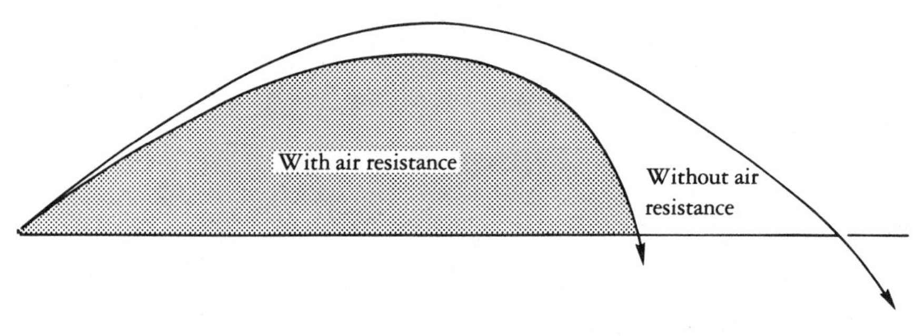

FIGURE 10.3
Air resistance causes a change in the trajectory of a projected object.

10.2 Projecting for Vertical Distance

An object may be directed in a predominantly vertical path. A vertical path may be upward, as when one attempts to throw a ball as high as possible, or it may be downward, as when one spikes a volleyball down to the opposite court. Examples of purely vertical projection do not occur frequently in sport activities. Even in jumping for height, such as high jumping, some horizontal component must be included for the jumper to move from the near side of the bar to the far side. Nevertheless, we will first consider those activities in which the purpose is to project an object as high as possible without the need for a horizontal component.

Projecting Objects for Vertical Distance

Newton's law of inertia says that an object will stay in a prescribed straight path and will move at its given speed unless some force acts to change it. Thus, an object projected vertically would continue at its given velocity (speed and direction) if it were not for the forces of gravity and air resistance. Because gravity must always act downward, it acts as a resistive force to slow an upward-directed object until it stops (at the peak of its vertical path); at that point, gravity becomes a motive force acting in a downward direction to increase the speed of the object until it encounters another force (an external object or the earth), which decelerates it to a stop. Gravity also alters the object's direction of motion when the object stops moving upward and begins moving downward.

Air resistance of still air is always directed against the motion of the object. Thus, on the way up, air resistance is directed downward and is decelerating the object, as is gravity; on the way down, the air resistance is directed upward and opposes the force of gravity. Because with a vertically moving object, air resistance is always opposing the acceleration due to gravity, the *net* effect is a downward acceleration less than 9.8 m/sec^2. Because the air-drag force is increased with increased velocity, the magnitude of the drag decreases as the object moves upward with a continuously slower velocity and increases as the object moves faster and faster downward. Remember that the direction and magnitude of the object's acceleration owing to the force of gravity are constant, while the direction and magnitude of the changes that are due to air resistance depend on the direction and speed of the object.

When calculating the maximum height to which a projectile will travel after release, it is readily apparent that the longer the object continues to move upward, the farther upward it travels. The object continues to move upward until the resistive forces of gravity and air decelerate it to zero vertical velocity. Thus, the more vertical velocity the object has to begin with, the longer it takes the resistive forces to slow it to a stop, and the farther the object travels upward. An object goes the highest when it is projected with the greatest vertical velocity. Remember that neither the downward acceleration that is due to gravity nor its upward deceleration are affected by the weight of the object. The weight, however, is a factor in calculating the net force acting on an object moving vertically through the air.

The vertical distance that a projected object will travel may be figured by dividing its squared vertical velocity component at release by 2 g.

EQUATION 10.1

$$D_{vert} = \frac{V_{vert}^2}{2\ g}$$

Note that this equation calculates the distance the object will travel vertically from its takeoff point.[3] The higher the takeoff point for a given projection, the higher the object will be *from the ground* at its peak. For example, an object that is projected from a height of 2 m from the ground and that travels 12 m upward from that point will be 14 m from the ground at its peak. The same object projected at 3 m from the ground and traveling 12 m upward will be 15 m from the ground at its peak.

For an object projected predominantly downward at a given velocity, such as a spiked volleyball, gravity will increase its initial projection vertical velocity at a rate of 9.8 m/sec for every second until the ball hits the ground. A volleyball spiked with an initial downward velocity of 7 m/sec will be traveling faster when it hits the ground because of the added acceleration that is due to gravity. Air resistance, however, will make the final velocity somewhat slower than the velocity calculated.

Projecting the Body for Vertical Distance

For most activities in which the human body is being projected vertically, the height of reach of the hand from the ground is the important factor.

Examples of such activities are spiking a volleyball, performing a vertical jump-reach test, or a jump ball or a rebound in basketball. Three factors determine the body's reach height: (1) the vertical velocity of the CG at takeoff, (2) the height of the CG from the ground at takeoff, and (3) the vertical distance of the fingertips relative to the CG at the peak of the jump.

Vertical Velocity of the Body's Center of Gravity at Takeoff

The same laws govern the motion of the body's CG that govern the motion of an object's CG. As in projecting an object vertically, the vertical velocity of the body's CG at takeoff determines how high the CG will travel. Unless an external force should intervene, nothing can be done to change the determined path of the center of gravity after the body is in the air.

Because the vertical takeoff velocity of a

[3]For trigonometric calculations of the vertical distance a projected object travels, see Appendix V, Section III, D. 3.

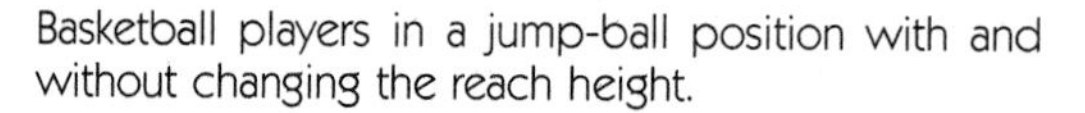

FIGURE 10.4

Basketball players in a jump-ball position with and without changing the reach height.

human body is usually relatively small and because the body is traveling through the air in a fairly streamlined position, the influence of air resistance is minimal. The air resistance force may be increased in activities in which the body is projected with the use of a spring (e.g., trampoline, diving board), because the body's vertical takeoff velocity is greater.

If horizontal displacement is unnecessary, then 100% of the resultant projection velocity should be in the vertical direction. The further the angle of projection from 90 degrees from the horizontal, the smaller the vertical component and the greater the horizontal component. (Refer to Figure 10.2b, p. 382)

Height of the Center of Gravity at Takeoff

The height of the CG of the body relative to the ground should be maximized before takeoff. Just as an object projected from a greater height will reach a higher peak, so the body's

CG will reach a higher peak relative to the ground if it is higher at takeoff.

Recall from Concept Module E.4 that while the body is in contact with a ground, the CG of the body may be moved relative to the body by relocating body segments. In anatomical position, the CG is located somewhere in the pelvic region. The CG may be raised within the body by flexing the arms overhead at the shoulder joints (flexing both arms overhead raises it more than a single arm). Raising the CG within the body while still supported also raises the CG relative to the ground. Thus, by leaving the ground with both arms flexed overhead, one may have the CG a few centimeters higher as it leaves the ground and therefore also when it peaks vertically. The technique of raising the segments to raise the CG must be employed before takeoff.

Center of Gravity to Fingertip Height

The third factor affecting the success of a vertical jump is the distance between the CG of the body at the peak of its flight and the fingertips of the reaching hand.

To maximize the CG height at takeoff, the performer should leave the supporting surface with both arms flexed overhead. If one arm is extended down to the side at the peak of the jump, the location of the CG *within the body* will be affected. The effect will be to raise the entire body upward in accordance with the fact that the CG must *conform to its predetermined location relative to the ground*. Thus, while free of support, a performer may rearrange the body parts around the CG but may not change the location of the CG in space. Such is the case with the basketball players shown in Figure 10.4, both taking off with equal vertical velocity and CG height. One player may be able to gain a slight reach-height advantage by dropping one arm to the side just before the peak of his jump. A successful performer may appear to "hang in the air" by manipulating body segments in this manner near the peak of the jump.

10.3 Projecting for Vertical Distance with a Horizontal Component

In many sport situations, an object that is projected for height also must have some horizontal velocity. While a vertical-jump test may prohibit the use of an approach run, a high jumper would not think of using a static vertical jumping technique for the high-jump event. The high jumper, the springboard diver, or the gymnast performing an aerial floor exercise stunt uses a running approach and thus has some horizontal velocity at takeoff. This horizontal velocity has been generated before the final push-off, and thus the actual direction of the ground reaction force of the push-off could be 100% vertical and the body will still have a horizontal component after takeoff.

High Jumping

As the jumper's body is projected through the air, the body's CG follows the prescribed parabolic path. Gravity begins to reduce the vertical velocity after takeoff and continues to do so until it is zero. (The relatively slow takeoff velocity makes the air-drag force of little consequence as compared with the body-weight force.)

The angle of takeoff determines the relative magnitudes of the vertical and horizontal velocity components. The jumper needs just enough horizontal velocity to move from one side of the bar to the other. If, in fact, the air-resistance force is an important factor (e.g., if there is a strong head wind), the horizontal velocity will decrease significantly during flight. In such a situation, the performer must increase the horizontal velocity component at takeoff to move from one side of the bar to the other during flight. On the other hand, if there is a strong tail wind, the performer will have to decrease the horizontal component at takeoff, assuming that the wind will help carry the body from one side of the bar to the other.

If we consider air resistance negligible for a high jumper, then no forces are acting in the horizontal direction to change the horizontal velocity component, and therefore we assume that it remains the same as it was at takeoff.

The high-jumping event requires that the performer jump only as high as the bar for each trial; however, the winner of the event is that jumper who can clear the highest bar during the competition. The ultimate performance of a single jumper is that in which **the CG is projected as high as possible**, while at the same time, **the individual body parts** are positioned so that **each clears the bar as it passes over the bar**. The highest point of the CG should be reached as it passes in vertical alignment with the bar (i.e., over, under, or through) if maximum effectiveness in the jump is to be attained. For maximum vertical height, the performer should have the vertical velocity at takeoff as large as possible. Two jumpers with the same projection speed may have different vertical velocity components because of the direction of takeoff. The one with the greater vertical component will raise the CG of the body farther during flight (see Figure 10.1c).

The intent of the performer is to attain the greatest takeoff velocity, to have the greatest vertical component of that velocity, and to maintain a large enough horizontal component to pass from one side of the bar to the other. In addition, with a given vertical velocity component, the jumper whose CG is the highest at takeoff projects the CG the highest from the ground. In this way, the taller jumper has a distinct advantage, even though the force of the jump and thus the takeoff velocity may be no greater than that of other jumpers.

Although the rules are set for this event, the truest test of vertical jumping ability would be to measure how high the jumper projects the CG from the takeoff height. A correction for the height of the CG of the jumper could be made as easily as handicaps are in golf and bowling. Classifications according to height would seem as appropriate as the weight classifications used in wrestling, boxing, and weight lifting. A second method of scoring would be to measure the jump as a percentage of the jumper's body height.

In a comparison of the paths of the CG of an elite male jumper and an elite female jumper, Atwater found that although the male jumper was reaching greater ground-to-peak height distances, both the male and female jumpers were, on the average, elevating their CG the same distances after takeoff. Furthermore, on the female's best jump, she raised her CG 5 cm higher than the male jumper raised his CG during any of his jumps (Atwater and Spray, 1975).

Hay has analyzed several high-jumping techniques to determine which technique best satisfies the mechanical laws in projecting the body the greatest vertical distance and still clearing the bar (Hay, 1975b). The factors of importance are the height of the CG at takeoff, the vertical velocity at takeoff, and the position of the body segments relative to the CG and to the bar. The two styles of jumping that seem to be the focus of much controversy are the back layout technique (flop) and the straddle. As Hay points out, the straddle technique has

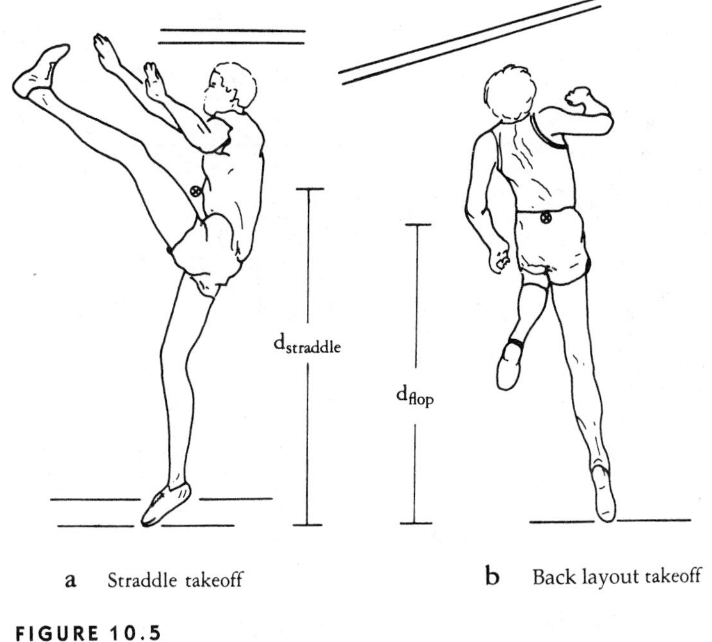

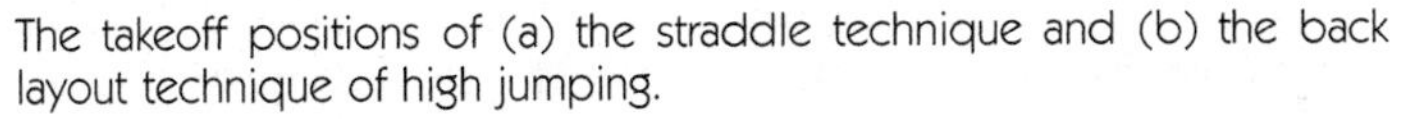

FIGURE 10.5
The takeoff positions of (a) the straddle technique and (b) the back layout technique of high jumping.

a definite advantage in that the CG at takeoff is higher, as can be seen in Figure 10.5.

Because of the high arms and high free leg of the straddle jumper at takeoff as compared with the low arms and the flexed knee of the free leg of the back layout jumper, the straddle jump style gives the user a few centimeters or so advantage in CG height. This means that if both jumpers have the same vertical velocity at takeoff, the straddle jumper will have the CG higher from the ground at the peak of flight. As each body part passes from one side of the bar to the other, the part must be above the bar. At any given time during flight, however, **the more body parts located below the bar, the higher will be the part passing over the bar**. While free of support, body parts can adjust themselves about the CG point. If one part moves downward, another part moves upward. Therefore, the most effective jump is one in which only the body part passing over the bar at a given instant is above the bar; the rest of the body parts are positioned as far below the bar as possible on either side. The position in the air is similar to the position of a snake crawling over a wall. As soon as any body part passes over the bar, it is dropped to the lowest possible position on the other side. Therefore, the technique that affords the jumper the best opportunity to do this yields an advantage over other techniques. The positions of the body during the bar clearance in the flop and straddle techniques are illustrated in Figure 10.6.

When the back layout technique became so successful for Debbie Brill and Dick Fosbury, one of the immediate points made was that it allowed the jumper to put all segments except the one passing over the bar in a lowered position. The success of this technique depends to a great extent on the flexibility of the back. Female jumpers find

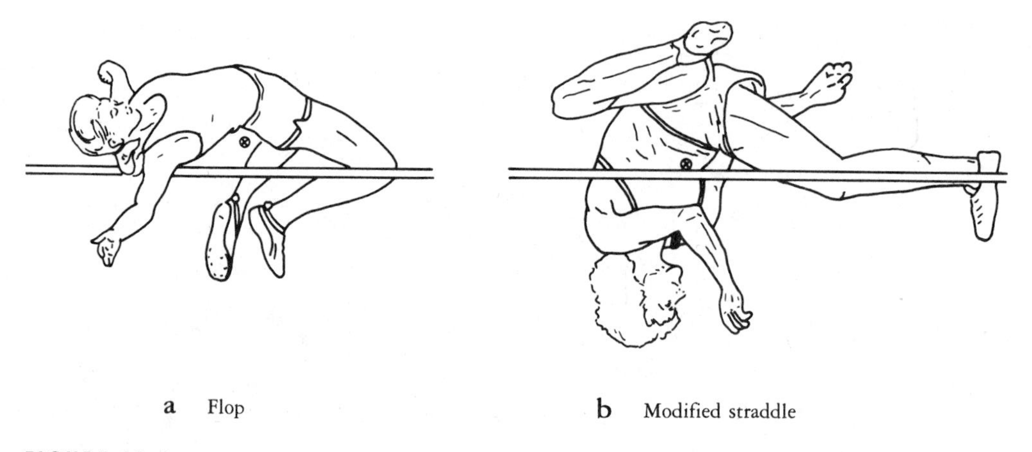

FIGURE 10.6

The body positions while crossing the bar in (a) a back layout and (b) a modified straddle technique high jump.

that the back layout technique is successful for them because of their generally greater back flexibility. Dapena (1980b) reports that for skilled back layout jumpers, the most effective bar clearance technique is one in which the jumper's CG passes through the bar (i.e., at its same level).

A straddle technique, however, enables the performer to use the flexion of the trunk and lead leg after they pass over the bar. Because one can flex the trunk to a greater extent than one can hyperextend it, there seems to be a distinct advantage to the straddle technique in this respect as well.

Hay suggests that a forward dive technique using a one-foot takeoff may well be the ultimate in high-jumping techniques. In such a jump style, one can maximize the height of the CG at takeoff and also curl the body over the bar, therefore keeping all body parts, except the one passing over the bar, as low as possible for the duration of the jump (Hay, 1975b).

Dapena (1980b) reports a horizontal velocity component of approximately 6 m/sec at touchdown and approximately 3 m/sec at takeoff for skilled back layout jumpers. The vertical velocity component was close to 4 m/sec at takeoff. The takeoff angle was between 40 and 48 degrees. One may be led to believe that a larger takeoff angle should be used (closer to 90 degrees from the horizontal) because this would increase the vertical velocity component and therefore the height of the jump. The resultant takeoff angle may be increased, however, by merely reducing the horizontal component, which would have no effect on the vertical component or the height of the jump.

In a detailed study of the takeoff motions of skilled back layout jumpers, Dapena and Chung (1988) report a ground reaction force at takeoff directed on the average of 66.8 degrees from the horizontal. At this angle, the vertical component of the ground reaction force is 8% less than if it were at 90 degrees. Complex linear and angular motions of the CG and body segments of the jumpers occur beginning at the plant and continuing through takeoff, however. Hay, Miller, and Canterna (1986) found that a decrease in horizontal velocity during the takeoff was seen with a corresponding increase in vertical velocity. Research has yet to determine whether, in an ideal situation, an athlete has the neuromuscular makeup to increase the vertical velocity component while maintaining the same horizontal component and also to control the eccentric and concentric muscular contractions.

Understanding Projections for Vertical Distance

1. Stand 10 m away from a partner. Throw a ball back and forth with a projection angle of approximately 45 degrees, noting the approximate amount of force used on the object during projection. Repeat the throw with a 60-degree and an 80-degree projection angle. Note the differences in the peak height and the distance that the object travels (its trajectory). In each case, explain the differences in terms of the vertical and horizontal velocities and the angles of projection.
2. Project a basketball vertically to approximate distances of 2, 4, 6, and 8 m. With a stopwatch, time the descent from the peak height until the ball hits the ground. Discuss the change in time that occurs at the various distances in terms of the relationship to (a) the velocity of projection, (b) the mass of the object, and (c) the height of projection.
3. Perform a vertical jump and reach, and mark the height of reach. Compare the difference in height attained when both arms remain overhead at the height of the jump and when one arm is extended downward quickly after takeoff. Explain the difference. How do these findings relate to the technique used for (a) a volleyball spike, (b) a basketball jump, and (c) a basketball rebound? What effect does the repositioning of the body segments have on the path of the CG of the body? on the path of the hand that remains above the head? How does this relate to clearing the bar in a high jump?

10.4 Projecting for Horizontal Distance

An object or the body is projected for maximum horizontal distance in many more sport activities than for maximum vertical distance. The object of all of the throwing events in track and field is to achieve maximum horizontal distance. Long jumping, triple jumping, golf drives, soccer placekicks, and punts are a few additional examples. An object or body projected for horizontal distance continues to travel horizontally until it is stopped from doing so by an external force. Usually that external force is provided by the ground. Gravity eventually halts the vertical movement of the object, and the ground friction eventually halts the horizontal velocity of an object and thereby stops it from traveling any additional horizontal distance. The longer an object is in the air with a horizontal velocity component, the longer it can continue to move in the horizontal direction. Time in the air is greater if the initial upward vertical velocity component is greater. For each second that the object is in the air, however, it travels a given number of meters horizontally, depending on the magnitude of that component. An object traveling horizontally 15 m/sec for 1 sec will travel 15 m; traveling for 2 sec, it will travel 30 m.

A trade-off has to occur between maximizing the time in the air, which is determined by the vertical velocity component, and maximizing the horizontal distance traveled in that time period, which is determined by the horizontal velocity component. To compare distances achieved with different projection angles, we will assume that the magnitude of the resultant projection velocity is constant. The magnitude of each component differs, however, as the angle of projection changes.

The horizontal velocity component is maximum at a 0-degree projection angle from the horizontal plane and minimum (zero) at a 90-degree angle. The vertical velocity component is the opposite: it is maximum at 90 degrees from the horizontal plane and minimum (zero) at 0 degrees. At 45 degrees, the two components are always equal in magnitude to each other at takeoff.

Horizontal Projections for Which Takeoff and Landing Heights Are Equal

Examples of sport skills in which maximizing horizontal distance is the primary mechanical purpose and in which the takeoff and landing heights are equal include long-distance placekicks and a long football pass or baseball throw to another person catching the ball at approximately the same height. Notice in Figure 10.7 that there are two landing heights. We will first consider landing height 1.

The greatest horizontal distance is achieved with a projection angle of 45 degrees when the takeoff and landing heights are the same. Thus, in this case, a 45-degree angle will give the greatest distance, as long as two or more

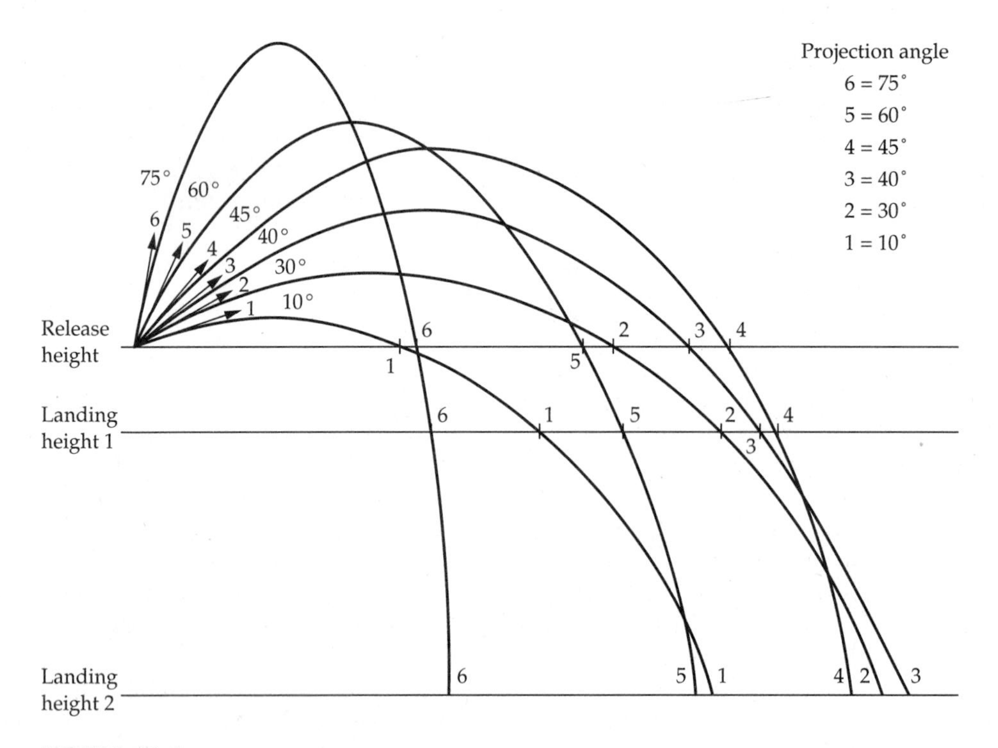

FIGURE 10.7

Horizontal distances covered by an object at various angles of projection.

projections with the same resultant velocities are compared. This angle will be optimum *regardless of the magnitude of the resultant velocity used*. This optimum angle is for activities in which the speed at which the object arrives at its target is inconsequential. For example, if a thrower is more concerned with how fast the baseball reaches the infield than how far it travels before it is caught, the thrower will decrease the vertical velocity component by decreasing the angle of projection to get the ball to its target quickly.

The general equation for calculating the horizontal distance of a projectile, for which the takeoff and landing heights are equal, is:

EQUATION 10.2

$$D_{horiz} = \frac{V_{horiz}}{t}$$

The greater the projection angle, the smaller the horizontal velocity but the greater the time in the air. The smaller the projection angle, the greater the horizontal component but the less the time in the air. Thus, a 45-degree angle *optimizes* these two factors to produce the *maximum* horizontal distance.[4]

Horizontal Projections for Which Takeoff and Landing Heights Are Unequal

The sport events in which objects are projected for maximum horizontal distance may be divided into two categories: events in which objects are thrown or struck (e.g., a shot, a discus, a baseball) and events in which the body is projected for horizontal distance (e.g., long jumping, racing dives, ski jumping).

[4]For trigonometric calculations of an object's projected horizontal distance, the reader is referred to Appendix II, Table II.2.

Projecting Objects for Horizontal Distance

Throwing or striking an object for distance involves projecting the object from a height. The height of the CG of the object depends on the height of the performer and on the place where the performer releases or impacts the object relative to the ground (e.g., underarm, overarm, sidearm). The landing point is usually the ground. Again referring to Figure 10.7, we will consider an object projected at various angles and reaching the ground at landing height 2. If we assume that air resistance is negligible, then the object will travel with the same horizontal velocity throughout its path.

The more time the object has to travel forward, afforded it by the extra vertical distance that it can travel before landing, the farther forward it will be when it lands. Each projection has traveled farther forward at landing height 2 than at landing height 1. The greater the difference in the two heights, the greater the time the object continues to move forward. Therefore, a tall performer has an automatic advantage over a shorter performer throwing with the same projection velocity.

Notice also that in Figure 10.7, the object projected at 45 degrees no longer "wins." In fact, in this case, the 40-degree projection travels the farthest before landing. The greater the difference between takeoff and landing heights, the smaller the *optimum* angle of release, given equal projection velocities.

The effect of varying the height of release on horizontal distance traveled is shown in Figure 10.8a. Given the same height and upward angle of projection, the object with the greatest velocity will go farther, as shown in Figure 10.8b. Thus, the performer able to project the object with the greatest speed will likely win. In fact, an increase in velocity provides a greater increase in horizontal distance than a similar increase in release height.

For each height and resultant speed, however, a unique *optimum* angle gives the object the greatest horizontal distance (Figure 10.8c). The

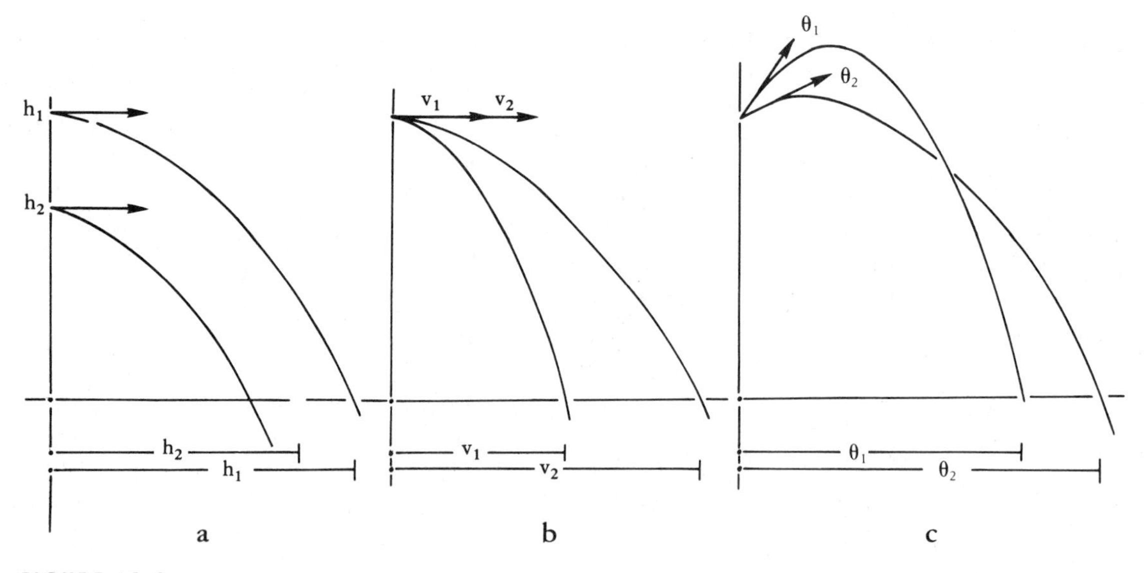

FIGURE 10.8

The effect of varying (a) the height of release, (b) the speed of release and (c) the angle of release on horizontal distance.

greater the difference between the takeoff and landing heights, the lower the optimum angle. The greater the projection velocity, the greater the optimum angle. If the performer alters the release height or changes the speed of projection, then a different angle is optimum. Each performer, therefore, has a unique angle of projection that gives the object the greatest distance. Note that when we talk about an **optimum** or an **ideal projection angle**, we are saying that with a *given release height and speed*, there is a unique projection angle that results in the object's traveling the greatest horizontal distance compared to that achieved at *all other angles*. (This angle may not be ideal for an object projected with a different velocity or height of release.)

The horizontal distance increases with increases in the velocity and with increases in the height of release, but the *ideal angle decreases* as the *height of release increases*. Thus, the taller the thrower, the less of an angle needed to produce the greatest distance for a given speed of release. In addition, the ideal angle increases (up to 45 degrees) as the speed of release increases. Therefore, a faster thrower needs a greater angle of release. The equation showing the relationship among the variables when takeoff and landing heights are unequal is in Appendix II, Table II.2.

The factors that determine the unique angle appropriate for various takeoff heights and speeds are important to consider when coaching athletes of various heights and muscular power. These differences are evident when working with athletes of different abilities, and they are particularly important when coaching athletes at different age levels, because heights, strengths, and muscular power change rapidly during the growing years.

Table 10.1 shows the variables of angle of projection, release heights, and speeds for a shot put and each variable's effect on the shot's horizontal distance.

In comparing three puts projected at different heights, A_1, A_2, and A_3, notice that the greater the height of release (A_1), the greater the horizontal distance. In terms of the most important factors in throwing for distance, the

TABLE 10.1

THE EFFECTS OF HEIGHT OF RELEASE (CASE A), SPEED OF PROJECTION (CASE B), AND ANGLE OF RELEASE (CASE C) ON SHOT PUT DISTANCE

Height of Release (m)	Angle of Projection With Horizontal (degrees)	Release Speed (m/sec)	Shot Put Distance (m)
C_2 2.44	$C_2$45	13.41	20.60
	45	13.11	19.79
	45	12.80	18.95
C_1 2.44	$C_1$40	13.41	20.73
	40	13.11	19.89
	40	12.80	19.08
2.44	35	13.41	20.29
	35	13.11	19.51
	35	12.80	18.72
2.44	30	13.41	19.43
	30	13.11	18.69
A_1	$A_1$30	12.80	17.96
2.13	45	13.41	20.37
	45	13.11	19.53
	45	12.80	18.72
2.13	40	13.41	20.42
	40	13.11	19.61
B_3	$B_3$40	12.80	18.80
2.13	35	13.41	19.94
B_2	$B_2$35	13.11	19.18
B_1	$B_1$35	12.80	18.39
2.13	30	13.41	19.08
	30	13.11	18.31
A_2	$A_2$30	12.80	17.60
1.98	45	13.41	20.24
	45	13.11	19.41
	45	12.80	18.57
1.98	40	13.41	20.29
	40	13.11	19.46
1.98	35	13.41	19.81
	35	13.11	19.02
	35	12.80	18.24
1.98	30	13.41	18.87
	30	13.11	18.14
A_3	A 30	12.80	17.40

Adapted from Dyson, G. (1978). *The mechanics of athletics* (7th ed.) (p. 219). New York: Holmes & Meier Publishers.

speed of release is the most significant. In comparing parts labeled B_1, B_2, and B_3, notice that a shot projected from a height of 2.13 m at 35 degrees and 12.80 m/sec travels 18.39 m (B_1). A gain of 0.41 m would be made by increasing the projection angle 5 degrees (B_3), but a gain of 0.79 m would be made by increasing the projection velocity 0.31 m/sec (B_2).

In comparing the parts labeled C_1 and C_2, notice that the shot projected at 40 degrees (C_1) travels farther than the same shot projected at 45 degrees (C_2).

Projecting the Body for Horizontal Distance

When the body is being projected for horizontal distance, as in the long jump or the triple jump, some additional factors play an important part in the total distance.

Long Jump

Figure 10.9 illustrates the factors that contribute to the total horizontal distance in a long jump. A horizontal distance between the takeoff board and the CG of body at takeoff is created by the forward lean of the body. This distance is designated as d_1. The d_2 distance is designated as the horizontal distance of a projectile when the takeoff and landing heights are equal. The d_3 distance designates the horizontal distance created by the extra time of projection afforded by the differences in takeoff and landing heights. It is during this time

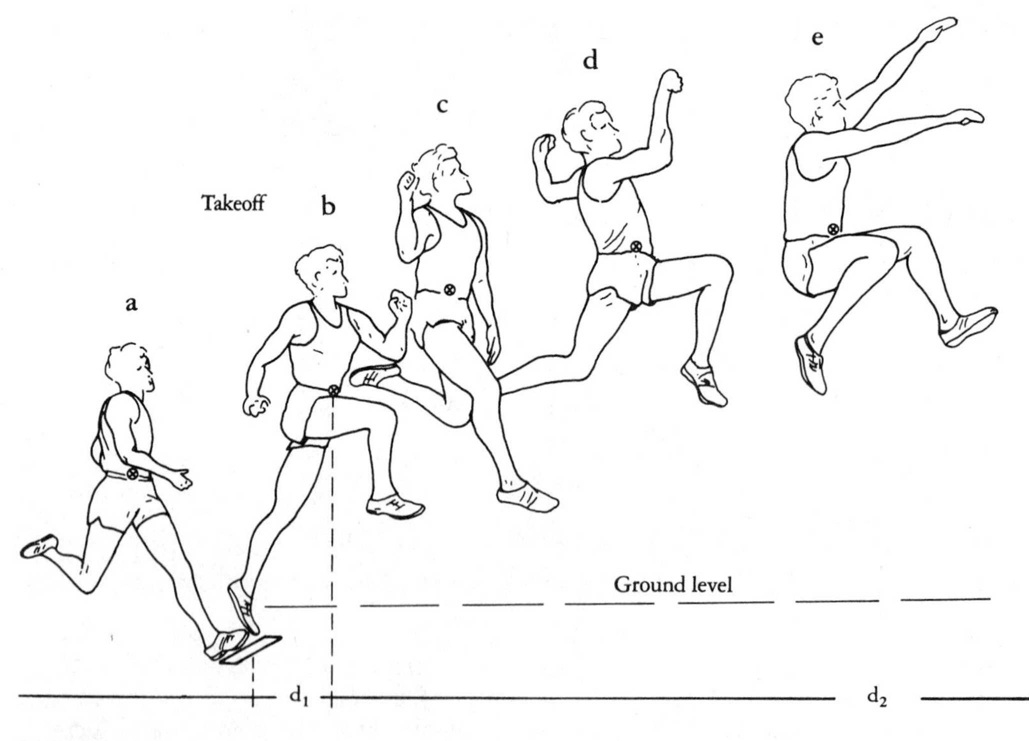

FIGURE 10.9
Jump distances involved in long jumping.

that the body is descending beyond the takeoff height. The d_4 distance designates the horizontal distance given to the jumper between the position of the CG and the heels when heel contact is first made with the ground.

The factors that increase d_1 distance are: the height of the performer (increases the CG height), positioning the arms and legs so that the CG is even higher relative to the body at takeoff, and decreasing the angle of takeoff from the horizontal.

Elliot and Newton (1987) studied the changes in the long jump projection when jumpers used an inclined takeoff board compared to the normal toe board. They found, not surprisingly, that an increase in vertical velocity and a slight decrease in horizontal velocity was associated with the use of an inclined board. Because of changes in the velocity components there was a decrease in the angular momentum and an increase in the angle of the body at takeoff.

Factors that increase the d_2 distance are the same as for projectiles in which takeoff and landing heights are equal. The d_2 distance is increased most markedly by increasing the projection velocity; it also is increased by increasing the takeoff height or by using the proper (unique) angle for the given speed of takeoff or both. Note that the actual d_2 distance designated here disregards the compounding effects of d_1, d_3, and d_4; the d_2 distance is for equal takeoff and landing points. Thus, the specific d_2 is increased by achieving a 45-degree angle and is not affected by height of takeoff from the ground.

The factor that increases d_3 is delaying the heel contact until the last possible instant to increase the time in the air.

Finally, the d_4 distance is increased by flexing the hip to position the lower extremity as close to the horizontal as possible to place the heels as far in front of the performer's CG as possible.

In practice, not all these factors can be used. Although a coach may seek a tall athlete who can be trained as a long jumper, the athlete must possess characteristics other than height,

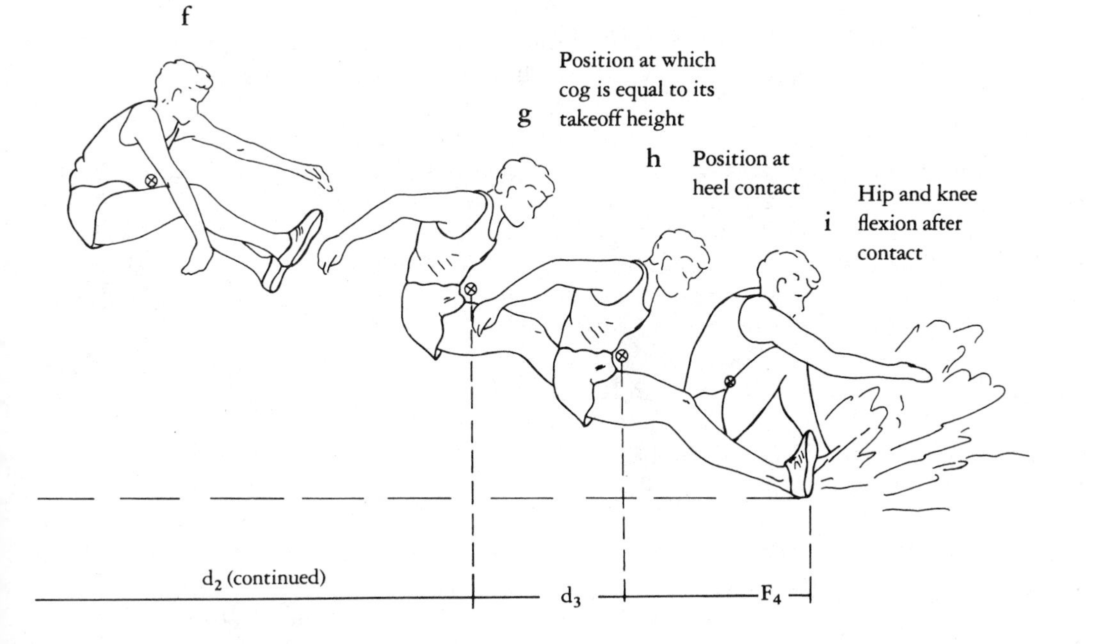

such as neuromuscular development and interest in the event. The total distance, then, depends on many inherent factors, about which a coach can do little.

To maximize the d_1 distance, a coach may teach a takeoff position that increases the height of the long jumper's CG by advising the performer to keep the arms and free leg high. The performer, however, must be careful not to sacrifice takeoff speed by taking too much time to get into this position.

Second, the performer can be coached to use a smaller projection angle, thus placing the CG farther ahead of the feet. This decreases the vertical velocity component of the CG, however, and if the body lean is too great, the reduction in the vertical component of the projection velocity reduces the d_2 and d_3 distances as well. Furthermore, the lower angle may not be ideal for the performer's takeoff speed.

Typically, jumpers take off at a projection angle far below that calculated as ideal. This is explained by the fact that the human body cannot maintain the magnitude of the resultant takeoff velocity if energy is taken to redirect the horizontally moving body in a more vertical path; that is, a reduction in the takeoff speed to achieve the ideal angle reduces the horizontal distance to a greater extent than the small increase gained by increasing the angle.

Most nationally ranked jumpers achieve a CG takeoff angle of 18 to 23 degrees. Widule (1974) calculated the theoretical projection distances of the CG for several takeoff angles of two jumpers, a female long jumper and Bob Beamon. The distances are shown in Table 10.2. Notice that the ideal angle for the given takeoff speed is calculated to be 41 and 43 degrees, respectively, although the actual angles used by the jumpers were 18 and 23 degrees, and the jump distances were 4.57 m and 8.90 m, respectively. The actual distances jumped include the d_4 distance, and the calculated CG distances are those that would have been produced if the centers of gravity had been allowed to fall to the ground. Optimum distances would have been reached had the ideal projection angles been used. Although the optimum projection angle would increase the time in the air, the human body has been incapable of achieving such a large angle without a great sacrifice in horizontal velocity. By increasing the angle of projection, one decreases the horizontal velocity component; thus, a greater projection angle requires a greater takeoff velocity to keep the horizontal component the same.

The delaying of heel contact is not difficult. Although some believe that muscular effort is required to hold the legs up toward the body, this is not the case. In freefall situations, gravity operates equally on all segments, pulling the trunk downward toward the legs as well, and therefore the body remains in a piked position without appreciable muscular effort to hold the legs up against gravity. (Slight muscular effort will be needed to counteract any elastic force caused by the stretching of the posterior hip and knee muscles.) To increase the d_4, one can position the body segments around the CG so that the heels are farther forward of the CG point. The most convenient technique is to throw the arms backward into hyperextension before heel contact. Leaning slightly backward with the upper body also projects the heels forward; care must be taken, however, because one must be able to recover over the feet once contact is made. Figure 10.10 illustrates the positional change and the advantage possible with this positional change.

The system has some forward angular momentum from the takeoff, and the jumper can effectively use this momentum to carry the body over the feet. The jumper can minimize the deceleration effect of the ground-to-heel reaction force by flexing at the hips and knees as soon as the heels contact the surface. A vigorous flexion of the shoulder joints after heel contact also helps maintain the angular veloc-

TABLE 10.2
THEORETICAL RANGES OF THE CENTER OF GRAVITY FOR TWO JUMPERS

Difference Between Height of CG at Takeoff and Landing (m)	Initial Velocity (m/sec)	Initial Angle (degrees)	Distance of the CG (m)
A. Theoretical Distances of the Body's CG for the Running Long Jump of a College Woman Participant on the Intercollegiate Track Team. Based on an Initial Velocity of 7.19 m/sec.[a]			
.19	7.19	25	5.48
(estimated)		30	5.81
		35	6.03
		40	6.11
		41	6.12[b]
		42	6.11
		43	6.11
		44	6.09
		45	6.07
B. Theoretical Distances of the Body's CG for the Running Long Jump of Bob Beamon. Based on an Estimate of Bob Beamon's Takeoff Velocity of 10.06 m/sec.[c]			
.19	10.06	25	9.52
(estimated)		30	10.30
		35	10.85
		40	11.14
		41	11.17
		42	11.18
		43	11.19[b]
		44	11.17
		45	11.15

[a] Actual angle of takeoff was approximately 18 degrees. Actual distance of the jump was approximately 4.57 m.
[b] Maximum distance.
[c] World record distance was 8.90 m (takeoff to landing of feet).
Adapted from Widule, 1974, pp. 168–69.

ity of the rest of the body forward about the heels. Figure 10.11 illustrates this application of the motion of the arms upon landing.

Triple Jump

The triple jump involves the mechanical factors of projectiles in each of three phases. During each phase—the hop, the step, and the jump—each projection is under the influence of the *same mechanical factors* as in the long jump.

Each successive phase in the triple jump is initiated at a reduced horizontal velocity. Bober (1974) states that "the competitors compensated for the drop in horizontal velocity by increasing the effectiveness of takeoff, thus resulting in comparatively high vertical velocity at the end of the takeoff and increase in the projection angle." The mean projection angles

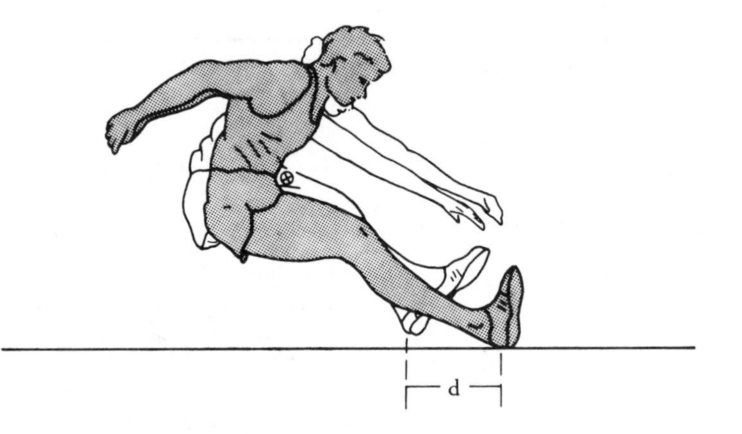

FIGURE 10.10
Increasing the heel distance by throwing the arms backward into hypertension and leaning the trunk back.

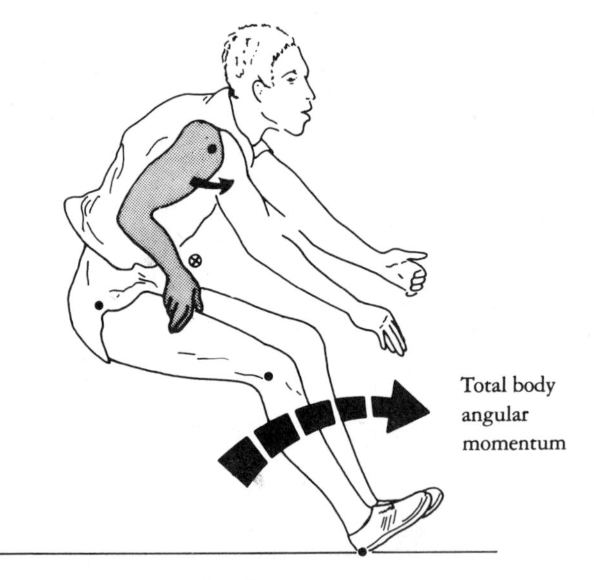

FIGURE 10.11
Application of the principle of conservation of angular momentum to landing in a long jump (hitch-kick technique).

reported were 18, 16, and 20 degrees for each phase, respectively. Application of the desirable qualities of projection—increased takeoff speed, high CG position at takeoff, the closest possible angle of takeoff to the ideal angle for the speed, and delaying the heel contact for as long as possible—must be combined in such a way as to maximize each factor without a substantial reduction in the distance achieved by another factor.

Fukashiro and Miyashita (1983) studied the parameters necessary for a triple jumper to attain a distance of 18 m. They found, not surprisingly, a significant positive correlation between run-up velocity and the total distance jumped. Second, they estimated that the jumper needs to apply a vertical force of 3.6 to 4.4 times his body weight during support for each phase of the jump. The hop takeoff angle was calculated to be 14.72 degrees; the step takeoff angle, 13.72 degrees; and the jump takeoff angle, 20.98 degrees. These angles follow a similar pattern to Bober's findings.

The 18-m triple jump model also indicated decrease in horizontal velocity with each successive phase (9.9 to 8.6 to 7.3 m/sec) and a vertical velocity that initially decreased from the hop to the step (2.6 to 2.1 m/sec) but then increased during the jump phase to a value greater than that at initial takeoff (2.8 m/sec).

In an attempt to gather kinematic data on world-class triple jumpers and to compare these data to other elite jumpers, Miller and Hay (1986) reported that with longer jumps there was a consistent decrease in the step phase of the jump. The average percentages of distances for the hop, the step, and the jump were 36.1, 29.6, and 34.35, respectively. While the mean horizontal velocities decreased from one phase to the next, 9.59, 8.44, and 6.93 m/sec, respectively, the lowest mean vertical velocity was the step phase with 1.88 m/sec. The highest vertical velocity was 3.89 m/sec for the hop, and the vertical velocity for the jump was 2.60 m/sec. The authors conclude that the best jumpers landed with a "pawing" action at the end of the hop and a blocking action at the end of the step. The "pawing" action would help to maintain the horizontal velocity from the hop to the step but would not help the vertical velocity. The blocking action at the landing of the step phase would not tend to help the horizontal velocity but would tend to increase the angle of takeoff for the jump as well as increase the vertical velocity component.

Racing Dives

The racing dive used in swimming starts is classified as a skill in which the primary mechanical purpose is horizontal distance; however, this is not strictly the case, as a result of numerous constraints placed on the dive relative to the total event. In a swimming race, the primary purpose is to move the body over a prescribed distance with the greatest speed. The dive section of that event plays a significant role in determining the total time. The following factors are considered here:

1. The human body can project itself horizontally through the air faster than it can propel itself through the water.
2. The angle of entry of the CG into the water partially determines the deceleration effect of the water on the resultant velocity of the body entering it.
3. The body configuration when entering the water has the greatest effect on how much horizontal deceleration the CG has on entering the water.

The grab start using two trajectories and two styles of entry is shown in Figure 10.12. Because the starting blocks are elevated from the water level, the difference in takeoff and landing heights suggests an optimum angle of projection less than 45 degrees. Maglischo (1982) reports that in the grab-start techniques (Figure 10.12a), the projection angle formed by a line from the hip to the toes should be 40 to 50 degrees from the top surface of the starting platform if the starting block is horizontal. Because the body's CG location is somewhat forward and upward relative to the hip joint during takeoff, the suggested angle of projection of the CG is somewhat less. Considering projectile motion parameters, Maglischo's suggested angles would produce a projection with excessive vertical velocity.

Wilson and Marino (1983) report on three types of starts: the swing technique in which the arms are swung forward to start, the grab

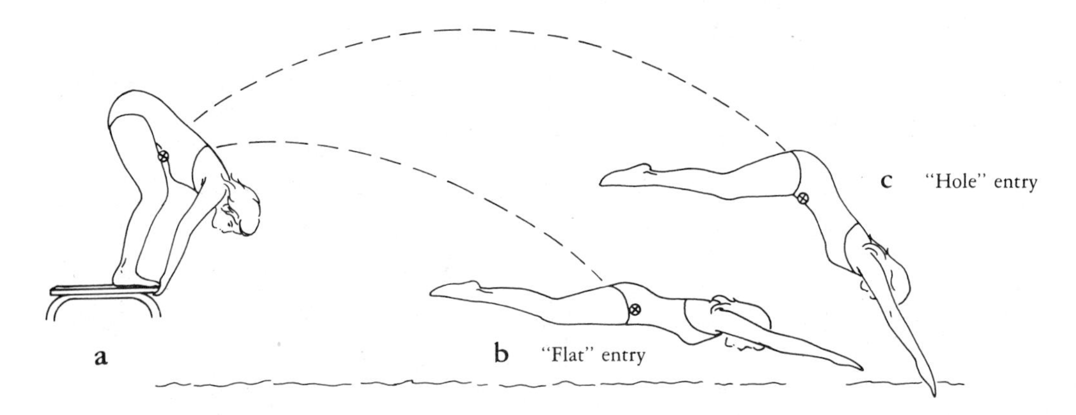

FIGURE 10.12
Trajectories of (a) the grab start with (b) the "flat" entry and (c) the "hole" entry techniques of a racing start.

start in which the body enters the water in a flat configuration (Figure 10.12b), and the whip start in which the body begins in the grab-start configuration and enters the water in a piked position (Figure 10.12c). They found no significant differences among the horizontal velocities of the styles at takeoff or entry. With respect to the angle of body lean at takeoff, a significant difference was noted between those subjects using the whip style (pike or hole entry) and those using the flat entry technique with a grab start. (The angle of body lean is defined as the angle made from a line between the shoulder and the point of contact of the feet on the block to the horizontal.) The angle of body lean in the grab with a flat-entry style was significantly smaller (19.85 degrees) than the grab with a pike-entry (whip) style (25.44 degrees). Furthermore, Stevenson and Morehouse (1979) found that varying the slope of the starting-block surface from 0 to 30 degrees had no effect on the projection angle for grab starts.

Generally speaking, the racing dive allows the swimmer to leave the starting block quickly. The resultant velocity of the swimmer's CG is therefore maximized and projected at an angle of approximately 19 to 26 degrees so that the horizontal velocity component is larger than the vertical component. The diver should increase the time in the air but not sacrifice too much horizontal velocity because the diver also should strive for the greatest horizontal distance before entering the water.

The pike entry position helps decrease the horizontal distance of the dive because the piking of the body causes the hand-entry spot to move backward slightly in space. The pike entry has the proposed advantage of allowing the hands to cut a "hole" in the water through which the rest of the body passes. This pike position also prevents the body from losing as much momentum as the flat position during entry. Maglischo (1982) reports an entry angle of the trunk of 30 to 40 degrees relative to the surface of the water. Hyperextension of the vertebral column and hyperflexion of the shoulder joints are needed to prevent deep plunging.

Understanding Projections for Horizontal Distance

1. Cut two pieces of heavy cardboard, 15 cm by 8 cm and 5 cm by 10 cm. As shown in Figure 10.13, pin the two pieces together so that the links are freely movable relative to each other.

   a. Hold piece A vertically and piece B horizontally, as shown in Figure 10.13a. Release piece B. What happens to it relative to A? What force is causing this motion?

   b. Place A and B perpendicular to each other, as shown in Figure 10.13b. Release both pieces at the same time, allowing them to fall to the ground. Is there any movement of B relative to A, as in the previous experiment? What force causes the movement of the system to the ground? Why does B not display the same movement relative to A as in the previous experiment?

   c. Relate the findings of these experiments to the example of a gymnast doing leg

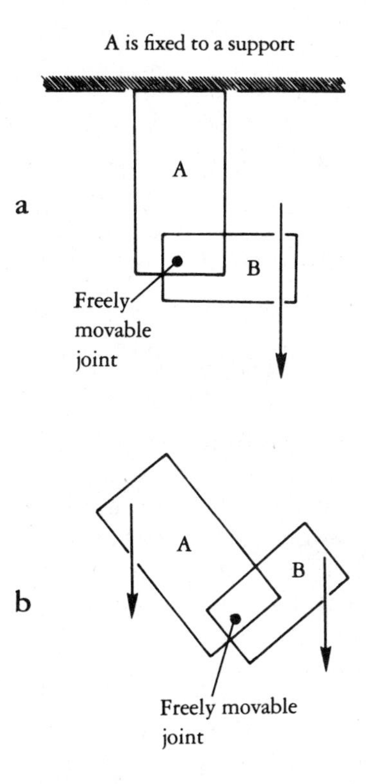

FIGURE 10.13

Cardboard model with freely movable joint (a) with segment A fixed and (b) with segment A and B free.

levers while hanging from a horizontal bar and to the example of a long jumper descending from a long jump (see Figure 10.9 or 10.10). (A represents the trunk; B represents the lower extremity.)

d. Why must the gymnast use muscle force to hold his or her legs up whereas the long jumper does not?

2. Draw a trajectory illustrating the path of a softball that is projected at 45 degrees with no air resistance (takeoff and landing heights are equal). Draw a trajectory illustrating the same object traveling with air resistance. What is the difference between the two trajectories? Why does the difference exist?

3. Draw an arbitrary trajectory for each of the following situations:

a. Projection velocities are equal at a 35-degree projection angle from a takeoff height of 1.5 m and a takeoff height of 2 m. Both land on the ground level.

b. Takeoff heights are equal at a 35-degree projection angle with projection speeds of 20 m/sec and 25 m/sec. Both land at the same level.

c. Takeoff and landing heights are equal, and speeds are equal. The projection angles are 30 degrees and 40 degrees.

For each of the situations, state a general principle that applies to the effects of altering the specifications of projection on the horizontal distance.

4. From a distant side view, observe the trajectories of a javelin and a discus. What are the differences between each trajectory and a parabolic trajectory? Why do these differences occur?

5. Imagine driving a badminton shuttlecock and a tennis ball horizontally as far as possible. Draw the trajectories. What are the differences? Why do they occur?

6. Perform a standing long jump, keeping the shoulders flexed forward on landing. Perform a standing long jump, hyperextending the shoulders during flight. Is there a difference in jump distances? To what could you attribute the difference?

10.5 Projecting for Accuracy

In many sport events, projecting an object or the body to an exact location is important. Maximum distance is not the primary mechanical purpose in these events; however, optimizing the distance so that the object or body reaches the target area may be necessary. For example, in throwing a dart one does not need to throw as fast as possible, but one must project it with enough velocity to get the dart to the board. Another example is the distance given to a basketball when it is shot; the velocity is *optimized* rather than *maximized* to place the ball in the exact location.

In sport events in which accuracy is the primary mechanical purpose, having the object travel at a high speed also may be desirable. Examples of such situations include pitching a baseball or a softball; serving a volleyball, tennis ball, or badminton shuttlecock; and most racquetball shots.

Projecting Objects for Accuracy

In several activities the primary concern is accuracy, and how fast the object is moving when it reaches the target is of little concern. These activities include slow-pitch

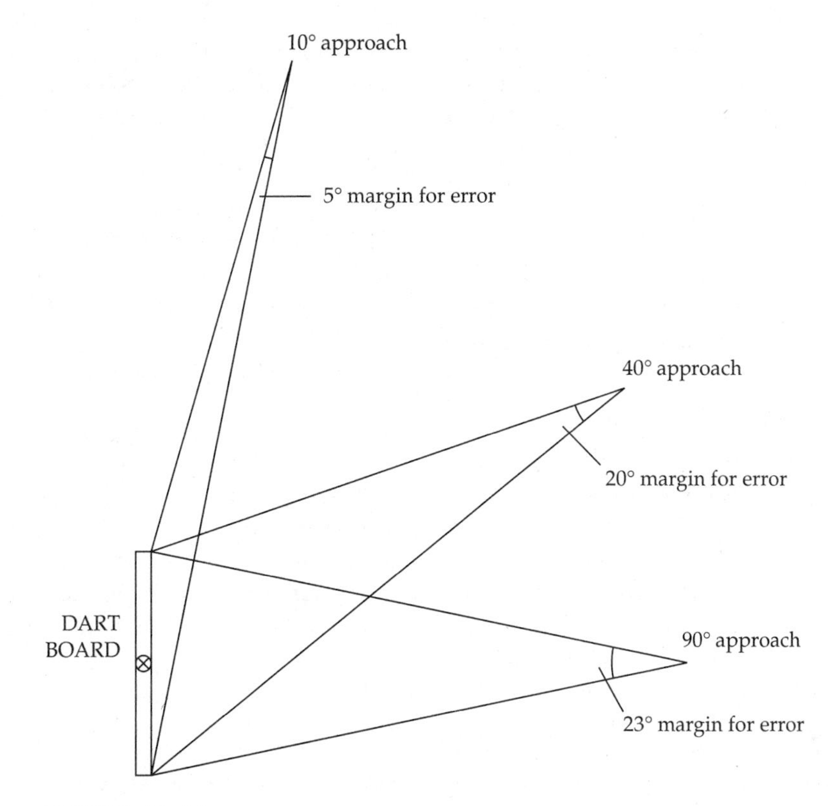

FIGURE 10.14

Vertical plane targets approached by a dart at three different angles of incidence.

softball pitching, horseshoe pitching, basketball shots, and approach shots in golf. The constraints under which each of these activities is performed help determine the projectile mechanics appropriate for analysis. One such constraint is where the plane of the target is located in space. There are vertical plane targets and horizontal plane targets.

Vertical Plane Targets

Targets located in a vertical plane are found in such activities as softball pitching, dart throwing, archery, and rifle marksmanship. Accuracy is easiest when projecting to a vertical plane target if one uses a trajectory that is perpendicular to the target as it coincides with the target face. Of course, the farther the projection distance, the more one must rely on a more curved parabolic path. Consider throwing a dart at a target on the wall from distances of 2, 5, and 10 m away. For the 2-m target, one must project the dart fairly straightforward; for the 5-m target, one must give the dart some vertical velocity so that the dart does not fall too much before reaching the target plane; and for the 10-m target, a considerable upward thrust must be made (Figure 10.14).

Now consider how each of the darts tossed in the aforementioned ways arrives at the target surface. The close toss probably arrives nearly perpendicular; the angle of projection could be a few degrees on either side of the horizontal, and the dart would land on target.

The 5-m toss arrives at the target at some angle as it travels downward. From the peak of its flight, it continues to accelerate downward in the vertical direction until it hits an external force—the target. This vertical acceleration is difficult to judge and places a constraint on the performer tossing from a distance. The 10-m toss involves a similar but more difficult problem. The toss must be given a greater projection velocity; it peaks farther from the target and thus has a greater downward-directed vertical velocity when it arrives at the target. The angle of incidence with the target also is smaller because the dart has a greater vertical velocity owing to spending more time in the air.

The smaller the angle of incidence with the target, the greater the chance of error. Generally, the more one must rely on vertical motion to get the object to a horizontal target, the greater the chance of error. The more horizontal velocity given to an object projected at a vertical plane target, the less one must rely on vertical motion to get the object to the target. For example, in archery the advent of the compound bow with its greater "weight" improved the performance of archers shooting at 50 m and greater distances.

In slow-pitch softball pitching, a peak is set on the trajectory. The pitcher not only wants the ball to pass through the target but also wants it to pass through in a particular manner. The more horizontal the trajectory as it passes through the target, the less effective it is as a pitch (i.e., a more vertically traveling ball is more difficult to contact than a more horizontally traveling ball). Thus, the less skilled pitcher probably sacrifices some effectiveness by projecting the ball more horizontally so that the accuracy is enhanced.

Horizontal Plane Targets

Accuracy activities in which the target is aligned in the horizontal plane are horseshoe pitching, basketball shots, clout shooting in archery, and a golf approach shot. In these activities, the object is projected upward to descend into the target area at close to 90 degrees. The more vertically the object arrives at the horizontal target, the more chance of success in hitting the target as well as minimizing the possibility of sliding or rolling out of the target area.

In horseshoe pitching, additional time in the air is given to the projectile by the difference between release and landing heights. The vertical component of the velocity is important. Because the object of the game is to place the horseshoe as close to the stake as possible, a high pitch with a more vertical landing is best; however, if the objective is to bump the opponent's shoe out of the way of the stake, then a more horizontal pitch provides the horizontal momentum for the necessary impulse at landing.

The pitch shot in golf is another example. A more vertical angle of projection and landing is beneficial for the accurate placement of the ball near the target. Along with the backspin, the enhanced vertical arrival on the green prevents the ball from rolling off the green into a sand trap or another unfortunate location (Figure 10.15).

Basketball shooting trajectories are often studied. By comparing the diameter of the basket and the diameter of the ball, one finds that there is a minimum angle at which the ball can approach the basket and still go through the hoop. This angle depends on the height of release and vertical distance of release from the height of the basket. The closer the release height to the height of the basket, the smaller the projection angle needed and also the less the velocity of projection. One would not want to make the projection angle too shallow, however, for an opponent could easily intercede in the ball's path.

There is some consensus that the height of release is important to the success of the shot. Hudson (1983) reports that successful shots are significantly related to a height ratio

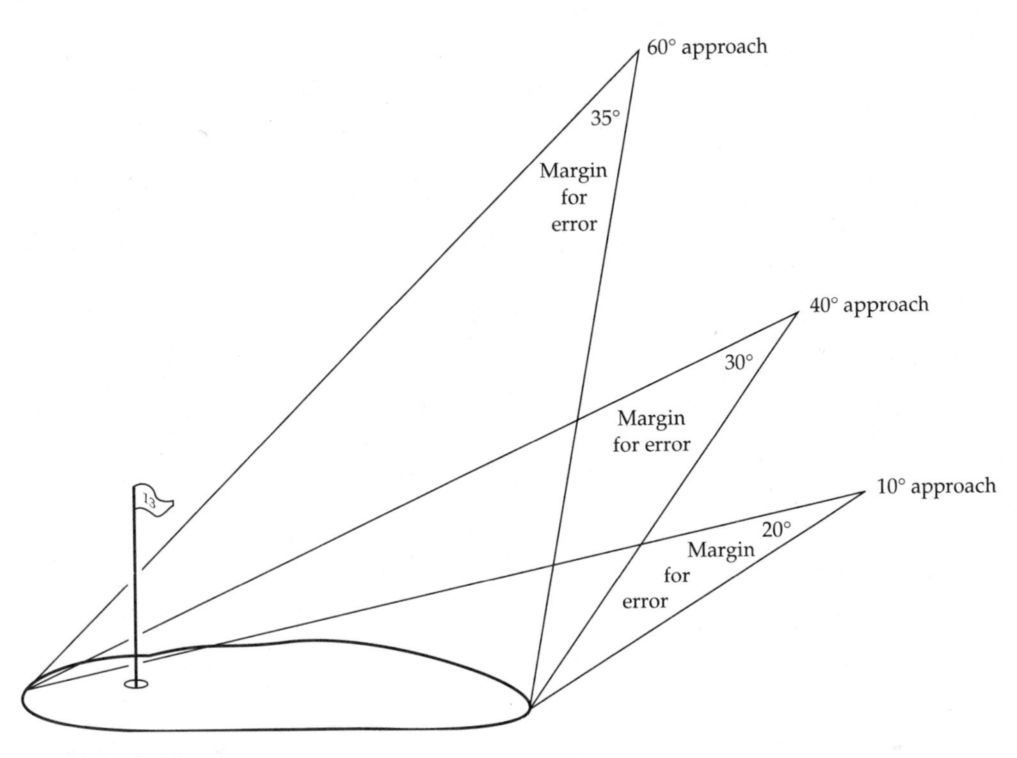

FIGURE 10.15
Horizontal plane targets approached by a golf ball at three different angles of incidence.

that was calculated by comparing the standing height of the shooter to the height of the shooter's release. Thus, the greater the release height for any given performer, the more successful the shot. This finding was supported by Yates and Holt (1983) when they found that successful shooters have greater shoulder flexion and elbow extension at release. This position gives the shooter a higher release point, and it gave the shooters in this study a more vertical angle of ball projection.

Maugh (1981) reports that "as the height at which a shot is released increases, so do the chances of scoring. . . . If the shooter jumps two feet, the width of this scoring band . . . increases by 18%." The scoring band to which Maugh refers is an arch in space, the diameter of which increases with the height of release. It is what Maugh calls the "margin of error" (p. 107). Any shot following within this arch is successful. The most common error in shooting as described by Maugh is using an angle of projection that is too small, thus decreasing the margin of error.

Projecting the Body for Accuracy

Projecting the body for accuracy occurs mainly in activities for which the primary mechanical purpose is to move or position the body or its segments or both in a prescribed pattern with the intent of achieving an ideal or model performance. Generally, the activities are judged or evaluated on the movement quality of the performance. Part of the quality of the performance depends on how effectively one projects the body into the air. Examples of activities include aerial tumbling, diving, trampoline, apparatus dismounts, ski jumping, and gymnastics vaulting.

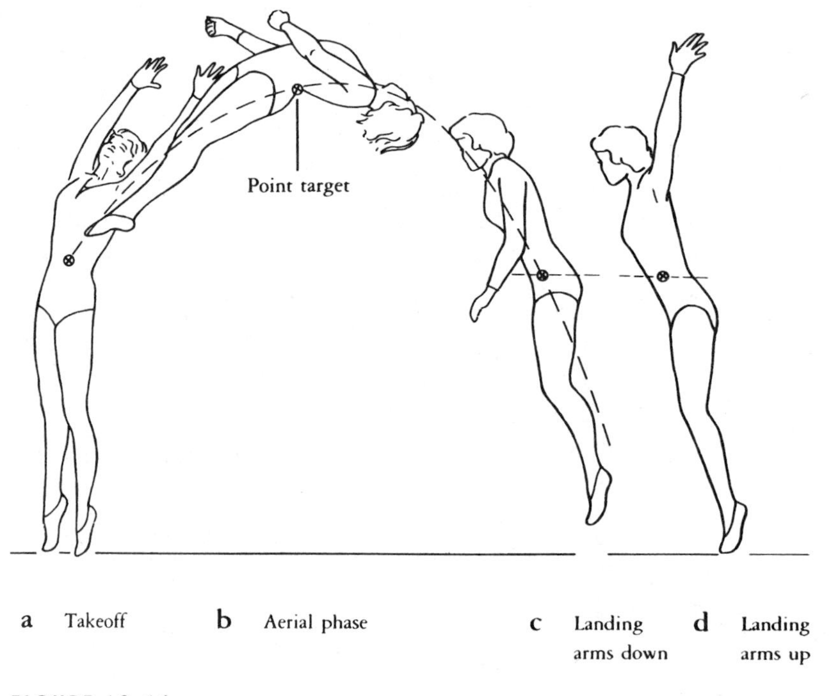

FIGURE 10.16

Positions of the feet relative to the ground (a and b) in landing from a back somersault when (c) the arms assume an extended position and (d) when they remain flexed overhead.

The target to which the body or objects are projected is an imaginary point in space and is called a **point target** in this text. This point target should be used as the target for projecting the body's CG as if one were projecting it to an ideal location in space. Placing the body's CG at this location assists in accomplishing some body maneuvers.

Gymnastics-Related Activities

AERIAL TUMBLING. Because aerial tumbling stunts progress from single turns in the air to multiple turns combined with twists, the performer must begin with a good takeoff so that multiple moves may be accomplished within the amount of available time in the air. Two principles govern this activity. First, the performance should maximize the time in the air so that the performer may have more time to turn and twist before landing. Second, the height of the CG should be maximized by lowering it within the body just before landing and by delaying foot contact so that additional time in the air is provided. As an example, a back layout somersault is examined, as shown in Figure 10.16. To maximize the time in the air, the performer must keep the angle of projection relatively large, producing a high vertical velocity. Therefore, the point target for the CG is as high as possible and peaks above the ground at a location approximately one half the performer's body length from the takeoff point. A greater vertical velocity means that the force of gravity requires more time to bring the body to a halt and more time to bring it back down again. Thus, the performer should be coached to project the body upward, rather than backward.

At *takeoff,* the body should be as extended as possible, with both arms fully flexed above the head, so that the performer may position the CG as high as possible (Figure 10.16a). There is a tendency to flex the arms upward and bring them partially downward before takeoff because the performer is anxious to assume the aerial position. If full body extension and arm flexion in the single back somersault are stressed by the teacher or coach, the performer usually progresses to the double back more easily. Once the body is free of support, the arms are brought down to the sides and should remain there to place the CG location more toward the foot end (Figure 10.16b). This serves to momentarily delay foot contact with the ground. This is illustrated by the two landing positions shown in Figure 10.16c and d. The gymnast in Figure 10.16c has time for a little more rotation before her feet contact the ground, whereas the gymnast in Figure 10.16d is already on the ground because her arms are overhead.

When the aerial stunt is preceded with a run and several linear-velocity generating stunts such as roundoffs and back handsprings, the projection of the body is less in need of horizontal velocity than vertical velocity. The projection direction in such cases is more in the vertical direction because the body already has horizontal velocity from the prior moves. In situations where a run-up precedes the aerial maneuver, the point target is close to vertical alignment from the takeoff point. The horizontal velocity combined with the vertical velocity from the vertical impulse provides the body with the proper trajectory. The point target for the performer's CG should be within 10 to 15 degrees from the vertical.

GYMNASTICS VAULTING. The takeoff characteristics of long-horse or side-horse vaulting are similar to those of aerial tumbling stunts in which run-ups or approach stunts precede the skill. In addition, a spring mechanism, the Reuther board, serves to enhance the vertical impulse. The approach run is usually longer in vaulting than in tumbling stunts, and therefore the horizontal velocity of the performer is greater. The takeoff angle is approximately 12 degrees from the vertical for skilled side-horse vaulters performing a handspring vault (Kreighbaum, 1974).

Because of the large horizontal velocity on initiating the hurdle step, the CG location at initial contact with the vaulting board is behind the foot contact position with the board. During the short interval of board contact, the CG moves forward so that it is located slightly ahead of the vertical. The point target for the vaulter is such that the CG is at its peak just before horse contact so that little vertical downward motion results onto the horse. The point target for the CG for handspring-related vaults is shown in Figure 10.17.

APPARATUS DISMOUNTS. In dismounting from an apparatus, the performer must consider that the body is taking off from a position higher than the landing height. This should not result in the performer adjusting the trajectory to a lower angle, as in projecting an object for the greatest horizontal distance. In these cases, it allows the performer to do a greater number of maneuvers in the air than can be accomplished from a ground takeoff. The point target for these stunts should be no more than one half the height of the performer in the horizontal direction and high as possible in the vertical direction. A beginning performer usually senses the need to "get around" immediately and therefore errs by projecting the CG too much in the horizontal direction. Second, the height of takeoff seems to have a negative effect in that it convinces the performer that the body gets to the ground sooner and that therefore the somer-

FIGURE 10.17
Point target for a long-horse vaulter.

saulting action must be done as soon as possible, which is frequently before the actual take-off.

Diving

Evaluation of diving performance includes such factors as the trajectory of the body in the air and the entry position of the body into the water. The trajectory determines the height to which the diver goes and how close the diver is to the diving platform or board.

The point target for high-platform diving is not as great as that used for springboard diving because the body is already quite high, and no spring can help the diver to attain greater heights. Therefore, platform divers use a point target as high as possible but horizontally far enough from the board to clear their bodies as they rotate forward or backward while passing the board on their descent. Because platform divers do not have a great deal of vertical velocity on takeoff, they do not have enough time in the air to move a great deal horizontally as they pass the platform on their way down. Consequently, platform divers must have greater horizontal velocity on takeoff than springboard divers.

Springboard divers have an extra elevation to which their CG moves owing to the reaction force of the elastic recoil of the board. Therefore, the horizontal component of the velocity vector need not be great, for by the time they move vertically upward and back downward to the level of the board, their CG is farther from the board than that of divers with less vertical velocity. The point target of platform divers, therefore, must be at a greater horizontal distance from the platform than that of springboard divers.

A diver should visualize the point target appropriate for the type of board or platform used. Also, the amount of vertical velocity obtained at takeoff helps determine the amount of horizontal velocity necessary to complete the dive safely.

As in the case with Greg Louganis's reverse tuck dive in the 1988 Olympic Games in which he hit his head on the board, the diver must also consider the time and place for opening up during the flight. Whereas, Louganis would have missed the board had he opened up earlier or later, the timing of his ill-fated dive resulted in disaster. Either the diver must increase the horizontal velocity vector so that he or she is farther from the board at opening or the opening must be at a different location relative to the board.

Projecting Objects for Accuracy When Speed May Enhance the Performance

Serving the ball in volleyball is an example of a situation in which the probability of success depends on the object's speed as well as its accuracy. Widule calculated the distance for a volleyball serve when it is projected at different speeds and angles (Widule, 1974). These distances are shown in Table 10.3. The impact (projection) height is 2.13 m (7 ft).

In the overarm volleyball serve, a spinning or floating serve, the movement is usually performed so that the ball travels at a high speed, passes close to the net, and is directed to land close to the back line; the server manipulates the projection variables so that these objectives are satisfied. The court is 9 m wide and 18 m long from service line to back line. The net is approximately 2.43 m high for men and 2.24 high for women.

At slower projection speeds, the angle must be increased so that the ball clears the net (note the case of the 10.67 m/sec [35 ft/sec] serve at 25 degrees). When using this slower serve, the server must increase the angle of projection to at least 30 degrees to clear the net. This results in a rather short serve (12.91 m). The ball descends at an angle greater than 30 degrees, which is undesirable because the opposing player can receive it with ease. Because the ball is of relatively low mass, air resistance helps decrease the resultant velocity, and thus the ball does not travel as far before it hits than would be the case with a more massive object. If the server selects a different type of serve (e.g., the sidearm or underarm), the angle of projection must be even greater to clear the net because the projection height is lower.

The server can achieve success at a lower angle by increasing the projection speed. At 12.19 m/sec (40 ft/sec), a 15-degree projection does not clear the net. At 12.19 m/sec (40 ft/sec), a 20-degree angle is the most desirable, but at this angle the ball only travels 13.85 m and falls 4.15 m short of the baseline. The angle has to increase to 30 degrees to come within 1.88 m of the baseline—still not a good mark. Of the types of serves listed in Table 10.3, the one producing the most desirable trajectory is that which is projected at 15.24 m/sec

(50 ft/sec) at an angle of between 15 and 20 degrees. The ball lands 0.72 m (approximately 28 in.) inside the baseline.

When considering players of various ability levels, the laws of projectiles can be applied to determine what type of serve is best for each individual. The height of the server must be taken into consideration as well as the speed with which the player can serve the ball in an underarm, sidearm, and overarm pattern. Some players cannot serve overarm well enough to impart sufficient velocity to the ball. Because such a player must increase the angle of projection to clear the net, the player may find that serving underarm elicits a better serve with regard to velocity, projection angle, and horizontal distance.

The tennis serve has restrictions similar to those of the volleyball serve. A few differences exist between some of the variables in these two serves: the tennis serve has a faster projection speed, a greater horizontal distance, and a greater projection height. The fact that the projection speed of the tennis serve is greater reduces the time that it takes the ball to get to the net. This reduces the time and, consequently, the vertical distance through which the ball drops because of the pull of gravity on it. (Note that this is because of a time reduction, not a reduction in the magnitude of gravity's force.) In addition, the tennis net is lower than the volleyball net, and the height of impact is higher so that the angle of projection is negative. In this case, the faster or taller server can decrease the angle of projection more than the slower or shorter server.

TABLE 10.3

DISTANCE FOR A VOLLEYBALL SERVE WHEN CONTACTED AT A HEIGHT OF 2.13 m

Velocity (m/sec)	Angle (degrees)	Height of Ball at Net	Distance (m)
9.14	45	1.46	—[a]
10.67	25	2.01	—[b]
10.67	30	2.60	12.91
10.67	35	3.16	13.38
10.67	40	3.66	13.56
10.67	45	4.07	13.44
12.19	15	1.62	—[a]
12.19	20	2.33	13.85
12.19	25	3.04	15.11
12.19	30	3.73	16.12
12.19	15	2.24	—[a]
13.72	20	2.99	16.65
13.72	25	3.74	18.35[b]
13.72	10	1.92	—[a]
15.24	12	2.23	—[a]
15.24	13.5	2.46	16.52
15.24	15	2.69	17.28
15.24	20	3.46	19.73[b]

[a] Ball fails to clear the net.
[b] Ball lands outside of the court.
Adapted from Widule, 1974.

Because the rear boundary of the service court is only 7 m from the net on the far side, the smaller angle of projection provides a distinct advantage in terms of directing the ball to land within the service court.

Understanding Projections for Accuracy and Speed

1. Consider the volleyball serve, given a court that is 9 m on either side of the net and a net height of 2.24 m. Using an underarm serve, draw the different trajectories that can be produced to have the ball clear the net and land within the opposite court. Specific factors that must be considered are height, velocity, angle of projection, net clearance, angle of approach to the opposing player, and horizontal distance.
2. Repeat the steps in question 1, using a sidearm serve and an overarm serve.
3. Consider three volleyball servers of various heights. What differences would there be in the ways they should be coached to serve? Draw the possible trajectories.
4. Consider three volleyball servers, one beginning, one intermediate, and one advanced. (All are of equal height, but ability may affect velocity of impact.) What factors should be considered in coaching each? Draw three possible trajectories.

References and Suggested Readings

Atwater, A. E., & Spray, J. A. (1975, May 22–24). Kinematic analysis of takeoff in the long jump and high jump. Research paper presented at the American College of Sports Medicine, 22nd Annual Meeting, New Orleans.

Ayalon, A., Van Gheluwe, B., & Kanitz, M. (1975). A comparison of four styles of racing start in swimming. In R. LeWillie & J. P. Clarys (Eds.), *Swimming II: Proceedings of the Second International Symposium of the Biomechanics of Swimming* (pp. 233–240). Baltimore: University Park Press.

Biesterfeldt, H. J. (1974). A comparison of selected factors relating to success of running front somersaults. *Gymnast, 16*, 32–33.

Bober, T. (1974). Investigation of the takeoff technique in triple jump. In R. C. Nelson & C. A. Morehouse (Eds.), *Biomechanics IV: Proceedings of the Fourth International Seminar on Biomechanics* (p. 150). Baltimore: University Park Press.

Bowers, J. E., & Cavanagh, P. R. (1975). A biomechanical comparison of the conventional & grab starts in competitive swimming. In R. LeWillie and J. P. Clarys (Eds.), *Swimming II: Proceedings of the Second International Symposium on the Biomechanics of Swimming* (pp. 225–232). Baltimore: University Park Press.

Dapena, J. (1974). Searching for the best straddle technique. *Track Technique, 55*(3), 1753–1756.

Dapena, J. & Chung, C. S. (1988). Vertical and radial motions of the body during the take-off phase of high jumping. *Medicine and Science in Sports and Exercise, 20*, 290–302.

Dapena, J. (1980a). Mechanics of rotation in the fosbury-flop. *Medicine and Science in Sports and Exercise, 12*, 45–53.

Dapena, J. (1980b). Mechanics of translation in the fosbury-flop. *Medicine and Science in Sports and Exercise, 12*(1), 37–44.

Denoth, J., Luethi, S. M., & Gasser, H. H. (1987). Methodological problems in optimization of the flight phase in ski jumping. *International Journal of Sport Biomechanics, 3*, 404–418.

Dyson, G. (1978). *The mechanics of athletics* (7th ed). New York: Holmes & Meier Publishers.

Elliot, B. C., & Newton, A. P. (1987). A cinematic analysis of the "inclined" board & "normal" long jump technique. *Journal of Human Movement Studies, 13*, 157–170.

Fukashiro, S., & Miyashita, M. (1985). A biomechanical study of the landing in triple jump. In D. A. Winter, R. W. Norman, R. P. Wells, K. C. Hayes, & A. E. Patla (Eds.), *Biomechanics IX-B* (pp. 454–457). Champaign, IL: Human Kinetics.

Fukashiro, S., & Miyashita, M. (1983). An estimation of the velocities of three takeoff phases in 18-m triple jump. *Medicine and Science in Sports and Exercise, 15*(4), 309–312.

Hay, J. G. (1974, January). Characteristics of the flop. *The Athletic Journal, 10*, 92–93.

Hay, J. G. (1975a, April). Straddle or flop? *The Athletic Journal*, 83–85.

Hay, J. G. (1975b, June). The Hay technique: The ultimate in high jumping style? *Coaches Digest*, 44–47.

Hay, J. G. (1978). *Biomechanics of sports techniques.* Englewood Cliffs, NJ: Prentice-Hall.

Hay, J. G. (1983). Springboard reaction torque patterns during nontwisting dive takeoffs. In H. Matsui & K. Kobayashi (Eds.), *Biomechanics VIII-B: Proceedings of the Eighth International Congress of Biomechanics* (pp. 822–827). Champaign, IL: Human Kinetics.

Hay, J. G. (1986). The biomechanics of the long jump. In K. B. Pandolf (Ed.), *Exercise and sport sciences review* (pp. 401–406.) New York, NY: Macmillan.

Hay, J. G., Miller, J. A., & Canterna, R. W. (1986). The techniques of elite male long jumpers. *Journal of Biomechanics, 19*, 855–866.

Herzog, W. (1986). Maintenance of body orientation in the flight phase of long jumping. *Medicine and Science in Sports and Exercise, 18*, 231–241.

Heusner, W. W. (1959). Theoretical specifications for the racing dive. *Research Quarterly, 30*(1), 25–37.

Hirata, T., Matsuo, A, Yuasa, K., & Fukunaga, T. (1987). Effect of takeoff velocity on long jump performance. In B. Jonsson (Ed.), *Biomechanics X-B* (pp. 745–748). Champaign, IL: Human Kinetics.

Hubbard, M., & Trinkle, J. C. (1985). Optimal Fosbury-flop high jumping. In D. A. Winter, R. W. Norman, R. P. Wells, K. C. Hayes, & A. E. Patla (Eds.), *Biomechanics IX-B* (pp. 448–453). Champaign, IL: Human Kinetics.

Hudson, J. L. (1983). A biomechanical analysis by skill level of free throw shooting in basketball. In J. Terauds (Ed.), *Biomechanics in Sports: Proceedings of the International Symposium* (pp. 95–102). San Diego, CA: Academic Publishers.

Kreighbaum, E. F. (1974). The mechanics of the use of the reuther board during side horse vaulting. In R. C. Nelson & C. A. Morehouse (Eds.), *Biomechanics IV: Proceedings of the Fourth International Seminar on Biomechanics* (pp. 137–143). Baltimore: University Park Press.

Lichtenberg, D. B., & Wills, J. G. (1978). Maximizing the range of the shot put. *American Journal of Physics, 46*(5), 546–549.

Luhtanen, P., Pulli, M., & Komi, V. (1987). A relative model of human movement with special reference to ski jumping. In B. Jonsson (Ed.), *Biomechanics X-B* (pp. 1145–1150). Champaign, IL: Human Kinetics.

Maglischo, E. W. (1982). *Swimming faster*. Palo Alto, CA: Mayfield.

Mann, R., & Herman, J. (1985). Kinematic analysis of olympic hurdle performance: Womens 100 meters. *International Journal of Sport Biomechanics, 1*, 163–173.

Maugh, T. H. (1981). Physics of basketball: Those golden arches. *Science 81*, 1(3), 106–107.

Miller, D. I. (1974). A comparative analysis of the takeoff employed in springboard dives from the forward and reverse groups. In R. C. Nelson & C. A. Morehouse (Eds.), *Biomechanics IV: Proceedings of the Fourth International Seminar on Biomechanics* (pp. 223–228). Baltimore: University Park Press.

Miller, D. I., Jones, I. C., & Pizzimenti, M. A. (1988). Taking off: Greg Louganis diving style. *Soma, 2*(4), 20–28.

Miller, D. I., & Munro, C. F. (1984). Body segment contributions to height achieved during the flight of a springboard dive. *Medicine and Science in Sports and Exercise, 16*, 234–242.

Miller, J. A., Jr., & Hay, J. G. (1986). Kinematics of a world record and other world-class performances in the triple jump. *International Journal of Sport Biomechanics, 2*, 272–288.

Phinizy, C. (1968, December 26). The unbelievable moment. *Sports Illustrated,* pp. 53–61.

Ridka-Drdacka, E. (1986). A mechanical model of the long jump and its application to a technique of preparatory and takeoff phase. *International Journal of Sport Biomechanics,* 2, 289–300.

Sanders, R. H., & Wilson, B. D. (1988). Factors contributing to maximum height of dives after takeoff from the 3m springboard. *International Journal of Sport Biomechanics, 4,* 231–259.

Stevenson, J. R., & Morehouse, C. A. (1979). Influence of starting block angle on the grab start in competitive swimming. In J. Terauds & E. W. Bedingfield (Eds.), *Swimming III: Proceedings of the Third International Symposium on the Biomechanics of Swimming* (pp. 207–214). Baltimore: University Park Press.

Toyoshima, S., Hoshikawa, T., & Ikegami, Y. (1981). Effects of initial ball velocity & angle of projection on accuracy in basketball shooting. In A. Morecki, K. Fidelus, K. Kedzior, & A. Wit (Eds.), *Biomechanics VII-B: Proceedings of the Seventh International Congress of Biomechanics* 525–530. Baltimore: University Park Press.

Widule, C. J. (1974). *An analysis of human motion* (pp. 156–176). La Fayette, IN: Balt.

Wilson, D. S., & Marino, C. W. (1983). Kinematic analysis of three starts. *Swimming Technique, 19.*

Yates, C., & Holt, L. E. (1983). The development of multiple linear regression equations to predict accuracy in basketball jump shooting. In J. Terauds (Ed.), *Biomechanics in Sports: Proceedings of the International Symposium* (pp. 103–109). San Diego, CA: Academic Publishers.

CONCEPT MODULE H

Fluid Forces

PREREQUISITES

Concept Modules C, D, E, F

CONCEPT MODULE CONTENTS

Why does a cyclist pedal behind another in a race? Why does a ski jumper lean toward a prone position in flight? Why do swimmers "spiral" their hands through their underwater stroking? And what makes a curveball curve? Answers to such questions are found through an understanding of forces created by fluids.

The two fluids we will focus on are air and water, for they are the ones we encounter in our movement activities. Air, a gaseous fluid, and water, a liquid fluid, behave according to the same mechanical principles. Therefore, the concepts that follow apply to both and are presented for fluids in general. Examples are given for activities done in air and in water. Chapter 11 presents sport activities affected by air forces (aerodynamics), and Chapter 12 presents the hydrodynamics of aquatic activities and buoyancy.

H.1 Fluid Drag Force

The term *fluid resistance* is commonly used to mean the type of force that is known technically as drag force. In air it is called aerodynamic drag force; in water, it is called hydrodynamic drag force. Several variables determine how much drag force is acting on a body. To study the interaction of a body with a fluid through which it moves, we must focus on the motion of the fluid flowing past the object. The fluid flow pattern is the same whether we look at the object moving through fluid or at the fluid moving past the object. The important condition is the *relative motion* between the object and the fluid. If a ball moves forward with a velocity of 30 m/sec through still air, the air-flow forces acting on the ball are the same as if the ball were stationary and a 30 m/sec wind were blowing backward past it. Figure H.1a illustrates the relative velocity of air particles moving past a ball as it moves through the air. Note that the air-flow velocity relative to the ball is of the same magnitude as the ball's velocity but is opposite in direction. When the air particles have movement of their own, as is the case on a windy day, the movement of the air past the projectile

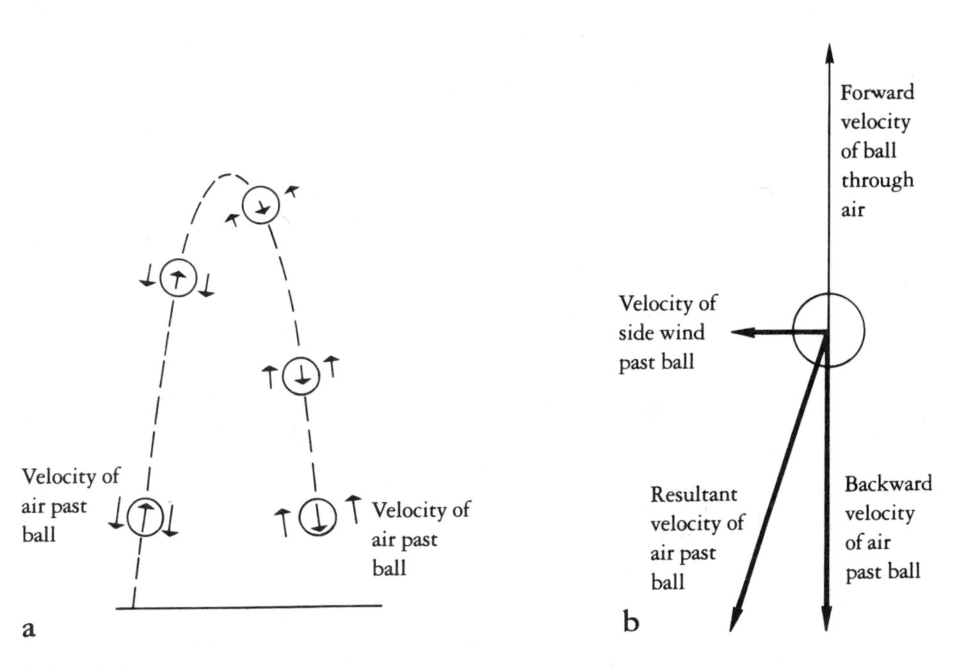

FIGURE H.1

(a) The same air-flow pattern results when the ball moves through stationary air as when the air moves past the stationary ball. (b) Vector composition of two air-flow velocity vectors relative to a thrown ball. The resultant air-flow velocity is a composite of the air flow backward past the ball (which is due to its forward motion through the air) and the air flow sideward past the ball (which is due to the side wind).

that is due to the projectile's motion must be combined with the movement of the air particles past the projectile *that is due to the wind* to find the resultant velocity of the air relative to the projectile. The *flow velocity* of the air past a ball thrown in a side wind is illustrated from an overhead view in Figure H.1b.

The ball can be pictured as experiencing two air flows: air flowing backward past it as it travels forward and air flowing sideward past it due to the side wind. The ball "feels" only one resultant air-flow velocity directed diagonally backward and to the side (recall linear vector composition). Similarly, if a ball is thrown directly into a head wind (a wind blowing in the direction opposite to the direction of the ball's motion), the wind velocity is added to the velocity of the air past the ball that is due to the ball's motion. For example, a ball thrown forward at a velocity of 20 m/sec into a head wind of 5 m/sec experiences an air flow past it of 25 m/sec (Figure H.2a). On the other hand, if the same ball were thrown with a tail wind (a wind blowing in the same direction as the ball's motion), the wind velocity would be subtracted from the velocity of the air flow

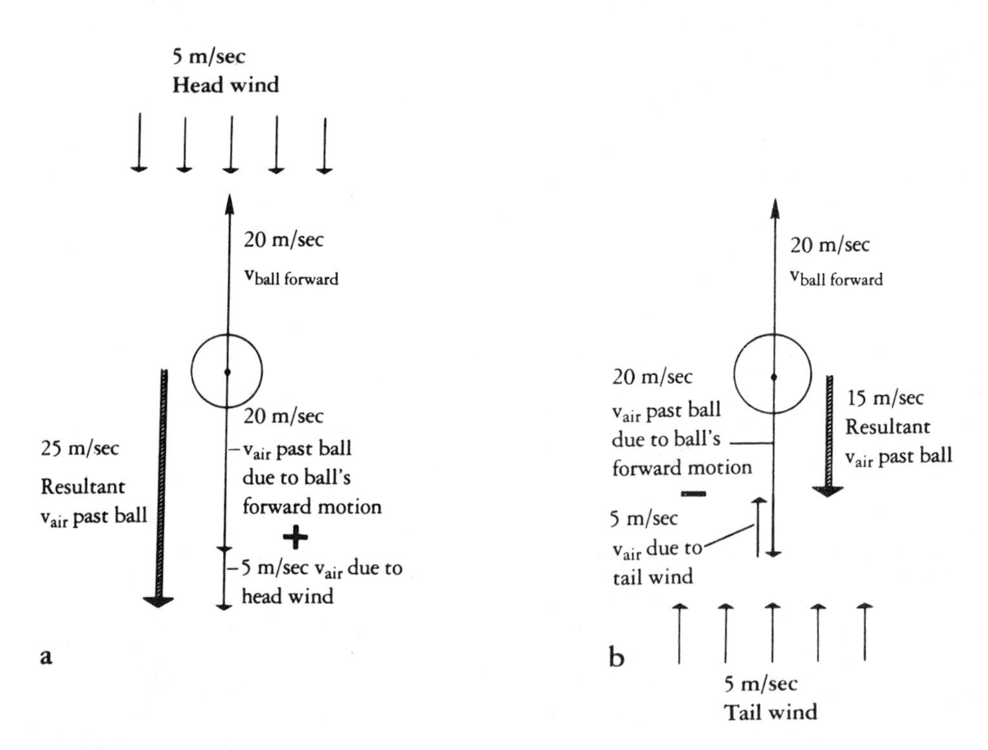

FIGURE H.2

Looking down on a thrown ball, the air-flow velocity relative to the ball moving through the air is determined by the velocity of the ball itself and the velocity of the surrounding air. (a) A head wind causes a greater relative air-flow velocity, and (b) a tail wind (following) causes a smaller relative air-flow velocity.

past the ball that is due to its motion (Figure H.2b), and the relative velocity of the air past the ball would be 15 m/sec.

The foregoing examples show how the direction and speed of fluid flow past a moving body are determined by the velocity of the body itself and of the motion of the surrounding fluid. No matter what direction a body moves, the relative flow that is due to its motion is opposite that motion. Thus, a falling body experiences an upward flow of air past it, and a rising body experiences a downward flow of air past it. It is the nature of the fluid flow around a body as it cleaves its way through the fluid that determines the amount of drag force against it.

Two types of drag force occur in both air and water sport activity: **skin friction** and **profile drag.** They act directly opposite to the body's direction of motion. Both are present and part of the total drag but are not of equal magnitude. Both depend on the size, shape, and position of the body, the velocity of fluid flow past it, and the density of the fluid (air is much less dense than water).

In swimming and other water sports, a third type of drag occurs: *wave drag.* Skin friction and profile drag are described here, and wave drag is described in Chapter 12.

Skin Friction

Skin friction, or *surface drag,* is the drag caused by the fluid tending to rub along the surface of the body. Actually, the thin layer of fluid in contact with the solid surface of the body does not slide; rather, it sticks to it and is carried along with the moving body. This layer in turn tends to tow along the adjacent outer layer of fluid, which then tugs on the next outer layer, and so on. Layer by layer, the differences in speed and in rubbing (shear) force become progressively less until the outer layers have no sliding tendency at all. The region of relative motion between adjacent layers of fluid particles is called the **boundary layer**. All this layer rubbing near the surface of the body is caused by the friction among the fluid particles. Such friction varies from fluid to fluid and can be characterized by the fluid's resistance to deformation, or flowing resistance; technically it is known as *viscosity* (honey and molasses are highly viscous fluids). Air has low viscosity and, therefore, causes little layer-rubbing friction. Water has a greater viscosity than air and causes greater layer-rubbing skin friction. Skin friction is also known as *viscous drag.*

Picture a body moving through a fluid so that the fluid flows around the body smoothly and with no wake or other disturbance. Such a smooth flow pattern is called *streamlined* or *laminar* (smooth, layered) flow because the layers of fluid around the body's surface remain smooth and are not forced into turbulence (Figure H.3).

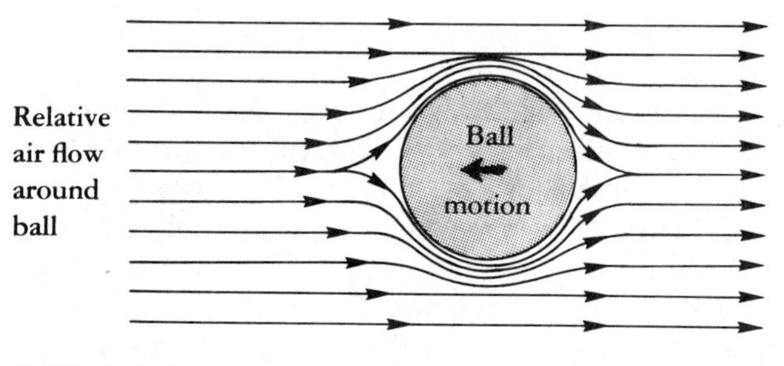

FIGURE H.3

A laminar or streamlined flow of air moving past a ball traveling very slowly through it. The layers of air pulled along with the ball's surface resist this pulling with the viscous friction forces and constitute surface drag (skin friction) against the ball's motion.

Skin friction drag will increase as the velocity of the flow increases, as the amount of surface area oriented parallel to the flow increases (more surface to tow more fluid), and as the roughness of the surface increases. The larger the body and the rougher the texture of the surface, the greater the skin friction drag. With such slow streamlined flow past a body, the most influential type of drag would, indeed, be skin friction. However, in the range of movement activities we encounter, such laminar flow is destroyed for most of them. The body and objects used in sport activities do not travel slowly enough through air or water to sustain the streamlined flow pattern, so it becomes turbulent; wakes and swirls are formed, even though they may be invisible. When that happens, the flow conditions are such that large pressure differences occur, making pressure (profile) drag predominant. Indeed, it has been shown that profile drag is the source of most of the drag encountered in sport activities.

Profile Drag

In addition to the skin-friction drag acting along the sides parallel to a body's motion, is the other type of drag, **profile drag**. Profile drag is the main air-resistance force operating on skiers, cyclists, runners, and all projectiles. It is the main water-resistance force acting on a swimmer's body and on body parts in aquatic resistance exercise. This type of drag also is known as *pressure drag* or *form drag*, each name being based on a factor that causes the drag. In most sports, the fluid particles are flowing past the body or projectile at speeds fast enough to produce a pattern of flow such that the fluid pressure on the leading surface of the body becomes greater than the pressure on the trailing surface. (Figure H.4a illustrates these pressure zones for a ball moving through air.)

The velocity of the air flow past the object is too fast for the air to follow the contour of the trailing side of the object as it cleaves its way through the air

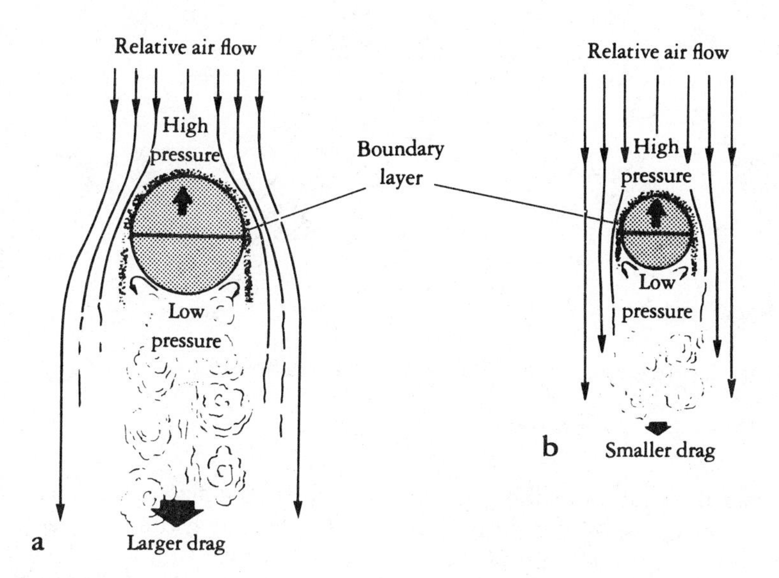

FIGURE H.4

A turbulent flow of air is created by most bodies moving through the air over a wide range of velocities. (a) An area of higher pressure exists on the leading surface, and an area of lower pressure exists on the trailing surface. (b) Profile drag is so named because the area exposed to the flow helps determine the magnitude of the drag force: A smaller ball experiences less drag than a larger one.

particles. When this happens, a "back flow" occurs at the surface of the ball, which causes the flow to separate from the surface contour. This **boundary layer separation** causes a large, turbulent low-pressure zone to be formed behind the ball. This region of low pressure is continually being formed as the object moves through the air. Such a flow pattern is called *turbulent flow*. Although it is invisible, turbulent air flow may be likened to turbulent water flow, which can be seen easily by sweeping the hand rapidly along the surface of water. The motion of the air or water that is "downstream" of the moving object is caused by the motion of the object relative to the surrounding fluid. The cause of the turbulent motion behind a moving body is the force applied by the object to the fluid as it pushes its way through. According to Newton's law of action and reaction, the fluid applies an equal and opposite reaction force against the object. This counterforce is the resistive drag against the object or body.

Profile drag is the name given to such drag because the *profile*, or area exposed perpendicular to the approaching flow, partly determines the magnitude of the drag. Thus, a small ball will experience less drag than will a larger ball moving at the same speed (Figure H.4b).

Profile drag is also called pressure drag, because it is the *pressure difference* between the leading and trailing surfaces of the object that creates a suctionlike

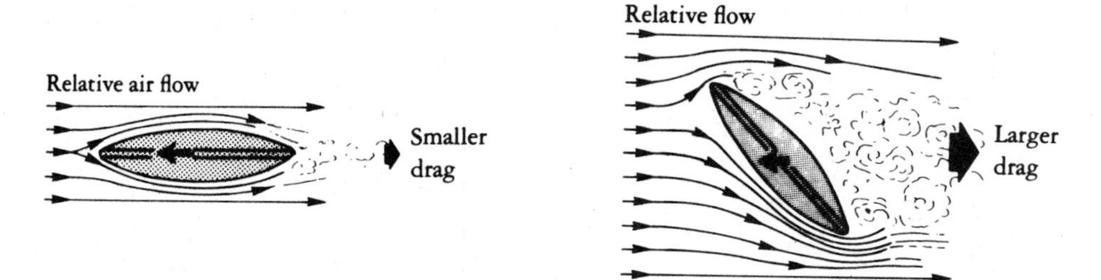

FIGURE H.5

A discus in two different orientations relative to the air flow. As the area facing the flow increases, the drag force increases.

effect acting against the motion of the projectile. Actually, the size of the low-pressure zone on the rear side of a body determines how much drag will act against it. The more abrupt the *change* in shape toward the rear of a body, the greater the magnitude of the suction effect. The orientation of a body relative to the fluid flow influences the area facing the flow, as well as the shape, such as when the air flow encounters the broad area of a discus or its edge. The illustration in Figure H.5 shows that the shape alone has no advantage in reducing drag unless that shape is positioned properly relative to the fluid flow.

Effects of Streamlining on Drag

Form drag is another name given to profile, or pressure, drag because the *form* or *shape* of the body determines how smoothly the object can cut through the fluid. Objects that are streamlined in shape are those that tend to create a streamlined flow pattern so that turbulent flow is minimized and laminar flow is sought to be maintained.

Figure H.6 illustrates the effect that streamlining an object has on reducing drag. Behind the sphere in Figure H.6a is a large zone of low pressure filling with turbulent fluid. But for the streamlined shape shown in Figure H.6b, this zone is filled with the object itself, and fluid is able to flow gradually backward and inward toward the tapered end. The less abrupt its contours, the more smoothly a body can cut through the fluid, without turbulent flow and an accompanying large pressure difference.

So important is the shape of a projectile in determining the drag that much attention is given to streamlining projectiles such as the discus, football, javelin, arrow, and bullet. Efforts also are directed to streamlining the body position when the body moves with such speed that air resistance is a major resistive

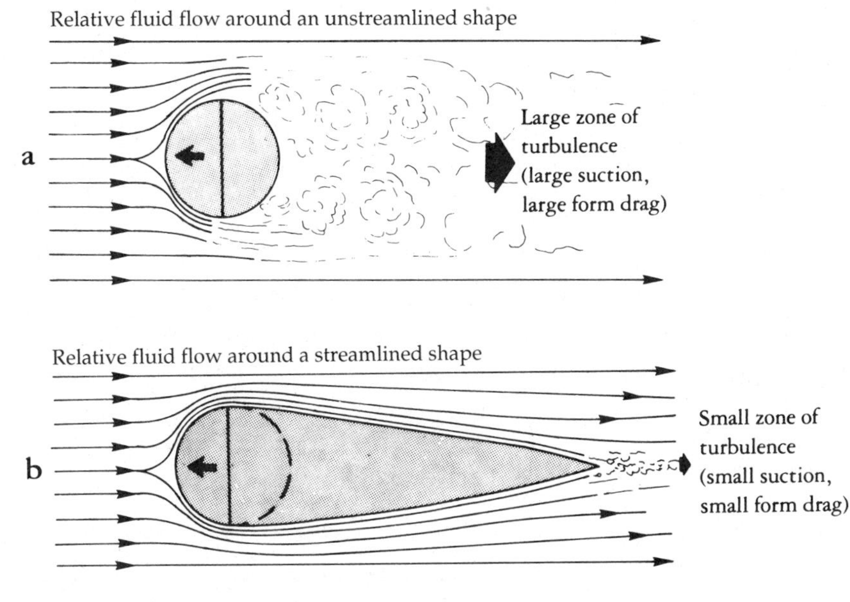

FIGURE H.6

Fluid flow (air or water) around (a) an unstreamlined shape creates much greater form drag on the body than it does around (b) a streamlined, or tapered, shape.

force as in speed skating, cycling, sky diving, and slalom skiing. Such streamlining efforts are guided by streamlined examples in nature. A tapering shape is characteristic of effective streamlining, such as the shape of a fish or the cross section of a bird's wing.

Recall, however, that a streamlined shape is streamlined *only* if it is positioned relative to the flow so that the fluid can pass around it with minimal disruption. If the streamlined object or body is moving through the fluid in an orientation, or position, that causes turbulence, it is no longer considered streamlined. Figure H.7 shows a streamlined javelin moving in a streamlined orientation and in a nonstreamlined orientation that increases the drag. Similarly, the human body may be considered streamlined with relatively low drag if it moves through a fluid from head to toe, but it is not streamlined if the flow passes front to back (e.g., a vertical dive water entry versus a belly flop). Downhill skiers and speed skaters try to perform in body shapes that are most streamlined in addition to trying to keep their frontal areas small. When a body is very streamlined, it is said to have a very small "coefficient of drag."

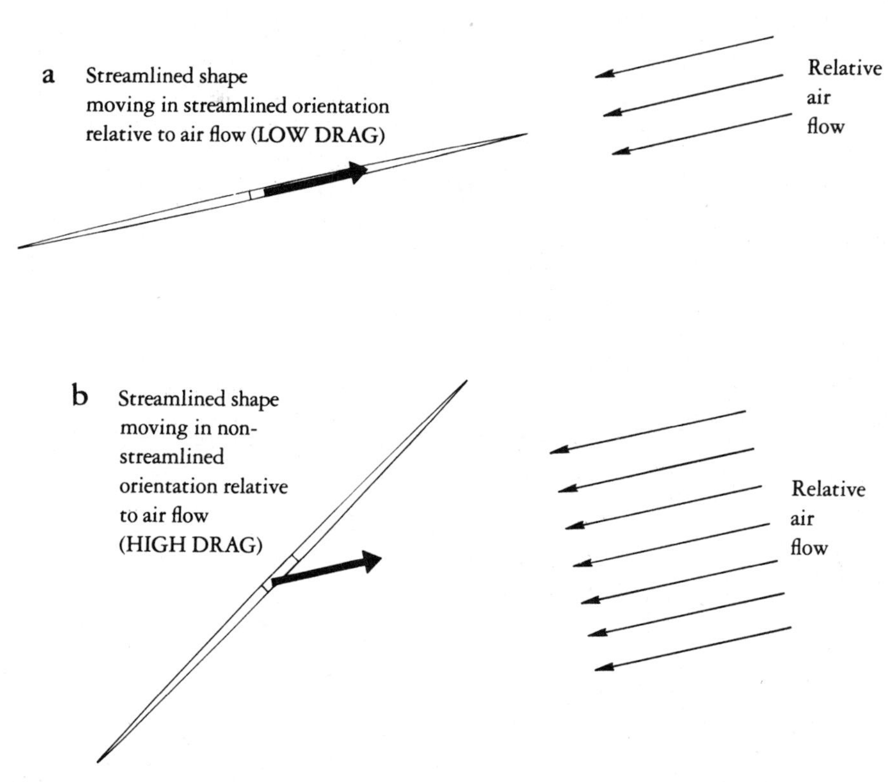

FIGURE H.7

(a) A javelin has a streamlined shape that allows for minimum drag only if it encounters the air flow in a streamlined position. (b) If it is moving at an angle to its motion direction, it is no longer in a streamlined position, and a large drag force results.

The Relative Influence of Factors Causing Drag

Examples have been given demonstrating how a number of factors contribute to fluid drag force. Putting them all together, the relationship of all the variables determining drag can be expressed as follows:

EQUATION H.1

$$\text{Drag force} = \tfrac{1}{2}\ \text{coefficient of drag} \times \text{area} \times \text{fluid density} \times \text{flow velocity}^2$$

$$F_D = \tfrac{1}{2}\, C_D\, A\, \rho\, v^2$$

where F_D is drag force, C_D is coefficient of drag (an index of "streamlinedness"), A is area facing flow, ρ (rho) is mass density of air, and v is relative flow velocity.

Coefficient of Drag

The coefficient of drag, C_D, for the human body and odd-shaped projectiles must be found experimentally in wind tunnels because it changes for every change in orientation of the body relative to the flow. A body's C_D tells how streamlined it is. Recall that whether a body or object is considered streamlined depends not just on its shape but on how that shape is presented to the passing flow. In tables based on experimental data, the C_D for a smooth-surfaced sphere is approximately 0.4. For a parachute, the C_D is approximately 1.2. The C_D for an adult male in running postures is estimated to be approximately 1.1 (Shanebrook and Jaszczak, 1976). For a discus, the C_D changes with the tilt of the discus relative to the flow. If it is inclined at an 11-degree angle, $C_D = 0.10$, but when it is tilted at 35 degrees to the flow, $C_D = 0.68$ (Ganslen, 1964). Similarly, as a javelin changes its inclination to the air flow, its C_D changes and is about 0.18 at a 20-degree angle (Ganslen, 1967).

Usually we can assume that an object's C_D will not change for a particular shape relative to flow and for speeds that the object travels in a particular event. In a given activity, however, when the human body moves within a wide range of possible speeds, such as in running, speed skating, cycling, and skiing, this C_D may change as the speed changes (Ingen Schenau, 1982; Shanebrook, 1974; Shanebrook and Jaszczak, 1976). The reason for this C_D difference is based on the changing dynamics of the fluid flow represented by a quantity called the Reynolds number.[1] Greater technical detail is not appropriate here, but it is important to recognize that the C_D can vary with body build, segmental movements, and for different ranges of speed of the body.

Area Facing the Flow

The C_D, which reflects the degree of streamlining of a body's *shape,* should not be confused with the area, A, of the body facing the flow. If the area facing the flow (frontal area) increases because of a change to an unstreamlined position, the drag is increased by the greater area *in addition to* the increased C_D. Increasing the area facing the flow increases the drag proportionately. For example, if the area is doubled, the drag is doubled.

Fluid Density

The density of an object or body is equal to its mass divided by its volume. Fluid density is the fluid's mass divided by a specific volume of it, such as a cubic foot or liter (mass/volume). It is a measure of how compactly the fluid's atoms and molecules are arranged to form the fluid. For example, fresh water

[1] For further technical information on flow dynamics, the reader is referred to Roberson and Crowe (1975).

(1 kg/liter) is almost a thousand times denser than air (.0013 kg/liter). Therefore, if a body were moving through air as compared to water, the drag would be extremely different. The density of water does not change, because it cannot be compressed. Air density does change, however, because it can be compressed by the weight of the atmosphere above it. Air density also decreases with an increase in temperature. It is about 12 percent less at 95 degrees F than at 30 degrees F. It is also decreased by increases in humidity but only slightly. The variability of air density with altitude, however, does not account for significant changes in drag when compared to the variability of the other factors.

Velocity

The C_D and the frontal area become more important at faster flow velocities because they are multiplied by the velocity *squared*. In the drag equation, the drag force increase can be seen to be proportional to the square of the velocity. To illustrate, suppose a skier goes downhill at 5 m/sec and encounters a drag force of 45 N (10 lb). If the velocity is *doubled* to 10 m/sec, the drag will be *quadrupled* to 180 N (40 lb). Such is the effect of the square of the velocity ($5^2 = 25$ and $10^2 = 100$).

Effects of Drag on Different Masses

Some objects and bodies cannot be streamlined because of the nature of the object or sport (e.g., running, gymnastics, a baseball, volleyball, tennis ball, and the badminton shuttlecock). A sphere is one shape that provides a great deal of air resistance; and in sports, the size and weight of each ball are standardized within certain limits by governing national and international organizations. We have seen how the size (profile) of a ball can influence the drag and thereby the nature of a particular game, but how is the mass of a ball important in air resistance? The mass of an object moving through air or water is important in determining only the amount of retardation, or deceleration, which is caused by whatever drag force is imposed on the object or body. Increasing or decreasing the mass of a body has no effect on the magnitude of the drag force acting against it. The mass of the body, however, determines, in part, how motion will be affected by that drag force.

Consider, for example, a Ping-Pong ball and a golf ball. The sizes are equal, but the masses are noticeably different. If you were to drop each ball to the floor from an elevation of 3 ft, each would take the same amount of time to hit the floor. The downward motion of the balls would serve to create an upward flow of air of a relatively low velocity because the balls would not have sufficient time to accelerate to some appreciable speed. If, however, the two balls were dropped from a higher elevation, say 10 ft, gravity would cause each to accelerate downward for a long enough time for the balls to gain appreciable speed. A difference in drop times would be evident: the lesser mass would be retarded more than the greater mass, even though the upward resistive drag force on each was the same. That is, the same drag force "seems" larger for the Ping Pong ball because its low mass makes it very susceptible to motion change. Its small weight is easily matched by the upward drag force, whereas the golf ball must fall much faster before its weight is matched by the drag force. This should not be surprising, for recalling Newton's law of acceleration, $a = F/m$, the acceleration of an object subjected to a given net force is inversely proportional to that object's mass; therefore, the smaller mass would be affected more by the same drag force.

A similar expression exists for the amount of retardation of a body moving horizontally, which is not affected by acceleration that is due to gravity: Deceleration of the body is equal to the resistive force of drag divided by the mass of the body. This expression means that the *slowing effect* is directly proportional to the drag force on the body (the greater the area facing the flow, the greater the drag force) and that the decelerating effect is inversely proportional to the mass of the body. The difference in mass explains why a basketball travels faster when thrown than a balloon or why a whiffle golf ball travels more slowly than a regulation golf ball. The difference in area and shape explains why most people are more likely to be pushed over when facing a high wind than if they are turned sideward to it.

Fluid Drag as a Propulsive Force

The preceding discussion of fluid drag has portrayed drag only as a resistive force acting on the body or sport implement. Drag, however, can act as a motive, or propulsive force, as well. The only requirement is that the drag force acts in the same direction that the body is moving. This happens in circumstances such as the following:

A sailboat "running before the wind" is pushed by air drag against the resistive water drag on the hull.

A dropped Ping Pong ball is pushed to the side as it falls in a sidewind.

The blades of oars pushing backward through the water experience a forward-directed drag force to serve as a "slipping" anchor point for the blades; the shaft of each oar, connected to the boat, forces the boat forward.

Standing at the top of a hill, a skier can get an early start down when the drag force from the blizzard wind acts on his or her body.

Understanding Fluid Drag Force

1. Under which condition would you experience more resistive drag?
   a. You're running at 10 mph in still air.
   b. You're running at 8 mph with a 2 mph tailwind.
   c. You're running at 7 mph with a 2 mph headwind.
   Explain your answer.
2. Which factor in the drag equation affects the magnitude of drag force more than the others? Why?
3. How do the coefficients of drag of a basketball and a softball compare? If both were moving at the same speed through the air, which would experience a greater drag force? Explain your answers.
4. Of the two types of fluid drag, which one accounts for the greatest part of the total drag acting on the body or implement in sport activities?
5. What is the boundary layer? What is meant by boundary layer separation?
6. Describe the difference between a fluid's density and its viscosity.
7. In what direction is air flowing past a ball at the instant it is projected at 45 degrees above the horizontal? What effect does the drag have on the ball's vertical and horizontal release velocity?
8. What is the direction of air flow past the ball in question 7 when it reaches the peak of its flight path? What is the effect of the drag force on the ball at the peak?
9. How does streamlining affect the air flow past a body so that drag is decreased?
10. Drop a racquetball and a badminton shuttle at the same time from a height of 2 ft and note which one lands first (the landings may be simultaneous). Repeat from heights of 4 ft, 6 ft, and 8 ft. Explain the results in terms of factors affecting acceleration due to gravity and aerodynamic drag force.
11. Repeat the experiment in question 7, using a tennis ball and a volleyball and then a small inflated balloon and a large inflated balloon. Explain the results.
12. Hold a piece of cardboard approximately 1 ft square, and move it quickly through a large arc so that it meets the air with its broadest surface. Repeat the movement with the edge of the cardboard meeting the air flow. Compare the difference in resistance felt. Repeat at different speeds and make comparisons of the resistance encountered.

H.2 Fluid Lift Force

Whenever a body or object moves through a fluid, it experiences a drag force opposing its motion. Under certain conditions a second type of force, **aerodynamic lift,** also will act on the object. Lift force, however, is always directed *perpendicular* to the oncoming flow, not parallel to the flow as is drag force.

Lift forces influence the trajectories of projectiles with particular shapes, such as the discus, javelin, and Frisbee, as well as ski jumpers and sky divers. The word *lift* suggests that all lift forces are directed upward; this is not necessarily true. *Horizontal lift* forces act to propel wind-dependent bodies such as sailboats and wind surfers. Fluid lift force is also used for propulsion by bodies whose parts move to produce lift force, such as propellers, canoe paddles, and human hands and feet in swimming.

In sport, lift force acts on bodies and objects whose *shapes* or positions influence the pattern of air flow around them. Lift force also acts on a projectile whose *spin* influences the pattern of air flow around it. Lift force created by spin of projectiles, such as seen in a pitched curveball, will be presented in Chapter 11 because it applies specifically to projectiles moving through air.

Lift Force Acting on Shapes and Surfaces

To understand how lift force is generated, the fluid flow relative to the moving object must be examined. The nature of air flow past the airfoil (wing-shaped object) in Figure H.8 demonstrates the concept of lift force. As air flows past the airfoil, it flows faster over the upper curved surface than the flatter surface underneath the object. The flow–velocity difference existing between opposite sides of an object causes a pressure difference between the two sides. The relationship between flow velocity and pressure is an inverse one, and is expressed in **Bernoulli's principle**[2]: *Where the flow velocity is fast, the pressure is low; where the flow velocity is slow, the pressure is high.* The existing pressure difference causes the wing shape to experience a force directed from the region of higher pressure to the region of lower pressure. Note that the direction of this lift force is *perpendicular to the flow direction* past the object, whereas the drag force is directed parallel to the flow direction.

Angle of Attack

If a body or projectile does not have the shape of a wing (curved on one side and flattened on the other), the difference in flow velocity (and pressure) may be achieved by tilting the object relative to the direction of flow past it. The angle

[2] Daniel Bernoulli (1700–1782) was a Swiss scientist who identified the inverse relationship between fluid flow velocity and pressure.

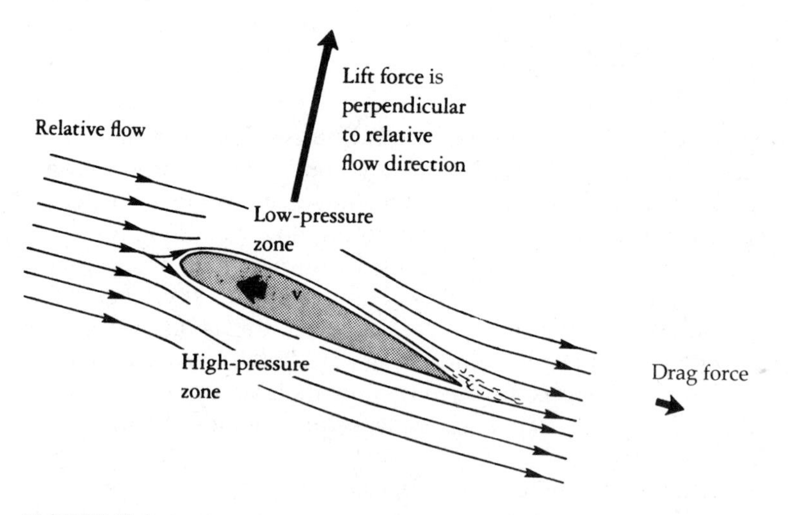

FIGURE H.8

Aerodynamic lift force acting on an airfoil (wing) is not necessarily upward. Lift force is perpendicular to the relative air-flow direction and is directed from the high-pressure zone toward the low-pressure zone. Drag force acts parallel to the relative air flow and against the object's motion.

formed between the main plane (chord) of the object and the flow direction is called the **angle of attack**. Figure H.9 shows a discus during flight, oriented so that its plane forms an angle of attack with the relative flow direction.

With no angle of attack, the flow velocity, and thus the pressure, is the same on both sides; all the air force is drag force, and no lift force is generated (Figure H.10a). As the angle of attack increases, the difference in flow velocity and pressure on opposite sides also increases, and lift force is generated (Figure

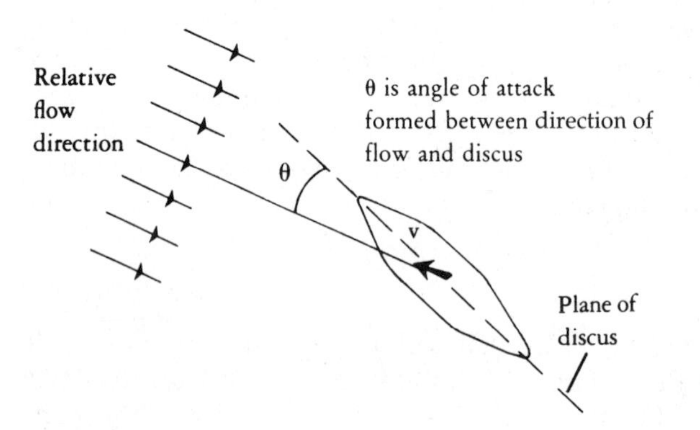

FIGURE H.9

The angle of attack, θ, is formed between the main plane (chord) of the discus and the relative flow direction.

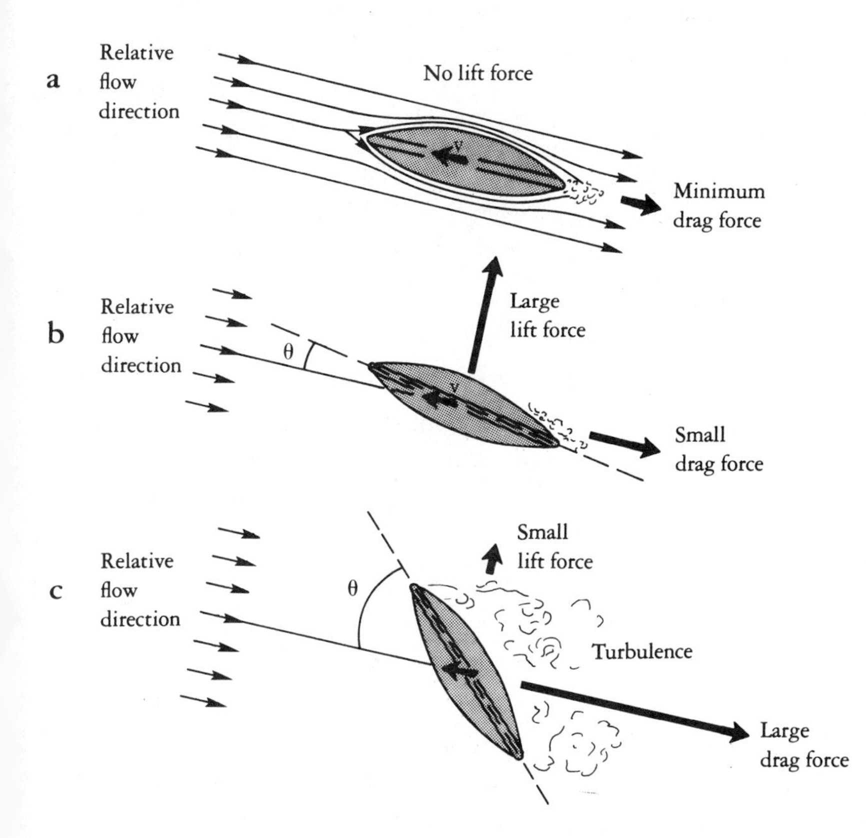

FIGURE H.10
Changes in the angle of attack cause changes in the lift and drag forces acting on a discus. (a) Zero angle of attack produces minimum drag force but produces zero lift force. (b) A large lift force is produced with a small angle of attack, and the drag force is increased slightly. (c) When the angle of attack increases beyond the optimum lift-producing angle, the lift force diminishes, and the drag force increases greatly, leading to stall.

H.10b). The increase in lift force will occur with a small angle of attack, but only up to a critical maximum angle, beyond which the lift force suddenly decreases and the drag force increases greatly due to extreme turbulence (Figure H.10c).

This critical angle, approaching 30 degrees for the discus (Ganslen, 1964), is called the **stall angle**. The object stalls as a result of increasing drag force and the loss of lift force that held it up against gravity, and it falls toward the ground. A Frisbee acts much like a lightweight discus, and if launched at even a small incline, it rotates to its stall angle very abruptly.

The best angle of attack for producing the greatest lift force with the least drag force depends on the shape of the object and the relative speed and direction of air flow. The angle of attack changes throughout the flight path as the relative air flow direction changes as the object ascends then descends.

The angle of attack must be considered in other activities using fluid lift force.

For example, a sailboard sail is positioned at an angle to the wind so that it acts as a wing, but the lift force is propulsive horizontally. A vertically positioned canoe paddle slices horizontally through the water with a small angle of attack; this produces a horizontal lift force on the blade as it "sculls" to move the canoe broadside. Large sculling movements of the hands are used in swimming strokes, treading water, and synchronized swimming skills. For all hand sculling, the angle of attack changes with changes in hand speed, size, and shape.

The Relative Influence of Factors Causing Lift

Just how much lift force can be produced depends on a number of factors. The variables that influence the lift force acting on a lift-producing surface are similar to those influencing drag. Their relationship is expressed as follows:

EQUATION H.2

$$\text{Lift force} = \tfrac{1}{2}\ \text{coefficient of lift} \times \text{area} \times \text{density} \times \text{velocity}^2$$

$$\text{or } F_L = \tfrac{1}{2}\, C_L\, A\, \rho\, v^2$$

where F_L is lift force, C_L is coefficient of lift (an index of the shape's "lift producibility"), A is area of the body on which pressure acts, ρ (rho) is mass density of air, and v is relative flow velocity.

The Coefficient of Lift

The lift coefficient, C_L, is an object's index of how well it can create lift force in fluid flow. For example, a wing has a much greater C_L than a book. The C_L changes, however, as the angle of attack changes; the C_L of any lift-producing body or implement depends on the shape characteristics and angle of attack of that body or implement. For example, a javelin traveling at 80 ft/sec with a 20-degree attack angle has a C_L of 0.25 (Ganslen, 1967). For a discus during flight, the C_L is about 1.2 at a 26-degree angle of attack; it decreases to $C_L = 0.1$ at a 2-degree attack angle, and to $C_L = 0.7$ (stall condition) at 35 degrees (Ganslen, 1964). These figures for the discus demonstrate the importance of *optimizing* the angle of attack for maximum C_L throughout the line of flight.

Velocity

As with drag force, lift force increases with the *square* of the flow velocity; that is, if the velocity is *doubled*, the lift force is *quadrupled*. For resistive drag force this relationship is a disadvantage. For lift force, which is usually desired, the squared effect is an advantage, however. Therefore, the faster a lift-producing shape moves through the air, the greater the lift force. The importance of fast relative flow velocity is demonstrated by pilots who take off against a wind rather than going with it. If no wind is present, the plane must accelerate to a speed fast enough to generate a lift force greater than the weight of the plane.

Area

As with drag force, an increase in area of the body will increase the lift force. This area, however, is *not* the area facing the flow (frontal area) but is the surface area that is angled to the flow to form the angle of attack. The dimensions of sport implements are usually specified to keep their areas constant. However, shapes and areas will vary among human beings seeking lift force in activities such as ski jumping, board sailing (wind surfing), and swimming. Depending on the hull and intended use, propeller blades for boats and aircraft will also vary.

Fluid Density

With all the other factors kept the same, the more dense the fluid, the greater the lift force. The effects of the lower density air at high altitude, therefore, can diminish the lift force and distance achieved by the same discus or javelin throw performed at sea level (unless the *advantage* of decreased drag cancels the disadvantages of decreased lift).

Lift/Drag Ratio

We have seen how drag increases with the area of a body facing the flow and how the C_D changes with the object's position relative to the flow. Also, we have seen how angling a lift-producing shape gives it an angle of attack necessary for lift force. The situation that arises, however, is that such tilting for an angle of attack creates a greater frontal area for greater drag and also increases the C_D. The target condition often is to try to have the body or object move through the fluid so that the lift force is *maximized* and the drag force is *minimized*. The object will always have drag force acting against it but not necessarily lift force. For each body, then, experimentation can determine the *lift/drag ratio* for different angles of attack. The value of this ratio is greatest when the angle of attack and lift force is the greatest it can be without overly increasing the drag force (i.e., when the lift force is *optimized*). As mentioned previously, the angle of attack of projected objects is constantly changing during their flight paths, and therefore, the lift/drag ratio changes also. The same is true for other lift-producing bodies whose angle of attack changes during a performance (sculling hands and oars, sails, etc.).

Understanding Fluid Lift Force

1. State Bernoulli's principle relating pressure to velocity of fluid flow.
2. In which direction relative to fluid flow direction past a wing-shaped body does lift force act?
3. Describe how a difference in pressure occurs on opposite sides of a lift-producing body.
4. Sketch a wing-shaped body moving through a fluid so that a lift force acts on it in a downward direction.
5. Sketch a paper plate that is traveling through air with a small angle of attack. Label the angle. Draw in the lift force.
6. Assemble a straw, a spool of thread, a card, and a thumbtack, as shown in Figure H.11. (A piece of corrugated cardboard, thick enough to hold the end of the straw, can be substituted for the spool of thread.) Place the card against the spool so that the tack prevents side-slipping of the card on the spool. While holding the card to the spool, blow through the straw, and then release the card while continuing to blow. The card stays on the spool while you are blowing and drops off when you stop blowing. Explain this in terms of Bernoulli's principle.

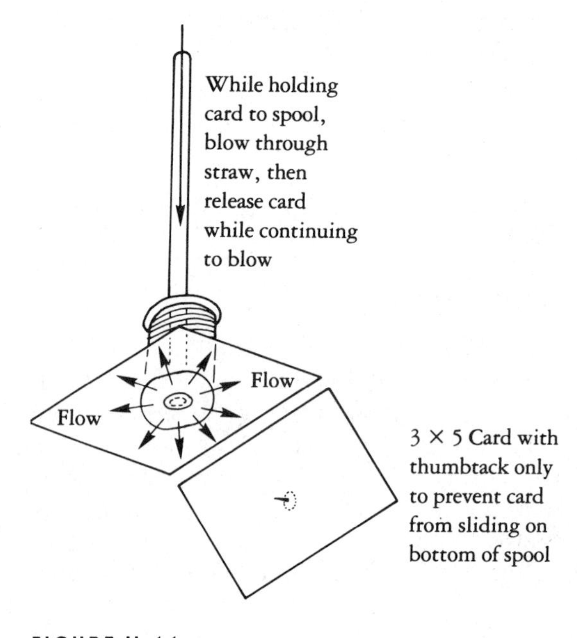
While holding card to spool, blow through straw, then release card while continuing to blow
Flow
Flow
3 × 5 Card with thumbtack only to prevent card from sliding on bottom of spool

FIGURE H.11

Bernoulli's principle explains the result of this experiment.

References and Suggested Readings

Alexander, R. M. (1968). *Animal mechanics.* Seattle: University of Washington Press.

Brancazio, P. J. (1984). *Sportscience.* New York: Simon & Schuster.

Daish, C. B. (1972). *Learn science through ball games.* New York: Sterling.

Ganslen, R. V. (1964, April). Aerodynamic and mechanical forces in discus flight. *Athletic Journal,* pp. 50–51, 68, 88–89.

Ganslen, R. V. (1967, December). Javelin aerodynamics. *Track Technique,* pp. 940–942.

Gustafson, D. R. (1980). *Physics: Health and the human body.* Belmont, CA: Wadsworth.

Roberson, J. A., & Crowe, C. (1975). *Engineering fluid mechanics.* Boston: Houghton-Mifflin.

Shanebrook, J. R., & Jaszczak, R. D. (1976). Aerodynamic drag analysis of runners. *Medicine & Science in Sports, 8*(1), 43–45.

Shanebrook, J. R. (1974). Aerodynamics of the human body. In R. C. Nelson & C. A. Morehouse (Eds.), *Biomechanics IV: Proceedings of the Fourth International Seminar on Biomechanics* (pp. 567–571). Baltimore: University Park Press.

Shapiro, A. (1961). *Shape and flow: The fluid dynamics of drag.* Garden City, NY: Doubleday.

Tricker, R. A. R., & Tricker, B. J. K. (1967). *The science of human movement.* New York: American Elsevier.

Walker, J. (1979a). Boomerangs: How to make them and also how they fly. *Scientific American, 240*(5), 162–172.

Walker, J. (1979b). More on boomerangs, including their connection with the dimpled golf ball. *Scientific American, 240*(4), 180, 183–186, 188–190.

CHAPTER 11

Applications of Aerodynamics in Sport

PREREQUISITES

Concept Modules B, C, D, E, F, G, H

CHAPTER CONTENTS

In many sport activities the effort used or the outcome of an event is influenced by *air forces*. Such activities are those in which the body or a projected object is moving fast through the air and a noticeable resistance is felt. Running, cycling, skiing, skating, surfing, and long jumping are some activities subjected to appreciable resistive air forces. In many slow activities, or those of brief duration, such as walking, tumbling, climbing, wrestling, dancing, and weight lifting, air resistance can be ignored or considered negligible. Some projectiles affected by air resistance are badminton shuttlecocks, baseballs, softballs, footballs, and javelins. For example, air resistance acting on punted footballs has been estimated to reduce the achieved distance to one half what it would be if the ball were kicked in a vacuum (Cunningham and Dowell, 1976). Air forces need not be only resistive to motion but can be used to advantage as propulsive and/or supportive forces.

11.1 Effects of Drag on the Body and Objects in Sport

Recall from Concept Module H that the aerodynamic drag force that acts on a moving body is influenced by its shape and surface qualities (coefficient of drag), the area exposed to the relative air flow, the density of the air, and the flow velocity relative to that body. These factors will have different amounts of importance depending on the particular activity examined.

Activities in Which One Approaches Terminal Velocity

When an object or body is dropped through air, water, or any other fluid, the force of gravity causes it to accelerate downward. At the instant the object is dropped, it has no velocity relative to the fluid and, consequently, experiences no drag force. If air is the fluid, gravity then causes the object to accelerate at a rate of 9.8 m/sec^2 downward, because the net motive force acting on the object is the pull of gravity, that is, the weight of the object. As the object starts to fall, however, it gains velocity under the accelerating force of its weight, and as it moves through the air faster and faster, it experiences a greater and greater drag force directed upward against its downward motion. Such a situation applies to the sky diver in Figure F.1 (p. 338). This drag force increases as the square of the velocity; therefore, by the time the falling object has tripled its velocity, the drag force has increased nine times. The drag force is an external resistive force, whereas the object's weight is the external motive force. As long as the drag force against the object is less than the weight of the object, the object continues to gain downward velocity, that is, to accelerate in the direction of the net force downward. The velocity of the falling object soon becomes so great, however, that the drag force acting upward on it increases to a value equal to the weight of the object. When such a velocity is reached, the net external force acting on the falling object is equal to zero, and no further acceleration occurs. The velocity at which the upward drag force is equal to the weight of the falling object is called the **terminal velocity**. Terminal velocity is approached gradually, because as soon as the object starts to experience some upward drag force, the *net* force acting downward to accelerate it is decreased (its weight minus the resistive drag), and the gain in downward velocity is less than 9.8 m/sec^2. The greater the drag force becomes, the smaller the downward acceleration becomes, until, finally, the terminal velocity is reached, and the remaining descent is at this constant velocity.

The concept of terminal velocity is useful in understanding the differences in falling times of various projectiles, including the human body. The lighter the object, the sooner the terminal velocity is reached if the other drag-determining variables are the same (velocity, shape, area, fluid). One would expect, therefore, a feather badminton shuttlecock to reach a terminal velocity sooner than a plastic shuttle of slightly greater weight. Similarly, a light sky diver will reach a terminal velocity sooner than a heavy diver who has the same body configuration and the same surface area facing downward (facing the air flow). The terminal velocity of a sky diver also varies with the nature of the maneuvering being done. A sky diver is able to regulate the falling speed by changing the orientation of the body relative to the air flow. A horizontal spread-eagle position, for example, exposes the greatest surface area to the approaching air flow and causes the most upward drag force, whereas a vertical streamlined body position causes the least. Thus, the latter body position needs a greater downward velocity to generate enough drag force to counteract the body weight. The greater the surface area, the smaller the terminal velocity. When skydivers work in groups, they may change the arrangement of body segments and positions to "wait" for or to overtake other members. Various arrangements of body segments also are used for maneuvers such as horizontal spins and somersaults. These maneuvers are accomplished by increasing drag on one area of the body and decreasing drag on another to create a torque on the body for rotational motion. For the final portion of descent, the diver opens the parachute to expose an extremely large surface area to the air flow so that the terminal velocity is reduced to only approximately 32 km/hr (20 mi/hr) for landing. With modern parachutes, the diver can guide himself horizontally for a less abrupt vertical landing.

For some common projectiles used in sports, even if time is inadequate to reach a terminal velocity (due to insufficient height from which to fall), the reduced rate of velocity gain downward (due to the resisting drag force) will influence the object's time in the air. A badminton shuttlecock, for example, does not fall to the ground from a high clear shot as rapidly as might be expected because of its low mass and high drag characteristics. The change in horizontal shuttle velocity in a drive or smash shot is readily apparent to the observer on the sideline. The high velocity with which it leaves the racket is more suddenly reduced than its downward velocity because of the absence of any horizontal accelerating force acting on it. Such a horizontal velocity change should not be confused with the attaining of terminal velocity, which occurs when the motive force equals the resistive force. The velocity decrease in the horizontal direction will continue as a result of the drag force, until the shuttle lands.

Terminal velocity can occur in any direction, not only vertically. The conditions for a body reaching a terminal velocity are merely those that specify linear equilibrium; that is, the motive forces must be equal to the resistive forces. In bicycling, boating, and other potentially high fluid-drag-force activities, therefore, a terminal velocity will be reached when the fluid-drag force and other resistive forces match the motive force being applied by the mover or machine. Note that for nonvertical terminal velocity to be reached, the moving object or body must be capable of generating continuous propulsive forces against the continuous resistive forces; otherwise, progressive deceleration that is due to a net resistive force will occur.

The Floater Volleyball Serve

A unique application of the effects of aerodynamic drag force on an object is illustrated by a floater serve in volleyball. The floater serve is identified by its lack of spin and its irregular pattern of deviations right and left during its flight. The large circumference of the ball with its relatively small mass makes it particularly susceptible to the influence of aerodynamic forces. A manufactured ball is never perfectly round; an air valve is located on one side of it, and the seams are arranged in a pentagonal pattern on the surface. A feasible explanation for the action of the floater serve is that since the surface texture and shape of the ball are so irregular, the presentation of a symmetrical surface on opposite sides of the ball at any one time is improbable. Because of the irregularity of the opposite surfaces as the ball travels forward, the air flow on opposite sides of the ball is irregular and turbulence is created asymmetrically on either side of the ball. The low-pressure zone created in the turbulent areas tends to shift around on the trailing surface of the ball, thus producing a drag force whose direction shifts accordingly and produces the erratic flight. For instance, if turbulence is induced earlier on the right side than on the left side as a result of the irregular surface on the right side, the pressure on the right side will be less than that on the left side, and the ball will be forced slightly to the right. This movement will cause a slightly different surface to be presented to the air flow. Depending on the surface irregularities on the new opposing sides, the ball will again move to the right, to the left, upward, or downward.

The Knuckleball Pitch

The aerodynamic forces on a good (erratic) knuckleball pitch are similar to those on a floater serve. A detailed experiment by Watts and Sawyer (1975) indicated that the best explanation for the midflight shifting of the baseball is that the ball spins very slowly and the location of the rough seams is changed relative to the flow. This causes a shifting of the turbulent wake from one side to the other and a consequent lateral force with each shift. Watts and Sawyer point out that too much spin would prevent these wake shifts and the resulting deflection; the most effective knuckleball rotates much less than one revolution on its way to the plate. To throw such a pitch, the ball is gripped with the first knuckles or fingernails so that little or no torque is applied to the ball at release to cause it to rotate.

The Golf Ball

The drag force on a driven golf ball would seem to be increased by the rough surface provided by the dimples. Actually, at the high speeds attained by a golf ball, the *total* drag *decreases* because of the dimples. This phenomenon is explained by the fact that dimpling, or roughening the otherwise smooth surface, causes the thin boundary layer to become turbulent earlier than it would on a smooth ball right next to the ball's surface. The result of this is a later separation of the boundary layer farther back on the rear portion of the ball. This creates a much smaller suction zone behind the ball, and the profile drag is drastically reduced. The skin friction drag is increased only slightly by the roughness. The *total* drag (profile + skin friction = Total) is therefore less.

The Hammer Throw

Maximum distance of projection is the purpose of the hammer throw in track-and-field competitions. The diameter (and frontal area) of the metal sphere on the end of the handle and wire influences the drag force acting on it. Rule revisions effective in 1981 changed the minimum and maximum diameter of the hammer head from 102 mm and 120 mm to 110 mm and 130 mm. With the larger heads, the distances would be expected to decrease because of greater air resistance. Dapena and Teves (1982) studied the problem and found that the 110-mm-minimum-diameter head (compared to the 102 mm) could indeed reduce the distance by 0.20 m as a result of the increase in drag. Use of the 130-mm head could reduce world-class distances of about 80 m by 0.5 m. The minimum mass (7.260 kg) and a C_D of 0.4 for the different heads were used in Dapena and Teves's computer prediction as well as release velocities between 20 m/sec and

32 m/sec. Other factors enter into the distance achieved, but the influence of drag is a factor worth considering in selecting the implement.

The Shot Put

Air resistance is usually neglected in the shot put because of the large mass of the shot and the relatively slow release velocity. After calculating the retarding effect of drag for a world-class put, however, Lichtenberg and Wills (1978) found that with a given speed of release, a put could be reduced by over 5 cm by air resistance. The slower the speed of release, the smaller the drag and the smaller the retardation. As in the hammer throw, the projected (frontal) area of the shot influences the drag. Thus, in selecting a shot, the putter would want to select the smallest-diameter shot so as to reduce the area. Lichtenberg and Wills state that a world-class putter who uses the smallest permissible shot will have a "two inch advantage over a putter who uses the largest allowed shot, all other things being equal" (p. 548). Thus according to calculations, air resistance may play an important role in the shot put, just as it does in the hammer throw.

The Long Jump

Although air resistance certainly reduces the distance of a long jumper, the jumper can do little to decrease that drag and still perform the movements necessary for balance and landing position. Decreased air density, however, has been used in attempts to explain Bob Beamon's 1968 record-breaking long jump of 8.90 m (29 ft 2.5 in.) in Mexico City. Some believe the high altitude and thin air of Mexico City gave Beamon a gravitational and aerodynamic advantage, enabling him to take more than a half meter off the record. According to the analysts, the drag force was decreased due to the 24% lower density air compared with sea level air. Based on Beamon's .87-m^2 area facing the flow (A) and his C_D of 1.0, the analysts estimate that the thin air could not have allowed for an increase of more than 1% in Beamon's horizontal speed; this would have given him only 20 cm more distance. The other variable was g, the acceleration resulting from gravity. (Recall that g decreases at higher altitudes.) The reduced g was estimated to have given Beamon an advantage of only one tenth of 1% over what would be expected at sea level—a negligible advantage. The conclusion of the analysts was that Beamon had simply "put it all together" for that one jump and that the aerodynamic and gravitational advantages were not totally responsible for his accomplishment (Phinizy, 1968).

"Putting it all together" is one way of summarizing the conditions described by Hay (1975) as being responsible for the distance achieved: a very fast approach run (aided by a tail wind), a high angle of takeoff, and "almost perfect coordination of body segment motions at takeoff."

Running

Aerodynamic drag acting on runners has been estimated from studies on models by Shanebrook and Jaszczak (1976) and Dapena and Feltner (1987). Because of the constantly changing segment positions, the area and the C_D are also changing. Shanebrook and Jaszczak estimated the continuous average drag force on a world-class male sprinter traveling at 30 m/sec to be about 31 N (6–8 lb), depending on the size of the sprinter. Such a constant resistive force requires significant energy expenditure to overcome it. This total added energy requirement is not as detrimental in short-duration sprints as in longer distances. In longer races the velocity is slower and the drag is less, but the time over which the drag acts is much greater. A runner at higher elevations will experience less drag, as a result of the lower air density factor, than a sea-level runner (given that they both have the same C_D, body area, and velocity).

The practice of "drafting" (running in the wake of leaders) by runners in groups gives the trailing runners less drag and thereby conserves their energy over a distance. Running in a tail wind decreases the relative air flow velocity past the runner and decreases the drag. In sprinting, the presence of a 2 m/sec tail wind for the full distance can decrease a 100-m sprint time by 0.18 sec (Ward-Smith, 1985). A head wind increases the flow velocity and resulting drag. For a runner traveling around a track, a tail wind on one side would be a head wind on the other. The conditions would seem to cancel each other. Because of the squared effect of the relative flow velocity, however, the head wind causes much more hindrance than the tail wind provides help. For example, a runner traveling at 6 m/sec into a head wind of 2 m/sec on one side of the track experiences a relative flow of 8 m/sec. On the other side, the tail wind of 2 m/sec for the runner decreases the relative flow to 4 m/sec. The drag increases with the velocity squared, so it is four times as great in the head wind ($8^2 = 64$ and $4^2 = 16$) than in the tail wind. A side wind would be better on the straightaway.

Wind fluctuations are reported to lead to worse times than constant wind (Dapena and Feltner, 1987). Results of studies confirm that the time hindrance produced by a head wind is greater than the time aid produced by a tail wind of the same speed.

Kyle and Caiozzo (1986a) have conducted wind tunnel tests to study the effects of clothing and hair on the drag of runners. They found that long hair was the worst culprit, producing about one third more drag than a loose cotton jersey. Their results showed that by covering or trimming the hair or making the clothing more aerodynamic (smooth and tight-fitting), a runner could reduce drag at least 2%. Such a reduction would amount to a 0.01-sec faster 100-m time and a 0.08-sec faster 400-m time. The skin suits seen in the 1988 Olympic track events are evidence of efforts made to gain every aerodynamic advantage possible. Some fabrics, however, may pose a disadvantage. For example, Kyle and Caiozzo found a 2.9% drag increase with a leg model covered with nylon spandex compared to a bare leg. A loose cotton jersey and a cotton crew sock caused increases in drag of 15.0% and 12.4% respectively. The only fabric that reduced the drag of a bare leg was a fine wool jersey covering, which led to a reduction of 0.6%. These results were in a wind tunnel velocity of 8.89 m/sec (comparable to 100-m sprint speed). At faster speeds, however, the boundary layer behaves differently and different fabrics yield different results. Controlled drag tests would be the only way to predict the best fabric to use in different situations.

Cycling

Aerodynamic drag accounts for over 90% of the resistive force encountered by a racing cyclist (Kyle and Burke, 1984). Because cycling speeds are much higher than running speeds, the drag is much greater. Faria and Cavanagh (1978) report some theoretical predictions that demonstrate the effect of body position on speed. Using the touring position (Figure 11.1a) as a reference for comparison, if the work effort were the same and the rider pedaled at 24.2 km/hr (15 mph), he could increase his speed by 1.13 km/hr (0.7 mph) if he assumed the low racing position (Figure 11.1b). An upright (bell and basket) position, shown in Figure 11.1c, would cause a decrease in speed of 0.97 km/hr (0.6 mph) compared to the touring position. Therefore, the speeds that could be maintained with the same work effort in the three different positions would be 23.2 km/hr (14.4 mph), upright; 24.2 km/hr (15 mph), touring; and 25.3 km/hr (15.7 mph), racing. To travel at the same speed as in the racing position, the work effort would have to be greater in the touring or upright position.

a Touring position

b Racing position

c Upright position

FIGURE 11.1

Drag force changes with the frontal area of a cyclist riding in three different body positions: (a) touring, (b) racing, (c) upright.

Other data presented by Faria and Cavanagh point out the differences in drag force at different speeds as well as the greatly increased work effort required for only small increases in speed. For example, cycling against a relative air velocity of 10 mph, the drag is 0.86 lb; however, at 20 mph the drag increases to 3.43 lb, and at 25 mph, the drag is 5.36 lb. The power (rate of work performed) required to pedal against 5.36 lb is 0.33 horsepower (HP), which is approximately equal to the *maximum* steady state exercise power output for average individuals working 10 min or more. For a nonexhausting 1-hr ride, 0.1 HP is realistic.

With the preceding data in mind, the importance of minimizing drag for greater speed with less effort is apparent.

In coasting downhill, differences in velocity may be observed for two riders of apparently the same frontal area and drag force. The rider with greater mass would be retarded less than the lighter rider with the same drag force. If the two wished to remain together, however, the lighter rider could assume a racing position while the heavier rider remained upright.

In the four-person team pursuit, an Olympic cycling event, the lead rider cuts the wind, thereby forming a low-pressure channel within which the other three can follow closely (Figure 11.2). The drag reduction is considerable, about 40%, and members take turns at the lead to allow the former lead rider to recover. When drafting, the power required to maintain the same speed is only about two thirds of that used as a lead rider. The closer the following distance, the better. Riding in a group provides the same kind of advantage (Gross, Kyle, and Malewicki, 1983).

Loose clothing also can increase the drag of a rider, and rubberized skin suits specialized for cycling are used by both competitors and recreational cyclists to minimize their drag. Aerodynamically shaped helmets, tapering to a point over the shoulders, reduce the drag of a ride with long hair and no helmet by about 7%. Spandex shoe covers are used, as well as streamlined shoes, to reduce drag by almost .5 lb at 30 mph. It is estimated that, with all the aerodynamic improvements made, a maximum of 1 lb of drag force can be eliminated at 30 mph; such a decrease could reduce a 4,000-m pursuit race time of 4 min 50 sec by 13 sec (Kyle and Burke, 1984).

Experimentation and wind tunnel tests by researchers in Italy resulted in a "new bicycle." It was constructed with disc-like spokeless wheels, the front wheel being smaller than the rear, and it was lightweight with different proportions than typical bikes. The use of such a bicycle, along with an aerodynamic helmet and suit, contributed to the gold medal successes of the 1984 Italian 4 × 100-km cycling team (Dal Monte et al., 1987).

Human-Powered Vehicles

The extremely fast speeds achieved by human-powered vehicles (HPVs) have been made possible by enclosing the rider and cycle in a streamlined shell (fairing) to reduce drag. Mainly because of changes in aerodynamic design in the last 15 years, speeds have increased to close to 70 mph. The most improvement has occurred since 1976, when the International Human-Powered Vehicle Association (IHPVA) was formed, and competition inspired improvement. The recumbent position of the rider on a long, low cycle allows for a low profile as well as a streamlined fairing shape. The drag coefficient of a crouched racing cyclist is 0.88, which is eight times greater than the fairing-enclosed recumbent cyclist (Gross, Kyle, and Malewicki, 1983).

Cross-Country Skiing

In cross-country skiing, the effects of drag can be considerable, especially when added over

FIGURE 11.2
Drag is reduced for the riders drafting in the low-pressure zone formed by the lead rider. (Photo courtesy of the Bozeman Daily Chronicle, Bozeman, Montana, Cliff Willis, photographer.)

a long period of time. Posture, varying from upright to semisquatting, has a large influence on drag, as do the positioning and movements of the hands and arms. Drafting in groups or behind another skier has advantages for both the lead and following skier. Spring et al. (1988) found that a lead skier experienced 3% less drag with another skier drafting close enough behind to prevent the full formation of the low-pressure zone: The drafting skier's drag was reduced about 25% in that situation. The distance between the skiers is important. They found that drag is decreased by about 16% on a skier drafting 2 m behind, but it was decreased by 23% when the distance closed to 1 m.

The clothing of cross-country skiers is important also for the reduction of drag. In the range of skiing velocities, the roughness of the skiers' clothes can change the drag. Van Ingen Schenau's work (1982) showed that a smooth, tight-fitting suit produced more drag than a rough woolen suit for skaters at speeds of less than 6 to 7 m/sec. However, at speeds greater than 7 m/sec (e.g., cross-country skiing speeds), the smooth suit produced less drag. In certain velocity ranges, therefore, the roughness of the suit can change the nature of the boundary layer behavior and, consequently, the drag force. Such a phenomenon would be similar to the influence of dimples on a golf ball on drag.

Downhill Speed Skiing

In downhill ski racing, the purpose is to travel as fast as possible with control. Therefore, to try to *maximize* the terminal velocity would be to the skier's advantage. Figure 11.3 shows the resistive air drag and ski–snow friction that limit the skier's velocity increase caused by the motive force of gravity (the downhill component of the skier's weight force).

The variables that are under the control of the skier to change are:

1. Ski–snow friction. Reducing this friction is accomplished by proper ski surface preparation.
2. Shape and area of the body facing the flow. Reducing the area facing the flow and possible reduction of C_D are accomplished by crouching and by keeping the arms in close and the poles aligned parallel to the flow. Racing suits fitting tightly help reduce the skin friction. Aerodynamic covers for the boot buckles can reduce drag enough to take tenths of seconds off the time of a downhill ski run.

A third variable, which can be changed from one day to the next but not within a given run, is body mass. Recall that the deceleration of the body is inversely proportional to its mass. Gros (1979) found that a 75-kg skier with an added mass of only 1 kg could increase his terminal velocity from 30 m/sec to 30.2 m/sec. Such an increase would make a difference of about 0.5 sec over 2,500 m. The greater mass effect means also that for the same body size, volume, and area, a higher density (large mass with small volume) skier would have an advantage. (Recall from Concept

FIGURE 11.3

Aerodynamic drag force is the major source of resistance to a downhill skier's speed. Snow-to-ski friction is the second source of resistive force.

Module H the analogy of terminal velocity for a golf ball and a Ping-Pong ball dropped simultaneously.)

Downhill speeds of two skiers in egg-shaped and upright positions have been measured by Watanabe (1978), who found differences of over 1 m/sec even at slow speeds. Wind tunnel experiments using the same two skiers in a relative air speed of 10 m/sec showed a drag force averaging 1.6 kg for the egg-shaped position and an 8.0-kg drag for the upright posture. At an air speed of 30 m/sec, however, the body in the egg-shaped position encountered a drag of 11.8 kg (Watanabe, 1981), over seven times the drag experienced at 10 m/sec for a threefold increase in speed. (The same subject may not have been used in both the 1978 and 1981 reports.)

The modern aerodynamic helmets provide for a smooth flow of air around the head, cutting drag noticeably at average speeds of 60 mph. At the top speeds near 90 mph, the effect is even more important (Brancazio, 1984).

Speed Skating

Wind tunnel experiments reported by van Ingen Schenau (1982) reveal how speed skaters' C_D changes as the air velocity changes as well as how C_D changes with body build. Figure 11.4 shows a typical posture of a speed skater at one instant during a stride cycle.

Van Ingen Schenau found the trunk angle to be an important factor influencing speed. The smaller this angle (0 degrees = horizontal trunk), the smaller the C_D (and frontal area). As skating velocity increased from ranges below 12 m/sec to ranges above 16 to 17 m/sec, the C_D decreased. The results demonstrate that since C_D depends on flow velocity and body build and since drag force depends on the C_D, the drag encountered at different velocities is not easily predicted for all skaters as a group.

The tight skin suit that skaters wear reduces the skin friction portion of the total drag. The hood is continuous with the rest of the suit so that the contours are as smooth as possible.

FIGURE 11.4

An almost horizontal trunk position is used in speed skating to reduce frontal area for minimum drag at high speeds.

Triathlons

The triathlete should be particularly aware of the factors influencing fluid drag. In swimming, the fluid drag must be minimized to cover the distance with reasonable speed and without undue fatigue (see Chapter 12). In cycling, a body position that exposes minimal area to the approaching flow is crucial in maintaining speed without undue fatigue (see Figure 11.1b). If the cyclist were to coast on long downhill runs, a terminal velocity might be reached if the speed became great enough to generate a resistive drag force to match the motive force of gravity. In both cycling and running, the effect of "drafting" in the wake of another can save valuable energy by reducing the drag.

Understanding Drag Force in Sport

1. Name three different skills or activities in which aerodynamic drag is significantly detrimental. If possible, how could the drag be minimized in each activity?
2. Tell which cyclist would encounter more resistive drag: (a) one riding at 20 km/hr into a head wind of 5 km/hr, or (b) one riding at 30 km/hr with a tail wind of 10 km/hr. Explain why.
3. How could the cyclists in question 2 minimize the air resistance without changing speed?
4. Describe how the air would flow around a volleyball that is hit with no spin.
5. What effect does dimpling a golf ball have on the drag force it experiences?
6. Why do speed skaters place their arms along their backs?

11.2 Effects of Lift in Sport

Variables Influencing Aerodynamic Lift Force

In Concept Module H fluid lift force was described. The factors that determine the amount of lift force applied by air or water on a body are the body's shape (coefficient of lift), the area of the surface past which the fluid flows, the fluid's density, and the velocity of the flow relative to the body. A body's angle of attack was described as the angle formed between the main plane or chord of the body and the direction of relative flow. A body's lift coefficient changes as its angle of attack changes; therefore, the angle of attack must be optimized for a given body to achieve the maximum lift coefficient and lift force.

When we examine the aerodynamic forces acting on certain projectiles in sport, we must first define several factors that affect the angle of attack when a body changes direction as it moves along its path. These factors are angle of projection, line of flight, attitude angle, and center of pressure.

Angle of Projection and Line of Flight

The angle of attack is not the same as the angle of projection (release) of an implement. The **projection angle** is the angle formed by the center of gravity's (CG) instantaneous projection *velocity vector* and the horizontal (Figure 11.5a). After release, the instantaneous velocity of the CG at any instant depicts its line of flight (i.e., the instantaneous direction in which the object's CG is traveling through the air) (Figure 11.5b).

Attitude Angle

The **attitude angle** is the angle formed between the main plane of the object (chord) and the horizontal ground. It represents the orientation of the implement in space and is the angle that is most obvious to the observer (Figure 11.6).

Angle of Attack at Release

In some events, namely discus and javelin, the wind speed and direction cannot be determined at the time of release of the implement. Therefore, it has become the practice in anal-

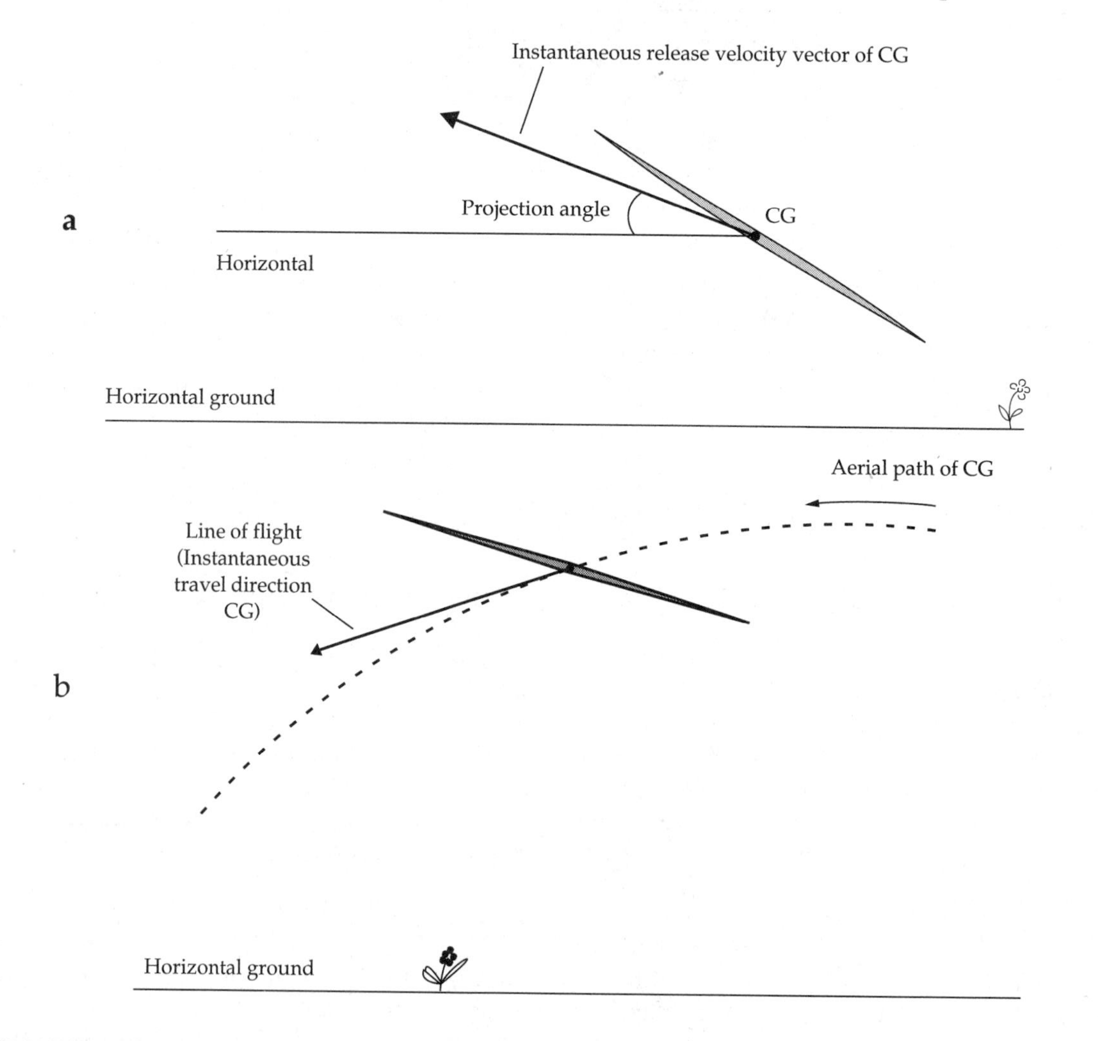

FIGURE 11.5

(a) The projection angle is formed between the horizontal and the instantaneous release velocity vector. (b) The line of flight is the direction the CG is traveling at any given instant, tangent to the curve of the path.

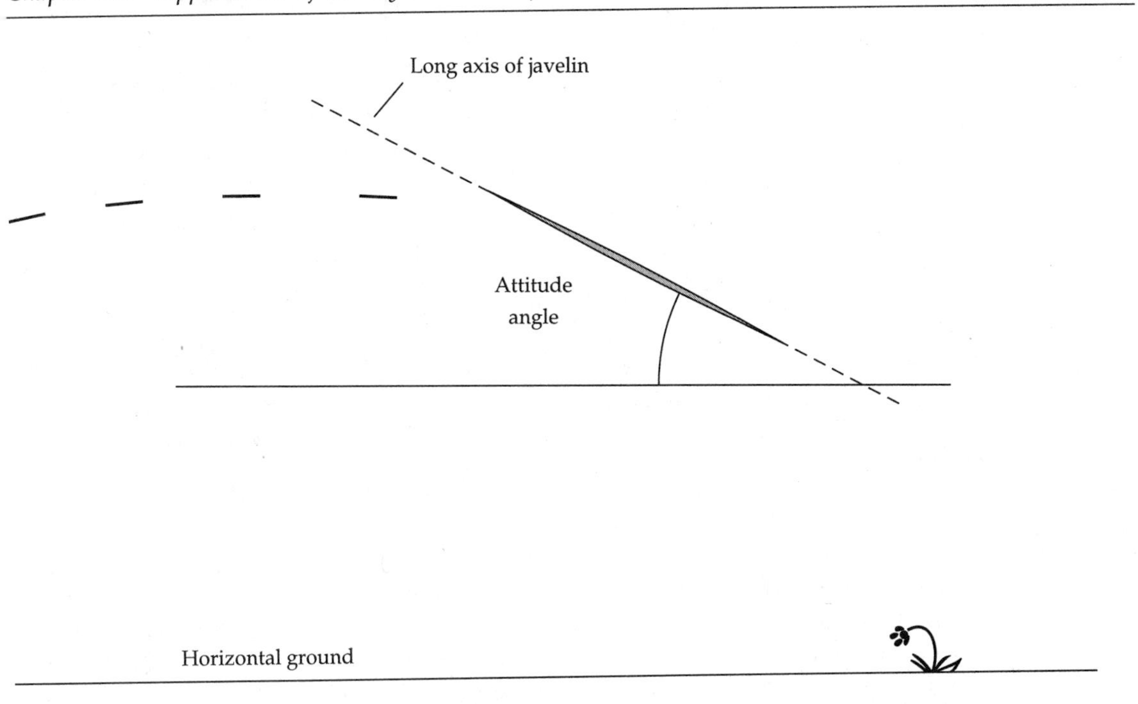

FIGURE 11.6
The attitude angle is formed between the main plane or long axis of the javelin and the horizontal ground.

ysis to define the angle of attack as that angle formed between the main plane or axis and the instantaneous projection velocity vector (Rich et al., 1984).

Center of Pressure

The point at which the resultant air force (lift and drag) acts on a projectile is called the **center of pressure.** Its location on a projectile changes with the changing angle of attack. For example, the center of pressure is located approximately one third the distance back from the leading edge of a wing, and as the wing's angle of attack increases beyond the optimum angle, the center of pressure moves forward on the wing. If the center of pressure is in front of the object's CG, the projectile experiences a torque (called a *pitching* moment) that rotates the leading edge upward, resulting in stall. Figure 11.7 shows the rotating effect on a discus, when the center of pressure is in front of the CG.

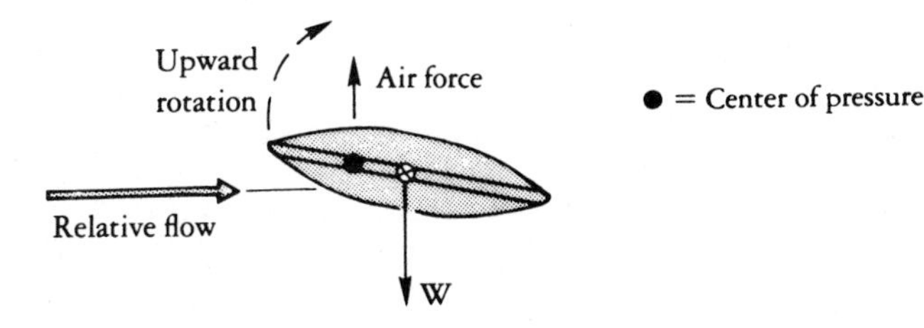

FIGURE 11.7
An upward pitching torque is produced on a discus whose center of pressure is in front of its CG.

The Discus Throw

Controlling the angle of projection of the discus' CG while at the same time controlling the release angle of attack and maximizing the release velocity requires practiced skill. The discus is relatively unstable in flight. It pitches upward throughout its flight, so the angle of attack at release must be small enough to prevent early stalling, which usually occurs when the relative flow forms an angle of attack between 26 and 29 degrees (Ganslen, 1964). (When the stall occurs, the lift/drag ratio is low.) The upward pitching of the discus may be delayed if the spin imparted at release is sufficient to give it some **gyroscopic stability**, or the stability against changing its orientation in space (like a spinning top). Measurements made from film indicate a common spin rate of about 8 revolutions per second (Ganslen, 1964). The angle of attack of a discus increases as a result of the line of flight, in addition to the natural pitching torque. Ganslen calculated that a distance of 228 ft (69.5 m) can be achieved by releasing the discus at 80 ft/sec (24.4 m/sec) at a projection angle of 40 degrees and an angle of attack of –14 degrees.

Figure 11.8 illustrates the changing angles for a discus at its release, at the peak of its trajectory, and during its descent. Note that the attack angle at release is slightly *negative*. This is desirable because the attack angle changes along the line of flight as the spinning discus tends to maintain its same *attitude* angle, while the relative flow direction changes as the discus travels up and then down.

Data from the gold and silver medal winners in the 1984 Olympics showed about the same projection velocities (25 m/sec) for men's distances of 66.6 m and 66.3 m with projection angles of 33.8 and 34.4 degrees, respectively (no attack angles were measured). The top two women's throws of 65.36 m and 64.86 m were achieved with about the same velocities (25 m/sec) and projection angles. Again, no angles of attack were measured (Gregor, Whiting, and McCoy, 1985).

Because of the difficulty in obtaining accurate measures of the angle of attack of the discus at release, it has not been studied adequately enough to make conclusive remarks about it. However, some recommend decreasing it in a head wind and increasing it in a tail wind. The lateral tilt of the discus is important in side winds. With the wind from the right,

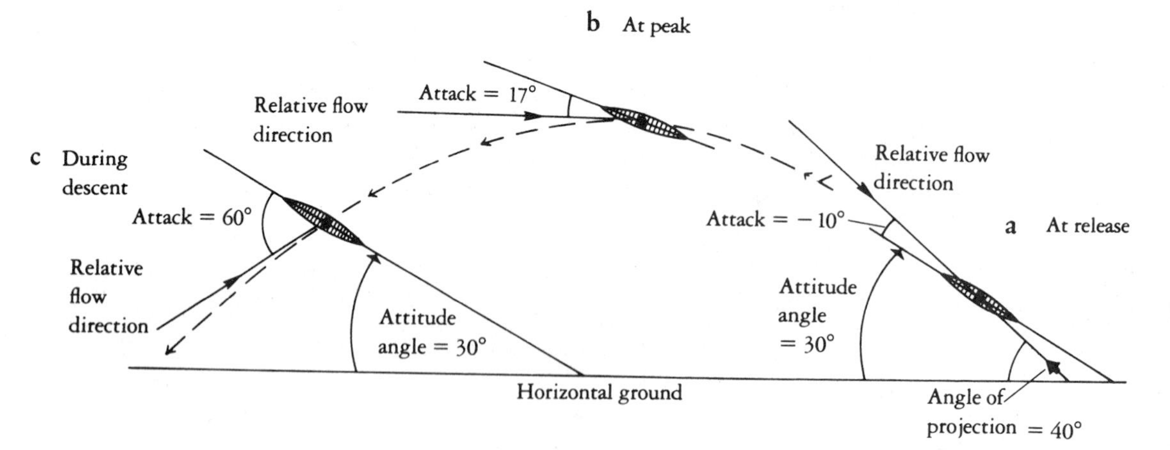

FIGURE 11.8
The angle of attack, angle of projection, and attitude angle of a discus (a) at the instant of release, (b) at the peak of its trajectory, and (c) during its descending path.

the right side must tilt upward to prevent a downward lift force; the opposite is true for a wind from the left. A *negative* angle of attack at release delays the increasing positive attack angle that develops through the flight, thereby delaying stall and producing more distance.

The Javelin

The angle and velocity of release and the angle of attack of a javelin greatly influence its behavior in flight and the distance achieved.

The relative positions of the CG and the center of pressure are particularly important in throwing the javelin. If the center of pressure lies ahead of the CG, a positive pitching moment (torque) is created. If the angle of attack is too great, it rises and then falls sharply as it stalls. Ideally, the javelin should be thrown at a projection angle and with an angle of attack that keep the center of pressure and the CG lined up vertically until just before landing. The change in direction of air flow resulting from the descending flight path causes the center of pressure to move backward behind the CG. The resulting torque rotates the leading edge downward slightly and produces the desired landing position as shown in Figure 11.9c.

Gregor and Pink (1985) analyzed Tom Petranoff's 1983 world record throw of 99.72 m. His projection velocity of 32.3 m/sec, one of the fastest ever recorded, probably played a major role in the distance achieved. The angles at release (32-degree projection and 36-degree attitude) gave the javelin a small positive 4-degree angle of attack at release. The

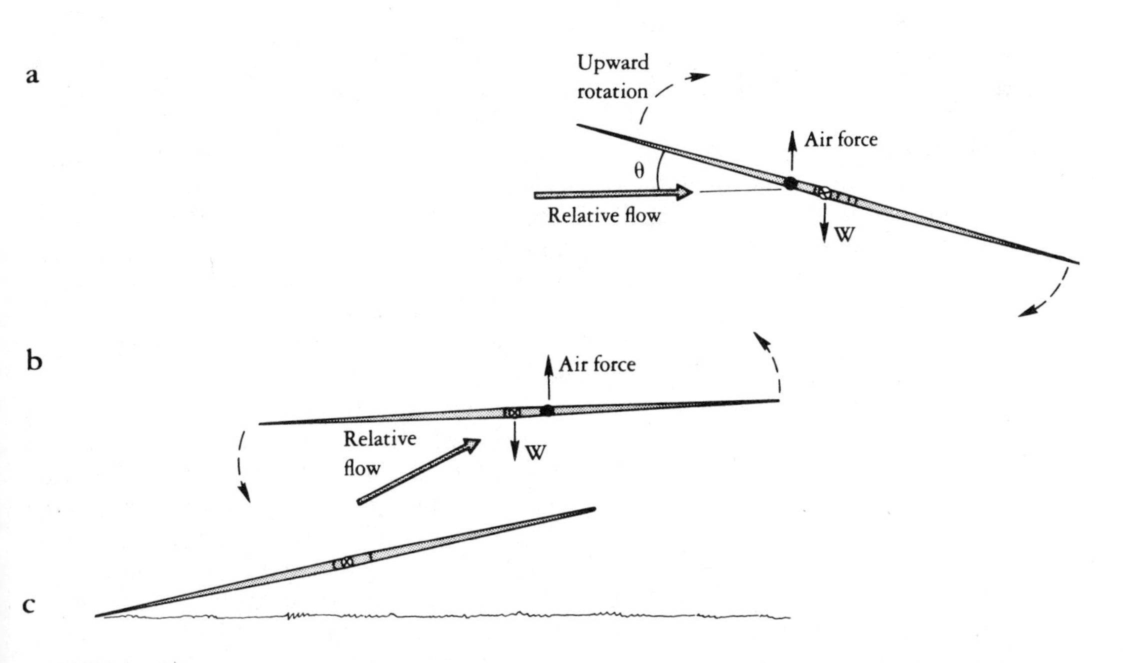

FIGURE 11.9
The torque produced on a javelin when the center of pressure and CG do not coincide. (a) The center of pressure is in front of the CG. (b) The center of pressure is behind the CG, causing the tip to rotate down before landing.

new javelin construction rules (1984) make the javelin less sensitive to release conditions and more responsive to release velocity (Hubbard and Alaways, 1987). Komi and Mero (1985) report data they collected from the 1984 Olympics. The top two male throwers achieved distances of 86.76 m and 85.74 m with projection velocities of 29.12 and 29.09 m/sec, respectively. The best throw was characterized by projection, attitude, and attack angles of 42, 29, and −13 degrees, respectively. The same angles for the second best throw were 36, 51, and +15 degrees. The variability in these initial conditions is evident. The top two women both achieved distances of 69.56 and 69.00 m with projection velocities of 20.73 and 23.62 m/sec, respectively. Their initial angles were 33, 35, and +2 degrees and 38, 38, and 0 degrees, respectively. Optimum values for the angles at release continue to be the focus of investigations. Rich et al. (1984) conclude from studying many elite throwers that "distance probably results from a complex interaction between several different parameters, as well as the unknown wind factor" (p. 59).

In general, Hubbard and Rust (1984) characterize a good throw as one with a small angle of attack producing low lift and low drag in the early stages of the trajectory; as the angle of attack increases through midflight, the lift and drag increase, and then decrease again near the end of its path. These characteristics are similar to those proposed by Remizov (1984) for ski jumping described in the next section.

Ski Jumping

In ski jump takeoffs the relative positions of the CG and the center of pressure on the jumper's leaning body are important for success. A biomechanical analysis of ski jumping by international competitors showed that the jumper would lean far forward so that the torque produced by the reaction force at take-off produced a forward rotation of the body around the CG. This forward rotation counteracted the lift and drag force acting on the center of pressure that otherwise would have caused a backward rotation of the jumper during flight (pitching moment). The use of these two opposing torques allowed the jumper to take advantage of the aerodynamic lift force on his trunk and he was kept in the air longer (Campbell, 1980). Figure 11.10 shows how the torque can be exerted on the jumper by the lift and drag forces acting on the trunk inclined to the relative air flow.

A ski jumper who wants to increase the horizontal distance by staying in flight longer wants to maximize the upward lift force and minimize the backward drag force; that is, maximize the lift/drag ratio. Some disagreement exists, however, and perhaps the lift/drag ratio needs to be optimized during different air phases. As soon as the body orients itself to form an angle of attack to the flow to obtain lift force, the area facing the flow increases to increase the drag force. Experimentation, often in wind tunnels, provides information about the best angle of attack (and body position) so that the lift/drag ratio can be adjusted. One such study on life-size models of ski jumpers revealed that the most effective arm position for increasing lift and decreasing drag is at the sides with the palms facing the relative flow (Watanabe, 1981), as shown in Figure 11.10.

Remizov (1984) conducted wind tunnel experiments to determine the best angle of attack for the longest flight distance. He presented some useful conclusions: the skier should lean well forward with a small angle of attack at take-off to reduce drag, not necessarily to maximize lift; then the angle of attack should be increased for maximum lift during the later stage of flight (more lift is of more benefit than decreased drag). He proposed that larger initial angles of attack (30 degrees)

FIGURE 11.10

Aerodynamic lift force acting on the ski jumper's body counters the forward rotation of the body and provides upward support for more time in the air.

are better for smaller jumping hills where take-off velocity is 20–25 m/sec, and smaller angles of attack (15–23 degrees) for larger hills giving a take-off velocity of 27–33 m/sec.

Understanding Lift Force in Sport

1. Explain the difference between a javelin's angle of projection and attitude angle.
2. Differentiate between the angle of attack at the instant of projection of a discus or javelin and how it is normally defined.
3. In which direction is an aerodynamic lift force applied to an airfoil mounted on the rear portion of a racing car if the airfoil has a negative angle of attack? What would be the function of such a lift force?
4. Experiment with throwing a Frisbee or other lightweight disc to see how it behaves with different angles of projection, attitude, and attack. Try throwing it upside down, and note the adjustments you need to make to have it travel where you want it to.
5. Fold a piece of paper into the shape of an airplane, and project it at various angles of release and at different speeds. For each,

visualize the angle of attack of the wing to air flow and the center of pressure location in relation to the CG location (in front or behind it). Where are the relative locations of the center of pressure and CG if the plane noses upward? if the plane noses downward? if the plane glides linearly forward? Explain.

11.3 Lift Force Produced by Spin: The Magnus Effect

Aerodynamic lift force was identified earlier in this chapter as a force acting perpendicular to the flow direction past a projectile that is inclined to the flow at some angle of attack to create the necessary pressure difference between two sides of the object. Aerodynamic lift force is also created on objects projected with spin. The lift force generated by a spinning object traveling through the air is responsible for the deviation in the flight path from that which is expected, such as with a curve ball. Such a lift force results from a pressure difference between opposite sides of the ball, and this pressure difference depends on the flow–velocity difference, in accordance with Bernoulli's principle. This particular mechanism by which lift force is generated on a spinning ball is called the **Magnus effect,** and the force of lift is sometimes referred to as the **Magnus force.**

To gain an understanding of this force, consider the motion of a ball thrown with top spin[1] in Figure 11.11. As the ball spins, it carries a thin layer of air, the boundary layer, around with it. The ball's forward linear motion creates an air flow *backward* past the ball. The boundary layer on the top surface of the ball is moving forward with the ball surface. This forward-moving top boundary layer encounters the backward flow of air past the ball. The mixing of these oppositely directed air layers results in decreased flow velocity relative to the top surface of the ball and a consequent increase in pressure. On the bottom side, the boundary layer is moving backward with the ball surface as a result of its spin, and the backward flow of air past the moving ball is reinforced, thus creating a lower pressure region on the bottom. The result is a net force on the ball directed downward from the region of higher pressure on the top to the lower pressure region on the bottom. The effect of this force is to cause the ball to accelerate downward at a faster rate than it would be accelerated by gravity alone. The net force acting downward on the ball is its weight plus the downward force of lift. Thus, top-spinning balls are forced to the ground sooner, or faster, than balls with either no spin or other spin directions. For this reason, top spin is used for projecting balls that must fall within a court boundary, as in tennis, table tennis, and volleyball. Figure 11.12 shows a trajectory of a top-spinning tennis ball and the Magnus force acting perpendicular to the flow direction at three different points in the path.

In the initial part of the flight, the Magnus force has a downward component that supplements the force of gravity's decelerating effect. The Magnus force also has a small forward component that tends to augment the forward speed of the ball. (If the initial ascent were

[1] Top spin may be defined as that spin during which the top surface of the ball moves in the same direction as the ball is traveling. In general, the types of spin are named according to the side that is moving in the same direction as the ball is traveling.

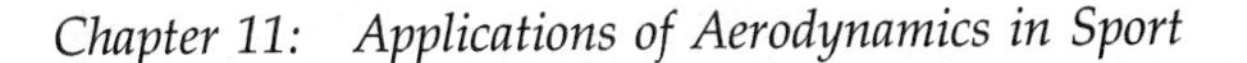

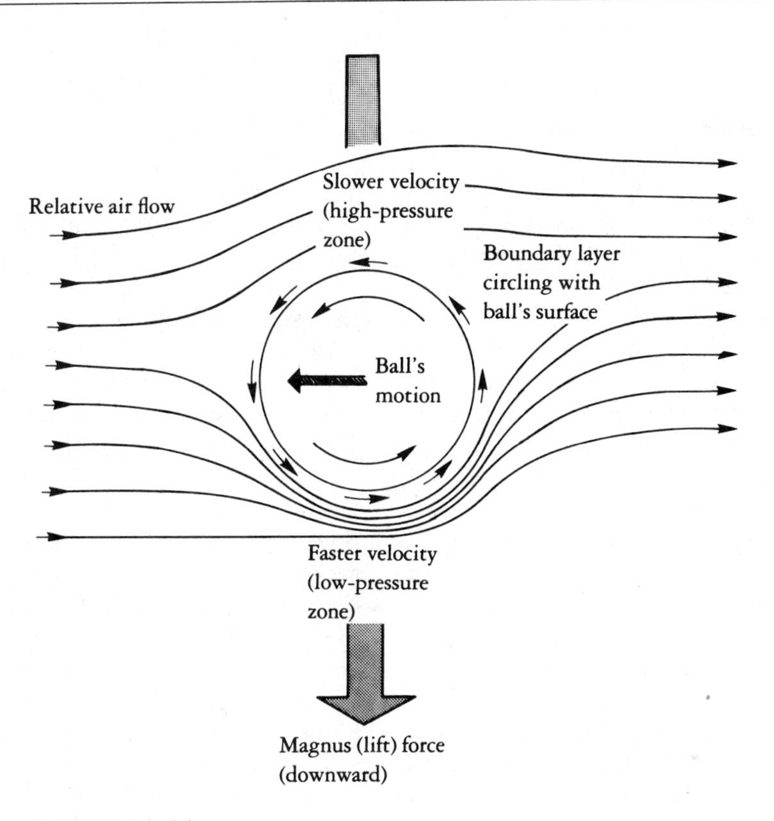

FIGURE 11.11

A ball thrown with top spin experiences a slower air-flow velocity and higher pressure on the top surface. It experiences a faster air-flow velocity and lower pressure on the bottom surface. A downward-directed Magnus (lift) force on the ball is produced as a result of the pressure difference.

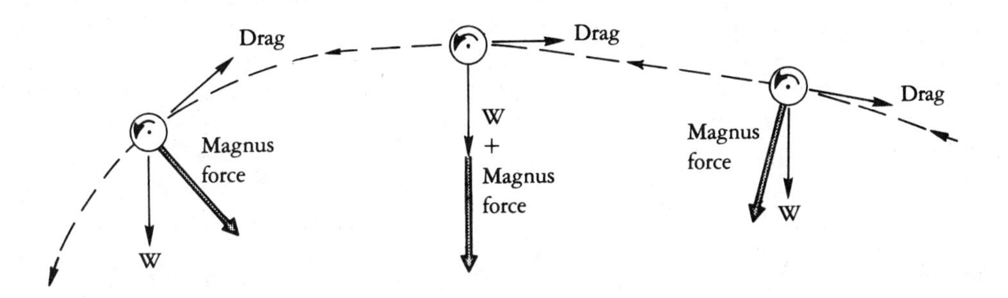

FIGURE 11.12

The Magnus force on a top-spinning ball has vertical and horizontal components that affect the flight of the ball differently at different points in its path. Note that the relative flow changes direction as the ball changes direction.

more vertically directed, this forward component of the Magnus force would be greater.) At the peak, where the air flow is horizontal past the ball, the Magnus force is entirely downward. During the descending portion of flight, the air flow is directed backward and upward past the ball, and the Magnus force is therefore downward and backward. This change in the direction of the air flow past the object is what causes the ball to take on a greater curve downward toward the end of its flight. The downward component continues to augment gravity's pull, accelerating the ball downward at a greater magnitude than would gravity itself. The backward component of this Magnus force resists the forward motion of the ball, causing it to fall at a steep angle of approach to the ground. It is readily apparent why top spin is an effective mechanism for restricting the horizontal distance of a ball that is traveling at a relatively high velocity: if the ball were projected with the desired horizontal velocity without top spin, it would travel too far before hitting the ground. Thus a tennis player able to impart top spin to the ball can project it with a higher velocity and still keep it in bounds. A top-spinning ball also strikes the ground at a more vertical angle than a non-spinning ball, because of the greater deviation downward near the end of its flight.

The Magnus effect is evident also on bottom-spinning balls and is illustrated in Figure 11.13a. The lift force is directed upward from the high-pressure region below the ball toward the low-pressure zone above. Bottom spin is used to project the ball for the purpose of maximizing the horizontal distance or time in the air or for accurate placement, as in golf, tennis, and soccer, because the upward Magnus force causes the ball to experience less acceleration downward due to gravity's pull than a ball with another type of spin. The net downward force experienced by a bottom-spinning ball is its weight minus the upward Magnus force. The trajectory of a bottom-spinning golf ball is shown as an example in Figure 11.13b (Daish, 1972).

Note the changing direction of flow past the ball from the initial part of the trajectory to the final portion. Just as in the previous example (the top-spinning ball), the Magnus force acts perpendicular to the flow direction past the ball, but, in this case, the ball is spinning in the reverse direction. During the ascent, therefore, an upward component of force tends to elevate the ball, and a backward component increases the resistance against its forward motion. Near the peak of the path, the air flow is horizontal, and the Magnus force is entirely upward. During the descent, the flow direction past the ball is such that the Magnus force is directed upward and forward. The upward component still resists the normal downward acceleration that is due to gravity, and the forward component acts to provide a forward propulsive force on the ball. The flight of a ball with backspin is prolonged as a result of the lift force. Therefore, the downward acceleration that is due to gravity acts over a greater period of time so that the ball strikes the ground at a more vertical angle than a ball with no spin, but not as steeply as a ball with top spin. The backspin on impact also tends to cause the ball to rebound at a steeper angle; this angle depends on the angle at which the ball strikes and the amount of skidding that occurs between the ball and rebound surface.

Because of the angle of the club face and the pattern of the swing, a golf ball hit for distance and hit straight usually has a backspin imparted to it. Consequently, the ball is lifted toward the tail end of its flight. Because the lift force acts to keep the ball in the air longer, there is less need to project it with a large vertical velocity component, which also serves to keep the ball in the air longer. Therefore, long shots such as drives should be projected with more horizontal velocity on club impact than shots with middle or short irons.

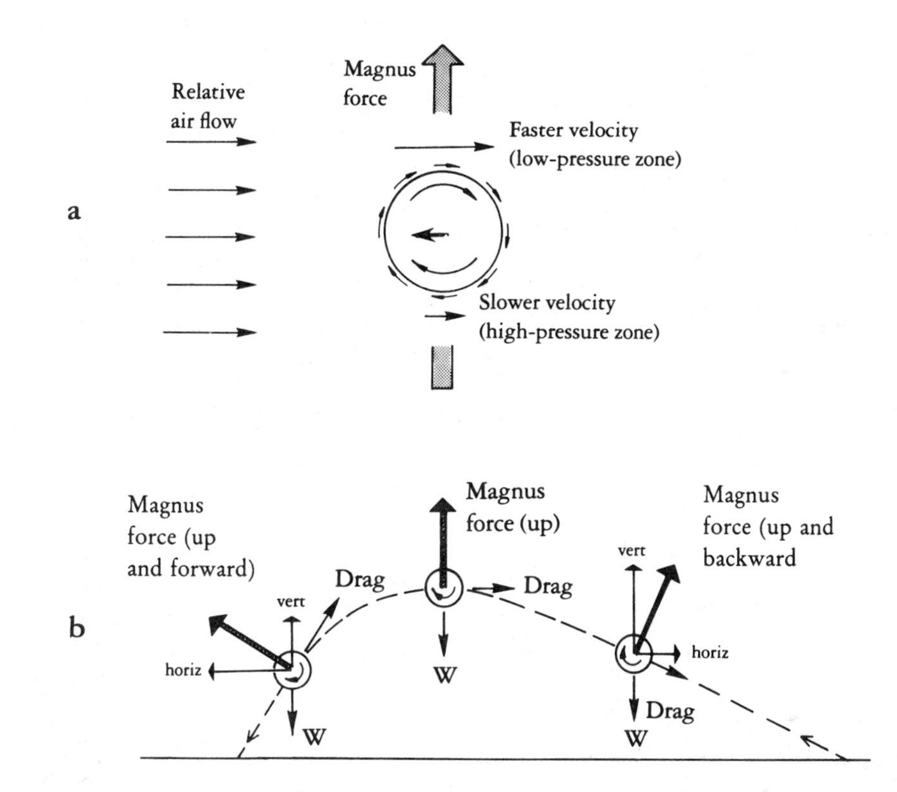

FIGURE 11.13

(a) An upward lift force on a ball projected horizontally with bottom spin. (b) The trajectory of a bottom-spinning golf ball experiencing a Magnus force at different points in its path. The vertical and horizontal components of the Magnus force change as the direction of the Magnus force changes.

The Magnus effect on side-spinning, or diagonally spinning, balls accounts also for the lateral curving of the ball's path commonly seen in softball and baseball pitches. Figure 11.14a illustrates a top view of a ball thrown with right spin (i.e., the right side of the ball moves in the same direction in which the ball is traveling). A right spin causes the ball to deviate from a forward direction toward the left, that is, toward the low-pressure side of the ball.

A common observation is that a side-spinning ball "breaks" to the side late in its flight path. An explanation for this abrupt direction change is twofold. First, the direction of air flow past the ball in the latter part of its trajectory is not only backward but also backward and toward the right (Figure 11.14b) because the ball's direction has already changed to some degree to the left during the initial and middle portion of the trajectory as a result of the right spin. The Magnus force is therefore directed toward the left and backward. The leftward component continues to force the ball toward the left, whereas the backward component, which has developed as a result of the ball's changing direction, serves to delay the ball's progress and allows

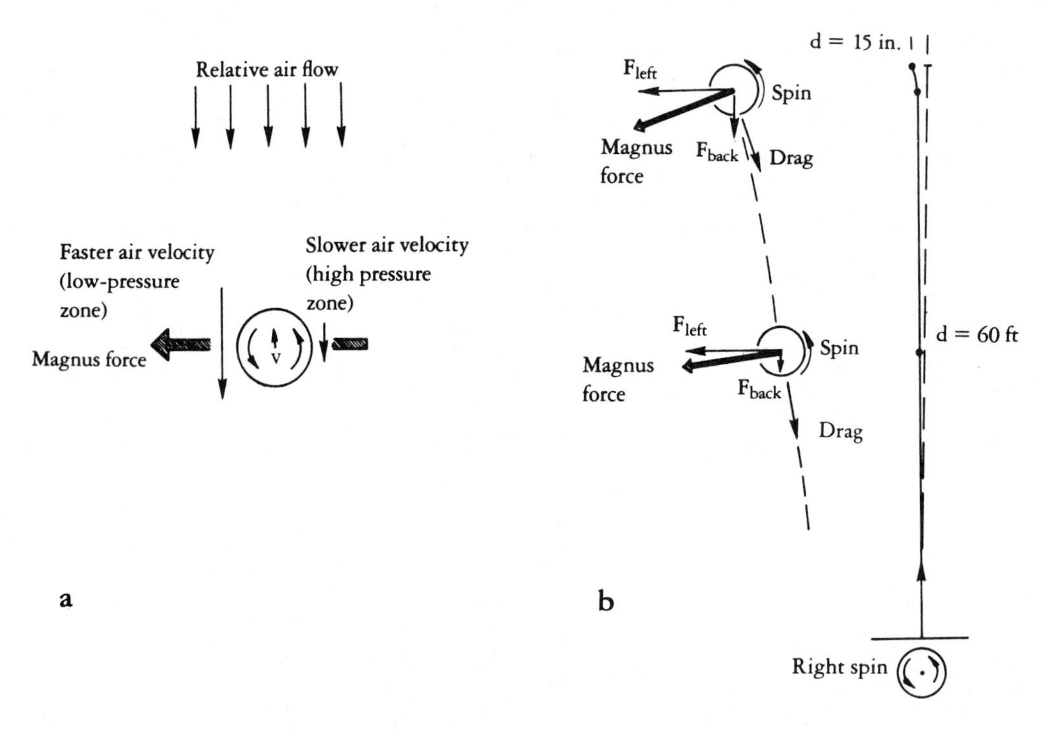

FIGURE 11.14

(a) The Magnus effect produces a lift force directed from right to left on a ball thrown with right spin. (b) A backward component of lift force develops as the ball continues to move toward the left and forward.

a greater time for the leftward component to act. Second, if the rate of ball spin is assumed not to slow appreciably during the time interval that it takes the ball to travel to the batter, the ball is being acted on by a *constant* Magnus force. The acceleration caused by a constant force applied for this length of time results in an appreciable deviation of the ball from a straight-line path. The actual distance the ball will travel toward the left increases as the *square* of the time the force is being applied (i.e., if the time of application is doubled, the ball will be moved four times the distance it moved in the previous time segment). Another way to picture this exaggerated curve is to liken the top view of the side-spinning ball's path to the second half of the parabolic flight path of a projectile seen from a side view; it is an *accelerated* change in direction. According to Daish (1972), a 5-oz baseball thrown with a side spin of 27 revolutions per second at 90 mph will experience a Magnus force of approximately 0.4 N. The resulting acceleration of the ball will be about one third the acceleration that is due to gravity (approximately 11 ft/sec^2). In the 0.5 sec it takes the ball to travel 60 ft to the plate, this Magnus force will cause a sideward movement of the ball of about 15 in.

The downward deviation from the force of gravity should not be overlooked, and the combined leftward and downward deviation results in a diagonally curving ball. In a diagonally spinning curve ball (which has side and top spin), a component of the Magnus force is directed downward, which augments the normal downward acceleration that is due to gravity, and it has a sideward component, which causes sideward displacement. The result of such a pitch is a "break" in the ball's path downward and to the side as it completes its path. From the point of view of the batter, who is expecting the ball to continue approaching in a straight line, the exaggerated displacement of the ball from a straight-line path seems even more exaggerated than it really is.

It is in just this way that the baseball pitcher produces various types of pitches (e.g., fastball, curve, slider). The release of the ball with a spin creates a lift force toward the side of the object with the higher velocity air passing by it. Thus, when a ball is released with bottom spin, the velocity of the air passing on the underside of the ball is relatively slower than the velocity of air passing on the top side of the ball. The force of the lift from bottom spin on the ball prevents the ball from dropping as quickly from the downward force of gravity as it would without any spin at all.

To summarize, Magnus force is applied on any spinning ball whose axis of rotation (spin) is not in the same direction as the flow past the ball. The lift force on the ball is maximum when the spin axis is oriented at 90 degrees to the air-flow direction and decreases to zero as the axis becomes more in line with the flow direction. For example, a spiraling football, whose line of flight is parallel to its long axis, does not experience any Magnus force that is due to its spin. As its axis is angled to its line of flight (flow direction), however, a Magnus force can be generated.

Effects of the Magnus Force on Projectiles

Golf Ball

In addition to the desirable bottom spin put on the ball for a primarily upward Magnus force, side spins may result from intentional or unintentional efforts. Just as the left lateral deflection of the baseball in Figure 11.14 was due to a right spin put on the ball, a similar but more dramatic sideward change in projection occurs in golf if the ball is hit in such a way that side spin is imparted to it. Because the golf ball is smaller, lighter in weight, travels faster and farther, and spins faster than a baseball, the hook, slice, fade, and draw are created under far more impressive conditions. These paths are shown in Figure 11.15.

The Curveball

For years, the curveball in baseball has been the object of controversy as to whether it breaks suddenly before it reaches the batter. The batter perceives the top-spinning ball as approaching in a fairly straight line and then dropping quite abruptly about 10 ft from the plate. Aerodynamic analyses of actual top-spinning curveballs reveal that the illusion of the sudden break downward is caused by the same forces that are shown in Figure 11.12. The constant pull of gravity, which accelerates the ball downward from the time of release, and the downward Magnus force acting at the same time, cause the descent to occur more rapidly during the latter part of the 0.4-sec flight. In addition, the backward component of the Magnus force serves to slow the forward speed during the last part of the curve, thereby allowing more time for the ball to descend before it reaches the batter. In the 0.4 sec a pitched ball is in the air, a nonspinning ball drops 0.15 m (6 in.) as a result of gravity alone, but a top-spinning pitch drops 0.43 m (17 in.) (Allman, 1982).

The deflection of a baseball that is due to the Magnus force is reported to be directly proportional to the square of the ball speed up to 45 m/sec (150 ft/sec), with spins up to 1,800 revolutions per minute (RPM) (Briggs, 1959). Measurements made on pitched balls indicate that a baseball rotates approximately 7 to 16

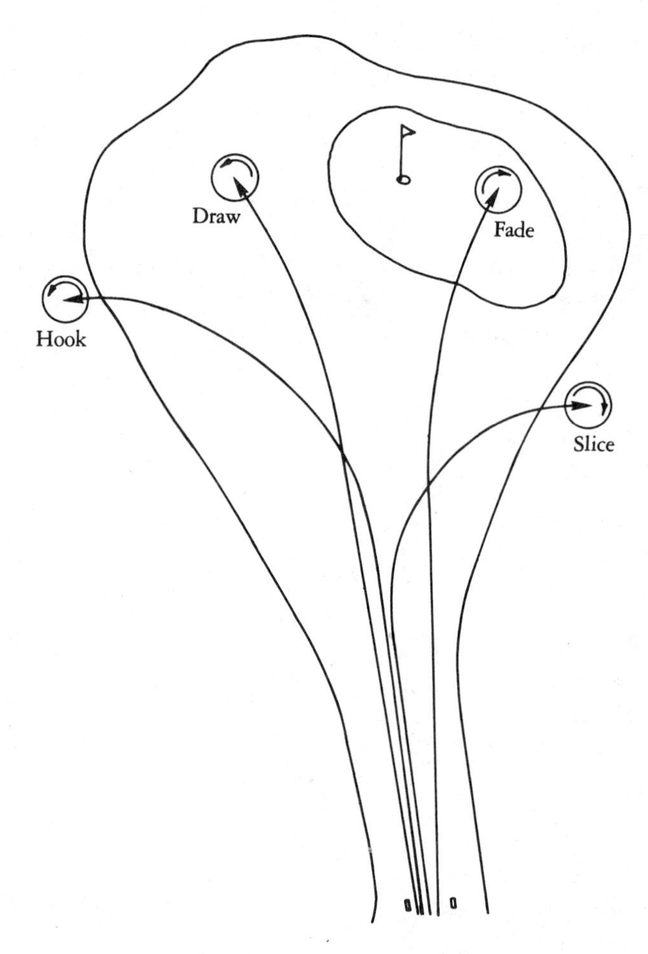

FIGURE 11.15
The flight paths in golf known as the hook, slice, fade, and draw.

turns on its way to the plate if released at 100 ft/sec. Thus, the maximum spinning rate would be about 1,600 RPM. These values can be considered target conditions for throwing curveballs.

The Discus

The Magnus effect may have some influence on a spinning discus as well as on a spinning ball. According to Dyson (1977), as the discus is thrown by a right-handed athlete, it is given a clockwise spin when viewed from the top so that the left side of the discus is seen to move forward and the right side to move backward, relative to the direction the discus is traveling.

Figure 11.16 shows the underside of a discus near the middle portion of its flight path, encountering the air flow at a small angle of attack. The axis of spin is near 90 degrees to the backward flow direction. The left side of the discus carries with it a forward-moving

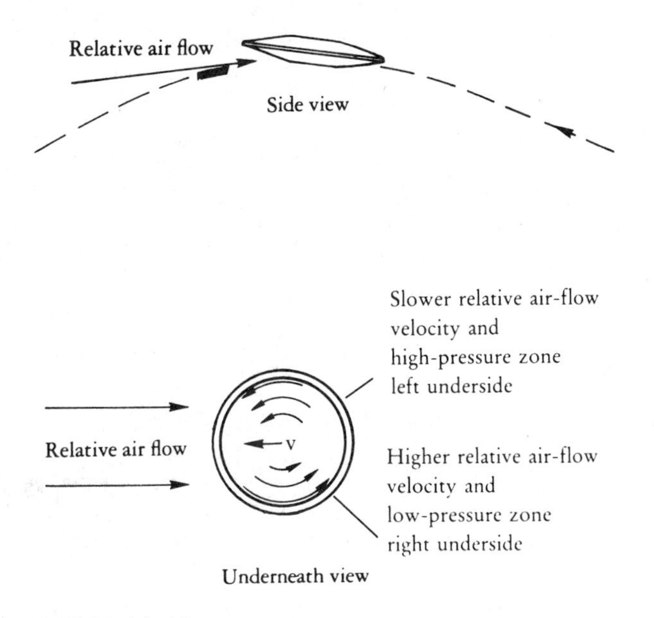

FIGURE 11.16
Differences in pressure between the right and left underside of the spinning discus tend to cause it to roll.

boundary layer that reduces the flow velocity of air past the left side. The right side of the discus carries a backward-moving layer that augments the air-flow velocity on the right side. Again, according to Bernoulli's principle, the pressure will be greater *under* the discus than on top. Because of the velocity differences between the right and left undersides, however, the center of pressure is shifted toward the left side. This tends to cause a rolling of the discus so that the right side lowers and the left side rises. This rolling tendency is not observed, as a result of the angular momentum of the spinning discus, which gives it a certain amount of gyroscopic stability (i.e., stability against a change in the orientation of the axis of spin of the discus) (Dyson, 1977).

Understanding the Magnus Effect

1. Identify three different skills in which putting backspin on a ball would be desirable. Do the same for top spin.
2. For balls thrown with different types of spin, discuss the effects of spin direction (axis of rotation), spin velocity, and velocity of projection on the spatial path of the ball. Discuss different kinds of baseball pitches relative to the same factors.
3. Draw a top-spinning ball during its ascent and descent in its trajectory. Draw in the Magnus force vector on each diagram. Then, resolve each Magnus force vector into its horizontal and vertical components. Compare the horizontal components on the two diagrams, and explain the effect of each one on the path of the ball.

4. Throw a large whiffle ball, Nerf ball, or Trak ball as fast as you can with back spin. Visualize the air flow past it caused by its forward linear motion, and visualize also the boundary layer of air on the spinning ball's surface. Have an observer note, from a distant side view, any deviations from a parabolic trajectory. Explain in terms of Bernoulli's principle and the Magnus effect.

5. Repeat the steps in question 4, imparting top spin instead of back spin.

6. Repeat the steps in question 4 with right and left spin. The observer must take a distant front view to best note the deviations in flight-path direction.

References and Suggested Readings

Allman, W. F. (1982). Pitching: The untold physics of the curve ball. *Science 82, 2*(10), 32–39.

Barkla, H. M., & Auchterlonie, L. J. (1971). The magnus or robins effect on rotating spheres. *Journal of Fluid Mechanics, 47*(3), 437–447.

Bartlett, R. M., & Best, R. J. (1988). The biomechanics of javelin throwing: A review. *Journal of Sports Sciences, 6*(1), 1–38.

Brancazio, P. J. (1984). *Sportscience*. New York: Simon & Schuster.

Briggs, L. (1959). Effect of spin and speed on lateral deflection (curve) of a baseball, and the magnus effect for smooth spheres. *American Journal of Physics, 27*, 589–596.

Campbell, K. R. (1980). Comparative analysis of take-off techniques for Olympic ski jumpers. In J. M. Cooper & B. Haven (Eds.), *Biomechanics: Proceedings of the Biomechanics Symposium* (pp. 92–100). Indiana State Board of Health.

Chapman, S. (1968). Catching a baseball. *American Journal of Physics, 36*(10), 868–870.

Cunningham, J., & Dowell, L. (1976). The effect of air resistance on three types of football trajectories. *Research Quarterly, 47*(4), 852–854.

Daish, C. B. (1972) *Learn science through ball games.* New York: Sterling.

Dal Monte, A., Leonardi, L. M., Menchinelli, C., & Marini, C. (1987) A new bicycle design based on biomechanics and advanced technology. *International Journal of Sport Biomechanics, 3*, 287–292.

Dapena, J., & Teves, M. A. (1982). Influence of the diameter of the hammer head on the distance of a hammer throw. *Research Quarterly for Exercise and Sport, 53*(1), 78–81.

Dapena, J., & Feltner, M. E. (1987). Effects of wind and altitude on the times of 100-meter sprint races. *International Journal of Sport Biomechanics, 3*, 6–39.

Dillman, C. J., et al. (1979). Biomechanical analysis of U. S. ski-jumpers. In J. Terauds & H. Gros (Eds.), *Abstracts of the International Congress of Sports Sciences*. San Diego, CA: Academic.

Dyson, G. (1977). *The mechanics of athletics.* New York: Holmes & Meier.

Erlichson, H. (1983). Maximum projectile range with drag and lift. *American Journal of Physics, 49*, 1125.

Faria, I. E., & Cavanagh, P. R. (1978). *The physiology and biomechanics of cycling.* New York: John Wiley & Sons.

Ganslen, R. V. (1964, April). Aerodynamic and mechanical forces in discus flight. *Athletic Journal*, pp. 50–51, 68, 88–89.

Ganslen, R. V. (1967, December). Javelin aerodynamics. *Track technique*, pp. 940–942.

Gregor, R. J. & Pink, M. (1985). Biomechanical analysis of a world record javelin throw: A case study. *International Journal of Sport Biomechanics, 1*, 73–77.

Gregor, R. J., Whiting, W. C., & McCoy, R. W. (1985). Kinematic analysis of Olympic discus throwers. *International Journal of Sport Biomechanics, 1*, 131–138.

Gross, A. C., Kyle, C. R. & Malewicki, D. J.

(1983). The aerodynamics of human-powered land vehicles. *Scientific American, 249*(6), 142–152.

Gros, H. (1979). Basic mechanics and aerodynamics applied to skiing. In J. Terauds & H. Gros (Eds.), *Science in Skiing, Skating, and Hockey: Proceedings of International Congress of Sport Sciences* (pp. 9–21). San Diego, CA: Academic.

Gutman, D. (1988, April). The physics of foul play. *Discover, 9*, pp. 70–76.

Hay, J. G. (1975). Biomechanical aspects of jumping. In J. H. Wilmore & J. F. Keogh (Eds.), *Exercise and sports sciences reviews* (pp. 135–161). New York: Academic Press.

Hopper, B. J. (1973). *The mechanics of human movement.* New York: American Elsevier.

Hubbard, M. (1984). Optimum javelin trajectories. *Journal of Biomechanics, 17*, 777–778.

Hubbard, M., & Alaways, L. W. (1987). Optimum release conditions for the new rules javelin. *International Journal of Sport Biomechanics, 3*, 207–221.

Hubbard M., & Rust, H. J. (1984). Simulation of javelin flight using experimental aerodynamic data. *Journal of Biomechanics, 17*, 769–776.

Kirkpatrick, P. (1963). Batting the ball. *American Journal of Physics, 31*, 606.

Komi, P. V., & Mero, A. (1985). Biomechanical analysis of Olympic javelin throwers. *International Journal of Sports Biomechanics, 1*, 139–150.

Kreighbaum, E. F., & Hunt, W. A. (1978). Relative factors influencing pitched baseballs. In F. Landry and W. A. R. Orban (Eds.), *Biomechanics of Sports and Kinanthropometry: Proceedings of the International Congress of Physical Activity Sciences* (pp. 227–236). Miami: Symposia Specialists, Inc.

Kyle, C. R. (1979). Reduction of wind resistance and power output of racing bicyclists and runners traveling in groups. *Ergonomics, 22*, 387–397.

Kyle, C. R., & Burke, E. (1984). Improving the racing bicycle. *SOMA, 1*, 34–45.

Kyle, C. R., & Caiozzo, V. J. (1986a). The effect of athletic clothing aerodynamics upon running speed. *Medicine and Science in Sports and Exercise, 18*, 509–515.

Kyle, C. R., & Caiozzo, V. J. (1986b). Experiments in human ergometry as applied to the design of human-powered vehicles. *International Journal of Sports Biomechanics, 2*, 6–19.

Lichtenberg, D. B. & Wills, J. G. (1978). Maximizing the range of the shot put. *American Journal of Physics, 46*(5), 546–549.

Luethi, S. M., & Denoth, J. (1987). The influence of aerodynamic and anthropometric factors on speed in skiing. *International Journal of Sport Biomechanics, 3*, 345–352.

Marino, G. W. (1983). Selected mechanical factors associated with acceleration in ice skating. *Research Quarterly for Exercise and Sport, 54*(3), 234–238.

Miller, D. I., & Munro, C. F. (1983). Javelin position and velocity patterns during final foot plant preceding release. *Journal of Human Movement Studies, 9*, 1–20.

Phinizy, C. (1968, December 26). The unbelievable moment. *Sports Illustrated*, pp. 53–61.

Raine, A. E. (1970). Aerodynamics of skiing. *Science Journal, 6*, 26–30.

Remizov, L. P. (1984). Biomechanics of optimal flight in ski-jumping. *Journal of Biomechanics, 17*, 167–171.

Rich, R. G., Gregor, R. J., Whiting, W. C., & McCoy, R. W. (1984). Kinematic analysis of elite javelin throwers. In J. Terauds, K. Barthels, E. Kreighbaum, R. Mann, & J. Crakes (Eds.), *Sports Biomechanics, Proceedings of the International Symposium of Biomechanics in Sports* (pp. 53–60). Del Mar, CA: Academic.

Roberson, J. A., & Crowe, C. (1975). *Engineering fluid mechanics.* Boston: Houghton-Mifflin.

Selin, C. (1959). An analysis of the aerodynamics of pitched baseballs. *Research Quarterly, 30*(2), 232–240.

Shanebrook, J. R. (1974). Aerodynamics of the human body. In R. C. Nelson & C. A. Morehouse (Eds.), *Biomechanics IV: Proceedings of the Fourth International Seminar on Biomechanics* (pp. 567–571). Baltimore: University Park Press.

Shanebrook, J. R., & Jaszczak, R. D. (1976). Aerodynamic drag analysis of runners. *Medicine and Science in Sports, 8*(1), 43–45.

Shapiro, A. (1961). *Shape and flow: The fluid dynamics of drag.* Garden City, NY: Doubleday.

Soong, Tsai-Chen. (1975). The dynamics of javelin throw. *Journal of Applied Mechanics, 42E*(2), 257–262.

Spring, E., Savolainen, S., Erkkila, J., Hamalainin, T., & Pihkala, P. (1988). Drag area of a cross-country skier. *International Journal of Sport Biomechanics, 4*, 103–113.

Technology and applications: Basic body for power. (1986). *SOMA, 1*, 62.

Technology and applications: Improving the javelin throw. (1987). *SOMA, 2*, 50.

Terauds, J. (1974). A comparative analysis of the aerodynamic and ballistic characteristics of competition javelins. In R. C. Nelson & C. A. Morehouse (Eds.), *Biomechanics IV: Proceedings of the Fourth International Seminar on Biomechanics*, (pp. 180–183). Baltimore: University Park Press.

Terauds, J. (1975). Some release characteristics of international discus throwing. *Track and Field Quarterly Review, 75*, 54–57.

Tricker, R. A. R., & Tricker, B. J. K. (1967). *The science of human movement.* New York: American Elsevier.

Unger, J. (1977). Aerodynamics and the discus. *Modern Athlete and Coach, 15*(4), 59–60.

van Ingen Schenau, G. J. (1982). The influence of air friction in speed skating. *Journal of Biomechanics, 6*(15), 449–458.

Walker, J. (1979). Boomerangs: How to make them and also how they fly. *Scientific American, 240*(5), 162–172.

Ward-Smith, A. J. (1983). The influence of aerodynamic and biomechanical factors on long jump performance. *Journal of Biomechanics, 16*(8), 655–658.

Ward-Smith, A. J. (1985). A mathematical analysis of the influence of adverse and favourable winds on sprinting. *Journal of Biomechanics, 18*, 351–357.

Ward-Smith, A. J. (1986). Altitude and wind effects on long jump performance with particular reference to the world record established by Bob Beamon. *Journal of Sports Sciences, 4*(2), 89–99.

Watanabe, K. (1978). Running speed of skiing in relation to posture. In F. Landry & W. A. R. Orban (Eds.), *Biomechanics of Sports and Kinanthropometry* (pp. 203–210). Miami: Symposia Specialists, Inc.

Watanabe, K. (1981). Skiing research in Japan. *Medicine and Science in Sports and Exercise, 13*(3), 202–209.

Watanabe, K. (1983). Aerodynamic investigation of arm position during the flight phase in ski jumping. In H. Matsui and K. Kobayashi, (Eds.), *Biomechanics VIII-B: Proceedings of the Eighth International Congress of Biomechanics* (pp. 856–860). Champaign, IL: Human Kinetics.

Watts, R., & Sawyer, E. (1975). Aerodynamics of a knuckleball. *American Journal of Physics, 43*, 960–963.

CHAPTER 12

Application of Hydrodynamics in Swimming

PREREQUISITES

Concept Modules C, D, E, F, G, H

CHAPTER CONTENTS

The discussion in this chapter focuses on some important aspects of human movement in water: The nature of water as a medium and forces affecting a person's ability to float and assume different floating positions (gravity and buoyancy); the roles drag and lift forces play in resisting or propelling a person in swimming and swimming-related skills; and factors that determine efficiency and speed.

In aquatic activities forces must be dealt with that normally are not encountered during movement on land. In land activities, body movements are facilitated because the mover feels at home in the environment and takes a number of things for granted. We learn from infancy to expect the firm support exerted by the ground or other surface supporting the body. We learn to predict what surfaces are slippery, sticky, or rough and to know that removal of support results in a fall. In a similar manner, the more an individual can explore movement in the water and develop a set of expectations to depend on, the greater is the sense of feeling at home in the water. As a result, developing skills for managing the body in water becomes easier. Increasing numbers of infants and young children are exposed to moving and playing in water, and they learn to expect different sensations and to predict the results of different body movements. A common practice, unfortunately, is the attempt to teach formal swimming strokes to those who have not yet developed a set of expectations that come from experience in moving the body and controlling breathing responses while in the water. The swimming teacher or coach is most qualified to expedite learning and to refine swimming skills if he or she (1) appreciates the importance of the learner developing a "feel for the water" and (2) understands the forces involved in floating and how forces are generated during movements in a water environment.

In swimming races, the primary mechanical purpose is to move the body over a prescribed distance as fast as possible. Recreationally, the purpose may be merely to interact with the environment. For competitive synchronized swimmers, the primary mechanical purpose is to move or position the body and its segments in a prescribed pattern to achieve an ideal or model performance. For any of these purposes, hydrodynamic principles govern the performance outcome.

A body's state of motion—whether it is still, accelerating or decelerating, or traveling at a constant velocity—depends on the forces acting on that body. Identifying the forces that influence the motion of a body in water is more difficult than identifying the forces acting on a body during land activities. On land, external forces affecting body motion are more easily visualized and measured than are the forces created by the "invisible" medium of water. When a body moves through water, the types of force created are the same as when a body or object moves through the air, that is, the forces of **drag** and **lift**. These forces were presented in Concept Module H. Water is a more dense medium than air and consequently generates greater drag and lift forces in response to movements through it. The nature of these forces is presented here with specific application to resistance and propulsion in swimming. The similarity of hydrodynamics and aerodynamics (Chapter 11) will be readily apparent.

12.1 Buoyancy and Flotation

The nature of water as a force-producing fluid must be considered for understanding how water provides support for a body or watercraft as well as how it reacts to movements performed during aquatic activities. This section on floating focuses on the human body, but the principles and concepts are the same for any object or body.

Static Water Pressure

Water pressure increases with depth, and its effects on the body should be understood and appreciated by the teacher or coach of swimming or diving. *Pressure* is defined as the amount of force a mass exerts against a surface divided by the area of the surface, that is, force per unit area (p = F/A). Air pressure on an object is greater at sea level (10.4 N/cm^2 or 14.7 $lb/in.^2$) than at higher elevations because of the weight of the atmosphere above it. Similarly, as one descends in water, the pressure on an object increases because of the weight of the water above it. Such an increase in pressure is felt by a descending diver as the increased water pressure compresses the volume of air in the middle ear and sinus cavities. Equalizing the pressure by "popping" the ears relieves the discomfort produced at a given depth and pressure. As the diver swims deeper, the pressure continues to increase; thus, the internal and external pressures must be equalized once again. Neoprene wet suits used by scuba divers contain many tiny air cells for insulation. These are compressed as the diver descends, and the suit becomes more dense as it loses volume. Consequently, during descent to greater depths, some inflation (expansion) of a flotation device (vest) may be necessary to compensate for this. Figure 12.1 shows the increase in water pressure with depth. A swimmer without compressible clothing is not compressed with depth; the body volume can be changed only by lung volume changes with breathing.

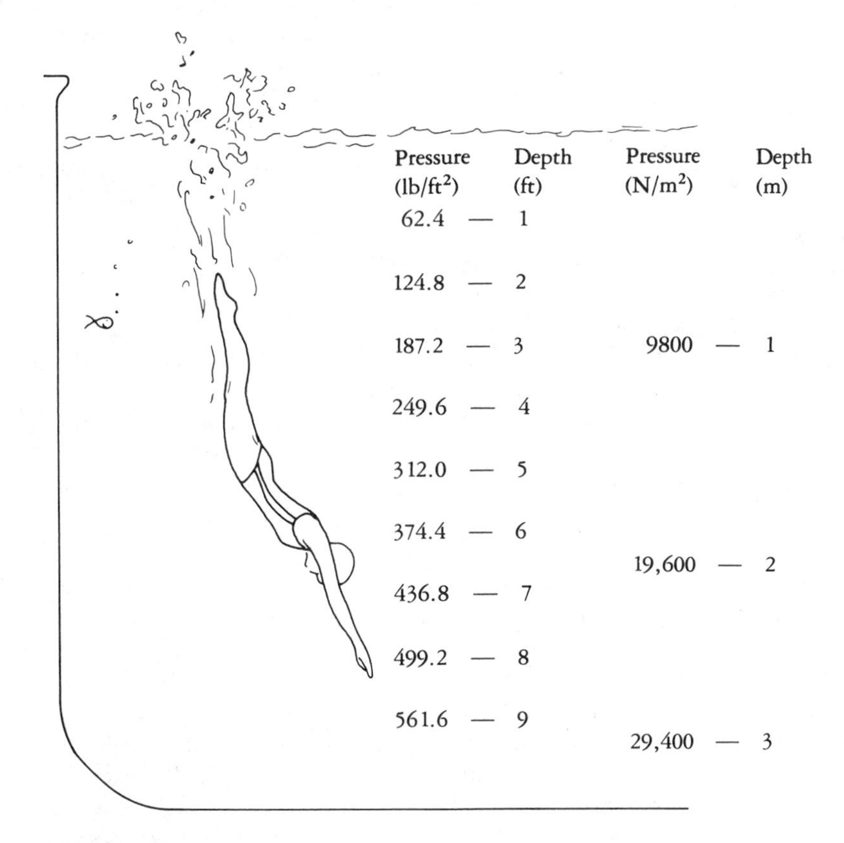

FIGURE 12.1

The water pressure experienced by a submerged swimmer increases with depth. The pressure at any level is exerted equally in all directions.

Volume and Density

Volume is the space a body occupies; it has length, width, and height. It is not necessarily true, however, that the larger a body, the greater its mass. The concept of mass is often confused with that of volume, or size. Consider, for example, the relative mass of a golf ball and of an inflated balloon. The amount of matter in the golf ball is contained in a smaller volume than is the matter in a balloon. In this case, the golf ball is more massive.

Recall that the **density** of an object or body is its **mass** per unit volume, or a measure of how compactly the body's matter is contained in its volume. **Weight** per unit volume is defined as the **weight-density** of a body.

EQUATION 12.1

$$\text{weight density} = \frac{\text{weight}}{\text{volume}} \quad \text{or} \quad WD = \frac{W}{V}$$

where W is weight and V is volume.

Because the weight of water depends on the volume chosen, its weight is expressed in terms of a given volume such as Newtons per liter (N/l) or pounds per cubic foot (lb/ft^3). Although *density* is the term used to express mass per unit volume, thinking in terms of water's weight per unit volume is more practical for us because weight is directly proportional to mass.

The weight density of fresh water is 9.8 N/l (62.4 lb/ft^3), whereas the weight density of salt water is closer to 10.1 N/l (64 lb/ft^3). The reason for the difference is that salt water, as a result of its dissolved salt content has more mass than the equivalent measure of fresh water, and thus has more weight per unit volume. The density of water does not change with depth. Water, as a liquid fluid, is practically incompressible, so that a greater mass of water cannot be squeezed into a given volume to increase its density. The human body, being mostly water, is also incompressible. Air, however, which is a gaseous fluid, can be compressed; sea-level air is more dense (0.01 N/l or 0.075 lb/ft^3) than air at high elevations because it is compressed by the weight of the atmosphere above it. Air also can be mechanically compressed into tanks for use in scuba diving.

The fact that the weight-density of the human body is approximately the same as that of water is an important consideration in studying the buoyancy of the body in water.

Because the human body is composed of a variety of materials, its density is not uniform; that is, each body segment has a somewhat different density. The large thorax volume is not as massive or heavy as it appears, which is due to the low-density air in the lungs. Bones, muscles, and body fluids are relatively high-density materials in the body, whereas fat and air have lower densities. Figure 12.2 illustrates the relative densities of air, fat, water, muscle, and bone. Air is the least dense; bone is the most dense. Adipose (fat) tissue is the least dense of body tissues.

Floating Ability of the Human Body

The rule that determines whether a body will float or sink is called **Archimedes' principle**, named after the ancient Greek scholar who is credited with identifying the concept. The principle states that *a body that is partially or totally immersed in a fluid will experience an upward buoyant force that is equal to the weight of the volume of fluid displaced by that body*. Another way of expressing this concept is the following: If the weight density of an immersed body is less than or equal to the weight density of water, the body will float; if its weight density is greater, it will sink. Wood, for example, has a weight density of 7.8 N/l (50 lb/ft^3) and will float. Aluminum, at 26 N/l (168 lb/ft^3), will sink; that is why air or foam-filled flotation

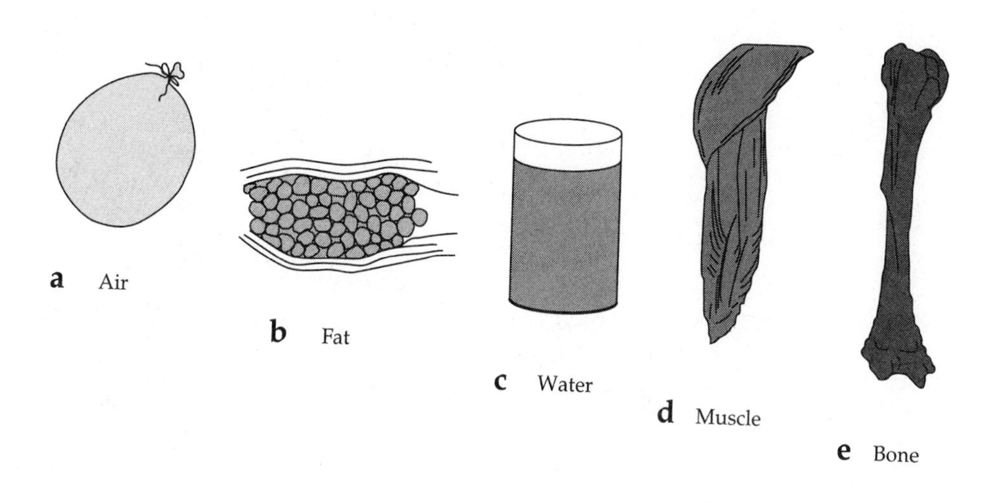

FIGURE 12.2
The relative densities of air, fat, water, muscle, and bone are indicated by the degrees of shading.

chambers are installed in canoes for buoyancy when they capsize and swamp. Plastics, also, are usually more dense than water.

When we speak of a person's weight density, we are speaking of average total body density, because the different body tissues, segments, and sections throughout the body vary in density. The skeletal framework is approximately 50% more dense than water, muscular tissue is only slightly more dense than water, and fat tissue is slightly less dense than water. Blood is slightly more dense than water.

The person in Figure 12.3a has the same weight density as water, meaning that her body weighs the same as water with the same volume. Therefore, when she is submerged, she displaces a volume of water that weighs the same as her body. She will float in the water with no part of her body above the surface. If she were placed 1 m under the surface and then released, she would remain suspended there because her weight force downward would be equal to the buoyant force upward. She would be called *neutrally buoyant*. Since zero net force is acting on her, she will neither sink nor rise. If she were lifted partially above the surface and then released, she would displace less water and the buoyant force would decrease. Consequently, her downward weight force would be greater than the upward buoyant force (weight of the water displaced), and a net downward weight force would exist, causing her to sink. She would descend into the water until her body displaces a volume of water whose weight is equal to her weight. Her downward momentum, however, might cause her to come to rest at some level below the surface.

Figure 12.3b shows a person whose higher fat content gives her a body density less than that of water. Assume that her body volume is the same as the floater in Figure 12.3a. With the same volume but less weight in that volume, she need not be completely immersed to float; that is, she need only displace a volume of water that weighs the same as her body. She can do that and still have part of her volume above the surface. Such a floater is said to be *positively buoyant*. If she is completely submerged, the net upward force (buoyant force minus weight) causes her to rise until the buoyant force is equal to her weight.

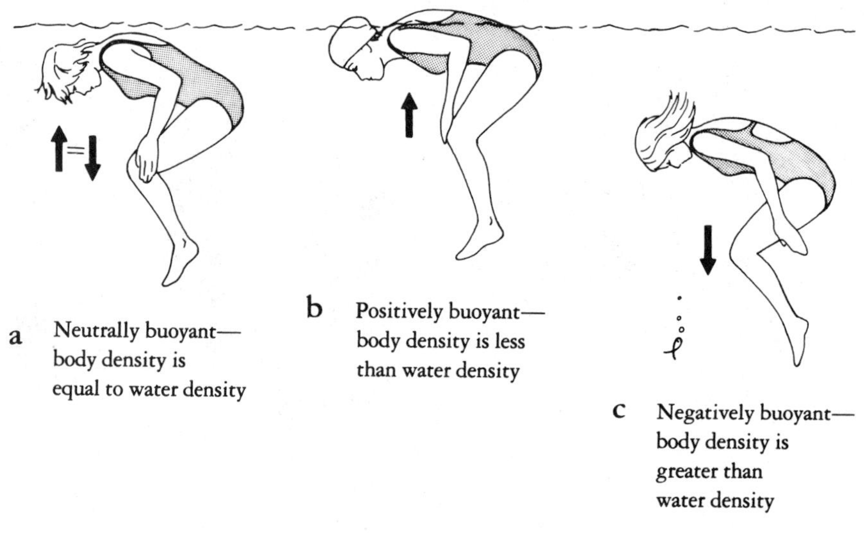

FIGURE 12.3
Three bodies whose densities differ will experience different degrees of support from the water.

Figure 12.3c shows a person immersed whose body composition is lean and has a higher percentage of muscle and dense bone. Her body volume is the same as each of our previous two floaters, but being made up of dense tissues, her weight density is greater than that of water. Consequently, her body volume cannot displace a volume of water great enough to obtain a buoyant force matching her weight; she will continue to sink as a result of the net downward force acting on her (weight minus buoyant force). She is called *negatively buoyant*.

Types of Floaters

In the foregoing examples, we assumed that all three swimmers had equal body volumes but different weights, and therefore, different *weight densities*. Changing body density by changing body's volume with breath control is possible, however.

Without air in the body to take up some of the body volume, the human body would have an average density greater than that of water. The thorax, however, can change its density through the breathing process. The volume of the thorax increases upon inspiration by about 4 liters (0.15 ft^3), whereas the weight of the inspired air is negligible, only about 0.04 N (.01 lb). Control of air volume in the lungs, therefore, influences average body density and is a key factor in floating ability. A person with a body composition that allows floating when the lungs are inflated but who sinks upon exhalation is called a *conditional floater*; that is, whether the body floats or sinks depends on the expansion volume of the lungs and thorax. A conditional floater, floating with the face up, can prolong the buoyant phase by establishing a breathing pattern in which a large volume of air is in the lungs most of the time: inhale quickly and deeply, hold the breath for a few seconds, exhale quickly, and inhale again to repeat the pattern. With such a breathing pattern, the period of time during which the body's density is being increased (exhalation) is quite brief and is not a sufficient time for the face to sink below the surface before the next inhalation is made. Performance of such a maneuver demonstrates how the average body density may be manipulated, but in actual practice, slight supportive movements of the arms or legs or both are usually employed for additional upward support.

Figure 12.4a shows an 800-N (180-lb) floater with a weight density of 9.5 N/l in the jellyfish float position, which is used to test floating ability. He holds his breath after a full inspiration (the volume of his body is now 84 l), and, upon total immersion in water, he displaces 832 N of water (84 l × 9.9 N/l). His body, therefore, experiences a net upward force of 32 N (832 N upward − 800 N downward), which raises his body so that some of his body is above the surface (Figure 12.4b). As part of the body rises above the surface, less water is displaced, and the buoyant force decreases until the weight of the water displaced is equal to 800 N (180 lb). The density of this particular swimmer with an inhalation is less than water density. In Figure 12.4c when he exhales, his volume decreases by approximately 4 liters and his average body density is 10 N/l. His body density is now greater than water density. The upward buoyant force exerted by the water displaced is 792 N (80 liters × 9.9 N/l), thus yielding a net force downward of 8 N, which acts to sink the swimmer.

Some people float no matter whether the lungs are inflated. These people are the *true floaters*—those whose average body density remains less than that of water even after exhalation. The body composition of such individuals typically includes a greater percentage of low-density tissues, such as adipose. Girls and women have a greater percentage of body fat on the average than do boys and men. This characteristic is an average one, however, and some women do have a lower percentage of body fat than many men. Obesity is a definite advantage in floating, whereas a person with a low percentage of fat or well-developed musculature is more likely to have difficulty floating without using supportive movements.

Although usually rare, the body density of an individual may be so great that even the chest expansion volume with inhalation is not sufficient to reduce the average body density enough for unassisted floating. Such a person is called a *sinker*. Blacks frequently experience greater difficulty in floating than others because of greater average body density, which is primarily due to higher-density bone tissue. Buoyancy characteristics of any given individual, however, should not be presupposed on the basis of population norms; individual variability often accounts for the exception to the rule. Some persons who claim to be sinkers are really conditional floaters who may lack information about adjusting body density with breath control.

In a normal population, most individuals—male and female—are *conditional* floaters, some are true floaters, and few are true sinkers. The jellyfish float position (Figure 12.4) is the most convenient test for classifying an individual according to the floating type. If a person can float in this test position, then that person can float in other positions as well; other floating positions are discussed in the next section. For learners to understand their buoyancy characteristics is helpful, and for the teacher to identify buoyancy problems in beginning and intermediate swimmers is imperative.

Different Floating Positions of the Human Body

The average density of a person's body determines if it will float or sink. The orientation of the body, or floating *position* in the water, is determined by where the buoyant force is applied on the body (**center of buoyancy [CB]**) and where the weight force is applied on the

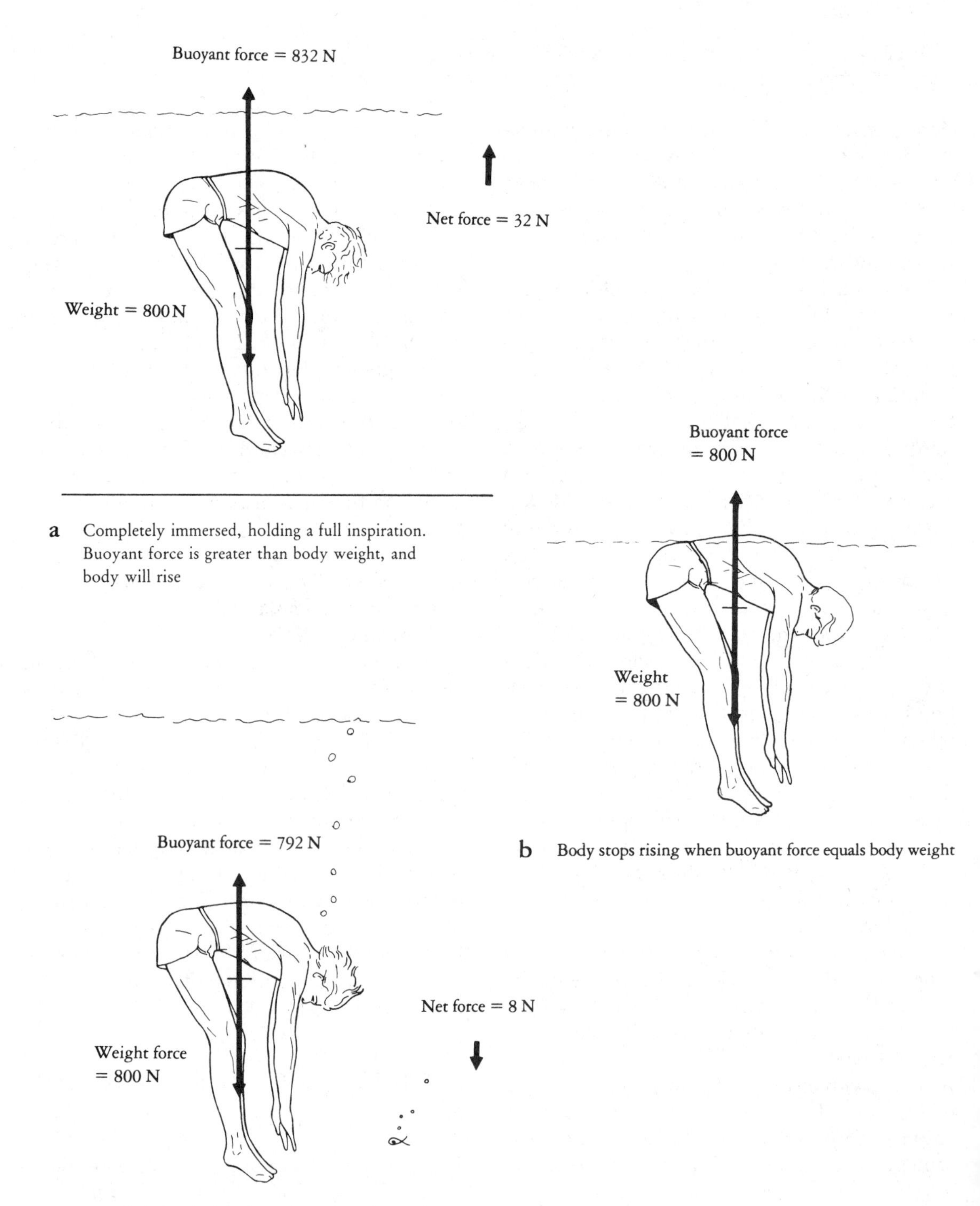

Buoyant force = 832 N
Net force = 32 N
Weight = 800 N
a Completely immersed, holding a full inspiration. Buoyant force is greater than body weight, and body will rise
Buoyant force = 800 N
Weight = 800 N
b Body stops rising when buoyant force equals body weight
Buoyant force = 792 N
Net force = 8 N
Weight force = 800 N
c On exhalation, body volume decreases, volume of water displaced decreases, and buoyant force decreases

FIGURE 12.4
Body density and flotation ability changes with thorax volume changes that accompany inhalation and exhalation.

body (**center of gravity [CG]**). Recall that the CG is that point about which all the mass of the body is balanced and is the point through which gravity exerts its resultant downward force on the body. Recall also that as body segments are repositioned, the location of the CG shifts in the direction of segment shift. The center of buoyancy of a body can be defined as the center of the volume of the body displacing the water.[1]

Look at the sealed can in Figure 12.5a, the CG of which is at the center of its volume. If the can is submerged (Figure 12.5b), it displaces a volume of water whose CG is also at its center. The center of buoyancy of the can is, therefore, in the same location as its CG. If, however, the bottom of the can is filled with cement or another dense material, the can's CG shifts to the position shown in Figure 12.5c. When this can is then submerged, it displaces the same volume of water as before, but now the CG of the can and the CG of the displaced water do not coincide (Figure 12.5d). Similarly, when the human body is submerged, its CG location is not the same as that of the displaced volume of water (Figure 12.5e and f). The CG of the displaced volume of water is the center of buoyancy of the person's body and is the point through which the water exerts its upward buoyant force on the body. The center of buoyancy normally is closer to the chest region (the region of greater water-displacement volume), and the CG is closer to the pelvic region. As Figure 12.6a shows, a person initially in a supine position experiences a downward force equal to the body weight acting through the body's CG.

The person also experiences an upward force equal to the weight of the displaced water acting through the center of buoyancy. The *line of action of gravity's force* and the *line of action of the buoyant force* are parallel, but the two forces act in opposite directions. If the lines of force do not coincide—that is, if the CG and center of buoyancy are not in vertical alignment—the forces cause the body to rotate. This rotation of the body occurs along a sagittal plane about a mediolateral axis located through the center of buoyancy, thereby causing the lower body to rotate downward (Figure 12.6b). A tendency also exists for the upper body to rotate upward; however, the elevation of the upper body above the surface is precluded by the required maintenance of water displacement for upward support. The main effect, therefore, is the downward rotation of the lower body to a position in which the CG is beneath (in vertical alignment with) the center of buoyancy. When this occurs, the line of gravity and the line of buoyancy are *colinear*; no further turning action exists, and the body is in equilibrium (Figure 12.6c).

The greater the horizontal distance between the CG and the center of buoyancy, the greater the resulting torque (turning force) and the more the lower body rotates before the forces are colinear. Also, if the lower body tends to sink (rotate downward) fast, the legs' downward momentum may be enough to pull the floater's face underwater, giving the immediate impression to a conditional floater that he or she is a sinker. A nearly vertical body orientation is usually the equilibrium position for conditional floaters who are located toward the "sinker end of the buoyancy continuum." The true floater, however, tends to remain in more of a horizontal position, because the body's CG and center of buoyancy tend to be closer together. Accumulation of adipose tissue in the abdomen or pelvic region displaces

[1]The center of buoyancy may be technically defined as the point that is the CG of a volume of water equal to the volume of the submerged body and having an identical shape.

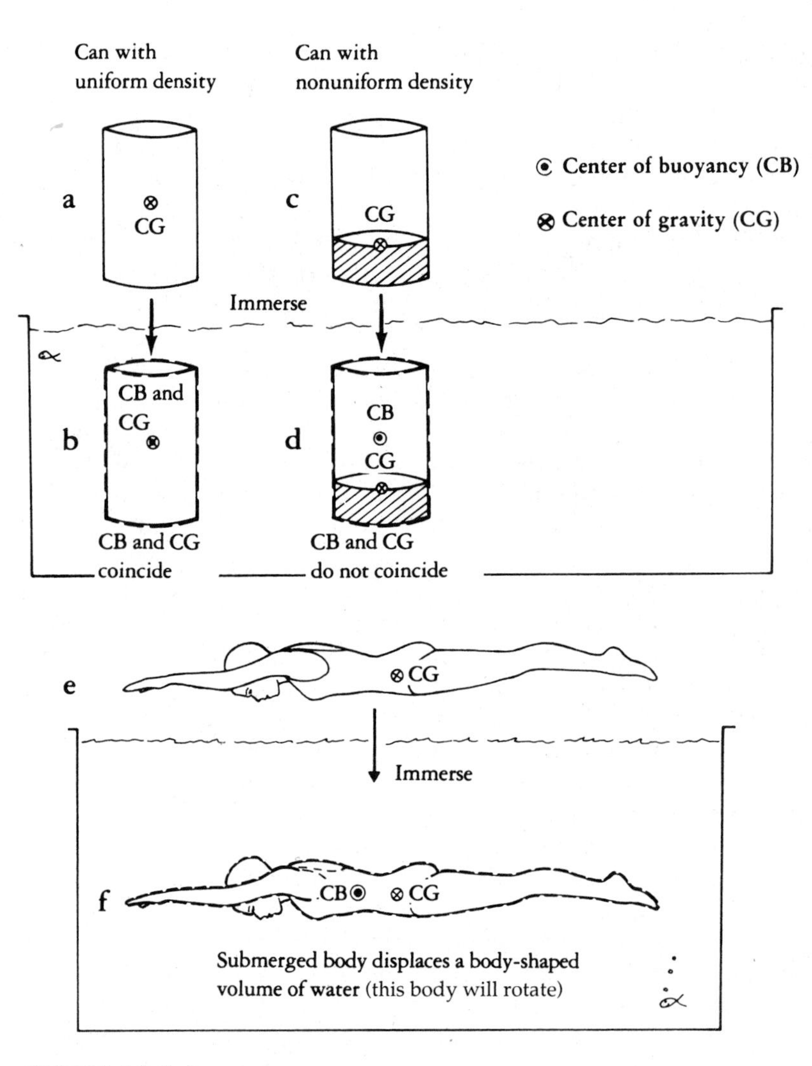

FIGURE 12.5
Relative locations of the CG and center of buoyancy (CB) of two cans and a human body.

more water there and shifts the center of buoyancy toward the pelvis, bringing it closer to the CG.

Floating positions can be altered by changing the positions of body segments to shift the CG closer to the body's center of buoyancy so that the torque and resulting body rotation are decreased. When body segments are repositioned, the CG has a greater change in location than does the center of buoyancy. Three

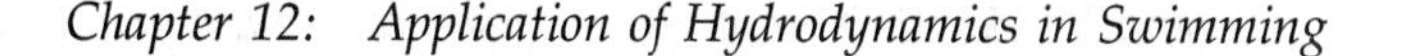

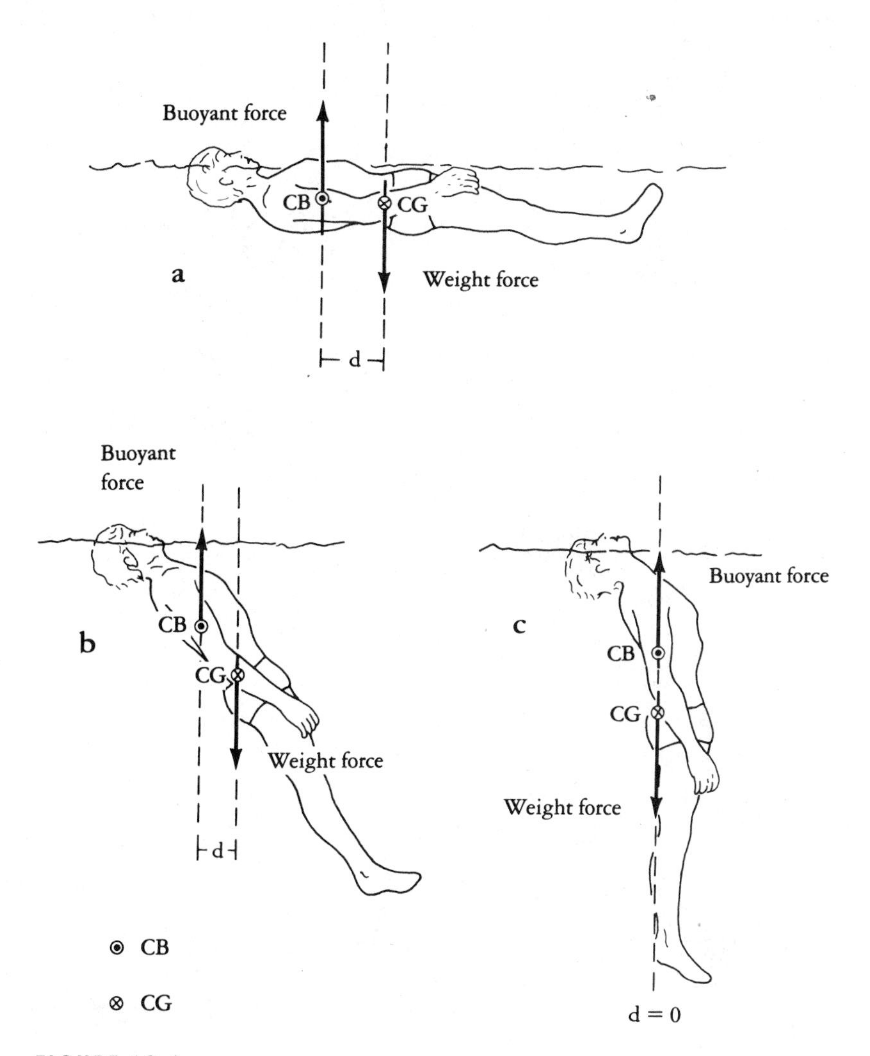

FIGURE 12.6

(a and b) The torque exerted by forces of gravity and buoyancy causes (c) the body to rotate to a position of equilibrium.

common segmental adjustments are shown in Figure 12.7.

The CG does not necessarily have to be beneath the center of buoyancy for a zero torque situation; it can lie above it. This is the case for an inverted body position, but it is unstable. In this situation, all that is needed for imbalance to occur is a slight rotation in a vertical plane to create a horizontal distance between the CG and center of buoyancy, thus creating an upsetting torque. Such an inverted position is used frequently in synchronized swimming skills, and balancing movements of the hands must continually be made to maintain the vertical body position. The body is in a stable float when the CG is lower than the center of buoyancy.

The body can be caused to rotate also in its

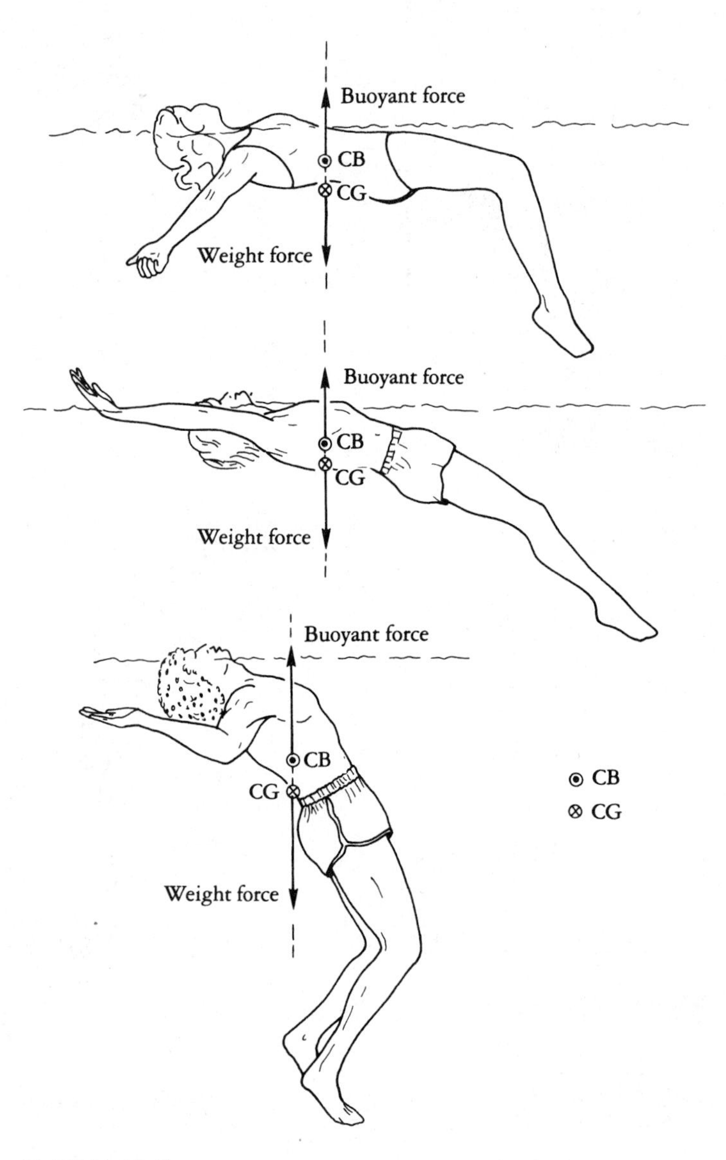

FIGURE 12.7

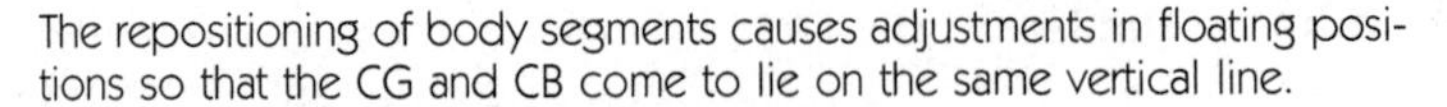
The repositioning of body segments causes adjustments in floating positions so that the CG and CB come to lie on the same vertical line.

transverse plane under the action of the two vertical forces as well as in its sagittal plane. Such a "log roll" can be executed on purpose as a stunt, or it can be caused by unintentional lateral flexion of the vertebral column (as is often done by a tense beginner) or by otherwise shifting the CG location laterally. Because rearrangement of body parts caused by tension can influence body position in the water, it should not be overlooked as a possible source of position problems in swimming skills.

Understanding Buoyancy and Flotation

1. What determines the magnitude of the buoyant force acting on a person in the water?
2. What happens to the average weight density of the body when a full inspiration is made? Why?
3. State what characterizes a conditional floater.
4. Who is more likely to float: a lean female weighing 668 N (150 lb) or an obese male weighing 801 N (180 lb)? Why?
5. Describe what is meant by center of buoyancy (CB).
6. Describe the relative locations of the CG and CB for a person whose legs drop when attempting to float in a horizontal position on the back. Describe the locations for a person floating horizontally.
7. In a swimming pool, working with a partner or in a small group, perform the following experiments and discuss the results.
   - **a.** Stand in chest-deep water, inhale deeply, hold the breath, and slide your hands down the front of your legs toward your feet; reverse the movement to regain the upright position.
   - **b.** Determine whether you are a floater, conditional floater, or a sinker by using the criteria discussed in the text.
   - **c.** Assume a prone float position and explain the changes in body orientation that occur as a result of (1) changing segmental positions, and (2) lateral flexion of the lumbar vertebrae.
   - **d.** Assume a prone float position with the neck fully flexed to "tuck the chin." Note the level of the body as it floats. Repeat the prone float position while hyperextending the neck; note the level of the body as it responds to this head position. Explain.
   - **e.** Attempt to use the body positions in Figure 12.6 and 12.7 to float. Explore other segmental arrangements, holding each for several seconds, and explain why your body moved to the position it did for a static float.

12.2 Resistive Forces in Swimming Skills

An understanding of the forces that resist movement of the body or segments in the water is a major factor contributing to one's ability to help others learn. Whenever the body or its segments move through water, the motions are resisted by the water. The present section discusses ways in which skin friction and profile drag forces on the body can be decreased to improve the performance and introduces a third type of fluid drag not covered in Concept Module H—wave drag. As we shall see, wave drag is an important source of hydrodynamic drag because it is created whenever a body is moving at or near the surface of the water, that is, at the air–water interface.

Skin Friction

Skin friction, or surface drag, is a resistive force caused by the water flowing backward along the surface of a body that is moving forward through the water. When compared with the other types of drag encountered in swimming, it is the *least* significant.

Skin friction is less along smooth-textured surfaces and is proportional to the amount of *surface area* of the body along which the water flows (parallel and opposite to the direction of body motion). If the skin or surface conditions are extremely rough (e.g., a loose-fitting swimsuit or an inordinate amount of hair on the head or body), then alteration of these conditions may produce a significant reduction in surface drag. The smoother the skin and swimsuit and the smaller the surface area of the body oriented parallel to the flow, the smaller this source of resistance. Changes in competitive swimsuit materials have been based on the desirability of reducing this skin friction, as has the practice of shaving the body, legs, and head if the amount of hair is excessive. If the body is not exceptionally hairy, shaving may serve more of a psychological than a mechanical purpose. Swimmers who shave their bodies often report that tactile sensitivity to the water is enhanced. Also conceivable is the assertion that the swimmer's tactile sensitivity is reduced, rather than enhanced, due to the removal of hairs, which, when moved, serve to stimulate receptors beneath the skin. In either case, the sensation is different with and without hair. The significance of this factor is probably negligible in light of the fact that skin friction is so small compared to profile and wave drag. Efforts would be better spent addressing the reduction of the significant drag forces.

Profile Drag

A significant type of resistance in swimming is that created by the swimmer's body "spreading the water apart" as it moves through the water. The amount of this *profile drag* depends on the size, shape, and speed of the swimmer as well as the orientation of the body relative to the flow. Recall that the speed factor is squared, so that by doubling the speed, the drag is quadrupled. Also, if the speed is doubled, the power output needed to maintain that increase is eight times as much.

As the body moves forward through the water, the water is "spread apart" to allow the body to pass through. The greater the area of the body and its parts that face the flow (i.e., the area you would see as a silhouette if you were the flow), the greater the drag. High-pressure zones are created on the leading surfaces of the body, and low-pressure zones are formed at the trailing surfaces where the water is moving faster and is turbulent, or full of eddies, thus creating a suction effect. The result is a net force backward against the forward-moving body and its parts.

Huijing et al. (1988) studied a large number of anthropometric characteristics of swimmers to learn which body features might affect drag. A highly significant feature contributing to drag was the maximum cross-sectional area of the body (transverse plane). Also found was that the greater the length/thickness ratio of the swimmer, the less the drag. This ratio also is referred to as the slenderness ratio, implying a long slender shape.

Figure 12.8a illustrates the relative flow past "swimmers" of different shape and the pressure zones caused by the different flow characteristics. As the velocity of the body increases (thereby increasing the velocity of water flow relative to the body), the pressure differential between the high- and low-pressure zones increases. The squared effect of increasing the velocity means that if the velocity of the body were tripled, the profile drag would increase by approximately nine times.

Recall that the shape of the body influences the way the water flows past it. The body's shape that is facing the flow determines its coefficient of drag (C_D), an index of how nonstreamlined it is relative to the flow. The influence on drag of the actual human body shape, however, may be noticed only when

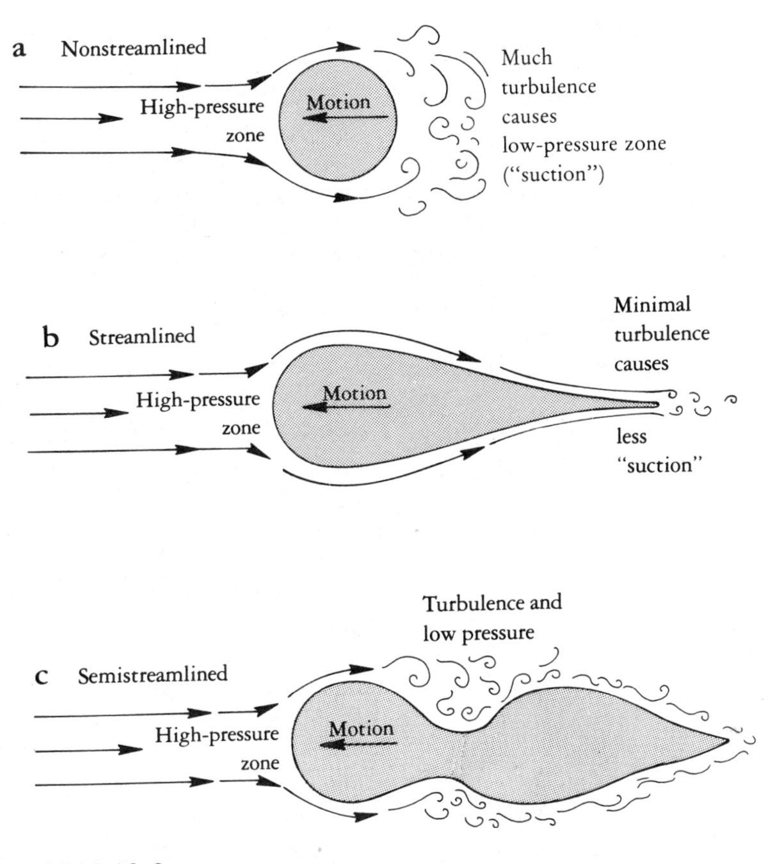

FIGURE 12.8

A high-pressure zone is formed at the leading surface of a body moving through the water, and a low-pressure zone is created on the trailing surfaces. The more abrupt the contour changes are, the greater the turbulence and low pressure in those areas.

the body is held in a given configuration, as in a glide from a wall push-off. Under such a condition, a more streamlined body shape may be beneficial. If the shape of the body is more streamlined as shown in Figure 12.8b (i.e., if the body tapers gradually from shoulders to hips to feet), the profile (form) drag will be less than it would be for a body with irregular contours (Figure 12.8c). Contours in the body cause water to flow at varying speeds around these shapes, and turbulence is created on the downstream sides, thus producing low-pressure, or suction, zones. Note that a *streamlined* shape means that the shape encountering the water flow allows the water to flow smoothly around and past the body without much turbulence. For example, a streamlined fish body would *not* be a streamlined shape if it were exposed to the flow sideward in a vertical orientation.

The streamlined shape, therefore, is a function of the swimmer's body build, but the shape and area exposed directly to the flow by virtue of body position are factors that influence the drag coefficient. The influence of actual body build or shape on drag, however, may be limited to a moving body that is *not* changing shape (Clarys, 1978). Clarys reported *no* relationship between shape and drag force measured *during* swimming. In addition, he calculated that the drag during swimming is twice as great as the drag on the body being towed in a prone position. These calculated values have since been challenged by researchers using film analysis (Schleihauf, Gray, and de Rose, 1983) and an underwater force-measuring pull-handle system (Hollander et al., 1986, 1987). They found that drag during actual swimming (active drag) may be just about the same as passive drag (during towing) which indicates that the arm and hand swimming movements of *skilled* freestyle

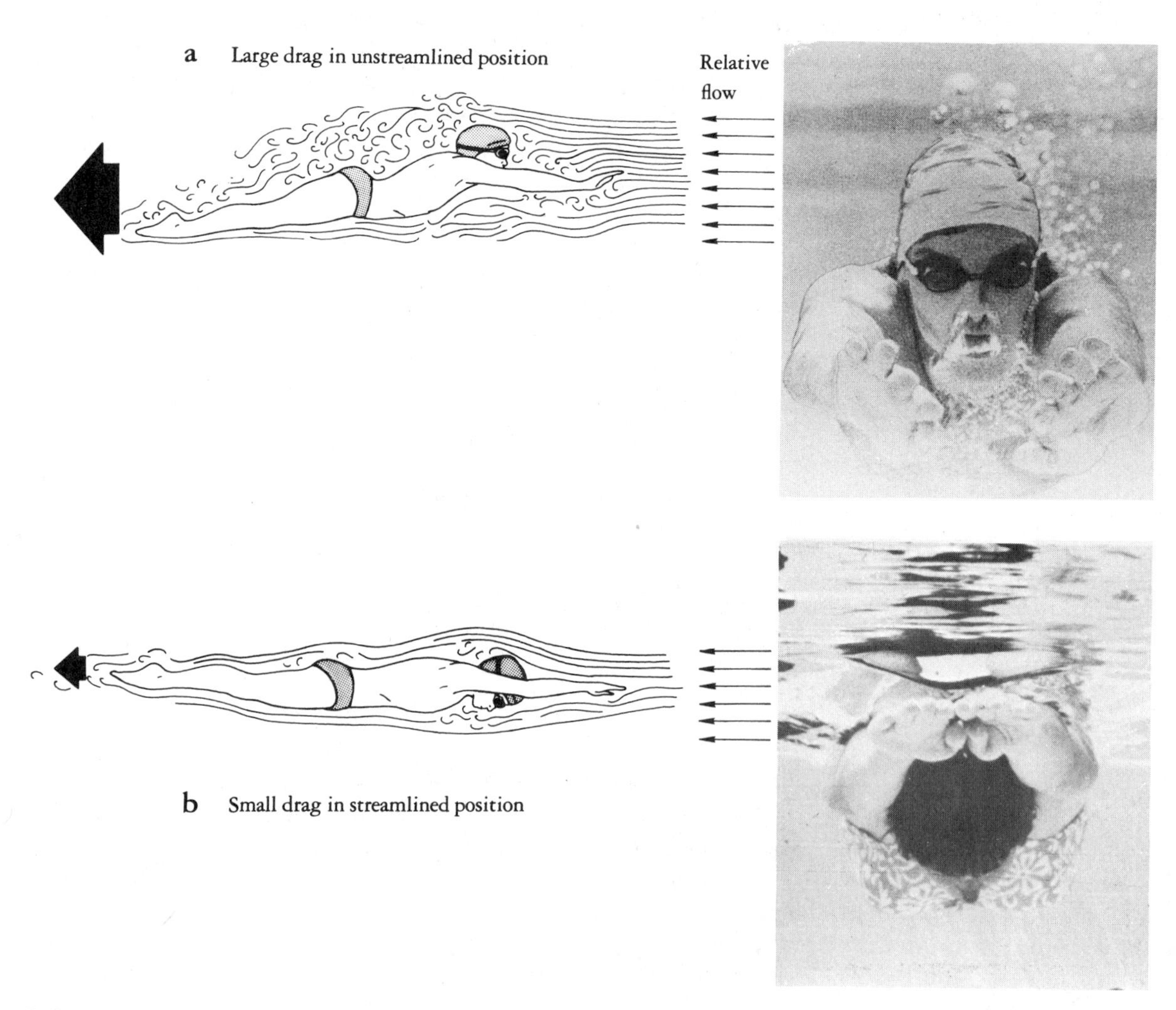

FIGURE 12.9

Although a person may have a streamlined body shape, the body position relative to the flow actually determines how streamlined the body is.

swimmers do not seem to contribute to the resistive drag. Whether, or how much, the arm movements of less skilled swimmers contribute to drag has not been studied, although one might speculate that they are a significant source of resistance.

Swimming movements produce changes in the relative flow patterns that account for drag forces. Although we would not stop the swimming movements to decrease the drag, we can control the body position influences. If the body is maintained in a horizontal position during a swimming stroke rather than one in which the legs are maintained in a position lower than the upper body, the drag coefficient and thus the profile drag can be effectively reduced, and the swimmer need not work as intensely to maintain a given speed.

Figure 12.9 shows the greater drag on an *unstreamlined* body *position* during a wall push-off glide compared with the more *streamlined position*. In Figure 12.9a, the swimmer is holding his head up, facing forward rather than downward toward the pool bottom. Such a push-off glide head position is common in competitive swimming practice for two reasons: (1) to see what is approaching is safer in crowded lanes, and (2) the water pressure holds the swimmer's goggles on the face. In the more streamlined head position (Figure 12.9b), the flow of water past the face frequently wipes the goggles off the eyes. Unfortunately, swimmers in practice usually avoid the streamlined "goggle wiper-offer" head position, and the high head position used in practice carries over into competition even when goggles are not worn.

Another way in which drag can be reduced during swimming is by trying to minimize drag-producing aspects of limb recovery movements. How the limbs are brought forward under water during recovery, or preparatory, movements in swimming strokes determines the amount of resulting profile drag acting on them. Figure 12.10a and b illustrates the underwater recovery movements of a breaststroker's arms and legs. Not only do these movements expose more *area* to the backward flow, but moving the limbs forward through the water *faster* than the body (trunk) is moving increases the resistance even more. The higher relative speed between the water and limbs on recovery is the source of drag that makes the deceleration of the body so evident in the breaststroke. For example, if the body is moving forward at 1.5 m/sec and the arms are moving forward relative to the body at 1 m/sec, the water flows backward past the body at 1.5 m/sec but past the arms at 2.5 m/sec. Such an increase in flow velocity past the arms increases the drag greatly, especially if the arms are in less streamlined positions during their forward movement.

In the back crawl stroke, a common tendency among swimmers is to lift the head upright above the surface of the water to look toward the feet (Figure 12.10c). This movement leads to some body sinking (recall the effect of displacing less water for support). The arm movements are then used for the purpose of keeping the head and upper body at the surface, that is, pushing downward toward the pool bottom. The result is usually the assumption of a slight pike, or sitting position, in the water, which increases the area exposed to the flow, undoubtedly increases the drag coefficient, and therefore increases the profile drag.

Wave Drag

Wave drag is the resistive force caused by any body moving along, through, or near the surface of the water and forming waves, or elevating some water in front of the leading parts. The force applied by the body or its segments to the water to form these bow waves generates an opposite reaction force applied by the waves against the body. The force of these swimmer-created waves increases with

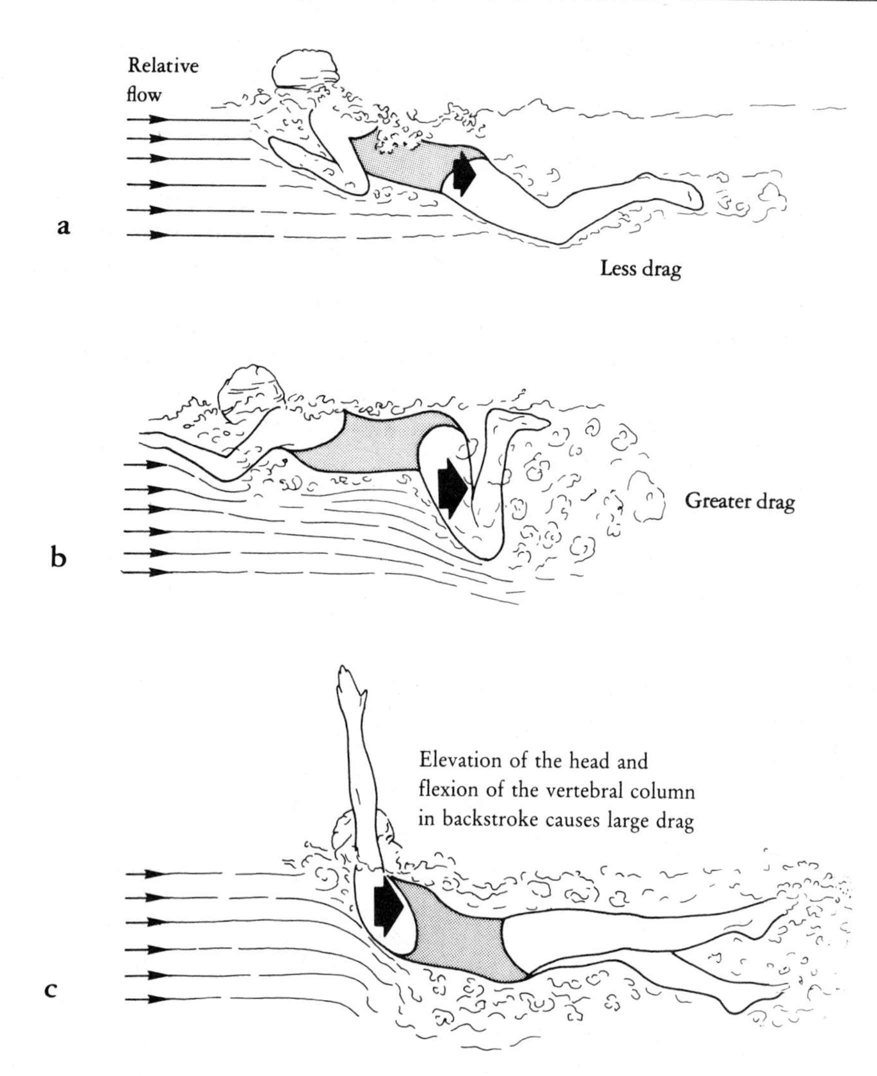

FIGURE 12.10
Profile drag on swimmers moving through the water is increased with increases in area facing the flow and with faster relative flow past the limbs.

the speed of the swimmer, just as the wave against a bridge piling exerts more force as the speed of flow increases in a river. Even if a moving object or body is completely submerged but is still within several feet of the surface, wave drag is not entirely eliminated as a source of resistance. Although at low speeds wave drag is not significant, it may be the most important source of resistance at the faster swimming speeds seen in competition. As the speed of the swimmer increases, the size of the bow waves increases and the "wall" of water presses backward against the swimmer. Bow waves are formed against the leading surface of any parts of the body that move through the air–water interface. The formation of bow waves also increases greatly with speed and with up-and-down movements of the shoulders or head found in many strokes. Consequently, more resistance is experienced by swimmers moving fast in a stroke, than swimmers moving more slowly. The skill level of a swimmer influences how much wave drag is produced. Highly skilled swimmers produce less wave drag than recreational swimmers moving at the same speed (Takamoto, Ohmichi, and Miyashita, 1985).

Figure 12.11 shows wave drag acting against swimmers as they move forward through the air–water interface. A large bow wave can be seen in front of the face and chest of a breaststroker as the breath is taken. The positioning of the body shown here is one of several "wave action" breaststroke techniques. These techniques utilize a "snaking" movement of the body integrated with well-timed limb movements to "weave" the body through the water, rather than "cleave" it through. As the body is raised out of the water, more wave drag is produced. This additional drag may be compensated for by decreased area and profile drag beneath the surface. Also, it is likely that the body movements taking the trunk through this position may provide greater propulsion than movements done with the body kept horizontal.

The bow wave formed around the head is used by skilled swimmers to facilitate breathing in the front crawl stroke. As the head is turned to the side, it need not be rotated very far for the mouth to be exposed to air, for behind the bow wave is a depression below the natural surface of the water, sometimes called a "pocket of air" (Figure 12.11d). If the swimmer is moving slowly and the bow wave by the head is small, this pocket barely exists and the head must be turned through a greater range of motion (ROM) for the mouth to be out of the water. This extra rotation may lead to a greater amount of rolling of the trunk in its transverse plane and subsequent changes in arm-stroke patterns. Many other instances of unnecessary resistance in swimming may be identified if the causes of drag are understood, for the principles may be generalized for any swimming skill.

The determination of actually how much drag force is encountered by skilled swimmers at different speeds has been a popular subject for research during the past 15 to 20 years. Drag is very difficult to measure directly, and estimates have been produced basically from three methods. One procedure is to measure the force necessary to tow a passive swimmer through the water at a constant speed (the motive towing force is equal to the resistive drag force). Drag values for this method range from about 25 N ($\approx$ 6 lb) to 37 N ($\approx$ 8 lb) at speeds of about 1.5 m/sec (Hollander, de Groot, and van Ingen Schenau, 1987). No matter what the actual magnitude of drag force is, if steps can be taken to reduce it by recognizing drag-producing features, swimming efficiency will be improved.

In summary, the total body drag is the resistive force that must be matched by the propulsive forces if the swimmer is to maintain a constant speed. The sum of the three types of drag constitutes the total drag force acting on a body in water. Profile and wave drag as sources of resistance in swimming are so large when compared to skin friction drag that the latter usually may be considered negligible. If the swimmer is to increase speed, the propulsive forces must be greater than the total body drag. If this drag is greater than the propulsive forces, the body will decelerate. Although often not noticeable to an observer, a swimmer's body is changing speed continually during every stroke cycle, which is due to the interplay of changing resistive and propulsive forces. The next section describes how propulsive forces are generated in swimming.

a

c

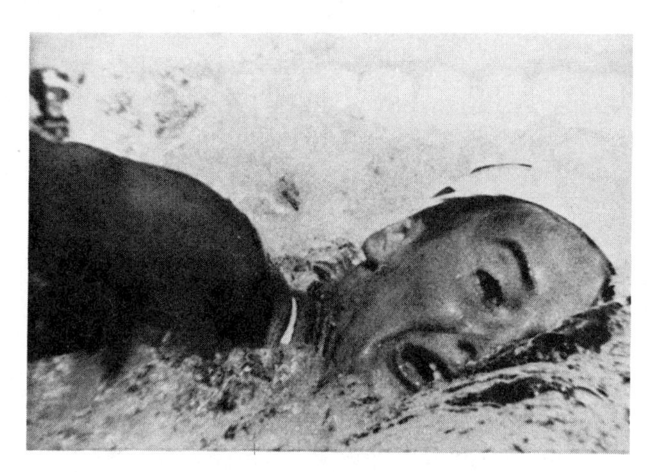

d

b

FIGURE 12.11

Wave-making drag acts against the body parts traveling forward through the surface of the water in (a) the breaststroke, (b) backstroke, (c) butterfly, and (d) freestyle.

Understanding Resistance in Swimming Skills

1. Name and describe each of the three types of drag force that can resist the forward body motion of a swimmer.
2. Why would a "sinker" sink faster in a vertical position than in a horizontal position?
3. In the swimming pool, perform these experiments and discuss the results:
   a. Lie along the edge of the pool and move your hand and forearm horizontally through the water at different speeds and with different hand shapes and positions relative to the water. Compare the forces you feel.
   b. Compare the speeds with which you can walk or run a given distance through hip-deep, waist-deep, and chest-deep water.
   c. Select a swimming stroke you can perform reasonably well. Swim a given distance with the head out of the water and count the number of strokes taken. Compare the number of strokes necessary to swim at the same speed with the head supported in the water. Explain why the numbers might be different.
   d. Force a kickboard under water to a depth of one width vertically on its edge and then in a horizontal broadside position. Compare and explain the relative difficulty of the two ways.
   e. Perform an underwater push-off glide with the same force of leg extension for a number of different arm, hand, and head positions. Explain the differences in resistance felt.

12.3 Propulsive Forces in Swimming Skills

Most explanations of swimming skills describe limb movements as they occur relative to the swimmer's body rather than how the limbs are moved through the water. The popularity of this viewpoint undoubtedly stems from the swimmer's perception of what the arms and legs *seem* to be doing in a stroke as well as what the limbs *appear* to be doing from the observer's viewpoint. As a result of such kinesthetic and visual impressions, the true nature of the propulsive hydrodynamic forces operating on the limbs usually remains obscure.

To identify these forces, we must examine the swimmer's limb movements relative to the water, not relative to the body. To do this we must focus on how the water flows past the moving part, just as we did in analyzing resistive forces, for this relative flow is responsible for the forces used for propulsion. Two types of propulsive force can be produced by the swimmer's body or limbs or both moving through water: **propulsive drag force** and **propulsive lift force**.

Propulsive Drag Force

Drag force was explained previously in terms of how it acts as resistance against the forward motion of the swimmer's body and its segments. The direction of drag force is always opposite to the direction of body or segment movement. Therefore, if a segment moves to the left, the drag force acts against it to the right; if the segment moves forward, drag acts backward. When drag force acts on *backward-moving* body segments, however, it is called **propulsive drag force** because it acts forward in the desired direction of body travel. Such a propulsive (forward-directed) drag force results from paddling or pushing movements of the hands back through the water, and it acts to resist the backward hand motion. As was shown for resistive drag acting on a body, a high-pressure zone is created on the leading surface (the palm) and a low-pressure zone is created on the back side of the hand. The faster the hand movement and/or the larger the hand area, the greater the forward drag force against the hand's backward movement. Using hand-arm movements that generate propulsive drag as the main source for propulsion, such as in

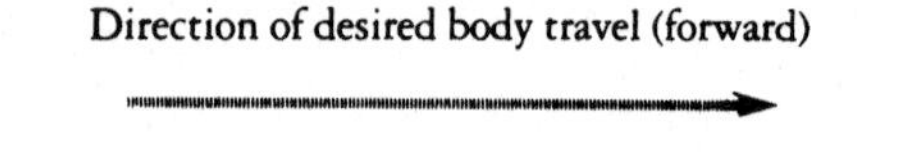

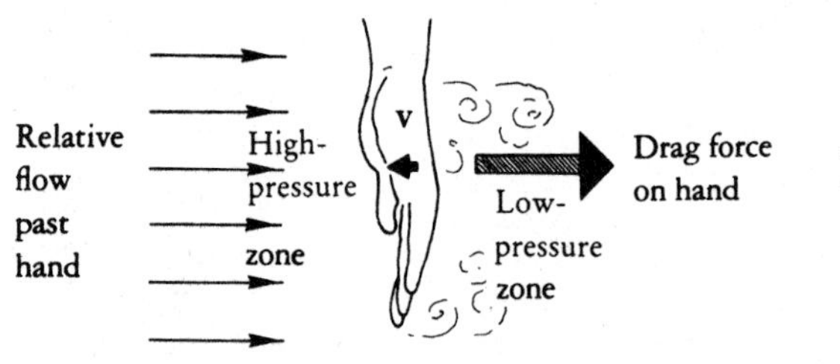

a View from side:
Hand is pushing straight backward through water

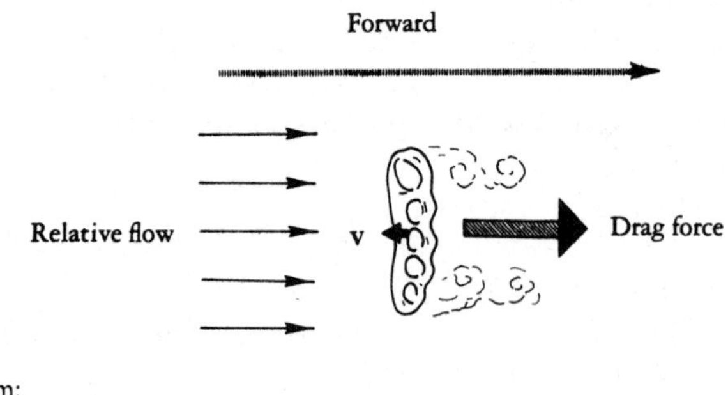

b View from bottom:
If the hand pushes straight backward through water; all of the drag force is directed forward

FIGURE 12.12

Propulsive drag force can be generated by paddling the hand backward through the water so that the water flows forward past the hand. This method of pulling is not efficient.

straight-back arm pulls, is inefficient, and the reasons for this inefficiency are discussed in a following section. Figure 12.12a and b shows how a drag force is produced on the hand if it moves backward through the water so that the flow is directed against the palm.

Propulsive Lift Force

In swimming activities, relative water flow past the moving body segments can provide not only drag force but also a second kind of force that plays an important propulsive function—**hydrodynamic lift force**. Recall that, although the term *lift* implies that it is an upward force, *it is not always upward*.

Horizontal lift force is used for propulsion by bodies whose blade-shaped parts move in a special way to produce horizontal lift force, just as propeller blades do.

To understand how hydrodynamic lift force can be a horizontal propulsive force in swimming, the water flow *relative to the moving body segment* must be examined. The nature of water flow past the hand (a wing-shaped object) in Figure 12.13a is responsible for the lift force that is caused by the hand's movement. The hand and forearm can be likened to a propeller blade. If the hand is

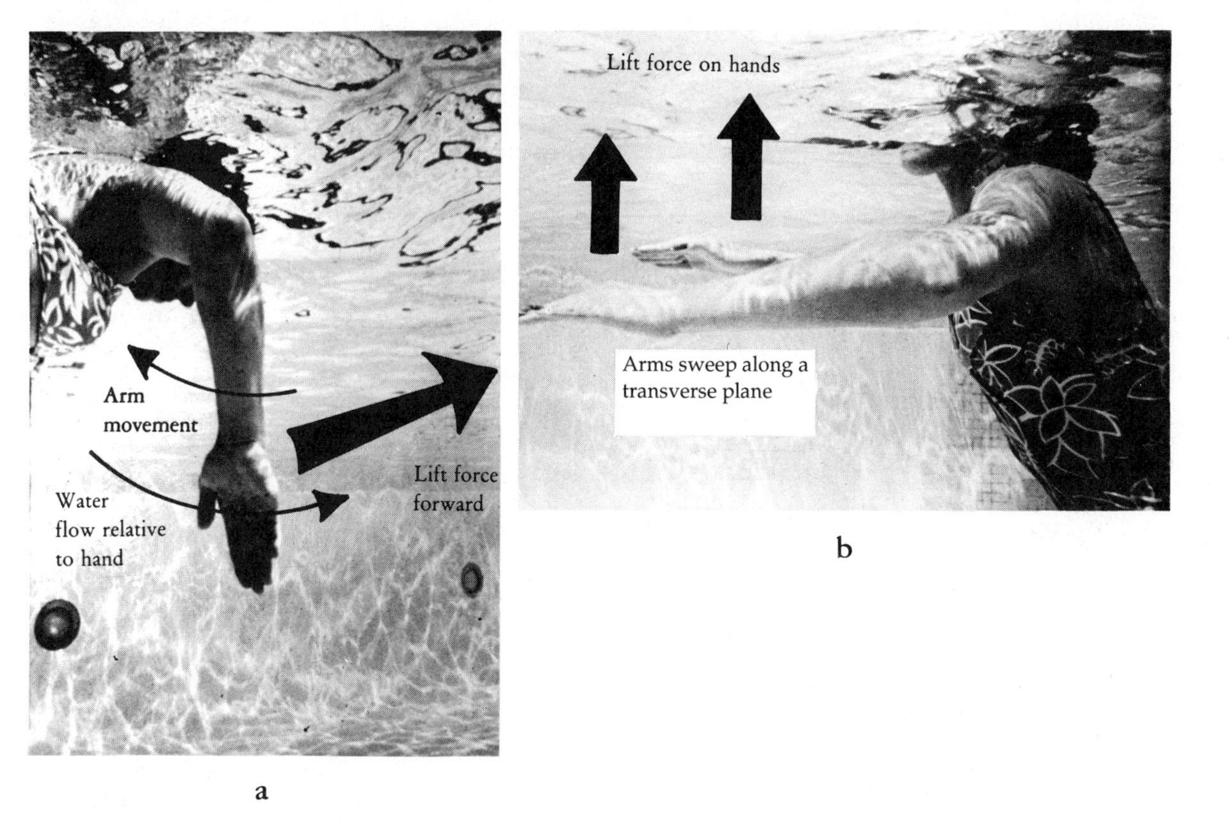

FIGURE 12.13

Blading (sculling) the hand through the water medially and laterally simulates a propeller to form a lift force on the hand. (a) A forward lift force on the hand is produced when the palm faces backward. (b) An upward lift force acts on a downward-facing palm.

positioned at an angle to the direction of flow, the water flowing past the moving hand flows faster over one side than the other. Such a flow–velocity difference existing between opposite sides of the hand causes a pressure difference between the two sides. The existing pressure difference causes the hand to experience a lift force directed from the region of higher pressure to the region of lower pressure. Note that the direction of lift force is *perpendicular* to the flow direction past the hand (and therefore perpendicular to the *path* of the hand), whereas drag force is directed parallel to the flow direction (opposite the hand motion). Therefore, lift and drag are always perpendicular to each other.

Lift force is not generated on all bodies moving through fluids; drag force is. For the flow patterns formed around the moving segment to generate lift force, the segment must have a shape and orientation to the flow that permits the formation of lift, such as an airplane wing (airfoil) or hydrofoils; the forelimbs ("wings") of a sea turtle, seal, or penguin; the pectoral fins and tails of fish, whales, and dolphins; and the human hands and feet. Lift force, directed perpendicular to the path of the "slicing" hand, is felt by the swimmer as pressure on the palm when the slightly tilted hand blades through the water, as in the sculling movements used in synchronized swimming skills and treading water. Figure 12.14 shows the hydrofoil-shaped hand oriented at an angle relative to the flow direction so that lift force can be generated. Such tilting of the hand is described by its **angle of attack**, which is the angle formed between the direction of flow and the main plane of the hand.

Each hand experiences lift force caused by a low-pressure zone across its knuckle side and a high-pressure zone on its palm. In this example the lift force is upward. Such lift-force

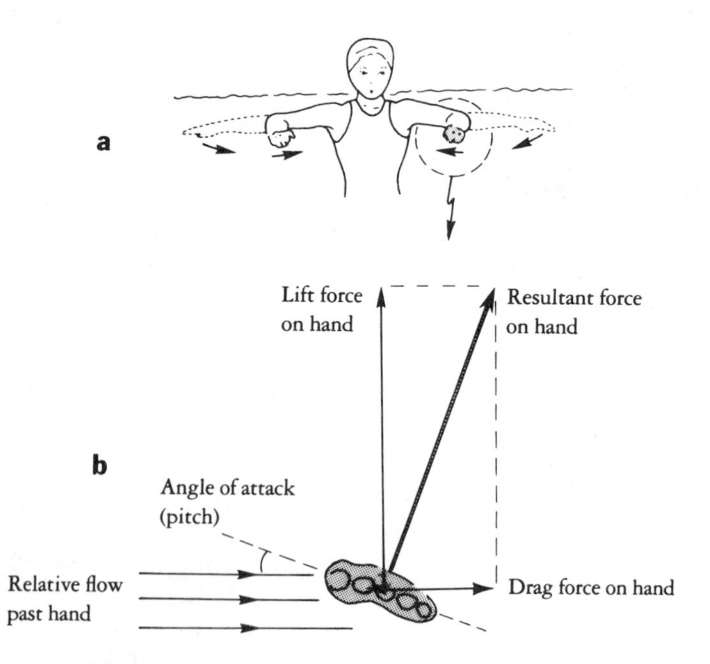

FIGURE 12.14

(a) Lift force on the hands is directed upward to support the weight of the head out of the water. (b) A close-up view of the blading hand in (a).

producing movements of the hands are used effectively for treading water without the use of the legs. The hands are used as hydrofoils, and the arms sweep horizontally so that broad blading, or sculling, movements are made to provide what might be pictured as a "layer of force" for the hands to use for supporting the weight of the head above the surface. Note that the hand's angle of attack must be small so that the palms are facing *primarily* toward the feet as they scull. Note also that the hand experiences drag force opposing its horizontal movements. This horizontal drag force contributes nothing to head support (it has no vertical component). To generate supporting lift force, the flow must encounter the palm side of the hand as it blades medially and laterally, so the lateral sweep demands more forearm pronation than does the medial sweep to produce the proper angle of attack for the hand. The upward lift force generated on the palms serves as a "handle" as the shoulder adductor muscles contract to keep the head supported above the surface. Thus, the body and arms are kept in their same relative position.

Although our attention is focused on lift force, note that drag force also acts on the hand. If the angle of attack becomes too large, the drag force increases, and the flow becomes so turbulent that lift force is lost. Obtaining an angle of attack that provides for maximum lift force for a given hand shape and blading speed is possible. Skilled swimmers may be able to feel for and sense the best angle as it changes throughout the stroke.

Note that the hands are not used as *pushing paddles*. They do not push downward through the water toward the feet seeking an upward *drag* force for support. Rather, the hands continue to blade inward and outward in a horizontal section, or layer, of water close to the surface. In this example, upward lift force is being used as supportive lift force; it is directed perpendicular to the path of the sideward-sweeping hands to counter the tendency of the body to sink downward. In treading water in this manner the hands will move downward only when they reverse direction. This downward motion is momentary and provides some upward drag force on the hands to make up for the lift force lost when the hand blading speed is reduced to zero for direction change.

Recall from Concept Module H that the magnitude of lift force that can be produced depends on the following factors:

EQUATION 12.2

$$F_L = \frac{C_L A \rho v^2}{2}$$

where F_L is hydrodynamic lift force, C_L is coefficient of lift of the hand and forearm for a given angle of attack, ρ (rho) is water density, v is flow velocity, and A is area of the palm-forearm unit.

Note that lift force, like drag, is a function of speed; the faster the hand moves through the water at an effective angle of attack, the greater the lift force. *The direction of a lift force is not necessarily upward* as it was in the treading water example. The direction of lift force is always perpendicular to the flow direction and to the drag force. Therefore, if the hand is positioned along a vertical plane with the fingers pointing downward and is then moved sideward through the water, the lift force will be directed horizontally *forward*. Visualization of the proper direction in which lift force is generated is important for understanding the role of lift force in propulsive movements in swimming. For example, if the treader in Figure 12.14 were to lie in a prone position in the water and then perform the same sculling movements ("canoe" scull), the lift force on the hands would be directed *horizontally* (Figure 12.15). This lift force would have a large forward component, since the palms would be facing primarily backward as they moved through the broad sculling motions along a vertical plane. The horizontal lift force would cause the whole body to move horizontally

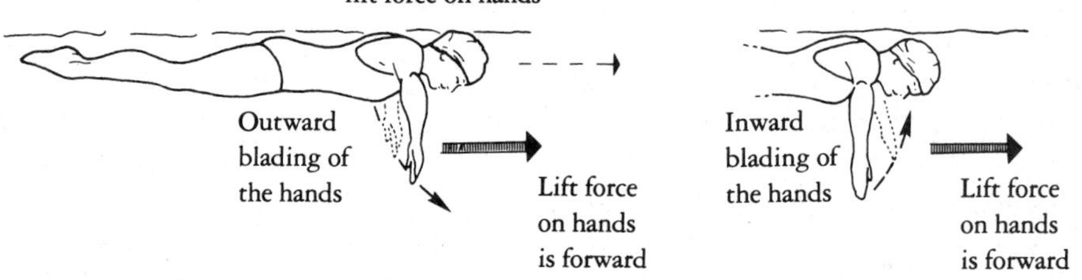

FIGURE 12.15
Lift force is directed forward on the hands as they scull to propel the body forward through the water.

forward through the water rather than cause a vertically directed force to support the head weight, as was done during treading water.

The hand shape as well as its size can affect the magnitudes of both lift and drag force generated. It is conceivable that the forearm and hand as a single unit may act more effectively as a hydrofoil, or wing, because of its dimensions. The *aspect ratio* is the length to width ratio of a winglike shape; and a large aspect ratio is associated with greater lift force. The hand alone has an aspect ratio of about 7:4, whereas the hand with the forearm has an aspect ratio of about 16:4, tending to favor a greater ability to generate lift force. The use of the hand and forearm as a single unit wing is seen not only in treading water but in synchronized swimming support sculling and in proficient breaststrokers. It is also seen in the early phases of the underwater stroking movements in the other styles of swimming.

Movements That Propel the Body

To understand the swimming "mechanism" responsible for effective propulsion of the body, we must recognize that the swimmer is a segmented body and that the purpose of the trunk, arm, and leg movements is to move the swimmer's whole body unit forward. Because of the differences of the trunk, arms, and legs in shape, joint structures, and movement capabilities, movements of these body segments are considered separately.

Propulsive Movements by the Hands and Arms

The two methods for moving the hand–arm unit through the water are basically the *paddling* method (using drag force) and the *sculling* method (using primarily lift force). These methods will be compared and evaluated in terms of their relative effectiveness.

To move the body forward most effectively we would want a situation in which the hand could grip an underwater "handle" that would not move backward as the hand pulled on it. With the hand pressing back against such a handle, the contraction of the shoulder extensor muscles would then cause the body to move forward toward and past this handle as the arm moved through its ROM. Such a hypothetical situation is illustrated in Figure 12.16a. If, however, this handle were not fixed solidly within a section of water, but slipped backward as the hand pressed against it, the body would not move forward through the water as far. With the arm moving through the same shoulder-extension ROM as before, the body would be moved forward a small distance as the hand holding the "slipping" handle moved backward a large distance (Figure 12.16b).

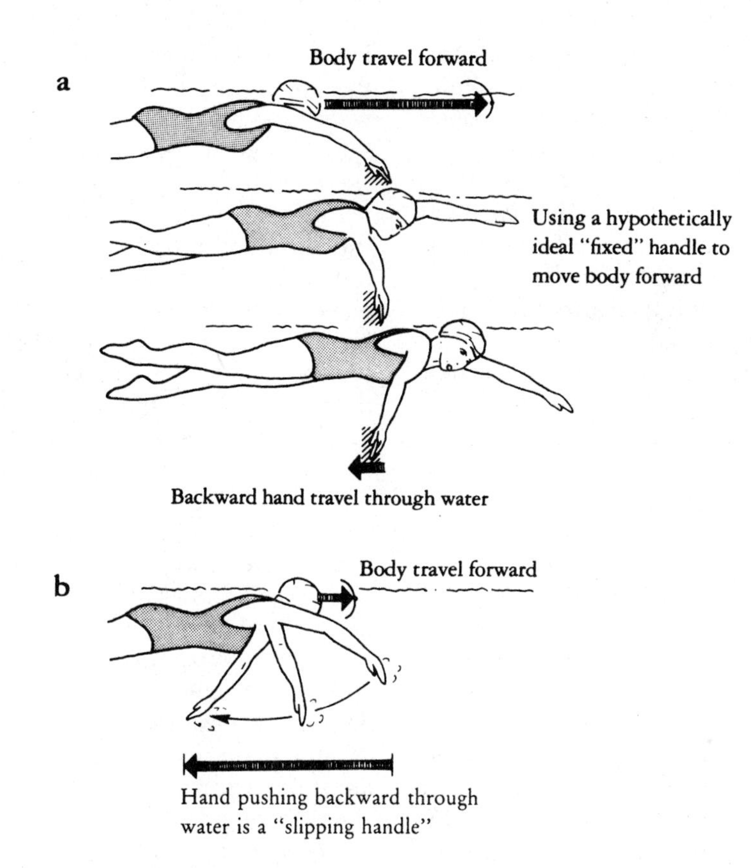

FIGURE 12.16

The shoulder extensors shorten to pull the body forward using an ideal "fixed" handle versus a slipping handle.

A similar analogy would be trying to climb a ladder with stable rungs to push against as opposed to trying to climb by pushing against rungs that slipped downward.

Keeping this analogy in mind, let us look at how the hand and arm can move through the water to provide a relatively *stable*, or *fixed*, "handle" rather than a *slipping* "handle." The more stable handle is created by the *sculling* method of propulsion and the slipping handle by the *paddling* method (Barthels, 1979).

For some time the belief was that to move the body directly forward, the swimmer must pull or push the hands, used as paddles, directly backward through the water. If using drag force alone were the most effective means for propulsion, straight backward pushing would be desirable. Great pressure is felt on the hands and arms when they are moved backward quickly through the water, which is probably why performers have assumed erroneously that propulsive drag is the major force responsible for body propulsion. If the hand is pushed backward slowly, less force is perceived. Therefore, the tendency for the swimmer desiring to swim fast is to pull the hands backward through the water as fast as possible to feel the most force against them in the forward direction. On face value, this may appear quite valid; however, the following reasoning will show the ineffectiveness of backward pushing of the hands relative to the water.

The purpose of a stroke is to move the body mass forward, and any *backward* hand movement *relative to the water* results in less forward motion of the body per arm stroke (see Figure 12.16b). But the hand *must* move backward through the water to create the flow that produces drag force on the hand. Therefore, if drag force on the hand is to be used for propulsion of the whole body, the imaginary handle on which the arm pulls must necessarily be a slipping handle. Moreover, if a greater drag force on the hand is desired, the hand has to move backward through the water even faster, thus yielding even *less* forward body motion by the time the arm reaches the end of its stroke. Such straight-back hand motion, however, is *not* characteristic of skilled swimmers' movements, even though it may feel as though the hands pull straight backward through the water. The ineffectiveness of using the hands solely as paddles to create a propulsive drag force by pushing straight back through the water is apparent from the ladder-climbing analogy. A more effective use of the hands for propelling the body that *is* exhibited by highly skilled swimmers involves using the hands to generate a *horizontally directed hydrodynamic lift force* with a large forward component.

When the hands are used to execute broad sculling movements within an arm-stroke pattern, the palms should form an angle of attack relative to the path of the hand so that the resulting lift force is directed forward. The function of such a horizontal lift force is to *prevent* backward hand motion relative to the water, that is, to help form a *stable* handle to be pressed against to move the body forward. Refer to Figure 12.15, in which the swimmer's body is being propelled forward by broad sculling sweeps of the hands and forearms. For faster body movement forward through the water, the body needs to be pulled forward past the sculling hand by contraction of the shoulder extensor muscles. If these muscles were to contract and shorten, as if to move the arm toward the feet, and at the same time the hand and forearm were to perform a fast, broad sculling motion within a vertical section of water, the *forward-directed* lift force on the hand would provide a fairly stable "handle." The body would be moved forward through the water with little *backward* hand motion relative to the water (Figure 12.17). Although, in such a pattern, the hands do not move backward through the water much at all, the swimmer has the impression that they do as the body moves forward past the sculling hand. The figure shows the swimmer blading the hand through still water along a plane that is primarily *perpendicular to the direction of body travel*; the resulting horizontal lift force on the hand provides a relatively slip-free handle, and the body can be moved over a greater distance per arm stroke by the contraction of the shoulder muscles. Such a mechanism is similar to propeller blades rotating in a vertical plane to propel a boat horizontally forward.

In highly skilled breaststrokers, the lateral, not backward, movement of the hands have been shown to increase body velocity (Tilborgh, Stijnen, and Persyn, 1987). Such evidence supports the use of vertical plane blading of the hands and forearms for effective propulsive forces. Observations of highly skilled swimmers confirm that these sculling motions are used in all four competitive strokes. Figure 12.18 shows a typical path of the hands of a butterfly swimmer as they move relative to the *water* (not relative to the body) from front, side, and bottom views (Barthels, 1974).

The hands are observed *not* to move directly backward through still water, although the swimmer may have the kinesthetic impression that they are doing so. Even though the swimmer may try to maximize the use

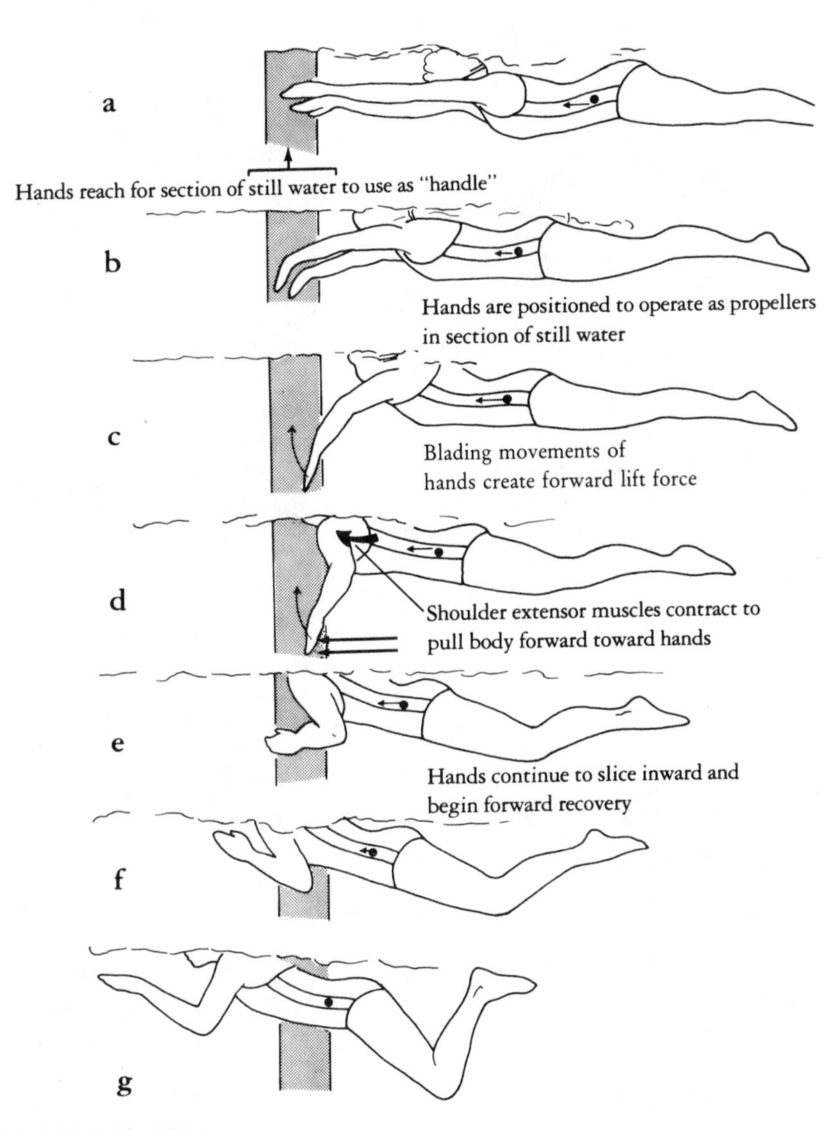
a
Hands reach for section of still water to use as "handle"
b
Hands are positioned to operate as propellers
in section of still water
c
Blading movements of
hands create forward lift force
d
Shoulder extensor muscles contract to
pull body forward toward hands
e
Hands continue to slice inward and
begin forward recovery
f
g

FIGURE 12.17

The shoulder extensor and adductor muscles contract to pull the body past the hand. If the forward-directed lift force on the hand is great enough, very little backward hand travel occurs, and more forward body motion past the hand results.

of the sculling movements to obtain propulsive lift force, unless the hand is very large or the body is very streamlined, some backward hand motion usually occurs during the stroke. The resulting movement produces a combination of propulsive lift force and propulsive drag force acting on the hand (Figure 12.19a); the forward component of the resultant force (of lift and drag) is the force that provides the "handle" for the swimmer (Figure 12.19b).

Most of the backward hand movement occurs when the hand changes blading direc-

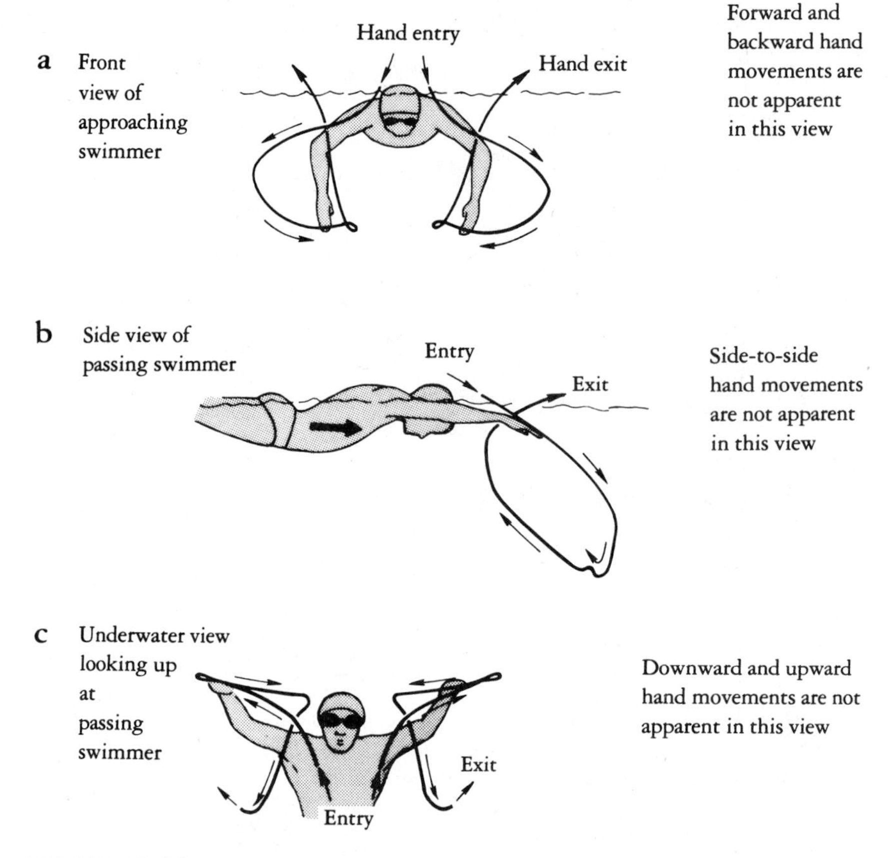
a
Front view of approaching swimmer
Hand entry
Hand exit
Forward and backward hand movements are not apparent in this view
b
Side view of passing swimmer
Entry
Exit
Side-to-side hand movements are not apparent in this view
c
Underwater view looking up at passing swimmer
Exit
Entry
Downward and upward hand movements are not apparent in this view

FIGURE 12.18

The underwater paths of the hands relative to the water they interact with. The hands are seen to operate more as propellers to create forward lift force, rather than as paddles to create forward drag force.

tion, because at this point, hand speed is lost as it stops and then reverses its direction (as it does when the elbow flexes, stops, extends, stops, flexes, and so on) (Barthels, 1974). When hand speed is lost, lift force is lost. When this lift force handle is lost, the shoulder muscles pulling on the arm force the hand back through the water a greater distance than they pull the body forward through the water. Swimmers with larger hands should not suffer as much lift force loss when blading hand speed decreases because lift force is proportional to the area of the hydrofoil. Swimmers with fast stroke rates (fast turnover) seem to use propulsive drag force more than do swimmers with slow stroke rates *traveling at the same body speed*.

Within a group of highly skilled swimmers tested, the more elite swimmers had greater forearm and hand area, arm length, and greater distance per stroke than the less successful competitors who were equal with regard to stroke turnover rate and physiological capacities (Toussaint et al., 1983). Grimston and Hay (1986) also reported that arm and leg length and hand and foot size influence the distance a swimmer attains with each stroke.

Clenched-fist swimming demands a very

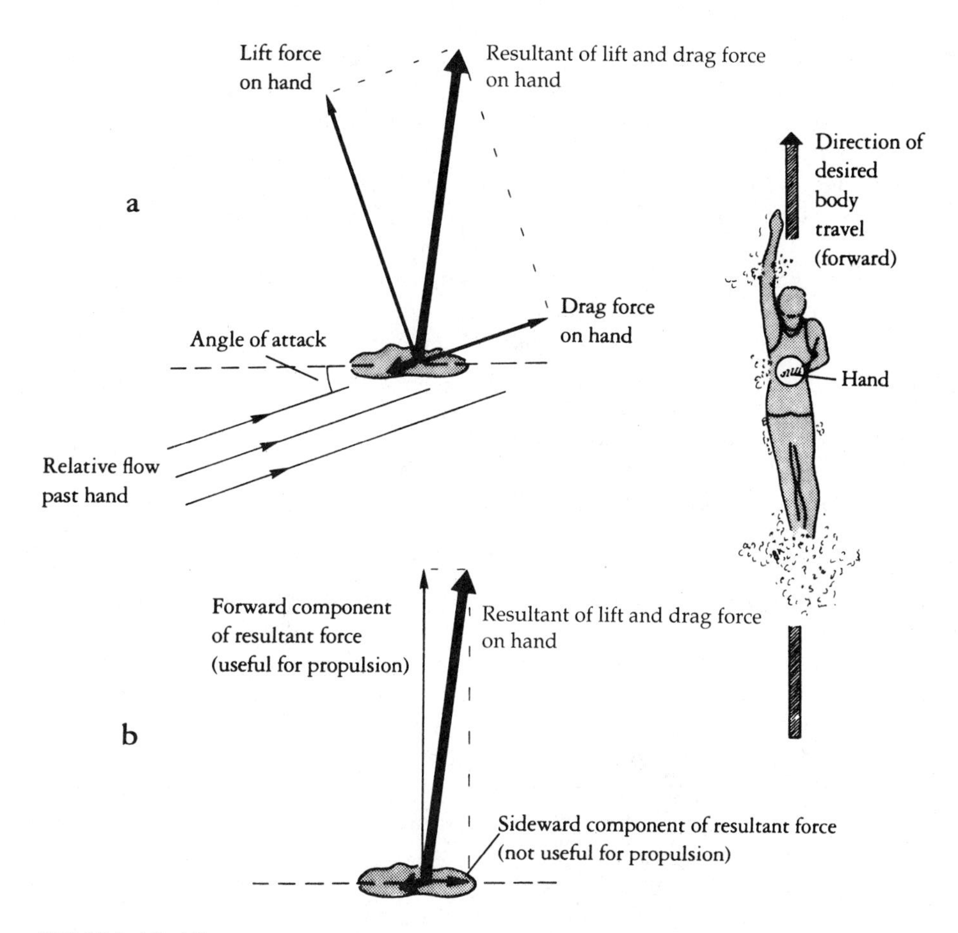

FIGURE 12.19

(a) The movement of the hand through the water creates lift force and drag force. (b) The forward-directed component of the resultant force is the force used as a "handle" by which to pull the body forward.

fast stroke rate to make the body travel fast. The hands must move extremely fast backward through the water to generate sufficient forward (propulsive) drag force, because the fist shape cannot create lift force. Attempts to identify the best angle of attack for the hand show that this best angle changes as the hand speed changes during its underwater path.

The effects of angle of attack were studied in a wind tunnel investigation by Remmonds and Bartlett (1981). They found that maximum lift force of about 12 N (2.7 lb) was obtained by the hand with the fingers closed and at an angle of attack of about 50 degrees. The thumb was positioned in the same plane of the hand.

Schleihauf (1979) reported that the optimal lift coefficient for the hand was 1.0 when the angle of attack was approximately 40 degrees. Of course, the greater the attack angle, the greater the drag force that resists the hand's sculling motion. According to de Groot and van Ingen Schenau (1988), the greater the drag force on the sculling hand, the more power the swimmer wastes moving water in a nonproductive direction. What a swimmer wants to do for greater propulsion efficiency is maximize the lift/drag ratio of the hand while still directing the resultant lift and drag force forward. Seemingly, a skilled swimmer can develop a feel for the best angle for the hand as it should change throughout the stroke to produce the most useful forward force on the hand.

The use of plastic hand paddles by competitive swimmers in training serves to increase the area of the hand so that a slower backward hand speed produces propulsive drag force. Consequently, the speed of the arm pull is not specific to the speed used in racing. Using hand paddles serves as a form of overload training for the shoulder muscles. Such paddles also may be effective overload devices for the shoulder muscles for the sculling arm pull because the greater area produces more lift force for a more stable handle.

If hand paddles are to be used in training, they should be carefully selected. The specificity of hand-paddle swimming depends greatly on the shape of the paddle. Of six different shapes commonly used in practice, the degree of similarity to actual stroking movements produced ranged from very different to highly similar. The best shape of the six for specific freestyle training was found by Bollens and Clarys (1986) to be a rectangle with the fingertip end rounded and bent slightly out of plane; this paddle also had the smallest surface area (closer to the actual area of the hand). The worst shape for specificity was found to be a long narrow rectangle with a much larger surface area.

Swimmers using hand paddles should be closely observed so that any changes in their underwater stroking patterns may be caught and corrected.

Propulsive lift force operates in all styles of swimming. For example, the leading hand in the sidestroke can produce lift force during elbow flexion as the hand blades up toward the face (a movement traditionally accepted as the beginning of the recovery movement of the hand). In the elementary backstroke, given an adequate ROM in the shoulder and shoulder girdle, it is possible to produce sculling arm movements similar to those in the back crawl that employ the "blading handle" mechanism.

Figure 12.20a–c shows the formation of propulsive lift force on the hand during elbow flexion in the backstroke, freestyle, and butterfly. Figure 12.20d–g is a head-on view of the propellerlike arm action of a breaststroker's underwater arm pull following wall push-off.

Propulsive Movements of the Legs and Feet

Traditionally, the forward thrust developed by the legs with various styles of kicking has been attributed to drag force acting in a forward direction against the feet and legs. If forward-directed drag force is to be exerted on the legs and feet for forward thrust, however, the legs *must* move backward relative to the water so that the water flows past them in a forward direction, similar to the way in which propulsive drag force is generated by the arms. Such conditions producing forward-directed drag force on the legs and feet occur to some extent in the whip (breaststroke) and scissors (sidestroke) styles of kicking at the beginning of the thrust phase when the body speed is relatively slow. However, backward leg movement relative to the water has not been observed in the flutter kick performed by skilled swimmers. Such observations indicate that drag force is not responsible for forward thrust. The different leg movements used in the breaststroke kick and the flutter and dolphin kicks suggest that propulsion may be achieved by two types of movement patterns: propellerlike and wavelike. The propellerlike breaststroke kick will be examined first because it is more like the hand-sculling patterns already described.

PROPELLERLIKE PATTERN. Although the arms and legs cannot move in a true revolving motion of a propeller, our joints permit the

Hand and forearm scull with elbow bending while body is pulled forward by shoulder muscles

a Backstroke

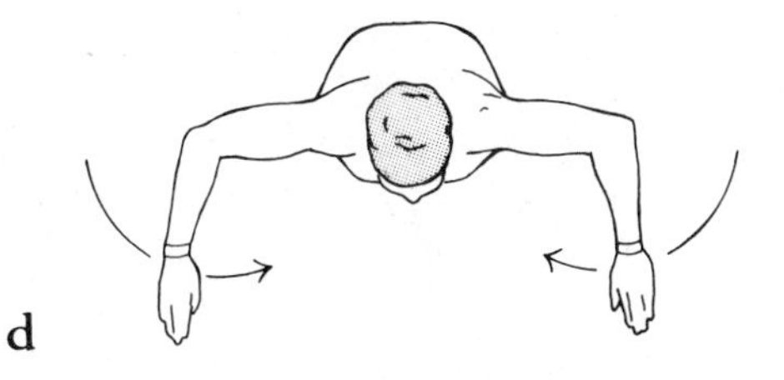

b Front crawl (freestyle)

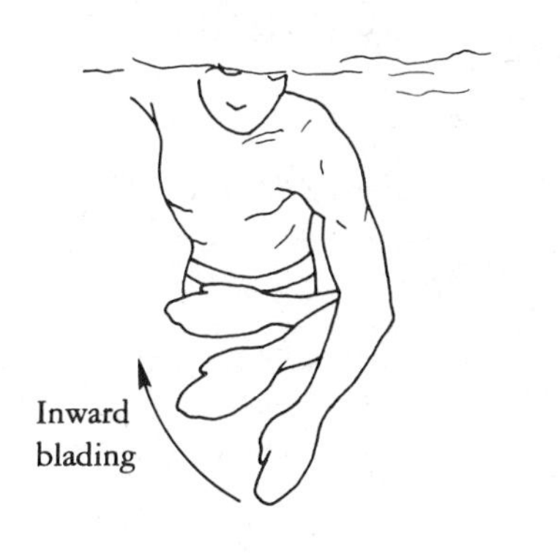

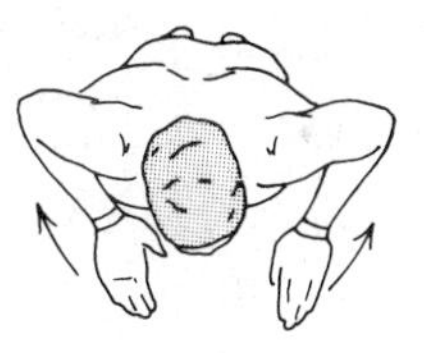

c Butterfly

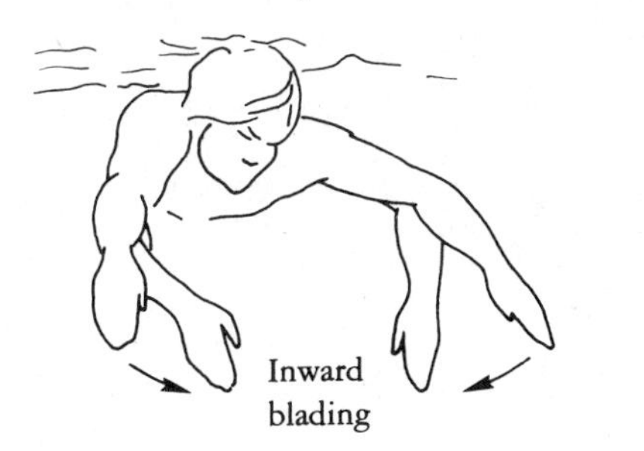

FIGURE 12.20

Propulsive horizontal lift force is created on the hands as they blade through the water in a vertical plane. From the swimmer's point of view, and from that of an observer, the hand seems to press backward through the water; however, the body actually is pressed forward past the hand as it blades through a three-dimensional, helical path. (a) Backstroke, (b) freesytle, (c) butterfly. (d–g) The sequence of a breaststroker's arm movements used for the underwater arm "pull" after a wall push-off.

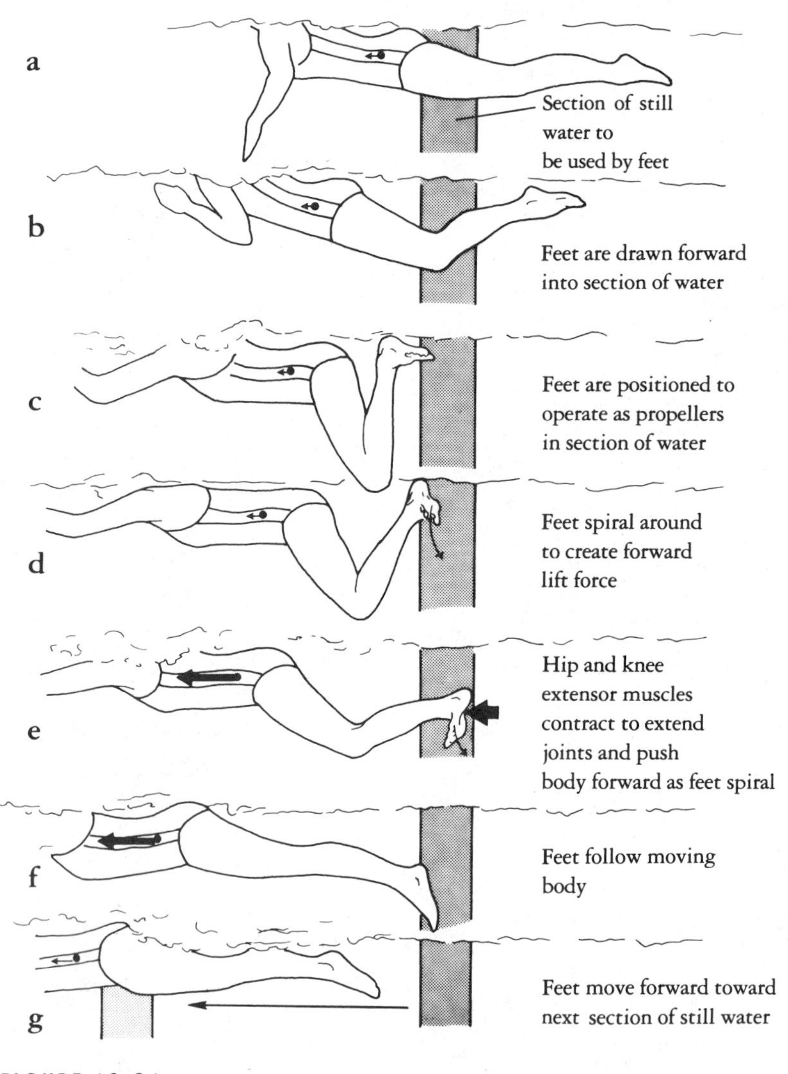
a
Section of still water to be used by feet
b
Feet are drawn forward into section of water
c
Feet are positioned to operate as propellers in section of water
d
Feet spiral around to create forward lift force
e
Hip and knee extensor muscles contract to extend joints and push body forward as feet spiral
f
Feet follow moving body
g
Feet move forward toward next section of still water

FIGURE 12.21

The forward component of lift force on the spiraling feet of a breaststroker forms a "force wall," against which the swimmer extends the legs to push the body forward.

positioning of the hands and feet so that they may act on the water in a fashion similar to propeller blades.

Evidence indicates that the propulsive force involved in the breaststroke whip kick is produced by the feet as they simulate the motion of a propeller. Such a propellerlike mechanism is similar to that involved in the formation of the lift force "handles" discussed earlier. The joint movements at the hip, knee, and ankle cause the feet to be circled in a spirallike path in a vertical plane behind the swimmer. The

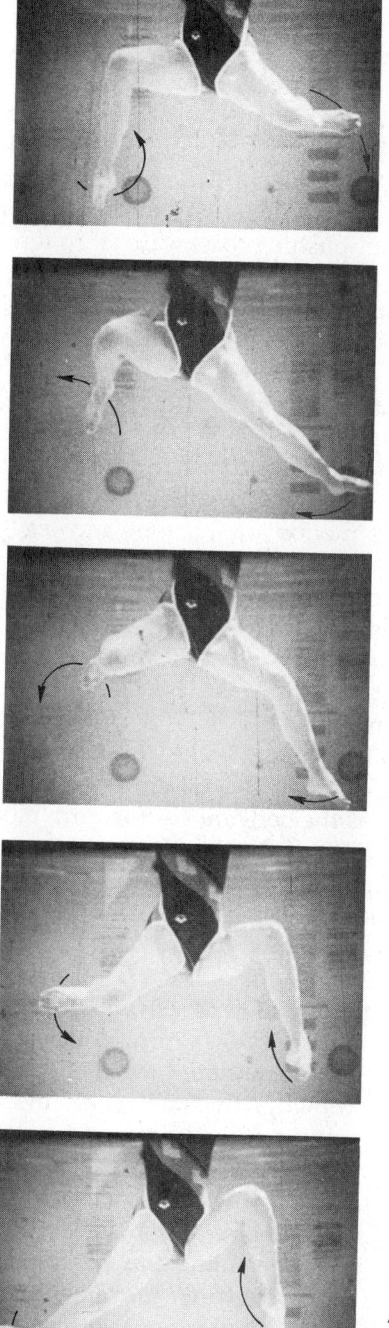

flow engages the sole of each foot at an angle of attack, and a lift force is generated on the foot. The forward component of this lift force provides what could be thought of as a "force wall" against which the legs can extend to push the body forward (Barthels, 1979). The swimmer senses that the feet and legs are pushing backward through the water, because the legs are extending and pressure is felt on the soles. Actually the feet are blading, or sculling, and remain in a vertical section of water while the body moves forward in response to leg extension. The greater the lift force produced on the feet in a forward direction, the less the backward movement of the feet relative to the water as the legs extend. Figure 12.21 illustrates the movement of the body forward relative to the swimmer's feet, which are spiraled through a vertical section of water to obtain lift force with a forward component for thrust.

Measurements taken on skilled breaststrokers by Tilborgh, Stijnen, Persyn (1987) showed that the lateral movements of the feet, rather than backward movements, caused body velocity to increase.

Synchronized swimmers and water polo players must use the continuous supportive lift force on their propellerlike feet and legs in the "eggbeater" kick. This kick is an alternating left-then-right breaststroke kick, as illustrated in Figure 12.22.

WAVELIKE PATTERN. The movements of the legs and feet in the propellerlike breaststroke kick are relatively "unnatural"; that is, they

FIGURE 12.22

A synchronized swimmer performing an "eggbeater" kick to support the head and arms above the surface. The feet and legs act as propeller blades to generate an upward lift force.

do not resemble natural segmental movement patterns used by humans on land. In contrast, the "flutter" kick used in freestyle (front crawl) and backstroke (back crawl) uses joint movements and timing similar to those used in normal walking. *Flutter* is an unfortunate term used to inaccurately portray the wavy motion of the legs used in this kick. As each leg "waves" from the hip, it simulates the motion of a fish or dolphin. In such an undulating pattern, the leading segment (thigh) rotates from the hip with a small amplitude (ROM), the leg rotates around the knee joint with slightly greater amplitude, and the foot rotates freely around the ankle with the ROM depending on ankle flexibility for plantar flexion. The timing of these segmental rotations produces an undulatory path of the foot through the water. Ungerechts (1983) reported similar wavy movement patterns of the leg and foot of butterfliers and the tail of dolphins. Because of the similarity of this undulatory foot path to the path of undulating dolphin flukes, it is likely that similar hydrodynamic forces are created for propulsion. Propulsive drag force must be ruled out if no backward movement occurs relative to the water to create a forward flow over the leg and foot. In dolphin and flutter kicking, Ungerechts (1987) has shown that the flow of water traveled along the dorsal surface of the foot and did not flow from the instep toward the sole, as it would have to if propulsion was caused by forward-directed drag force. To clarify this point, consider this: When the flutter kick is observed for a kicker whose body is not moving forward (e.g., while braced against the pool wall), the foot motion would be backward and downward during knee extension and would indeed cause a forward-directed drag force on the leg and foot. For a moving swimmer, however, the legs move forward along with the body, as they wave up and down. The direction of relative flow past the leg and foot, then, is backward and upward during the downbeat and backward and downward during the upbeat. The relative flow, therefore, has a backward component that creates resistive drag.

Figure 12.23 illustrates how this relative flow past the leg might look during the downbeat of a flutter kick. The legs are actually snaking their way forward through the water rather than pushing backward through the water. If the resultant flow of water past the anterior surface of the moving foot and leg is examined during the downbeat, the water flow can be seen to be directed backward and upward along the shin and foot. The backward component of flow results from the leg being towed forward by the moving body, while the upward component results from the downward motion of the foot and leg. The *resultant* flow direction creates a small angle of attack with the foot surface, and the conditions for generating lift force may be present (Figure 12.23b). The forward component of the lift force (thrust) is useful for body propulsion. This same type of flow condition is created on the tail flukes of dolphins and other whales when the tail undulates in a vertical plane as the body moves forward. Note in Figure 12.23b that the drag force on the foot is directed upward and backward; the backward component of the drag force cancels part of the thrust created by the forward component of lift force.

The magnitude of propulsive force depends on the relative values of propulsive lift and resistive drag encountered by the particular leg that is moving, the rate of kicking, and the forward speed of the body and legs. Some investigators have questioned the role of flutter kicking in swimming, proposing that it may produce little or no propulsive force but that it acts to streamline the body position or prevent excessive trunk movement (Counsilman, 1968; Watkins and Gordon, 1983). Hollander and others (1988), however, found that the legs actually did contribute to propul-

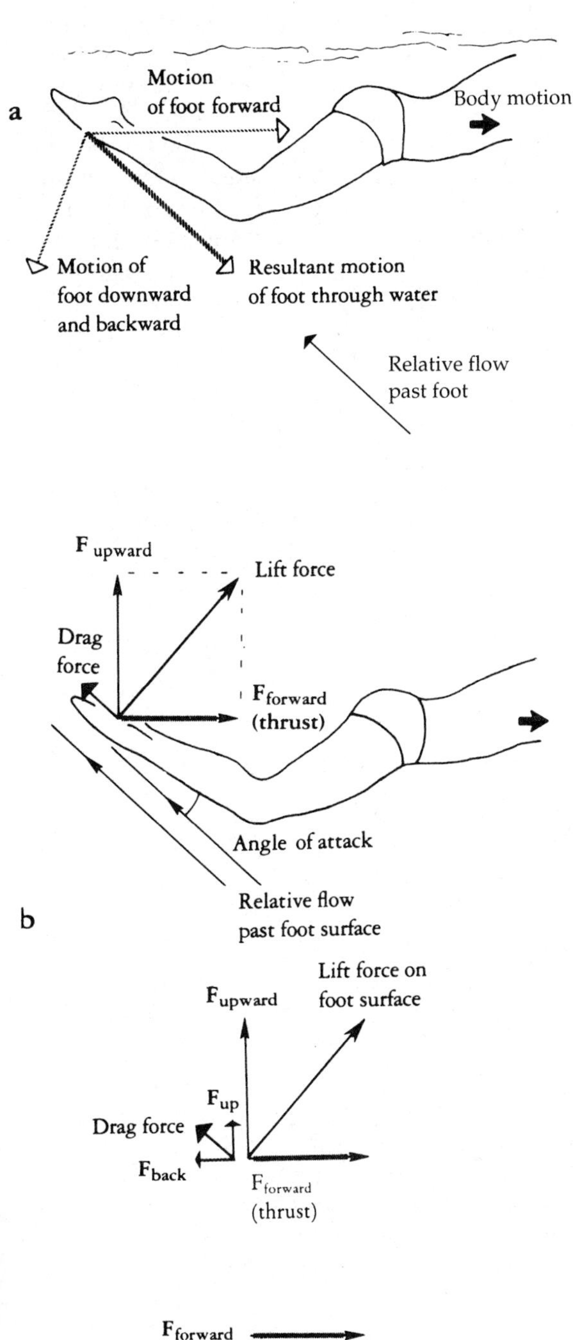

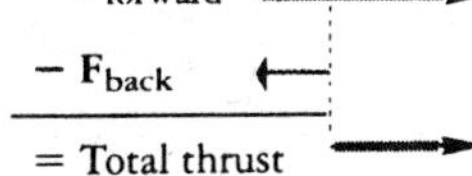

sive force but indicated that the percentage of total power output was lower for the legs than for the arms. Because of the relatively low efficiency of the flutter kick compared to the arm stroke for producing propulsive force, it seems reasonable that longer distance swimmers, who need to be more energy efficient to last the race, should reduce the effort used for kicking (as seen in 2- and 4-beat kicking) if the oxygen transport system cannot maintain a kick demanding greater oxygen consumption (6- or 8-beat kicking). Sprinters, however, who do not need to conserve energy over a long time can benefit from the added propulsion of a higher beat kick.

Propulsive Movements of the Total Body

In most styles of swimming performed by humans, the arms and legs work as separate units to move the whole body forward through the water. The trunk is usually stabilized, and most intervertebral movement is restricted, with the exception of some lateral flexion and transverse rotation that occur in response to alternating arm and leg movements or breathing to one side. The trunk, as a segment, usually is not considered a propulsive agent. In the dolphin butterfly stroke, however, the intervertebral movements of flexion and extension actually contribute to body propulsion, as do the arm and leg movements. In the preceding section, the flutter kick and the dolphin kick were illustrated as being isolated from the rest of the body. In the flutter kick used in the front and back

FIGURE 12.23

(a) The resultant foot motion and (b) the relative flow direction past the foot in a flutter or dolphin kick during the downbeat of the legs while the swimmer moves forward. (c) The backward-directed drag force on the leg cancels part of the propulsive thrust.

crawl strokes, each leg works as a single unit attached to a fairly stable trunk base. In the dolphin kick, however, both legs move together as though they were one "tail" unit, which does not work independently of the trunk but functions more as an extension of the vertebral column to continue the waving of the body along its sagittal plane.

The dolphin kick, therefore, may be viewed as a kick that is initiated from the thoracic vertebrae rather than from the hip joints. The intervertebral flexion and extension capabilities, along with the flexion and extension capabilities of the lower extremity segments, permit an eellike movement of the total body. Figure 12.24 shows the total body weaving its way through the water. Such undulations result in flow patterns conducive to the generation of lift forces, with the forward-directed components useful for thrusting the body forward. The arm stroke in the dolphin butterfly stroke contributes intermittent addi-

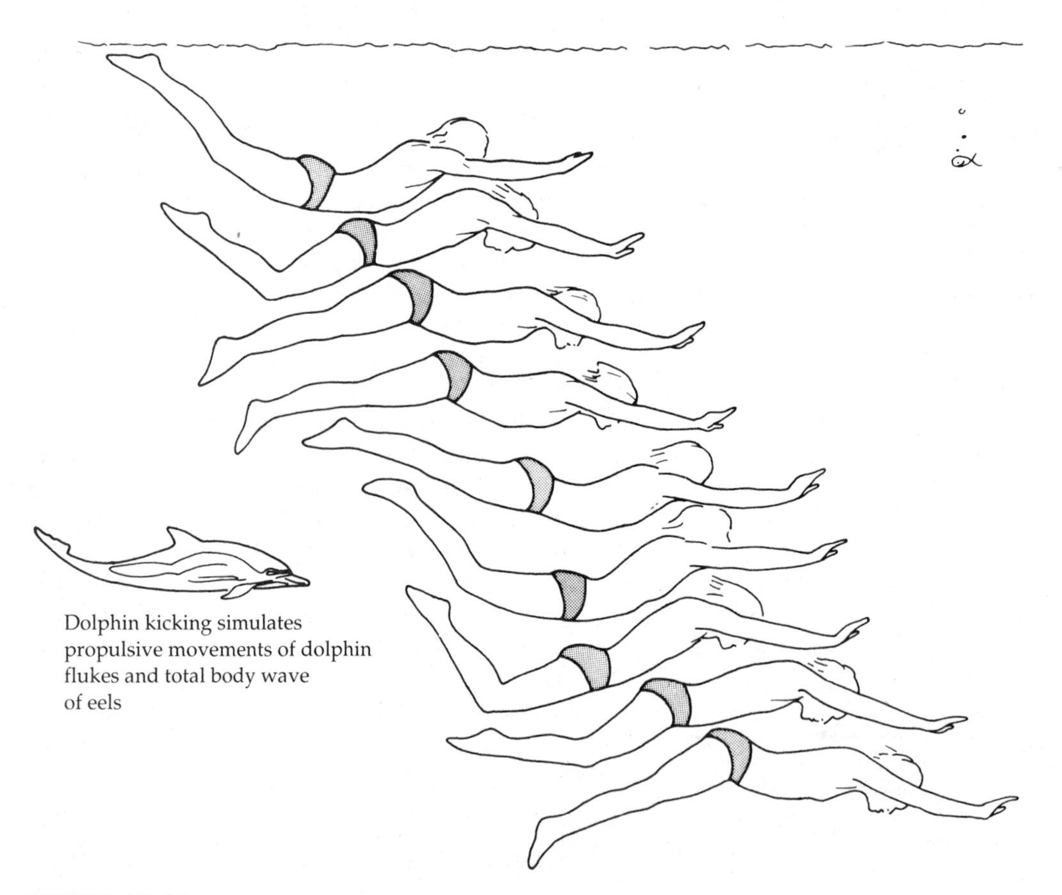

Dolphin kicking simulates propulsive movements of dolphin flukes and total body wave of eels

FIGURE 12.24

The waving movement of the legs is a continuation of the intervertebral movements. This pattern results in propulsive undulatory motion of the whole body.

tional propulsion and also provides for the positioning of the head for breathing. The propulsive nature of this total body weave or wiggle is demonstrated by the progress made by the swimmer who uses only trunk and leg movements in an underwater "dolphin" action.

Although we call it the dolphin kick, certain dissimilarities are present between the human movements and dolphin movements because of differences in structure and the use of the arms in the full butterfly stroke. The full body wave seen in eels and water snakes is not seen in the fastest fish, dolphins, and whales. The latter have evolved tail flukes (caudal fins) so that the undulation of the body is greatly reduced, and the wavy movement is confined to the low mass tail end of the body. Humans, without benefit of large flukes, incorporate the use of the eel's body wave for propulsion. Wearing swim fins, however, the dolphin's kick can be more accurately simulated and body weave can be reduced. Almost immediate success is experienced by learners using fins for butterfly kicking; because of this, fins are often used in the learning process. Difficulty, however, may arise when the fins are removed and more of the body must be "waved" for success, rather than merely flexing and extending the knees to wave the fins.

Another feature in butterfly swimming that leads to more eellike propulsion is the use of the arms and head. The overwater recovery movements of both arms simultaneously lead the shoulders through an undulatory path, establishing the wavelike pattern for the rest of the body.

Because of the arm stroke timing, the two "kicks" used in the full stroke are of different forms (Barthels and Adrian, 1971). One major and one minor kick usually are observed for each stroke cycle. The perceived size of the kick is not the distinguishing feature between the two kicks; one kick involves total body movement, the other, only the legs. As the arms enter the water in front of the head, the elevated shoulders follow in a forward and downward path. The hips move forward and upward, following the shoulders, as the thighs and legs follow in a downward-forward and upward-forward "swoop." This sequence of full-body weaving is eellike. When the hands are blading through the last part of the stroking, the head and shoulders travel forward and upward (for the breath) while the hips remain lower. A second downbeat is delivered at this time, but it does not correspond to hip elevation as in the full body weave. It occurs through knee extension and is dolphinlike because it involves only the "tail" end. These two kinds of kicks may be pictured as a body kick and a leg kick. The amount of body weave and the ROM at the knee vary considerably according to the swimmer's anthropometrics. Because the fastest vertebrate swimmers use the tail-end portion of their bodies to move their propulsive flukes (mammals) or tails (fish), swimmers whose feet are large and flexible could benefit more from the dolphin kicking pattern with reduced body weave and unnecessary energy expenditure. Most swimmers must depend on a combination of the eellike and dolphinlike patterns for effective propulsion.

Breaststrokers also use vertebral movements, coordinated with the whip kick, so that the body contributes to propulsion, rather than remaining a rigid drag-producing unit. Figure 12.25 illustrates the weaving of the body during the kick and the beginning of the arm pull in this contemporary breaststroke technique.

In summary, propulsion of a swimmer depends on the way the water flows past the body and its parts; this flow is determined by the way in which the body and

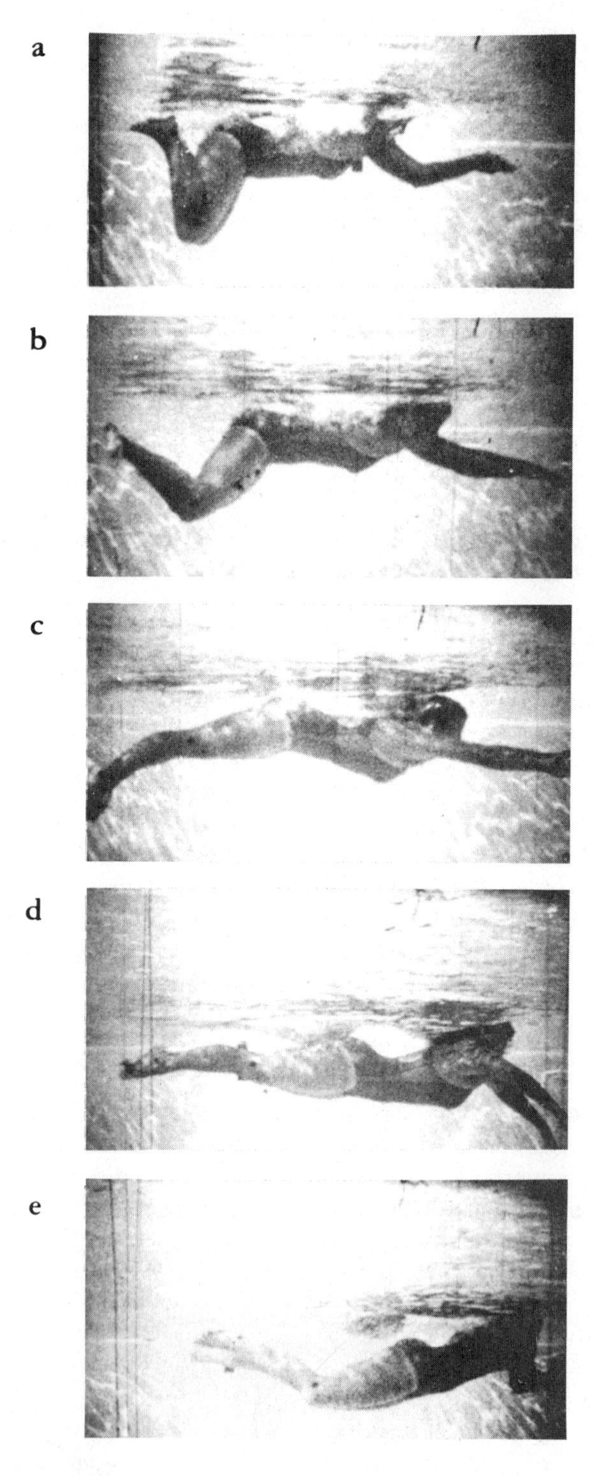

FIGURE 12.25

Intervertebral flexion and extension is incorporated into the breaststroke so that the body is not merely a rigid unit connecting the arms and legs.

parts move through the water. Flow patterns and directions differ depending on whether the body is moving forward by means of kicking only, by arm stroking only, or by using both the arms and legs; or not moving forward at all. The teacher or coach who understands the propulsive process can best help the learner-performer, and coaching cues can be developed to focus the swimmer's attention on how the movements should feel and appear. Research focusing on the exact mechanism used for best propulsion continues.

Understanding Propulsion in Swimming Skills

1. Explain how drag force can serve as a propulsive force in swimming.

2. Explain how lift force is produced on the hand moving through water.

3. Explain how lift force on the hand can be used to support weight above the surface of the water.

4. Explain how lift force can prevent the hand from moving backward through the water in a front crawl stroke. Is this desirable? Why?

5. In the swimming pool, working with a partner or in a small group, perform the following experiments and discuss the results.

a. Tread water (without using the legs) by using broad horizontal blading sweeps of the hands to obtain upward lift force, as shown in Figure 12.14. Repeat with the hands clenched into fists. Repeat with hand paddles. Experiment with different hand and finger shapes to determine which shape produces the greatest lift force.

b. Tread water as explained in 5a; continue the horizontal blading movements, but hyperextend the wrist about 45 degrees. What is the direction of lift force relative to the plane of your palm? Identify the two rectangular components of the lift force and the result of each directional component. Repeat with a flexed wrist position.

c. Perform the same treading movements in a prone position, as shown in Figure 12.15. Can you propel your body forward without a backward paddling motion? Note the orientation of the plane of the hand for maximum upward support (for instance, when the head is held above the surface of the water in a prone position). Repeat in a supine position, blading the hands alongside the body.

d. In a supine position, use horizontal blading movements with your palms facing the pool bottom to obtain upward lift force on the hands. Compare the sculling speeds necessary for generating enough lift force to (1) hold one knee out of the water, (2) hold one leg out, and (3) hold two legs out.

e. Using no kick for propulsion, compare the number of arm strokes necessary to swim (front crawl) one length of the pool when (1) using no flexion of the elbow during the underwater phase, and (2) maximizing the blading mechanism to obtain propulsive lift force.

Repeat two or three times, maintaining the same time per length.

f. Have a partner hold your ankles as you perform front-crawl arm movements in a prone position, using (1) straight-back pushing motions with your arms and (2) blading movements with the arms as you stroke at the same stroke rate. Which movements, according to your partner, produce the greatest force for pulling the body forward? Experiment with different styles of arm pulls, hand shapes, stroke rates, and head positions to identify what factors increase the force (according to your partner) for moving the body forward. (Note that the direct force acting to pull your body forward is the shoulder muscle contraction force acting on the trunk; the movements of the hands and arms through the water serve to form the handles that stabilize the arm so that trunk movement will occur.)

g. Compare the propulsive effectiveness of the legs moving in (1) a bicycling pattern, (2) a wavy flutter pattern, (3) an "up-out-together," wedge-type breaststroke kick, and (4) a spiraling, whip-type breaststroke kick.

h. Compare the propulsive effect of the whip kick if the ankles are (1) kept in dorsiflexion throughout, (2) kept in plantar flexion throughout, and (3) moved from dorsiflexion to plantar flexion as the legs near the end of their drive phase.

12.4 Swimming and Speed Efficiency

Factors Influencing Swimming Velocity

The total time a swimmer takes to complete a race depends on swimming velocity, the time from the starting signal to when the swimming actually begins (starting time), and the time used for the turns and wall push-offs.

The most important part of the race, however, is swimming velocity. As we know, the forward velocity fluctuates within each stroke cycle; therefore, the *average* stroke velocity is the value that is analyzed.

Swimming velocity can be calculated in terms of how far body travels per stroke cycle (stroke length) and the "turnover" rate at which the stroke cycles are performed (stroke frequency):

EQUATION 12.3

$$\text{swimming velocity} = \text{stroke length} \times \text{stroke frequency}$$

$$v = SL \times SF$$

where v is body velocity, SL is distance covered per stroke cycle (stroke length), and SF is number of stroke cycles per second (stroke frequency).

An example of the use of this relationship would be a body velocity of 1.4 m/sec obtained with a stroke length of .95 m per stroke and a stroke rate of 1.5 strokes/sec. We can see that an increase in stroke length (SL) or an increase in stroke frequency (SF) or an increase in both would increase the swimmer's body velocity. A large increase in stroke length obtained at the expense of reducing the stroke frequency too much, however, would result in *slower* body velocity. For example, if a very long "glide" is held with each entering hand in a freestyle (front crawl) stroke, the stroke frequency would be low. Moreover, if the arms were "resting" in this fashion every stroke, there would be a large time interval of body deceleration resulting from the resistive force being greater than the propulsive force. Such a low stroke frequency leading to body acceleration-deceleration-acceleration would be inefficient as well.

Similarly, if the stroke frequency is increased too much, the stroke length will be sacrificed. Such is the tendency among many competitive swimmers; they attempt to increase their stroke frequency so much that the distance traveled per stroke is reduced greatly, resulting in slower times. Therefore, neither the stroke length nor stroke frequency should be maximized; each should be *optimized* to yield the fastest body velocity.

A likely explanation for the decrease in distance per stroke with a stroke frequency increase beyond optimum is that the underwater stroke mechanics change to produce less propulsive force; that is, the hand and arm "slips" backward through the water as the swimmer attempts to increase the stroke frequency, thereby eliminating some of the sculling movements that provide the handle for pulling the body forward.

Looking a 200-m distances, Hay and Guimaraes (1983) found that improvements in butterfly, breaststroke, and freestyle speeds during a season were due to increasing SL without increasing SF. Possible explanations for an individual's increase in SL with no change in SF could include improved propulsive lift force, because use of propulsive drag stroking technique would necessarily increase SF. The way an already skilled swimmer improves propulsive lift might include slight changes in hand path, hand angle of attack (pitch) or less local muscular fatigue that would lead to hand path deviations (hand "slipping"). Decreased body drag by improved head and body positioning also would be a likely reason. A combination of both, of course, would account for greater improvement.

In 100- and 200-m freestyle races, Toussaint et al. (1983) found a significant difference in SL between Olympic caliber and other highly trained female swimmers. In this cross-sectional study, the faster swimmers, with greater SL but same SF, had larger hands (compared to body length) than the slower group. Other anthropometric features that distinguished the faster from the slower group were greater arm length and forearm-and-hand frontal area.

Body height, weight, and proportions also seem to be important characteristics affecting swimming performance. From 1964 through 1980, Olympic swimming data for males reveal increases in height without corresponding increases in weight (Montpetit and Smith, 1988). Body type may also distinguish among swimmers suited best for certain events (sprint vs. distance or type of stroke). Results of research attempting to establish beneficial or detrimental anthropometric characteristics is contradictory in some cases. However, the basic features that theoretically reduce drag and favor propulsion offer a basis for a natural initial advantage for some participants. Although anthropometric variables may not determine maximum attainable speed, some characteristics, such as arm and leg length and hand and foot size, do influence the SL and SF that a swimmer combines to reach a given speed (Grimston and Hay, 1986).

In most reports, group averages are presented for SL and SF and swimming velocity. Group averages can show trends or general differences between groups but often mask the specific combinations of SL and SF used by individuals to achieve a given swimming speed. A report by Letzelter and Freitag (1983) helps "unmask" these relevant individual variables by giving, at least, the *range* of values for the different groups. Using their data on fast and slower male and female swimmers, extreme differences in possible individual SL and SF yielding about the same velocity can be seen. For example, one group's average velocity was 1.82 m/sec (the average SL of 1.28 × the average SF of 1.42). The range of SL was 1.00–1.52 m. The range of SF was 1.18–1.84 strokes/sec. It is possible that an individual swimmer using the minimum SL of 1.00 times the maximum SF of 1.84 could achieve a velocity of 1.84 m/sec. Another could achieve a velocity of 1.79 m/sec by using the maximum SL of 1.52 times the minimum SF of 1.18. It seems clear that for any given swimmer achieving a given velocity in a given stroke for a given distance, the optimum SL and SF will be determined by that swimmer's anthropometric characteristics, movements used for propulsion, and drag encountered. For an interesting account of split times and how SL and SF vary over the course of a 100-m race, the reader is referred to Letzelter and Freitag (1983).

Craig, Boomer, and Gibbons (1979) and Craig and Pendergast (1979), report the stroke frequency and stroke length of swimmers in the 1976 Olympic swimming trials. Their data show that faster stroke frequencies were seen in the shorter events, while the longer event swimmers were characterized by longer stroke lengths. Craig et al. (1985) report the data they obtained from swimmers in the 1984 Olympic trials. They note the changes in SL and SF associated with the faster times than in 1976. For most men's events, faster speeds were attained by increasing distance per stroke, often even when stroke frequency *decreased*. The finalists of all men's and women's events had greater distance per stroke than the slower swimmers. In the men's backstroke, SL also increased, but SF stayed the same. The point was made by the authors that these changes in SL and SF were indicative of improvements in stroke mechanics. They conclude with the following remark:

> *For the past 20 (years) training for swimming has been characterized by long hours spent swimming long distances. Such programs are aimed at increasing the swimmers' capacities to produce energy. It is suggested that training programs which emphasize improving the biomechanics of swimming and using these changes during competition might be a better use of time in the pool. (Craig et al. 1985, p. 634)*

Factors Influencing Swimming Efficiency

The meaning of efficiency incorporates the concepts of energy expenditure and work production. Specifically, **swimming efficiency** is the ratio of the mechanical work performed by the swimmer (propulsive force × the distance the body is moved) to the energy expended by the swimmer to do this work (determined by oxygen consumption). If the performance is examined for a given time interval, the efficiency can be expressed in terms of mechanical power (force × distance/time) divided by the rate of energy expenditure (oxygen consumption/time). Multiplying the result by 100 gives efficiency in terms of percentage.

In recent years the work performed by the swimmer has been shown to include not merely the force used to move the body forward but also the force used to move the surrounding water. Some of the power (energy) used by the swimmer is used for propulsion; the rest is "wasted" on giving the water kinetic energy as a byproduct of propulsive and nonpropulsive movements.

Greater propulsion efficiency results when the limb movements do not cause useless turbulence (kinetic energy given to the water). When the power applied by a swimmer is used to move the hand through a large pathway to create propulsive force ("spiraling" hand path rather than straight back), while minimizing the nonpropulsive work done on the water, efficiency is increased.

Research has shown that highly skilled swimmers display a greater distance per stroke than the less skilled. This is characteristic of high propelling efficiency. More energy is wasted by using short arm pulls that generate high peak forces than by using longer arm strokes that deliver more uniform force (de Groot and van Ingen Schenau, 1988). A major factor in determining efficiency is the power ($F \times v$) required by the swimmer to travel at a given speed. Because resistive drag force is proportional to the velocity squared, and power is force × velocity, the power needed to overcome drag force is proportional the velocity *cubed* (Toussaint et al., 1988). The greater the power used, the greater the rate of energy (work) expenditure. Therefore, the more power, or effort, that goes into propulsion and the less that is wasted on giving the water unnecessary motion, the greater the efficiency.

Because the evidence shows that skilled swimmers swim more efficiently than the less skilled, the movement patterns exhibited by elite competitors represent the critical features of efficient swimming. An analysis of swimming efficiency must involve consideration of biomechanical and physiological factors in the performance. Lack of research instrumentation and a number of uncontrollable variables, however, have made swimming efficiency estimates difficult to determine. Values up to only approximately 5%–7% have been reported for moderate-speed freestyle, which seems to be the most efficient of the strokes performed, followed by backstroke and then breaststroke (Karpovich and Millman, 1944; Holmer, 1972). No value has been reported for the efficiency of the contemporary butterfly stroke.

The economy of any given stroke performance is influenced by how the body speed fluctuates within each stroke cycle. More energy is required in variable-speed swimming than in constant-speed swimming, just as more fuel is required for driving a car that is continually accelerating and decelerating during a trip. For efficiency in swimming, therefore, a relatively constant body velocity should be sought. If sources of resistive force acting on the swimmer's body can be identified and reduced as much as possible, less propulsive force will be needed to maintain any given speed.

Speed fluctuations are quite apparent in the sidestroke, elementary backstroke, and breaststroke because of the intermittent propulsive movements of the limbs and the large drag of the underwater recovery movements. To keep the resistive drag on the limbs to a minimum, the speed of the underwater recovery movements should be relatively slow, and the surface area of the limb facing the flow should be minimized. To take full advantage of the propulsive movements of the limbs, the rest of the body should be in as streamlined a position as possible for minimum drag. Less apparent, but still present, are speed fluctuations in the butterfly stroke. The overwater arm recovery in the butterfly eliminates the drag caused by the underwater recovery in the conventional breaststroke, and the wavelike body and leg movements in the dolphin motion provide more or less continuous propulsion. A more constant body speed is observed in the front and back crawl strokes, because the limbs operate to provide some propulsive force throughout the total stroke cycle.

A general statement that can be made is that swimming efficiency depends on swimming speed, the stroke used, and the skill of the swimmer. For each stroke and swimmer, there seems to be an *optimum* swimming speed for the highest efficiency. At very slow speeds, efficiency is low; at moderate speeds, efficiency improves; at higher speeds, efficiency decreases again.

If the mechanical purpose of swimming is merely to interact with the environment, efficiency may be of no concern. If the mechanical purpose is to cover a given distance in the fastest time possible, however, the efficiency of the swim will be compromised by the speed (pace) demanded by the distance; that is a sprint at maximum speed would not be *efficient*, but it would be *effective*.

In competitive swimming *maximum* efficiency is not the purpose. *Optimal efficiency* for *maximum effectiveness* is the goal. The effectiveness of the swimming performance is measured by the speed with which the distance is covered.

Understanding Speed and Efficiency

1. Identify the forces that act *directly* on the swimmer's trunk to cause it to move forward relative to the water with one arm cycle.
2. Count the number of stroke cycles performed by a swimmer for one pool length (25 yd or 50 m), and record the time for the swim. Calculate the body velocity and average distance covered per arm-stroke cycle. Calculate the stroke rate.
3. Repeat the procedure in question 2, but have the swimmer hold the hands in fists. Compare and explain the differences in body velocity, stroke length, and stroke rate.
4. Observe a 100-yd or 200-yd race. Count the number of strokes taken during the second 25-yd lap and time the lap. Do the same for the last 25-yd lap. Calculate the swimming velocity, stroke rate, and stroke length for each of these laps, and compare them. Explain the differences.

References and Suggested Readings

Adrian, M. J., Singh, M., & Karpovich, P. V. (1966). Energy cost of leg kick, arm stroke, and the whole crawl stroke. *Journal of Applied Physiology, 21*, 1763–1766.

Alexander, R. M. (1968). *Animal mechanics*. Seattle: University of Washington Press.

Andersen, P. (1976). The use of hand paddles, overload training, and after-effects. *Swimming Technique, 13*(2), 58–62.

Azuma, A. (1983). Biomechanical aspects of animal flying and swimming. In H. Matsui & K. Kobayashi (Eds.), *Biomechanics VIII-A: Proceedings of the Eighth International Congress of Biomechanics* (pp. 35–53). Champaign, IL: Human Kinetics Publishers.

Bachman, J. C. (1983). Three butterfly pulls. *Swimming Technique, 20*(1), 23–25.

Barthels, K. M. (1974). *Three dimensional kinematic analysis of the hand and hip in the butterfly swimming stroke*. Unpublished doctoral dissertation, Washington State University, Pullman, WA.

Barthels, K. M. (1979). The mechanism for body propulsion in swimming. In J. Terauds & E. W. Bedingfield (Eds.), *Swimming III: Proceedings of the Third International Symposium on the Biomechanics of Swimming* (pp. 45–54). Baltimore: University Park Press.

Barthels, K. M. (1982). Biomechanical research in swimming: Past, present, and future. In J. Terauds (Ed.), *Biomechanics in Sports: Proceed-*

ings of the International Symposium of Biomechanics in Sports (pp. 381–390). San Diego, CA: Academic.

Barthels, K. M., & Adrian, M. J. (1971). Variability in the dolphin kick under four conditions. In L. Lewillie & J. P. Clarys (Eds.), *First International Symposium on Biomechanics in Swimming, Waterpolo and Diving Proceedings* (pp. 105–118). Brussels: Universite Libre de Bruxelles Laboratoire de L'effort.

Bollens, E., & Clarys, J. P. (1986). Front crawl training with hand paddles: A telemetric EMG investigation. In M. Adrian & H. Deutsch (Eds.), *Biomechanics. The 1984 Olympic Scientific Congress Proceedings* (pp. 271–277). Eugene, OR: University of Oregon Microform Publications.

Bucher, W. (1975). The influence of the leg kick and the arm stroke on the total speed during the crawl stroke. In J. P. Clarys & L. Lewillie, (Eds.), *Swimming II* (pp. 180–187). Baltimore: University Park Press.

Campbell, W. A. (1978). Relationship between buoyancy of the black male and learning the crawl stroke. In F. Landry & W. A. R. Orban (Eds.), *Biomechanics of Sports and Kinanthropometry: Proceedings of the International Congress of Physical Activity Sciences* (pp. 149–155). Miami: Symposia Specialists.

Ciccone, C. D., & Lyons, C. M. (1987). Relationships of upper extremity strength and swimming stroke technique on competitive freestyle swimming performance. *Journal of Human Movement Studies, 13*, 143–150.

Clarys, J. P. (1978). An experimental investigation of the application of fundamental hydrodynamics to the human body. In B. Eriksson & B. Furberg (Eds.), *Swimming Medicine IV: Proceedings of the Fourth International Congress on Swimming Medicine* (pp. 386–394). Baltimore: University Park Press.

Clarys, J. P. (1979). Human morphology and hydrodynamics. In J. Terauds & E. W. Bedingfield (Eds.), *Swimming III: Proceedings of the Third International Symposium on the Biomechanics of Swimming* (pp. 3–41). Baltimore: University Park Press.

Costill, D. L., King, D. S., Holdren, A., & Hargreaves, M. (1983). Sprint speed versus swimming power. *Swimming Technique, 19*(1), 20–22.

Counsilman, J. (1968). *The science of swimming.* Englewood Cliffs, NJ: Prentice-Hall.

Craig, A. B., Jr. (1984). The basics of swimming. *Swimming Technique, 20*(4), 22–27.

Craig, A. B., Boomer, W. L., & Gibbons, J. F. (1979). The use of stroke rate, stroke length, and velocity relationships in training competitive swimmers. In J. Terauds & E. W. Bedingfield (Eds.), *Swimming III: Proceedings of the Third International Symposium on the Biomechanics of Swimming* (pp. 265–274). Baltimore: University Park Press.

Craig, A. B., Jr., Boomer, W. L., & Skeehan, P. L. (1988). Patterns of velocity in competitive breaststroke swimming. In B. E. Ungerechts, K. Wilke, & K. Reischle (Eds.), *Swimming Science V* (pp. 73–77). Champaign, IL: Human Kinetics.

Craig, A. B., Jr., & Pendergast, D. R. (1979). Relationships of stroke rate, distance per stroke, and velocity in competitive swimming. *Medicine and Science in Sports, 11*, 278–283.

Craig, A. B., Jr., Skeehan, P. L., Pawelczyk, J. A., & Boomer, W. L. (1985). Velocity stroke rate and distance per stroke during elite swimming competition. *Medicine and Science in Sports and Exercise, 17*(6), 625–634.

diPrampero, P. E., Pendergast, D. R., Wilson, D. W., & Rennie, D. W. (1974). Energetics of swimming in man. *Journal of Applied Physiology, 37*, 1–5.

Erbaugh, S. J. (1986). Effects of body size and body mass on the swimming performance of preschool children. *Human Movement Science, 5*(4), 301–312.

Francis, P. R., & Welshons-Smith, K. (1982). A preliminary investigation of the support scull in synchronized swimming using a video motion analysis system. In J. Terauds (Ed.), *Biomechanics in sports* (pp. 401–407). Del Mar, CA: Research Center for Sports.

Gray, J. (1974). How fishes swim. In *Animal Engineering: Readings from Scientific American* (pp. 29–35). San Francisco, CA: W. H. Freeman.

Grimston, S. K., & Hay, J. G. (1986). Relationships among anthropometric and stroking characteristics of college swimmers. *Medicine and Science in Sports and Exercise, 18*(1), 60–68.

Groot, G. de, & Ingen Schenau, G. J. van (1988). Fundamental mechanics applied to swimming: Technique and propelling efficiency. In B. E. Ungerechts, K. Reischle (Eds.), *Swimming Science V* (pp. 17–29). Champaign, IL: Human Kinetics.

Hay, J. G., & Guimaraes, A. C. S. (1983). A quantitative look at swimming biomechanics. *Swimming Technique, 20*(2), 11–17.

Hollander, A. P., Groot, G. de, & Ingen Schenau, G. J. van (1987). Active drag in female swimmers.

In B. Jonsson (Ed.), *Biomechanics X-B* (pp. 717–720). Champaign, IL: Human Kinetics.

Hollander, A. P., Groot, G. de, Ingen Schenau, G. J. van, Kahman, R., & Toussaint, H. M. (1988). Contribution of the legs to propulsion in front crawl swimming. In B. E. Ungerechts, K. Wilke, & K. Reischle (Eds.), *Swimming Science V* (pp. 39–43). Champaign, IL: Human Kinetics.

Hollander, A. P., et. al. (1986). Measurement of active drag during crawl arm stroke swimming. *Journal of Sports Science, 4*, 21–30.

Holmer, I. (1972). Oxygen uptake during swimming. *Journal of Applied Physiology, 33*, 502–509.

Holmer, I. (1974). Propulsive efficiency of breaststroke and freestyle swimming. *European Journal of Applied Physiology, 33*, 95–103.

Holmer, I. (1978). Time relations: Running, swimming, and skating. In B. Eriksson & B. Furberg (Eds.), *Swimming Medicine IV: Proceedings of the Fourth International Congress on Swimming Medicine* (pp. 361–366). Baltimore: University Park Press.

Huijing, P. A., et. al. (1988). Active drag related to body dimensions. In B. E. Ungerechts, K. Wilke, & K. Reischle (Eds.), *Swimming Science V* (pp. 31–37). Champaign, IL: Human Kinetics.

Jensen, R. K., & Tihanyi, J. (1978). Fundamental studies of tethered swimming. In F. Landry & W. A. R. Orban (Eds.), *Biomechanics of Sports and Kinanthropometry: Proceedings of the International Congress of Physical Activity Sciences* (pp. 135–142). Miami: Symposia Specialists, Inc.

Jiskoot, J., & Clarys, J. P. (1974). Body resistance on and under the water surface. In J. P. Clarys & L. Lewillie (Eds.), *Swimming II* (pp. 105–110). Baltimore: University Press Park.

Karpovich, P. V., & Millman, N. (1944). Energy expenditure in swimming. *American Journal of Physiology, 142*, 140–144.

Kornecki, S., & Bober, T. (1978). Extreme velocities of a swimming cycle as a technique criterion. In B. Eriksson & B. Furberg (Eds.), *Swimming Medicine IV: Proceedings of the Fourth International Congress on Swimming Medicine* (pp. 402–407). Baltimore: University Park Press.

Letzelter, H., & Freitag, P. (1983). Stroke length and stroke frequency variations in men's and women's 100-m freestyle swimming. In A. P. Hollander, P. A. Huijing, & G. de Groot (Eds.), *Biomechanics and medicine in swimming* (pp. 315–322). Champaign, IL: Human Kinetics.

Lighthill, M. J. (1969). Hydromechanics of aquatic animal propulsion. *Annual Review of Fluid Mechanics, 1*, 413–445.

Loetz, C., Reischle, K., & Schmitt, G. (1988). The evaluation of highly skilled swimmers via quantitative and qualitative analysis. In B. E. Ungerechts, K. Wilke, & K. Reischle (Eds.), *Swimming Science V* (pp. 361–367). Champaign, IL: Human Kinetics.

Lugt, H. J. (1983). *Vortex flow in nature and technology*. New York: John Wiley & Sons.

Maglischo, E. W. (1982). *Swimming faster*. Palo Alto, CA: Mayfield.

McGrain, P., Rose, D., & Davison, R. (1984). Temporal analysis of competitive male swimmers. In J. Terauds, K. Barthels, E. Kreighbaum, R. Mann, & J. Crakes (Eds.), *Sports biomechanics* (pp. 149–160). Del Mar, CA: Research Center for Sports.

Midtling, J. (1974). *Swimming*. Philadelphia: W. B. Saunders.

Miller, D. I.(1975). Biomechanics of swimming. In J. H. Wilmore & J. F. Keogh (Eds.), *Exercise and Sports Science Reviews* (Vol. 3, pp. 219–248). New York: Academic Press.

Miyashita, M. (1975). Arm action in the crawl stroke. In R. LeWillie & J. P. Clarys (Eds.), *Swimming II: Proceedings of the Second International Symposium on the Biomechanics of Swimming* (pp. 167–173). Baltimore: University Park Press.

Miyashita, M. (1978). Method of calculating overall efficiency in swimming crawl stroke. In F. Landry & W. A. R. Orban (Eds.), *Biomechanics of Sports and Kinanthropometry: Proceedings of the International Congress of Physical Activity Sciences* (pp. 135–142). Miami: Symposia Specialists, Inc.

Montpetit, R. M., Cazorla, G., & Lavoie, J. M. (1988). Energy expenditure during front crawl swimming: A comparison between males and females. In B. E. Ungerechts, K. Wilke, & K. Reischle (Eds.), *Swimming Science V* (pp. 229–235). Champaign, IL: Human Kinetics.

Montpetit, R. M., & Smith, H. (1988). Built for speed. *Swimming Technique, 24*(4), 30–32.

Mutoh, Y. (1978). Low back pain in butterfliers. In B. Eriksson & B. Furberg (Eds.), *Swimming Medicine IV* (pp. 115–123). Baltimore: University Park Press.

Nelson, R. C., & Pike, N. L. (1978). Analysis and comparison of swimming starts and strokes. In B. Eriksson & B. Furberg, (Eds.), *Swimming Medicine*

IV: Proceedings of the Fourth International Congress on Swimming Medicine (pp. 347–360). Baltimore: University Park Press.

Nursall, J. R. (1958). The caudal fin as a hydrofoil. *Evolution, 12,* 116–120.

Ohmichi, H., Takamoto, M., & Miyashita, M. (1983). Measurement of the waves caused by swimmers. In A. P. Hollander, P. A. Huijing, & G. de Groot (Eds.), *Biomechanics and medicine in swimming* (pp. 103–107). Champaign, IL: Human Kinetics.

Pai, Y. C., & Hay J. G. (1988). A hydrodynamic study of the oscillation motion in swimming. *International Journal of Sport Biomechanics, 4,* 21–37.

Pendergast, D. R., diPrampero, P. E., Craig, A. B., & Rennie, D. W. (1978). The influence of selected biomechanical factors on the energy cost of swimming. In B. Eriksson & B. Furberg (Eds.), *Swimming Medicine IV: Proceedings of the Fourth International Congress on Swimming Medicine* (pp. 376–378). Baltimore: University Park Press.

Remmonds, P., & Bartlett, R. M. (1981). Effects of finger separation. *Swimming Technique, 18*(1), 28–30.

Saito, M.(1982). The effect of training on the relationships among velocity, stroke rate and distance per stroke in untrained subjects swimming the breaststroke. *Research Quarterly for Exercise and Sport, 53*(4), 323–329.

Schleihauf, R. E. (1979). A hydrodynamic analysis of swimming propulsion. In J. Terauds & E. W. Bedingfield (Eds.), *Swimming III: Proceedings of the Third International Symposium on the Biomechanics of Swimming* (pp. 70–109). Baltimore: University Park Press.

Schleihauf, R. E., Gray, L., & Rose, J. de (1983). Three dimensional analysis of hand propulsion in the sprint front crawl stroke. In A. P. Hollander, P. A. Huijing, & G. de Groot (Eds.), *Biomechanics and Medicine in Swimming* (pp. 173–184). Champaign, IL: Human Kinetics.

Sharp, R. L., Troup, J. P., & Costill, D. L. (1982). Relationship between power and sprint freestyle swimming. *Medicine and Science in Sports and Exercise, 14*(1), 53–56.

Stoner, L. J., & Luedtke, D. L. (1979). Variations in front crawl and back crawl arm strokes of varsity swimmers using hand paddles. In J. Terauds & E. W. Bedingfield (Eds.), *Swimming III: Proceedings of the Third International Symposium on the Biomechanics of Swimming* (pp. 281–288). Baltimore: University Park Press.

Takamoto, M., Ohmichi, H., & Miyashita, M. (1985). Wave height in relation to swimming velocity and proficiency in front crawl stroke. In D. A. Winter, R. W. Norman, R. P. Wells, K. C. Hayes, & A. E. Patla (Eds.), *Biomechanics IX-B* (pp. 486–491). Champaign, IL: Human Kinetics.

Tilborgh, L. V., Daly, D., & Persyn U. (1983). The influence of some somatic factors on passive drag, gravity, and buoyancy forces in competitive swimmers. In A. P. Hollander, P. A. Huijing, & G. de Groot, (Eds.), *Biomechanics and medicine in swimming* (pp. 207–214). Champaign, IL: Human Kinetics.

Tilborgh, L. V., Stijnen, V. V., & Persyn, U. J. (1987). Using velocity fluctuations for estimating resistance and propulsion forces in breaststroke swimming. In D. A. Winter, R. W. Norman, R. P. Wells, K. C. Hayes, & A. E. Patla (Eds.), *Biomechanics IX-B* (pp. 779–784). Champaign, IL: Human Kinetics.

Tilborgh, L. V., Willems, E. J., & Persyn, U. (1988). Estimation of breaststroke propulsion and resistance resultant impulses from film analyses. In B. E. Ungerechts, K. Wilke, & K. Reischle (Eds.), *Swimming Science V* (pp. 67–71). Champaign, IL: Human Kinetics.

Toussaint, H. M., et. al. (1983). A power balance applied to swimming. In A. P. Hollander, P. A. Huijing, & G. de Groot (Eds.), *Biomechanics and medicine in swimming* (pp. 165–172). Champaign, IL: Human Kinetics.

Toussaint, H. M., et. al. (1987). Active drag related to velocity in male and female swimmers. *Journal of Biomechanics, 21*(5), 435–438.

Toussaint, H. M., et. al. (1988). Measurement of efficiency in swimming man. In B. E. Ungerechts, K. Wilke, & K. Reischle (Eds.), *Swimming Science V* (pp. 45–52). Champaign, IL: Human Kinetics.

Toussaint, H. M., et. al. (1989). Effect of a triathlon wet suit on drag during swimming. *Medicine & Science in Sports and Exercise, 21*(3), 325–328.

Ungerechts, B. E. (1983). A comparison of the movements of the rear parts of dolphins and butterfly swimmers. In A. P. Hollander, P. A. Huijing, & G. de Groot (Eds.), *Biomechanics and medicine in swimming* (pp. 215–221). Champaign, IL: Human Kinetics.

Ungerechts, B. E. (1985). A description of the reactions of the flow to acceleration by an oscillating flexible shark model. In D. A. Winter, R. W. Norman, R. P. Wells, K. C. Hayes, & A. E. Patla (Eds.), *Biomechanics IX-B* (pp. 492–496). Champaign, IL: Human Kinetics.

Ungerechts, B. E. (1987). On the relevance of rotating water flow for the propulsion of swimming. In B. Jonsson (Ed.), *Biomechanics X-B* (pp. 713–716). Champaign, IL: Human Kinetics.

Ungerechts, B. E. (1988). Peak body accelerations and phases of movements. In B. E. Ungerechts, K. Wilke, & K. Reischle (Eds.), *Swimming Science V* (pp. 61–66). Champaign, IL: Human Kinetics.

Valiant, G. A., Holt, L. E., & Alexander, A. B. (1982). The contributions of lift and drag force components of the hand/forearm to a swimmer's propulsion. In J. Terauds (Ed.), *Biomechanics in sports* (pp. 391–400). Del Mar, CA: Research Center for Sports.

Van der Vaart, et. al. (1987). An estimation of drag in front crawl swimming. *Journal of Biomechanics, 20*, 543–546.

Watkins, J., & Gordon, A. T. (1983). The effects of leg action on performance in the sprint front crawl stroke. In A. P. Hollander, P. A. Huijing, & G. de Groot (Eds.), *Biomechanics and medicine in swimming* (pp. 310–314). Champaign, IL: Human Kinetics.

Webb, P. W. (1984). Form and function in fish swimming. *Scientific American, 251*(1), 72–75, 76–82.

Wood, T. C. (1979). A fluid dynamics analysis of the propulsive potential of the hand and forearm in swimming. In J. Terauds & E. W. Bedingfield (Eds.), *Swimming III: Proceedings of the Third International Symposium on the Biomechanics of Swimming* (pp. 62–69). Baltimore: University Park Press.

CONCEPT MODULE I

Rotary Movement Responses to Applied Torques

PREREQUISITES

Concept Modules C, D, E, F, G

CONCEPT MODULE CONTENTS

How a system responds to forces and torques applied to it depends on the magnitude of the net force and the magnitude of the net torque. The response also depends on the inertial characteristics of the body in question, which differ between rotary and linear motion. To explain how torques affect rotary movement responses, we will begin with some descriptions of rotary motion.

I.1 Angular Speed and Velocity

A system in rotary motion is represented by a straight line connecting the axis of rotation to a selected point in the system. The point selected might be the center of gravity (CG) of the rotating system (e.g., when describing the rotation of a gymnast swinging from a horizontal bar). The point may be some part of the body, such as a point on the head if the body is rotating about an axis through its CG, as it does when it is rotating free of support.

A body that is rotating around some axis has **angular speed**, which tells us how fast the body is changing its angular position. To determine angular speed, we measure the angular distance through which the body has rotated (in degrees or radians) and divide by the time it took to move through that arc:

EQUATION I.1

$$\text{angular speed or velocity} = \frac{\text{angular displacement}}{\text{time}}$$

$$\omega = \frac{\theta}{t}$$

where ω (omega) is angular speed (or velocity) of the body (radians/second or degrees/second), θ (theta) is angular displacement of the body (radians or degrees), and t is time during which the displacement occurred (seconds).

In cycling, for example, the angular speed of the thigh rotating around the hip axis during a down stroke might be 5 rad/sec. Even though the thigh does not rotate a full 5 rad, its rate of rotation for the time it is rotating is 5 rad/sec.

If a direction is specified for the rotating body, the body is said to have **angular velocity,** a vector quantity; that is, it has speed and direction. The direction of rotation is specified by the right-hand thumb rule if we wish to draw a vector diagram, or we can specify the direction as positive (counterclockwise [ccw]) or negative (clockwise [cw]). Recall that the axis or rotation is always perpendicular to the system's plane of rotation and that direction can be specified by the spatial orientation of the axis or the plane of rotation. A phonograph record, for example, rotates with an angular velocity of $33\frac{1}{3}$ rpm around a vertical axis and in a horizontal plane. The state of motion of the second hand on a clock is described as having an angular speed of 360 degrees per minute. The forearm may be rotating about a horizontal axis at the elbow at 100 degrees per second at some specified time during its movement. The angular velocity of the leg rotating in a vertical plane about a horizontal axis through the knee is about 700 degrees per second as a skilled runner swings the leg forward (Cooper and Glassow, 1976).

I.2 Linear Velocity of a Point on a Rotating Body

We can describe a system's rotary motion in terms of how far (θ) and how fast (ω), and we can specify the direction of its rotation. Angular displacement (θ) and angular velocity (ω) also can be used to determine how far some point on the rotating system travels along its curved path (linear distance) and how fast it travels (linear speed). The relationship between the angular velocity of a rotating object or segment and how fast the end of it is moving linearly has direct application in understanding throwing, kicking, and striking skills.

The linear distance and speed of some point on a rotating segment depends on the distance that point is from the axis of rotation, that is, the point's **radius of rotation**. A point further from the axis moves a greater linear distance than a point closer to the axis. The greater the radius, the greater the distance the point moves as the segment rotates. The linear distance, d, that a point moves may be found by multiplying its radius of rotation, r, by the angular displacement, θ, of the segment in radians:

EQUATION I.2

$$d = r\theta$$

where d is linear distance of a point on a rotating segment (meters), r is radius of rotation of the point (meters), and θ is angular displacement of the point (radians).

Whenever any calculations are performed to relate angular quantities to linear quantities, radians must be used for angular measure, not degrees or revolutions. Therefore, to change degrees to radians, divide the angular displacement in degrees by 57.3 degrees (57.3 degrees = 1 radian).

We may also calculate the instantaneous linear velocity of a point rotating with some given angular velocity. The linear velocity may be calculated from either the point's linear distance or its angular velocity. The relationship between the angular and linear velocities is given as follows:

EQUATION I.3

linear speed or velocity = radius of rotation × angular velocity

$$v = r\omega$$

where v is linear speed or instantaneous velocity of a point on a rotating segment (meters/second), r is radius of rotation of the point (meters), and ω (omega) is angular velocity of the segment (in radians/second).

Notice that v stands for speed or velocity. It means speed if the point is traveling along its curvilinear path. It means velocity if we look at its instantaneous value with its straight line direction at that instant.

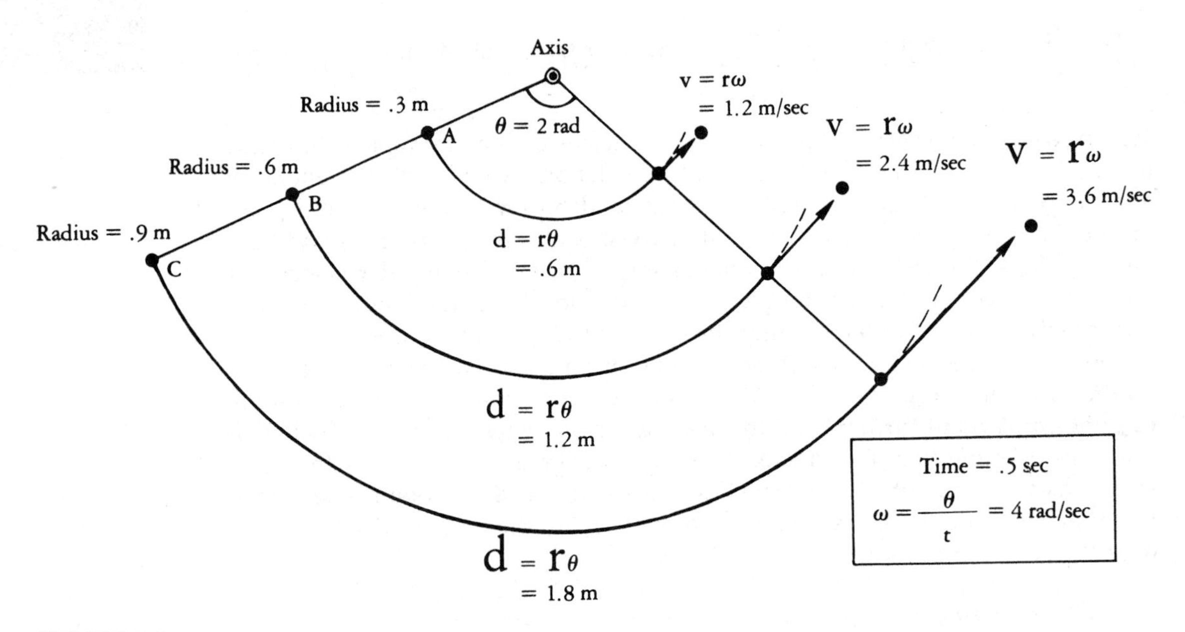

FIGURE I.1
The linear distance traveled and the instantaneous linear velocity of a point on a rotating segment increase with an increase in its radius of rotation.

The greater the angular velocity and/or the greater the radius of rotation, the greater the rotating point's linear speed or velocity.

Figure I.1 shows the relationship between the angular displacement and linear distance of three points on a rotating segment. The figure also shows how the linear velocities of the three points are related to the angular velocity of the segment. Note in the figure that the angular displacement for all three points, no matter how far the points are located from the axis, is the same. Notice, however, that the farther the point is located away from the axis, the greater the linear distance it moves for a given angular displacement. Thus, point C moves linearly farther than point B, which moves farther than point A. Recall that the distance of each point from the axis is called its radius of rotation, which is designated by a lower-case r and measured in linear units, such as meters, centimeters, feet, or inches. Thus, if the radius of rotation of point A is 0.3 m, it travels $d = r\theta = 0.3(2) = 0.6$m. Point B, with a radius of 0.6m, travels 1.2m. Point C, located even farther from the axis, travels 1.8m.

To determine the angular velocity, ω, of the segment we must know the amount of time the segment took to rotate through the angle of 2 rad. If the time measured is 0.5 sec, the angular velocity will be $\omega = \theta/t = 2$ rad/0.5 sec = 4 rad/sec.

Just as the radius of rotation of each point determines the linear distance it travels, the radius also determines the linear velocity of each point. If we assume that the segment rotates around at the same angular velocity, $\omega = 4$ rad/sec, then point A will have a linear velocity of $v = r\omega = 0.3\ (4) = 1.2$m/sec; point B, a linear velocity of $0.6(4) = 2.4$ m/sec; and point C, with the greatest radius of rotation, has a linear velocity of $0.9(4) = 3.6$ m/sec. Usually, linear velocity of any point on a rotating segment is determined for some given instant during the rotation. At the instant selected for calculating the linear velocity of the point, the direction of the linear velocity (a vector) is tangent to the arc; that is, it is perpendicular to the point's radius of rotation. Such an instantaneous linear velocity is often called the **tangential velocity**, a term that indicates its direction relative the the point's path.

The model in Figure I.1 could represent a leg swinging to kick a ball, and the linear velocity of a point on the foot would be of interest because a fast foot applies more force than a slow foot. Therefore, with a given angular velocity for two different legs, a longer leg will have a faster foot. In a following section, we will see, however, that the angular velocity of a body segment or implement could be more important in producing fast linear velocity than the radius of rotation. The influence of both factors, r and ω, on linear velocity of a point should be observed.

I.3 Angular Acceleration

How far, θ, and how fast, ω, a rotating body travels are two ways to describe its motion. Rarely, however, does a human body segment or implement rotate with the same angular velocity throughout its range of motion (ROM); it changes its state of rotary motion by speeding up or slowing down or changing direction. The rate at which a body's angular speed or direction is changed is defined as its **angular acceleration**. More precisely, acceleration is the time rate of change in a body's velocity; that is, how fast it is changing its speed or direction or both.

We can show the relationship of velocity change and acceleration as follows:

EQUATION I.4

$$\text{angular acceleration} = \frac{\text{change in angular velocity}}{\text{time it took for the change}}$$

$$\alpha = \frac{\omega_2 - \omega_1}{t} \text{ or } \frac{\Delta\omega}{t}$$

where α (alpha) is angular acceleration of the system, ω_2 is final angular velocity of the system, ω_1 is initial angular velocity of the system, t is time during which the angular velocity was changed, and Δ is "change in."

If a large change in angular velocity occurs in a short period of time, the angular acceleration will be large. If the arm starts to rotate downward and forward to release a fast softball pitch and it reaches an angular velocity of 20 rad/sec in 0.2 sec, its angular acceleration is $\alpha = \Delta\ \omega /t = 20/0.2 = 100$ rad/sec per second. In slow pitch the arm may rotate to only 5 rad/sec in 0.2 sec. Its angular acceleration then is $\alpha = \Delta\ \omega /t = 5/0.2 = 25$ rad/sec per second. Note that time is mentioned twice in expressing acceleration values. The change in velocity includes a time unit in its measurement, and then the rate of change for acceleration adds the second time unit.

A rigid body, segment, or implement experiences an angular acceleration (or deceleration) only during the time a net external torque is being applied. At the instant the net torque ceases, the system has reached its new velocity and maintains it until another torque is applied. The acceleration is always in the direction of the net torque acting. The greater the torque, the greater the angular acceleration. (We will see later that a body that can change shape during rotation can speed up or slow down without an external torque.)

When a body gains angular velocity, we say it is accelerating, and when it loses angular velocity, we say it is decelerating. Acceleration is also referred to as positive (+) acceleration, and deceleration, as negative (−) acceleration. Angular directions also are given + and − labels: ccw is + and cw is −. Therefore, a positive (+) acceleration value may be assigned to an angularly decelerating segment that is still rotating in the negative direction. (In essence, slowing down in the negative direction is mathematically interpreted as speeding up in the positive direction. We will be using the more qualitative terms, *acceleration* and *deceleration*, and specify the direction when necessary (cw or ccw).

Examples of angular acceleration are readily found among segmental movements of the human body. Every time a movement of a body segment is initiated or terminated, it has been angularly accelerated or decelerated. Almost all of our movements are reciprocal, or back-and-forth movements. Rarely, if ever, do the body segments move with a constant angular speed or direction. During the course of one rotary movement, the angular velocity of the segment is variable from one instant to the next. Consider, for example, the movement of the arm in a volleyball spike (Figure I.2). The arm segment is angularly accelerated to start the backswing, or preparation phase of the spike, and then it is decelerated at the end of the ROM of the shoulder joint (Figure I.2a). Angular acceleration and deceleration of the arm occur at the shoulder joint near the ends of the ROM. The forward rotation of the arm begins, and the arm is accelerated to move the hand toward the ball rapidly forward in an arc. The ball is struck during the arm's rotation when its angular velocity is large (Figure I.2b). The arm's follow-through after ball contact is a period of angular deceleration; this deceleration must be great enough to prevent contact with the net. To some, the term *acceleration* connotes fast movement, but this is not an accurate interpretation of the concept. In fact, when velocity is at maximum value, acceleration is zero. In general, the angular velocity of a rotating segment or implement is the quantity of ultimate importance, and the magnitude of the acceleration or deceleration determines how fast that desired velocity is attained. Consider, for example, a bat accelerated from 0 degrees per second to 10 degrees per second

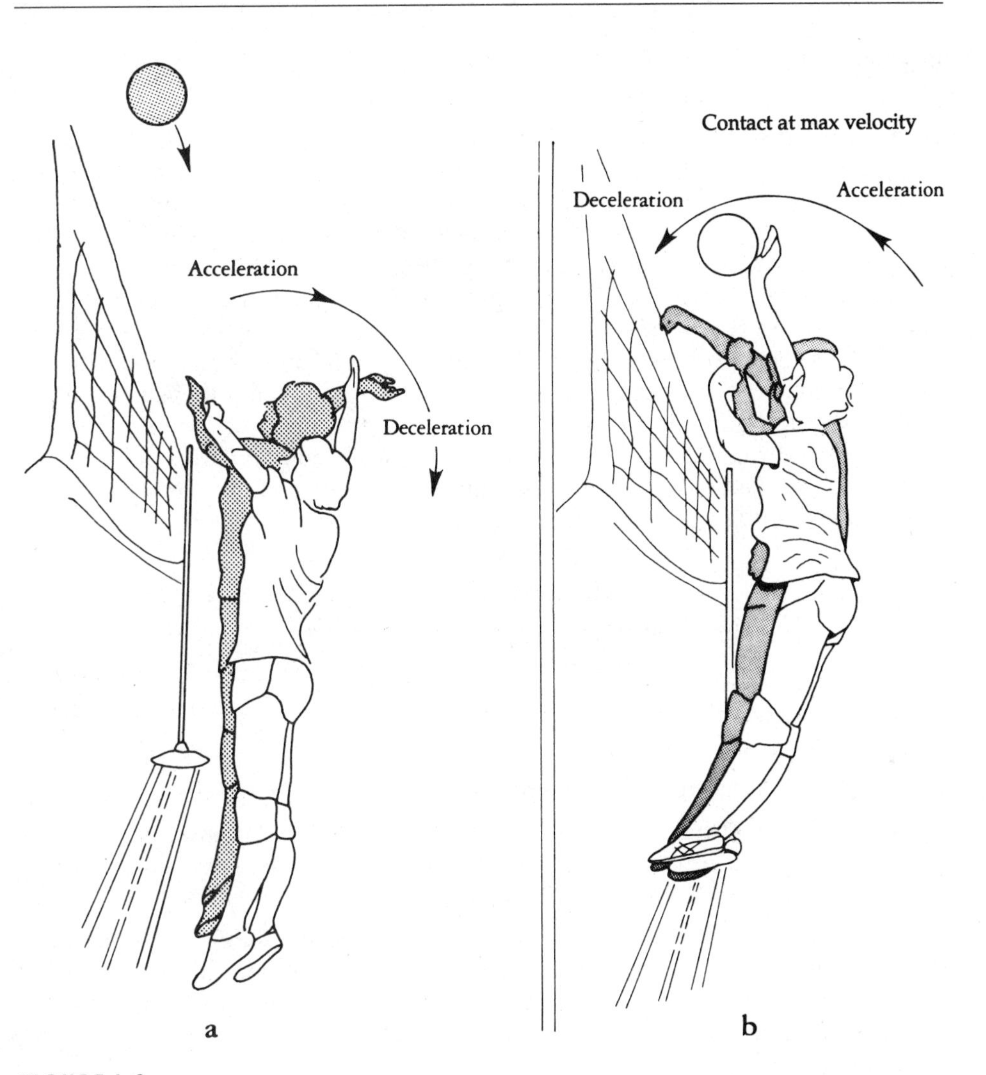

FIGURE I.2
A volleyball spiker showing the angular acceleration then decleration of the arm in (a) the backswing and in (b) the execution of the hit.

in 0.1 sec. The final velocity is 10 degrees per second, and the acceleration is 100 degrees per second per second. Now consider a bat being accelerated from 490 degrees per second to 500 degrees per second in 0.1 sec. The bat's angular acceleration is still 100 degrees per second per second; however, the angular velocity is 500 degrees per second, and such a velocity will cause the bat to have a much greater impact force than a bat rotating at 10 degrees per second.

The magnitude of the acceleration of a body segment depends on the purpose of the movement and the constraints imposed on the performer. For example, the forward angular acceleration of the lower extremity during locomotion is considerably less in walking than it is in running.

Variable Angular Velocity: Average and Instantaneous Values

When segments or implements are rotated, the velocity varies during the ROM. In examining Figure I.1, we assumed that the angular velocity was constant when the linear tangential velocity of a point was determined. When we look at a rotation of varying velocity we could describe its motion by giving its **average angular velocity,** which is calculated by dividing its ROM, θ, by t (the time it took to cover that ROM). Such a gross measure, however, does not have much use in sport analysis; most rotations are performed with the purpose of angularly accelerating a segment or implement up to some angular velocity at a particular instant (e.g., swinging a bat to reach maximum angular velocity at ball contact, swinging the arm to pitch a horseshoe at an optimal angular velocity at release, and rotating the leg segments to a maximum angular velocity during a high jump takeoff). **Instantaneous angular velocity** is the velocity used for calculating the instantaneous linear velocity ($v = r\omega$).

Understanding Rotary Motion

1. What distinguishes angular velocity from angular speed? Give an example for each.
2. If a body is rotating with constant angular velocity, what is its acceleration?
3. Which segment has the greatest angular acceleration: (a) an arm that gains an angular velocity of 300 degrees per second in 0.5 sec, or (b) an arm that gains an angular velocity of 200 degrees per second in 0.2 sec?
4. Perform ten different segmental movements (one at a time). For each, identify when angular acceleration and angular deceleration are taking place. Can you rotate a body segment with a constant angular velocity somewhere in its ROM?
5. What is the instantaneous release velocity of a softball released by an upper extremity, 0.5m long, that is rotating at 573 degrees per second at the time of release? (Remember to use radians in $v = r\omega$.)

I.4 The Relationship of Torque, Rotational Inertia, and Angular Acceleration

Mechanically, the magnitude of the angular acceleration of a system depends on the net torque applied and the resistance of the system to angular motion change. When external torque is applied to a system to cause its angular acceleration, the greater the torque, the greater the acceleration. In addition to this condition, however, the system's resistance to angular acceleration also plays a part in how the system responds to the torque.

Newton's second law, the law of acceleration, is expressed here specifically for rotary motion: the angular acceleration of a body is directly proportional to the net torque applied, and the acceleration is inversely proportional to the **rotational inertia**[1] of the body.

This interrelationship of three variables is expressed by the following:

EQUATION I.5

$$\text{angular acceleration} = \frac{\text{torque}}{\text{rotational inertia}}$$

$$\alpha = \frac{T}{I} \quad \text{or} \quad T = I\alpha$$

where α is angular acceleration of the segment or object, T is net torque applied to the segment, and I is rotational inertia of the segment.

A detailed description of the term *rotational inertia* is necessary for understanding its significance and for appreciating its effect in all rotational movement. Recall that a body's mass is the inertial resistance to a change in that body's *linear* motion (resistance to linear acceleration). The mass of a body or segment, however, makes up only part of the body's resistance to angular acceleration. A system's resistance to a change in its state of *rotary* motion also depends on how far the mass is distributed away from the axis of rotation. The combination of these two factors, the system's *mass* and its *mass distribution* (i.e., arrangement relative to the axis of rotation), determine the system's **rotational inertia (I).** The rotational inertia is the measure of resistance to angular acceleration of a body. Since this resistance is specific to a given axis, a system's rotational inertia about one axis may be quite different from that same system's rotational inertia about another axis. The greater a system's mass or the farther from the axis of rotation the mass is distributed or both, the greater the system's rotational inertia. This mass-distribution factor is more influential than is the quantity of mass in terms of determining the magnitude of the resistance to rotary motion change. In fact, the resistance to changes in the state of rotary motion of a system is directly proportional to increases in its mass, but it increases as the square of the distance of the system's mass relative to the axis of rotation. We can express this concept as follows:

EQUATION I.6

$$\text{a body's rotational inertia} = \text{mass} \times \text{radius}^2$$

$$I = mr^2$$

[1] Rotational inertia is also called "moment of inertia"; however, the term *rotational inertia* will be used in this text because the name implies its meaning.

where I is rotational inertia of the segment or body, m is mass of the segment or body, and r is radius (distance) of the mass from the axis of rotation.

To appreciate better the importance of the mass-distribution factor in the determination of rotational inertia, imagine trying to swing a baseball bat while holding it at the grip end and then trying to swing that same bat while holding it at the barrel end. The relative ease with which the bat can be swung while held at the barrel end is evident, yet the mass of the bat has not changed. Examine the mass distribution relative to the axis of rotation for each swing, and a difference will be apparent. Using the wrist area as the axis in both cases, more of the bat's mass can be seen to be farther from the axis when it is held at the small end. More effort is required to start the bat rotating in this case because the bat's rotational inertia is larger. If it were necessary to swing the bat faster in a game situation, one would not see the player gripping the barrel end of the bat to permit a faster rotation, but one would see the player choking up on the bat, or moving the axis of rotation (the wrist) closer to the barrel of the bat. Such a shift in handgrip serves the purpose of bringing the more massive part of the bat closer to the axis of rotation (wrist).

Figure I.3 illustrates the effect on a system's rotational inertia of doubling its mass, compared to the effect of doubling the distance its mass is distributed from the axis of rotation. Note that in this model, all the mass is considered to

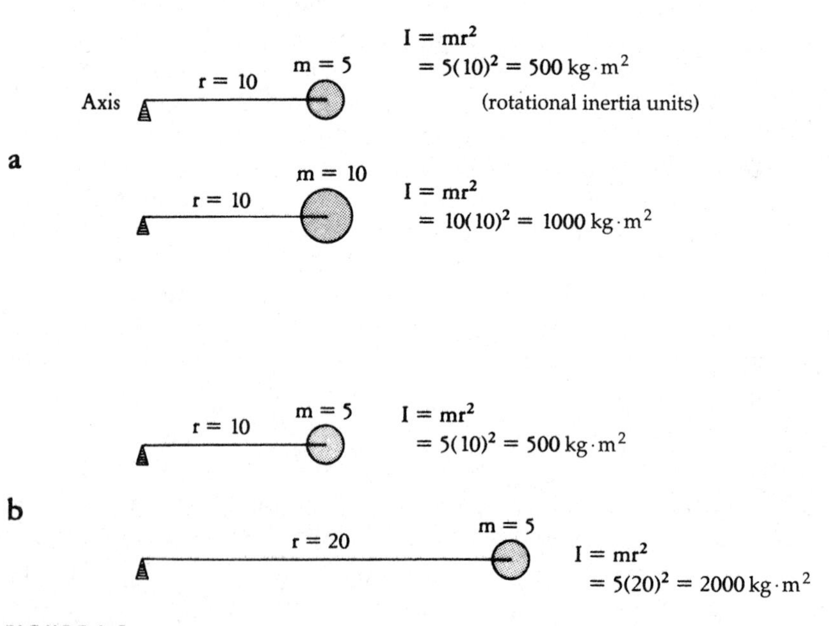

FIGURE I.3

(a) Doubling the mass of a rotating object doubles its rotational inertia. (b) Doubling the distance of an object's mass from the axis quadruples its rotational inertia.

be packed into a single point. Because the mass is doubled in Figure I.3a, an increase in the rotational inertia of the system is doubled. In Figure I.3b, however, a doubling of the distance of the mass from the axis of rotation causes a fourfold increase in the rotational inertia of the same system.

In the real would of rotating systems are bodies or segments or implements whose masses are not condensed into single points as in Figure I.3; the mass is spread out along the radius and has some kind of shape. For real bodies, then, the rotational inertia is the mass times the distance that would be the radius if all the body's mass were concentrated at a single point. This distance that represents how far away from the axis of rotation a body's mass is distributed is technically called the **radius of gyration (k).** A rotating body whose mass is spread away from the axis has a large radius of gyration (e.g., a somersaulting body in a layout position). If the mass is packed in close to the axis, its radius of gyration is small, as in a tuck somersault. A body with a small radius of gyration has a smaller rotational inertia than that body would have with a larger radius of gyration.

The radius of gyration distance should not be confused with the distance of the CG from the axis of rotation. The radius of gyration is always a distance greater than the distance of the CG of a system from the axis of rotation.

Instead of using $I = mr^2$ for a single condensed mass, therefore, we can express the rotational inertia for real bodies as follows:

EQUATION I.7

$$\text{rotational inertia} = \text{mass} \times \text{radius of gyration}^2$$

$$I = mk^2$$

where I is rotational inertia of the system, m is mass, and k is radius of gyration.

The units are designated as "rotational inertia" units. A body's resistance to change in rotary motion, then, is the product of a body's mass and its radius of gyration squared and is specific to a given axis of rotation.

For our purposes, calculating the numerical values of a body's rotational inertia is unnecessary. What is important, however, is understanding fully the concept of rotational inertia and its influence on the angular acceleration of any body or segment that can rotate about an axis. Every time a body segment is moved (rotated), the rotational inertia of that segment determines how the segment responds to the torque applied to it. The shape of each of the segments forming our upper and lower extremities produces a relatively small rotational inertia about an axis at its *proximal* end, because the more massive part of each segment is situated close to the proximal joint, thus decreasing the amount of torque that needs to be applied to the segment to angularly accelerate the segment around the proximal joint. The rotational inertia of these segments about axes through their *distal* joints is greater because of the mass

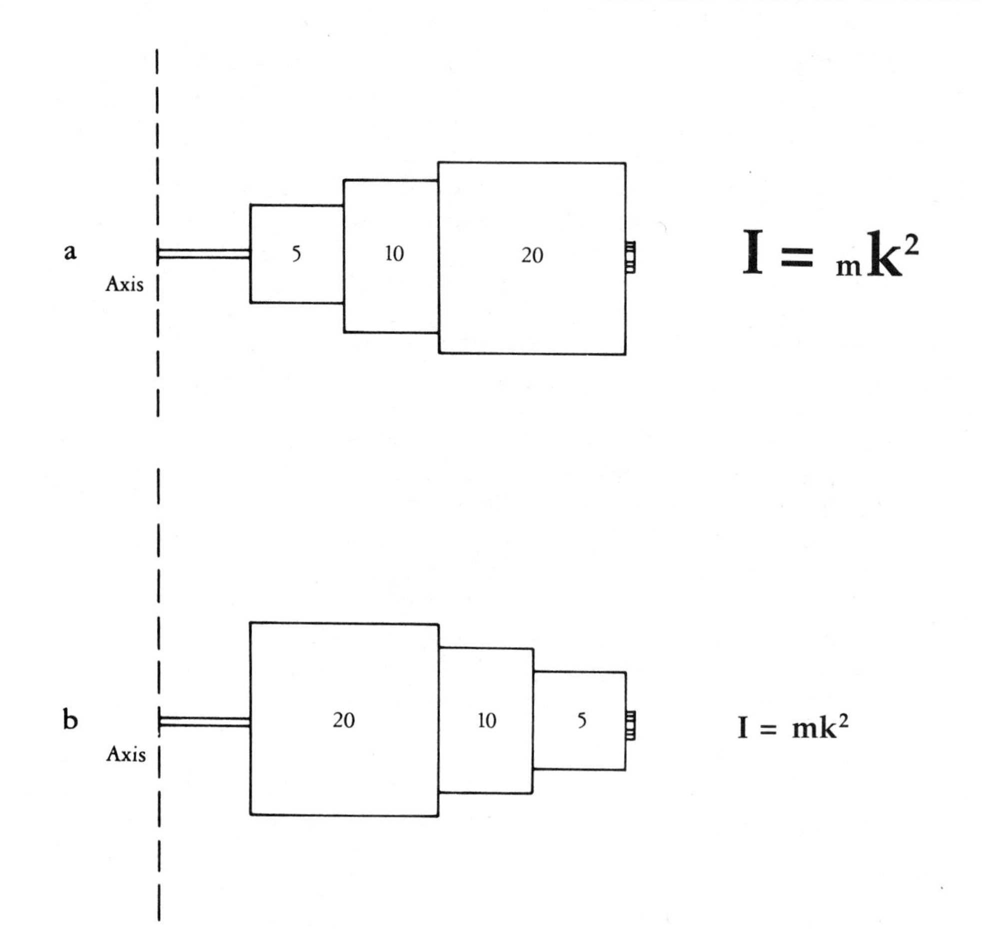

FIGURE I.4

Two segments with the same masses but with the masses distributed differently from the axis. Much more torque would be necessary to angularly accelerate the segment in (a) than the one in (b).

distributions of the segments. The arrangement of a body segment's mass close to its axis of rotation permits rapid limb rotation.

Consider the two segments in Figure I.4, each to be rotated around the same axis. Each has the same total mass, but the distribution of the mass relative to the axis is reversed. The mass arrangement in Figure I.4a, would require a much larger torque to angularly accelerate it than would the arrangement in Figure I.4b.

When the whole human body is rotated in spins and aerial twists and somersaults, the torque necessary to rotate it varies according to the body's rotational inertia around the specific axis used. The body's rotational inertia is the least when it rotates about its longitudinal axis; its radius of gyration is small because the body's mass is packed in close along the axis. The body's rotational inertia is the greatest when somersaulting in a layout position about the body's mediolateral axis because that same mass is distributed far away from the mediolateral axis.

Now that we have identified rotational inertia as the resistive property of a system to angular acceleration, we can return to the expression for Newton's second law as applied to rotary motion: $\alpha = T/I$. The interrelationship of the three variables in the expression is well worth understanding. The meaning is similar to that for linear motion. The angular acceleration of the segment is directly proportional to the applied net torque and is inversely proportional to the segment's rotational inertia. The relationship of any two of the variables can be expressed graphically while the third factor is kept the same. Figures Figure I.5, I.6, and I.7 illustrate the effect of holding one of the three variables constant to note how the other two change in relation to each other.

Case I: Varying the motive torque applied by the hip flexors to the lower extremity with its rotational inertia kept constant, and observing the magnitude of the resulting angular acceleration (Figure I.5).

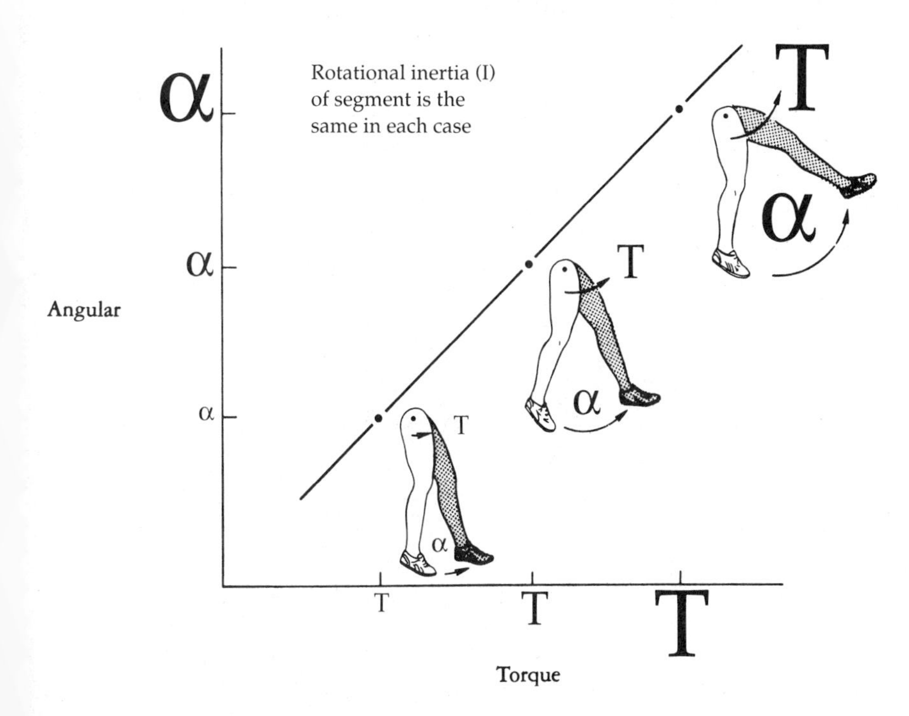

FIGURE I.5
Increasing the torque applied by the hip flexor muscles increases the angular acceleration of the segment if the segment's rotational inertia is not changed.

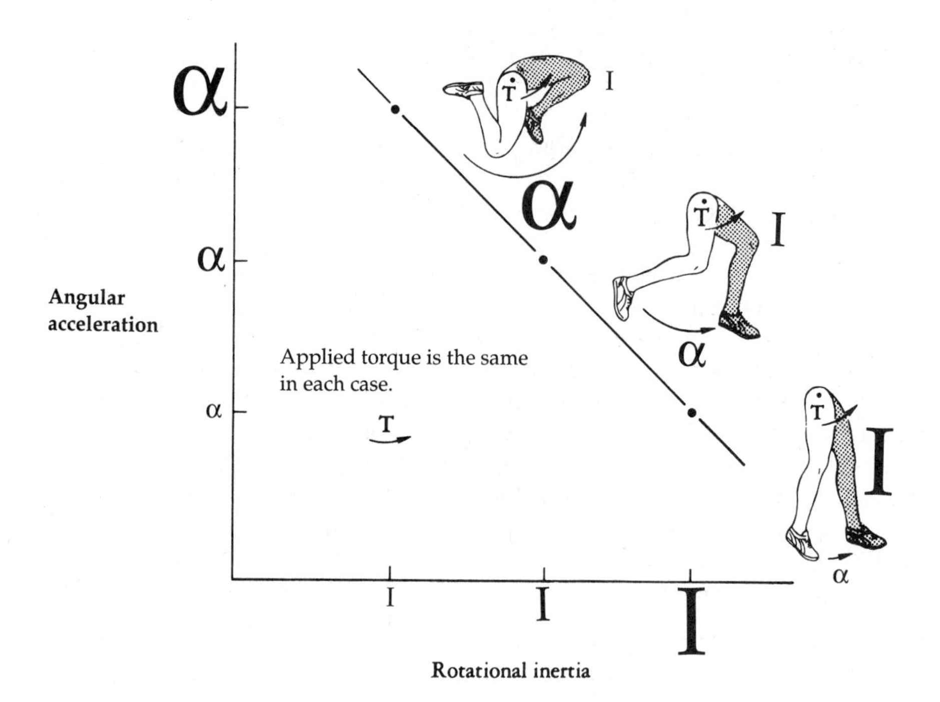

FIGURE 1.6

As the rotational inertia of a segment is increased, it experiences less angular acceleration from the same amount of torque applied by the hip flexor muscles.

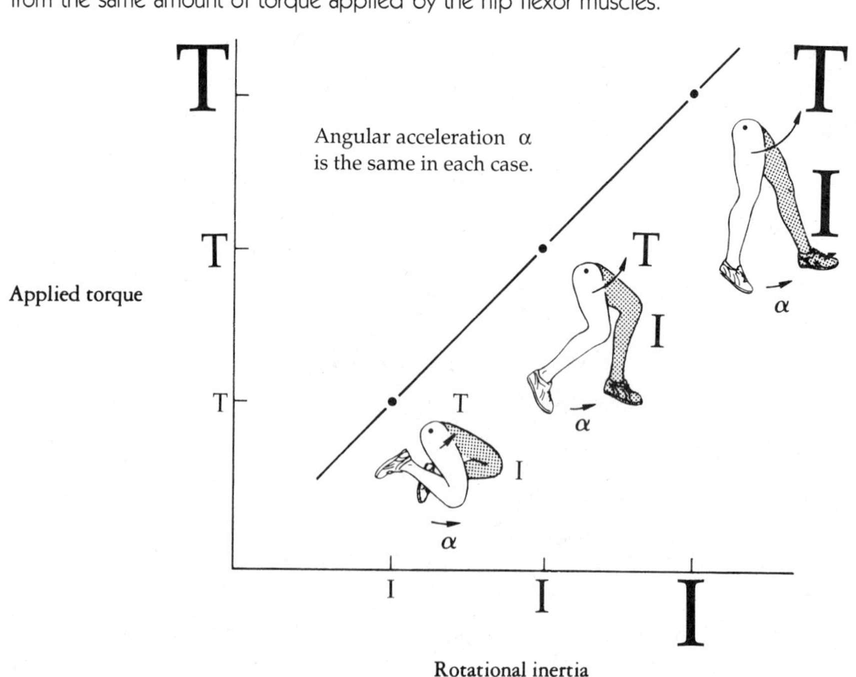

FIGURE 1.7

As the rotational inertia of a segment is increased, the applied muscle torque must be increased to produce the same angular acceleration.

Case II: Varying the rotational inertia of the lower extremity (knee fully extended to fully flexed), applying the same amount of muscle torque, and observing the resulting angular acceleration (Figure I.6).

Case III: Varying the rotational inertia of the lower extremity, specifying that the angular acceleration of the extremity is to be the same each time it is swung forward, and observing the magnitude of the applied muscle torque required to produce that angular acceleration (Figure I.7).

Understanding the concept of rotational inertia of individual body segments and how this property influences resulting rotary motion and necessary muscular effort is a valuable tool in exercise and skill analysis. The ability to vary the rotational inertia of the whole body and body parts relative to specific axes of rotation is a factor of primary importance in the performance of somersaults, twists, and suspension swinging skills, as will be seen in the chapters concerning rotations of the body.

We will now examine the familiar movement of swinging the leg forward in the recovery phase of running, as illustrated in Figure I.8. The system for which $\alpha = T/I$ will be represented by the total lower extremity (thigh, leg, and foot). The axis of rotation is through the hip joint. The mass of the lower extremity remains constant; however, the rotational inertia depends on the extremity's mass distribution relative to the axis of rotation at the hip joint. The measure of mass distribution is the radius of gyration (k). In Figure I.8, this measure of the distribution from the axis of rotation is indicated by the shading. When the knee is extended, the rotational inertia is the greatest, as shown for the left leg in Figure I.8a. When the knee is fully flexed, the mass is brought as close as possible to the hip axis; therefore, the rotational inertia is the least, as shown for the right leg in Figure I.8a and b. The resistance of the extremity to angular acceleration about the hip joint in this case is the least. The external motive torque (external to the segmental system) is the contraction of the hip flexor muscles, which cross over the hip joint to pull the femur forward.

The preceding example explains why more knee bend is observed during fast running, when the leg must be brought forward more quickly than is seen in walking. The reduction of the rotational inertia of the lower extremity produces a greater angular acceleration of the extremity for a given muscle torque.

Understanding Torque and Motion Relationships

1. Describe the difference between the mass (inertia) of your arm and its rotational inertia.
2. Grip a baseball bat and rotate it back and forth horizontally. Grip it at the opposite end and rotate it as before. Compare the difficulty with which the

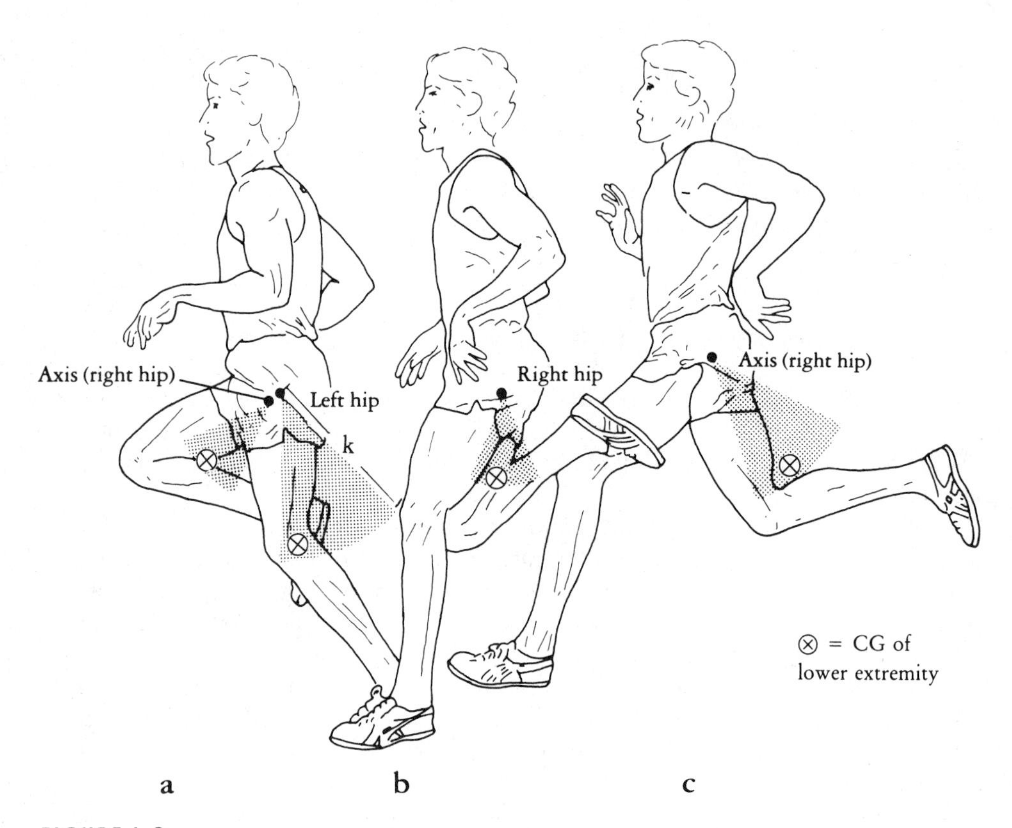

FIGURE I.8

The rotational inertia of the lower extremity around the hip joint is changed by knee flexion and extension, which alters the radius of gyration, k.

same mass may be rotated depending on how the mass is distributed from the axis of rotation. Explain.

3. With one hand, grip each of the objects listed below at one end. Using wrist motion only, rotate each back and forth through approximately a 100-degree ROM along a horizontal plane of motion. Compare the degree of difficulty experienced with each object and rank them from the greatest to least in terms of rotational inertia. The objects are: (a) a racquetball racket, (b) a fencing foil, (c) baseball bats of various weights and shapes, (d) tennis rackets of various weights and shapes (if possible), (e) a badminton racket, (f) a yardstick, and (g) a book.
4. Hold your arms horizontally out to your sides and rapidly swing them up and down in a vertical plane several times. Next, bend the elbows to place your fingers at the front of your shoulder joints. Repeat the sweeping motion of the arms in a vertical plane. Explain the difference in difficulty (you have not altered the mass of the arm).

5. Why would a front-crawl (freestyle) swimmer choose to bend the elbow as the arm is brought forward over the water to begin another stroke?
6. Why is it easier to perform jumping jacks with bent elbows?
7. Examine the masses and shapes or mass distribution of the following body segments and make a conclusion regarding the amount of muscle torque necessary to rotate these segments: (a) upper arm, (b) forearm, (c) hand, (d) fingers, (e) thigh, (f) lower leg, and (g) foot.
8. Describe an example in which a body segment is angularly accelerated by a given muscle torque. How could that segment be accelerated at a faster rate?

References and Suggested Readings

Barham, J. (1978). *Mechanical kinesiology.* St. Louis: C. V. Mosby Co.

Cooper, J. M., & Glassow, R. B. (1976). *Kinesiology.* 4th ed. St. Louis: C. V. Mosby Co.

Miller, D. I., & Munro, C. F. (1985). Greg Louganis' springboard takeoff: II. Linear and angular momentum considerations. *International Journal of Sport Biomechanics, 1,* 288–307.

Plagenhoef, S., Evans, F. G., and Abdelnour, T. (1983). Anatomical data for analyzing human motion. *Research Quarterly, 54*(2), 169–78.

Tichonov, V. N. (1976). Distribution of body masses of a sportsman. In P. V. Komi (Ed.), *Biomechanics V-B: Proceedings of the Fifth International Symposium on Biomechanics* (pp. 103–108). Baltimore: University Park Press.

Tricker, R. A. R., and Tricker, B. J. K. (1967). *The science of human movement.* New York: American Elsevier.

Wheeler, G. F., and Kirkpartick, L. D. (1983). *Physics: Building a world view.* Englewood Cliffs, NJ: Prentice-Hall.

CONCEPT MODULE J

Angular Momentum

PREREQUISITES

Concept Modules C, D, E, F, G, I

CONCEPT MODULE CONTENTS

J.1 Angular Momentum

To understand how and why the body moves as it does in rotary motion about an axis, angular momentum must be considered. We know that a body's resistance to change in its state of angular motion is called its rotational inertia and is designated by **I**.

We have seen how the angular acceleration of a system is determined by the net external torque applied and the rotational inertia of the system. Angular acceleration of a system by a torque produces a certain amount of angular velocity, or speed of rotation in a given direction. Were it not for the application of an external torque, the system would have remained as it was, either at rest with zero angular velocity or rotating with some preexisting angular velocity.

Just as a system moving linearly possesses a certain amount of linear momentum, a rotating system has **angular momentum (L)**. The magnitude of a system's angular momentum is the product of its rotational inertia (**I**) and its angular velocity (ω):

EQUATION J.1

$$L = I\omega \quad \text{or} \quad L = mk^2\omega$$

where L is angular momentum, m is mass, k is radius of gyration, I is rotational inertia, and ω is angular velocity.

Although **I** is the most frequently used symbol for rotational inertia, remember that **I** is a function of both the mass, **m**, and the *square* of the radius of gyration, **k**.

Calculations of angular momentum are not used here, but the relationship of the factors that influence angular momentum are addressed.

The significance of angular momentum of body segments and of the whole body in a multitude of movement activities cannot be overestimated. It is a primary factor involved in swinging, jumping, throwing, kicking, twisting, somersaulting, swimming, and locomotion. Angular momentum, or the quantity of angular motion, is created originally from an externally applied torque. After some angular velocity (and angular momentum) is achieved, the body continues to rotate after the removal or cessation of the motive torque, and it loses or gains angular momentum only upon receiving another external torque.

In Figure J.1a, the follow-through movements of the arm and racket after ball contact are the result of angular momentum of these segments caused by applied muscle torque. The extremity and racket continue to rotate forward and upward on their own angular momentum until resistive torque of the shoulder and arm muscles and joint ligaments causes angular deceleration of the segment. In Figure J.1b, once the gymnast executes the push-off to rotate the body in vaulting, the body continues to rotate because of its angular momentum. The body continues to rotate with this angular momentum until acted on by the external torque provided by the floor at landing.

J.2 Angular Impulse

To angularly accelerate a body, a net external torque must be applied. If a large angular velocity of the body is desired, the torque must be large or the torque must be applied for a longer period of time to continually accelerate the body. Whether a torque acts for a short time or a long time, the product of torque multiplied by its time of application is called the **angular impulse**. The magnitude of the angular impulse acting on the body determines the *change* in the body's angular momentum ($I\omega$ before to $I\omega$ after the impulse acts). Quantitatively, angular impulse is expressed as follows:

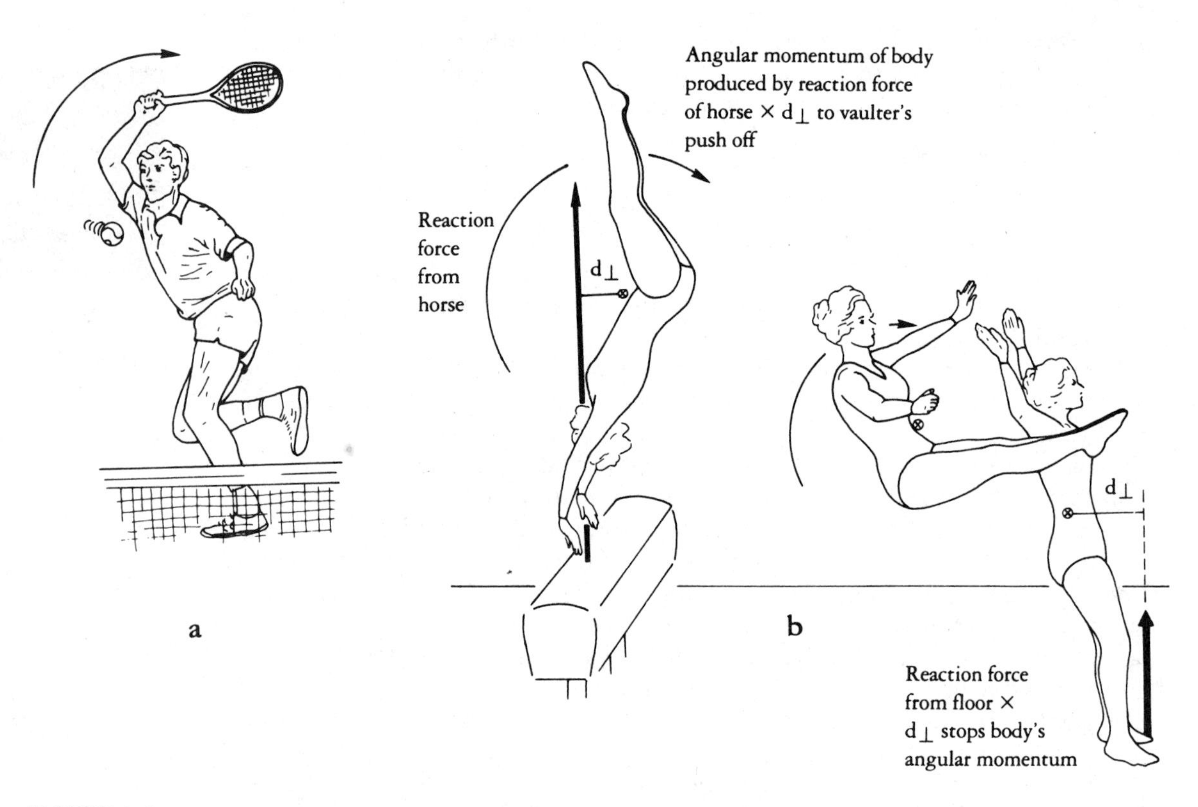

FIGURE J.1

Angular momentum of (a) segments and (b) the whole body is produced by external torque applied to the system. If the rotating system does not encounter an external torque, it will continue rotating with the same angular momentum.

EQUATION J.2

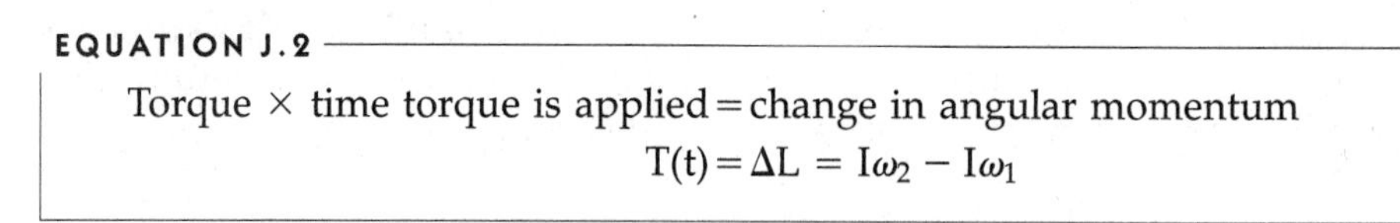

$$\text{Torque} \times \text{time torque is applied} = \text{change in angular momentum}$$
$$T(t) = \Delta L = I\omega_2 - I\omega_1$$

where T is external torque applied, t is time during which the torque is applied, and ΔL is change in angular momentum.

Thus, a body projected into the air by a force, which acts for a period of time before takeoff and acts eccentric to the body's center of gravity (CG), will create a linear motion of that body's CG and a rotation of that body about its CG. The magnitude of the velocity of rotation will be determined in part by the magnitude of the force, the perpendicular distance of the line of force from the CG axis, and the length of time the force is applied.

The diver in Figure J.2 depresses a springboard, giving it elastic potential energy. The upward force of the board causes the diver's upward linear motion,

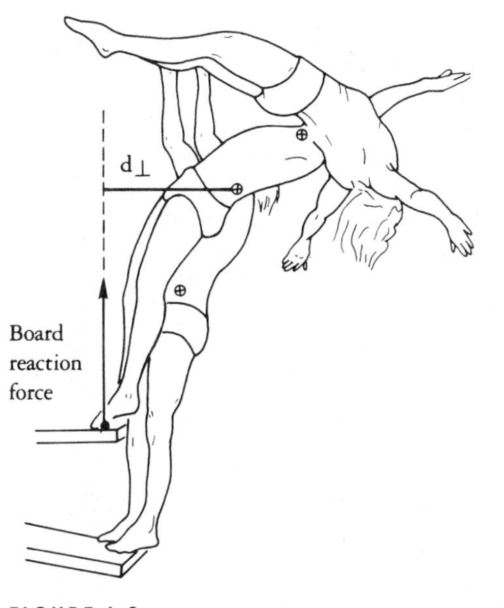

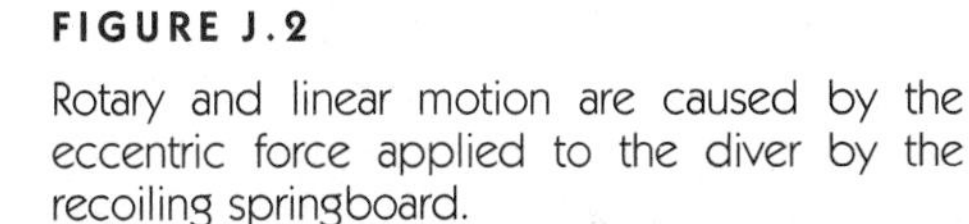

FIGURE J.2

Rotary and linear motion are caused by the eccentric force applied to the diver by the recoiling springboard.

and because the line of action of the force passes eccentric to the diver's CG ($d_{\perp}$), it is torque producing and causes him to rotate. The greater the force, the $d_{\perp}$, or the time of application of the torque, the greater the angular impulse and consequent angular momentum.

Another example of an impulse creating rotation of the body about an axis is a figure skater who wants to spin about her longitudinal axis. She initiates the spin by pushing against the ice with the toe of the skate (Figure J.3). The push is directed opposite to the spin direction, and its line of action is eccentric to the body's longitudinal axis. Figure J.3a indicates the skater's position during the force application, and Figure J.3b is a view from the top showing the perpendicular distance the force is from the longitudinal axis of rotation. The longer the time she can generate an accelerating reaction torque, the greater her angular momentum when the push ceases.

Once angular momentum is established in a body, it can be decreased by a resistive angular impulse, such as an off-center contact of the fingers receiving a top-spinning volleyball. The angular momentum of an object also can be increased by a motive (accelerating) angular impulse. For example, in catching a Frisbee creatively, the glancing blows of the fingers on the rim of the spinning disc increase the spin velocity before it is caught. Such fast spinning of the Frisbee, like the spinning of a toy top, gives stability to the object; that is, it is resistant to changing its axis orientation, or direction. In fact, the faster

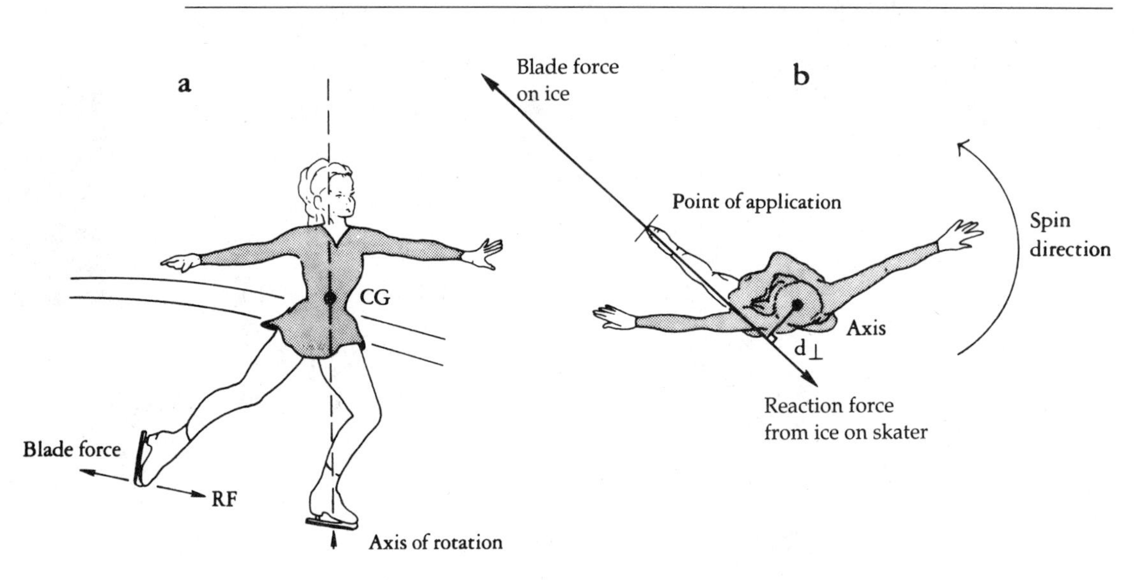

FIGURE J.3

(a) A figure skater initiating a spin about the longitudinal axis. (b) A view from the top showing the line of action of the reaction force and the force arm distance $d_\perp$ from the axis of rotation.

the spin (angular velocity), the greater the stability of the body in its particular position or orientation. Such resistance to change in position is called **gyroscopic stability**. It accounts for the directional stability of a spiraling football pass or spinning discus, of a cyclist maintaining an upright position when the wheels of the bicycle are rotating rapidly, or of an ice skater spinning about a vertical axis. Such a phenomenon is a by-product of the body's angular momentum.

J.3 Conservation of Angular Momentum Within a System

We have seen that once an angular impulse, T(t), has given a system some angular momentum, that angular momentum continues in the absence of external torques.

Newton's first law about the constancy of a body's angular momentum is the basis for the principle of **conservation of angular momentum**. This principle declares that the angular momentum of an isolated system will remain the same regardless of any movements or torques that are produced *internal* to the system, as long as no *external* torque is applied to the system.

The term *isolated* means that the rotating body is free from an external source that could apply a torque (e.g., a supporting surface or object, contact with another body, or air resistance or water resistance that is strong enough to create a torque). Conservation of angular momentum of the human body is evident in a number of activities in which the body is isolated, as in diving and

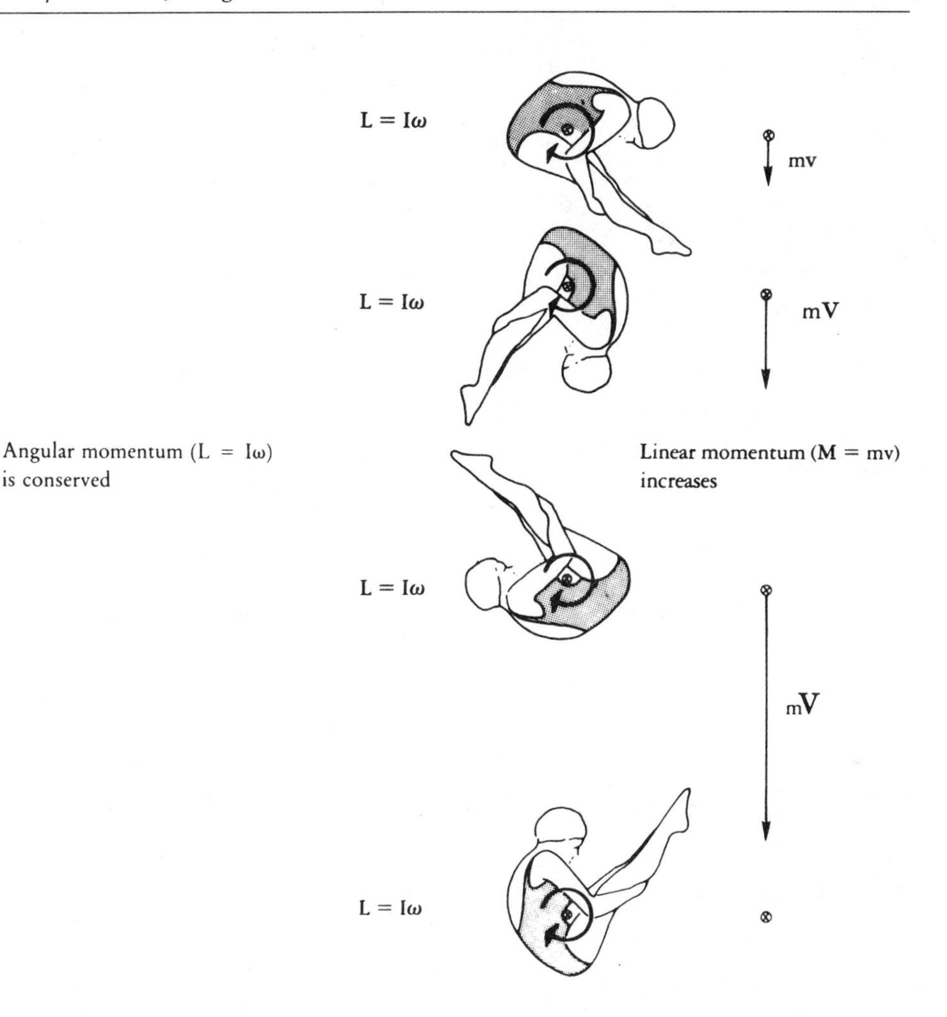

FIGURE J.4
After angular momentum is given to a body that is then free from external torques, its angular momentum ($I\omega$) will stay the same. The body's linear momentum will change under the influence of the external force of gravity.

aerial gymnastics, and also in situations in which the external resistive torque is negligible, as in spinning about on the tip of an ice skate or on a smooth shoe surface.

The diver in Figure J.4 illustrates the constancy of angular momentum during descent after leaving the board with angular momentum ($I\omega$) about the mediolateral axis. Once the body leaves the supporting surface, the only external force acting on it is gravity. Gravity acts through the CG of the body, which serves as the location of the axis of rotation during free flight. The gravitational force, therefore, does not have any perpendicular distance from the axis of rotation and thus can produce neither torque nor angular impulse. Consequently, dur-

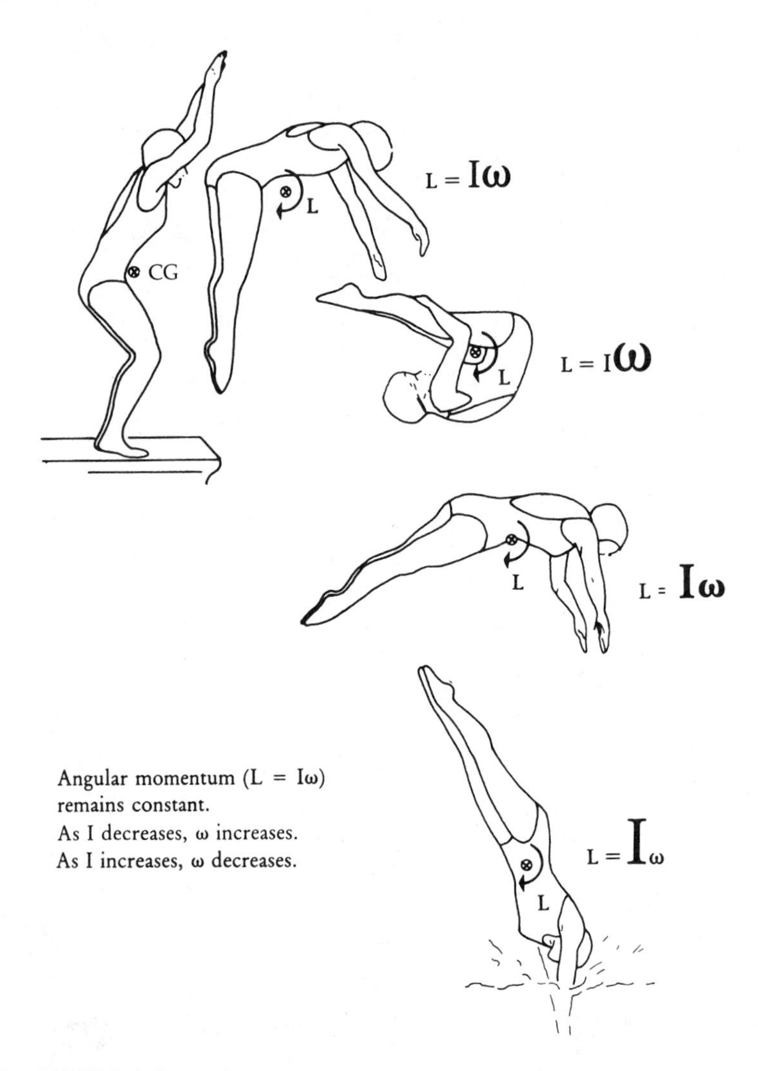

FIGURE J.5

The angular momentum of a diver is conserved throughout the aerial phase, during which she receives no external torque (according to Newton's first law).

ing free flight, gravity cannot cause any change in the angular momentum of the body. Because there are no external torques, there will be no change in the diver's angular momentum until she lands or collides with another object. For simplicity, air resistance is considered negligible.

Note that gravity is accelerating the diver's CG linearly downward and causes her to fall faster and faster but has no effect on her angular momentum.

Now consider the diver who initiates a forward somersault as she pushes off from the board (Figure J.5). Her initial angular velocity in a layout position is quite small, and her rotational inertia relative to the axis passing through her CG is large. Her angular momentum is the product of these two factors; once isolated in air, her angular momentum, **I**ω, will not change. If she then goes into a pike position, she brings her body mass close in toward her axis of rotation and thereby decreases her radius of gyration and thus her rotational inertia. Such a decrease in rotational inertia causes her angular velocity to increase so that the product of **I** and ω remains the same value as it was before she changed body position. Her angular momentum is conserved.

Before entering the water she moves out of the pike position into the layout, thereby slowing her angular velocity so that she can enter in a straight position. Once her hands enter the water, she encounters a resistive torque from the water, stopping her rotation and allowing the rest of her body to enter vertically. Such changing of segment positions while rotating changes the body's rotational inertia and leads to an opposite type of change in the angular velocity, so that the total angular momentum of the body remains unchanged. The principle of conservation of angular momentum forms the basis for controlling rotary movements of the body and its segments.

Such a situation in which an isolated rotating body can change its shape (and rotational inertia) with internal forces and thereby change its angular velocity can be depicted by the symbols shown in Figure J.6.

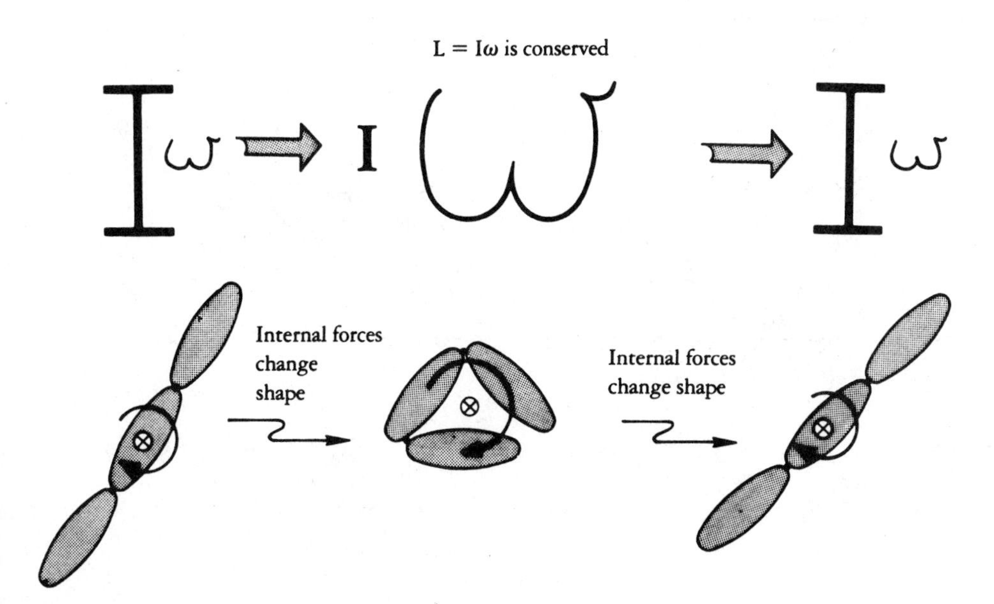

FIGURE J.6

Internal forces can change the shape of a body to change its mass distribution (k), which changes its rotational inertia ($I = mk^2$). Angular momentum is conserved; therefore, angular velocity must change as the shape is changed.

J.4 Vector Resolution of Angular Momentum

The resolution of an angular momentum vector is important in understanding one method of initiating rotation in the air, a common maneuver performed in diving and gymnastics. The process set forth here is like that used for resolving a body's linear velocity, except that the angular directions must be indicated by the right-hand thumb rule that was described previously for determining torque direction. The following method demonstrates the resolution of a body's angular momentum into two perpendicular angular momentum components. For example, in Figure J.7a, a gymnast is in the process of simultaneously somersaulting and twisting around a horizontal axis in space, indicated by AB; the body's spatial angular momentum vector is directed toward B.

To resolve the body's spatial angular momentum vector into a twisting component and a somersaulting component, the graphic vector resolution method (completing the rectangle) is used. The two directions sought are along the axis for somersaulting (the mediolateral axis, which runs through the body from side to side) and along the axis for twisting (the body's longitudinal axis, which runs from toe to head). These directional lines are drawn perpendicular to each other to form two sides of the rectangle that will be completed. Figure J.7b shows the completed rectangle, the vector direction and magnitude (lengths) of the twist-

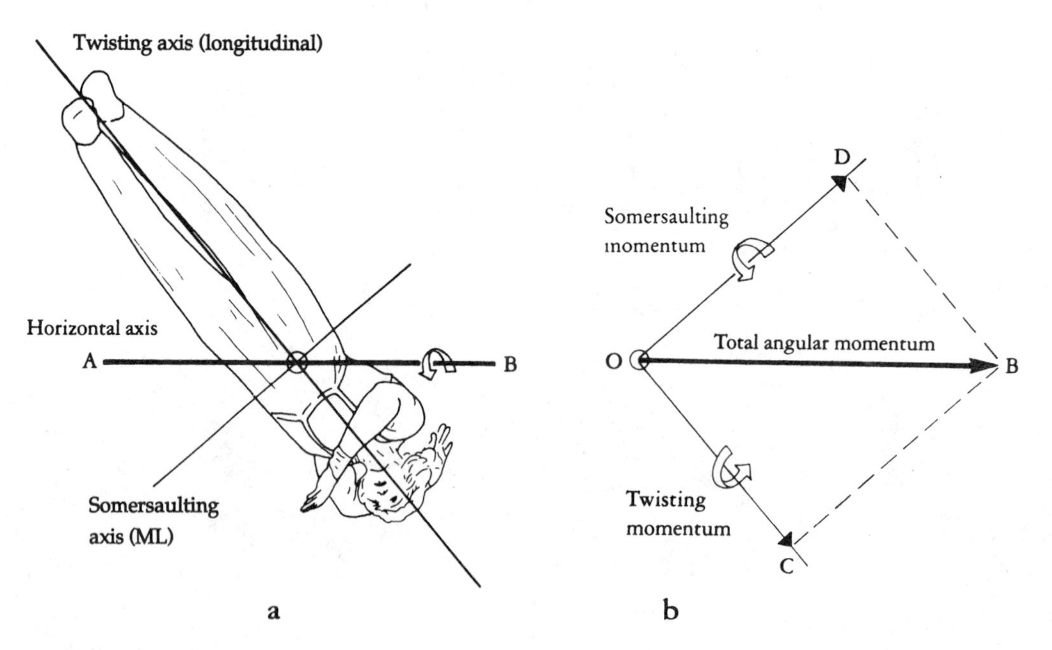

FIGURE J.7

A somersaulting and twisting gymnast. (a) The right hand thumb rule indicates the total angular momentum of the body around a horizontal spatial axis (A-B). (b) The body's angular momentum is resolved into its somersaulting component and its twisting component.

ing momentum (OC), and the direction and magnitude of the somersaulting momentum (OD). The component vector direction determines the direction of twisting and somersaulting according to the right-hand thumb rule.

Understanding Angular Momentum

1. Give an example in human movement of a system that has angular momentum; tell what factors comprise its angular momentum.
2. Define angular impulse.
3. Identify the angular impulse that causes a body segment to rotate and therefore gain angular momentum. Identify the angular impulse that is responsible for diminishing a segment's angular momentum to zero.
4. Hold a book in each hand with the arms held horizontally out from your sides. Turn and spin on one foot. While spinning, bring both books to your chest. Explain the body's motion response in terms of the conservation of angular momentum.
5. Repeat the steps in question 4, but return the arms out to the sides just after you bring the books to your chest. Explain the response.
6. Describe several examples in sport and dance in which the angular momentum of the body is conserved although the angular velocity changes.
7. The following experiments involve using a turntable and various pieces of external equipment. The table should be made level in the horizontal plane, have a small rotational inertia, but be sturdy enough and large enough to hold an adult. The joint between the table and the base should be as frictionless as possible. An external wheel should be available, which is weighted on the outside rim to increase its rotational inertia. It may be rotated around its axis while a person is holding onto the extended axle (Figure J.8a).

   For all the experiments, the following should be noted:

   The axis of rotation of the system and of the parts
   The direction of rotation of the system and the parts
   The direction of the angular momentum vectors involved
   The relative rotational inertias (masses and radii) of the parts
   The relative speeds of rotation of parts
   The description of what happens and the reason why the experiment operates as it does

Turntable Experiments

a. Stand on the motionless turntable, holding a baseball bat in both hands. Swing the bat overhead in large circles.

FIGURE J.8

(a) Standing on a turntable isolates the body from external torques around a vertical axis. Holding a bicycle wheel by its extended axle creates changes in rotary motion of the body, depending on the spin of the wheel. (b) Holding weights in the hands enables the user to easily alter the body's rotational inertia around a vertical axis.

b. Perform the same motion as in experiment **a** without holding the bat in your hands. Compare.

c. Perform the following arm actions while standing on the motionless table:

i. Abduct the right arm to shoulder level, and flex the left arm to shoulder level. Swing both arms vigorously to the left.

ii. Perform the same experiment as in **i**, while holding two heavy books in the hands. Compare the results (see Figure J.8b).

iii. Perform the same experiment as in **i**, with the legs spread to the side as far as they will go.

iv. Hold both arms in abduction at shoulder level. Transversely flex them simultaneously.

v. Perform the same experiment as in **iv**, holding a book in the right hand.

d. Stand on the motionless turntable and swing a baseball bat vigorously while gripping the handle. Turn the bat around so that you are now gripping the hitting end and swing it vigorously again. Compare the two reactions.

e. Hold two heavy books in the hands and abduct the arms to shoulder level. Have a partner spin the turntable. As you are spinning, pull the books in toward your chest. (Be careful, as the spin velocity may increase appreciably.) Put the arms back out again.

Wheel Experiments

The following experiments involve the turntable and the bicycle wheel:

f. Stand on the motionless turntable. Your partner spins the free wheel as fast as possible and hands it to you in a position so that the wheel is spinning horizontally; that is, the axis of the wheel is vertical. Turn the wheel to a vertical spinning position (axis horizontal).

g. Repeat experiment **f**, but have your partner hand you the wheel in a vertical spinning position (axis horizontal).

h. Repeat experiment **f**, with the person on the table spinning the wheel. Compare.

i. Stand on the motionless turntable. Your partner hands the vertically spinning wheel to you, and then spins the turntable. Turn the wheel axis to the right, return the wheel to the original position, and turn the wheel to the left.

j. Repeat experiment **i**, altering the turntable speeds to a slow spin, a medium spin, and a fast spin. (Be careful!)

References and Suggested Readings

Barham, J. (1978). *Mechanical kinesiology*. St. Louis: C. V. Mosby Co.

Dyson, G. (1973). *The mechanics of athletics* (6th ed.). London: Hodder & Stoughton.

Frohlich, C. (1979). Do springboard divers violate angular momentum conservation? *American Journal of Physics, 47*(7), 583–592.

Hopper, B. J. (1973). *The mechanics of human movement*. New York: American Elsevier.

Resnick, R., & Halliday, D. (1977). *Physics: Part I*. New York: John Wiley & Sons.

Tricker, R. A. R., & Tricker, B. J. K. (1967). *The science of human movement*. New York: American Elsevier.

Wheeler, G. F., & Kirkpatrick, L. D. (1983). *Physics: Building a world view*. Englewood Cliffs, NJ: Prentice-Hall.

CHAPTER 13

Analysis of Activities in Which the Body Rotates Free of Support

PREREQUISITES

Concept Modules B, C, D, E, F, G, I, J
Chapters 1, 9, 10

CHAPTER CONTENTS

During the discussion of projectiles in Chapter 10, principles were applied to the human body as an object moving through space. The focus of attention for projectiles was the center of gravity (CG), and its motion was curvilinear and parabolic. In most projectile activities, the primary mechanical purpose is to give the greatest distance or speed or accuracy or all three to the object being projected. During activities in which the human body is projected, however, important rotational movements may occur.

The principles of rotary motion apply to the movements of the body during the air phases of such activities as long jumping, high jumping, and pole vaulting. Most of the body's rotational activities, however, are judged events in which the overall performance objective is to achieve an ideal or model performance rather than to attain the greatest horizontal or vertical distance. The OPO of most of these rotationally based activities is to move or position the body or its segments in a predetermined pattern with the intent of achieving an ideal or model performance. The quality of movement is then judged. Gymnastics, diving, dance, and trampoline activities are obvious examples.

13.1 The Human Body in Rotary Motion

The Center of Gravity and Axes of Rotation

When the total body is rotating and free of support (aerial movements), body rotation takes place around one or more of the three principal axes of the body, the mediolateral (ML), anteroposterior (AP), and longitudinal axes. The three principal axes must pass through the CG of the total body whether the center is located within the body or outside the body.

When a body is free of support, the path in space of the total body's CG cannot be changed by segmental movements. During free flight, the body parts move relative to the CG (i.e., the body parts realign themselves around the CG). The realignment of body parts, however, does not change the path of the CG from what it would have been without segmental movement.

Figure 13.1 shows the path and position of the CG of the postflight of a long-horse vaulter performing a forward somersault around the body's ML axis. Notice that the performer takes off in a layout position, moves into a tuck position, and lands in a layout position. Although the body parts realign themselves relative to the body's CG, the path is not altered; it is the predetermined parabolic path that was established at takeoff. (The path would be the same had the performer remained in the layout position.)

Rotation of the body around any of the body's principal axes may be accomplished when the body is in a layout, pike, arch, or even a tuck position. The body may leave the ground in one of these positions and move *symmetrically* into any of the other positions without changing the principal axis of rotation.[1]

Movements of the entire body about each of the principal axes of rotation may be reviewed from Concept Module B. Rotations about the ML axis are called **turns** or **somersaulting** motions, and rotations about the longitudinal axis are called **twists**.

[1]Note that the movements must be symmetrical; if not, quite a different situation occurs, as will be seen later.

FIGURE 13.1
A vaulter performing a front somersault vault in a tuck position.

The aerial cartwheel is one of the few examples of a movement taking place around the body's AP axis. AP-axis motion in free flight is probably the least common of body movements, and often appears as the result of turning and twisting rotations being performed simultaneously.

Angular Momentum and Human Body Rotations

The angular momentum of the human body in rotational motion consists of three factors: the mass of the performer, the radius of gyration of the performer's mass relative to the axis of rotation, and the angular velocity of the performer about the axis of rotation. Because the mass of a performer does not change, it is considered a constant factor in the body's angular momentum.

Recall that the radius of gyration is a distance that represents the distribution of the mass relative to the axis of rotation. The radius of gyration of the performer's mass relative to an axis of rotation depends on which axis the body is rotating about—ML, AP, or longitudinal—and on the position of the body segments relative to that axis. As segmental movement occurs, the distribution of the mass changes, and thus its radius of gyration changes. Figure 13.2 illustrates body positions that change the distribution of the body's mass about each of the three principal axes. In Figure 13.2a, one can see that the radius of gyration about the ML axis is the largest when the performer assumes a layout position with the arms overhead and is the smallest when her body is in a tuck position. During free flight, the radius of gyration is measured from the ML axis, which passes through the CG whether the CG is located within the body (as in the layout position) or is located somewhat in front of the abdomen (as in the pike position). The body mass in this case is constant (barring any loss of clothing), no matter how the person has changed the distribution of the mass about the axis of rotation. In a tuck position, the body has a smaller radius of gyration because the mass is closer to the ML axis about which it is rotating.

In Figure 13.2b, three positions are shown relative to the body's longitudinal axis. In a large radius of gyration position, the upper and lower extremities are positioned away from the longitudinal axis, whereas in the small radius position, the extremities are packed in close to the axis. Similarly, changes in the distribution of the body's mass may be made about the AP axis by assuming the various positions in Figure 13.2c.

The two factors, then, that determine the body's resistance to changes in rotary motion are the body's mass and the distribution of that mass around the axis of rotation in question. When both these factors are considered, the resistance to rotation or rotational inertia (I)

FIGURE 13.2
A change in the body's configuration will change the radius of gyration of the body relative to the three principal axes of rotation: the (a) mediolateral, (b) longitudinal, and (c) anteroposterior axes.

b Longitudinal axis

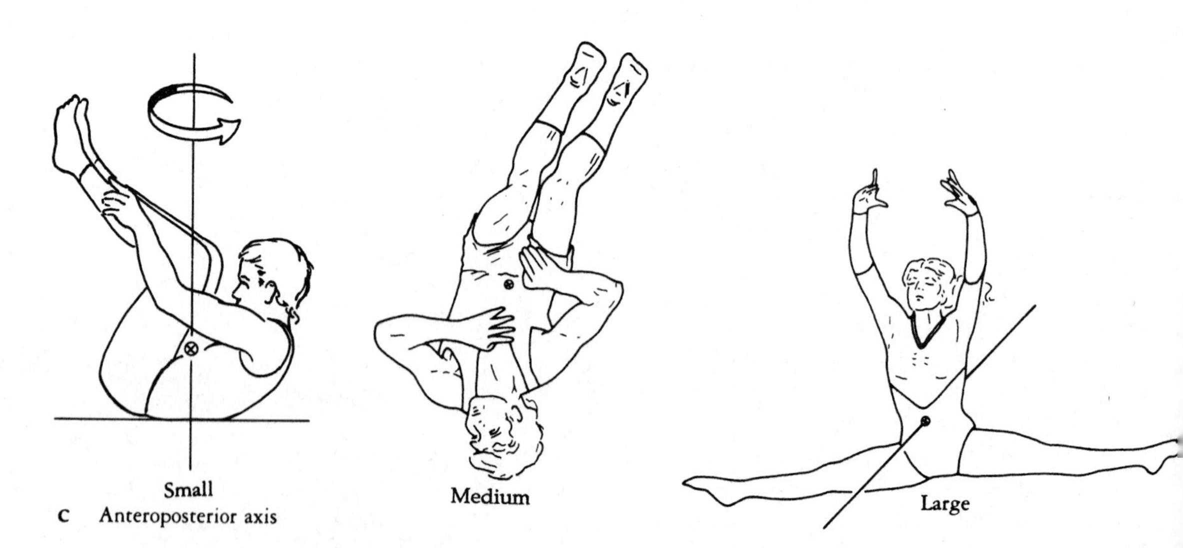

c Anteroposterior axis

FIGURE 13.2
(continued)

can be calculated. Recall that because of the particular way in which the radius of gyration is calculated for a given body position, the term is squared and multiplied by the mass (mk^2). Because we are considering the mass of the body to be constant, the *mass distribution* becomes the major focus of attention when analyzing free-rotation activities.

The body is a single system made up of individual pairs. The angular momentum that is established at takeoff is established for the system and cannot be changed unless an external torque is applied. The performer is not rigid, however, and individual body segments are free to rotate about individual axes of rotation passing through the joint centers. Any segments caused to rotate about their own axes have angular momentum produced by *internal* muscle torques; that is, they act on the segments within the system but do not act as external torques on the total system.

The segments are part of the system, however, and their angular momenta become part of the total angular momentum of the system. Because the angular momentum of the system must be conserved (cannot change), the segmental momenta cannot add or subtract from the established total momentum of the body, but the segments may "assume" or redistribute the momentum within the system. As an example, consider the performer in Figure 13.3, whose body and extremities are represented by cylinders projected into space by a force acting through the CG. Because the projection force acts through the CG, the body has linear momentum but no angular momentum at any time during the flight. If the arm segments are then circled backward around their ML axis so that the sum of their angular momenta equals three angular momentum units, then the trunk segment must take on enough forward momentum about *its* ML axis so that the algebraic sum of all the angular momenta in the system remains zero. This can be shown by solving the equation:

EQUATION 13.1

$$\sum L = 0$$
$$L_{arms} + L_{legs} + L_{trunk} = 0$$
$$I_1\omega_1 + I_2\omega_2 + I_3\omega_3 = 0$$
$$(+3.0) + 0 + L_{trunk} = 0$$
$$L_{trunk} = -3.0 \text{ angular momentum units}$$

Note that the solution to the equation indicates that the rotation of the trunk is negative, and, therefore, the trunk is rotating clockwise (cw). Knowing the mass and the radius of gyration of the trunk, we can then calculate how fast the trunk will be rotating. The mass and radius of gyration of the trunk will be much greater than the masses and radii of the arms; consequently, the trunk angular velocity will be much slower than the angular velocity of the arm rotation. Remember that the *sum* of the angular momenta of the system during flight is, and must remain, zero in this case.

The summation technique also can be used when the magnitude of the angular momentum is *not* zero at takeoff. For example, if the body has an angular momentum of +30 angular momentum units (circling counterclockwise [ccw] about the body's ML axis) at takeoff, the arms and legs then circle during flight so as to generate +2.5 angular momentum units *each* (also circling ccw) about their segmental ML axes ($4 \times 2.5 = 10$). Thus, the trunk will be left with +20 angular momentum units. It will have its angular velocity decreased enough to let the product of $mk^2\omega$ of the trunk equal +20 angular momentum units. The calculation would be as follows:

$$L = +30 \text{ angular momentum units}$$

$$L_{arms} + L_{legs} + L_{trunk} = +30 \text{ angular momentum units}$$

$$2(+2.5) + 2(+2.5) + L_{trunk} = +30 \text{ angular momentum units}$$

$$L_{trunk} = +30 + (-5.0) + (-5.0)$$

$$L_{trunk} = +20 \text{ angular momentum units}$$

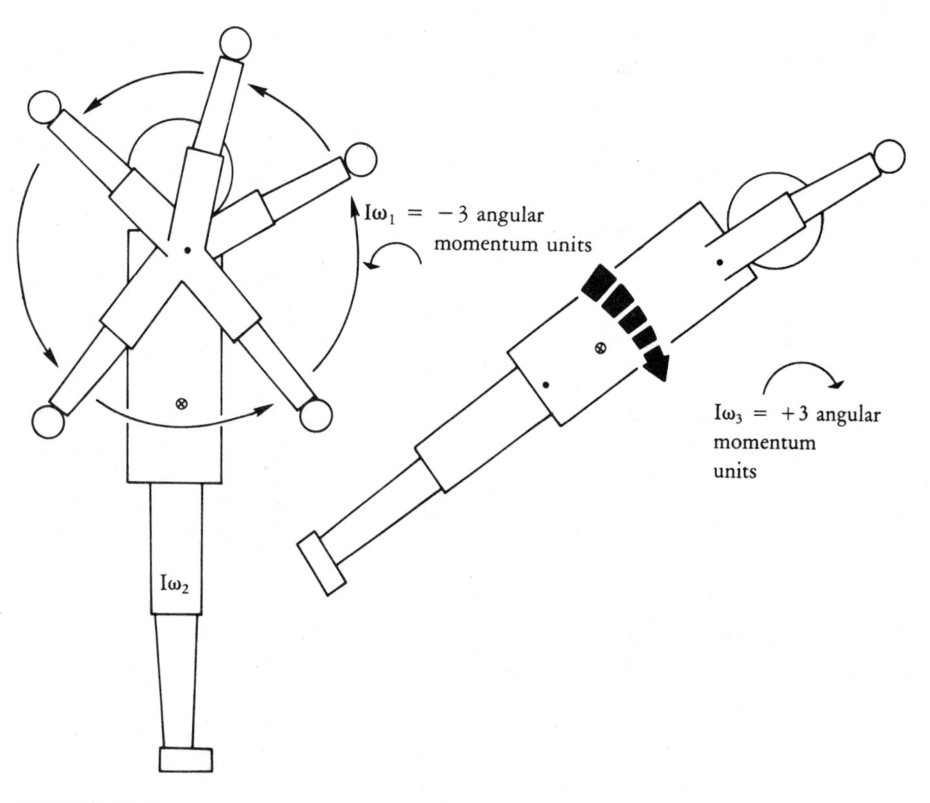

FIGURE 13.3
The body represented by a cylinder having zero angular momentum. Circling the arms backward will cause the trunk and lower extremities to circle forward, thus allowing the body to maintain zero net angular momentum.

If enough angular momentum can be generated from the circling of the arms and legs, the body will cease rotating completely. Remember that the summation always must be around axes that are parallel to each other, for momentum generated around one axis in the air does not alter the momentum established around other nonparallel axes.

Application of the Right-Hand Rule for Determining Angular Momentum Vectors

To determine the magnitude and direction of the angular momentum vector for a body rotating around an axis, one may use the right-hand rule, which is described in Concept Modules E and J. To determine the vector direction of the body rotating around an axis, one wraps the right hand around the axis of rotation so that the fingers are flexed in the direction of the rotation. The vector direction of that particular angular velocity (and angular momentum) is the direction in which the right thumb points. For example, if one were looking down at a gymnast performing a pirouette in the air about the longitudinal axis (Figure 13.4), she would appear to be rotating ccw, and the right hand would be wrapped around the longitudinal axis as shown. The right-hand thumb would be pointing toward her head.

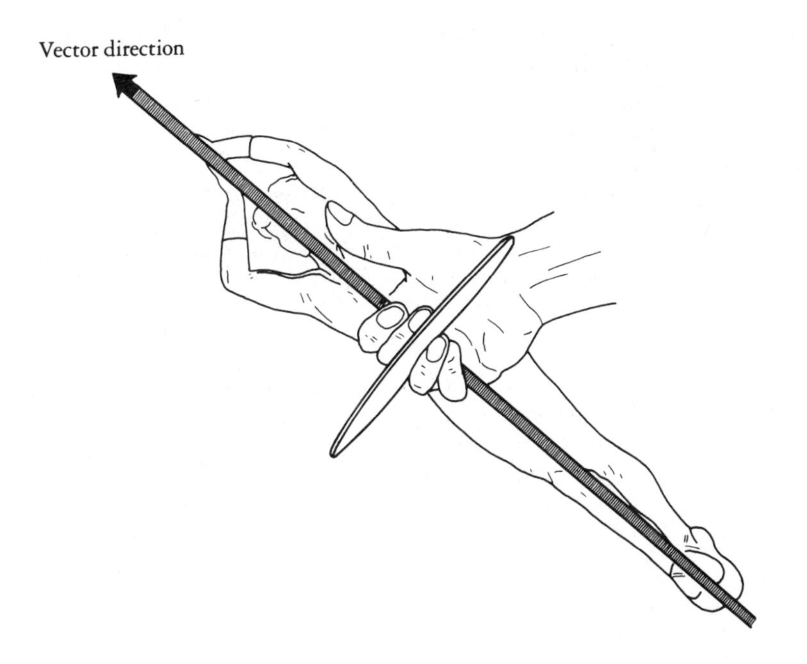

FIGURE 13.4
A gymnast doing a pirouette, with the right-hand thumb rule applied to show the angular momentum vector.

The vector associated with angular motion is represented by an arrow that has magnitude and direction and may be added or resolved just like any other vector. Vector addition could be used to solve the problem of the performer circling the arms and legs in Figure 13.3. Resolution of a resultant angular momentum vector is important in understanding one method of *initiating* rotation in the air.

Understanding the Body Rotating Free of Support

1. For the following segmental movements, state the segmental axes about which the segment is rotating; what effect (increase, decrease, remain the same) does the segmental movement have on the body's radius of gyration about each of the three body's axes?
   a. arm abduction to 90 degrees
   b. arm flexion to 90 degrees
   c. knee flexion to 90 degrees
   d. transverse flexion (adduction) from 90 degrees abduction
   e. lateral trunk flexion 70 degrees
   f. trunk hyperextension of 40 degrees

2. How many mass units of resistance to changes in *horizontal* motion will a body weighing 400 N have?
3. How many mass units of resistance to changes in *vertical* motion will a body weighing 400 N have?
4. If a body has 2 mass units and a radius of gyration of 5 cm, what is the rotational inertia? if 2 mass units and a 10- cm radius of gyration?
5. What is the *change* in resistance to rotation of a body with an increase of the radius of gyration to three times what it was initially? a decrease of the radius of gyration to one fourth of what it was?
6. If angular momentum is equal to zero at takeoff, indicate about which axis and in which direction the trunk will move with the following limb movements: (a) circumduction (circling) of the upper extremities forward, (b) circumduction of the upper extremities backward, (c) abduction of the right upper extremity and adduction of the left upper extremity.
7. If the same body has 50 angular momentum units forward at takeoff, would the trunk's rotational velocity increase, decrease, or stay the same when the movement indicated in question 1 takes place?
8. Answer question 2 for the same body having 50 angular momentum units backward around the ML axis.

13.2 Initiating Rotations

Somersaults or twists may be initiated from the ground using the ground reaction force, and they also can be initiated in the air. Rotation may be begun in the air when the body has no angular momentum about the axis of rotation or when the angular momentum that a body has gained from takeoff is redistributed or used for rotation around a second axis. There are three basic methods of influencing rotation while in the air. The method of rotation used may be recognized from the body position prescribed in performing the stunt. These positions are similar for diving, trampoline, tumbling, and aerial planes of freestyle skiing and apparatus routines.

Ground-Reaction Rotations

Figure 13.5 shows a performer initiating rotation around her ML (somersaulting) and longitudinal (twisting) axes by using the reaction force of the beam during a dismount.

As the gymnast leaves the beam, her CG is ahead of the line of action of the ground reaction force (GRF) of the beam. Consequently, external torque is created about the ML axis through the performer's CG. The gymnast leaves the beam with angular momentum about the ML axis. In this illustration, the angular momentum is directed cw. Using the right-hand rule for determining direction, the vector would be pointing along the gymnast's ML axis and to her left.

If a person rotates some body segments in one direction around a joint axis while the feet are still in contact with the ground, the friction force of the ground on the feet produces a ground reaction force which results in the momentum of those segments being transferred to a parallel total body axis as the performer leaves the supporting surface.

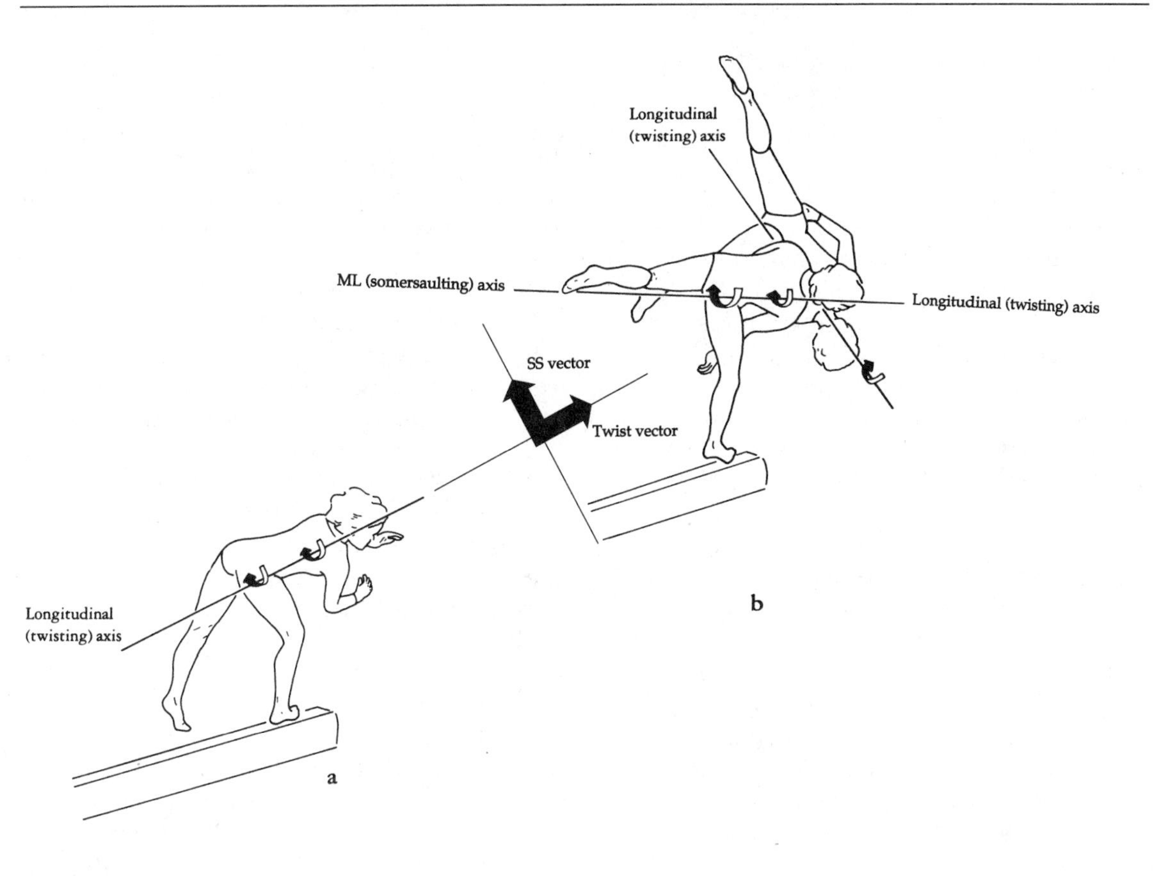

FIGURE 13.5
A gymnast initiating rotation around her longitudinal axis from the supporting surface to produce a twisting somersault dismount.

In the gymnast's case, her trunk and free leg are rotated around a longitudinal axis passing through the supporting hip joint. In effect, the movement is transverse abduction of the pelvis at the right hip. After takeoff, the angular momentum created by this segmental motion during beam contact causes the body to rotate around its longitudinal axis, thus providing the twisting momentum. The vector direction using the right-hand rule is directed headward. Therefore, the ground reaction force provides an external torque that acts to give the gymnast's body angular momentum about her ML and longitudinal axes. Once the gymnast leaves the ground with her established angular momentum, she may change segmental positions relative to her angular momentum axes. Comparisons are then made between the size of the rotational inertia about an axis at takeoff and the size of the rotational inertia in the new position. Because of the conservation of angular momentum principle, if the gymnast increases her rotational inertia about an axis after takeoff, her angular velocity about that axis will decrease; if she decreases her rotational inertia about an axis, her angular velocity about that axis will increase.

Ground-reaction torques are used in establishing momentum on takeoff in long jumping and high jumping as well. In the back layout high-jump technique, the performer leaves the ground with his or her CG at approximately a 40-degree angle from the horizontal. Depena and Chung (1988) estimate that the ground-reaction force (GRF) at takeoff is 66.8 degrees from the horizontal. Thus, the GRF is eccentric to the jumper's CG therefore applying a torque to the body. The jumper then leaves the ground with some angular momentum. Figure 13.5b illustrates the position of takeoff of a high jumper using the flop technique.

Two spatial axes of rotation are useful in the back layout technique. The x spatial axis passes through the jumper's CG and is parallel to the bar; the z spatial axis passes through the jumper's CG and is vertical and perpendicular to the bar. A third axis, the y axis, is perpendicular to the bar, through the jumper's CG, and horizontal. The angular momentum about the y axis should be minimized. All three of these axes should be imagined to move along with the jumper's CG as it moves through the air.

The angular momentum about the x axis is produced by the jumper's CG being to the bar side of the line of action of the reaction force. This external torque causes the jumper to rotate in a backward somersaulting motion. Angular momentum is produced about the jumper's z axis when the jumper plants the takeoff foot and medially rotates her pelvis about the supporting hip and rotates her trunk in the same direction. The friction force produced between the ground and the supporting foot provides the external torque responsible for giving the jumper angular momentum about the z axis. The z axis rotation is a twisting motion.

In the straddle technique, the GRF gives the body angular momentum that is produced in a similar fashion to the back layout technique. The takeoff position places the CG inside the line of force of the push-off and the bar. Because the jumper is facing the bar, angular momentum is established about the x axis which is through the jumper's CG and parallel with the bar. Because the jumper is not square to the bar, the axis is aligned from the near shoulder to the far hip joint. The jumper is in effect somersaulting forward about this diagonal axis relative to his or her body.

Finally, ground-reaction torques are used to establish angular momentum about somersaulting and twisting axes during platform and springboard diving. Hamill, Ricard, and Golden (1986) report that actions of the arms while the diver is in contact with the platform can contribute as much as 74% to the total angular momentum. For example, in a forward somersaulting dive, the quick extension of the flexed arms to the side while the feet are still in contact with the board generates torque at the feet by the friction force there. This friction-force torque is directed backward at the feet, thus causing forward rotation of the body. Miller and Munro (1985b) state that Greg Louganis' arm action contributed approximately one third of his total angular momentum. The arm action adds to the angular momentum generated by the reaction force of the board acting behind the diver's CG.

Understanding Ground-Initiated Momentum

A long jumper taking off from the board extends the hip, knee, and ankle joints to push off into the air. Draw a diagram of a long jumper taking off. Show the body position relative to the takeoff foot, the CG position at takeoff, and the direction and line of action of the (GRF). Draw arrows indicating the linear and rotational characteristics of the jumper after leaving the ground. (Using Figure 10.9 on p. 394 will be helpful.)

Initiating Rotation in the Air

Once the performer leaves the supporting surface, no increases or decreases can be made in the *magnitude* of the system's angular momentum. The performer, however, can rearrange or redistribute the angular momentum within the system so that rotations that were not evident at takeoff are possible.

Angular momentum, once established, may be assumed by individual body parts within the system, and the velocity of rotation can be increased or decreased by changing the radius of gyration about the axis of momentum. The following section deals with techniques that are used to initiate rotation of the total body about an axis around which it was not initially rotating or to change the velocity or axes of existing rotation. Initiating rotation does not change the total angular *momentum* of the system while free of support; therefore, it does not contradict the conservation principle.

The advantage of air-initiated rotations, particularly in judged activities such as gymnastics and diving, is that in each of the methods the performer can easily stop any rotation that has been created by ceasing the movement that is creating it. Timing the initiation and cessation of the rotation is crucial to successful performance. Generally, rotation may be initiated by three methods while the body is free of support: reaction rotation, cat rotation, and twist from a somersault.

Reaction Rotation

The reaction rotation technique is so named because the movement of the arms, legs, or trunk about an axis causes a reaction or movement response of the rest of the system in the opposite direction about a parallel body axis. If the performer is free in the air while the segmental rotations are executed, the rest of the body must react in the opposite direction so that the sum of the angular momenta of the system remains constant.

For illustration, visualize a person rebounding vertically off a trampoline with the shoulder joints flexed forward to 90 degrees. The system's axis of rotation is a spatial vertical (z) axis that corresponds to the longitudinal body axis. The arms move in a transverse plane around their respective z axes that pass through the shoulder joints; for the total body, the axis is longitudinal and passes through the body's CG. If the performer's arms are thrown to the left (ccw as viewed from the top) during flight, the rest of the body will react by moving to the right in a cw direction so that the angular momentum generated by the arms will equal the angular momentum response of the rest of the body. Note that the angular momentum is equal and opposite, and therefore angular velocities may be different as a result of the differences between the masses and radii of gyration of the arms and the rest of the body. In this case, even with the arms having greater radii of gyration than the trunk and legs, the greater amount of mass in the trunk and legs results in the body moving only about one fourth of the distance that the arms move. The performer has now twisted only about 45 degrees.

If the performer wants to prevent a rotation of the trunk and legs back to the original position, the arms cannot be brought back to the right (cw), because that would cause a return movement of the rest of the system and cancel out the trunk twist that was gained. To prevent that motion, the arms must be dropped down to the side, a rotation that is *not* around an axis parallel to the vertical axis. (The dropping of the arms to the side and flexing them to the front also causes reactions around other axes; however, a slower angular velocity of movement and the relative amounts of masses and radii result in a smaller response of the body about those axes.)

To perform a full turn around the longitudinal axis, the performer must repeat this series of steps eight times. Clearly, this technique is quite tedious and time-consuming and not practical for anything but a free fall for a great distance or for playing in a gravity-free chamber! In fact, the reaction rotation method has been explored as an effective way in which astronauts may realign their bodies during periods of weightlessness without the use of cumbersome mechanical propulsion devices (Kane and Scher, 1970). Mathematically, a 70- degree reaction twist is derived from a complete circle action of the legs moving in a semiconical fashion about a vertical axis, and a 30-degree reaction twist is derived from a complete circle of the arms over the head.

Similarly, a backward somersault of the body may be produced by circling (circumducting) the upper extremities forward around an ML axis that passes through the shoulder joints. Derived values for a forward or backward body somersault per arm cycle were 12 degrees with the legs straight and 24 degrees with the legs tucked (Kane and Scher, 1970). An adaptation of the technique may be performed while suspended by the hands from a single gymnastics ring. Beginning in a slight pike position, swing the legs in a conical fashion to describe a circle on the mat below (circumduction of the lower extremities at the hip joints). The reaction of the trunk will be in the opposite direction; it will continue to turn as long as the legs are circumducting. Biesterfeldt (1974a) notes that a male gymnast's body will react approximately 300 degrees for every leg circle; however, this depends on how far the legs are held away from the longitudinal axis during the circumduction.

Usually in the execution of airborne skills, the reaction rotation method is employed for error correction; that is, if a performer has too much or not enough angular velocity to land properly. A discussion of error correction is given under the analysis section later in this chapter.

Cat Rotation

The cat rotation is so-called because it is the technique cats use to land upright if dropped from an upside-down orientation. A cat falling and twisting is shown in Figure 13.6. The technique also has been called the modern twist (Biesterfeldt, 1974b) and the counter-rotation twist (Hery, 1975).

The cat rotation is a complicated version of the reaction rotation previously described, and it can bring about greater rotational displacement with less effort on the part of the performer. In addition, the body position required for the execution of a cat rotation is dictated in some gymnastics and diving skills. The cat rotation can be recognized by seeing a performer in a body position in which the longitudinal axis of the lower part of the body is at some angle to the longitudinal axis of the upper part of the body. Effective cat-rotation positions are the pike, the side arch, and the back arch. These are illustrated in Figure 13.7.

The longitudinal axis of the upper body is at some angle relative to the longitudinal axis of the lower body. The closer this angle is to 90 degrees, the more effective is this method of rotation. The twisting axis for the entire body (system) is the body's longitudinal axis through the CG, and the angular momentum about this axis is usually, but not always, zero. The cat-rotation method is commonly seen in gymnastics stunts, trampoline stunts, dives, and apparatus dismounts in which a 180-degree or 360-degree turn is required and a bending of the body is acceptable. Usually, the rotation a 180-degree body rotation may be completed in three steps. Once the flexed or pike position is assumed, the sequence begins.

Through *internal* muscle torques, the upper body may be made to rotate (twist) around its

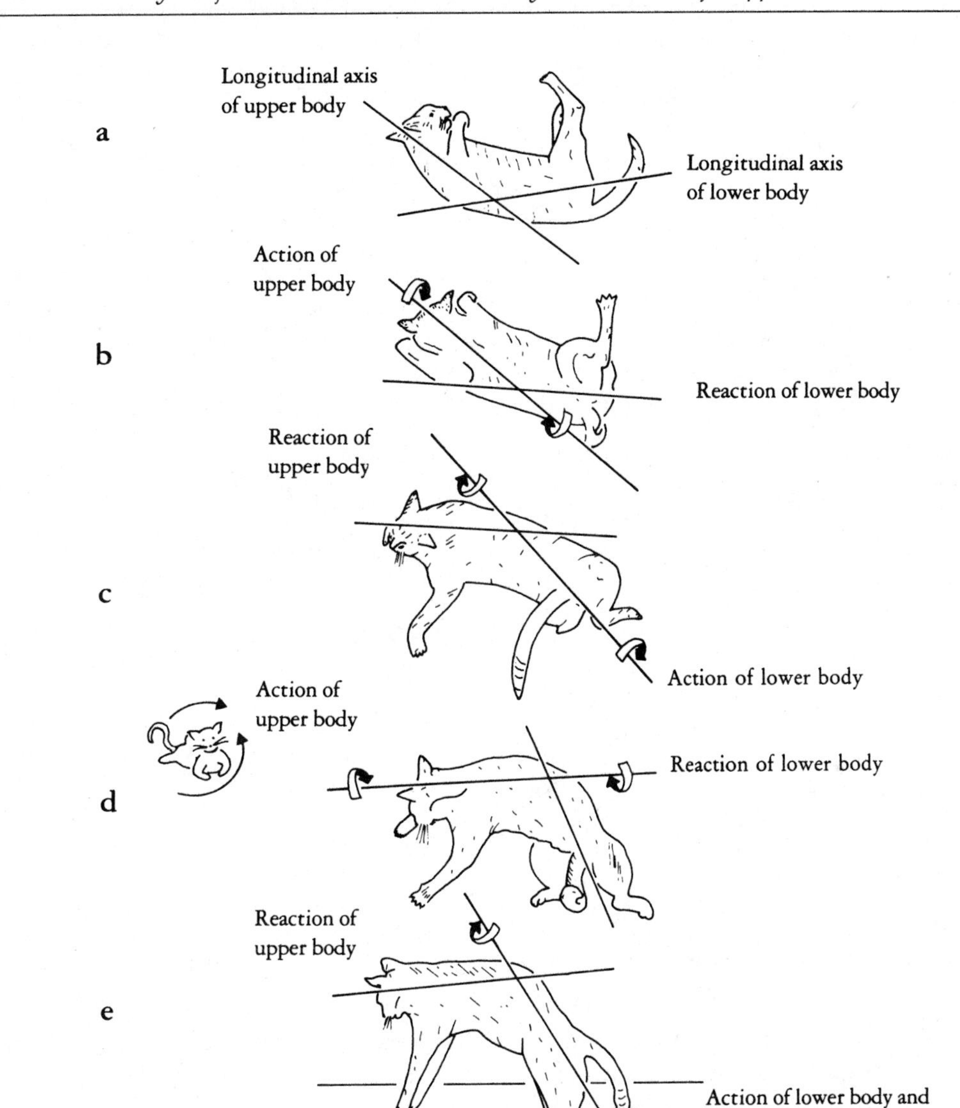

FIGURE 13.6

A cat rotating to its feet when it is dropped from an upside-down position.

longitudinal axis to the left. In keeping with the reaction rotation principle, a reaction must occur in the rest of the system (the lower half), and that reaction must be around the *same* axis as the action, but in the opposite direction. The radius of gyration of the lower body about the longitudinal axis of the upper body is quite large, and, therefore, the response is quite small. Immediately, the lower body rotates by internal muscle torques about its longitudinal

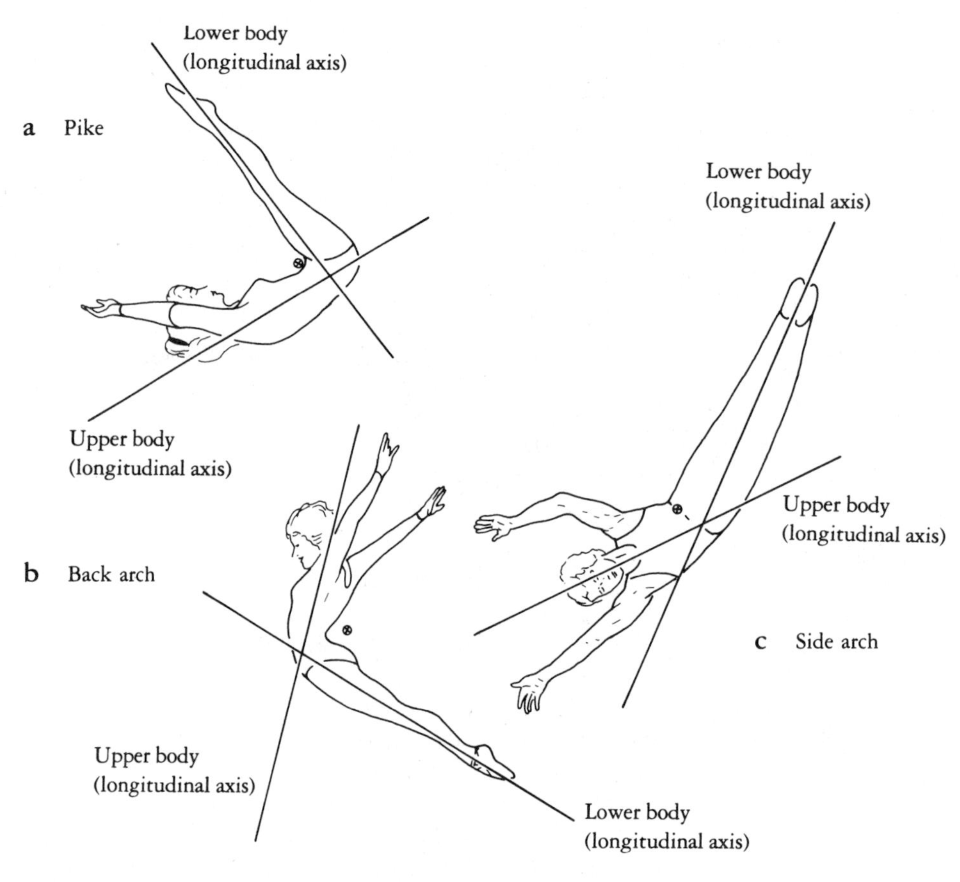

FIGURE 13.7

Positions effective for initiating the cat rotation. (a) Pike. (b) Back arch. (c) Side arch.

axis in a similar direction as the upper body rotated. A similar response occurs in the upper body in the opposite direction when the lower body twists by muscular effort about its longitudinal axis; the response of the upper body about the longitudinal axis of the lower body is quite small. If the twisting motions of the upper and lower halves are rotated alternately, the body moves from a pike into a back arch position, facing in the opposite direction. The total system has, in effect, twisted 180 degrees.

Figure 13.8 shows the cat-rotation method used by a gymnast performing a vault. Notice that the performer leaves the horse without any twisting, or rotation, about the body's longitudinal axis (Figure 13.8a). By moving into a pike position (Figure 13.8b), the gymnast may use the cat-rotation method by rotating the upper and lower body portions alternately about their respective longitudinal axes. The body moves from the pike to a side-arch to a back-arch position. By flexing the hips, the gymnast can land on her feet, after a one-half twist, facing the horse.

This same technique is used for a swivel hips on the trampoline. The performer leaves the trampoline in a hip-flexed (pike) position,

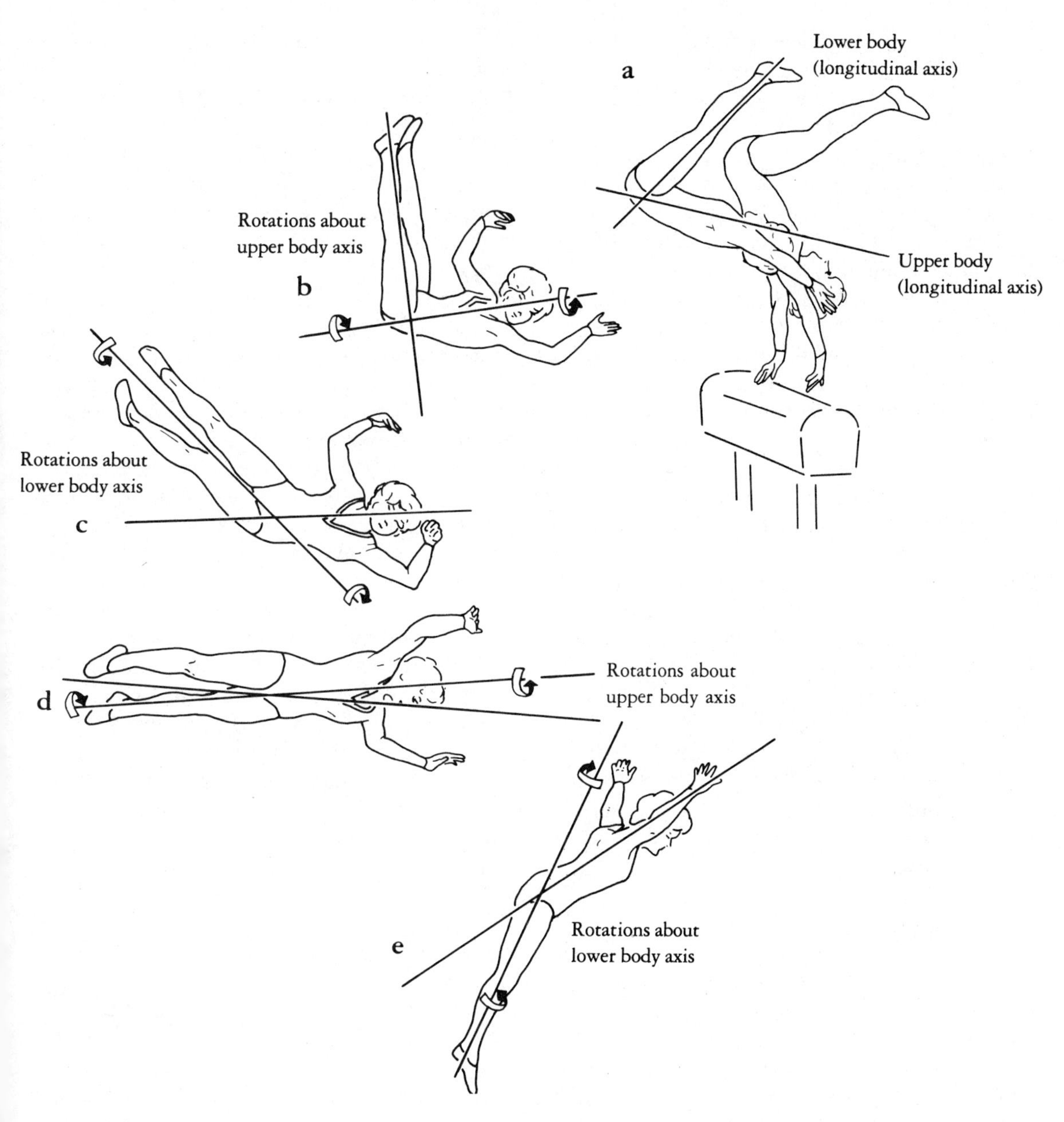

FIGURE 13.8
A gymnast employing the cat rotation during a vault.

rotates the upper and lower halves of the body cw to a side-arch and then to a back-arch position (Figure 13.8c and d). When a one-half turn has been completed, a pike at the hips allows the performer to land in a seat drop again. The same technique can be used in performing a front drop one-half twist to a back drop, or a back drop one-half twist to a front around the longitudinal axis (Figure 13.8e).

To perform a full 360-degree twist using the cat-rotation technique, the performer must be extremely flexible in the vertebral column to maintain a large enough angle between the upper and lower sections of the body. A simplified way of testing this technique is underwater, where the body is suspended and free of support without the dangers of falling to the ground. A second method is to have the gymnast hang by both hands from a single gymnastics ring. Beginning with a slight pike, the performer should twist the lower body to the right, around the longitudinal axis of the lower extremity, and the upper body to the right, around its longitudinal axis, so that the body ends in a back-arch (hyperextended) position facing the opposite direction. (This is not the same movement pattern described for the reaction twist, in which the legs are circled in a conical fashion [circumducted] to create an opposite reaction of the upper body.

Twist From a Somersault

Twisting from a somersault is probably the most common technique used to initiate rotation while the body is free of support because it is the most efficient of the three techniques. Unlike the other two types, it does not require a series of segmental movements, since once the twist has been initiated, the body continues to twist until the performer returns to the original position before twisting began. The body *must* have angular momentum established about some principal axis at takeoff. The axis about which the angular momentum is established at takeoff is called the *axis of momentum*.

The momentum is usually established around the performer's ML (turn) axis causing the performer to somersault either forward or backward. The ML axis is thus the axis of momentum, and the resultant angular momentum established around it does not change in magnitude or direction. The axis of momentum can be thought of as fixed in space, regardless of the change in position of the body and the body's principal axes. Figure 13.9 illustrates the axis of rotation of a performer doing a *backward* somersault who has taken off with angular momentum about her ML axis.

The initial body position is symmetrical; that is, both arms are held above the head. During flight, the performer drops her right arm to the side, as shown in Figure 13.9b. This asymmetrical position reduces the radius of gyration of the right side of the body relative to the ML axis. This smaller radius of gyration on the right side forces the body to tilt to the left. The tilt in the body position alters the body's principal axes to those shown in Figure 13.9c. In addition, by dropping the right arm cw around its AP axis (adduction), the performer causes a reaction of the body ccw around its AP axis, thus *enhancing* the tilt.[2]

The axis of angular momentum that was initially established does not change. It should be thought of as an x spatial axis along which the body's ML axis is aligned at takeoff. The original momentum is now resolved into two perpendicular components, one along the body's now tilted ML axis and the other along the body's longitudinal axis. Thus, the somersaulting motion around the ML axis continues

[2]The tilting of a body's principal axis of rotation from its original position is called **nutation** and also may be referred to as gyration or wobbling. Nutation occurs when angular momentum is established about a spatial axis that is the same as one of the body's principal axes. Changing the position of the body segments during the flight causes the principal axes to move away from their original positions.

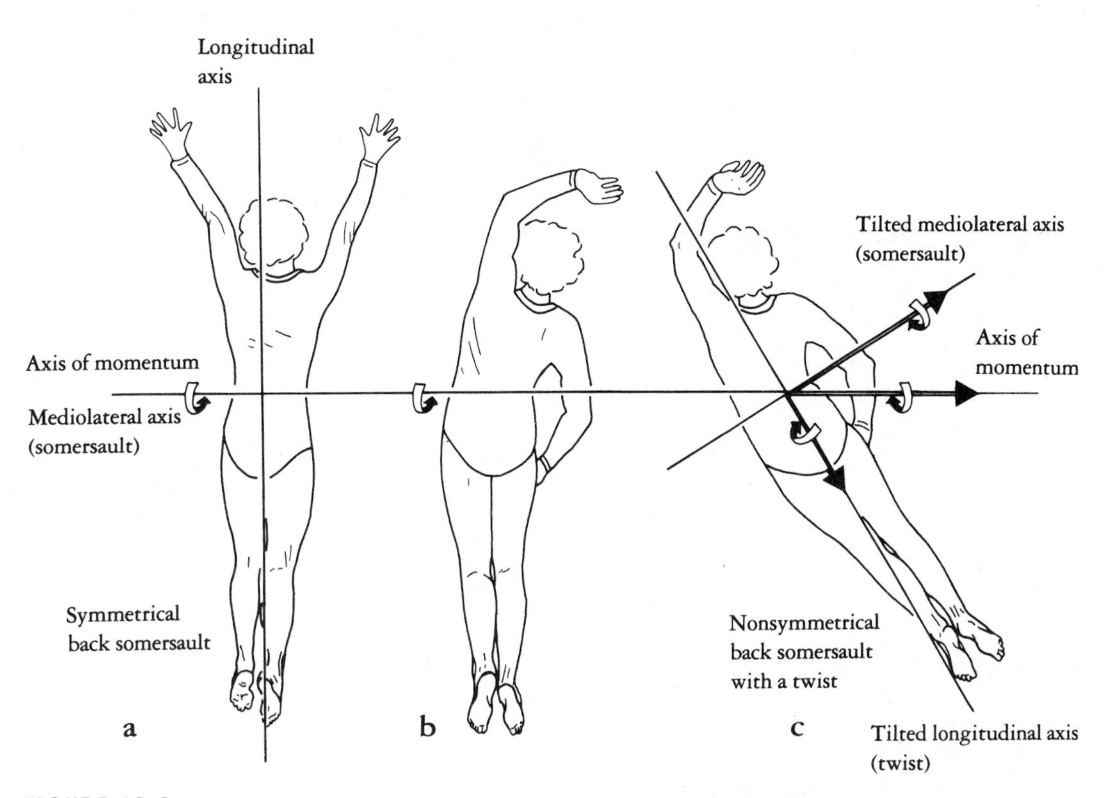

FIGURE 13.9

(a and b) A tumbler somersaulting backward about the mediolateral axis. (c) A tumbler tilting the body's principal axes in executing a nonsymmetrical back somersault with a twist.

even though the ML axis is now tilted from its original position, and in addition, the body begins to twist about its longitudinal axis (the other component).

The resolution of the initial angular momentum vector into two components may be seen more clearly by using the right-hand rule for vectors. The arrow L_m in Figure 13.10 represents the vector direction and magnitude of the original turning momentum about the axis of momentum for the gymnast in Figure 13.9a.

L_f represents the vector of the realigned principal ML axis, and L_1 represents the vector of the realigned principal longitudinal axis. Because the radius of gyration about the twisting axis (longitudinal) is small, the velocity of the twist will be large, in spite of the smaller vector magnitude as shown. Recall that for a body with a given magnitude of angular momentum, a reduction in the radius of gyration will result in an increase in angular velocity.

As soon as the performer's body assumes a symmetrical position again, the momentum vectors will compose (combine) into the original resultant vector, the axis of which will be the original axis of momentum. The angular momentum has been conserved even though it was resolved into two components. The principal ML axis of the performer and the axis of momentum will again coincide. The per-

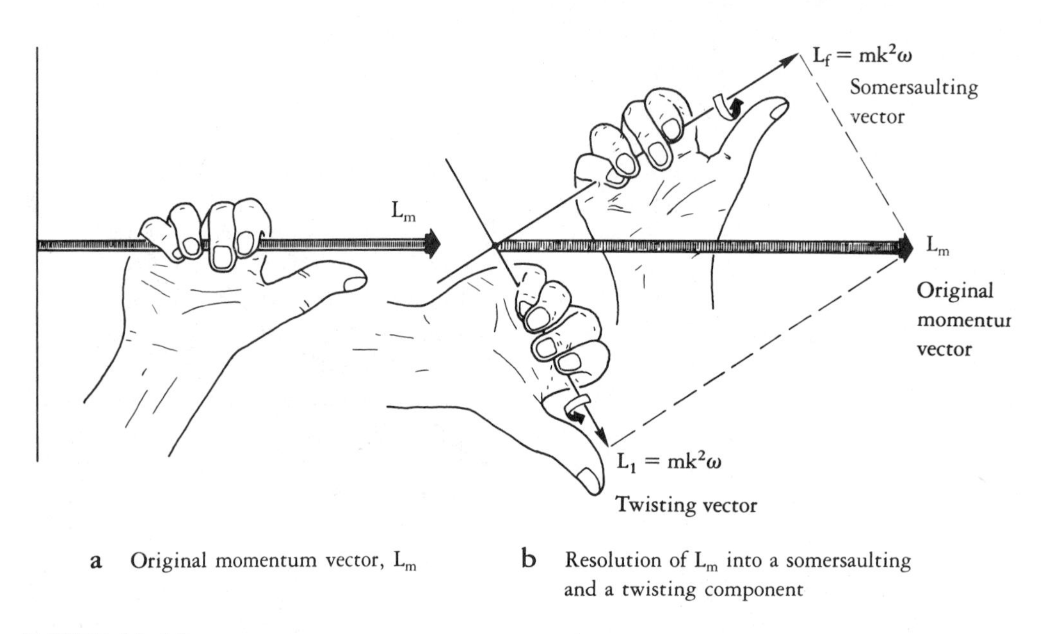

FIGURE 13.10
The resolution of the angular momentum vector into two components using the right-hand rule.

former now will be somersaulting backward again without any twist.

An easier way of thinking through this technique may be employed. Recall Figure 13.9b in which the performer has dropped one arm. The radius of gyration is decreased on the right side of the body, and thus its velocity backward increases about the ML axis. The velocity of the left side does not change because its radius of gyration has not changed. As the right side speeds up, it "pulls" the right shoulder backward to the left hip, thus initiating the twist to the right. If the performer were somersaulting forward, the turning speed of the right side about the ML axis would increase forward and pull the performer's right side forward. As viewed from the top, a somersault with a right arm drop will cause the body to move into a twist represented by a headward-directed vector; a back somersault with a right arm drop will cause the body to move into a twist represented by a footward-directed vector. Figure 13.11 shows a vaulter using this technique in her postflight.

Note that the body must simply resume a symmetrical position again to stop the twisting; it does not have to resume the initial position. For example, if the initial position is a layout and the performer drops the right arm to the side to initiate a twist, the body may resume a symmetrical position again and stop twisting by also dropping the left arm to the side.

Usually, the gymnastics, trampoline, and diving stunts that require a single or partial twist can be accomplished by using only one of the methods of initiating the twist in the air. The accomplishment of multiple twists, however, frequently demands the employment of more than one method. Gaining twisting momentum from the ground and enhancing this momentum with a twist initiated in the air will give the performer the greatest twisting velocity. Such a technique is used for stunts requiring double and triple twists.

FIGURE 13.11
A gymnast initiating a twist from a somersault during vaulting.

Summary of Principles of Body Rotations

Angular momentum may be initiated from the ground by providing a force that acts on the body for a period of time, the line of which is eccentric to the body's CG. The greater the force, the time, or the distance from the axis, the greater the angular momentum generated.

The amount of angular momentum that a body has at takeoff cannot be changed once the body is free of support (assuming no other forces act on the body during flight). Angular momentum is not altered by the force of gravity, because gravity acts directly through the CG of the body and, consequently, has no effect on the body's turning.

The sum of the angular momenta of the body and its parts must equal the angular momentum with which the body left the ground.

Factors that constitute angular momentum may change in magnitude. In body rotation activities, the mass usually does not change throughout the execution. The body's radius of gyration may be altered in the air by moving body segments closer or farther away from the axis of rotation. The angular velocity of a rotating body increases or decreases, according to the change in the body's radius of gyration, so that the angular momentum always remains constant. The change in the angular velocity is inverse to the change in the radius and is squared in magnitude.

Body rotation about an axis may be initiated in the air without changing the body's total angular momentum. Three methods may be employed: reaction rotation, cat rotation, and the method of twisting from a somersault. In the first two methods, the body does not need any angular momentum at takeoff. In the third method, angular momentum must be established about a single axis at takeoff, and it is resolved into two perpendicular components that coincide with two of the body's principal axes.

Understanding the Initiation of Rotation in the Air

1. What three body positions identify the cat-rotation method of twisting?
2. If one wants to cause forward rotation of the body during flight, what directions would you rotate your upper extremities to cause a forward reaction rotation of your body?
3. What position of the body is required for one "twist from a somersault" technique?
4. Which of the three twisting techniques produces the greatest amount of rotation with the least amount of effort? Which produces the least twisting for the greatest effort?

13.3 Analysis of Rotations While Airborne

Altering the Velocity of Rotation

Once rotation has been initiated from the ground, the radius of gyration about the axis may be altered to slow down or speed up the existing velocity of rotation. Increasing the angular velocity of spin by decreasing the radius is probably the most common technique used in rotational moves. As with the roller skater in Figure 13.12, angular momentum is established by pushing the skate against the ground in one direction to create a torque about the vertical axis, thus initiating the spin in the opposite direction.

Usually the performer has the arms abducted while applying the force. Once the spin has begun, the skater draws the arms and free leg close to the vertical axis and spins at increased speed. The reduction of the radius of gyration about the longitudinal axis causes the angular velocity to increase, but the angular momentum remains the same. Note that in the case of an ice skater performing a spin, the friction of the supporting skate on the ice retards the angular momentum slightly, but friction force is relatively insignificant in this example.

Because every *change* in the radius of gyration is squared, the change in the angular velocity is squared also. For example, if the radius of gyration is decreased one third, the angular velocity increases 9 times; if k

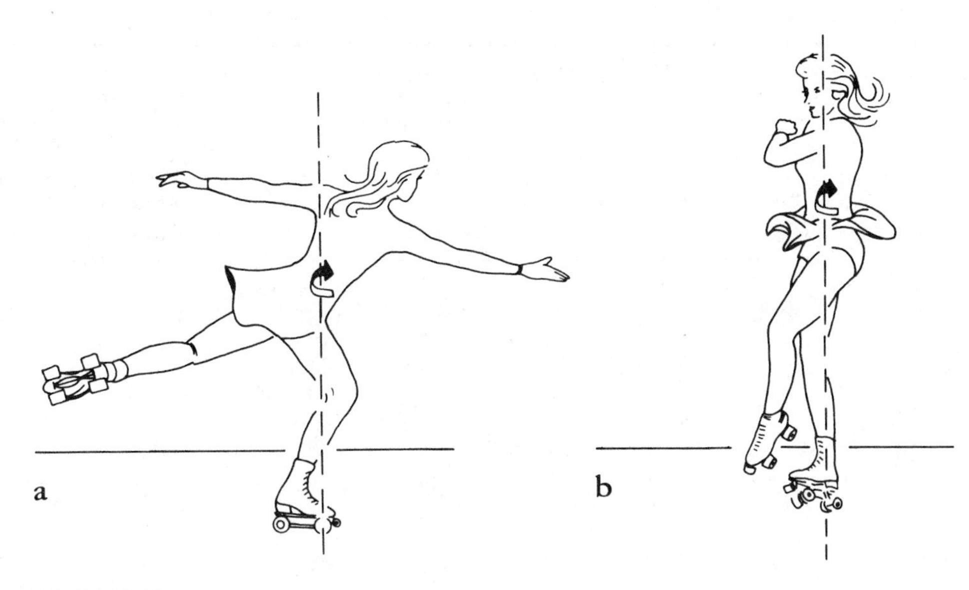

FIGURE 13.12

(a) A roller skater initiating a spin about the longitudinal axis with the arms and one leg abducted.
(b) The position of the arms and legs for increasing rotational velocity.

is decreased to one half, the angular velocity increases 4 times; if one fourth, 16 times; and so forth. The same relationship exists for increasing the radius of gyration and decreasing the angular velocity. Changes in the velocities of rotation when starting from one position and then assuming another position in the air are presented as ratios in Table 13.1. The body positions used in formulating the table data are shown in Figure 13.13a–f. The data were calculated for a gymnast performing a somersault forward or backward about the ML axis. The table indicates the ratio of revolutions from the starting position, listed along the left side of the table, to the assumed position, shown along the top of the table. For instance, moving from position f in the left column, layout with arms overhead, to position a in the top row, tight tuck position, would produce a 4.67 increase in the velocity of rotation; moving from position d to position e would reduce the velocity of rotation to 0.58 of original velocity. These ratios were calculated for an average male, aged 18 to 22 years.

Preventing or Delaying Unwanted Rotation: Error Correction

In performing some activities, a person may obtain some angular momentum from the takeoff that is not wanted or that needs to be delayed until later in the performance. Because the body works as a system of parts, the angular momentum acting on the body may be prevented or delayed by letting the extremities "assume" or take up some of that angular momentum. Mainly, these movements are used for correcting errors in the takeoff that could prove disastrous if not corrected.

Ski Jumping

The principle of conservation of angular momentum is applied to prevent unwanted rotation in ski jumping, as shown in Figure 13.14. If the takeoff position results in an angular impulse that causes the jumper to rotate forward too fast in the air, the performer can

TABLE 13.1

RATIOS OF RELATIVE SPEEDS OF ROTATIONS BETWEEN BODY POSITIONS

	Assumed Position					
Starting Position	**a. Tight Tuck**	**b. Medium Tuck**	**c. Tight Pike**	**d. Medium Pike**	**e. Layout Arms Down**	**f. Layout Arms Overhead**
a. Tight tuck	1.00	0.69	0.69	0.47	0.28	0.21
b. Medium tuck	1.45	1.00	1.00	0.70	0.41	0.31
c. Tight pike	1.45	1.00	1.00	0.70	0.41	0.31
d. Medium pike	2.08	1.47	1.47	1.00	0.58	0.46
e. Layout arms down	3.56	2.48	2.48	1.72	1.00	0.76
f. Layout arms overhead	4.67	3.22	3.22	2.24	1.31	1.00

Modified from Biesterfeldt, H. J. (1975, March). Salto mechanics I: Moment of inertia. *International Gymnast,* p. 45.

rotate the arms forward to assume some of that forward momentum. The magnitudes of the masses and the radii of the body parts must always be compared to determine how much angular velocity must be created in the extremities to alter the angular velocity of the trunk; that is, because the arms are less massive than the entire trunk, lower extremity, and skis, the arms must rotate much faster to affect the rotation of the rest of the system.

If the ski jumper, on takeoff, pushed slightly harder with the right foot so that a torque was set up about the performer's AP axis, the momentum established would tend to rotate the skier to the left. The skier's angular momentum cannot change during flight, but angular velocity may be decreased. By increasing the radius of gyration about the AP axis (e.g., by abducting the arms at the shoulder joints), the angular velocity of rotation about that axis could be decreased to a point where it would not be detrimental to the jump.

A second error-correction method would be to abduct only the left arm, thus creating an aerodynamic force on the left side that would result in a torque causing the skier to rotate to the right.

Gymnastics

A gymnast who dismounts from a piece of apparatus with one side of the body coming off a little before the other side will have forced some twisting rotation in the air. If the performer does not want the twist around the body's longitudinal axis, the twist may be controlled by spreading the arms or legs to the side, thus showing the velocity to a point where it may be unnoticed. Often gymnasts assume what is called a "cowboy" tuck position in the air to stop unwanted twisting

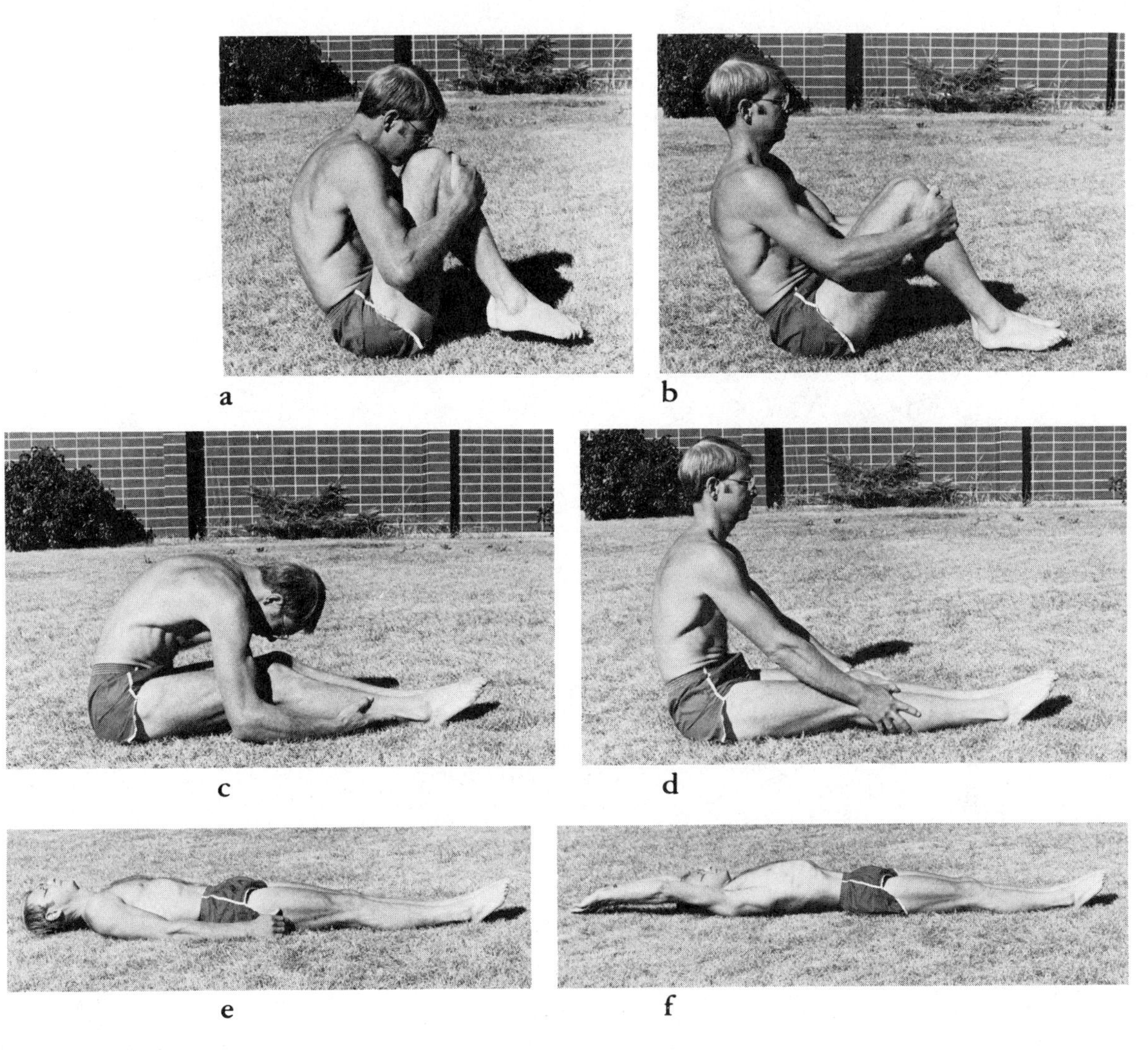

FIGURE 13.13

Body positions with (a) the smallest to (f) the largest radius of gyration about the mediolateral axis. (a) Tight tuck. (b) Medium tuck. (c) Tight pike. (d) Medium pike. (e) Layout with arms at the sides. (f) Layout with arms flexed overhead.

action (Figure 13.15). In the cowboy tuck position, the knees are spread, which increases the radius of gyration about the longitudinal axis and slows the undesirable twisting action. (The position also may allow performers to tuck tighter, which will decreased the radius of gyration about the ML axis and thus increase the speed of the somersaulting action.)

Increasing Rotational Inertia in Striking Skills

As a volleyball spiking movement is initiated, the body is free in the air. Little, if any, angular momentum is in the system. The spiking movement consists of transverse rotation of the trunk about the vertebral column axis. Because there is no momentum about the lon-

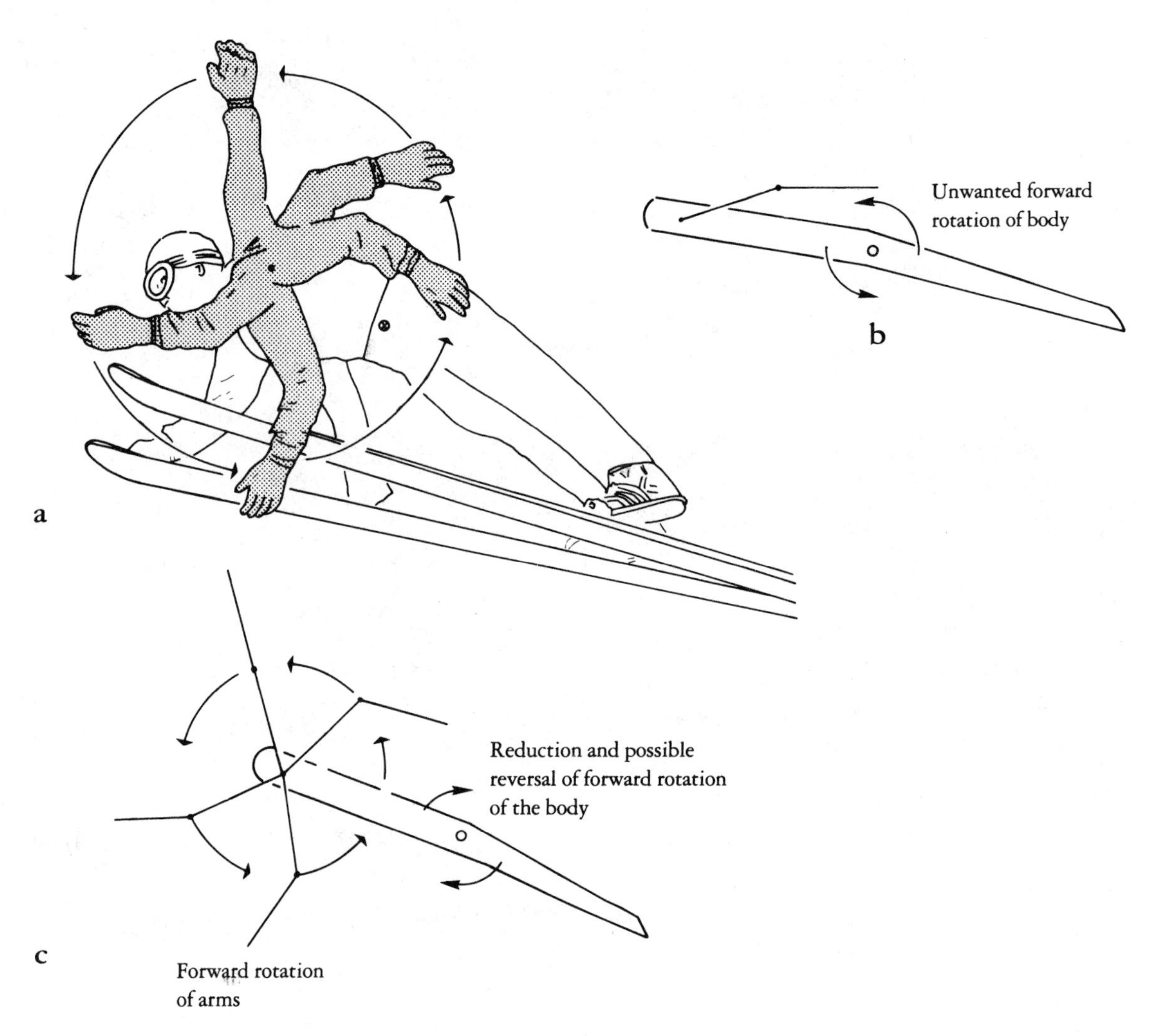

FIGURE 13.14

A ski jumper rotating the upper extremity around a mediolateral axis through the shoulder joints to prevent unwanted forward rotation of the total body.

gitudinal axis, the lower part of the body must "react" with an equal and opposite momentum to keep the angular momentum at zero. The volleyball spiker in Figure 13.16 has positioned his legs to increase the radius of gyration about the lower body's longitudinal axis. This decreases the lower body's reaction to the upper body's rotation around the longitudinal axis during the spiking movement. Thus, by abducting the hips or flexing one hip and extending the other, the radius of gyration of the lower part of the body is increased. This reduces the necessary movement response of the lower part.

Similar positional changes are made in tennis serves in which the feet are off the ground.

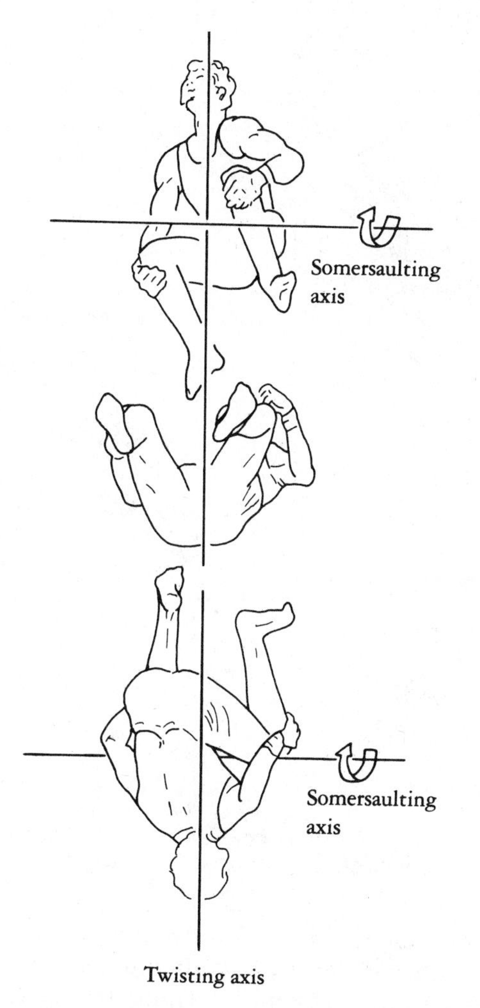

FIGURE 13.15

A gymnast assuming a "cowboy" tuck position, which increases the radius of gyration about his longitudinal and anteroposterior axes but results in a form deduction.

Using Angular Momentum in the Long Jump

The angular momentum generated from the takeoff in a long jump may be used to enhance the success of the jump if the rotation of the body can be delayed until landing.

In the case of a long jumper performing the hitch-kick technique (Figure 13.17), the line of force of the GRF at takeoff passes slightly behind the jumper's CG and thus gives angular momentum to the system in cw(–) direction around the jumper's ML axis. Because of the angular momentum his body is rotating on takeoff and continues to rotate in that fashion until landing. Forward rotation is not ideal for any long jumper not wanting to do a front somersault in the air. The jumper can delay the rotation of the total body by using inter-

FIGURE 13.16
A volleyball player increasing the radius of gyration of the lower extremity about the longitudinal axis to reduce the lower body's rotation, which occurs in reaction to the upper body's spiking movement.

nal torques to rotate his arms and legs in a cw (–) direction around their segmental ML axes through the shoulder and hip joints. As the segments rotate in the same direction as the body's angular momentum, they assume some of the ccw angular momentum of the body (system).

If the limbs rotate to generate enough cw angular momentum to equal that which the body has from the takeoff, the trunk ceases rotation entirely. While the forward rotation is detrimental to the jumper during flight, such rotation is helpful when he wants to rotate forward over the heels in landing. When the arms and legs slow or stop rotating, the trunk again assumes the momentum. Likewise, if the arms and legs are rotated in the opposite direction (ccw) to the body's momentum (cw), the trunk rotates faster cw, enough to keep the sum of the angular momenta the same. Thus, the conservation of angular momentum principle is used in this case to delay unwanted body rotation until landing.

Herzog (1986) supports this theory in a study of maintenance of body orientation during the flight phase of the sail and the hitch-kick technique in long jumping. For the jumpers using the sail technique, a 180-degree rotation of the takeoff arm and a full circle of the lead arm is required with the lead arm taking up 50% of the angular momentum of the entire body. The hitch-kick technique requires a 1 1/2 circle of the takeoff arm and a full circle from the lead arm. Because it is impossible to circle the lower extremities in the same fashion as the upper extremities, the legs are

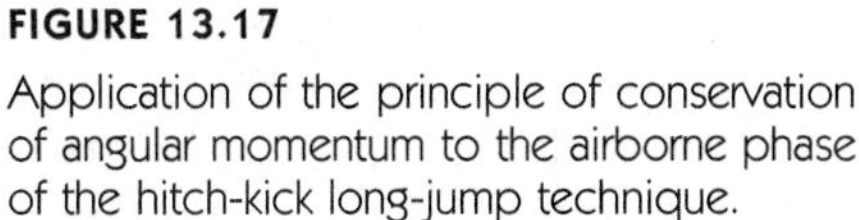

FIGURE 13.17

Application of the principle of conservation of angular momentum to the airborne phase of the hitch-kick long-jump technique.

only useful in taking up momentum of the body during their backward motions. Herzog says, "It is a distinct advantage to execute the forward swing of the legs out of the sagittal plane, like the trail leg in hurdling, rather than in the sagittal plane, like the recovery leg in sprinting. . . ." (1986, p. 239). For the sail technique, Herzog found that the lead leg did not take up any of the forward angular momentum of the body, whereas the takeoff leg served to take up forward angular momentum. In the hitch-kick technique, however, both the lead and takeoff legs rotate backward during the flight phase, and thus both served to take up angular momentum of the system.

Using Angular Momentum in the High Jump

The conservation of angular momentum and the reaction rotation are useful in analyzing the motion of a high jumper during the airborne phase. Once the jumper is in the air, the angular momentum about the x axis helps the jumper rotate over the bar. Trunk hyperextension and knee flexion reduce the radius of gyration of the total body and thus increase the velocity of rotation. Two factors are important to the success of the rotation about the x axis. First, the jumper must not be too close or too far from the bar at takeoff. If she is too close, she will be unable to rotate in time for the hips to clear the bar; if she is too far at takeoff, she will rotate too much before the shoulders pass in vertical alignment with the bar and they will not clear. The second factor involves using the reaction rotation method to clear body segments as they pass down over the bar. The lowering of body parts immediately after they pass the bar will raise the next segments as they pass over the bar. For example, the head, shoulders, and lead arm should lowered immediately after they have passed the bar so that the pelvis will be raised as it passes in vertical alignment with the bar. The pelvis should be dropped immediately after it passes the bar (a sitting motion) so that the knees, legs, and feet will be raised. Timing is critical in this use of the conservation of angular momentum because segmental motions performed too early or too late will cause the segment to displace the bar.

The z axis rotation is used mainly for squaring the shoulders and pelvis relative to the bar. If this component were not available, the body would not be parallel to the bar as it passed over, and the right or left side probably would displace the bar as segmental movements were made.

In the straddle technique, the conservation of angular momentum is used in a fashion similar to that employed in the back layout technique. Rather than hyperextending over the bar as in the back layout, however, the jumper flexes the trunk and drops the lead arm over the bar after their clearance. This serves to increase the velocity of rotation and to raise the lumbar and pelvis region as they pass the bar.

To clear the thighs, legs, and feet, the jumper must use a reaction rotation. Dapena (1974) suggests swinging the lead arm in transverse adduction across the body so that a reaction rotation in the lower part of the body allows the lead leg to clear. (If the right arm is swung to the left, the right leg swings to the right, thus allowing it to pass over the bar.) As soon as the right leg clears, the right and left arms are swung to the right and the lower body responds by rotating to the left, thus clearing the trailing left leg.

In both of these high-jump techniques, angular momentum is established at takeoff and then rearranged during flight to clear the segment passing over the bar at any given time. The reaction rotation method of initiating rotations in the air is used to further reposition the body segments to enhance the success of the jump.

Skill Progressions Related to the Principle of Conservation of Angular Momentum

Progressions for teaching many tumbling, trampoline, and gymnastics skills are based on manipulating the size of the radius of gyration to increase or decrease the angular velocity of rotation. One of the obvious applications of the principle of conservation of angular momentum to skill progression is the use of the tuck, pike, and layout positions while airborne.

To successfully execute one or more rotations before landing, the performer must have a large enough vertical linear velocity component to provide the necessary height that allows greater time in the air. Second, the body must have enough angular velocity to complete the desired number of turns in the time available.

The greater the difference in the takeoff height compared to landing height, the less angular momentum needed to complete the desired rotations. Also, the greater the vertical velocity component, the less angular momentum needed to complete the desired number of rotations.

For the skilled performer, vertical velocity is optimized, and therefore the height and the time in the air are sufficient to complete the desired number of rotations. Beginning performers, however, usually cannot achieve the necessary height and time in the air. This may be due to several factors. The beginning performer is apprehensive about being able to complete the rotation before landing and thus becomes anxious to start rotating as soon as possible. Therefore, the beginner may start the rotation before takeoff.

If a front somersault is being done, the beginner may attempt to rotate the upper body forward as the push-off is being performed. This decreases the angle of takeoff and the vertical impulse, which in turn decreases the vertical velocity, the peak height, and the time in the air. The given angular momentum does not allow enough time to complete a full rotation.

Establishing angular momentum at takeoff while the body is in a layout position is particularly important so that the radius of gyration in that position will be large. Once angular momentum has been established, the body has the possibility of decreasing the radius of gyration by moving into a tuck position to complete the necessary number of rotations before landing. The tuck position in the air enables the body to make the maximum number of rotations, because moving from a lay-

FIGURE 13.18
An (a) advanced, (b) intermediate, and (c) a beginning gymnast performing a back handspring.

out to a tuck reduces the performer's radius of gyration the most and thus increases the angular velocity (Table 13.1, p. 564).

As the performer progresses and becomes more accomplished at the skill, the linear takeoff velocity increases, which increases the time in the air. As the time in the air increases, the angular velocity needed to rotate one revolution before landing decreases, and the performer can complete a full rotation in a pike and finally in a layout position. Even in the layout position, a performer may reduce the radius by bringing the arms from a fully flexed position overhead at takeoff to an extended position at the sides during flight.

The teacher-coach is encouraged to pay particular attention to the takeoff position of a performer for that is where the stage is set for unsuccessful attempts. In tumbling activities, the takeoff time is so short that a slow-motion or stop-action film is helpful. For diving, trampoline, and apparatus dismounts, the difference in takeoff and landing heights makes the takeoff position less critical to success. The same factors influence the quality of any performance, however, and should be the focus of attention of a teacher or coach in any of these areas.

Handspring progressions illustrate a similar change in the radius of gyration as a performer becomes skilled. These differences can be observed in a front or back handspring vault or handspring performed on the floor. Figure 13.18 illustrates a back handspring on the floor. The progression of difficulty from least to most difficult is seen by observing a beginner (Figure 13.18c), who springs backward in a layout position and brings the body over the hands in a tuck position; an intermediate performer (Figure 13.18b), who brings the body over the hands in a pike position; and an advanced performer (Figure 13.18a), who brings the body over the hands in a near layout position. By bringing the body over the supporting hands in a tuck or pike position, the performer causes a reduction in the body's rotational inertia, making it easier to successfully complete the rotation before landing on the feet.

The amount of vertical and angular velocity the performer has on takeoff from the feet and the hands determines whether the body has enough time to remain in the layout position during the rotation or whether the body must reduce the radius of gyration to increase the angular velocity and carry the body over the hands and to the feet.

Understanding Airborne Rotations

1. For the following changes in the radius of gyration, k, state the change in the angular velocity that would be effected: (a) 1/3 k, (b) 1/5 k, (c) 6 k, (d) 12 k.
2. If a performer can do two somersaults in a layout position, how many somersaults can be done in a tight tuck? in a medium pike? Use Table 13.1, p. 564.
3. To effect as much change as possible in the angular velocity, a performer should leave the support in what position? Why?
4. Discuss the problems a freestyle skier would have in attempting to initiate a reaction rotation about the longitudinal axis, as compared with a gymnast attempting the same rotation.
5. Figure 13.17, p. 569 shows a person initiating a twist from a somersault. Imagine that you are standing on the landing mat observing the performer coming over the horse toward you. Draw the resultant momentum vector for rotation about the

axis of momentum and the two vector components. Use circles about the axes to indicate directions.

6. Figures 10.5 and 10.6 (pp. 387–388) show high jumpers using a straddle and a flop-style jump. Indicate and discuss the axis of rotation about which the jumper will be turning in the air, and discuss how the takeoff created this momentum. How could each jumper make the body position more effective while moving over the bar?

7. Kane and Scher (1970) estimated that circumducting the arms forward once would cause a 12 backward rotation of the body with the legs extended and a 24 backward rotation of the body with the legs tucked. Why is there a difference?

References and Suggested Readings

Bajin, B. (1978, June). Three Tsukahara vaults. *International Gymnast*, 58–59.

Bangerter, T. (1968). Comparison of the twisting somersault in diving and trampoline. *Scholastic Coach, 38*(1), 68–69.

Bedi, J. F., & Cooper, J. M. (1977). Takeoff in the long jump—angular momentum considerations. *Journal of Biomechanics, 10*(9), 541–548.

Biesterfeldt, H. J. (1975, March). Salto mechanics I: Moment of inertia. *Gymnast*, pp. 44–45.

Biesterfeldt, H. J. (1974a, April). Twisting mechanics I. *Gymnast*, pp. 28–30.

Biesterfeldt, H. J. (1974b, June–July). Twisting mechanics II. *Gymnast*, pp. 46–47.

Brown, J. R. (1974, December). A comparison of selected factors relating to success of running forward somersaults. *Gymnast*, pp. 32–33.

Bruggemann, P. (1983). Kinematics and kinetics of the backward somersault takeoff from the floor. In H. Matsui & K. Kobayashi (Eds.), *Biomechanics VIII-A: Proceedings of the Eighth International Congress of Biomechanics* (pp. 793–800). Champaign, IL: Human Kinetics.

Dapena, J. (1974). Searching for the best straddle technique. *Track Technique, 55*(3), 1753–1755.

Dapena, J. (1978). A method to determine the angular momentum of a human body about three orthogonal axes passing through its center of gravity. *Journal of Biomechanics, 11*(5), 251–256.

Dapena, J. (1980). Mechanics of rotation in the Fosbury-flop. *Medicine and Science in Sports and Exercise, 12*(1), 45–53.

Dapena, J., & Chung, C. S. (1988). Vertical and radical motion of the body during the take off phase of high jumping. *Medicine and Science in Sports and Exercise, 20*, 290–302.

Dillman, C. J., Cheetham, P. J., & Smith, S. L. (1985). A kinematic analysis of men's olympic long horse vaulting. *International Journal of Sport Biomechanics, 1*, 96–110.

Dyson, G. (1978). *The mechanics of athletics* (7th ed). New York: Holmes & Meier.

Eaves, G. (1971). Recent developments in the theory of twist dives. In L. LeWillie & J. Clarys (Eds.), *Biomechanics in Swimming: Proceedings of the First International Symposium* (pp. 237–242). Brussels: University of Brussels.

Frohlich, C. (1979). Do springboard divers violate angular momentum conservation? *American Journal of Physics, 47*(7), 583–592.

Goehler, J. (1977, October). The mechanical effect of the forward leg snap. *International Gymnast*, pp. 56–59.

Hamill, J., Richard, M. D., & Golden, D. M. (1986). Angular momentum in multiple rotation nontwisting platform dives. *International Journal of Sport Biomechanics, 2*, 78–87.

Hery, G. (1975). Mechanics of rotational movements. Paper presented at the Southwest Gymnastics Clinic, Las Vegas.

Herzog, W. (1986). Maintenance of body orientation in the flight phase of long jumping. *Medicine and Science in Sports and Exercise, 18*, 231–241.

Hopper, B. J. (1973). *The mechanics of human movement*. New York: American Elsevier.

Igaraski, H. (1983). The prediction of the quadruple backward somersault on the horizontal bar. In H. Matsui & K. Kobayashi (Eds.), *Biomechanics VIII-*

B: Proceedings of the Eighth International Congress of Biomechanics (pp. 787–792). Champaign, IL: Human Kinetics.

Kane, T. R., Headrick, M. R., & Yatteau, J. D. (1973). Experimental investigation of an astronaut maneuvering scheme. *Journal of Biomechanics, 5* (3), 313–320.

Kane, T. R., & Scher, M. P. (1970). Human self rotation by means of limb movements. *Journal of Biomechanics, 3*(1), 39–49.

Kinolik, Z., Garhammer, J., & Gregor, R. J. (1981). Biomechanical comparison of front aerial somersaults performed from different starting positions. In A. Morecki, K. Fidelus, K. Kedzior, & A. Wit (Eds.), *Biomechanics VII-B: Proceedings of the Seventh Congress of Biomechanics* (pp. 531–536). Baltimore: University Park Press.

Knight, S. A., Wilson, B. D., & Hay, J. G. (1978, March). Biomechanical determinants of success in performing a front somersault. *International Gymnast*, pp. 54–56.

LaDue, F., & Norman, J. (1967). *Two seconds of freedom*. Romford, Essex, Great Britain: Glover House.

Liu, Z. C., & Nelson, R. C. (1985). Analysis of twisting somersault dives using computer diagnostics. In D. A. Winter, R. W. Norman, R. P. Wells, K. C. Hayes, & A. E. Patla (Eds.), *Biomechanics IX-B* (pp. 401–408). Champaign, IL: Human Kinetics.

Miller, D. I., & Munro, C. F. (1985). Greg Louganis' springboard takeoff: I: Temporal and joint position analysis. *International Journal of Sport Biomechanics, 1*, 209–220.

Miller, D. I., & Munro, F. (1985b). Greg Louganis' springboard takeoff: II. Linear and angular momentum considerations. *International Journal of Sport Biomechanics, 1*, 288–307.

Mohan, J., & Hiltner, W. (1976). *Freestyle skiing*. New York: Winchester Press.

Nelson, R. C., Gross, T. S., & Street, G. M. (1985). Vaults performed by female Olympic gymnasts: A biomechanical profile. *International Journal of Sport Biomechanics, 1*, 111–121.

Page, R. L. (1976, February). The mechanics of swimming and diving. *The Physics Teacher*, pp. 72–80.

Payne, H. A., & Barker, P. (1976). Comparison of the takeoff forces in the flic flac and the back somersault in gymnastics. In P. V. Komi (Ed.), *Biomechanics V-B: Proceedings of the Fifth International Congress on Biomechanics* (pp. 314–321). Baltimore: University Park Press.

Ramey, M. R. (1973). Significance of angular momentum in long jumping. *Research Quarterly, 44*(4), 488–497.

Robertson, E. A., Paul, J. P., and Nicol, A. C. (1985). Investigation into the biomechanics of certain men's olympic gymnastics. In D. A. Winter, R. W. Norman, R. P. Wells, K. C. Hayes, & A. E. Patla (Eds.), *Biomechanics IX-B* (pp. 377–382). Champaign, IL: Human Kinetics.

Sanders, R. H., & Wilson, B. D. (1987). Angular momentum requirements of the twisting and nontwisting forward $1\frac{1}{2}$ somersault dive. *International Journal of Sport Biomechanics, 3*, 47–62.

Stroup, F., & Bushnell, D. (1969). Rotation, translation, and trajectory in diving. *Research Quarterly, 40*(4), 812–817.

Takei, Y. (1988). Techniques used in performing handspring and salto forward tucked in gymnastic vaulting. *International Journal of Sport Biomechanics, 4*, 260–281.

Tichonov, V. N. (1978, November). Improvement of vaulting technique. *International Gymnast*, pp. 56–57.

Van Gheluwe, L. (1981). Computer simulation of an airborne backward twist somersault. In A. Morecki, K. Fidelus, K. Kedzior, & A. Wit (Eds.), *Biomechanics VII-A: Proceedings of the Seventh International Congress of Biomechanics* (pp. 200–207). Baltimore: University Park Press.

Wilson, B. D. (1977). Toppling techniques in diving. *Research Quarterly, 48*(4), 806–811.

Yeadon, M. R., and Atha, J. (1985). The production of a sustained aerial twist during a somersault without the use of asymmetrical arm action. In D. A. Winter, R. W. Norman, R. P. Wells, K. C. Hayes, & A. E. Patla (Eds.), *Biomechanics IX-B* (pp. 395–400). Champaign, IL: Human Kinetics.

CHAPTER 14

Analysis of Activities in Which the Body Rotates While Supported

PREREQUISITES

Concept Modules B, D, E, F, G, I, J
Chapters 1, 9, 10, 13

CHAPTER CONTENTS

The concepts of rotary motion, which were discussed and applied to a system free of support, also may be applied to a system while it is supported or suspended. While supported, the total system appears to rotate around the point of support and possesses the same variables of mass, radius of gyration, angular velocity, and angular momentum. A net external torque is necessary to rotate the system. Although the principles remain the same, slight differences occur in their application when the body is supported.

14.1 The Human Body in Supported Rotary Motion

During supported rotation, the body is subjected to two forces: the reaction force of the external support and the weight force acting at the center of gravity (CG) of the body. As long as the line of gravity falls within the base of support, the body remains in equilibrium. As soon as the line of gravity falls outside the base of support, an unopposed torque is created by the body's weight force and produces a rotation of the body. The weight force at the CG functions as a potential torque-producing force in the case of a supported body.

Figure 14.1 illustrates the lines of action of forces acting on a body when it is supported and nonsupported. W represents the weight force, and RF, the reaction force at the supporting surface. The magnitude of the weight force does not change as the performer changes positions relative to the axis of rotation, but the perpendicular distance of that line of force from the axis of rotation may be changed by altering the segmental positions. Thus, the torque produced by the weight force can increase from zero, when the weight force is in vertical alignment above or below the base of support, to its greatest magnitude, when the weight force is at its greatest horizontal distance from the base of support.

Similarly, the reaction force at the point

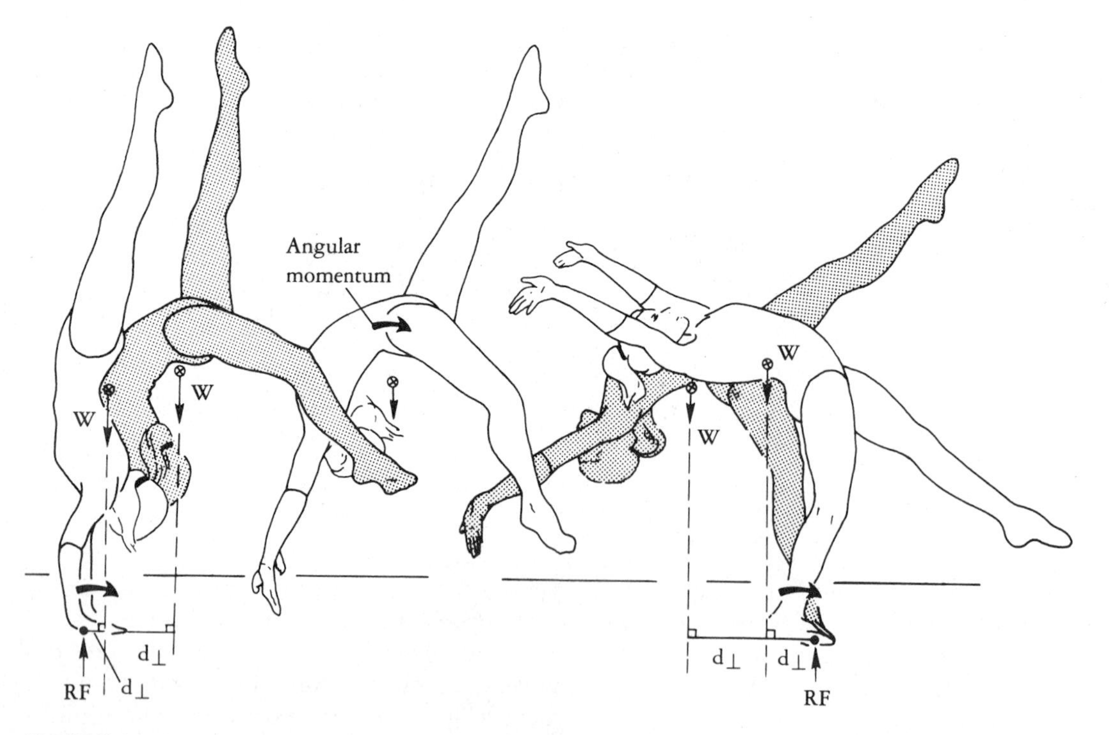

FIGURE 14.1
Motive forces in a handstand walkout.

of support also may be a torque-producing force when the axis of rotation is considered as being through the system's CG. This is particularly evident when the performer is being projected by a spring mechanism such as a diving board, Reuther board, or trampoline. With a springboard diver, the upward-directed force of the spring is easier to comprehend as a torque-producing force about the diver's CG. Also possible is the production of torques by the reaction forces of other supporting surfaces such as the horizontal bar, the parallel bars, tumbling mats, and Reuther boards.

The body mass, its radius of gyration, and the resulting rotational inertia are as important in supported rotation as they are in nonsupported rotation. The axis of rotation in supported body rotation may change during the execution of a series of stunts. For instance, on the uneven bars, a gymnast may swing around the high bar as an axis until body contact is made with the low bar. A release of the high bar allows the body to begin to rotate around a new axis at the low bar.

The radius of gyration of the body varies according to which axis of rotation the body is using and the position of the body relative to that axis. Figure 14.2 shows a body rotating around a horizontal bar. In positions 1, 2, and 3, the shoulders and hips are flexed, and the knee joints are fully extended; the ankles are plantar flexed. The radius of gyration is shown by the perimeter of the shaded area. In positions 4, 5, and 6, the body is pulled closer to the axis by

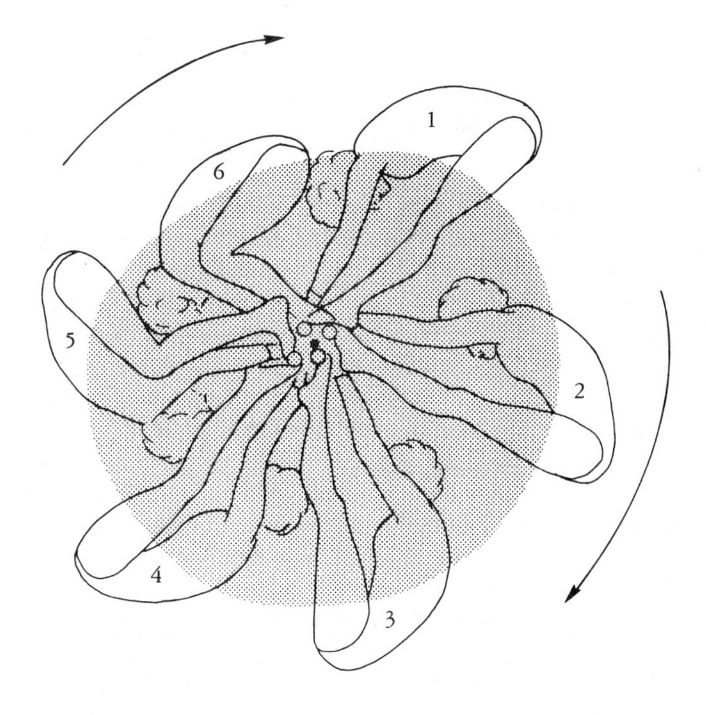

FIGURE 14.2

A gymnast in various positions while swinging around a horizontal bar to show changes in the body's radius of gyration while rotating.

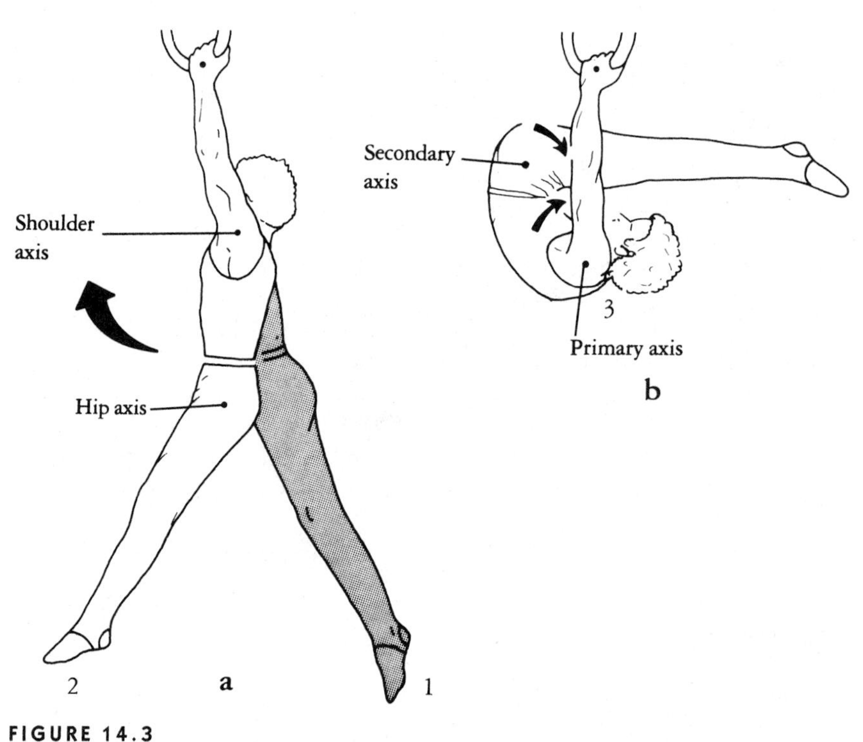

FIGURE 14.3
A gymnast performing (a) a swing on the rings and (b) moving into a pike hang.

flexion of the knees, dorsiflexion of the ankles, and increased flexion of the hips. An elevation of the shoulder girdle also serves to pull the body closer to the bar. Notice that the radius of gyration increases (positions 1, 2, 3) and decreases (positions 4, 5, 6) according to the body's position relative to the rail.

A gymnast moving from a swing to a pike hang on the rings provides another example of changing the distribution of the body's mass of rotation (Figure 14.3). The gymnast begins the swing from an extended layout position (positions 1 and 2) and then changes the body position to that of a pike (position 3). As he moves from the layout position to the pike, he decreases his radius of gyration (distribution of his mass) from the axis of rotation, which in this case is the shoulder joint during the upswing. Note that as he performs the piking action, he is additionally rotating his legs about a secondary axis, the mediolateral (Ml) axis through the hips.

14.2 Conservation of Angular Momentum in a Supported System

The principle of conservation of momentum applies to bodies rotating around a support; however, because support exists, additional factors must be considered. The primary axis of rotation is located at the point of support, and the force of friction acts on the point of contact between the body and the supporting surface.

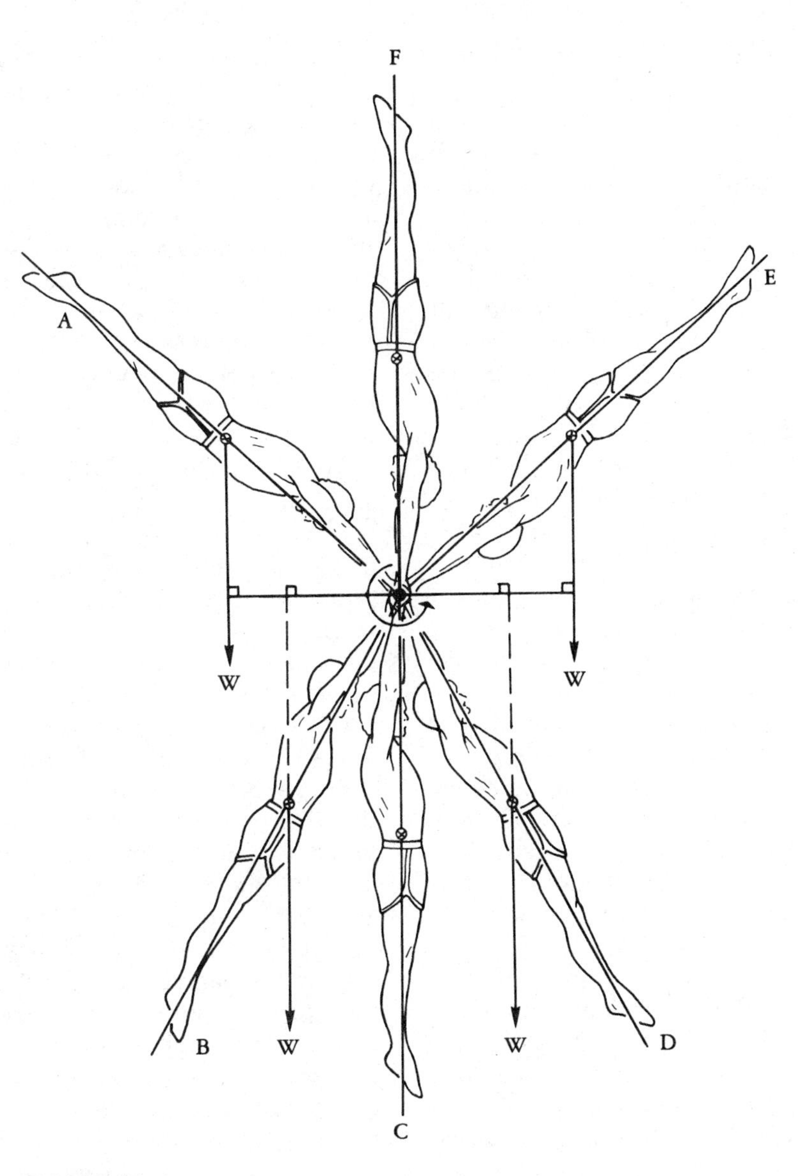

FIGURE 14.4
A rigid body circling a horizontal bar.

Figure 14.4 shows a rigid body rotating about an axis. One can see that the body begins to fall as soon as its line of gravity falls outside the base of support. The weight of the body, represented as a vector, is a line of force that does not pass through the axis of rotation. As the performer begins his downswing, the weight force vector continually progresses farther and farther away from the axis of rotation. The farther the line of force of the body's weight from the axis of rotation, the greater the torque it produces. The increase in the perpendicular distance from the axis of rotation to the line of action of the force continues until the body is horizontal.

From the horizontal until the body is vertical under the bar, the force arm for the weight force decreases until it is zero and no torque is produced.

The motive torque produced by the weight continues to act through 180 degrees until the body's CG is again in vertical alignment with the support; that is, the body has a motive torque acting on it from the time that its line of gravity passes outside the base until the bottom of the swing.

The force of gravity is acting in the direction of body movement during the downswing and against the direction of body movement on the upswing; thus, the torque changes from a motive torque during the downswing to a resistive torque during the upswing. During the first half of the upswing the force arm for the weight torque increases. During the last 90 degrees of upswing, the resistive torque decreases until it is again zero at the top.

Certain positions of the body on the downswing may be compared with similar positions on the upswing. Similar positions are defined as two positions, one on each side of the swing where, neglecting friction and air resistance, the angular momentum is equal.

Figure 14.4 can be used to compare various body positions during the swing. If the angular impulse acting until position A has produced a magnitude of 10 angular momentum units in the body, and if the body has 30 angular momentum units by the time it reaches the bottom of the swing, then the angular impulse acting as a resistive force during the upswing will cause the angular momentum of the body to be reduced to 10 angular momentum units again when it reaches position E and zero units when it reaches the top. When one considers the total swing, one can understand that a perfectly rigid body acting without friction or air resistance will come to rest at the top of the bar where it started.

Activities such as the giant swing would be simple if one could perform them without friction or air resistance, but such is not the case. The direction of the resistive forces of friction and air resistance depends on and is opposite to the direction of the moving body. Figure 14.5 indicates the directions of the forces of friction on the hands and the air resistance on the body during the downswing and the upswing phases of a giant swing.

On the downswing, both of the resistive torques are acting against the torque produced by gravity and thus are retarding the increase of the angular momentum. Similarly, on the upswing, the friction and air resistance torques are acting against the direction of movement and along with the now resistive torque of gravity to retard the angular momentum. Thus, with the directions of the motive and resistive forces as they are, a body could not possibly return to the top of the bar at the end of the swing. The angular momentum would truly not be conserved because of the external forces of friction and air resistance.

By making use of the principle of conservation of momentum, however, and remembering that, at any given time, the angular velocity of the body may be increased by decreasing the body's radius of gyration, one can compensate for the resistive torques involved. The loss of angular velocity, from what it would have been without the resisitive forces at any given point during the upswing, may be compensated for by decreasing the body's radius of gyration. Because the radius of gyration is squared in the angular momentum formula, a decrease in the radius of gyration may satisfactorily compensate for a reduction in the angular momentum, due to the resistive torques, by increasing the body's angular velocity. Therefore, the body in position E, Figure 14.4, p. 579, has a smaller angular momentum than that in position A; however, the angular velocity of position E may be increased by a reduction of the body's radius of gyration. This increase in the body's angular velocity allows the body to return to the top of the bar.

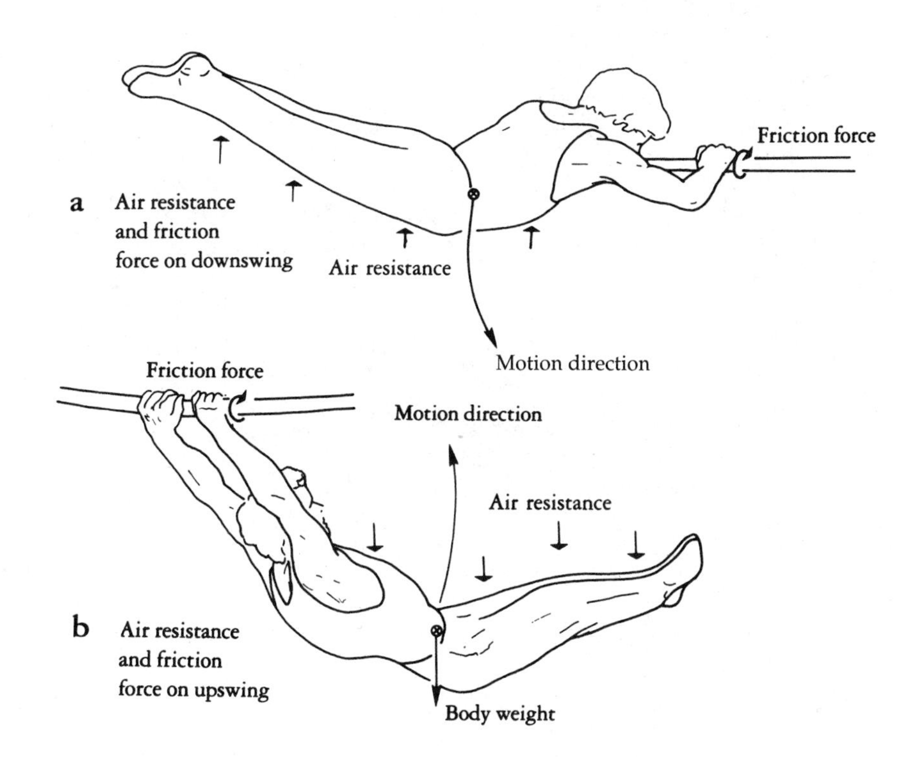

FIGURE 14.5
Directions of the resistive forces acting on a rotating body.

14.3 Applications of Angular Momentum Principles to a Supported Body

The principle of conservation of angular momentum may be applied to sport situations in which the body is supported in one of four general categories: (1) altering the radius of gyration to change rotational inertia, (2) altering the radius of gyration to change angular velocity, (3) altering the radius of gyration in learning progressions, and (4) dealing with angular momentum about multiple parallel axes.

Altering the Radius to Change Rotational Inertia

During the execution of some activities, the body must reduce the radius of gyration to reduce the resistance of a body segment or of the total body (i.e., its rotational inertia). The axes of rotation would be the anatomical axes through the shoulder, elbow, hip, and knee-

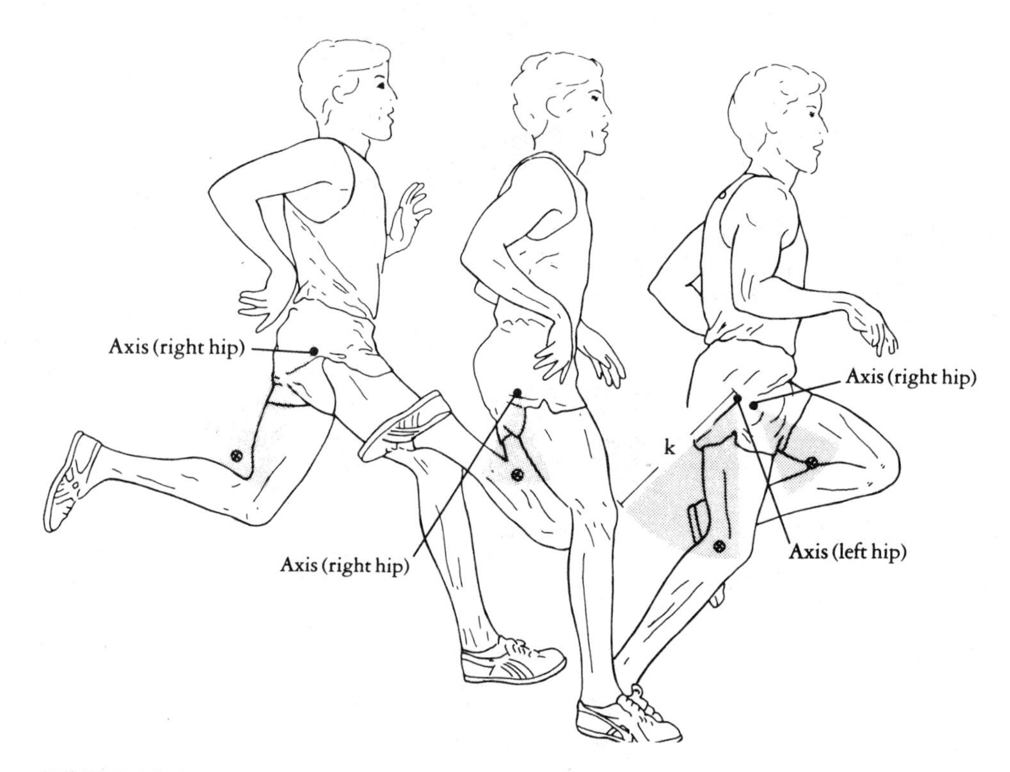

FIGURE 14.6
The changing radius of gyration of the extremities during running.

joint centers, and the radii of gyration of the segments would be measured from those particular axes.

The change in the radius of gyration during the recovery phase in running is shown in Figure 14.6. A runner also must change the amount of flexion that occurs in the arms and legs during the recovery phase so that it matches the pace of the run. During the recovery phases of the upper and lower extremities, a sprinter must use considerably more flexion than a distance runner so that the segments recover faster and thus speed up the entire pace of the run. As the pace of the run slows, the amount of necessary flexion decreases so that a long-distance runner may be able to keep the lower extremity at a fairly large angle during the forward swing; the recovery phase of leg motion may be just enough to clear the ground with the toe.

A redistribution of the body's mass relative to an axis is experienced by the pole vaulter shown in Figure 14.7. By planting the pole in the vaulting box, the vaulter gains angular momentum from the linear momentum developed during his approach. The plant fixes the end of the pole and establishes the first axis of rotation. The system is defined as the pole and the vaulter's body as both rotate about the fixed end.

As the pole–body system rotates, the vaulter repositions his body relative to the axis. This repositioning changes the radius of gyration of the system relative to the fixed end, and because the body is internal to the pole–body system, it "assumes" some of the angular momentum of the entire system as the body rotates relative to the pole. Because the rotations of the vaulter's body and its segments are in the opposite direction, ccw, of

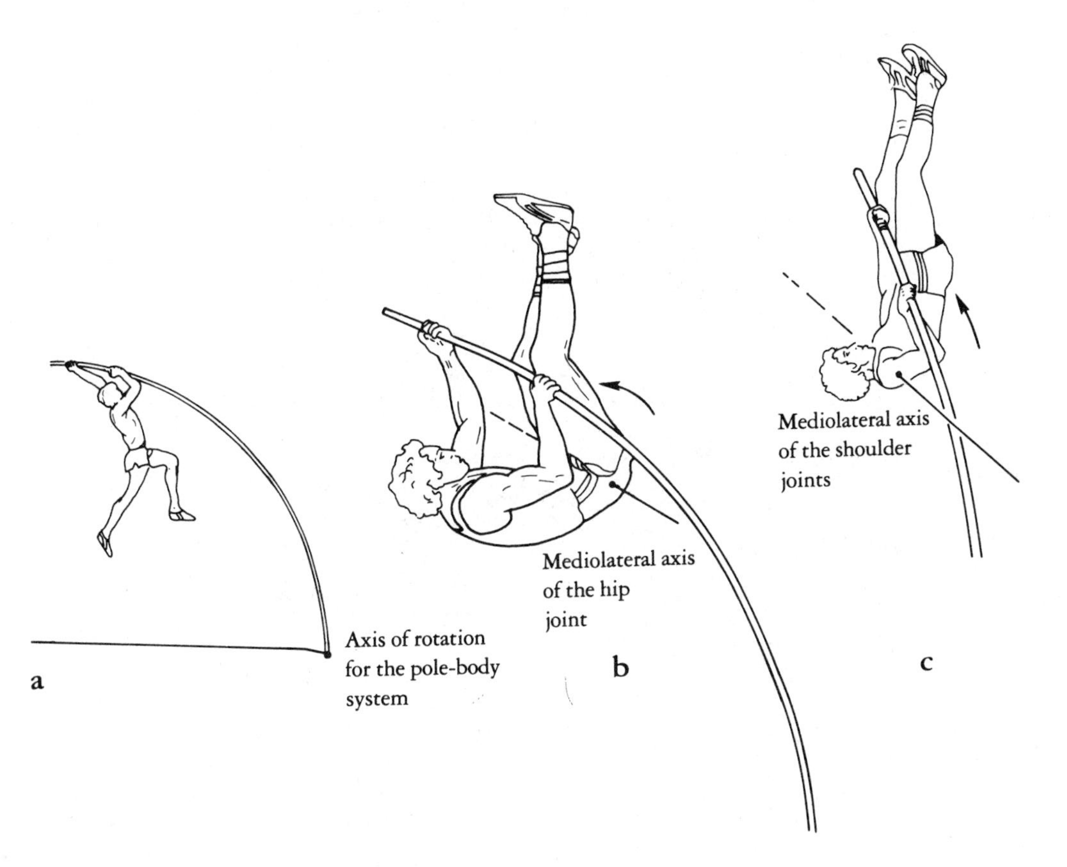

FIGURE 14.7
A pole vaulter during the hang phase of the vault.

the system's rotation, cw, the vaulter's movements tend to increase the system's angular velocity about the fixed end.

The vaulter swings counterclockwise (ccw) about an axis at his top hand as the pole rotates clockwise (cw) (Figure 14.7a). The rock back is initiated by extending the shoulder joints, which serves to pull the trunk ccw and by flexing the vertebral column and hips, which serves to pull the lower extremities ccw (Figure 14.7b). McGinnis and Bergman (1986) found that the arm extensor torques acting about the shoulder joints must be strong in order for a vaulter to excel in the pole-vaulting event; whereas the lower extremity torques required were not great. The actions of the hands help bend the pole. By the time the pole is ready to recoil, the vaulter should be in a vertical position, ready to receive the impulse, and the pole should have lost most of its angular velocity about its fixed end (Figure 14.7c).

A third example of altering the timing of a movement is in the gymnast's position during vaulting. The vaulter must gain enough velocity on the approach to carry the body from the board through the preflight, horse contact, postflight, and landing. The CG is moving linearly during the hurdle, but as soon as the feet contact the board, the friction force at the feet fixes the contact point and establishes an axis of rotation about the feet (Figure 14.8). The excursion of the CG during board contact is critical to the result-

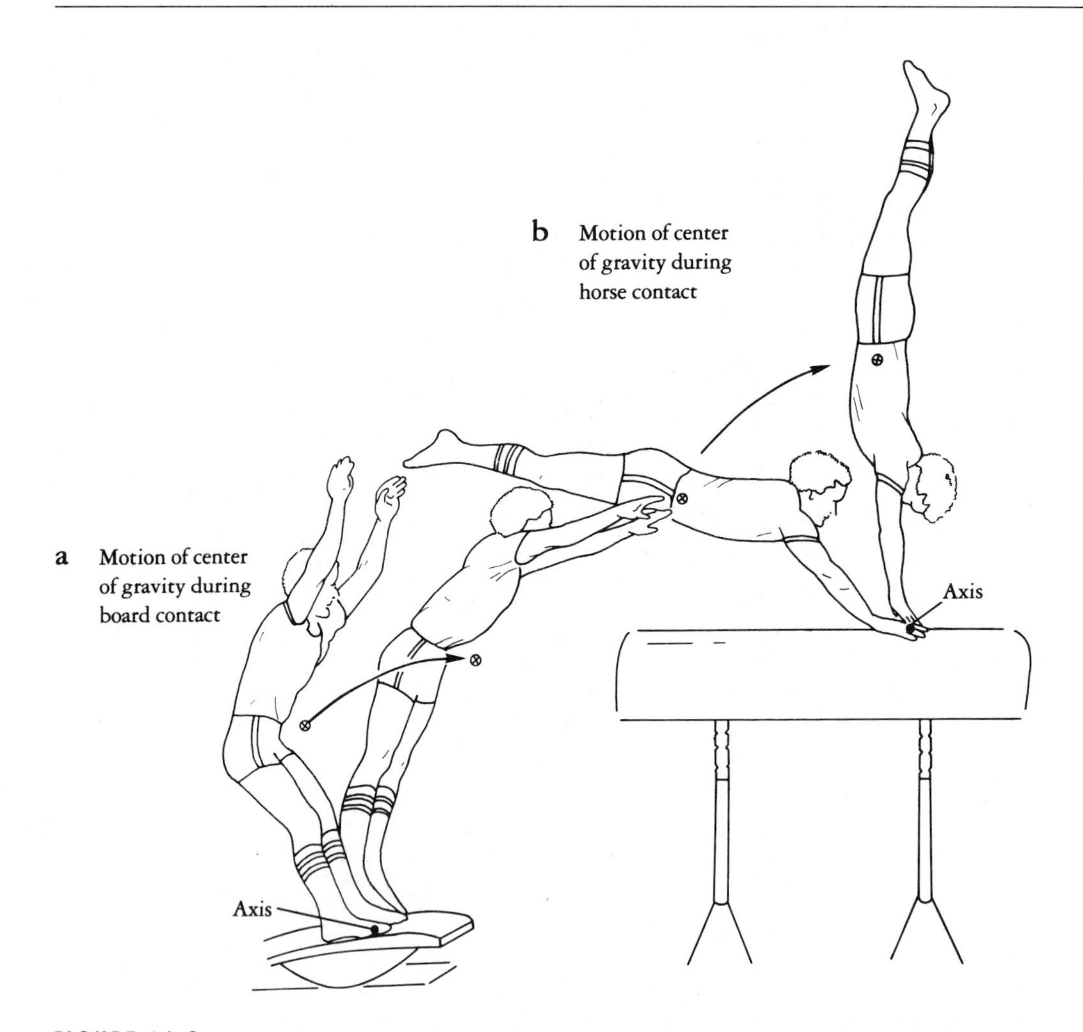

FIGURE 14.8
Two supported rotations used in vaulting (a) on the Reuther board and (b) on the horse.

ing body motion. The velocity of this rotation depends on the velocity of the approach run and on the position of the body during board contact. The speed of the run determines how far the CG travels during the time that the vaulter is in contact with the board. All other things being equal, a very fast runner contacting the board at the same angle as a slower runner will have the CG farther ahead of the feet than the slow runner when each leaves the board. Having the CG farther ahead of the feet during takeoff provides the gymnast with a greater torque and resulting angular momentum at takeoff. The accompanying reduction in the angle of takeoff, however, changes the linear path of the CG during preflight, the trajectory of which will be a flatter arc and may not be beneficial to the vault being performed.

A body landing on the board with the arms positioned overhead rotates more slowly around the support axis than a body with the arms held at waist level. Slower rotation results in a more vertical angle of takeoff.

Vaulters should have the arms overhead on takeoff. Vigorously flexing the arms during contact (accelerating the arms upward) increases the depression of the board, which enhances the board reaction force and thus the linear and rotary velocity of the body at takeoff. Because this arm action must occur during a time interval of no more than 0.1 to 0.2 sec, the vaulter may not be able to complete the arm action in that length of time.

Similar considerations are appropriate for the vaulter during horse contact (Figure 14.8b). Enough angular momentum must be conserved to carry the gymnast over the point of support at the hands. If he arrives at the horse at a low angle, he has to reduce the body's radius of gyration relative to the hands so that the resulting increase in the velocity carries him over. Reducing the radius of gyration means a form break, because it is usually accomplished by flexing the hips or flexing the knees or both. A more skillful technique would be to increase the angular displacement enough during the preflight so that enough of a turn around the ML axis is executed in the air before horse contact. By controlling the angular momentum, the gymnast is able to maintain a straight body position throughout the vault.

During the contact with the board and contact with the horse during vaulting, the gymnast loses horizontal velocity and gains vertical velocity (Dillman, Cheetam, and Smith, 1985; Nelson, Gross, and Street, 1985; Takei, 1988). The forces encountered in these support phases have been estimated to be 0.8 to 1.3 times body weight on the board and 2.0 to 5.2 times body weight on the horse. Takei (1988) calculated correlations between numerous independent variables in a vault and the judges' score. He found that the following variables were related to high scores: higher vertical velocity at takeoff from the horse, vertical force exerted on the horse, a sharp quick blocking action, and a greater height of the center of gravity at takeoff. Associated with higher points was also a smaller height of the CG at landing due to more backward lean. Clearly, the support phases of vaulting contain critical components for effective performance.

Altering the Radius of Gyration to Change Angular Velocity

In the example of the performer on the high bar performing a giant swing, the torque produced by the force of gravity acted to reduce the angular momentum on the upswing. To slow this reduction of the angular momentum, the radius of gyration could be reduced.

Figure 14.9 illustrates a beginning gymnast performing a giant swing on the horizontal bar. Notice that just after passing the bottom point in the swing, the gymnast flexes in two places, the hips and the elbows, while allowing the shoulder joint to extend from a fully flexed position overhead to midposition in front of the body. These actions bring his mass closer to the bar (reduce the body's radius of gyration) and thus tend to counteract the reduction in the body's angular momentum during the upswing. By bringing the CG closer to the bar on the upswing, the gymnast also decreases the force arm of the gravitational force and thus decreases the magnitude of its resistive torque. The reduction of the resistive torque produced by gravity allows his body to maintain its angular velocity for a longer period of time; and, consequently, as he approaches the top of the bar, he moves faster than he would had he remained in the rigid position. The performer is able to help counteract the resistive forces of friction and air resistance by reducing the radius of gyration and thus increasing the velocity of rotation.

As a second example, a gymnast attempting to perform a forward stride circle on the low bar of the uneven parallel bars is shown in Figure 14.10.

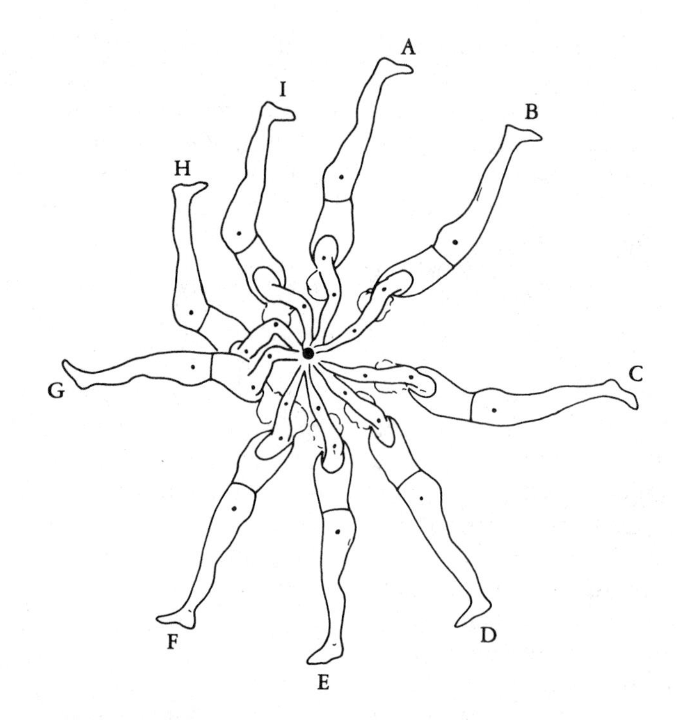

FIGURE 14.9

A beginning gymnast performing a giant swing on the horizontal bar.

The initial positions (2 and 3) are used to gain as much angular momentum as possible on the downswing. This is accomplished by increasing the distance between the CG of the body and the axis of rotation (the bar), thus maximizing the motive torque. A coach may cue the performer to keep the head up, and the chin, chest, and front leg out as far as possible away from the bar. Once the performer begins to fall, the extended, stretched position should be maintained for as long as possible until the bottom of the circle is reached (position 4). On reaching the bottom, the gymnast assumes a more flexed position; that is, she flexes the cervical and thoracic vertebrae and pulls the body into the bar with an elevation of the shoulder girdle (positions 5 and 6). The reduction of the radius of gyration in this manner increases the angular velocity on the upswing, decreases the resistive torque of gravity, and allows the performers to partially overcome the air resistance and friction on the upswing so as to come to rest on the top of the bar. Most beginners have trouble with two parts of this skill. First, a beginner frequently does not maintain the extended position for more than a split second after starting to fall forward, and therefore the motive torque on the downswing is minimized. This results in a reduction in the possible angular momentum that may be gained during the downswing. Second, in fear, the beginner usually grasps the bar tightly, which increases the friction component and also reduces the angular momentum to a magnitude that is less than that necessary to complete the circle.

As a skilled performer approaches the top of the bar (position 1), she again must extend as much as possible to reduce the angular velocity that may carry her forward into another

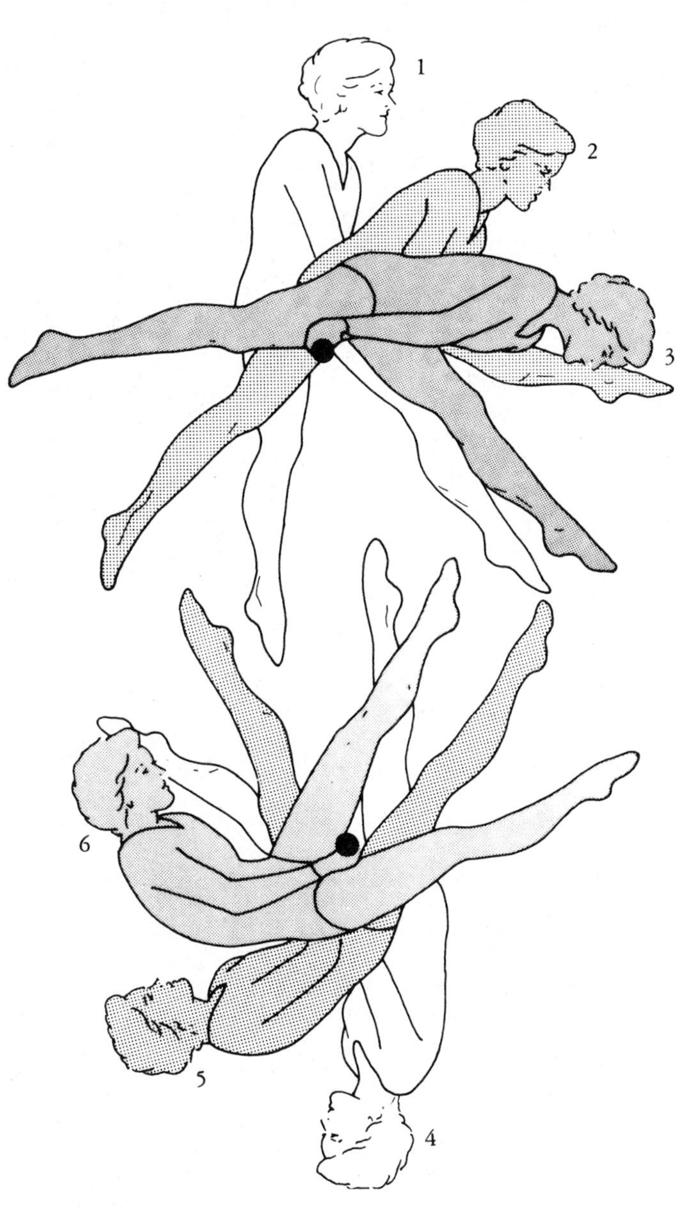

FIGURE 14.10

A gymnast performing a forward stride circle on the uneven parallel bars.

circle. A similar analysis may be made with a forward or backward seat circle on the low bar or horizontal bar.

A final example is a gymnast performing a cast back hip circle on the uneven bars (Figure 14.11). On initiating the cast, the performer lifts her CG up to the horizontal (position 1 to 2) and pushes it outward away from the high bar (position 3). This increases the distance of the weight force from the axis of rota-

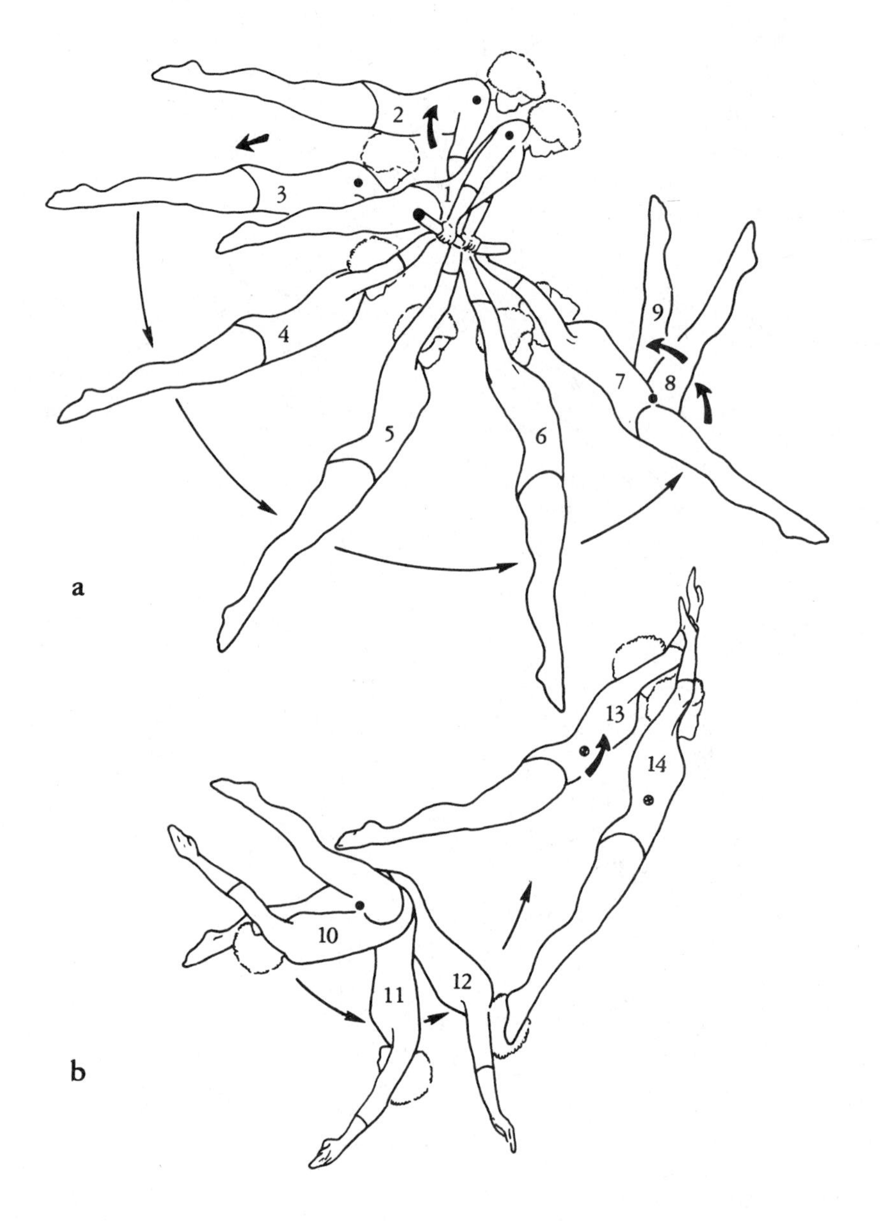

FIGURE 14.11
A cast back hip circle on the low bar.

tion (the rail) and thus increases the motive torque. A straight and extended body on the downswing has greater angular momentum (positions 4 and 5). On contact with the low bar (position 7), the gymnast releases the high bar and initiates a rotation around a new axis at the hip. The velocity of rotation increases as the body moves around the new axis because the performer's radius of gyration relative to this new axis is smaller. The velocity shows a greater increase if the performer flexes at the hips and knees during positions 8 and 9.

Altering the Radius of Gyration in Learning Progressions

In many instances, motor skills are taught in a learning progression that relies on the reduction of the body's radius of gyration to help the beginning performer through attempts at a new skill. Examples of aerial stunts were used in the previous chapter to demonstrate these progressions. Similar progressions are used when a body is rotating around an external support. Generally, the ideal performance of the most difficult stunts is that in which a body maintains a position of greatest rotational inertia, or full extension with the arms usually flexed over the head (a layout position). The intermediate level of a skill uses a pike position (flexion of the hip), and the beginning level uses a tuck position (flexion of the hips, knees, and vertebral column).

Beginning performers use obvious deviations from the ideal (layout) body position, such as flexing the hips and knees to reduce the radius of gyration to increase the angular velocity. For example, the performer depicted in Figure 14.11 would have increased her angular velocity about the low bar if she had flexed her hips and knees during the wrap. As performers become skillful in the intermediate and advanced stages, they perfect the timing of segmental movements and the positioning of the body parts and also reduce the resistive friction force in cases of hand-held support. These improvements result in greater initial angular velocity so that less reduction in the radius of gyration is necessary to complete the stunt. In fact, in judged activities, the measure of a skilled performance is that it displays as little obvious deviation as possible from the ideal or model positions.

An example of a learning progression is a cast back hip circle (Figure 14.12). In the beginning stages of learning the stunt (Figure 14.12c), the gymnast gets a small cast; she grips the bar tightly, does not place the upper body away from the bar on the downswing as much as she should, and, therefore, must tuck as tightly as possible on the upswing to complete the circle. Her hips, knees, and vertebral column are flexed, and her ankles are dorsiflexed. In the intermediate stages of learning (Figure 14.12b), the performer increases the distance of the CG from the bar during the back drop, and decreases the tightness of her grip. Consequently, she can make it to the top by merely flexing the hips. A skilled gymnast (Figure 14.12a) performs the stunt with little deviation from the rigidly extended body position. Just before reaching the bottom, she assumes a slight flexion of the cervical and the thoracic vertebrae. During the upswing she keeps her body's CG close to the bar by not allowing her arms to flex much from their extended positions. At the top of the bar, the gymnast extends her body again to prevent the added velocity from carrying her body into another circle backward.

Angular Momentum About Multiple Parallel Axes

Few examples of supported rotational skills concern movements about only one axis. These include the giant swing on the horizontal bar or some simple rotational moves on the low bar of the uneven parallel bars, such as a forward stride circle. Understanding the movements about multiple parallel axes is desirable when one is attempting to teach coordination and timing of a stunt that involves a series of body movements, such as the kipping movements.

A glide kip mount on the low bar of the uneven parallel bars serves as a good example.

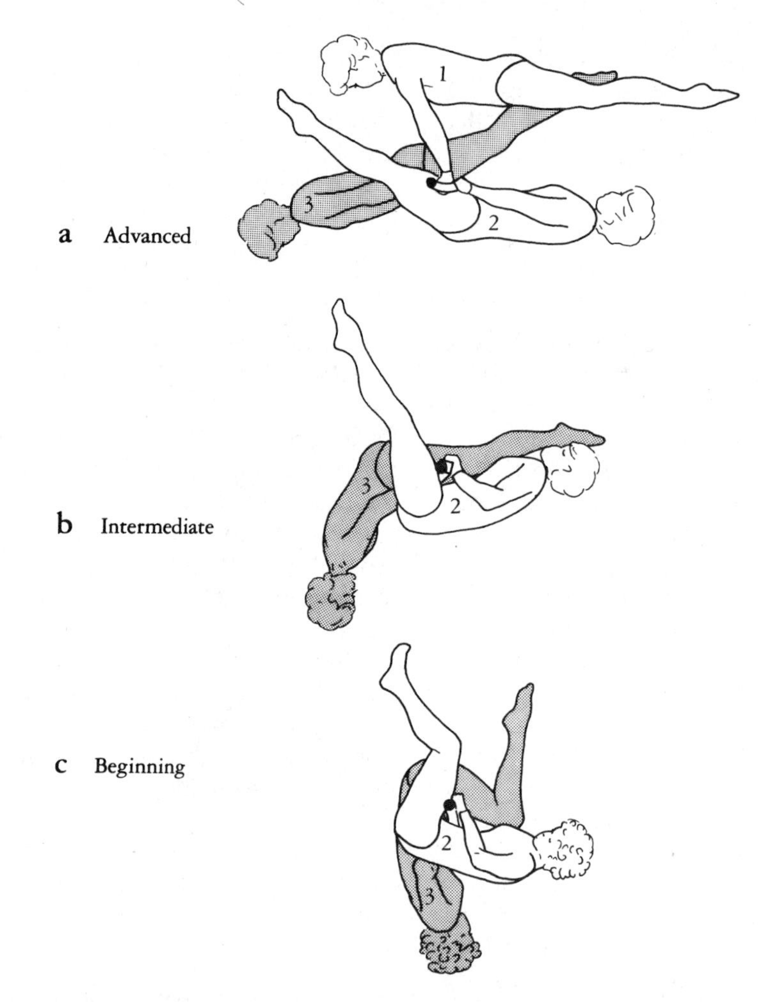

FIGURE 14.12

(a) An advanced, (b) an intermediate, and (c) a beginning gymnast performing a cast back hip circle on the horizontal bar.

This move is shown in Figure 14.13. Sequentially, the first and most obvious rotation is the body rotation about the rail; the feet are removed from the floor support, and the torque produced by gravity rotates the body from position 1 to 2. The upper body position should be kept as rigid as possible during this phase because the angular momentum gained by the system on the downswing may be lost by allowing the shoulders to flex (i.e., the body system should be rotating about only one axis at the rail until the arms are perpendicular to the floor). The second axis used is the ML axis, parallel with the rail and through the hip joints. It is used as the hips extend during the glide phase, from positions 2 to 3, and as they flex during positions 4, 5, and 6. The vertebral column should be stabilized in a position of slight flexion throughout.

The next axis used is the ML axis through the shoulder joints (shoulder joint extension) as the performer brings her pelvis up to the rail from positions 7 to 8. The momentum gained on the downswing from the outside of the rail

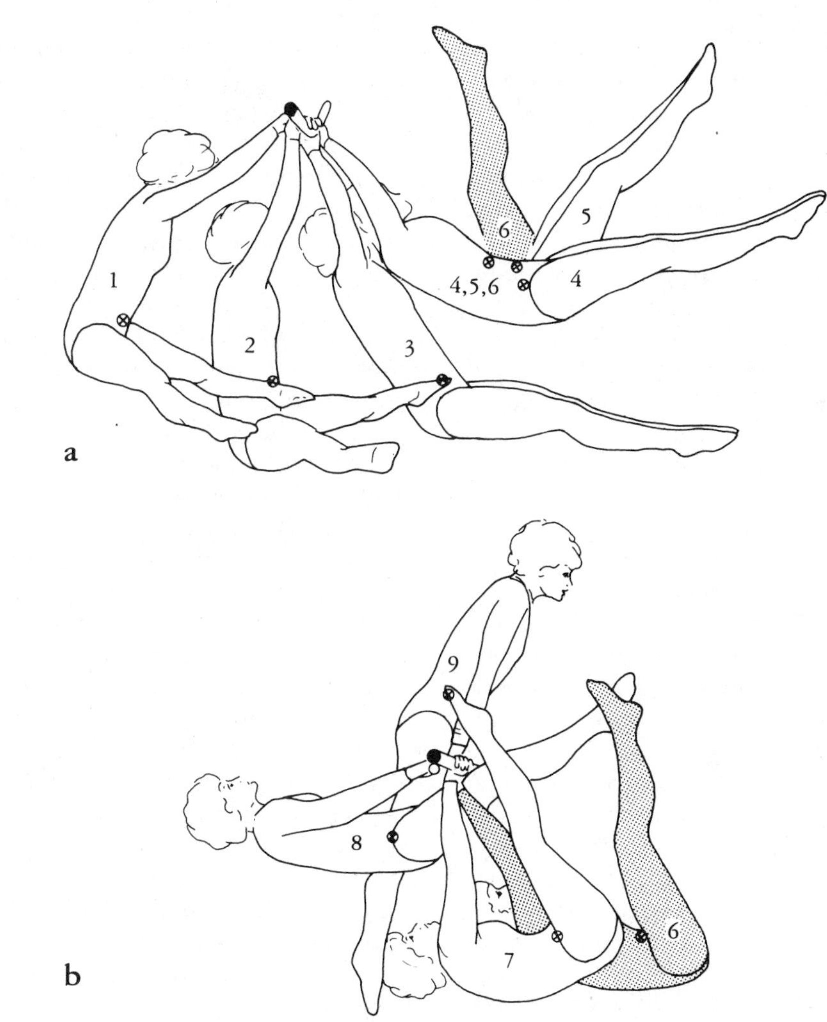

FIGURE 14.13

A gymnast performing a glide kip mount on the uneven parallel bars.

generates momentum to carry the performer up on the other side. From positions 8 to 9, the performer may have to flex the vertebral column and elevate the shoulder girdle to pull the body inward toward the axis so that the body's angular velocity increases. The proper sequence and timing of rotations about these axes are crucial to the performance because the momentum gained or maintained within the system during one phase may be lost by improper timing of the next movement.

Initially, as the body rotates downward, it gains angular momentum from the torque produced by the force of gravity. The body's CG should be placed and carried as far as possible from the rail to increase the perpendicular distance of that line of force from the axis. As the body reaches the bottom of the swing and begins to move upward, it must not be allowed to extend too far outward from the rail. The correct position can be achieved by preventing the shoulders from fully flexing and by not allowing the lumbar vertebrae to hyperextend during the extension of the hip joints. Although the hyperextended position is a popular one, it has little mechanical advantage because it only increases the resistive torque on the upswing, thereby enhancing the reduction of the body's angular momentum and its upward displacement. As the body reaches the mid-phase of the upswing on the outside of the bars, the hips are flexed. This motion takes place about a parallel axis to the rail axis and temporarily assumes some of the positive (ccw) angular momentum of the body about the rail. (The motion of the legs is in the same direction as the momentum about the rail.)

As the body reaches the end of the upswing, the lumbar and hip flexion bring the legs near the rail. The body is in a position to gain negative (cw) angular momentum on the far side as the torque from the weight acts on the body to accelerate it downward again. The angular momentum gained in this phase is gained by the torque applied by the force of gravity on the CG. As the body reaches the bottom of the arc, the angular velocity may be increased by extending the shoulder joints, keeping the hips close to the rail, and further flexing the vertebral column, which brings the body's CG close to the rail and decreases the body's radius of gyration. (A retraction and elevation of the shoulder girdle also will aid in the upswing.) The body now has an increased velocity to bring it up on the near side. Important at this point is keeping the body close to the rail because any separation of the two would increase the radius of gyration of the body, causing the body to slow to the point that it would not be able to reach the top. Beginners often have to "muscle" their way up the rail, because the angular velocity has been decreased to zero. A slight flexion of the neck and thoracic vertebrae on reaching the top of the rail also decreases the radius and aids the upward rotation.

As one teaches or coaches skills in which the concepts of angular momentum are important to the successful completion of stunts, one should watch carefully for rotation about axes other than the intended ones. Many times, unwanted rotation occurs because the body cannot stabilize its joints in the desired positions. Consequently, the body segments are allowed to rotate about joint axes while the body system rotates about an external axis. Moving about more than one axis can be undesirable if the movement about the second axis assumes some of the angular momentum needed for the successful performance of rotation about the first axis.

Figure 14.14 shows the proper (Figure 14.14b) and improper (Figure 14.14c) rotations that may occur during performance of a backward seat circle on the parallel bar. To perform this stunt correctly, the body must be stable at the shoulder joints, as in Figure 14.14a and b. The basis for completion of the circle comes from the gaining of angular momentum on the downswing by holding the body out as far away from the bar as possible to increase the motive torque (Figure 14.14a). Once the body's CG reaches its lowest position at the bottom of the swing, an elevation of the shoulders brings the body in closer to the bar, as does a flexion of the thoracic and cervical vertebrae. The decrease in the radius of gyration and resistive torque provided by the force of gravity on the upswing is enough to increase the velocity and bring the body to the top (Figure 14.14b). An extension of the body with a tighter grip at the completion of the stunt stops the body from falling backward or forward again. In Figure 14.14c, however, the performer has not held the shoulder joints stable at position 4, and the cw angular momentum gained by the system during the downswing is assumed by the trunk and leg segments rotating cw about a parallel axis through the shoulder joints. Thus, the body rotation about the bar is halted. The performer ends with her feet hanging downward and with her shoulder joints in a hyperex-

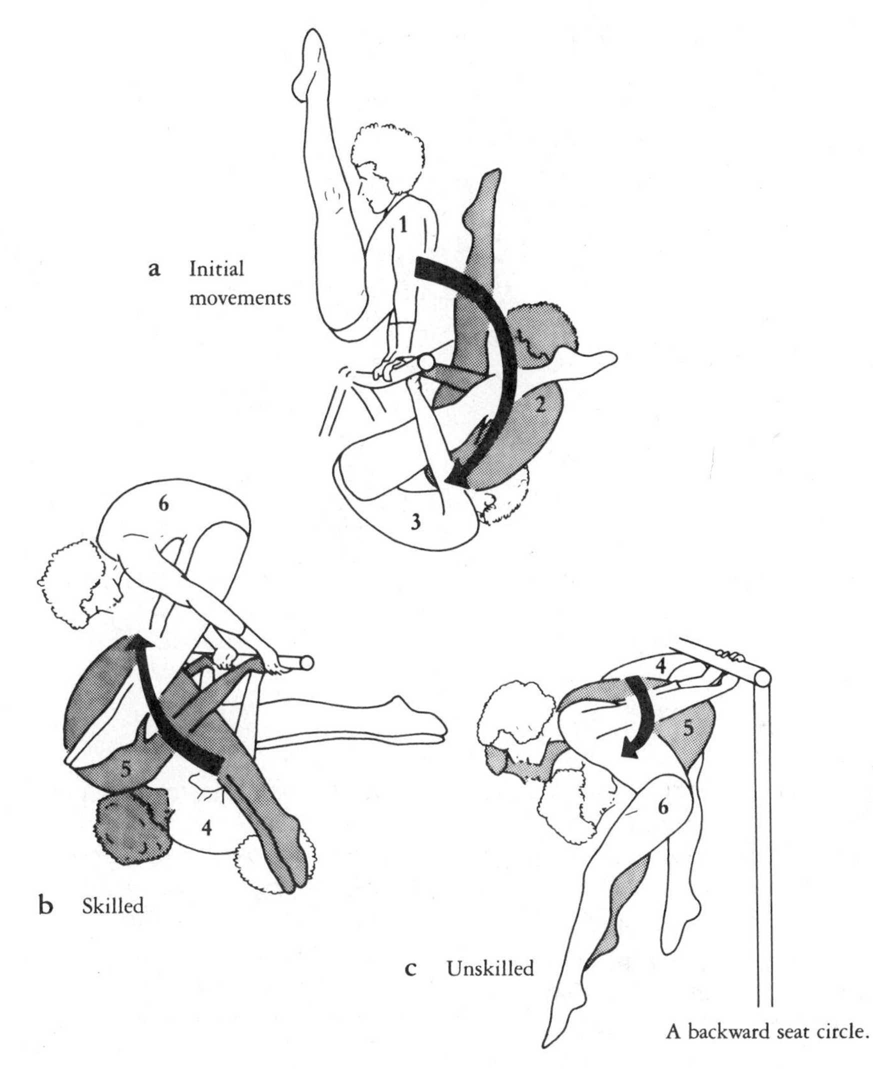

FIGURE 14.14
A backward seat circle.

tended position. The second rotation about the axis through the shoulder joints must be prohibited. The angular momentum of the system must remain about the axis at the rail and not be assumed by rotation of the body about the shoulder joints.

Understanding the Rotation of a Body Supported

1. Figure 14.15 illustrates a system falling around an axis of rotation. The body's weight is equal to 450 N. If 1 cm = 0.1 m, what is the torque of the body's weight at points a, b, c, and d?
2. Ignoring friction and air resistance, in

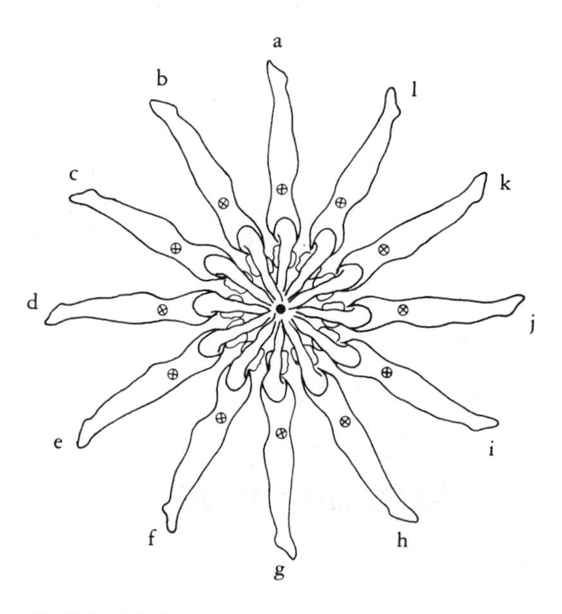

FIGURE 14.15

which positions in Figure 14.15 should the angular velocities be identical? In which positions should the angular momenta be identical? Explain why.

3. In Figure 14.15, considering friction forces at the axis of rotation, the positions on the upswing (g through l) will have an angular velocity less than, equal to, or more than the velocity of the corresponding (opposite) position on the downswing (a through f). Explain why.

4. Observe a runner jogging slowly. Note the degrees of elbow, hip, knee, and ankle flexion during recovery. Repeat for a medium-speed run and a sprint. What differences are noticed in the degree of flexion? Explain the differences in terms of the rotational inertia and muscular torques.

5. Observe gymnasts performing on the horizontal bar, the uneven parallel bars, or the parallel bars. While the body is supported by one of these pieces of apparatus, identify the axis of rotation as the gymnast performs. Can you identify when the apparatus is the axis of rotation? when the axis of rotation is located through a joint (e.g., shoulder joint, hip joint)? Observe how the axis of rotation changes from the apparatus to the joint centers; at times, the body and its segments may be rotating around both.

6. Observe a beginning or intermediate gymnast. Identify three skills in which the performer had to flex the hips or knees or extend the shoulder joints to reduce the rotational inertia so that the skill could be completed and where ideally the movements should have been prevented. Explain the relationships.

References and Suggested Readings

Bajin, B. (1986, October). Kasamatsu vault: A comparison between one and two-arm push off. *International Gymnast*, p. 46.

Borms, J., Moers, R., & Hebbelinck, M. (1976). Biomechanical study of forward and backward giant swings. In P. V. Komi (Ed.), *Biomechanics V-B: Proceedings of the Fifth International Congress on Biomechanics* (pp. 309–313). Baltimore: University Park Press.

Cheetham, P. J., & Mizoguehi, H. (1987, April). The gymnast on rings—A study of forces. *SOMA*, pp. 30–35.

Dillman, C. J., & Nelson, R. C. (1968). The mechanical energy transformations of pole vaulting with a fiberglass pole. *Journal of Biomechanics, 1*(4), 175–181.

Ganslen, R. V. (1974, March). Penetration—A new concept in pole vaulting efficiency. *The Athletic Journal*, pp. 36–39.

Goehler, J. (1977, October). The mechanical effect of the forward leg snap. *International Gymnast*, pp. 56–59.

Goehler, J., & Wieman, K. (1978, June). Swings—Snaps—Straddles. *International Gymnast*, pp. 51–52.

Hopper, B. J. (1973). *The Mechanics of Human Movement*. New York: American Elsevier.

Kaufman, D. A. (1973, December). A biomechanical block diagram of the pole vault. *Track Technique*, pp. 1732–1734.

Kopp, P. M., & Reid, J. B. (1979). Kinetic analysis of the giant swing on the horizontal bar. In J. Terauds (Ed.), *Abstracts of the International Congress of Sport Sciences*, p. 66. San Diego, CA: Academic.

McGinnis, P. M., & Bergman, L. A. (1986). An inverse dynamic analysis of the pole vault. *International Journal of Sport Biomechanics, 2*, 186–201.

Miller, D. I., & Munro, C. F. (1985a). Greg Louganis' springboard takeoff: I: Temporal and joint position analysis. *International Journal of Sport Biomechanics, 1*, 209–220.

Miller, D. I., & Munro, C. F. (1985b). Greg Louganis' springboard takeoff: II. Linear and angular momentum considerations. *International Journal of Sport Biomechanics, 1*, 288–307.

Nissinen, M. Preiss, & Bruggemann, P. (1985). Simulation of human airborne movements on the horizontal bar. In D. A. Winter, R. W. Norman, R. P. Wells, K. C. Hayes, & A. E. Patla (Eds.), *Biomechanics IX-B* (pp. 373–376). Champaign, IL: Human Kinetics.

Nissinen, M. A. (1983). Kinematic and kinetic analysis of the giant swing on the rings. In H. Matsui & K. Kobayashi (Eds.), *Biomechanics VIII-B: Proceedings of the Eighth International Congress of Biomechanics* (pp. 781–887). Champaign, IL: Human Kinetics Publishers.

Payne, H. A., & Barker, P. (1976). Comparison of the takeoff forces in the flic flac and the back somersault in gymnastics. In P. V. Komi (Ed.), *Biomechanics V-B: Proceedings of the Fifth International Congress on Biomechanics* (pp. 314–321). Baltimore: University Park Press.

Prassas, S. G. (1986). Biomechanic model of the press handstand in gymnastics. *International Journal of Sports Biomechanics*, 326–341.

Robertson, E. A., Paul, J. P., & Nicol, A. C. (1985). Investigation into the biomechanics of certain men's Olympic gymnastics. In D. A. Winter, R. W. Norman, R. P. Wells, K. C. Hayes, & A. E. Patla (Eds.), *Biomechanics IX-B* (pp. 377–382). Champaign, IL: Human Kinetics.

Sprinings, E. J., & Watson, L. G. (1985). A mathematical search for the optimal timing of the armswing during spring-board diving take-offs. In D. A. Winter, R. W. Norman, R. P. Wells, K. C. Hayes, & A. E. Patla (Eds.), *Biomechanics IX-B* (pp. 389–394). Champaign, IL: Human Kinetics.

Stepp, R. D. (1976, March). Mechanical analysis of the pole vault. *Track Technique*, pp. 2017–2018.

Takei, Y. (1988). Techniques used in performing handspring and Salto forward tucked in gymnastic vaulting. *International Journal of Sport Biomechanics, 4*, 260–281.

Tichonov, V. N. (1978, November). Improvement of vaulting technique. *International Gymnast*, pp. 56–57.

Zinkovsky, A. V., Vain, A. A., & Torm, R. J. (1976). Biomechanical analysis of the formation of gymnastic skill. In P.V. Komi (Ed.), *Biomechanics V-B: Proceedings of the Fifth International Congress on Biomechanics* (pp. 322–325). Baltimore: University Park Press.

CONCEPT MODULE K

Throwlike and Pushlike Movement Patterns

PREREQUISITES

Concept Modules A, B, E, I, J
Chapters 1, 3, 9

CHAPTER CONTENTS

K.1 Introduction and Terminology

In sport skills in which the performer is throwing or pushing an object, skills have one of the following overall performance objectives: (1) To project the object the greatest vertical or horizontal distance or (2) to project an object primarily for accuracy with the speed of projection enhancing the projectile's effectiveness. The most common throwlike skills are projections for horizontal distance, such as javelin, discus, and shot put, or skills in which an object is projected primarily for accuracy with the speed enhancing its effectiveness, such as volleyball, tennis, and racquetball serves, baseball pitches, and infield throws. In the latter groups, accuracy is primary, because without it, the projection is quite ineffective. In other words, speed may have to be compromised so that accuracy is not sacrificed.

Within each of these groups of skills are biomechanical factors, biomechanical principles, and critical features governing how the body's segmental movements best produce the desired accuracy and/or speed necessary to perform effectively, that is, to achieve the overall performance objective. The ability to recognize misapplications of the principles by performers, or to see errors in performance, is a key element in effective teaching and coaching. Because the biomechanical principles are similar for all skills in which an object is projected for the same performance objective, developing the ability to analyze performances in terms of these principles and to correct mechanical errors is relatively easy. To discuss throwing, kicking and striking, a few terms need to be reviewed from Chapter 9.

Movement Pattern

The term **movement pattern** refers to a general series of anatomical movements that have common elements of spatial configuration, such as segmental movements occurring in the same plane of motion. Within a movement pattern, the individual segmental movements may vary slightly in their ranges of motion (ROMs), velocities, and planes of motion. Throwing, kicking, and pushing are all general movement patterns. These patterns may be further defined according to where the movements occur relative to the body (e.g., *overarm, sidearm,* or *underarm* throwing patterns of the upper extremity or a kicking pattern with the lower extremity). Atwater (1977) distinguishes an *overarm* pattern as one in which the trunk laterally flexes away from the projecting arm and a *sidearm* pattern as one in which the trunk laterally flexes toward the projecting arm.

The terms *overhand* and *underhand* technically refer to the grasping position of the hand; that is, an underhand—or undergrip—for grasping when the forearm is in a supinated position, and an overhand—or overgrip—for grasping when the forearm is in a pronated position.

Skill

When a general movement pattern is adapted within the constraints of some particular movement activity or sport, it is called a **skill**. For instance, an overarm pattern is used in reference to sport skills such as a tennis serve, a baseball pitch, a badminton smash, and a shot put. Table K.1 offers one possibility for the classification of skills under general movement patterns.

If individual skills within a similar movement pattern are examined, the skills may be seen as slightly different in spatial and temporal terms. For instance, on first consideration, a tennis serve and a badminton smash are observed to be of a pattern similar enough to be identical. A critical analysis of each skill, however, shows that although the gross pattern is the same, slight variations exist between individual segmental movements in the two performances. The uniqueness of the two skills is due to the difference in the masses of the two rackets, the difference in the forces generated by each racket on the projectile, and the difference in the masses of the two projectiles.

Constraints

Factors that influence the time and space variables to a movement pattern are called **constraints**. Examples of common variable factors influencing the observable characteristics of a performance include the mass of the implement, the mass of the object to be projected, the size of the target, the size of the sport environment or playing area, and the strength or ability of the performer. Thus, the mass of a medicine ball compared to a baseball becomes a constraint to one

TABLE K.1

SPORT SKILLS CLASSIFIED UNDER FOUR GENERAL MOVEMENT PATTERNS

Underarm Patterns	Sidearm Patterns	Overarm Patterns	Kicking Patterns
Softball pitch	Baseball throw	Baseball pitch	Football punt
Bowling ball delivery	Discus throw	Shot put	Soccer placekick
Horseshoe pitch	Hammer throw	Javelin throw	Swimming flutter kick
Volleyball serve	Volleyball serve	Volleyball serve	Walking or running, swing phase
Volleyball bump	Baseball batting	Volleyball spike	Dolphin kick
Badminton underarm clear	Tennis drive	Basketball one-hand shot	
Badminton serve	Badminton drive	Tennis serve	
Field hockey drives and passes	Racquetball shots	Tennis smash	
	Squash shots	Badminton smash and around-the-head shot	
	Handball shots	Cricket "pitch"	

who wishes to use the high-velocity baseball throwing pattern for projecting the medicine ball.

For example, consider the overarm throwing pattern used for projecting a basketball and a baseball for horizontal distance. The basic pattern is the same for both activities, but the positions of the upper extremity segments may be quite different. If the hand is not large enough to hold the basketball, then the hand segment must remain behind the ball so that the ball does not slip from the hand as the ball is pushed forward. The baseball can be easily grasped with the fingers. If the thrower is not strong or powerful enough to accelerate the ball when it is positioned away from the midsagittal plane of the body, then the thrower must bring the ball in closer to the shoulder. Basketballs are thrown close to the midsagittal plane more often than are baseballs because basketballs are larger and greater in mass.

Understanding Segmental Movement Concepts

1. List three general movement patterns. List three sport skills under each pattern.
2. How does one distinguish between a movement pattern and a sport skill?
3. List three sport skills displaying an overarm, an underarm, and a sidearm pattern.
4. List two sport skills in which an underhand grip is used. List two sport skills in which an overhand grip is used.
5. How does Atwater distinguish between an overarm throw and a sidearm throw?

K.2 Throwlike Patterns: Sequential Segmental Rotations

The term *throwlike,* as used in this text, is characterized by movements used to project an object that is allowed to lag back behind the proximal segments that have finished their backswings and are now moving forward. The term throwlike refers to the characteristics of the general movement pattern of a throw and not to a throw *per se*. For example, a kick or a strike is classified as throwlike. In the case of striking activities, such as kicking or batting, the body part or the implement that will contact the object lags behind the other segments until just before contact, at which time it "whips" forward to catch up to its proximal segments.

The overall performance objective of these throwlike activities is either to project an object for the greatest horizontal or vertical distance, or to project an object for accuracy where the velocity of the object enhances its effectiveness. Regardless of which overall performance objective is the appropriate one, the velocity of the object at release is a very important variable. The velocity that the object will have at release depends on the velocity of the contact point that the object has with the hand, the foot, or the implement being used. So, in throwlike activities, we are attempting to achieve high "end-point" velocity.

The Kinetic Link Model

To illustrate how one may generate high end-point velocity with a system of segmented links, a model is shown in Figure K.1. The three segments are labeled **A, B,** and **C,** and the three axes of rotation for these segments are labeled **a, b,** and **c.** The arrows on segment **A** represent muscle torques applied to segment **A** and to a fixed "ground" segment. Because the muscle torques applied to segment **A** are *external* torques (i.e., external to the system of links), they can effect changes in the system's angular momentum. In the figure, the muscle torques between segments **A** and **B** and the muscle torques between segments **B** and **C** function as *internal* torques (they are internal to the system).

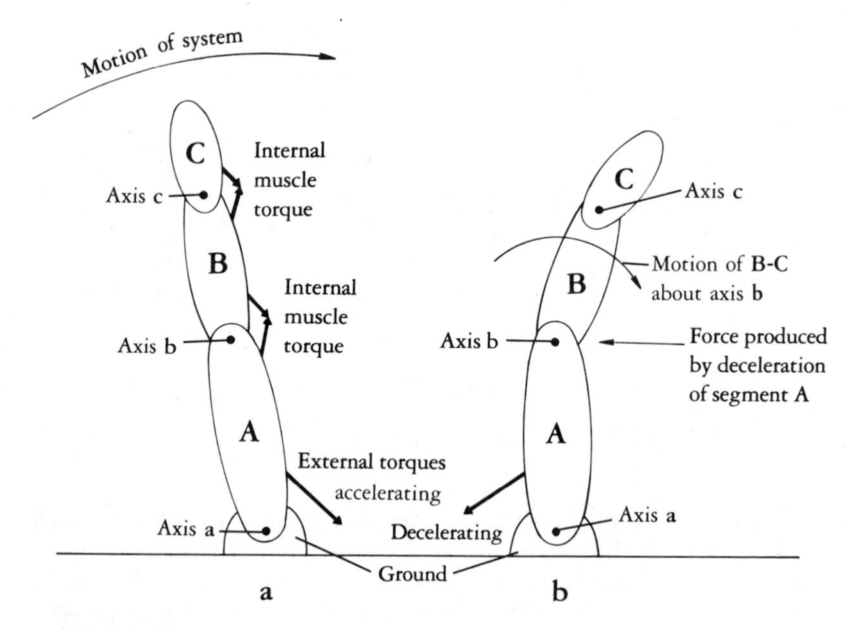

FIGURE K.1

A three-segment model of the kinetic link concept with external and internal torques influencing its motion.

The way such a model functions can be understood by examining the following situation:

Suppose that the muscles to the right of segment **A** apply an external torque large enough to accelerate segment **A** in a clockwise (cw) direction. The intersegmental muscle torques between segments **A** and **B** and **B** and **C** work to prevent lagging back, or counterclockwise (ccw) motion of the distal segments relative to the proximal segment as the system accelerates cw. If these internal stabilizing torques were not applied, segments **B** and **C** would "lag back" as segment **A** rotates forward. Thus, the external torques applied to segment **A** accelerate the entire system and give the entire system angular momentum.

Now suppose that a second external torque, the torque to the left of segment **A** and axis **a**, acts to decelerate segment **A** (Figure K.1b). The deceleration of segment **A** places a left-directed force on axis **b**, tending to "fix" that axis in space. Because segments **B** and **C** are free to move cw about axis **b**, **B** and **C** continue to rotate with the same angular *momentum* about axis **b**.

Because the axis of rotation for segments **B** and **C** has moved from axis **a** to axis **b**, the radius of gyration and thus the rotational inertia for the remaining moving system, segments **B** and **C**, have decreased. Consequently, because the momentum of the systems is conserved and segments **B** and **C** are smaller masses and have a smaller radius of gyration about axis **b**, the angular *velocity* of segments **B** and **C** increases.

Note that the rotational inertia of segments **B** and **C** decreased because the axis of rotation moved closer to these two segments, resulting in a decrease in the radius of gyration as well as eliminating the mass of segment **A** from the system. Second, no *external* torque (force × zero distance) was applied to segments **B** and **C**, only to segment **A**. Therefore, the angular momentum of segments **B** and **C** as a system is being conserved, but the angular momentum of the original system, which included segment **A**, is not being conserved.

Figure K.2 illustrates the application of the previously described model to a woman moving in a similar fashion. Each body segment moves by rotating around an imaginary axis of rotation that passes through the articulation of the segment. Recall that the segment moves along a plane that is perpendicular to the axis around which it is rotating. Figure K.2 illustrates (1) the movement of the hand in space caused by flexing the pelvis and trunk at the hip joint around its mediolateral (ML) axis, position 1 to position 2; (2) the movement of the hand in space caused by the medial rotation of the abducted humerus around a longitudinal axis through the shoulder joint, position 2 to position 3; and (3) the hand movement caused by flexing the hand at the wrist around an ML axis, position 3 to position 4.

In this human body application of the model, segment **A** is accelerated by an external torque, which would be the hip flexors. This torque would give the entire system including segments **B** and **C** angular momentum. The external torque used to decelerate segment **A** would be the hip extensors. The deceleration of the pelvis and trunk segment's deceleration would apply a force on axis **b**, the longitudinal axis of the humerus through the shoulder joint. This force would thereby cause axis **b** to be "fixed" in space. The distal segments would

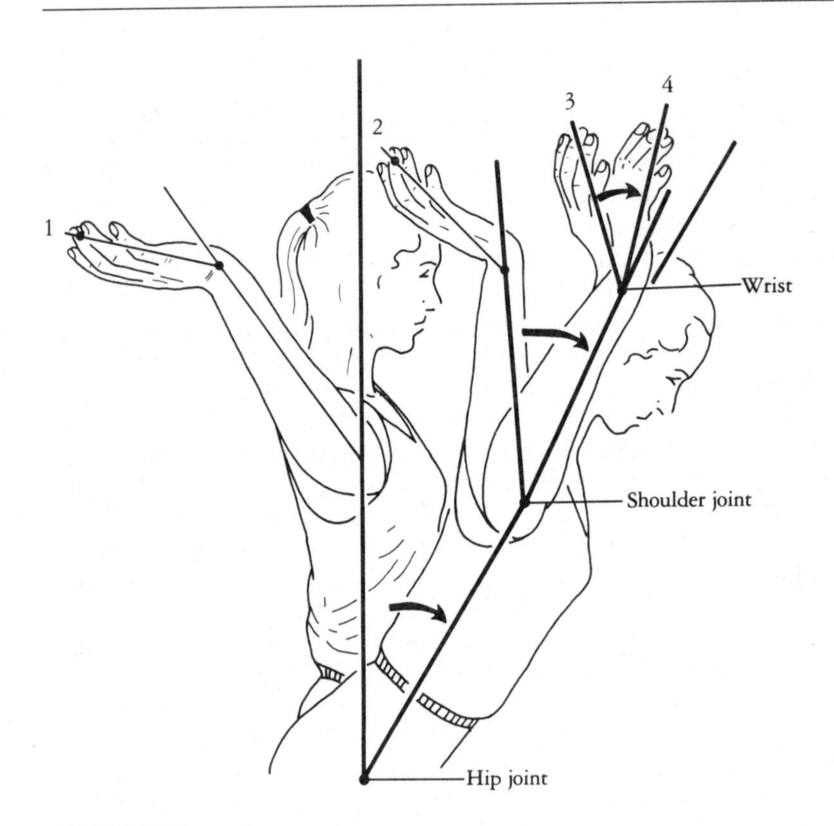

FIGURE K.2

Flexion of the hip, medial rotation of the shoulder joint, and flexion of the wrist with all distal segments of the upper extremity fixed.

freely rotate around axis **b** and do so with increased angular velocity because the rotational inertia of these two distal segments is smaller relative to axis **b** than it is relative to axis **a**.

Linear Motion Parameters of a Rotating Point

Recall that the velocity of the point of release on the body or implement is the critical factor in giving high velocity to the object being projected. In the previous two models, we were attempting to generate high angular velocity. The *linear* velocity of a given point on a rotating segment is directly proportional to both its angular velocity and its radius of rotation ($v = r\omega$). The **radius of rotation** for any point on a rotating system is defined as the perpendicular distance between the contact point of the object being projected or struck on the segment or implement and the axis of rotation of that segment. The radius of rotation is that to which people refer when they say that a longer implement, such as a baseball bat, gives greater leverage and hence greater velocity.

If a projectile is to be given a large *linear* velocity, the distal segment of the body must have a large linear velocity when the projectile is released (as in throwing) or contacted (as in striking or kicking). When the extremity is acting as an open chain, the movement of any proximal segment affects the motion of all segments distal to it. To illustrate this concept in the simplest manner, we will stabilize all segments in the chain and allow them to move one at a time.

The radius of rotation for the fingertips extends from the fingertips to the axis where the rotation is taking place—the hip. In Figure K.2, the linear velocity of the fingertips that is due to the flexion of the hip may be determined by multiplying the angular velocity that is due to hip flexion by the distance of the fingertips from the ML axis of the hip joint. The greater the radius of rotation of a point or the angular velocity of a point or both, the greater is the point's linear velocity.

The angular displacement for each of the motions shown in Figure K.2 is 30 degrees, or 0.524 radians. Thus, if we assume that these segments move the 0.524 radians in 0.5 sec, then each segment's angular velocity would be 1.048 radians/sec (0.524 radians ÷ 0.5 sec). Further, if we measure the perpendicular distances from the hip, shoulder, and wrist axes to the middle of the fingers (assuming that is the location of the center of gravity of an object such as a baseball), then we can calculate the linear velocity of the ball that is due to each rotation. If we say that the perpendicular distances from the hip, shoulder, and wrist axes are 95, 46, and 14 cm, respectively, then the linear velocity that is due to the hip motion may be calculated as follows:

$$V = r\,\omega$$
$$V = 95\text{cm} \times 1.048 \text{ radians/sec}$$
$$V = 99.56\text{cm/sec}$$

linear velocity that is due to the shoulder motion:

$$V = 48\text{cm} \times 1.048 \text{ radians/sec}$$
$$V = 50.304\text{cm/sec}$$

linear velocity that is due to the wrist motion:

$$V = 14\text{cm} \times 1.048 \text{ radians/sec}$$
$$V = 14.672\text{cm/sec}$$

Thus, one may increase the linear velocity of the end point by increasing the perpendicular distances between the object and the axes of rotation being used as well as increasing the angular velocity of the rotating segments.

To achieve an effective throw, we use a combination of many segmental movements, with each segment moving around its own particular axis. Consequently, the throw, or any other open kinetic link activity, is produced by the coordinated (well- timed) motion of all segments in that chain. For the throw, the velocity of the fingertips is affected by the displacement of the pelvis, the trunk, the shoulder girdle, arm, forearm and hand segments.

Sequencing Segmental Rotations: The Kinetic Link Principle

The model of throwlike movements of an open kinetic link system just described has the following characteristics: (1) The system of links has a base, or fixed, end, and a free, or open, end; (2) the more massive segments are at the proximal end and the less massive segments are at the free end; and (3) an external torque is applied to the base segment to initiate the system's motion and give the entire system angular momentum. Intersegmental torques are internal to the system and, when applied, change the angular velocities of individual segments in the system.

The previous example was a hypothetical situation in which muscle torques were acting either within or on a system of segmental links. Because of the complexity of the human link system, predicting exactly which muscle torques might be operating at any given time is impossible. For example, the accelerating external muscle torque acting on the base segment may operate all the way through the motion, most of the way through the motion, or part of the way through the motion. The intersegmental torques may be voluntarily controlled bursts of muscular contraction, muscular contractions evoked by the stretch reflex response, torques produced by the elastic recoil of the muscles and surrounding tissues, or any combination of these. The sequence of segmental rotations used for throwing, kicking, or striking patterns may be likened to the motion of a whip, a fly-fishing rod, or the lashing action of a fish's tail. Figure K.3 illustrates this motion for a fisherman casting a fly.

If segmental rotations are free to occur at each of the distal joints, the body's system of links behaves in a manner similar to a flexible chain of links, with the earth and base segment acting like the handle of a bullwhip. Just as the tip of the whip can be made to travel so fast that it has supersonic speed (hence the sonic boom heard as the crack), the small distal segments of the hand or foot can be made to travel extremely fast by the sequential acceleration and deceleration of the body segments. The general pattern of movements is one in which the initial rotation occurs in the base segment, which is the most stable part of the system, and is followed by the forward rotation of the next distal segment. Each segment comes forward as the movement of its proximal segment reaches its greatest angular velocity (the distal end of the proximal segment has its greatest linear speed at this time). The sequencing procedure continues to occur from proximal to distal, one segment's movement overlapping the other until the most distal segment finally comes forward.

Such a pattern includes a "lagging back" of the distal segments as the more proximal segments are accelerated forward. The lagging back of the distal segments can be seen on the tip of the fly rod and line as the initial movements of the arm and hand occur. Like the action of the fly line or the action of a bullwhip, the lagging back of the distal segments of the body may be due to the backward momentum developed during the original backward motion of the system (commonly known as the backswing or windup), or it may be due to the proximal segment moving out from under the distal segment, in which case it is described as *inertial lag*.

Sometime during the backswing, muscle torques cause the proximal or base segment to begin to accelerate forward. In a system of freely rotating segments, as the proximal link angularly accelerates in one direction, the next distal link in the kinetic chain lags back in the opposite direction. The lagging back of the distal segments persists until one or more of the following occurs: (1) the acceleration of the proximal segment decreases or ceases, (2) the elastic or structural limit of a joint's ROM is reached, or (3) the stretch reflex is activated. If the stretch reflex is evoked, elastic recoil occurs, or voluntary intersegmental muscular torques are activated, the lagging segment's forward rotation will be initiated.

Ideally, as the limit of this backward motion is reached, the proximal segment begins to decelerate. This should occur sometime after the midrange of the proximal segment's motion and is the place at which the base segment has its maximum angular velocity (zero acceleration). This deceleration of the proximal

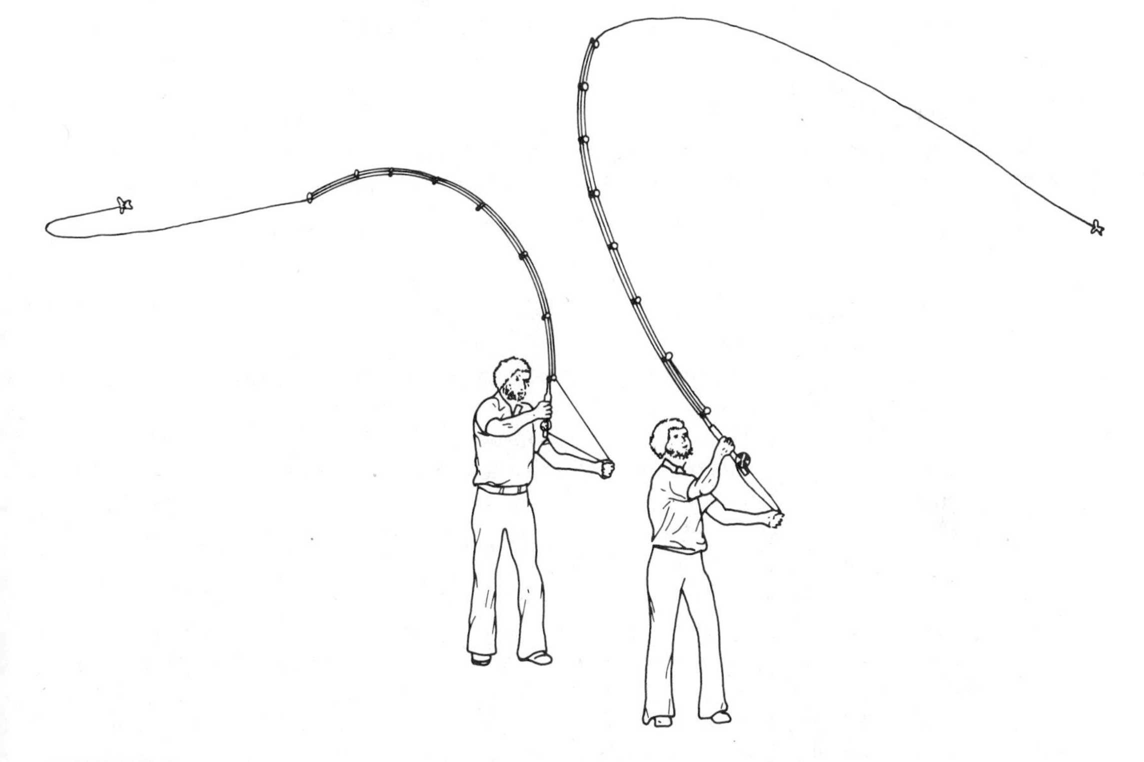

FIGURE K.3

A fisherman casting a fly. The rod and line may be likened to the whiplike action of the segmental system.

segment and the next distal segment may be caused by either or both of the following: (1) by the muscles that are antagonists to the motion occurring in the proximal segment, and (2) by the intersegmental muscles between the proximal segment and the next distal segment. As the proximal segment decelerates, the adjoining distal segment accelerates.

Similarly, when the second segment reaches the midrange of its rotation, it too decelerates and puts a force on the next distal axis, which increases the velocity of the next adjacent distal segment in the chain. Such a sequence occurs link by link, from base to end, proximal to distal, until the end of the chain is reached.

Just as a fly rod or bullwhip is tapered, the more distal parts of the human body are composed of successively smaller and smaller masses. By virtue of the smaller mass, the rotational inertia of each successive distal segment becomes progressively less (e.g., the pelvis compared to the thigh, the thigh to the leg, the leg to the foot; the trunk compared to the arm, the arm to the forearm, the forearm to the hand, and the hand to the fingers).

The tapering of the distal masses is significant when the proximal segment is decelerated by an internal torque between it and the next distal segment. A given magnitude of muscle torque, when applied to a larger mass such as a proximal segment, will cause that segment to lose momentum. The same muscle torque applied to the smaller adjacent segment will cause that smaller segment to gain momentum. As a result of its smaller mass, the distal segment will gain a larger velocity from the applied muscle torque. Thus, the timed successive segmental rotations that originate from the stable base segment and terminate with the most distal segment serve to produce extremely fast linear speed on the distal end of the kinetic chain.

A Model of a Kinetic Link System

To illustrate the simplified effects of the acceleration and deceleration of the proximal segments on the distal segments, a conceptual model is shown in Figure K.4. The presence of muscle torques acting on segments whose velocities are continuously increasing and the sequence and timing of a mechanically effective link system are shown. The model is adapted from that of Morehouse and Cooper (1950) and modified to account for the kinetic contributions of muscle torques which are indicated.

Segments **A, B, C, D,** and **E** represent an arbitrary number of body segments used sequentially to perform a skill. Applying the model to a kicking motion, the segmental movements would be **A**, the pelvis rotating in flexion about a lumbar intervertebral axis; **B**, the thigh segment rotating in flexion about the hip axis; and **C**, the leg and foot segments rotating in extension about the knee axis. In the case of kicking, three segments are used. Applying a throwing motion to the model, the segments used are (**A**),, the pelvis medially rotating about the longitudinal hip axis; (**B**), the trunk rotating about the longitudinal vertebral axis; (**C**), the shoulder girdle protracting about the longitudinal sternoclavicular axis; (**D**), the arm medially rotating about the longitudinal shoulder axis; (**E**), the forearm extending about an ML elbow axis; and (**F**), the hand and ball as a single segment flexing about the ML wrist axis. In this case six segments are used. The model may be shortened or lengthened to fit the skill being analyzed.

The muscle torques are represented in Figure K.4 by the squiggly lines. The angular velocity is represented by the symbol ω, shown along the left side of the model. Time is represented along the baseline. The rotational inertia (I) of the system distal to the joint axis being used at different times during the sequence of the skill is shown to the left of the angular velocity lines and is sized to show differences in its magnitude.

This is a conceptual model and is not to be interpreted as representing actual magnitudes. It has been developed to show the relationships of various mechanical quantities with an *external* muscle torque applied at the base segment and *internal* muscle torques applied between segments thereafter. The model represents a theoretical, ideal pattern in which achievement of high angular and

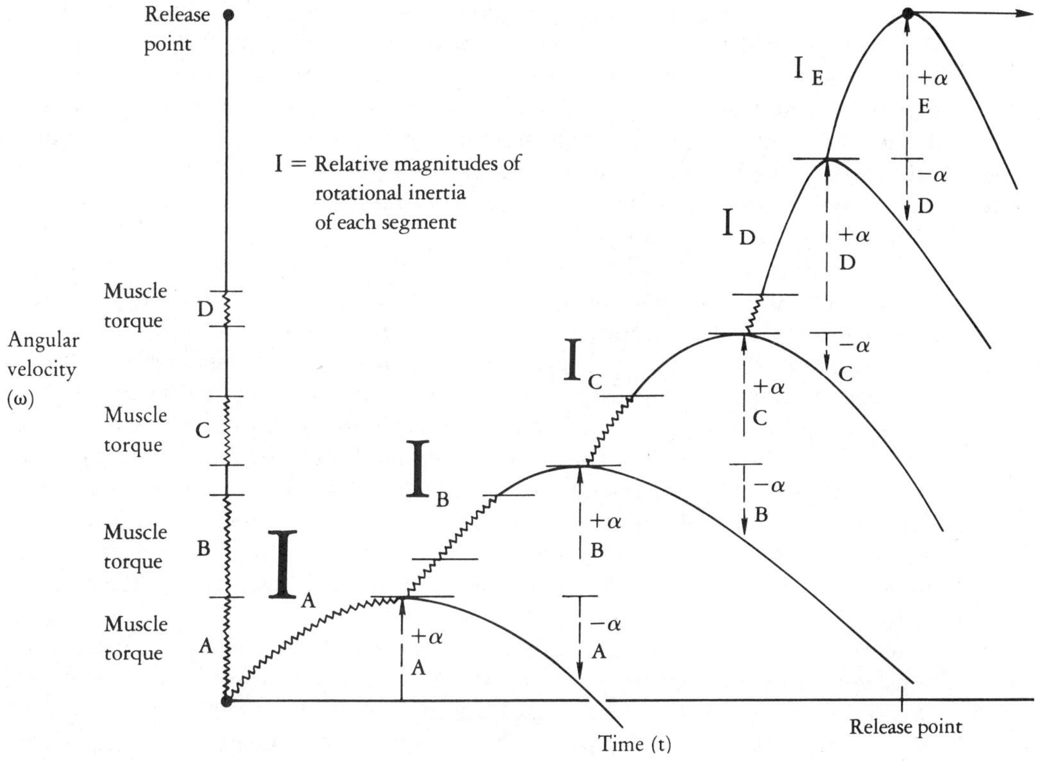

FIGURE K.4

A conceptual model of the kinetic link principle.

linear velocity of the "endpoint" on the distal segment is desired so that an object to be thrown, kicked, or struck will travel at a high speed.

Using the kinetic link model, we will describe the skill of punting. Segment **A** represents the pelvis; segment **B**, the thigh; and segment **C**, the leg and foot. From a position of slight hyperextension of the lumbar vertebral column, the vertebral column flexors contract to flex the pelvis about an axis through the lumbar spine. Because the supporting extremity is fixed to the ground, the initial muscular contractions to initiate the forward motion of the pelvis and free extremity are *external* to the system and therefore can give momentum to the system. In this case the distal segments lag back, as would be exemplified by hip hyperextension and knee flexion. After the initial flexion of the pelvis about the lumbar axis, the hip flexors contract, causing deceleration of the pelvis relative to the vertebral axis and thigh acceleration relative to the hip axis. The rotational inertia of the thigh and leg–foot segments rotating about the hip is less than that of all three segments rotating about the vertebral column axis, and thus the thigh and leg–foot segments rotate faster than they were rotating previously.

Sometime during the midrange of hip flexion, the knee extensors contract to accelerate the leg–foot segment about the knee axis. The knee extensors simultaneously decelerate the thigh about the hip axis. Because the leg–foot segment is less massive than the previous segments and because the radius of gyration of the leg–foot segment is less about the new knee axis, the rotational inertia of the leg–foot system is considerably less, and consequently, the angular velocity is considerably more. We are now at the end of the link system. At the maximum linear velocity of foot motion, the foot should contact the ball to impart the greatest velocity to the ball.

For the intersegmental torques to play a part in the more distal segments' acceleration, the muscular contractions must be fast enough not only to keep up with this already fast-moving distal segment but also to apply force to it. Consequently, the faster the distal segment is traveling, the faster the muscular contraction must be to exert a force on it. Remembering the force–velocity relationship, we know that the faster the muscle must shorten, the smaller the force that can be applied on the attached segment (and the smaller the angular impulse it can apply to the segment). After the initial contraction of the internal torque associated with the stretch reflex and the elastic recoil of the muscle, the velocity of the distal segments may be so great that the muscle contraction on the distal segments may not be able to apply a torque to accelerate them further.

In addition, considering the length–tension relationship of muscular contraction, we know that muscles can exert less tension as they become shorter. Determining the length of the muscles at any given time is difficult; however, the lagging back of the distal segments helps place the muscles in a lengthened state at the time they are needed to apply the internal torque and thus enhances their effectiveness.

Throwlike Movements Performed While in the Air

Several throwlike skills involve the body initiating a rotation while it is in the air. Examples include a volleyball spike, a tennis serve, and a badminton smash. In these cases, the performance is similar to gymnastics and diving rotations that are initiated in the air in that a reaction rotation is caused by the rotation of the trunk and upper extremity in one direction, resulting in an equal and opposite rotation of the pelvis and lower extremities. These situations may be shown by a model illustrated in Figure K.5.

After the body is off the ground, the backswing (preparation phase) and the force phase (acceleration phase) cause responses of the lower body in the opposite direction. In Figure K.5a, a three-segment model is shown in which the initial rotation of the system occurs before the system leaves the ground; thus, the system leaves the ground with some angular momentum. In Figure K.5b, contraction of the internal torques between segments **B** and **C** each result in a movement response in segment **A** resulting in the motion of that shown in Figure K.5c. In addition to the angular velocity given to it by the external force on segment **A** before the system left the ground, the distal segment has the angular velocity that is due to its own internal segmental torque's action on it.

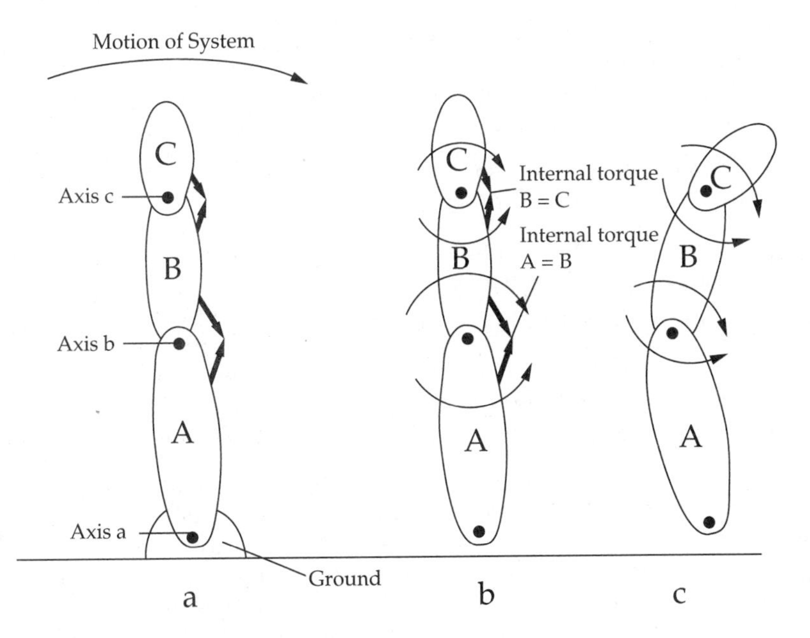

FIGURE K.5

A three-segment model of the kinetic link concept with internal torques influencing its motion.

The movements of the "responding" segments may be minimized by increasing their rotational inertias about parallel axes such as the volleyball spiker illustrated in Figure 13.16, p. 568. By abducting his lower extremities, he maximizes the lower extremities' resistance to the rotational response. This increase in the resistance helps to establish a more stable base on which the upper body segments can rotate around a parallel axis.

Understanding Sequential Segmental Rotations

1. Of the following activities, circle those to which the kinetic link principle applies: a soccer punt, a golf swing, a right jab in boxing, a back handspring takeoff, hammering a nail, a hammer throw, pushing a table across the floor, a football pass for distance, bowling.
2. Observe the masses of the arm, forearm, and hand segments. Discuss the similarities and differences in terms of the relative masses of the segments (greatest to least) and the distribution of those masses (closer to the proximal or distal end). Repeat for the thigh, lower leg, and foot segments. Refer to Appendix III to find mean values of the segmental weights and radii of gyration. List those values. Discuss the relative masses of the segments and their relationship to the kinetic link principle.
3. State three principles that relate to the use of body segments in generating velocity on the open end of the open kinetic chain.
4. Define the radius of gyration. How does this concept relate to throwing an object or swinging an implement?
5. Define the radius of rotation. How does this concept relate to swinging an implement?
6. Describe the differences in the radius of gyration and the radius of rotation in relation to throwlike movements. Relate these to linear and angular velocity.
7. Attend a beginning class in badminton, tennis, or golf. Observe an unskilled performer and a skilled performer in the class, and compare the segments that are used and the sequencing (or lack of sequencing) of the segments in a similar skill. Does the beginning performer use the same segments and movements? Does the beginner sequence the movements, or are some of the movements used simultaneously? What differences do you notice about the speed of the object projected by each? the accuracy?

K.3 Lever Versus Wheel–Axle Rotations

Two machines used in segmental movements are associated with the kinetic link model. These mechanisms, discussed in Chapter 3, are the lever and the wheel–axle. The most common segmental motions are *lever-type* motions such as flexion–extension, abduction–adduction, and protraction–retraction. The wheel–axle motions are those motions that occur around the longitudinal axis of the segment and include all of the joint movements that have the word *rotation* in the name (e.g., medial and lateral rotation, right and left rotation). Protraction and retraction of the shoulder girdle occur around the longitudinal axis of the sternum and are *lever-type* motions and are exceptions to this rule. In addition, pronation and supination of the radioulnar joint and inversion and eversion of the subtalar joint are wheel–axle movements. In wheel–axle motions, the bone that rotates as a result of direct application of muscle torque serves as the axle segment, and an adjoining distal segment, or segments, serves as the "wheel."

Lever and wheel–axle systems are both rotational systems, and therefore, when either is used for throwing, striking, or kicking skills in which a high linear velocity is desired, two aspects of that rotational motion are important: the angular velocity of that system and the radius of rotation of the contact point of the object. The point of concern in throwing, kicking, and striking skills is that point at which the projectile makes, or will make, contact with the body part or with an implement. If one wants to adjust the radius of rotation of the point of contact on a lever, one selects a shorter or longer implement or contacts the object at a different location on the lever. For example, the leg and foot in a kicking skill serve as a lever system during knee extension. To decrease the radius of rotation for that leverlike movement, one must contact the ball on the tibia, an unlikely choice.

Unlike the lever system, the wheel–axle mechanism has an adjustable radius of rotation. By flexing or extending the distal "wheel" segment relative to the rotating axle segment, one can increase or decrease the radius of rotation for the point of contact on the wheel. Figure K.6 illustrates how, through the use of radial or ulnar flexion of the wrist, one can change the radius of rotation for the center point on a racquetball racket relative to the axis of rotation for forearm pronation and supination. In this example, the racket and hand serve as the wheel segment. To achieve high linear velocity at the sweet spot of the racket, one should position the racket so that the radius of rotation is as large as possible.

A second factor to consider in achieving high contact-point velocity is the angular velocity of the segment on which the contact point is located. The amount of angular velocity that a segment can achieve depends on its rotational inertia (I = mass × radius of gyration^2) and the torque applied:

$$(\alpha = \frac{T}{I} \text{ or } \frac{T}{mk^2})$$

If the muscle torques are at their maximum, then the muscle torques are constant. However, changing the segment's radius of gyration will change the segment's resistance to those torques' accelerating effects. A smaller radius of gyration will result in a larger acceleration for a given torque; a larger radius of gyration will result in a smaller acceleration for a given torque.

This situation leaves somewhat of a dilemma. If one increases the radius of rotation to increase the linear velocity of a rotating point ($V = r\omega$), one also increases the radius of gyration, which increases the system's rotational inertia (resistance to acceleration) and thus the torque's accelerating effect ($\alpha = \frac{T}{I}$). There is a way to theoretically solve the problem.

Figure K.7 shows a model of a lever arrangement of three segments and a wheel–axle arrangement of the same three segments. The lever arrangement has a greater radius of rotation (2.75 units); the wheel–axle arrangement has a smaller radius of rotation (1.72 units). Thus, if both systems were rotating with the *same angular velocity*, the end of the lever would be moving twice as fast linearly (double the radius of rotation, $v = r\omega$). In this example, however, the rotational inertia of the system used as a wheel–axle is approximately five times smaller than the rotational inertia of the same system used as a lever. ($I_L = .44$; $I_{WA} = .09$).

Because the radius of gyration is squared in determining rotational inertia, it has a squared effect on the resistance of the system to acceleration by muscle torques ($I = mk^2$). Thus, for a situation in which equal muscle torques are acting for the same time period on the two systems, the smaller rotational inertia

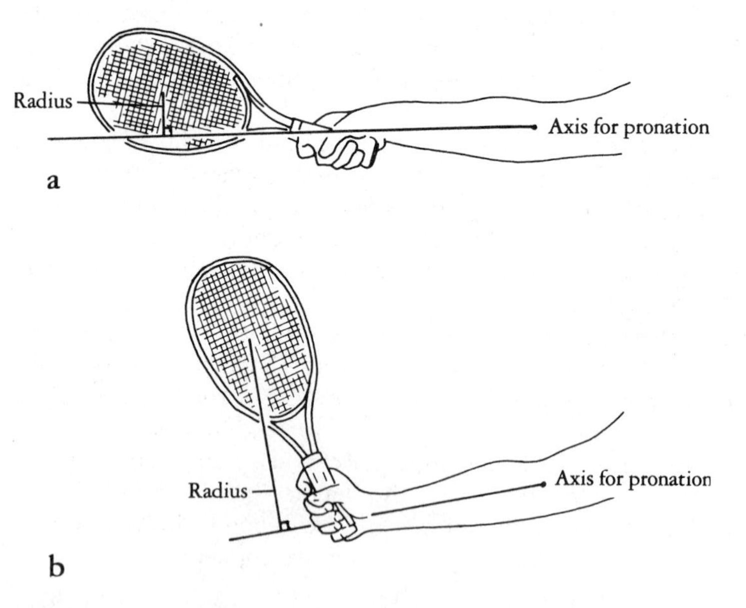

FIGURE K.6

(a) Small-radius "wheel" segment. (b) Large-radius "wheel" segment, used in the wheel–axle motion of radioulnar pronation.

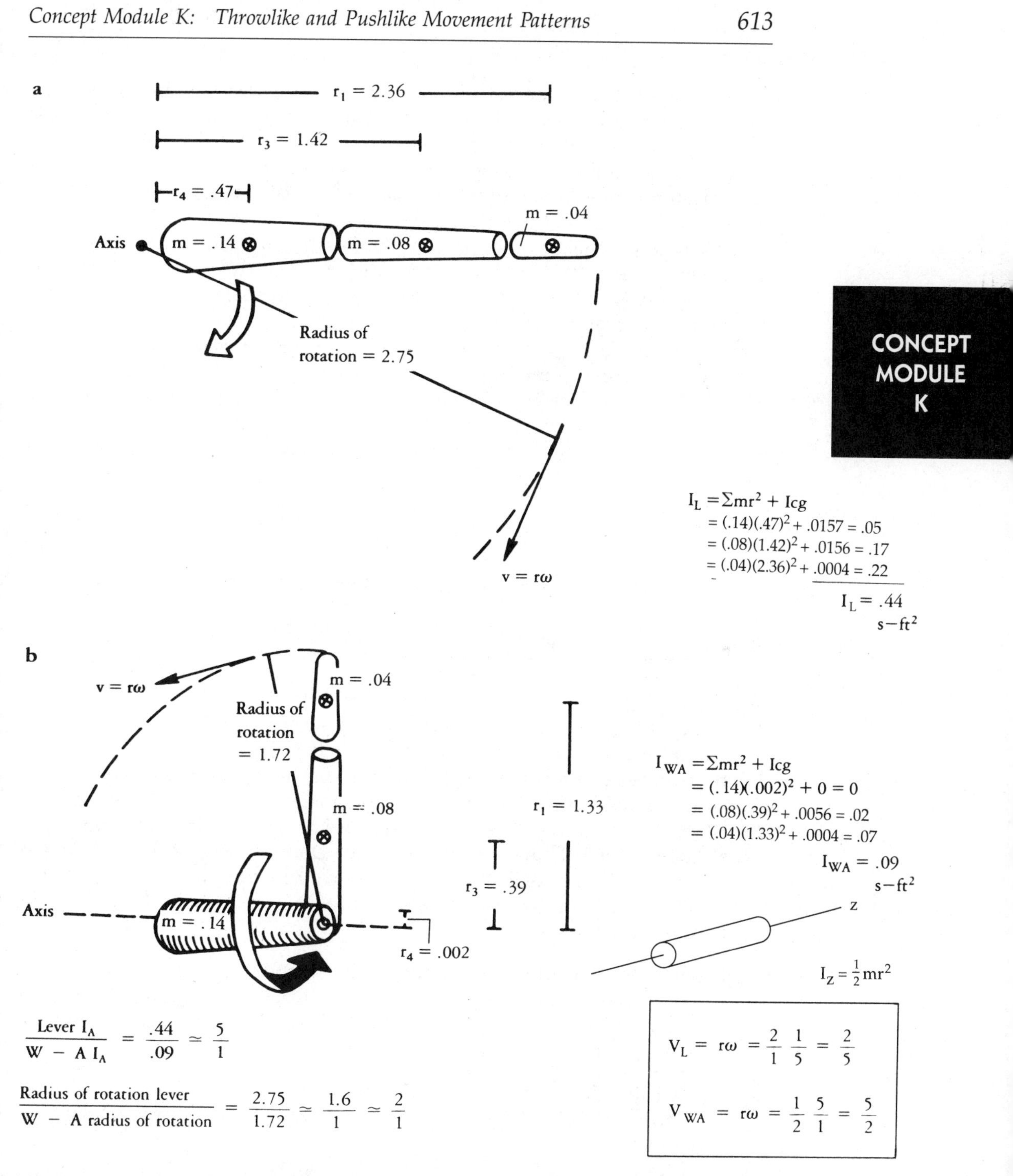

FIGURE K.7

The rotational inertia of a segment in (a) a lever arrangement and (b) in a wheel–axle arrangement. Icg designates the rotational inertia of a segment about its own center of gravity.

of the wheel–axle mechanism allows a five times greater angular acceleration and thus greater angular velocity. The greater acceleration in the wheel–axle system gives this system two and one half times the *linear* velocity of the lever system, although the lever system has a larger radius of rotation. Probably for this reason the wheel–axle mechanism is used prevalently in throwing-striking skills in which high end-point velocity is desired.

The torque-producing capabilities of the muscle groups involved in each of the mechanisms must be considered; that is, the relative strengths and speeds of contraction of the lever-system muscle groups compared to the wheel–axle system muscle groups.

In summary, the lever system has the potential of giving a faster linear velocity to a distal point by virtue of the contact point's greater radius of rotation, but the wheel–axle system has the potential to give considerably more linear velocity to a distal point by virtue of the wheel's smaller rotational inertia, which therefore allows a greater angular acceleration for a given muscular torque.

Understanding Lever and Wheel–Axle Rotations

1. With a badminton racket, assume a grip such that the radius of rotation for forearm pronation is maximized. By changing the grip and the wrist position relative to the forearm, assume four more positions that progressively decrease the radius of rotation for forearm motion. The last position should place the center of the racket's face along the axis of rotation of the radioulnar joint. Draw the five positions. Discuss these positions in terms of the forearm's potential linear velocity contribution to the swing. Which position allows the greatest potential linear velocity, and which allows the least potential linear velocity?

2. Discuss why it is impractical or impossible to position the hand using radial or ulnar flexion such that one could effectively use forearm (radioulnar) motions for added velocity in a baseball pitch, a tennis forehand stroke, or a basketball throw for distance.

3. Discuss the phrase "a longer lever gives more velocity." Why is this statement misleading? In what kind of situations would a longer lever not provide more velocity?

4. List the advantages of using a wheel–axle motion rather than a lever motion when one wants to generate high end-point velocity on an implement.

K.4 Pushlike Patterns: Simultaneous Segmental Rotations

The previous section described the sequence of movements that most effectively produces a throwlike motion for which the overall performance objective is to generate high velocity on the end of the distal link. These movements include throwing, kicking, and striking skills.

For some skills, however, the overall performance objective is maximum accuracy of projection, such as a basketball throw; for others, a large motive force must be applied to an object to overcome the resistive forces acting on that object, such as weight lifting; and for others the body projects itself, as in a takeoff. Further examples of high-accuracy skills are darts, horseshoe pitching, archery, the tennis volley, and the volleyball set. Examples of skills in which the primary purpose is to overcome the resistive forces are those used in weight lifting and wrestling. All jumping activities are examples of pushing the body.

Figure K.8 illustrates the path of the hand in a throwlike motion and the path of the hand in a pushlike motion. The throwlike pattern is used for generating high linear velocity on the distal end of the series of links; therefore, the point of contact of the object on the distal link moves in a *rotary* or a *curvilinear* path. In activities involving accuracy or the overcoming of a resistive force, moving the point of contact with the object in a *rectilinear* path is important. The rectilinear path is a movement that may be likened to a fencing lunge. The segmental rotations are taking place simultaneously to control the hand and to cause it to move in a straight line. The importance of the rectilinear path for high accuracy is shown in Figure K.9.

If an object is traveling in a curved path, as in Figure K.9a, it must be released within a narrow space along the path between points A and C to hit the target.

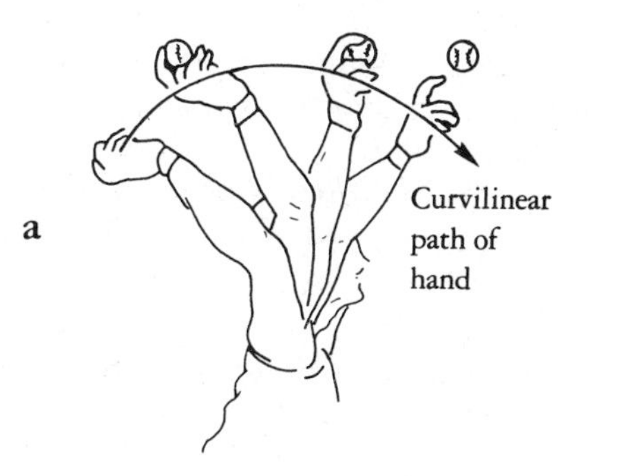

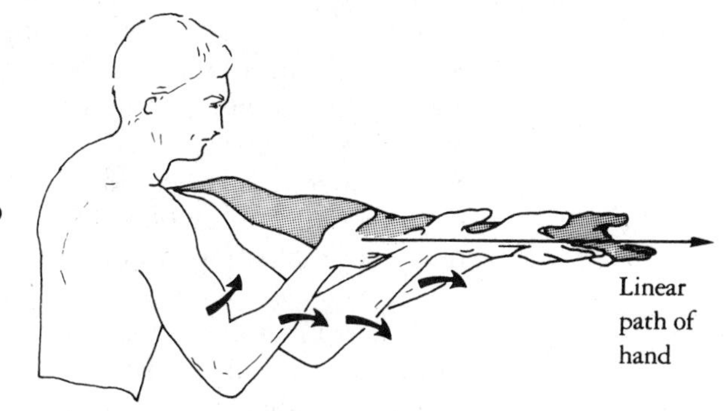

FIGURE K.8
(a) An overarm throw. (b) A fencing lunge.

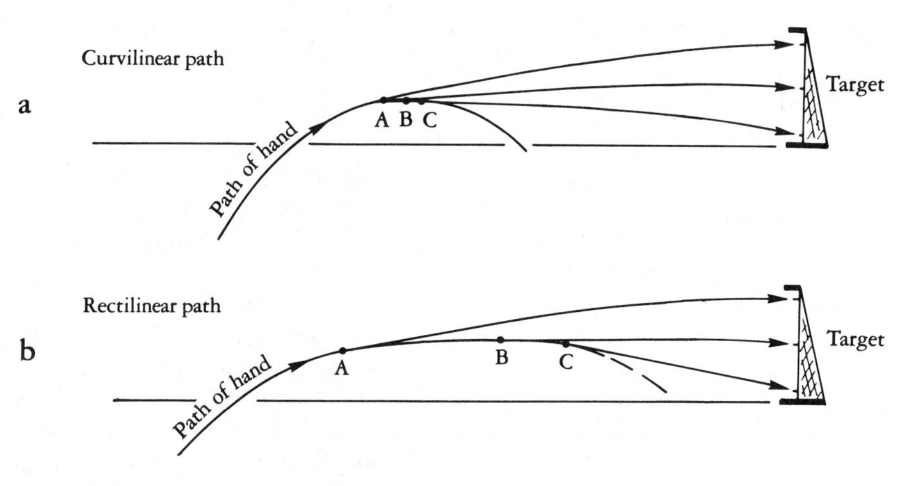

FIGURE K.9

Accuracy of projection at a target when using (a) a curvilinear path and (b) a rectilinear path.

Little time would be available within the total performance for the performer to contact or release the object and be successful. In Figure K.9b, notice that the object could be released within a wide range of space between points A and C and still hit the target. The more linear the path surrounding the contact or release, the more *time* the performer has to project the object accurately. A rectilinear path allows the performer more time to align the release direction with the target than does a curvilinear path. A curvilinear or rotary path requires that the performer release or strike the object along an extremely short section on the curved path so that the tangent is directed to the target.

As a practical example, picture the students in a beginning tennis class learning and practicing the serve. During the learning stages, the students may attempt to reproduce a fairly skilled sequential serving pattern. Knowing that speed as well as accuracy are important in a good serve, the students practice the skill in an attempt to develop both of these components. If the students are required to perform in a test situation, however, in which success depends to a greater extent on accuracy than on speed of serving (e.g., in counting the number of serves landing within the service area out of 10 attempts), then the sequential throwlike striking pattern, designed for greater speed, usually reverts to a pushing pattern, designed for greater accuracy. A beginner's pushlike serve is shown in Figure K.10.

The pushlike pattern of segmental movements is easily distinguished from a throwlike pattern. These four differences are noted:

1. For throwlike movements, the segment–object contact point "lags back" as the proximal segments "move out from under" the distal segment, and eventually the distal end will catch up to the proximal segments at release.

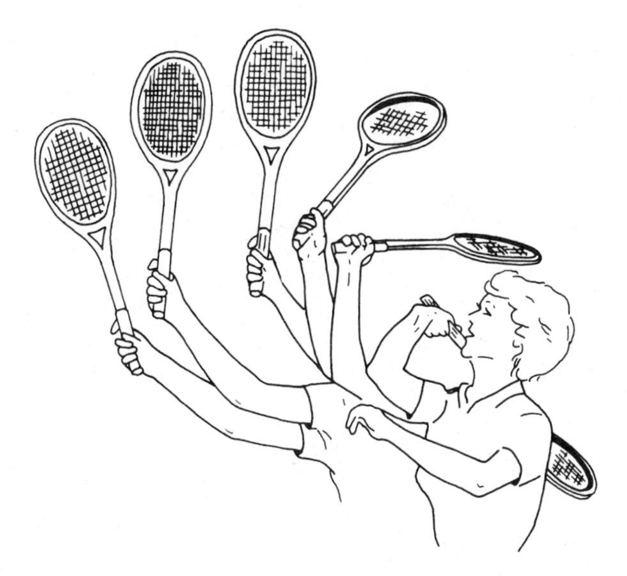

FIGURE K.10
A simultaneous serve, designed for accuracy.

In a pushlike motion, the segments are positioned either behind the object to be projected, such as in projecting darts, or the object or implement is pulled along, such as the oar in rowing.

2. In throwlike movements, segmental rotations occur sequentially to produce high velocity. In pushlike movements, segmental rotations occur simultaneously to produce high accuracy or force or both.
3. In a throwing pattern, the object moves along a curvilinear path or a rotary path before contact or release; whereas in a pushing pattern for accuracy, the object is kept in a rectilinear path before contact or release.
4. Throwlike patterns have a predominance of wheel–axle movements, and pushlike patterns have a predominance of leverlike movements.

A comparison of the throwlike and pushlike patterns may be seen by observing a volleyball spike and a volleyball set pattern in Figure K.11. Figure K.11a illustrates the motion of the hand as the player performs a spiking motion. The motion of the hand is curvilinear; it may be close to purely rotational and involves the sequential rotations of segments. Figure K.11b shows the hands moving in a rectilinear path during the setting skill. Accuracy of placement is important in this skill. The setting pattern, however, involves the *simultaneous* rotations of segments. The set is for accuracy; the spike is for high velocity.

A pushing pattern also is used in situations in which the center of gravity (CG) location of the object being moved must be kept in a precise location for

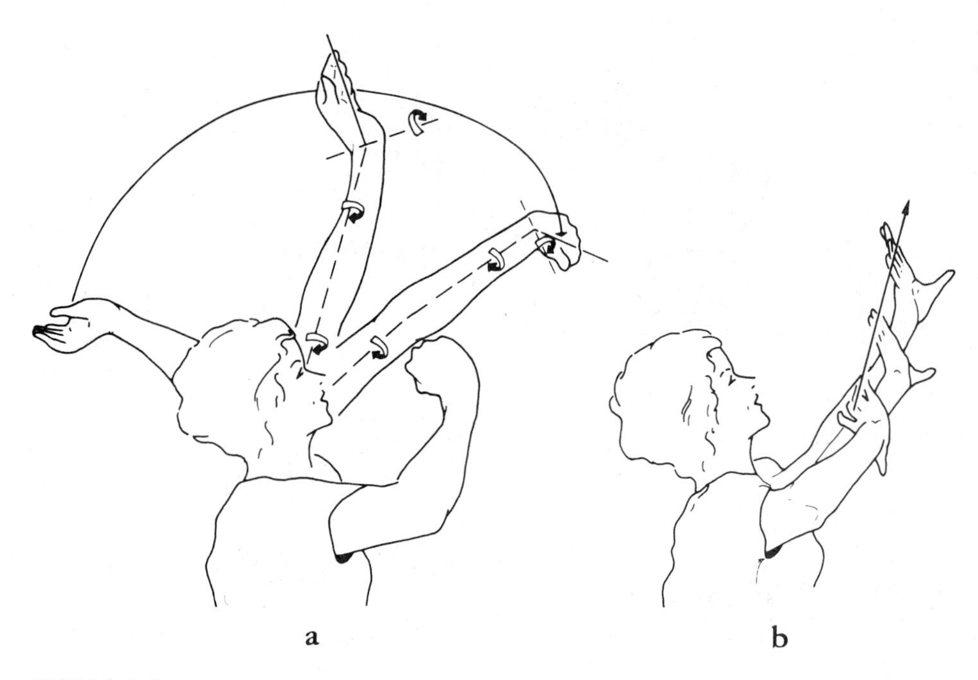

FIGURE K.11

(a) Curvilinear motion and (b) rectilinear motion caused by the spatial and temporal characteristics of the segmental movements.

balance. An example is lifting free weights in weight training. The CG of the barbell must be kept in vertical alignment with the base segments. If the barbell moves horizontally from its vertical rectilinear path, other muscle groups must be used to bring it back into alignment, or it will rotate over and fall to the floor. The use of a weight-lifting machine eliminates the necessity for the performer to maintain the vertical rectilinear direction of the resistance and hence makes the lifting task somewhat easier and safer.

Whether a skill ideally should be a throwlike or a pushlike motion depends primarily on the overall performance objective of the skill; that is, whether an object must be projected at a high velocity (used in projecting for distance), projected or moved for accuracy, or manipulated as a large resistance. Examples of skills in which one projects primarily for distance include throwing the javelin or discus. Skills in which one projects primarily for accuracy are the volleyball set, a horseshoe pitch, and the basketball free throw.

The choice of pattern also depends on the constraints of the activity and the performer, for the ideal pattern may have to be compromised as a result of the physical environment, the unique competitive situation, or the physical attributes of the performer.

K.5 The Throw–Push Continuum

The kinetic link principle applied to the throwlike pattern may be said to govern the performance of events that demand high velocity of the object. Similarly, the pushlike pattern may be said to govern the performance of accuracy events that require precision in the direction of projection. Most sport skills, however, cannot be classified as being purely one form or the other. Therefore, we must consider a set of constraints in the activities that require *blending* of the throwlike and pushlike patterns into a combination of the two. For this purpose, skills can be placed conveniently along a **throw–push continuum**, with skills performed with entirely sequential rotations on one end of the continuum and skills performed with entirely simultaneous rotations on the opposite end of the continuum. In between are skills in which a sequential throwlike pattern occurs in the first part of the performance and a simultaneous pushlike pattern occurs in the latter part of the performance. Keep in mind that the throwlike patterns are characterized by the sequencing of the segmental rotations from most massive to least massive and from most stable to most free; pushing patterns are characterized by the simultaneity of segmental rotations (i.e., all rotating at the same time to produce a straight line motion of the distal link).

These are the constraints that influence the location of a skill on the continuum:

1. The massiveness of the object to be moved or projected (shot or basketball)
2. The size of the object to be moved or projected (balloon ball or volleyball)
3. The shape of the object (football or waterpolo ball)
4. The strength of the performer (conditioned athlete or intramural participant)
5. The skill of the performer (beginner or advanced)

To apply this concept, we will work through several examples. If high speed of the object is important for achieving distance of projection, then the most sequential pattern possible should be used. For a skilled player to throw a baseball from the outfield using sequential rotations entirely is both possible and desirable. In the shot put, however, in spite of the fact that the performer works to achieve the same overall performance objective as the baseball thrower (i.e., distance of projection), the shot would not be thrown because of distal musculoskeletal limitations in manipulating a massive object. One should see the sequential rotations of the massive segments (the pelvis, trunk, and shoulder girdle) followed by simultaneous rotations of the less massive distal segments of the upper extremity. The shot put therefore is a combination activity that is placed in the middle of the continuum; it is partly sequential, partly simultaneous. It is an activity in which constraints on the performer demand the use of less than the ideal pattern.

An ideal pattern for projecting the shot is a throw because the objective is that of projecting for distance, which depends mainly on release velocity. The constraint in the case of the shot put is the large mass of the object to be projected. A similar adaptation is necessary if the strength of the performer is decreased. A weak performer attempting to put the shot may use all simultaneous motions without any evident sequencing. Thus, an immature or weak performer may turn an otherwise sequential movement into a more simultaneous one by ordering the initiation of the movements of the segmental links at the same or nearly the same time.

Second and third constraints are the size and shape of the object to be thrown. A javelin and a football, because of their somewhat unwieldy shapes, are thrown for distance with a small component of a push at the distal end, the javelin because of the importance of the placement of the tip and the football because of the possibility of wobble.

One identifying characteristic of a throwing pattern that has been adapted to a partial pushing pattern owing to the previously listed constraints is that the object to be projected is brought in closer to the longitudinal body axis because the rotational inertia of the distal segment and object is too great for the torques produced by the smaller muscles to stabilize or accelerate. As the mass is brought closer to the body, the *sequential* segmental rotations become more *simultaneous* segmental rotations and combine to move the distal end of the kinetic chain in a rectilinear rather than a curvilinear path before release. When a less massive object is used, it can be held farther from the midline of the trunk as it is being accelerated. Figure K.12 illustrates the patterns used for projecting three objects of varying mass and size.

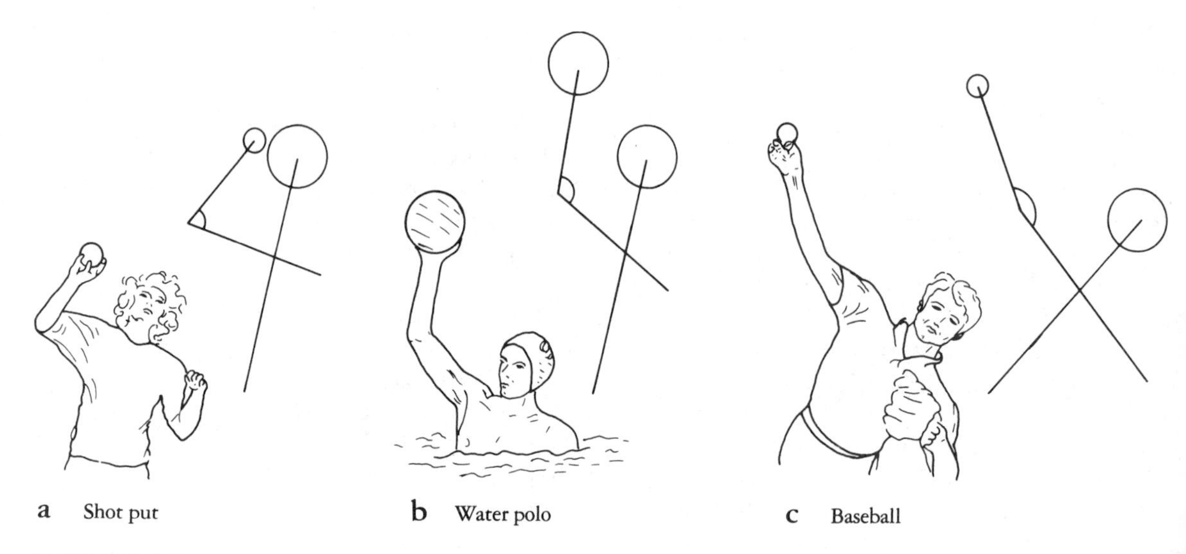

FIGURE K.12

Relative positions of body segments and objects for three different masses and sizes of objects being projected.

Understanding Pushlike Patterns

1. List the primary mechanical purposes associated with pushlike patterns. For each of the mechanical purposes, list five skills that have that purpose.
2. What four factors distinguish a pushlike pattern from a throwlike pattern?
3. List five constraints of the activity, equipment, or performers that would have the effect of putting a pushlike element into an otherwise ideal throwlike pattern.
4. For each of the three paths shown in Figure K.13, draw an arrow to indicate the direction that an object moving along the path would travel if released at point 1, 2, or 3.

   a. Which path is rotary?

   b. In which path is the timing of release the most critical to the accuracy of the projectile?

   c. In which path is the timing of release the least critical?

   d. Draw these paths:

      i. The path of the head of a golf club

      ii. The path of the head of a tennis racket during a forehand drive

      iii. The path of the CG of the body just before and after leaving the horizontal bar during a dismount

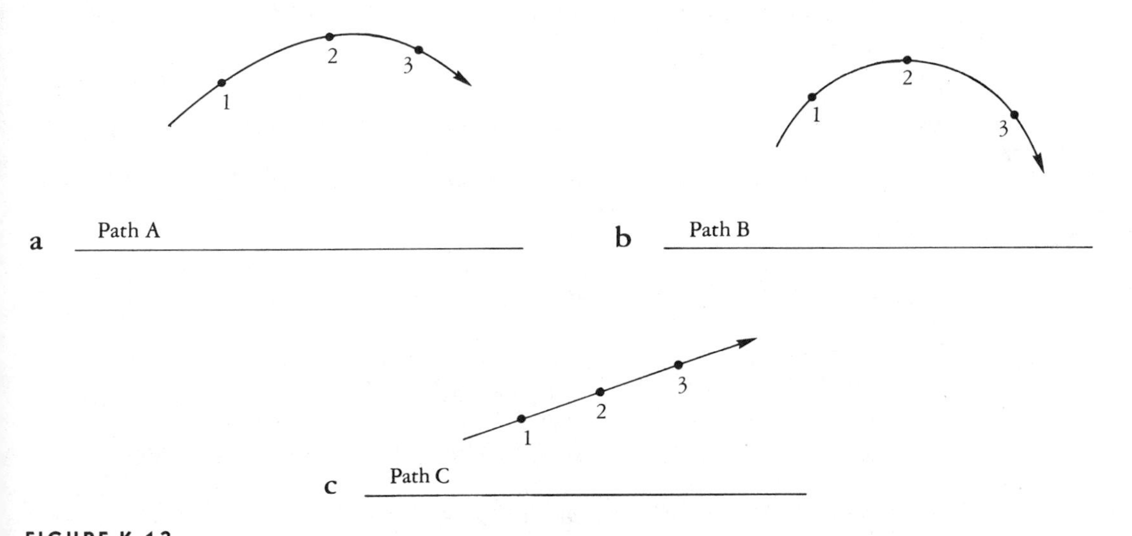

FIGURE K.13
Three trajectories.

iv. The path of the discus before release

v. The path of a basketball during a jump shot, before release

5. Within the following groups of activities, (a) list several examples of skills within each group, (b) list the sequence of the body segment movements used, (c) determine what influence an object's size or weight or both has on the segments used and on the sequence of the segments used, and (d) list the similarities and differences among individual skills within groups.

   Group I: Overarm striking skills in handball, volleyball, and all racket sp

   Group II: Overarm throwing or pushing skills for distance

   Group III: Kicking

   Group IV: Throwing underarm for accuracy

   Group V: Implements used in an underarm pattern to strike

   Group VI: Sidearm skills to throw or strike

6. Place all the skills listed in question 5 on a throw–push continuum from most throwlike to most pushlike. What effect does the object's size or mass or both have on the place at which the skill is located on the continuum?
7. List three skills that are not throws per se but that use the upper extremity in a throwlike pattern.
8. List three skills that use the upper or lower extremity in a pushlike pattern that are not mentioned as examples in the text.

References and Suggested Readings

Atwater, A. E. (1970, April). Overarm throw patterns: A kinematographic analysis. Paper presented at the AAHPERD Convention, Seattle, WA.

Atwater, A. E. (1977, January). Biomechanics of throwing: Correction of common misconceptions. Paper presented at the Joint Meeting of NAPECM and NAPECW, Orlando, FL.

Atwater, A. E. (1980). Biomechanics of overarm throwing movements and of throwing injuries. In R. S. Hutton & D. I. Miller (Eds.), *Exercise and sport science reviews*, (Vol. 7, pp. 43–85). Franklin Institute Press.

Bober, T., Putnam, C. A., & Woodworth, G. G. (1987). Factors influencing the angular velocity of a human limb segment. *Journal of Biomechanics, 20*, 511–522.

Daish, C. B. (1972). *Learn science through ball games*. New York: Sterling.

Dyson, G. (1978). *The mechanics of athletics* (7th ed.). New York: Holmes & Meier.

Lindner, E. (1971). The phenomenon of the freedom of the lateral deviation in throwing. In J. Vredenbregt & J. Wartenweiler (Eds.), *Biomechanics II: Proceedings of the Second International Seminar on Biomechanics* (pp. 240–245). Baltimore: University Park Press.

Morehouse, L. E. & Cooper, J. M. (1950). *Kinesiology*. London: Kinipton.

Phillips. S. J., Roberts, E. M., & Huang, T. C. (1983). Quantification of intersegmental reactions during rapid swing motions. *Journal of Biomechanics, 16*(6), 411–417.

Putman, C. A. (1983). Interaction between segments during a kicking motion. In M. Matsui & K. Kobayashi (Eds.), *Biomechanics VIII-B: Proceedings of the Eighth International Congress on Biomechanics* (pp. 688–694). Champaign, IL: Human Kinetics Publishers.

Roberts, E. M., Zernicke, R. F., Youm, Y., & Huang, T. C. (1974). Kinetic parameters of kicking. In R. C. Nelson & C. A. Morehouse (Eds.), *Biomechanics IV: Proceedings of the Fourth International Seminar on Biomechanics* (pp. 157–162). Baltimore: University Park Press.

Roberts, E. M., & Phillips, S. J. (1977). Angular acceleration in free segment motion. In C. Dillman & R. Sears (Eds.), *Kinesiology: (Proceedings of) A National Conference on Teaching* (pp. 328–334). Champaign-Urbana, IL: University of Illinois.

Widule, C. J. (1983, May 24). Contiguous segment velocity: Acceleration characteristics. Personal Communication.

Zernicke, R., & Roberts, E. (1978). Lower extremity forces and torques during systematic variation of nonweight bearing motion. *Medicine and Science in Sports, 10*(1), 21–26.

CHAPTER 15

Performance Analysis of Throwlike Movements

PREREQUISITES

Concept Modules A, B, E, I, J, K
Chapters 1, 3

CHAPTER CONTENTS

The kinetic link principle applies to numerous skills. An application of the kinetic link concept may serve the following purposes for the teacher or coach:

1. Understand and analyze a particular skill.
2. Help recognize performance ability levels.
3. Help gain a basic understanding of the sim-

ilarities and differences of skills that use the same general movement pattern.

4. Gain a better understanding of the developmental patterns that are used by children as they progress through stages of learning to a mature form of a movement pattern.

15.1 Biomechanics of Throwlike Patterns

In previous sections, throwlike skills are defined as those in which the mechanical purpose is to develop high linear velocity on the end or contact point of the segmental link system. These skills may or may not include the use of an implement such as a bat or a racket. If an implement is used, it serves as an additional link in the body's link system.

The specific links used and the motions of each link in the system depend on which skill is being performed. For example, major differences occur in the segmental movements used in performing a golf swing and baseball batting or between overarm throwing and a tennis backhand drive. Slight differences occur in more similar skills, those that belong to a common pattern. Specific skills within the overarm, sidearm, or underarm patterns are slightly different. To better understand how one can begin to analyze a pattern, we can examine an overarm throwing pattern. Figure 15.1 illustrates the sequence of body positions used in this pattern.

With the first step forward, the thrower begins to initiate the end of the windup. The end of the windup is first noticed when the base segments, the pelvis and the trunk, begin to move forward. The distal segments continue to lag back until their turn to come forward. In keeping with the ordering of the segments, the movements proceed from base to free end, from proximal to distal, and from most massive to least massive.

Six segments are *in a position* to be active during the throw. They are the pelvis, trunk, shoulder girdle, arm, forearm, and hand. Each has its own movements relative to its own proximal articulation. These articulations are the hip, intervertebral, sternoclavicular, shoulder, elbow, radioulnar, and wrist joints. More than one movement *could* occur in several of these articulations, and exactly which ones are used depends primarily on which specific sport skill is being performed. Because we are examining the general pattern of overarm throwlike movements, we will consider the movements common to most overarm patterns.

As the leading heel approaches the ground, the pelvis begins its medial rotation about longitudinal axis of the hip joint. The next distal segment is the trunk, which moves in transverse rotation about a longitudinal axis through the vertebral column. The distal segments and their possible movements are the protraction of the shoulder girdle about the longitudinal axis through the sternoclavicular joint; the transverse adduction of the arm about an anteroposterior (AP) and mediolateral (ML) oblique axis through the shoulder joint; the medial rotation of the arm about the longitudinal axis through the shoulder joint; the extension of the forearm about an ML axis through the elbow joint; the pronation of the forearm about the longitudinal axis through the radioulnar joint; and finally, the flexion and/or ulnar flexion of the hand about an ML axis or AP axis, respectively, through the wrist joint.

Such a listing provides two types of information to the teacher or coach and performer: (1) the segments, the movements, and the muscle groups that *may be* used to produce an overarm pattern, and (2) the ordering or sequencing of the movements used in an overarm pattern. After a general list has been made, the teacher or coach adapts this list to the specific skill that is being analyzed.

The kinematic and kinetic descriptions of the actual sport skill being analyzed depend on two other variables: the rotational inertia of the group of segments that rotates at any given time and the radius of rotation of the contact point relative to each of the axes of rotation. The rotational inertia for each combination of segments is difficult to assess; however, a general idea can be gained if the following are considered:

1. The greater the rotational inertia (mass times radius of gyration squared), the smaller the angular acceleration for a given muscle torque.
   a. The base segmental muscle torque must accelerate the mass of all the segments distal to its axis of rotation and therefore must accelerate a system with a greater rotational inertia.
   b. The distal segmental masses are located farther from the proximal axes of rotation being used, and therefore the radius of gyration for proximal segmental rotations is larger.
2. As the sequence of movements progresses to the distal axes, the rotational inertia of the system formed by the remaining rotating segments becomes less.

FIGURE 15.1

A baseball pitcher, showing the step and the movements of the hip, vertebral column, shoulder girdle,shoulder joint, forearm, and wrist.

a. The axis of rotation for each successive distal motion is located farther out on the link system, hence it moves closer and closer to the end.

b. The mass of each successive segment is progressively smaller.

c. The radius of gyration for a segment moving in a wheel–axle fashion around a longitudinal axis is smaller than that for a leverlike movement of the same segment (Module K).

Consequently, the link system tends to move faster as the movements proceed to the distal end of the link system, resulting from the continual reduction of its rotational inertia.

A second variable in the system is the radius of rotation (r) used in calculating the linear velocity of a point on a rotating system ($v = r\omega$). Figure 15.2 illustrates how to visualize the radius of rotation for several possible segmental rotations for an overarm pattern.

Note the larger radii of rotation for the base segments and the smaller radii of rotation for the proximal segments. Keep in mind that each axis of rotation is associated with its own radius of rotation. The repositioning of segments distal to the axis of movement will increase or decrease the radius of rotation for that axis. For example, the extension of the elbow during the overarm throwing skill increases the radius of rotation relative to the hip axis, the vertebral axis, and the sternoclavicular axis, but does not change the radius of rotation relative to the elbow axis itself.

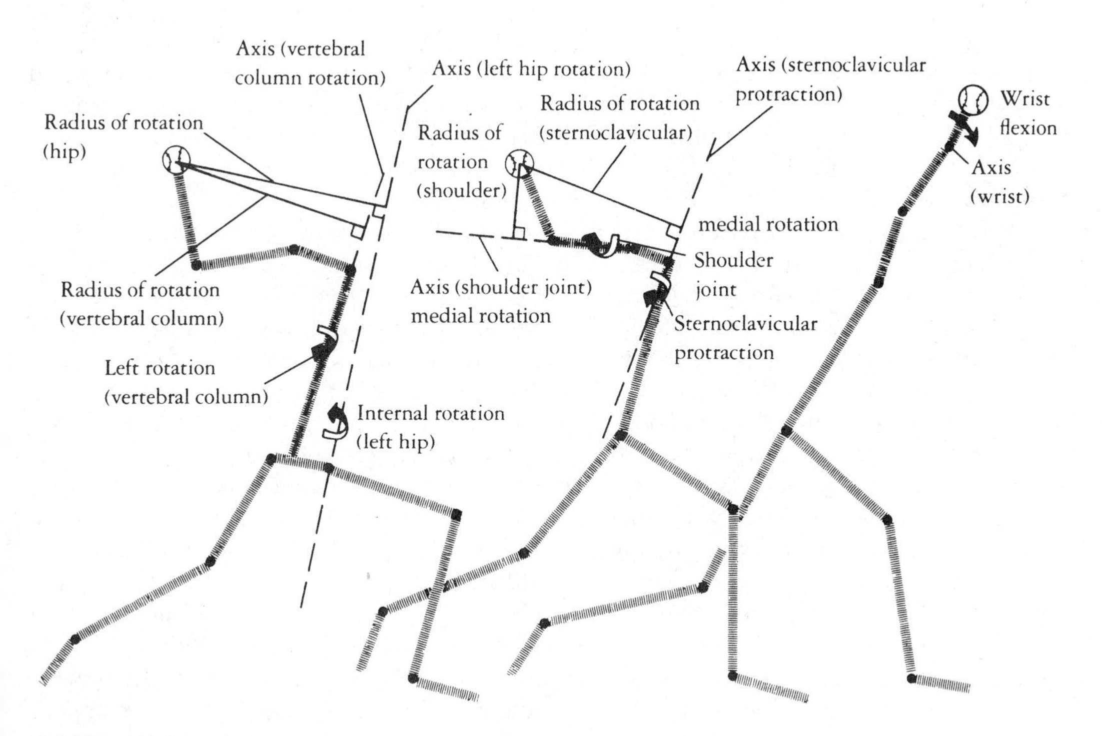

FIGURE 15.2
Axes of segmental rotations used in pitching a baseball.

15.2 Analysis of Sport Skills Using the Kinetic Link Principle

Baseball Pitch

The segmental movements described by researchers who have observed high-speed films of skilled baseball pitchers are consistent with the kinetic link model. That the wheel–axlelike medial rotation of the pelvis about the hip joint axis initiates the forward action of the pitch after the stride foot plant is fairly well established. The pelvic movement continues as the trunk begins transverse rotation about the vertebral column. During this time, the shoulder joint is abducted to 90 degrees, and the elbow is flexed to 90 degrees. These latter two movements serve to position the ball in space in preparation for the second part of the throwing motion rather than to contribute any forward velocity to the ball.

The transverse rotation of the trunk is associated with extreme lateral rotation of the shoulder joint as the upper extremity and ball lag back, which is typical of the throwlike pattern. The second part of the pitching motion consists of continuing trunk transverse rotation, accelerated elbow extension, accelerated medial rotation of the shoulder joint, radioulnar pronation, and wrist flexion. Note that wrist flexion occurs from hyperextension of the wrist to full extension of the wrist; the wrist does not "snap forward" into flexion at any time before release. The sequencing of these motions follows the kinetic link principle (i.e., from proximal to distal with the exception of elbow extension occurring before shoulder joint medial rotation).

Researchers have basically ignored the protraction of the shoulder girdle in studies of the kinematics of pitching. The motion is difficult to measure in that it is indistinguishable from trunk transverse rotation on videos and films regardless of the filming speed. Electromyography has provided useful information on the muscle action surrounding the shoulder girdle complex. Gowan et al. (1987) mention that the protraction of the shoulder girdle about the sternoclavicular axis occurs during maximal lateral rotation and before the medial rotation of the humerus at the shoulder joint and extension of the elbow. The serratus anterior has been found to be integral to the overarm throwing motion. It is unclear whether this muscle is acting in eccentric tension to decelerate the retraction of the shoulder girdle, is acting in static tension to stabilize the shoulder girdle to provide a stable base on which the shoulder joint muscles pull, or is acting in concentric tension, pulling the shoulder girdle into protraction. The serratus anterior muscle, however, may function in all three ways and thus would provide another link in the body's kinetic link model (Gowan et al., 1987; Jobe et al., 1984; Pappas, Zawacki, and Sullivan, 1985).

The medial rotation of the shoulder joint begins from extreme lateral rotation and is noticeable only during a brief time period prior to release. Feltner and Dapena, 1986, reported the humeral medial rotation as taking place in 0.005 sec). Recall that medial rotation is a wheel–axle type of movement and thus has the advantage of a relatively small rotational inertia. Because the elbow is not *fully* extended at release, a small radius of rotation is associated with the medial rotation motion of the shoulder joint.

The third segmental movement listed in the general order of upper extremity throwlike movements is shoulder joint transverse adduction. This movement does not appear in skilled pitchers throwing at near maximum speeds, but it has been reported as a part of the throwing movements of the unskilled. As will be discussed in a future section, placing the elbow joint ahead of a line connecting the shoulder joints before release is characteristic of an unskilled overarm pattern.

Because the hand and ball are closely aligned with the axis of rotation for radioulnar pronation, the radioulnar motion has no radius of rotation unless the type of pitch calls for holding the ball off axis by using radial or ulnar flexion. Thus, the motion usually serves to position one hand on the ball at release and influences the amount of spin imparted to it. For producing a fastball, the palm is facing forward at release with the forearm pronated. For a slider, the forearm is not pronated quite as much as in delivering a fastball. The curveball delivery is somewhat surprising. For years the curveball was associated with supination of the forearm before release. High-speed films disclaimed this idea, showing that the forearm is in midposition and that the fingers roll off the *forward outer* part of the ball during release in a motion of ulnar flexion of the wrist (Atwater 1980; Elliott, Marsh, and Blanksby, 1986). The forearm then pronates and the shoulder joint medially rotates throughout the follow-through. This sequence of motions produces a curve that has top spin and "breaks" downward and outward. A more detailed discussion of the deflections of a spinning ball may be found in Chapter 11.

The amount and location of muscle torques within the throwing motion have been studied by several researchers in sports medicine in attempts to explain rotator cuff injuries to the shoulder joint (Gowan et al., 1987; Jobe et al., 1984). In addition, a particularly thorough discussion of the relationships between the kinematics of segmental movements and the kinetics of muscle torques is given by Feltner and Dapena (1986). The action of the latissimus dorsi, teres major, and pectoralis muscles actively decelerate the lateral rotation of the shoulder and accelerate the medial rotation.

Forearm extension occurs before the medial rotation of the humerus. The elbow ceases extension at approximately 140–160 degrees of elbow angle. This move, however, results in a small radius of gyration for the shoulder joint medial rotation and probably allows the medial rotation to show great acceleration over a very short period of time. By the time elbow extension occurs, the link system has generated a considerable amount of angular velocity. Whether the elbow extensor muscles can shorten fast enough to keep up with or surpass the acceleration of the extending forearm to cause an increase in its angular velocity is unknown.

Studies have shown that the forearm extends considerably faster when preceded by proximal segmental rotations than it can extend without those rotations using only muscle extensor torque. Further support of this concept is given by Feltner and Dapena (1986). They reported very large angular velocities for elbow extension with very small elbow extension muscle torques. "This strongly suggests that the extension of the elbow is not due primarily to the action of the triceps but to the resultant joint force exerted by the upper arm on the forearm at the elbow" (p. 256). Electromyograph recordings substantiate the minimal activity of the triceps during elbow extension.

Finally, two researchers report percentage contributions of various segmental motions to pitching velocity (Elliott et al., 1986; Vaughn, 1985). Vaughn (1985) studied medial rotation of the shoulder joint and elbow extension as they contributed to wrist velocity. He found that the humeral rotation contributed approximately 54% and elbow extension contributed approximately 17.5% to wrist forward velocity. Elliott et al. (1986) reported that the hand itself moving at the wrist joint contributed 26.5% to ball velocity in a fast-ball pitch with the forearm moving at the elbow, contributing 35%; the humerus moving at the shoulder, 8.8%;

and the trunk moving around the vertebral column, 19.4%. This latter percentage may include shoulder girdle protraction as well as trunk rotation.

Javelin Throw

The segmental rotations used to project a javelin are similar to those used in pitching a baseball. Any slight differences in position are due to the angle of projection at which the javelin is thrown, which is estimated to be ideal at approximately 35 degrees above the horizontal. The body's position at release is surprisingly similar to that of throwing a baseball. The sequencing of the segmental rotations is the same; the proximal segments accelerate and then decelerate as the distal segments accelerate, resulting in maximum velocity of the hand at release.

Because a javelin is heavier than a baseball, one would anticipate the necessity of greater muscle torque involvement in projecting a javelin than in pitching a baseball. The release velocity of the javelin is approximately two thirds that of a baseball pitch, however. The lagging back of the distal segments places the distal segmental muscles on stretch, and consequently, these muscles, through the use of the stretch reflex and their elastic properties, may apply some accelerating muscle torque to the distal segments, which may not be possible in the baseball throw because of the speeds at which they are moving.

Menzel (1987) in studying good and poor javelin throwers found that the velocity of release is associated with a reduction of velocity of the hip and elbow and that the reduction of the velocity of the elbow depends on the maximum velocity of the shoulder. The velocity of release and the reduction of hip, shoulder, and elbow velocity were significantly higher for the good throwers. The good throwers also reduce the speed of the hip earlier. Menzel (1987) states that these data confirm that distal segments reach their maximum velocities after a more proximal segment. Supporting this theory, Komi and Mero (1985) found that high elbow extension velocity at javelin release was associated with poorer performances.

Football Pass

A football is heavier than a baseball but lighter than a javelin. The shape of a football requires that the nose of the ball be projected forward and that as little wobble as possible be imparted. Throwing a long forward pass with a football is close to the same overarm throwing pattern as the baseball throw in terms of segmental rotations and sequencing. The elbow is not extended quite as much at release, however, and this may be due to the necessity of imparting spin to the football while holding its nose forward. The last part of the throw resembles a slight pushlike motion to prevent ball wobble. More transverse adduction of the humerus allows elbow extension to contribute to ball speed, while ulnar flexion at the wrist along with medial rotation of the shoulder joints imparts spin to the ball as it rolls off the fingertips. Because the hand cannot grasp a football in the same manner as a baseball, the hand cannot easily hold a football in the lag-back position.

Badminton Smash

The general overarm throwing pattern also is used in performing overarm skills with the use of implements. The pelvic rotation initiates the forward motion while the next distal segment lags back; intervertebral transverse rotation follows. Gowitzke and Wadell (1979b) state that "it gave the impression of the body 'moving out from under the arm' with the hand staying almost motionless in space behind the head" (p. 6).

The trunk rotation is followed in sequential order by elbow extension, medial rotation of the shoulder joint, and forearm pronation. As was reported for baseball pitchers, elbow extension, a more distal motion, begins before a more proximal motion at the shoulder.

In the badminton smash, the elbow extension possibly serves to position the racquet head in space rather than to provide high velocity. Elbow extension and transverse rotation of the trunk *cease* before contact with the shuttle. Gowitzke and Waddell (1979a) report that forearm pronation and shoulder joint medial rotation continue to occur through contact and follow-through. These wheel–axlelike motions provide fast racquet-head velocity. Forearm pronation displays a greater range of motion (ROM) than shoulder medial rotation.

In the badminton smash performed by elite players, the hand movement at the wrist joint did not appear to contribute any velocity to the racquet head, because the hand either was held in radial flexion through contact or was moved from radial flexion to a neutral position before contact. Radial flexion was observed again after contact. Figure 15.3 illustrates an advanced pattern and a beginner's pattern for performing the overarm smash. Note the excessive wrist flexion occurring in the beginning performer (Fig. 15.3a). This common error is not surprising when one recalls the prevalent but misleading teaching cue, "snap the wrist." For achieving a high end-point velocity with a relatively light implement and object, the speedy wheel–axle movements of medial rotation and pronation are most effective (Fig. 15.36). Elite badminton

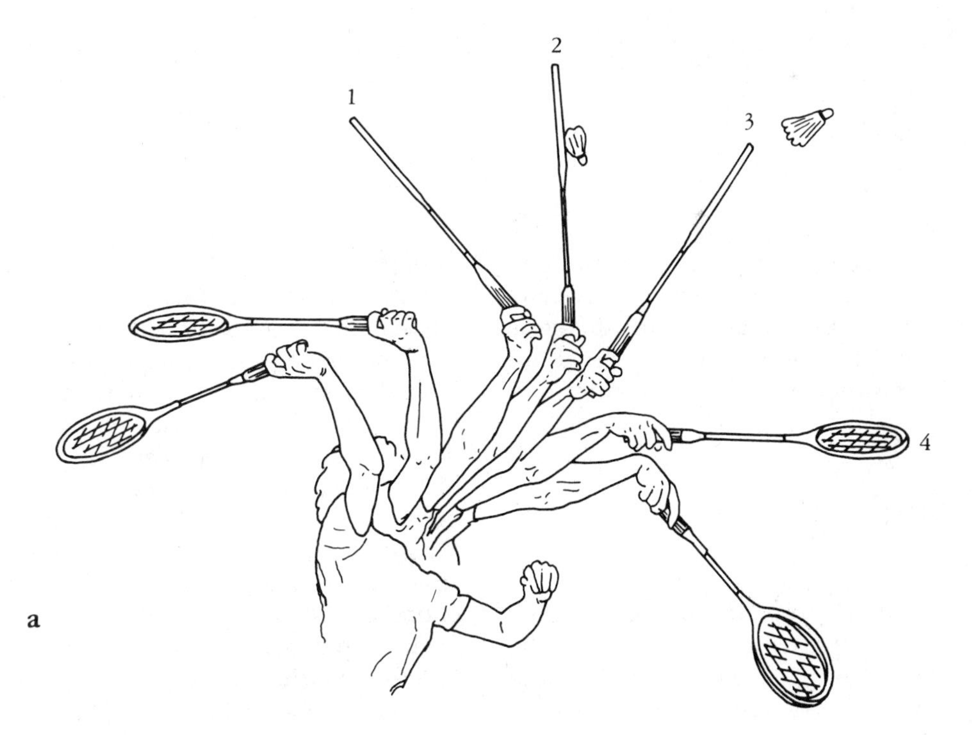

FIGURE 15.3

(a) A beginning and (b) an advanced badminton player, demonstrating the forearm and wrist contributions to a smash.

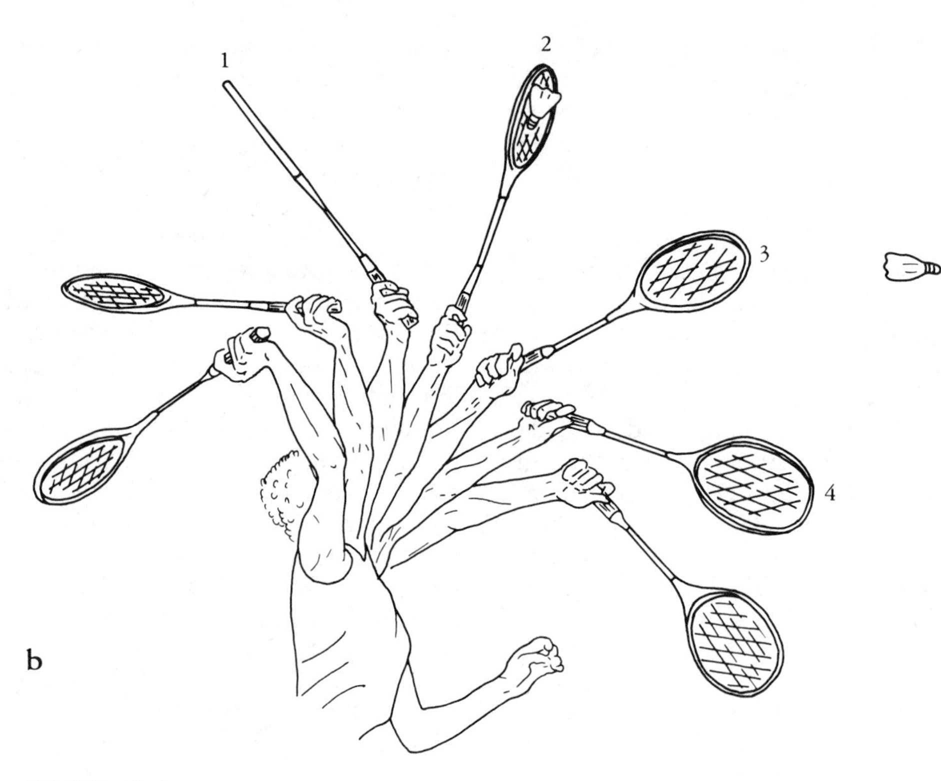

FIGURE 15.3
(continued)

players reposition the racquet face through the use of radial flexion, which serves to increase the radius of rotation and therefore the linear velocity of the contact point for these two movements. (Recall Figure K.6, p. 612.)

Golf Swing

The underarm pattern of a golf drive or long iron shot incorporates the kinetic link principle. The downswing is initiated by the base segments (rotation of the pelvis about the hip joint and the trunk about the vertebral column), followed by the underarm pattern of the upper extremities across the body. An investigation into the generation of high club head velocity was conducted by Milburn (1982). The researcher concluded that the greatest acceleration of the club head occurred when a delay wrist action allowed the arm acceleration to take place as long as possible, that is, allowed the arm to reach a maximum velocity before the action of the wrists. The wrist motion, when it did occur, acted to decelerate the arm motion; thus, an early "uncocking" of the wrists would serve to decrease the angular motion of the entire system.

The rapid acceleration of the club head, Milburn concluded, was due to the wrist's high acceleration, which was "summed with the pushing backwards of the arm about the shoulder to bring about the greatest angular velocity of the distal segment" (p. 63).

The Kick

The kicking pattern is a throwlike pattern in which the proximal segments initiate the movement forward, causing a lagging back of the distal foot segment (Figure 15.4). The knee extensor muscles and surrounding tissues prevent the leg and foot from lagging back too far. The pattern of motion is consistent with the link principle. The acceleration of the thigh flexion about the hip joint should take place as long as possible because the acceleration of knee extension tends to decelerate the thigh segment. Thus, the longer the thigh can accelerate, the greater the angular velocity of the foot when it reaches the ball. As the leg accelerates, the thigh either begins to decelerate or its acceleration is reduced. The knee becomes the new axis of rotation for the leg and foot motion. Thus, the rotational inertia of those segments is effectively reduced, and their angular velocity is increased.

Robertson amd Mosher (1985) concluded in studying lower extremity muscles during soccer kicking that the hip muscles are the most important muscles because they supply 90% of the work done by the leg muscles. The hip muscles were "responsible for both the thigh's motion and the knee's extension" (p. 537). These researchers did not find any knee extensor activity just before ball contact, and in fact, knee flexor torques dominated. This finding was supported also by Bollens, De Proft, and Clarys (1987). The dominant knee flexor torques before ball contact were speculated to be either a hyperextension (injury) prevention tactic or that " . . . the knee is extending so rapidly that the knee extensors are incapable by [sic] exerting a forceful contraction due to

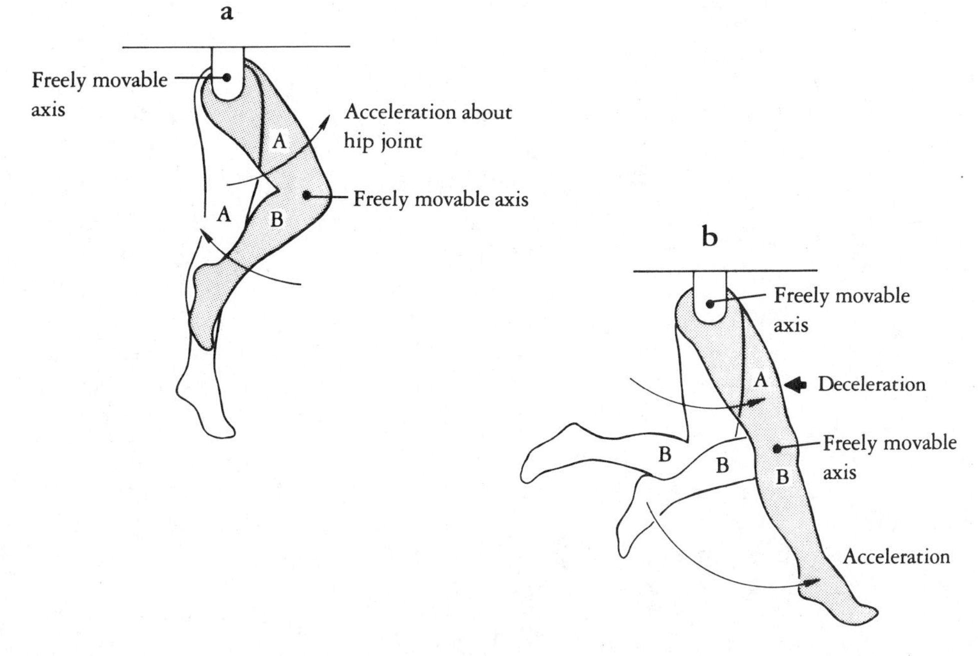

FIGURE 15.4
The effects of acceleration and deceleration of segments in a two-segment model. (a) Acceleration of segment A counterclockwise and acceleration of segment B clockwise. (b) Deceleration of segment A and acceleration of segment B.

their inherent force–velocity limitations" (p. 537). This conclusion supports the hypothesis in this text that while effectively applying the kinetic link principle, the distal segments are traveling so fast that the muscles cannot keep up, let alone accelerate the segment further.

Putnam (1987) compared variables in a punt and a place kick. She found no significant difference in ankle speed at ball contact, the thigh rotated more slowly in the place kick, and increased leg moments were required for the same leg angular acceleration in the place kick. These last two factors mean that if the thigh rotates more slowly, greater knee muscle moments are required to produce the same angular acceleration of the leg. Similarly, if greater angular acceleration of the thigh is wanted, then resultant hip joint muscle moments are called on to do this. Clearly, the kinematic variables of one segment influence another segment in the link system. The exact effects are still unclear. As is stated by Dunn and Putnam (1987), "Although peak thigh angular velocity and the initiation of knee extension do not occur at the same time, they appear to be related regardless of the speed at which the movement is performed" (p. 157). "There is a temporal relationship among the start of forward shank angular velocity, time of maximum thigh velocity, time of knee extension and ball impact" (p. 157).

15.3 Comparisons of Similar Skills Within the Same Pattern

Whenever a comparison is to be made of different sport skills within a similar pattern, the skills can be placed on a continuum according to the purposes of the projection and the relative sizes and masses of the objects and implements used. These general principles govern the differences that occur within similar patterns:

1. The smaller and lighter the object, the greater the number of segments that may be used.
   a. Heavy-object projections rely on the larger muscle groups surrounding the hip and vertebral column to form a movement base for the remaining segments. For example, one would not try to throw an object as heavy as a medicine ball by relying solely on radioulnar, wrist, or finger movements.
   b. Some precision of movement used for projection is lost by using the larger groups. Accuracy skills tend to use the smaller muscle groups.
2. The smaller and lighter the object, the greater the radius of rotation that may be used for a given segmental movement. For example, the forward motion of the shot put is initiated with the shot carried close to the body, and the forward motion of the baseball pitch is initiated with the ball carried out to the side of the body with an abducted and laterally rotated shoulder.
   a. Heavy objects tend to be projected with a motion that is initiated and carried out close to the midline of the body.
   b. Lighter objects may be thrown or struck along a more curvilinear path that may be characterized by a lengthening of the radii of rotation of the links used. Increasing the radius of rotation also

increases the rotational inertia of the distal links around the proximal axis. This increase in rotational inertia resists the desired acceleration and thus increases the muscle torque necessary to maintain or increase the angular velocity of the object. Thus, it may be effective only with lighter implements and objects.

3. The greater the mass of the object, the greater the contribution made by the larger segments (e.g., segments rotating about the hip, vertebral column, and sternoclavicular articulations). In the badminton smash, the medial rotation of the shoulder joint and forearm pronation are noticeable because the racquet is light enough to be manipulated by the smaller muscle groups. The thrower in water polo, however, has a problem in trying to grasp the ball while throwing it. To accomplish this, the basic pattern must be adjusted.
4. When an implement such as a tennis racquet or badminton racquet is used, the contact point (the face of the racquet) can be positioned eccentric to the longitudinal axis of the shoulder and radioulnar joints. Thus, these joints may become effective contributors to the end-point velocity.

Badminton Smash and Tennis Serve

The uniqueness of each individual skill within a given pattern is evident if the overhand pattern of a badminton smash performed by an advanced badminton player is compared with a badminton smash performed by an advanced tennis player. Figure 15.5 shows these patterns.

The tennis racquet and ball are heavier than the badminton racquet and shuttlecock; thus, although the general pattern is the same, the relative importance of the segmental movements in the two skills is slightly different.

The tennis player, being used to heavier equipment, uses greater motion of the larger-muscled segments (pelvis and trunk) and lesser motion of the smaller segments (arm and forearm) than the badminton player. Van Gheluwe, Ruysscher, and Craenhals (1987) reported that early pronation in the tennis serve is followed by medial rotation acceleration of the humerus.

In a study of the volleyball overarm serve, the tennis serve, and the badminton smash, Adrian and Enberg (1971) reported that "for the subject who was very highly skilled in one sport in particular . . . , the position at contact in the other skills greatly resembled the contact position in the best sport." The reverse situation may be seen by observing a skilled badminton player attempting a shot with a tennis racquet. The tendency to maintain the badminton pattern could produce injury because of the use of the smaller segments with the heavier equipment and greater impact.

In executing the tennis serve, the performer uses vertebral rotation and flexion and shoulder medial rotation. Forearm pronation is used mainly to position the racquet face for accuracy of ball placement (Van Gheluwe et al., 1987). Elliott et al. (1986) report a fairly large (33.3 m/sec) linear velocity for wrist flexion before impact.

In the badminton smash, however, the shoulder medial rotation and forearm pronation are maximized, while the hip and vertebral rotations are reduced or cease before impact. The muscle force is provided mainly by the larger muscle groups in the tennis serve. The lateral flexion of the trunk and elbow extension mainly serve to increase the radius of rotation distance between the hip, vertebral column, and the shoulder axes of rotation and the point of impact. In the badminton smash, however, 41% of the shuttle velocity is achieved from the medial rotation of the shoulder joint and forearm pronation,

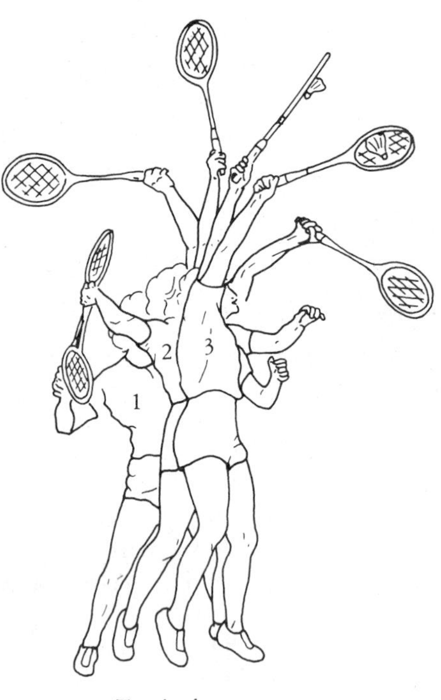

a Tennis player

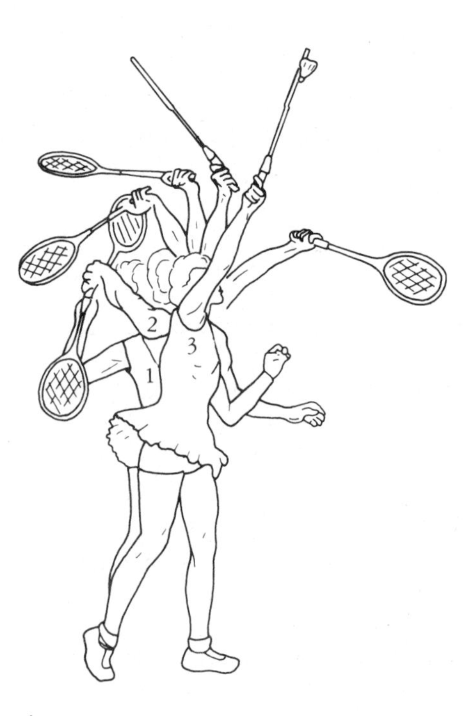

b Badminton player

FIGURE 15.5

(a) An advanced tennis player and (b) an advanced badminton player executing a badminton smash.

two of the fastest movements of the body (Gowitzke, 1979). Because of the smaller muscle groups associated with the arm and forearm segments, these motions are enhanced in activities in which the body is using fairly light objects and implements, such as the badminton racquet and the shuttlecock.

Consider also that in spite of the fact that the hip and vertebral movements may not occur at impact in badminton, they do move to initiate the forward "force phase" of the stroke and probably serve to accelerate the system and to increase the velocity of the upper extremity. The light weight of the racquet and shuttlecock in badminton precludes the necessity of using the larger segment rotations for developing racquet speed. In addition, in higher speed activities the performer has little time to employ a full-body link system during competitive situations.

15.4 Performance Errors: Teaching and Coaching Applications

Errors displayed by performers of different ability levels may be identified and placed into one of three categories: segmental movement errors, timing and sequencing errors, and errors owing to lack of strength or power.

Segmental Movement Errors

The position of body segments during a performance often provides a clue to performance errors. Keeping in mind that one identifying characteristic of the use of the kinetic link principle is the lagging back of more distal segments, teachers and coaches should watch for segments that are positioned ahead of a more proximal segment before release or contact. A beginner attempting to perform a kinetic link activity often positions a more distal segment ahead of a more proximal segment. A common example is that of the location of the wrist relative to the elbow. While the wrist is noticeably ahead of the elbow, the beginner may hold the racquet or the projectile back by hyperextending the wrist joint.

A second common and easily identified segmental position error that is frequently seen in the beginning performer involves the location of the elbow joint relative to a line connecting the shoulders. The elbow is positioned ahead of the shoulders before contact or release. Both of these errors may be seen in the illustrations of the beginning performers in Figures K.10 and 15.3a (pp. 617, 631).

As was mentioned in a previous section, transverse adduction of the shoulder joint is not observed in a skilled baseball throwing pattern but is observed in a less skilled one. The transverse adduction of the shoulder joint before release places the elbow ahead of the shoulder line. In cases of gross errors, an entire throw or strike pattern begins and ends with elbow extension from an elbow located almost directly in front of the same shoulder joint. This error is common to most beginning or immature throwing and striking skills. It is also found in throwing objects in which the front end or "mass" of the object must be maintained in forward-facing position such as a javelin or football.

Sequencing and Timing Errors

The second area of focus for the teacher or coach is that of observing sequencing and timing errors. Segmental movements are difficult to distinguish from one another without the aid of a video or movie film. If the teacher or coach has a mental picture of the proper sequencing, however, then comparing what *is* seen with what *should be* seen is easier.

The segmental sequencing in any skill that is governed by the kinetic link concept should generally proceed from proximal to distal, more massive to less massive, fixed end to free end. The initial forward movement of a distal segment should occur at the point of maximum velocity of the preceding segment. Although *seeing* velocities and accelerations is impossible, the teacher or coach should come to recognize, through experience, where in a pattern a proximal segment contributes its acceleration and when the next distal segment motion should begin. As a general rule, the proximal segment uses about one half to two thirds of its ROM to reach its peak velocity.

A second comparator that may be used by the teacher or coach is a mental image of whiplike timing. One may notice that the performer gets stuck during what should be a smooth flow of segmental movements. Often the beginner moves segments in distinct *blocks* rather than overlapping each with the adjacent segments. In this pattern the segments move simultaneously rather than sequentially.

At the novice level, insufficient relaxation of the muscle groups antagonistic to the desired movement can inhibit the free rotation of one segment relative to another. As the distal links are used, the internal muscular torques that act to bring those distal segments forward may be ill timed. If adequate lag back of distal segments is not noticeable, the distal segments may not be in a position to elicit the stretch reflex in the muscle tissues or to make use of the muscles' elastic properties.

Sometimes having the performer exaggerate each segmental movement in the proper order gives the performer the idea of the proper sequencing. The exaggeration of each segmental movement is particularly helpful to beginners who are having trouble focusing their attention on individual segmental movements.

A second technique is to hold back the contact point (hand, racquet, bat, or club) while the performer attempts to finish the skill. While the end point is held, the proximal segments move out from under the distal segments. The performer should feel the sequential nature of the skill.

A novice performer typically displays a more simultaneous pattern (e.g., with the contact or release point ill timed, too early or too late). Improper timing not only reduces the projection velocity but also alters the direction of the projectile. On the other hand, an intermediate performer tends to display *erratic* timing of the segments: some segmental movements may be initiated too early, some too late, and some just at the right instant.

At the intermediate level of performance, the critical analysis of the teacher or coach is necessary for identifying errors. The analyst should attempt to focus on the timing of one or two segments at a time. Unfortunately, without the aid of a video or film, spotting the error or errors in one viewing is difficult, and therefore, the performer is required to repeat the skill several times.

Errors Owing to Lack of Strength or Power

The skill, strength, or power of the performer may alter the location of any skill on the continuum from that of the ideal. For instance, an immature performer may turn an otherwise sequential movement into a more simultaneous one by ordering the initiation of two or more links at the same or nearly the same time. A person who lacks the necessary power to sequentially accelerate an object with the distal links carries the object closer to the midline of the body, so that a simultaneous pushing motion is observed. The basketball throw for distance may demonstrate this concept if performers of various strengths and hand sizes are observed. Those who have hands large enough to grasp the ball and have enough upper body and arm strength demonstrate a throwlike pattern. As the sizes of hands and strengths of arms decrease in the subjects, the ball is carried closer to the body during the throw. The sequential rotation of proximal segments is maintained while the distal links either will be stabilized or will display a nearly simultaneous pushing motion. A performer who is not able to manipulate either the size or the mass of the ball demonstrates a totally simultaneous pattern.

A second example of the effect of the size and weight of the equipment on throwlike patterns is observed in a tennis player performing a forearm stroke who does not have the strength to hold the racquet in a neutral position at the wrist. Eventually, the weight of the racquet causes the wrist to fall into an ulnar-flexed position. In addition, to reduce the inertial resistance of the racquet and the torque of the impact, the performer adducts the shoulder joint by pulling the elbow into the side, thus increasing elbow flexion (Ward and Groppel, 1980). The increased elbow flexion is an attempt to compensate for the racquet head being lowered vertically, which is due to the ulnar-flexed position of the wrist.

The effects of the size and mass of an object or implement on the throw-like pattern has several implications for the selection of equipment for people of differing sizes, strengths, and abilities. If the performer is to demonstrate a mature pattern with all of its segmental movement and timing requirements, then the equipment may have to be made smaller or less massive or both.

15.5 Developmental Patterns: Teaching Implications

Developmental kinesiology is the area concerned with studying the movement patterns of children during their developmental stages of growth. Children's throwing, kicking, and striking patterns are identified in the literature.

In many ways, the developmental movement patterns are similar to the beginning, intermediate, and advanced skill levels discussed previously; that is, the developmental patterns are described by the segments being used, the positions and movements of those segments, and the sequencing and timing of the segmental movements.

Kicking

Deach (1950) identified several developmental stages in kicking patterns of 2- to 6-year-old children. These stages are as follows:

Stage 1. The performer kicked the ball using a small amount of hip flexion. Little forward movement of the leg occurred after impact as well as little coordinated effort with the rest of the body. The knee remained relatively straight during the impact.

Stage 2. In addition to the hip motion, a flexion of the knee produced a backswing of the leg before impact. The arms were used in abduction for sideward balance.

Stage 3. The ROM of the hip extension and knee flexion was increased. The arms began to show a flexion motion as a balance mechanism for the increased forward–backward motion of the lower extremity.

Wickstrom (1983) states, "Overall the stages show gradual change from a relatively straight pendular leg action with little body movement to a sweeping, whiplike action with gross body movement" (p. 202). "The children . . . almost invariably tended to retract the kicking leg after the completion of the kick. They did not allow leg momentum to carry the rest of the body forward in a follow-through motion. This tendency to withdraw the kicking leg is a clearly identifiable aspect of developmental form in kicking" (p. 202).

In Deach's (1950) study, the subjects were not allowed to incorporate an approach. Bloomfield et al. (1979) found that the first three stages of kicking were similar to Deach's subjects for boys 2 to 12 years of age. They proposed three additional stages that incorporate a run-up (Stage 4), a lagging back of the leg on a forward accelerating thigh (Stage

5), and a complete run-up backswing, thigh hyperextension, and knee flexion (Stage 6).

In the early stages of kicking, little, if any, participation by the pelvis and little participation by the leg occur because of the lack of knee flexion before impact. Thus, the flexion of the thigh at the hip is the single link utilized by the kicker. As the kicker becomes more developed in the kicking pattern, more segments are used; that is, the leg and then the pelvis are added. In addition, an increase in the ROM of the distal links occurs because of its lagging back in compliance with the open kinetic link principle. This lag back gives the segments a greater distance over which to accelerate for the forward movements and thus allows them to generate more velocity to the free end segment, the foot. The greater knee flexion during the forward swing of the thigh provides the added advantage of reducing the rotational inertia of the lower extremity for hip flexion and therefore allows greater acceleration of the thigh during the hip-flexion phase. The momentum of the lower extremity is assumed by the leg (a smaller rotational inertia segment) as the thigh decelerates.

Similar developmental progressions occur in throwing and striking, and all follow similar trends (i.e., the less mature performance generally incorporates fewer segments, displays a smaller ROM for those segments used, and uses more simultaneous rather than sequential timing).

The progression of developmental patterns in kicking is similar to the beginning, intermediate, and advanced aspects of any of the throwing, kicking, and striking skills. Mechanically, a skilled pattern is characterized as a multilink system, with each link contributing its own particular movement to the production of velocity at impact with the object. Beginning from the most hyperextended position of the free hip, the links move forward in a prescribed order from the largest, most stable, and proximal link to the smallest, most freely movable, and distal link.

Throwing

Developmental stages of throwing were proposed initially by Wild (1938) as reported by Wickstrom (1983). The stages of throwing reflect an increased use of sequential segmental rotations as the mature pattern develops.

Stage 1. The elbow is located forward of the shoulder joint. The ball is thrown primarily with elbow extension. No rotation of the trunk is visible (Figure 15.6a).

Stage 2. A trunk rotation to the right (right-handed thrower) accompanies the backward motion of the arm. The throw is initiated by the arm swing forward accompanied by trunk rotation left. The elbow extends at variable times during the forward arm swing (Figure 15.6b).

Stage 3. A step is taken with the ipsilateral (same side) foot. The step is followed by trunk rotation and the forward arm swing. The elbow extension occurs later than in Stage 2 (Figure 15.6c).

Stage 4. A step is taken with the contralateral foot, the trunk rotates to the left, a transverse adduction of the arm and elbow extension ends the force phase (Figure 15.6d).

Roberton (1978) further identified segmental movement components or categories using a more kinesiological and biomechanical basis. These are (1) the action of the humerus, (2) the action of the forearm, and (3) the action of the pelvis-spine. In all categories, the pattern develops from less mature forms characterized by:

1. Block (simultaneous) rotation of adjacent segments with little or no differentiated sequencing
2. Little evidence of lag back

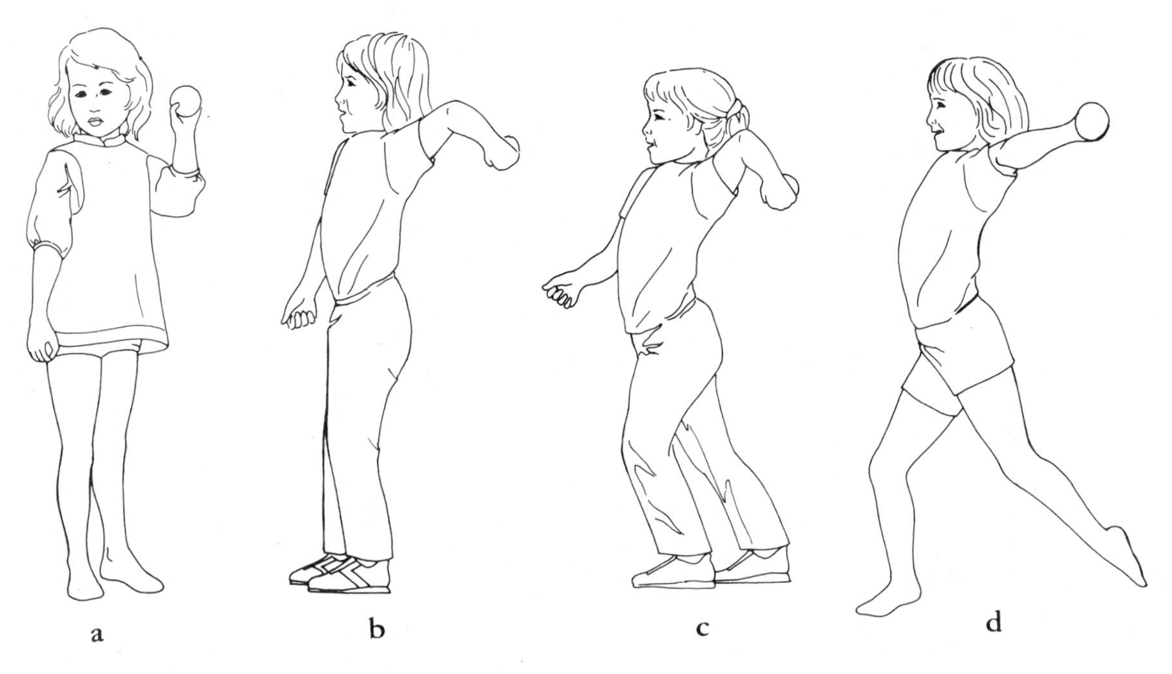

FIGURE 15.6
The four developmental stages of throwing.

3. Inappropriate segmental movements (e.g., transverse adduction of the humerus and hyperextension-flexion of the trunk)

Eventually these immature forms give way to more mature forms as described previously in this chapter and in keeping with the kinetic link concept.

Of interest are Anderson's findings (1977) as reported by Roberton (1978). In less-skilled throwers, pectoralis major activity continued through ball release, whereas in skilled throwers, the muscle activity diminished sharply before ball release. This finding implies that muscle torque is not necessary to increase all the segmental velocities in an open kinetic chain activity.

While watching children of different developmental stages perform the same pattern, some basic differences may be observed in the sequencing and timing of the segments within the same pattern. Considering kinesiological and biomechanical as well as motor development and motor learning factors, it is questionable as to which factors are most important in helping to develop the best performance possible. If one looks to the potential mature performance, one can approach the teaching of that skill in one of two ways.

One may ask the physically immature performer, no matter what size or strength, to train using regulation equipment. In skills in which maximum distance or velocity or both are important, the ideal projection angle has to be adjusted according to the height of release and speed of projection that the individual is able to produce. (A slower speed of release dictates a higher release angle.) Alternatively, one can reduce the size and weight of the equipment in proportion to the size of the performer so that the performer may pattern his or her movements closer to the ideal mature performance.

Teachers and coaches must consider (1) whether the child-athlete gains more by training with smaller, lighter equipment and thus is enabled to closely approximate the mature movement pattern, or (2) whether training should be a continual adjustment of the movement pattern because of growth (height) and development (strength) factors.

Understanding Applications of the Kinetic Link Principle

1. Assemble these objects: a tennis ball, a softball, a volleyball, a soccer ball, a basketball, and a medicine ball. For two observers, perform an overarm throw for maximum distance with each of the objects in the order listed. Answer the following:
   - **a.** Can any differences be detected among throws in terms of the simultaneous or sequential nature of the joint movements?
   - **b.** In what order are the segments moving?
   - **c.** Does one joint movement stop before another begins?
   - **d.** Do two or more segments start and stop moving simultaneously?
   - **e.** Do two or more segments start at the same time but not stop at the same time?
2. Throw and catch a softball with a partner. Identify the sequence of movements occurring when one is throwing as hard as one can. Perform the same throw with the nonpreferred arm. What differences are noted in the motion of the segments? What differences are noted in the sequencing of the segmental movements?
3. Throw and catch a softball with a partner in an overarm, underarm, and sidearm pattern. For each pattern, identify and list the sequential joint movements in the order of their occurrence and those that occur simultaneously. Compare the movements of each throw with the movements observed using the same object and the nonpreferred arm.
4. With the assistance of an unskilled partner (using the nonpreferred arm), attempt to identify the type of error that the unskilled partner is making from those listed in the section entitled Performance Errors: Teaching and Coaching Applications (pp. 637-639). Attempt to correct those errors with the help of the applications given in the same section.
5. Select one of these skills: baseball batting, volleyball spike, tennis one-hand backhand, tennis two-hand backhand, badminton round-the-head shot, racquetball forehand, golf drive, cricket pitch, hurling, soccer-style place kick, water polo shot, and javelin throw.
   - **a.** In groups of three, analyze the selected skill in terms of the following:
     - **i.** The segments and articulations that are doing the moving
     - **ii.** The movement that the articulation is doing
     - **iii.** The axis of rotation used in each of the movements

b. Draw an illustration (stick figure) of the release or contact position of a person performing the skill. Place dots or lines where appropriate to indicate the axes of rotation used by the segments as they move into the release or contact position. Draw the radius of rotation for each movement. You may need more than one view. (See Figure 15.2 for an example.)

c. Discuss how the radius of gyration and the mass of the system change as each segmental movement takes place in sequence. Consider the position of all the segments distal to the segment you are considering. How do the mass and radius of gyration changes affect the angular velocity of the moving segment?

6. Attend a Little League game, go to a playground, or observe a physical education class at an elementary school. Watch the children catch, throw, and kick. Identify performance errors that are related to the developmental stages. Can you relate these performance errors to the errors listed under the section Performance Errors: Teaching and Coaching Applications?

References and Suggested Readings

Adrian, M., & Enberg, M. L. (1971). Sequential timing of three overhand patterns. In C. J. Widule (Ed.), *Kinesiology review* (pp. 1–9). Reston, VA: AAHPERD.

Atwater, A. E. (1970, April). Overarm throw patterns: A kinematographic analysis. Paper presented at the AAHPERD Convention, Seattle, WA.

Atwater, A. E. (1977, January). Biomechanics of throwing: Correction of common misconceptions. Paper presented at the Joint Meeting of NAPECM and NAPECW, Orlando, FL.

Atwater, A. E. (1980). Biomechanics of overarm throwing movements and of throwing injuries. In R. S. Hutton & D. I. Miller (Eds.), *Exercise and sport science reviews* (Vol. 7, pp. 43–85). Philadelphia: Franklin Institute Press.

Bloomfield, J., Elliot, B. C., & Davies, C. M. (1979). Development of the soccer kick. *Journal of Human Movement Studies, 5*, 152–159.

Bollens, E. C., De Proft, E., & Clarys, J. P. (1987). The accuracy and muscle monitoring in soccer kicking. In B. Jonsson (Ed.), *Biomechanics X-A* (pp. 283–288). Champaign, IL: Human Kinetics.

Campbell, K. L., & Reid, R. L. (1985). The application of optimal control theory to simplified models of complex human motions: The golf swing. In D. A. Winter, R. W. Norman, R. P. Wells, K. C. Hayes, & A. E. Patla (Eds.), *Biomechanics IX-B* (pp. 527–532). Champaign, IL: Human Kinetics.

Dapena, J. (1984). The pattern of hammer speed during a hammer throw and influence of gravity on its fluctuations. *Journal of Biomechanics, 17*, 553–560.

Dapena, J. (1985). Factors affecting the fluctuations of hammer speed in a throw. In D. A. Winter, R. W. Norman, R. P. Wells, K. C. Hayes, & A. E. Patla (Eds.), *Biomechanics IX-B* (pp. 499–503). Champaign, IL: Human Kinetics.

Dapena, J. (1986). A kinematic study of center of mass motion in the hammer throw. *Journal of Biomechanics, 19*, 147–158.

Deach, D. (1950). *Genetic development of motor skills in children two through six years of age*. Unpublished doctoral dissertation, University of Michigan.

Deutsch, H. (1970, April). A comparison of women's overarm throwing patterns. Paper presented at the AAHPERD Convention, Seattle, WA.

Dillman, C. J. (1974). Effect of leg segmental movements on foot velocity during the recovery phase of running. In R. C. Nelson & C. A. Morehouse (Eds.), *Biomechanics IV: Proceedings of the Fourth International Seminar on Biomechanics* (pp. 98–105). Baltimore: University Park Press.

Dunn, E. G., & Putnam, C. A. (1987). Kicking speed and lower extremity kinematics. In J. Terauds, B. Gowitzke, & L. Holt (Eds.), *Biomechanics of Sports III & IV* (pp. 154–160). Del Mar, CA: Academic.

Dyson, G. (1978). *The mechanics of athletics* (7th ed.). New York: Holmes & Meier.

Elliott, B., Grove, J. R., & Gibson, B. (1988). Timing of the lower limb drive and throwing limb movement in baseball pitching. *International Journal of Sport Biomechanics, 4*, 59–67.

Elliott, B., Grove, J. R., Gibson, B., & Thurston, B. (1987). A three-dimensional cinematographic analysis of the fastball and curveball pitches in baseball. *International Journal of Sport Biomechanics, 2*, 20–28.

Elliott, B., Marsh, T., & Blanksby, B. (1986). A three-dimensional cinematographic analysis of the tennis serve. *International Journal of Sport Biomechanics, 2*, 260–271.

Feltner, M., & Dapena, J. (1986). Dynamics of the shoulder and elbow joints of the throwing arm during a baseball pitch. *International Journal of Sport Biomechanics, 2*, 235–259.

Gowan, I. D., Jobe, F. W., Tibone, J. E., Perry, J., & Moynes, D. R. (1987). A comparative electromyographic analysis of the shoulder during pitching. *The American Journal of Sports Medicine, 15*, 586-590.

Gowitzke, B. (1979). Biomechanical principles applied to badminton stroke production. In J. Terauds (Ed.), *Science in racquet sports* (pp. 7–16). San Diego: Academic.

Gowitzke, B. & Waddell, D. B. (1979a). Techniques of badminton stroke production. In J. Terauds (Ed.), *Science in racquet sports* (pp. 17–42). San Diego: Academic.

Gowitzke, B. & Waddell, D. B. (1979b). Qualitative analysis of the badminton forehand smash as performed by international players. In J. Groppel (Ed.), *The racquet sports* (pp. 1–15). Champaign: University of Illinois Press.

Gregor, R. J., Whiting, J. C., & McCoy, R. W. (1985). Kinematic analysis of Olympic discus throwers. *International Journal of Sport Biomechanics, 1*, 131–138.

Groppel, J. L. (1986). The utilization of proper racquet sport mechanics to avoid upper extremity injury. In F. A. Pettrone (Ed.), *American Academy of Orthopaedic Surgeons Symposium on Upper Extremity Injuries* (pp. 30–35). St. Louis: Mosby.

Hirano, Y. (1987). Comparative study of pitching motions between skilled and Little League baseball pitchers. In B. Jonsson (Ed.), *Biomechanics X-B* (pp. 649–654). Champaign, IL: Human Kinetics.

Jobe, F. W., Moynes, D. R., Tibone, J. E. & Perry, J. (1984). An EMG analysis of the shoulder in pitching. *The American Journal of Sports Medicine, 12*, 218–222.

Joris, H. J. J., Edwards van Muyen, A. J., van Ingen Schenau, G. J., & Kemper, H. C. G. (1985). Force, velocity & energy flow during the overarm throw in female handball players. *Journal of Biomechanics, 18*, 409–414.

Komi, P. V., & Mero, A. (1985). Biomechanical analysis of Olympic javelin throwers. *International Journal of Sport Biomechanics, 1*, 139–150.

Kulig, K., Mowacki, Z., & Bober, T. (1983). Synchronization of partial impulses as a biomechanical principle. In H. Matsui and K. Kobayashi (Eds.), *Biomechanics VIII-B: Proceedings of the Eighth International Congress on Biomechanics* (pp. 1144–1151). Champaign, IL: Human Kinetics.

Lindner, E. (1971). The phenomenon of the freedom of the lateral deviation in throwing. In J. Vredenbregt & J. Wartenweiler (Eds.), *Biomechanics II: Proceedings of the Second International Seminar on Biomechanics* (pp. 240–245). Baltimore: University Park Press.

Menzel, H-J. (1987). Transmission of partial momenta in the javelin throw. In B. Jonsson (Ed.), *Biomechanics X-B* (pp. 643–648). Champaign, IL: Human Kinetics.

Messier, S. P., & Brody, M. A. (1986). Mechanics of translation and rotation during conventional and handspring soccer throw-ins. *International Journal of Sport Biomechanics, 2*, 301–315.

Milburn, P. D. (1982). Summation of segmental velocities in the golf swing, *Medicine and Science in Sports and Exercise, 14*, 60–64.

Morehouse, L. E., & Cooper, J. M. (1950). *Kinesiology*. London: Kimpton.

Neal, R. J., & Wilson, B. D. (1985). 3D kinematics and kinetics of the golf swing. *International Journal of Sport Biomechanics, 1*, 221–232.

Pappas, A. M., Zawacki, R. M., & Sullivan, T. J. (1985). Biomechanics of baseball pitching. *The American Journal of Sports Medicine, 13*, 216–222.

Pettrone, F. A. The pitching motion. In F. A. Pettrone (Ed.), *American Academy of Orthopaedic Surgeons Symposium on Upper Extremity Injuries* (pp. 59–63). St. Louis: Mosby.

Phillips, S. J., Roberts, E. M., & Huang, T. C. (1983). Quantification of intersegmental reactions during rapid swing motions. *Journal of Biomechanics, 16*(6), 411–417.

Putnam, C. A. (1983). Interaction between segments during a kicking motion. In M. Matsui & K. Kobayashi (Eds.), *Biomechanics VIII-B: Proceedings of the Eighth International Congress on Biomechanics* (pp. 688–694). Champaign, IL: Human Kinetics.

Roberton, M. A. (1978). Longitudinal evidence for the developmental stages in the forceful overarm throw. *Journal of Human Movement Studies, 4*, 167–175.

Roberts, E. M., & Phillips, S. J. (1977). Angular acceleration in free segment motion. In C. Dillman & R. Sears (Eds.), *Kinesiology: (Proceedings of) A National Conference on Teaching* (pp. 328–334). Champaign-Urbana, IL: University of Illinois.

Roberts, E. M., Zernicke, R. F., Youm, Y., & Huang, T. C. (1974). Kinetic parameters of kicking. In R. C. Nelson & C. A. Morehouse (Eds.), *Biomechanics IV: Proceedings of the Fourth International Seminar on Biomechanics* (pp. 157–162). Baltimore: University Park Press.

Robertson, D. G. E., & Mosher, R. E. (1985). Work and power of the leg muscles in soccer kicking. In D. A. Winter, R. W. Norman, R. P. Wells, K. C. Hayes, & A. E. Patla (Eds.), *Biomechanics IX-B* (pp. 533–538). Champaign, IL: Human Kinetics.

Tarbell, T. (1974). Some biomechanical aspects of the overhead throw. In R. C. Nelson & C. A. Morehouse (Eds.), *Biomechanics IV: Proceedings of the Fourth International Seminar on Biomechanics* (pp. 180–183). Baltimore: University Park Press.

Toyoshima, S., & Miyashita, M. (1973). Force–velocity relation in throwing. *Research Quarterly, 44*, 86–95.

Tricker, R. A. R., & Tricker, B. J. K. (1967). *The science of human movement*. New York: American Elsevier.

Tsarouchas, E., & Klissouras, V. (1981). The force–velocity relation of a kinematic chain in man. In A. Morecki, K. Fidelus, K. Kedzior, & A. Wit (Eds.), *Biomechanics VII-A: Proceedings of the Seventh International Congress on Biomechanics* (pp. 145–151). Baltimore: University Park Press.

Van Gheluwe, B., & Hebbelinck, M. (1985). The kinematics of the service movement of tennis: A three-dimensional cinematographical approach. In D. A. Winter, R. W. Norman, R. P. Wells, K. C. Hayes, & A. E. Patla (Eds.), *Biomechanics IX-B* (pp. 521–526). Champaign, IL: Human Kinetics.

Van Gheluwe, B. V., Ruysscher, I. D., & Craenhals, J. (1987). Pronation and endorotation of the racquet arm in a tennis serve. In B. Jonsson (Ed.), *Biomechanics X-B* (pp. 667–672). Champaign, IL: Human Kinetics.

Vaughn, R. E. (1985). An algorithm for determining arm action during overarm baseball pitches. In D. A. Winter, R. W. Norman, R. P. Wells, K. C. Hayes, & A. E. Patla (Eds.), *Biomechanics IX-B* (pp. 510–515). Champaign, IL: Human Kinetics.

Ward, T. & Groppel, J. (1980). Sport implement selection: Can it be based upon anthropometric indicators? *Motor Skills: Theory Into Practice, 4*, 103–110.

Whiting, W. C., Puffer, J. C., Finerman, G. A., Gregor, R. J., & Maletis, G. B. (1985). Three-dimensional cinematographic analysis of water polo throwing in elite performers. *The American Journal of Sports Medicine, 13*, 95–104.

Wickstrom, R. (1983). *Fundamental motor patterns*. Philadelphia: Lea and Fibiger.

Widule, C. J. (1983, May 24). Contiguous segment velocity: Acceleration characteristics. Personal communication.

Wild, M. (1938). The behavior pattern of throwing and some observations concerning its course of development in children. *Research Quarterly, 9*(3), 20.

Yoshihuku, Y., Ikehami, Y., & Sakurai, S. (1987). Energy flow from the trunk to the upper limb in Tsuki motion of top-class players of the martial arts, Shorinji Kempo. In B. Jonsson (Ed.), *Biomechanics X-B* (pp. 733–738). Champaign, IL: Human Kinetics.

Zernicke, R., & Roberts, E. (1978). Lower extremity forces and torques during systematic variation of nonweight bearing motion. *Medicine and Science in Sports, 10*(1), 21–26.

CHAPTER 16

Performance Analysis of Pushlike Movements

PREREQUISITES

Concept Modules B, F, G, I, J, K
Chapters 3, 9, 10, 15

CHAPTER CONTENTS

The term *pushlike* is characterized by the body parts moving simultaneously or nearly so. There are three mechanical purposes related to pushlike patterns: to manipulate a resistance, to generate maximum power, or to maximize the accuracy of projection. To effectively manipulate a resistance, the performer is not concerned with creating the highest possible velocity of the object being manipulated. Rather, the performer wishes to move a large external load from one place to the other such as in "power" weight-lifting events. Those are the squat, the dead lift, and the bench press. Unfortunately, the term "power lifting" is a misnomer, for the events performed in this category are not associated with power *per* se but are associated with force production throughout a range of motion without regard to the velocity of moving the weight. It is the Olympic lifts that incorporate power to a great extent; those lifts are the clean and jerk and the snatch. Nevertheless, the squat, the dead lift, and the bench press are associated with strength or force production.

To generate maximum power, the performer must produce great force to accelerate the resistance in a limited amount of time. In maximum power activities, the limited time available for the application of force is crucial. There are four categories of activities for which the generation of power is paramount: jumping activities, takeoffs for aerial events, starts in events that have the overall performance objective of "to move the body over a prescribed distance as fast as possible," and continuous events in which there is an intermittent application of force such as rowing.

Last, pushlike patterns are used for activities in which the overall performance objective is the maximum accuracy of projecting an object. These pushlike patterns are at the pushlike end of the throw–push continuum and were introduced at the end of Module K.

16.1 Force Activities

Maximum force activities can be represented by competitive weight-lifting events called "power lifting." Power lifting events are the bench press, the squat, and the dead lift. Unfortunately, power lifting incorporates very little "power" per se in that the velocity of the lift appears to contribute little to the effectiveness of the performance. Maximum strength seems to be paramount.

Power Lifting Events

The power lifting events do not require much power. That is, a performer in the dead lift, the bench press, and the squat is required to move the weight from A to B. How fast he or she lifts the weight is really insignificant as long as the weight is moved through the required displacement. In studying skilled and unskilled adolescents performing the dead lift, Brown and Abani (1985) found that the skilled lifters took a longer time from lift off to knee passing than did the unskilled.

Thus, the magnitude of force application is highly important. The performer is required to summate the segmental torques to achieve the maximum force against the resistance. In maximum force activities, such as those shown in Figure 16.1a–c, simultaneous segmental rotations are necessary.

Whereas sequential segmental rotations produce curvilinear or rotational displacement of the end point, simultaneous segmental rotations cause the end point to move in a straight line. Straight-line movements are important for activities such as weight lifting, in which balance is extremely important as well as maximum force against a resistance.

The movements in a squat exercise involve slight extension of the intervertebral joints, extension of the hip and knee joints, and slight plantar flexion of the ankle joints. All of these movements should occur simultaneously so that the weight or resistance moves in a linear fashion vertically and rarely deviates horizontally from its vertical path. Additional examples of this type of movement occur in a maximum accuracy type of activity.

a

b

c

FIGURE 16.1

The three lifts included in power lifting competition. (a) The squat. (b) The bench press. (c) The dead lift.

16.2 Power Activities

Power activities are similar to force activities in that both ask for a high amount of force. Power activities require in addition a high velocity. We know from the force–velocity relationship that when velocity increases, force decreases and vice versa. In power activities, the velocity of the projectile or the velocity of the object being moved must be accomplished in a relatively short period of time.

As was introduced earlier in this text, power events may be placed on a continuum, from those in which the force application is more important to success to those in which the velocity of moving an object is more important. Force-dominated power events take on some of the characteristics of the force activities previously discussed in this chapter such as rectilinear or curvilinear movement of the object. These events do have time limitations, however, and the object or the body must be accelerated in a short amount of time. Shot putting is an example. The mass of the shot makes the magnitude of the force application important. However, the putter has a limited amount of time to apply this force, and the velocity of the shot at release is very important to the distance that it travels. Thus, shot putting is a force-dominated power event.

Speed-dominated power events are on the other end of a continuum. The object being manipulated is not as massive as the force-dominated objects, and the performer's speed of muscular contraction determines the success of the attempt rather than high generation of force.

Power activities are presented under the headings of jumping events, punching events, lifting events, and continuous events that require the steady application of force.

Jumping Events

The vertical jump has long been used as a measure of power. The performer must apply force to the body's mass to accelerate it as much as possible while it is still in contact with the ground. The vertical velocity of the body at takeoff is primary to the height achieved. Jumping events include the jumps that are discrete skills in themselves, such as high jumping, long jumping, and triple jumping, and jumps that are part of a larger skill, such as the jump preceding a volleyball spike, the jump preceding a jump shot in basketball, and takeoffs such as in gymnastics tumbling and diving.

Whenever the body is projecting itself, the movement pattern resembles that of pushing the body into space by sequential segmental rotations. The segmental rotations act to move the body's center of gravity (CG) in a rectilinear path. A series of rotational movements producing a linear path of the body's CG is illustrated by the basketball player in Figure 16.2.

In contrast to other models, the jumping model shows the most massive segment, the trunk, located at the open end of the link system and the least massive segment, the feet, located at the fixed end. This configuration is called a **closed kinetic chain**. The segmental movements are in reverse order; that is, beginning at the most free segment and ending at the most fixed segment, the feet. This is a slightly different order from that seen in throwing, striking, and kicking skills. To project the body, one must have an external force; that external force is the ground-reaction force (GRF) that the jumper creates by pushing against it. The greater the GRF, the greater the upward acceleration of the body's CG. The

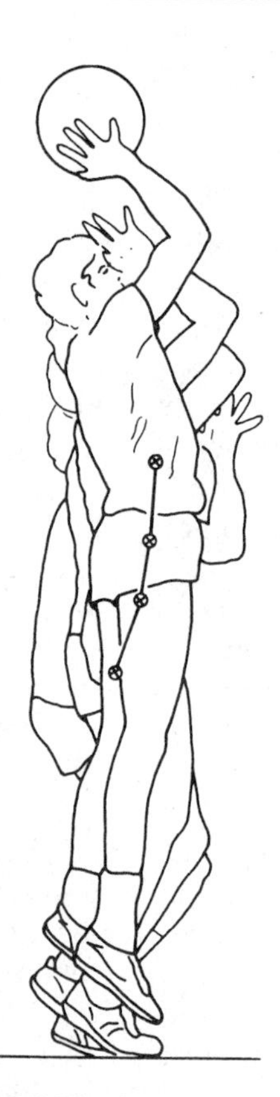

FIGURE 16.2
Segmental rotations producing a linear path of the center of gravity.

greater the upward acceleration of the segments, the greater the force downward so that the GRF increases. In the pushing pattern, the sequential application of segmental forces is used to produce a greater GRF on the body.

The jumping motion consists of a partial sequential contraction of the antigravity muscles in an attempt to summate the segments' forces as in pushing patterns involving power components. This sequencing of some of the segmental rotations is done to accommodate the neuromuscular properties of the muscles involved.

The initial upward motion of the sequence is flexion of the arms at the shoulder joints. During their upward acceleration, the arms exert a downward reaction force on the segments below them. This "squeezing" of the lower segments enhances the negative work or eccentric action of the antigravity muscles. Sometime during the midst of their flexion, the arms begin to decelerate. This deceleration unweights the lower segments. During the deceleration phase of the arms, the optimum time to begin extension of the trunk about the hip joints begins. Thus, the performer can use the elastic recoil of the lower segments.

The lower and less massive segments are

subjected to a large downward reaction force during the upward acceleration of the large trunk segment above them. Because the lower segments are acting under such influence, they may be unable to begin their extensions simultaneously with the trunk, as would be expected in such a pushing type of movement. Even though the muscles of the lower segments may be attempting to produce the force for upward acceleration, they are so heavily loaded with the acceleration of the more massive segments above them that they may not be able to produce the force necessary to counteract the load. Thus, these lower segments may in fact be squeezed into a greater range of flexion before their extensions. Because the extensor muscles exert tension during their joints' flexions, they are functioning with eccentric tension and therefore can use the elastic recoil action. Furthermore, as Bobbert and van Ingen Schenau (1988) point out, the hamstrings, which are functioning concentrically during the hip extension, are also knee flexors. The flexion torques of the hamstrings at the knee during this time may also prevent the extension of the knee although when the foot is fixed as it is in jumping, the hamstring torques at the knee may actually cause its extension.

Bobbert and van Ingen Schenau (1988) provide further evidence of the sequential nature of the muscular action in jumping. In noting the time history of peak maximum muscle activity of the uniarticulate muscles of the lower extremity, these researchers found that the gluteus maximus achieved its peak at the start of the push-off, whereas the uniarticulate extensors of the knee were at 62% of their peak at initial push-off and peaked at 190 msec before leaving the ground. Similarly, the plan-

a Initial foot plant position
b Arms and free leg accelerate upward
c Support leg extends, arm and free leg decelerate

FIGURE 16.3

A high-jump takeoff showing (a) upward acceleration of the arms and the free leg and (b) extension of the support leg as (c) the arms and free leg decelerate.

tar flexors of the ankle were at 26% of their peak at initial push-off and peaked at 100 msec before leaving the ground.

There are two important advantages to the sequential nature of the jumping action. First, the lower segments are allowed to extend through a greater range of motion (ROM). Second, and more importantly, the extensor muscles of the lower segments are stretched to a greater extent before their contractions. The stretch reflex is evoked, which enhances the muscle's contractile force, and it allows a greater use of the series elastic component of the tissues surrounding the joint. Note that merely increasing the ROM during flexion does not necessarily produce a greater extension force. The extensor muscles must be tensed during the joint flexion phase for the recoil of the elastic component to contribute to the extension of the joints.

The acceleration and deceleration of the arms in a vertical jump or of the arms and free leg during a high jump are commonly referred to as the transfer of momentum from the arms or the leg or both to the total body. The momentum of the arms or of the leg (created by their upward acceleration during ground contact) is conserved within the entire system; therefore, as these segments decelerate, an increase occurs in the velocity of the body's CG upward. Consequently, the force resulting from the final extension of the supporting knee and ankle joints is used to further increase the momentum of the already upward-traveling CG. Figure 16.3 illustrates the timing of the arms and of the free-leg and support-leg movments in the high-jump takeoff.

Dyson (1977) calls the impulse provided by free-swinging arms and legs, as in the case of a high jumper, a *transmitted impulse*, as opposed to a *controlled impulse*, which is produced by the direct muscular effort against the supporting surface.

Understanding Jumping Mechanics

1. Perform a vertical jump with the nonreaching arm held at the side throughout the jump and the reaching arm held above the head. Measure the jump. Repeat the jump, swinging both arms upward just before takeoff.
   a. Explain any difference between the two techniques in regard to the height of the projection.
   b. What is the acceleration–deceleration pattern during the ROM of the arms relative to the extension of the hips?
   c. Does a sequence of segmental rotations occur during the jump, or are the rotations simultaneous?
2. Stand on a scale in anatomical position. From this position, perform the following actions while watching the weight indicator: (a) flex the elbows quickly, (b) flex the shoulder joints slowly, (c) flex the shoulder joints rapidly, (d) quickly flex the hips, knees, and ankles, (e) from a half-squat postion, extend the hip, knee, and ankle joints rapidly. Explain the reactions of the scale in this experiment in terms of acceleration and deceleration of the segments. Relate this to the weighting and unweighting of the lower segments and the upward reaction force created at the feet.
3. Complete this sentence: Whenever the free end of a kinetic link system is more massive than the fixed end, the system is called a ________ kinetic chain.

Punching Motions

Punching events include boxing, karate, and taekwondo. The force of impact is important as is the speed with which the punch is "thrown." In boxing, a right-hand punch is performed most effectively by using the hips, trunk, and shoulder girdle segments as wheel and axle mechanisms about the supporting opposite hip joint. The right upper extremity is used as a lever system consisting of shoulder joint flexion and elbow extension. Figure 16.4 illustrates a right-hand punch.

The importance of maintaining a straight-line movement of the fist is understood when one considers the speed with which the punch must be thrown. In a straight-line movement, the fist will have a smaller linear displacement and hence arrive at the intended target sooner. Whiting, Gregor, and Finerman (1988) provide evidence that the jab is performed in a more straight-line motion than the hook, which is more angular.

The importance of the acceleration of the hand cannot be overemphasized when one considers that the impact is determined by the sudden deceleration of the hand on the opponent's face. The larger the velocity of the hand on initial impact, the larger the deceleration of that hand must be. Thus, the effectiveness of the punch in part relies on the instantaneous power at impact. The other part of the effectiveness relies on the accuracy of the punch.

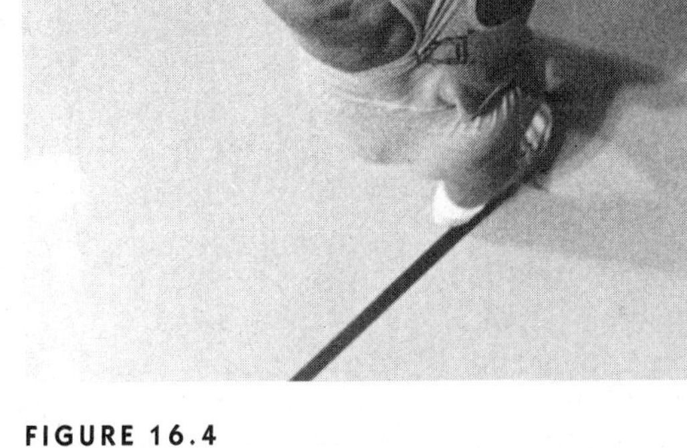

FIGURE 16.4
Top view of a right-hand punch in boxing.

a

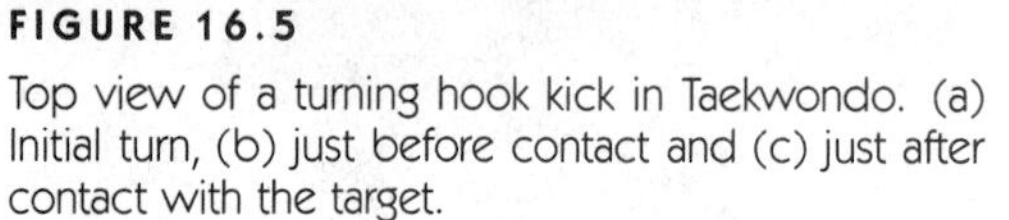

FIGURE 16.5

Top view of a turning hook kick in Taekwondo. (a) Initial turn, (b) just before contact and (c) just after contact with the target.

As discussed in the last section, rectilinear motion provides the best possibility of accurate movement.

The turning hook kick in taekwondo serves as a second example of a punching pushlike movement. The turning hook kick is shown in Figure 16.5a-c.

While the performer is rotating in a clockwise direction, as viewed from above, about the supporting left lower extremity, the actual kick is a pushlike punch performed near the end of the rotation. Figure 16.5a and b illustrate the fighter just before and at contact. As the fighter nears the target, the hip, knee and ankle joints are quickly extended so that the foot arrives at the target as quickly as possible. The speed of the foot at the instant of contact is crucial for the same reasons that the speed of hand arrival is important in the boxing motion. A higher velocity at impact results in a greater deceleration of the foot and a harder blow and of course less time for the opponent to dodge the blow. In taekwondo the rules state that points are scored when the punch is delivered with sufficient force to cause "trembling shock."

b

c

Olympic Lifts

As was stated previously, those competitive lifts called power lifts are not truly powerful in nature. The Olympic lifts, the clean and jerk and the snatch, have a high component of power in their performance. Garhammer (1985) states "High power output capacity was the most distinguishing characteristic . . . and is likely necessary for successful participation . . . at the elite level" (p. 122).

The Olympic lifts require the application of force to the bar such that the lifter causes the bar to be moving upward as fast as possible. Both lifts have several discrete parts. The first discrete part of the clean and jerk begins with the initial lifting of the bar from the floor (Figure 16.6a). In the clean and jerk, the body must drop under the upward-moving bar so that the bar can be held at the shoulder level (Figure 16.6b). The faster the bar is moving upward during the lift, the more time the lifter has to drop under it and the less range of motion the lifter must use in flexing the hips and knees.

FIGURE 16.6

(a) The bottom and (b) the top of the clean lift. (c) The bottom and (d) the top of the jerk in performing the clean and jerk in Olympic lifting.

FIGURE 16.6
(continued)

During the second half of the lift, the bar must be accelerated upward from the shoulder area such that the lifter can drop under the upward-moving bar so that the bar comes to rest on the arms, which are fully flexed overhead (Figure 16.6c-d). The same power moves are required, that is, a large application of force to achieve the largest possible accelerations. The application of these forces is limited to a very short period of time.

In the snatch lift, the lifter must bring the bar to the overhead position during the first part of the initial lift shown in Figure 16.7a-c. The lifter does this by using a greater range of motion for hip and knee flexions and dropping under the upward-moving bar and ending in a full-squat position (Figure 16.7c). A large power requirement is necessary to apply a large force to the initial lift so that the bar is moving as fast as is possible before the body's dropping under it. The second half of the snatch may be classified as a force activity with little power required. The lifter rises to a standing position while maintaining the bar overhead (Figure 16.7c-d). The speed with which he does this is of little importance to the success of the lift.

Continuous Power Events

Whenever there is a limited amount of time to accelerate an implement or an object, power is involved, because power includes force and velocity of movement. Whereas the previous examples were discrete skills, many examples of continuous power events exist in sports. Two examples are an oarsperson applying force to an oar in rowing, wherein there is a limited amount of time to apply a large force, and the sprinter attempting to produce a large amount of force against the ground in a very short period of ground contact.

After the initial start-up period, very little if any slippage occurs between the oar and the water. Thus, the athlete has a limited

FIGURE 16.7

(a) The beginning, (b) the lifting, (c)the squatting, and (d) the ending of the snatch in Olympic lifting.

FIGURE 16.7
(continued)

amount of time to apply a large amount of force against the oar to displace the shell maximally as fast as possible during each stroke. Nelson and Widule (1983) state that power training is more important than sheer strength training and should provide greater knee and trunk accelerations.

The rowing motion involves two blocks of segmental movements—the extension of the trunk, the hips, knees, and ankles and the retraction of the shoulder girdle, extension-transverse extension of the shoulder joint, and the flexion of the elbow. While the movements within the blocks occur somewhat simultaneously, the blocks themselves are initiated in an overlapping sequential fashion. The first block, consisting of the larger, more massive segments (lower extremities and trunk), initiates the action. In skilled rowers, the knee extension occurs before trunk extension, whereas the hip and trunk extend somewhat simultaneously (Nelson and Widule, 1983).

Powerful trunk and lower extremity angular accelerations are used to accelerate the oar. The arms must serve as rigid linkages from the legs to the oar. The trunk and lower extremity joints experience angular decelerations during the latter half of their ranges of motion, and at this time the upper body actions occur. These muscles contract fast enough to keep up with the oar; whether they can contract faster to surpass the speed of the oar to apply force is questionable. Wrist flexion and extension will follow to free the blade from the water on the recovery and to catch the water on entry.

After the start, the oar moves at the same speed as the shell. Therefore, the oarsperson must generate enough force (must contract the muscles fast enough) to surpass the aerodynamic and hydrodynamic drag forces and to apply additional force if possible. The start-up may be thought of as a force-dominated power event while the rest of the race is thought of as a speed-dominated power event.

In sprint running, the runner experiences the same power aspects as in rowing. Both are continuous events in which after the start, the body is moving at a speed that is under the influence of aerodynamic drag forces. In addition, the runner must continually support the body weight. The power surge comes during the short period of time that the runner's foot is in contact with the ground. It has been well established that the speed of the runner is correlated with the stride length and the stride frequency. Thus, the greater the power during push-off (the greater the force applied in that short period of time), the greater will be the stride length. In addition, the faster the runner can get the foot back down to the ground surface, the more frequently the power can be applied. The pushlike simultaneous extensions of the hip, knee, and ankle joints provide the power required in this skill.

16.3 Accuracy Activities

Accuracy activities include those events in which an object is projected to a target such as archery, shooting, field goal kicking; basketball shots and field hockey shots for the goal; pushing an implement or an object to place it in a specific location such as a fencing lunge or a volleyball set; and parts of events in which the body must project itself to a specific location in space to accomplish a total body movement. The latter category was discussed in Chapters 13 and 14. The specific location or target in space was called a "point target." In the following section, projecting an object to a target and pushing an object to a specific location will be discussed.

Projecting an Object to a Target

In all activities in which one is projecting an object to a target, consistency of movement is of primary importance. Consistency of movement depends on stability of the body during projection; maintaining the end point, a body part, or an implement in a straight line just before and just after release; and assessing the angle of release for distance projections. The angle of release for projectiles was discussed in Chapter 10 and will not be repeated here.

Pushlike activities are characterized by the simultaneous rotations of two or more body segments so that the end point moves in a rectilinear path. In shooting a basketball, the shoulder is flexed, followed by simultaneous extension of the elbow and flexion of the wrist. The two rotational motions allow the player to move the ball in a rectilinear path before release and thus increase the precision of the release direction.

In tossing a dart, the performer stabilizes the body with a wide base of support. The trunk and legs remain relatively motionless while the shoulder is moved along some diagonal plane between full flexion and full abduction. Simultaneously, the elbow is extended to produce a straight-line motion of the thrower's hand and dart. Although there may be some wrist flexion or ulnar flexion, the third segmental rotation produces a rotational motion of the hand. Thus, more successful is the toss in which the wrist is stabilized.

Similar simultaneous rotations are seen in a volleyball set in which the setter must place the ball in a precise location. While the ball is touching the fingertips for a very short period of time, the rectilinear path of the fingers preceding the touch ensures that the ball is placed exactly where the setter wants it.

During activities in which the overall performance objective is distance of a projectile but in which accuracy is also a factor, creating a "flat space" in the path of the rotating implement is an important part of the skill. Such flat spaces occur in a golf swing, a tennis forehand and backhand, and somewhat in a baseball swing. Releasing or striking an object while it is moving in a rotational path may be accurate; however, the performer has an extremely short time in which to do so. The rectilinear path surrounding the release allows the performer a longer time in which to release or strike the object and still maintain accuracy to the target.

Finally, the more the performer is in a state of static (in the case of archery and shooting) or dynamic (in the case of moving basketball shots) stability, the easier it is for the performer to perform the movements consistently. In accuracy, consistency enhances the effectiveness of the movement.

References and Suggested Readings

Ae, M., Shibukawa, K., Toda, S., & Hashihara, Y. (1983). A biomechanical analysis of the segmental contribution to the takeoff of the one-leg running jump for height. In H. Matsui & K. Kobayashi (Eds.), *Biomechanics VIII-B: Proceedings of the Eighth International Congress of Biomechanics* (pp. 737–745). Champaign, IL: Human Kinetics.

Ae, M., Sakatani, Y., Yokoi, T., Hashihara, Y., & Shibukawa, K. (1986). Biomechanical analysis of the preparatory motion for takeoff in the Fosbury flop. *International Journal of Sport Biomechanics, 2*, 66–77.

Antonio, P., & Renato, R. (1987). Evaluation of biomechanical motor patterns in ski jumpers during simulation of takeoff. In B. Jonsson (Ed.), *Biomechanics X-B* (pp. 679–684). Champaign, IL: Human Kinetics.

Bates, B. T., & Lander, J. E. (1985). Variability

assessment of a submaximal skill for accuracy. In D. A. Winter. R. W. Norman, R. P. Wells, K. C. Hayes, & A. E. Patla (Eds.), *Biomechanics IX-B* (pp. 443–447). Champaign, IL: Human Kinetics.

Baumann, W., Gross, V., Quade, K., Galbierz, P., & Schwirtz, A. (1988). The snatch technique of world class weightlifters. *International Journal of Sport Biomechanics, 4*, 68–89.

Biesterfeldt, H. J. (1975, October). Salto mechanics II: Energy and height. *Gymnast*, pp. 54–56.

Bobbert, M. F., & van Ingen Schenau, G. J. (1988). Coordination in vertical jumping. *Journal of Biomechanics, 21*, 249–262.

Bolourchi, F., & Hull, M. L. (1985). Measurement of rider induced loads during simulated bicycling. *International Journal of Sport Biomechanics, 1*, 308–329.

Brooke, J. D., & Goslin, B. R. (1985). Effect of extended habituation on variability patterns of within-movement forces applied in pedaling. In D. A. Winter, R. W. Norman, R. P. Wells, K. C. Hayes, & A. E. Patla (Eds.), *Biomechanics IX-A* (pp. 388–392). Champaign, IL: Human Kinetics.

Brown, E. W., & Abani, K. (1985). Kinematics and kinetics of the dead lift in adolescent power lifters. *Medicine and Science in Sports and Exercise, 17*, 554–556.

Dapena, J., & Chung, C. S. (1988). Vertical and radial motions of the body during the take-off phase of high jumping. *Medicine and Science in Sports and Exercise, 20*, 290–302.

Davis, R., Ferrara, M. & Byrnes, D. (1988). The competitive wheelchair stroke. *National Strength and Conditioning Association Journal, 10*(3), 4–10.

de Groot, G., de Boer, R. W., & van Ingen Schenau, G. J. (1985). Power output during cycling and speed skating. In D. A. Winter, R. W. Norman, R. P. Wells, K. C. Hayes, & A. E. Patla (Eds.), *Biomechanics IX-B* (pp. 555–559). Champaign, IL: Human Kinetics.

de Boer, R. W., Schermerhorn, P., Gademan, J., de Groot, G., & van Ingen Schenau, G. J. (1986). Characteristic stroke mechanics of elite and trained male speed skaters. *International Journal of Sport Biomechanics, 2*, 175–185.

de Groot, G., & van Ingen Schenau, G. J. (1987). Biomechanical aspects of push-off techniques in speed skating the curves. *International Journal of Sport Biomechanics, 3*, 69–79.

de Boer, R. W., Ettema, G. J. C., van Gorkum, H., de Groot, G., & van Ingen Schenau, G. J. (1988). A geometrical model of speed skating the curves. *Journal of Biomechanics, 21*, 445–450.

Dufek, J. S., & Bates, B. T. (1987). Temporal gait characteristics of cross-country skiers. In B. Jonsson (Ed.), *Biomechanics X-B* (pp. 729–732). Champaign, IL: Human Kinetics.

Dyson, G. (1977). *The mechanics of athletics*. New York: Holms and Meier.

Enoka, R. M. (1988). Load and skill-related changes in segmental contributions to a weightlifting movement. *Medicine and Science in Sports and Exercise, 20*, 178–187.

Freivalds, A., Chaffin, D. B., Orr, R. B., & Radin, E. L. (1984). A dynamic biomechanical evaluation of lifting maximum acceptable loads. *Journal of Biomechanics, 17*, 251–262.

Garhammer, J. (1985). Biomechanical profiles of Olympic weightlifters. *International Journal of Sport Biomechanics, 1*, 122–130.

Gerber, H., Jenny, H., Sudan, J., & Stuessi, E. (1987). Biomechanical performance analysis in rowing with a new measuring system. In B. Jonsson (Ed.), *Biomechanics X-B* (pp. 721–724). Champaign, IL: Human Kinetics.

Gervais, P., & Wronko, C. (1988). The marathon skate in nordic skiing performed on roller skates, roller skis, & snow skis. *International Journal of Sport Biomechanics, 4*, 38–48.

Gregor, R. J., Cavanagh, P. R., & LaFortune, M. Knee (1985). Flexor moments during propulsion in cycling—A creative solution to Lombard's paradox. *Journal of Biomechanics, 18*, 307–316.

Hay, J. G. (1967). Pole vaulting: A mechanical analysis of factors influencing pole-bend. *Research Quarterly, 38*(1), 34–40.

Hay, J. G. (1975). Biomechanical aspects of jumping. In J. H. Wilmore & J. F. Keogh (Eds.), *Exercise and sports science reviews* (Vol. 3, pp. 135–162). New York: Academic Press.

Hay, J. G. (1986). The biomechanics of the long jump. In K. B. Pandolf (Ed.), *Exercise and sport sciences reviews* (pp. 401–446). New York: Macmillan.

Hay, J. G., Miller J. A., Jr. (1985). Techniques used in the triple jump. *International Journal of Sport Biomechanics, 1*, 185–196.

Hay, J. G., Miller, J. A., & Canterna, R. W. (1986). The techniques of elite male long jumpers. *Journal of Biomechanics, 19*, 855–866.

Hopper, B. J. (1973). *The mechanics of human movement*. New York: American Elsevier.

Hudson, J. L. (1986). Coordination of segments in the vertical jump. *Medicine and Science in Sports and Exercise, 18*, 242–250.

Hull, M. L., & Gonzalez, H. (1988). Bivariate optimization of pedalling rate & crank arm length in cycling. *Journal of Biomechanics, 21*, 839–850.

Hull, M. L., & Gonzalez, H. K., & Redfield, R. (1988). Optimization of pedaling rate in cycling using a muscle stress-based objective function. *International Journal of Sport Biomechanics, 4*, 1–20.

Hull, H. L., & Jorge, M. (1985). A method for biomechanical analysis of bicycle pedalling. *Journal of Biomechanics, 18*, 631–644.

Iizuka, K., Kobayashi, T. & Miyashita, M. (1985). Ski robot for parallel turning: Comparison with skiers movement. In D. A. Winter, R. W. Norman, R. P. Wells, K. C. Hayes, & A. E. Patla (Eds.), *Biomechanics IX-B* (pp. 543–548). Champaign, IL: Human Kinetics.

Jorge, M., & Hull, M. L. (1986). Analysis of EMG measurements during bicycle pedaling. *Journal of Biomechanics, 19*, 683–694.

Klinger, A., & Adrian, M. (1987). Power output as a function of fencing technique. In B. Jonsson (Ed.), *Biomechanics X-B* (pp. 791–796). Champaign, IL: Human Kinetics.

Klissouras, V., & Karpovich, P. (1967). Electrogoniometric study of jumping events. *Research Quarterly, 38*(1), 41–47.

Knutzen, K. M., & Schot, P. K. (1987). The influence of foot position on knee joint kinematics during cycling. In B. Jonsson (Ed.), *Biomechanics X-A* (pp. 599–604). Champaign, IL: Human Kinetics.

Komi, P. V. (1985). Ground reaction forces in cross-country skiing. In D. A. Winter, R. W. Norman, R. P. Wells, K. C. Hayes, & A. E. Patla (Eds.), *Biomechanics IX-B* (p. 18). Champaign, IL: Human Kinetics.

Komi, P. V. (1987). Force measurements during cross-country skiing. *International Journal of Sport Biomechanics, 3*, 370–381.

Kreighbaum, E. (1974). Mechanics of side horse vaulting. In R. C. Nelson & C. A. Morehouse (Eds.), *Biomechanics IV: Proceedings of the Fourth International Seminar on Biomechanics* (pp. 137–43). Baltimore: University Park Press.

Kumar, S., Chaffen, D. B., & Foulke, J. (1987). Methodology for the measurement of dynamic lifting strength. In B. Jonsson (Ed.), *Biomechanics X-B* (pp. 1077–1080). Champaign, IL: Human Kinetics.

Lander, J. E., Bates, B. T., & DeVita, P. (1986). Biomechanics of the squat exercise using a modified center of mass bar. *Medicine and Science in Sports and Exercise, 18*, 469–478.

Leonardi, L. M., Komor, A., & Dal Monte, A. (1987). An interactive computer simulation of bobsled push-off phase with a multimember crew. In B. Jonsson (Ed.), *Biomechancis X-B* (pp. 761–766). Champaign, IL: Human Kinetics.

Luhtanen, P., Pulli, M., & Komi, V. (1987). A relative model of human movement with special reference to ski jumping. In B. Jonsson (Ed.), *Biomechanics X-B* (pp. 1145–1150). Champaign, IL: Human Kinetics.

Madsen, N., & McLaughlin, T. (1984). Kinematic factors influencing performance & injury risk in the bench press exercise. *Medicine and Science in Sports and Exercise, 16*, 382–388.

McLaughlin, T., Lardner, T., & Dillman, C. (1978). Kinetics of the parallel squat. *Research Quarterly, 49*(2), 175–189.

McMahon, T. A., & Greene, P. R. (1978, December). Fast running tracks. *Scientific American*, pp. 148–163.

Miller, J. A. Jr., & Hay, J. G. (1986). Kinematics of a world record & other world-class performances in the triple jump. *International Journal of Sports Biomechanics, 2*, 272–288.

Nadeau, M., Cuerrier, J. P., Allard, J., Tardif, M., & Brassard, A. (1985). The power of young active men and women on a bicycle ergometer. In D. A. Winter, R. W. Norman, R. P. Wells, K. C. Hayes, & A. E. Patla (Eds.), *Biomechanics IX-B* (pp. 560–564). Champaign, IL: Human Kinetics.

Nelson, R. C., & Martin, P. E. (1985). Effects of gender and Load on vertical jump performance. In D. A. Winter, R. W. Norman, R. P. Wells, K. C. Hayes, & A. E. Patla (Eds.), *Biomechanics IX-B* (pp. 429–433). Champaign, IL: Human Kinetics.

Nelson, W. N. & Widule, C. J. (1983). Kinematic analysis and efficiency estimate of intercollegiate female rowers. *Medicine and Science in Sports and Exercise, 15*, 535-641.

Nicol, A. C. & Watkins, J. (1987). Biomechanical analysis of somersault activities. In B. Jonsson (Ed.), *Biomechanics X-B* (pp. 673–678). Champaign, IL: Human Kinetics.

Norman, R. W., & Komi, P. V. (1987). Mechanical energetics of world class cross-country skiing. *International Journal of Sport Biomechanics, 3*, 353–369.

Ozolin, N. (1973). The high jump takeoff mechanism. *Track Technique, 52*, 1668–1671.

Pierce, J. C., et. al. (1987). Force measurement in cross-country skiing. *International Journal of Sport Biomechanics, 3*, 382–391.

Redfield, R., & Hull, M. L. (1986). On the relation between joint moments & pedalling rates at constant power in bicycling. *Journal of Biomechanics, 19*, 317–330.

Ridka-Drdacka, E. (1986). A mechanical model of the long jump and its application to a technique of preparatory and takeoff phase. *International Journal of Sport Biomechanics, 2*, 289–300.

Sagawa, K., Tsuruta, H., & Ishii, K. (1985). Starting and braking of squat exercise. In D. A. Winter, R. W. Norman, R. P. Wells, K. C. Hayes, & A. E. Patla (Eds.), *Biomechanics IX-B* (pp. 424–428). Champaign IL: Human Kinetics

Sanderson, B., & Martindale, W. (1986). Towards optimizing rowing technique. *Medicine and Science in Sports and Exercise, 18*, 454–468.

Smith, G. A., Mcnitt-Gray, J., & Nelson, R. C. (1988). Kinematic analysis of alternate stride skating in cross-country skiing. *Journal of Sport Biomechanics, 4*, 49–58.

Spaepen, A. J., Peters, J., & Van Leemputte, M. (1987). Simulation of a standing long jump with imposed action at the joints. In B. Jonsson (Ed.), *Biomechanics X-B* (pp. 1187–1190). Champaign, IL: Human Kinetics.

Tsarouchas, E., & Klissouras, V. (1981). The force–velocity relation of a kinematic chain in man. In A. Morecki, K. Fidelus, K. Kedzior, & A. Wit (Eds.), *Biomechanics VII-A: Proceedings of the Seventh International Congress on Biomechanics* (pp. 145–151). Baltimore: University Park Press.

Vagenas, G., & Hoshizake, T. B. (1987). Optimization of an asymmetrical motor skill: Sprint start. *International Journal of Sport Biomechanics, 3*, 29–40.

van Soest, A. J., Roebroeck, M. E., Bobbert, M. F., Huijing, P. A., & van Ingen Schenau, G. J. (1985). A comparison of one-legged & two-legged countermovement jumps. *Medicine and Science in Sports and Exercise, 17*, 635–639.

van Ingen Schenau, G. J., de Boer, R. W., & de Groot G. (1987). On the technique of speed skating. *International Journal of Sport Biomechanics, 3*, 419–431.

van Ingen Schenau, G. J. (1987). Push-off force in speed skating. *International Journal on Sport Biomechanics, 3*, 103–109.

van Ingen Schenau, G. J., Bobbert, M. F., Huijing, P. A., & Woittiez, R. D. (1985). The instantaneous torque-angular velocity relation in plantar flexion during jumping. *Medicine and Science in Sports and Exercise, 17*, 422–426.

van Ingen Schenau, G. J., de Groot, G., & de Boer, R. W. (1985). The control of speed in elite female speed skaters. *Journal of Biomechanics, 18*, 91–96.

Whiting, W. C., Gregor, R. J. & Finerman, G. A. (1988). Kinematic analysis of human upper extremity movements in boxing. *The American Journal of Sports Medicine, 16*, 130–136.

Wielki, C. Z., & Dangre, M. (1985). Analysis of jump during the spike of volleyball. In D. A. Winter, R. W. Norman, R. P. Wells, K. C. Hayes, & A. E. Patla (Eds.), *Biomechanics IX-B* (pp. 438–442). Champaign, IL: Human Kinetics.

Wilkerson, J. D. (1985). Comparative model analysis of the vertical jump utilized in the volleyball spike with the standing vertical jump. In D. A. Winter, R. W. Norman, R. P. Wells, K. C. Hayes, & A. E. Patla (Eds.), *Biomechanics IX-B* (pp. 434–437). Champaign, IL: Human Kinetics.

Wilson, B. D., McDonald, M. & Neal, R. J. (1987). Roller skating sprint technique. In B. Jonsson (Ed.), *Biomechanics X-B* (pp. 655–660). Champaign, IL: Human Kinetics.

Yoshihuku, Y., Ikehami, Y., & Sakurai, S. (1987). Energy flow from the trunk to the upper limb in tsuki motion of top-class players of the martial arts, shorinji kempo. In B. Jonsson (Ed.), *Biomechanics X-B* (pp. 733–738). Champaign, IL: Human Kinetics.

CHAPTER 17

Introduction to Biomechanics Instrumentation

PREREQUISITES

Chapters 1, 9
Concept Modules A, B, C, D, E, F, G

CHAPTER CONTENTS

17.1 Research In and Out of the Laboratory

Teachers and coaches who are involved with helping people move better are required to qualitatively evaluate movement based on observations of performances during class or practice. When precise quantitative information about a performance is required, however, the use of measuring devices is necessary. Movement is studied with a variety of instruments to unveil details of performance that cannot be discerned by the human eye alone. Information about skilled or normal movement patterns gained with these tools is used to form better models of performance that are used in such applications as improving industrial work tasks, rehabilitating the impaired, teaching sport and dance skills, and refining the elite athlete. When teachers and coaches combine their talents with research personnel and their instrumentation, a greater opportunity exists for conducting research to answer relevant, practical questions, the answers to which can be used to help solve movement problems.

The purpose of this chapter is to introduce to the student the most commonly used instrumentation for studying movement. Some instrumentation is high tech; some is not so high tech but serves the evaluator's purpose. Although some devices must be used within a laboratory setting, most can be used fairly easily in the field and even during competition.

17.2 Overview of Instrumentation and Its Uses

Instrumentation used for measuring the many aspects of human movement has been changing rapidly over the past several years. The development and widespread use of the microcomputer has played the most important role in facilitating biomechanical research. For example, by connecting (interfacing) some measuring devices to a computer with appropriate software (programmed instructions), signals can be measured, interpreted, and printed almost immediately.

Instruments most commonly used with a computer for data analysis include force platforms, force and pressure sensors, muscle activity sensors (electromyography), accelerometers, dynamometers, and joint angle sensors (goniometers). Points plotted from film or videotape also can be read by a computer for analyzing changing joint angles and selected points on a body or even to determine the body's center of gravity (CG) in any position. Mechanical work, power, and energy also can be determined from film or video records using segmental mass values appropriate for the subject.

Computer-assisted data analysis also allows for the output of very informative graphics (graphs, charts, and pictures). With data taken from film, videotape, or videodisc, the computer's screen images of the moving body or segments can be animated (put in motion) and, with three-dimensional data, viewed from any perspective the viewer chooses. Figure 17.1 shows an animated stick figure displayed on a computer monitor.

When a quantitative analysis of a movement is to be performed, the most meaningful quantities must be identified; then the appro-

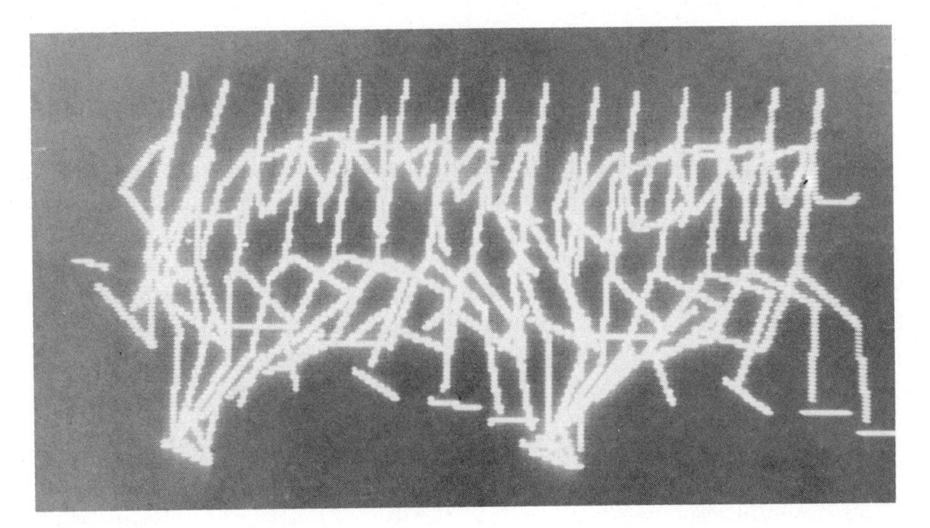

FIGURE 17.1

Animated stick figure of a runner displayed on a computer monitor. The figure was produced by a computer program using points traced from a series of still film frames.

priate tools for measuring can be selected. One must resist the temptation to measure something simply because the tool is available and the measurement would be convenient to obtain. The information sought should dictate the type of instrument selected, and the data should have the potential for answering the predetermined questions. The availability of computerized equipment designed for measuring specific things easily can lead to the practice of measuring only those things that one's equipment can measure.

Although very sophisticated equipment and devices are being used in biomechanical testing and research, some older methods and equipment also can provide quality information, but their use usually takes more time. For instance, the same signals that are sent to a computer for analysis could be examined by a person, but the computations involved would require days, weeks, or longer to complete.

Sometimes a research question dictates the use of a "customized" device to provide the necessary kinds of data. In such instances, the investigator should feel free to devise ways of obtaining the necessary measures even though the method may be nontraditional. A major consideration in the use of any measuring device is the extent to which the device may inhibit or change the real, or normal, movement performance of the subject. Often the subject will try to be careful not to damage or dislodge some piece of equipment or will try to make contact with a device in such a way that the normal movement is changed. Such interference may not be readily detectable by the observer or the subject, so the investigator must make every effort to ensure that the subject feels free to move normally without restriction. An advantage that film or video techniques have is that data can be obtained during nonlaboratory conditions or during competition while the subject is concentrating on the activity rather than on the fact that he or she is being studied.

17.3 Clocks and Timers

If the purpose of the study requires information about the time for something to move from one location to another, simple clocks or timing devices could be used. Clocks and timers can be activated by on and off switches that can be triggered by the investigator or by the performer during movement. Those that are triggered by the performer provide more accurate data because the response time of an observer is eliminated. Figure 17.2 shows one type of timing device that can be configured to provide time durations for whole-body displacements or segmental movements.

Figure 17.3 shows a series of light beams and photocells. When a single beam is interrupted by passing the body, segment, or implement through it, a clock is triggered to start, split-stop, or stop, depending on the wire connections between the light sensors and the clock. The clocks can record times in milliseconds (one msec = 0.001 sec).

Photocells such as those in Figure 17.3 are highly versatile in that they can be positioned very close together for sensing small movements of segments, or they can be spaced far apart for recording such things as running velocities through intervals.

With such timing devices, the position of the moving body is predetermined by the location of the switches or light beams, so that a known distance is covered between the starting and stopping or split-stopping of the clock. With the displacement and time known, the average velocity can be calculated. If more

FIGURE 17.2

Three contact-sensitive mats are wired to a Dekan timing device to time the rate of side-stepping from one to the other.

Top: Three light sensors spaced 0.5m apart connected to msec clock.
Passing an implement through each light beam triggers the clock to start, split-stop and stop.
Bottom: Three light sources direct beams toward sensors.
Clock

FIGURE 17.3

Photocells and clock apparatus. By swinging an implement through the space between the upper light-sensitive cells and the lower light beams, the clock is triggered to start, split-stop, and stop. Implement velocity can be determined for the first and second intervals as well as for the total distance.

split-stop triggers are activated, and the average velocity calculated for each interval of the total displacement, variable velocity can be determined. The smaller the distance between the trigger devices, the more "instantaneous" the calculated velocity; that is, the smaller the chance that the velocity changed (accelerated or decelerated) within that interval. Richards (1987) describes his method for connecting a photocell system to a microcomputer and includes the computer program used to drive the photocell circuit.

Time and position data are obtained through photographic methods, as well. Three common techniques are stroboscopy, cinematography, and videography. All are used to produce images of the performer from which measurements can be made.

17.4 Stroboscopy

Stroboscopy is a technique that produces superimposed images on a single photograph. The subject must perform in the dark, and a strobe (flashing) light is set so that the performer is illuminated, say 60 flashes/sec (60 Hz), and an open-shutter still camera records each lighted image on top of the previous image. Measurements can be taken from the photograph if each image is clear enough, and velocities can be calculated if the time between each image is known. The clarity of the image is assured by having a very short exposure time; industrial quality stroboscopes have flash durations of about 10 microseconds (0.000010 sec), which is much shorter than used for most photographs. Because the subject must be in complete darkness except for the flashing light, this technique is limited to movements performed in place, such as a golf swing, vertical jumping, calisthenics, or throwing.

17.5 Cinematography and Computer-Assisted Analysis

Cinematography (motion picture photography) is probably the most popular and productive method used for studying movement. Its level of sophistication has grown steadily over the past few decades. Motion pictures of a subject's performance provide a record that can be viewed repeatedly for a qualitative analysis as well as for obtaining quantitative data. Measurements can be taken from each still frame as it is projected (like a slide) on a flat surface.

Cameras

Most analysis cameras used today are 16 mm and are driven by rechargeable battery packs (Figure 17.4). Some super-8 cameras also are used. The cameras are specially built to operate at very fast filming rates so that a large number of visual "samples" of a movement may be captured. With these high-speed cameras, more detail can be examined than is possible with ordinary movie cameras that take only 18–24 pictures per second. Some older, spring-driven cameras have a maximum rate of 64 frames per second (fps) and must be rewound frequently to keep the rate up to maximum. The modern cameras can be run at very low speeds (1–10 fps) to over 10,000 fps. An internal timing light registers a dot of light at a selected frequency (e.g., 100 Hz) on the film border so that the actual frame rate may be verified when it is developed. For a camera without an internal timing light, an external timing device must be filmed to determine the exact frame rate.

Filming Speed Considerations

The selection of frame rate (film transport speed) depends on the nature of the movement and the information that is needed

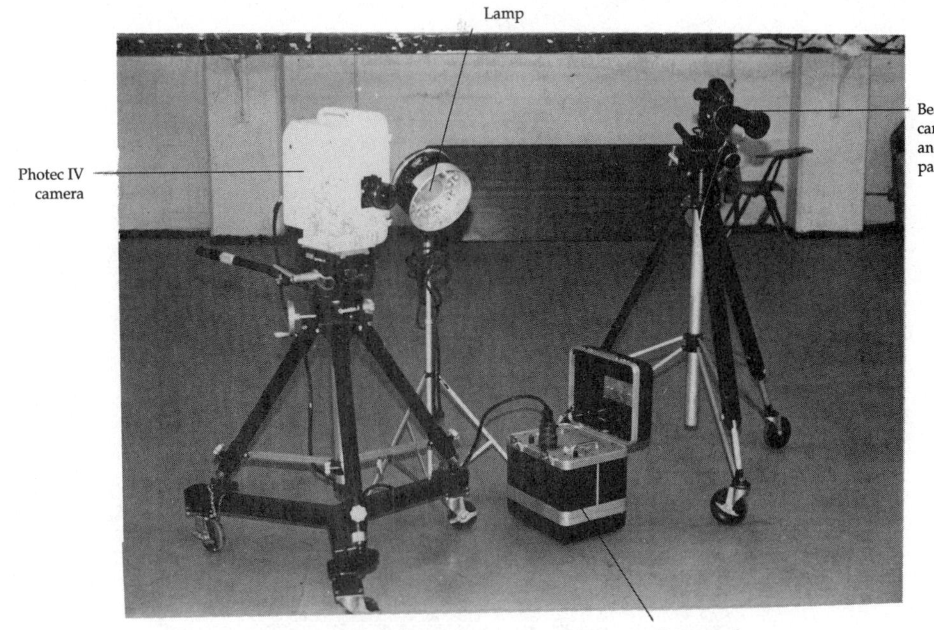

FIGURE 17.4
A Photec IV 16-mm high-speed camera (100–10,000 frames/sec) is shown with its battery pack and a Pallite VIII lamp. A motor-driven 16-mm, 64-frame/sec, Beaulieu camera with battery pack.

from the film. Very fast movements that may involve an important instant, such as contact of two objects (a golf drive, racquetball shot, a runner's foot plant, or a karate blow) would require many samples (200–500 fps, or more) to increase the chance of capturing the contact event. With fast frame rates, the time between each frame becomes very small. For example, at 100 fps the time between frames is 0.01 sec. At 200 fps it is 0.005 sec. (5 msec). Even faster rates are required if impact deformation and reformation of a ball, implement, or body part is to be revealed. With very fast frame rates, however, another factor must be considered: the time the camera's shutter is open to allow light to strike the film. If it is not long enough, the film is not exposed enough and the image is too dark. If it is too long, fast movement occurs while the shutter is open and the image is blurry.

The exposure time is determined by the setting of the camera's adjustable shutter opening size and by the frame rate. With a very fast frame rate and small shutter opening, the exposure time is very small. With a small exposure time, bright light on the subject is necessary. A light meter is used to measure the light reflected by the subject and is then used to determine the camera's lens opening (f-stop) based on the known exposure time. Given enough light, the exposure time can be set to a very small value so that the image of a fast-moving subject is not blurry. If the movement is fast but no particular instant needs to be captured, the frame rate can be reduced, but the exposure time still needs to be small to prevent blur.

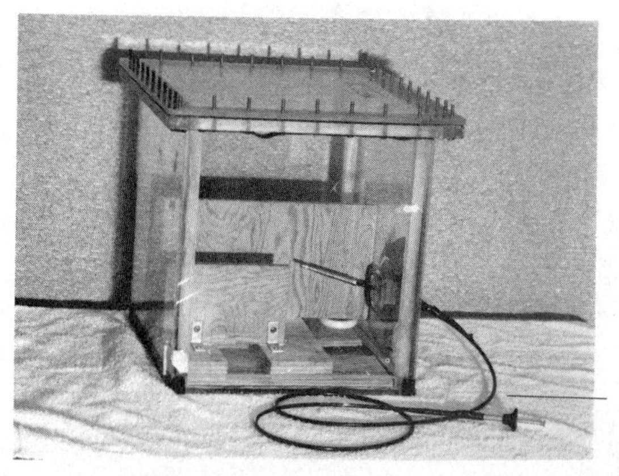

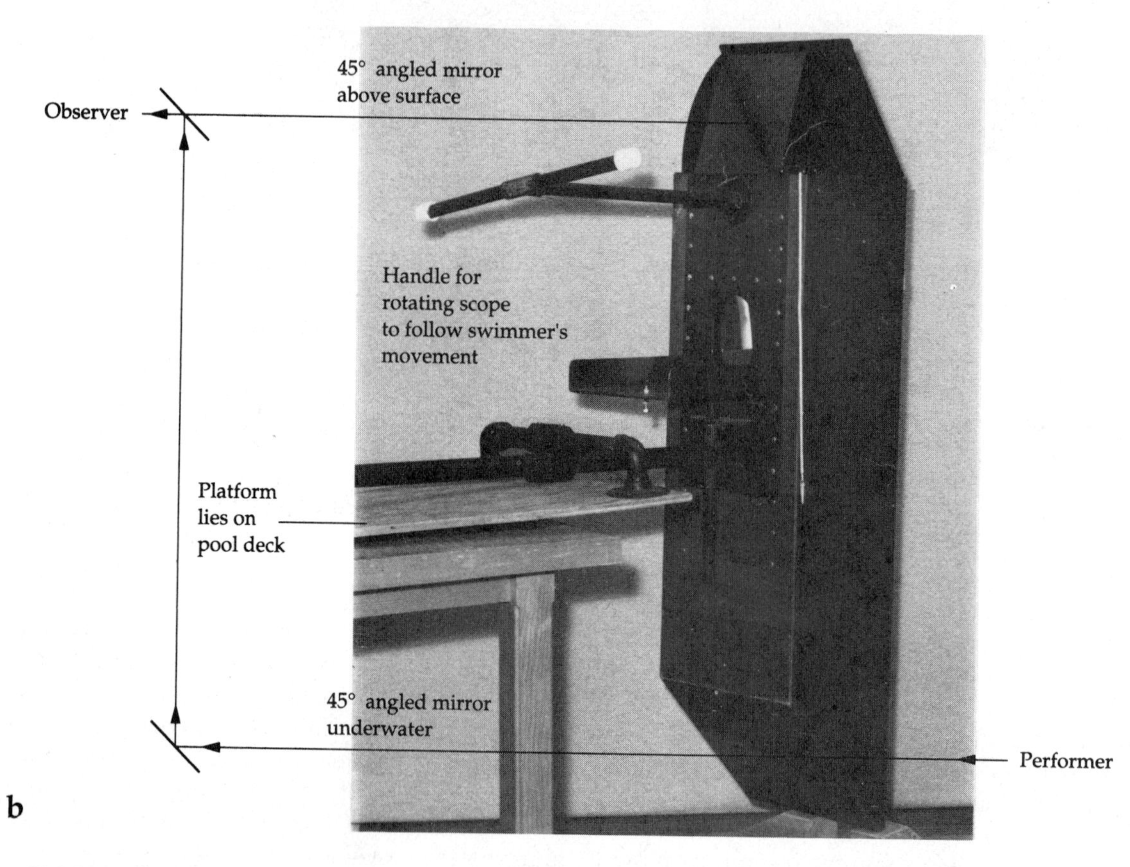

FIGURE 17.5

(a) A waterproof camera housing with a cable camera trigger contains a frame to support a camera for underwater filming. (b) A periscope used for viewing underwater from a pool deck. (c) The periscope allows the positioning of a camera to film or videotape.

FIGURE 17.5
(continued)

Lighting and Films

Outdoor filming in sunlight permits very fast frame rates and very small exposure times, even for underwater filming through underwater viewing windows, waterproof camera housings, or periscopes (Figure 17.5).

Indoor filming at fast rates and small exposure times requires the use of photographic flood lamps to provide enough light to expose the film in the short time it is exposed to the light. The selection of a more light-sensitive film (with a high ASA or ISO number) alleviates the lighting problem to some extent, but such film produces a grainier, less clear image. Black and white film typically is used in 16-mm cameras because it is more light sensitive than color film and it costs less to buy and process. In the high-speed super-8 cameras, however, color film cassettes are purchased and processed at a very low cost. An advantage to color is that more image detail can be detected and often more accurate differentiation of portions of the image can be made. The disadvantage of the super-8 format is its reduced image size, which could lead to a greater measurement error than that made with a 16-mm image of more than twice the area. However, given other sources for error that are present for filming as a technique, this source should not restrict its use for obtaining credible data.

Lenses and Camera Positioning

When high-speed cinematography is used, careful planning is necessary to ensure clear, measurable images. The subject must be marked in locations appropriate for the desired data. An object of known length must be in the image so that image measurements can be converted to actual measurements when the film is analyzed. Known vertical or horizontal references are also necessary. It is important to position the camera far enough away from the subject so that the total movement may be captured without using a wide-angle lens, which distorts the real dimensions of the subject. Therefore, a distant camera with a good-quality telephoto (magnifying) lens to enlarge the small distant subject is necessary. Camera lenses that can be adjusted for subject distance (zoom lenses) provide flexibility for the positioning of the camera relative to the subject.

The camera is mounted on a stable tripod, its height adjusted to the level of the center of the subject's activity, and leveled. The lens is adjusted to provide an image of the subject as large as possible without losing parts of the performance as a result of a small field width. If possible, several practice trials of the actual performance should be performed so the camera operator can determine the best field width but also to determine when the camera is to be switched on and off. In high-speed filming, starting the camera too soon (when the subject is far away from the camera's view) uses film unnecessarily. The same waste occurs if the camera is not switched off following the performance.

Two-Dimensional (2-D) Cinematography

When a single camera is used to film a subject's performance, the image produced is two dimensional; that is, the movement can be measured in only one plane. (Recall that a plane has two dimensions; a solid, or volume of space, has three dimensions.) Measurements taken from a two-dimensional image can be accurate only for movements that move along the plane that is perpendicular to the optical axis of the camera lens (parallel to the plane of the film in the camera). Unfortunately, most movements of the human body are three dimensional, and only a few segmental movements may be suitable for measuring from a 2-D picture. For example, if normal walking is viewed from a side view, the thigh and leg appear to move along the subject's sagittal plane. If viewed from the front, however, the legs are observed to move also slightly inward along the walker's frontal plane for foot plant. If only a 2-D side view is used to describe the kinematics of the lower extremity, the true three-dimensional (3-D) nature of the movement is not revealed. Filming from a front or back view is necessary. Movements that occur only in one plane can be fully described by a single-plane analysis, and it is possible that a 2-D analysis would be appropriate in some circumstances. For movements that the investigator is unsure of, the only way to tell is to observe and possibly measure the movement from two or more different views.

Three-Dimensional (3-D) Cinematography

To record a performance from more than one view, two cameras are needed. Although a performance can be filmed from one view and then filmed from another view with the performance repeated, a valid three-dimensional representation of the act is not obtained. The reason is that a single performance is not likely to be repeated without some measurable variation in some aspect. Therefore, simultaneous filming of the subject from two or more views is necessary. Before the availability of relatively recent computerized analysis techniques, it was necessary to position the cameras perpendicular to one another. For example, one camera was directed toward the side, another toward the front, and if a third camera was used, from overhead. Points and pictures could then be traced on an x–y plane, an x–z plane, and a y–z plane, each plane being taken from its respective camera. All three spatial coordinates of each point (x, y, z) could be determined, and the three-dimensional motion of points could be described. Major limitations to this procedure are the difficulty with which filming could be conducted and the potential for error in setting up the site.

With the demand for easier methods for obtaining 3-D data came a valuable tool that allowed for the positioning of two cameras at some angle more convenient to the investigator. This less restrictive requirement for camera positioning allows greater flexibility for setting up the filming site. The procedure, described by Shapiro, 1978, is called the *direct linear transformation method* (DLT). It is based on a series of equations that relate the x, y, z

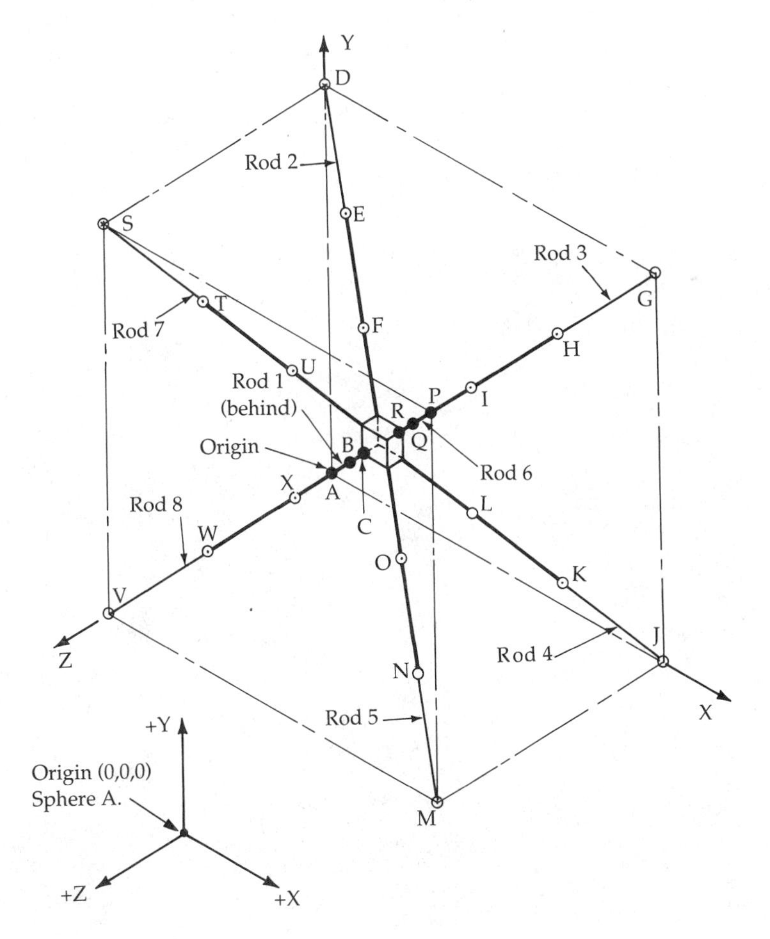

FIGURE 17.6

A calibration frame used to establish measurement coordinates in the x, y, and z directions. (Courtesy of Peak Performance Technologies, Inc.)

coordinates of known markers in the subject's movement field to the x, y, z coordinates of the point marked on the subject's body (or implement). It requires an apparatus constructed to resemble a giant tinker-toy arrangement, 1–2 m in length, height, and width (Figure 17.6).

Before filming the performer, the apparatus is placed in the activity area and filmed. The x, y, z values of the points on this apparatus are plotted and hand measured. The hand-measured values are fed into the computer for use during the subject's film analysis.

When using two or more cameras, it is essential to synchronize them; that is, there must be a way to ensure that the film frames from one camera exactly match the same instants in time recorded on the frames of the other camera. An external "phase-lock" triggering device can be used with some cameras to ensure the synchronous operation of both cameras set at the same frame rate. If such a device is not available, one cannot assume that the shutters of both cameras were open at exactly the same brief instant, consequently, the simultaneity of the images in the two planes cannot be assured. An event marker of some type (e.g., a ball hitting the ground) can be the basis for synchronizing the film images when they are viewed. If the frame rates are not the same, one camera should operate at at least twice the speed of the other so that its film can be used to match the frames of the slower camera.

Projectors

Standard movie projectors do not have the options necessary for taking measurements from single-film frames, so specially built "an-

FIGURE 17.7
Layfayette Analyzer projector for projecting film one frame at a time, slow motion, or normal speed.

alyzer" projectors must be used. The options necessary include a manual film-threading pathway, a high-intensity lamp with a cooling fan, a single-frame advance and flicker-free still frame, variable-speed forward and reverse drive, a frame counter, a remote control box, and a good-quality lens. Figure 17.7 shows a typical film-analysis projector that is made for projecting films frame by frame.

Computer-Assisted Data Analysis

Before computers were available and affordable for the biomechanics researcher, film analysis was performed by hand tracing, measuring, and calculating point displacements and angles. The time consumed by such efforts suppressed much of the research that otherwise would have been conducted. Now, instead of projecting a still frame on paper for hand-tracing points and lines, each frame is projected on a surface (or rear-projected through a frosted glass plate) so that points in the image can be *digitized*. To digitize a point means to plot its x and y coordinates. This is done electronically by the x and y sensor devices of an image digitizer. By positioning an electronic stylus or cross-hair cursor on a point, for example on the hip joint of a runner, the x, y position of that point on the screen (graphic tablet) is detected and displayed on a control unit panel (Figure 17.8).

By advancing the film to the next frame of the runner, the hip joint can be digitized again, producing its x, y coordinates in that frame. The change in the x value is equal to the displacement of the hip in the x direction, and the

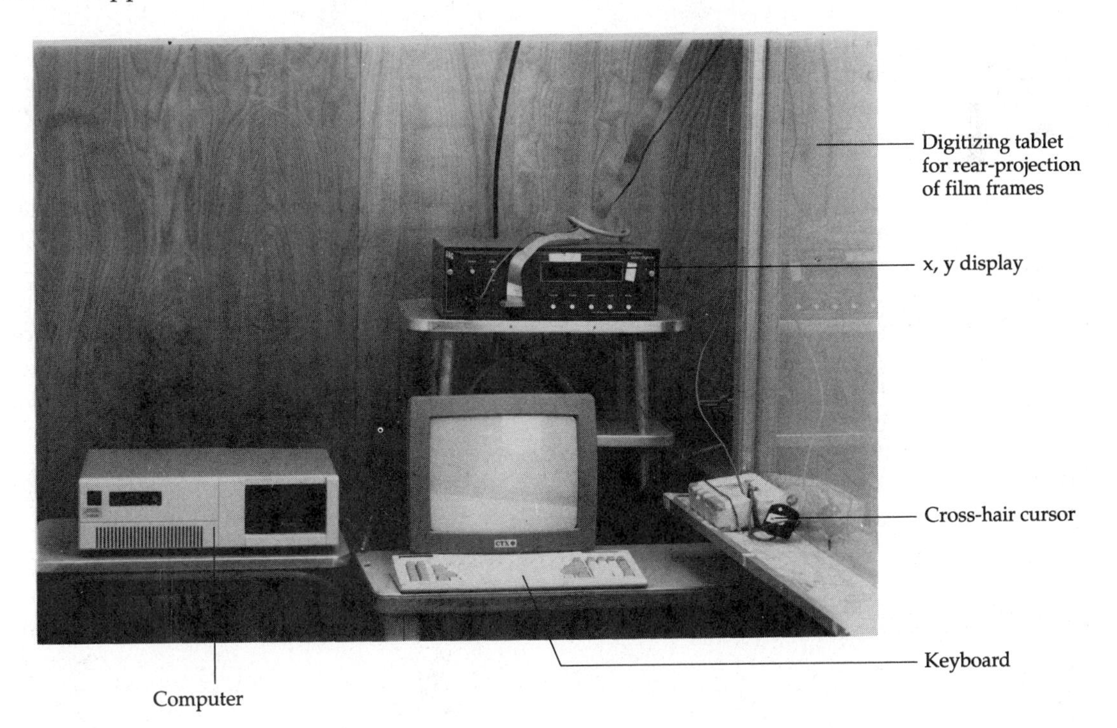

FIGURE 17.8

A computer coupled with an image-digitizing system. The cursor is placed on each desired point on an image projected on the digitizing tablet, and the x, y coordinates of the point are sensed and stored for analysis.

change in the y value is equal to the displacement in the y direction. Use of the Pythagorean theorem ($\sqrt{x^2 + y^2}$ = displacement) will yield the total displacement of the runner's hip from the first frame to the second. Knowing the frame rate of the camera, the time for that displacement is also known. Dividing the displacement by the time will give velocity. If a joint angle is wanted, three points must be **digitized** for the data required to calculate it (e.g., the hip, knee, and ankle, for calculating the angle at the knee).

If the digitizer is not interfaced with a computer, each x, y position must be hand written by the investigator and used later with subsequent coordinates; hand calculations would be necessary. Such a procedure eliminates most of the advantages of digitizing. Therefore, digitizing units are interfaced with a computer. Some are on-line to a central computer (main frame) via a modem (transmitting device to a distant computer); however, most are now connected directly to a microcomputer. In addition, the microcomputer is connected to a printer or plotter for producing a paper copy of the results.

Computer programs have been developed to use the x, y coordinates input by the digitizer in equations that produce linear and angular displacement, velocity, and acceleration of points and angles. If the motion of the body's CG needs to be determined, the software can use input segmental mass data and body joint coordinates to calculate the CG position in each frame and perform the same calculations to quantify its motion. The user determines exactly which parameters (quantities) are output by the computer.

The validity and reliability of the x, y position of the point depends on a number of potential errors that could be made in the filming process, the placing of the markers on the subject, the hand steadiness of the investigator, and other variables. Because some error always exists, computer programs have been created to decrease the effect of error by treating the incoming raw data with some kind of error-smoothing or data-filtering technique. Smoothing of the position data is necessary for the mathematical determination of velocities by the computer program. It becomes even more critical for the determination of acceleration values because the original error is magnified greatly. Many types of smoothing procedures exist, and the selection of one over another depends on the nature of the movement studied and the data produced (Sprigings, 1988).

Computer programs allow for the printing of data in tabular form, giving frame by frame displacements, velocities, and accelerations, as well as the x and y components of these parameters. These values also may be graphed on a coordinate system produced by a printer or plotter. Stick figures can be generated on the computer's monitor and printed on paper if desired. With three-dimensional data taken from two or more films, kinematic and kinetic parameters in x, y, z directions also can be generated (Figure 17.9). No doubt, with the continued refinements in computer software, faster and easier film data analysis will result.

17.6 Videography and Computer-Assisted Analysis

Videotaping performances during practice or competition has been commonplace for a number of years. For some time it was considered a technique to be used only for qualitative analysis because quantitative positional information taken from the relatively low-resolution (non-

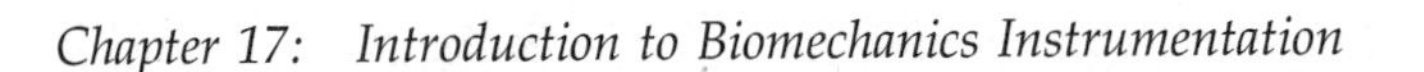

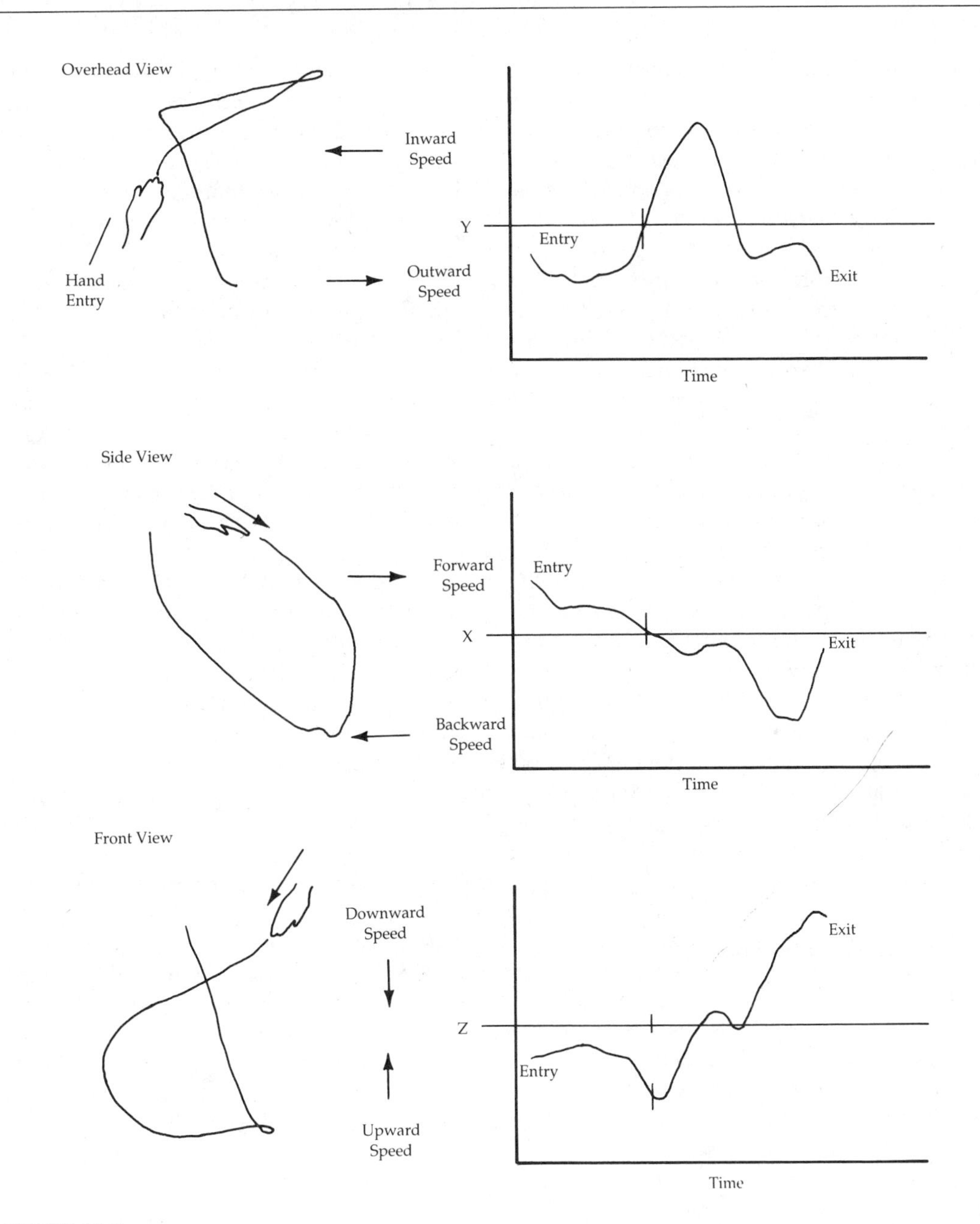

FIGURE 17.9
The underwater path of a butterflyer's hand is displayed from three perspectives. The velocity of the hand in each of the three directions is plotted against time.

sharp) and "shaky" images contained large errors. Error also was introduced if measurements were taken by hand from the glass screen over a curved tube. The scan ("frame") rate of standard cameras was (and still is) limited to 30 (60 fields/sec). Two fields interlace to combine to form a full picture. The exposure time from one full picture to the next amounted to 1/60 sec, as the image was "painted on across the screen." The lack of control of exposure times meant that the freeze frame image of a mover was always blurry.

Within recent years, however, the improved technology in television and videotaping, along with its relatively low cost and immediate results in color, has made it the rapidly growing competitor of cinematography. One appreciable advantage of using video is that the user can check the lighting brightness and contrast and focus with a trial run and then change it before actual data collection begins. With film, the user calculates the appropriate settings to use for filming, but never knows what the product looks like until after processing some days later. Other advantages include less expense, immediate results, and much longer recording time.

Cameras and Playback Units

Video cameras that have a mechanically or electronically shuttered lens feature are capable of producing very fast exposure times, most of which are faster than those used in most filming research. These exposure times can be as fast as 0.0001 sec and probably will become faster for even sharper images if adequate lighting sources are also available (Cheetham and Scheirman, 1988). Although the standard shuttered camera is still restricted to a relatively low frame rate, its other advantages make it a popular analysis technique for movements other than those that require fast frame rates for capturing a critical instant.

The *camcorder* is a camera that has the video recorder built into it and uses a rechargeable battery that is also contained within the camcorder. Camcorders are very lightweight and contain features that can be manually or automatically controlled to produce clear, sharp images. With the shuttered lens feature, the freeze frame contains no "noise" to obscure the image. Zoom lenses are used to the same advantage as they are for cinematography. The single-frame advance feature of a playback unit (videocassette recorder, VCR) gives a viewer the possibility of taking frame-by-frame measurements in the same way that film is viewed.

The clarity of the image depends also on the number of video recording heads in the camcorder and VCR. It varies from two, for ordinary use, to three, four, or more. The VCR selected will determine the options available for analysis use. For example, analysis requires a freeze-frame option and a single-frame advance both forward and backward. If no single-frame reverse capability is present, quick access to a recent previous frame is restricted. Locating frames can be facilitated by superimposing an image of a timer on the image while the performance is being recorded.

Because of the way the full image is formed by 60 fields/sec to form 30 pictures/sec, the standard VHS format VCR will display every second field, giving a maximum of 30 pictures/sec. Newer digital VHS VCRs are capable of advancing the tape field by field, yielding 60 pictures/sec, which is adequate for many human-motion studies if the exposure time is fast enough.

Image Measurements

The disadvantage of hand measurements made directly over a monitor screen is still present. Taking measurements on a video image that is projected on to a flat surface is possible, but the image distortion that may occur with some video projectors must be eliminated or corrected for. As is true for cinematographic analysis, digitization of the image with computer-assisted data analysis is required for an analysis to be completed within a reasonable amount of time.

Computer-Assisted Data Analysis of Video

Digitizing Projected Video Images

If the image distortion produced by projecting a video image on a flat surface or through a rear-projection digitizing tablet can be eliminated, video images can be treated as film images projected for digitization. The same procedures would apply, except for the limitation placed on the freeze-frame time. Film-analysis projectors have fans to cool the film frame being projected by a hot lamp and can be held for long periods of time. VCRs typically limit the time of freeze frame to 2 to 3 minutes before automatically advancing to the next frame to prevent damage to the heads. Unless the number of points being digitized is very large, such a time limit would not be an inconvenience.

Video Overlay Devices

To eliminate the error and time consumption of hand measuring on a video screen, a device known as a *video overlay* can be used relatively inexpensively with a microcomputer. It allows the display of a video image on the computer's monitor screen, which then can be digitized by placing on each point a cross-hair cursor, controlled by keyboard, joystick, or a "mouse." The computer program uses the x, y coordinates of each point for calculations of the parameters specified. Shapiro, Blow, and Rash (1987) reported such a procedure to be comparable to that used for cinematography digitization.

Similar in concept to the video overlay technique is an inexpensive technique described by Abraham, 1987. Using standard video equipment, the use of a special effects generator (SEG) enables the mixing of the performance video with the image of the digitizing cursor tablet ordinarily used for film analysis. The image of the cursor is superimposed on the subject on the monitor as the cursor is moved by hand on the tablet. The digitizer is interfaced with a microcomputer for data analysis and output.

Computerized Video Analysis Systems

Optoelectronic Cameras

Optoelectronic cameras operate on light-sensitive units and interface directly to a microcomputer. The camera records the coordinate locations of luminous markers placed on the subject's body. The digitization of the location of the markers is automatic, requiring no operator digitization of the markers with a cursor. The markers can be *passive* or *active*. Passive markers are reflective and adhere to the subject and allow the most freedom of movement, whereas the active markers are infrared light-emitting diodes (LEDs). Some cameras are useful for only slow movements, such as gait, in rehabilitation studies because of the low frame rate and image resolution (Vaughn, Smith, and du Toit, 1987). Others can record up to 200 frames/sec with high resolution. The VICON (Oxford Medilog, Inc.) is one system that digitizes up to 30 passive reflective markers from two to seven cameras at 50 frames/sec, stores the coordinates on computer disk, and performs the data analysis. The cameras use special infrared-sensitive tubes along with infrared strobe lights, invisible to the subject. The output data are in the form of graphs or stick figures.

The SELSPOT System (Selective Electronic, Inc.) uses LEDs attached to the subject, and up to 30 points are registered by the cameras simultaneously. The three-dimensional data are output to an interfaced computer for analysis. Macellari, Rossi, and Bugarini (1987) present the use of another system (CoSTEL) and discuss the features of other optoelectronic systems.

High-Speed Computer/Video Systems

Since the early 1980s, high-speed video recording and analyzing systems have become available. Higher "effective" frame rates are made possible by "screen splitting" and units called "frame grabbers." They store designated frames in memory and separate each frame into its odd and even fields (every other field is odd or even). The even field is filled in by interpolation to reconstruct a full, higher resolution image. The odd field is then retrieved from memory and treated the same way. With such computer-controlled frame splitting, the effective frame rate of a standard camera is 60, rather than 30, frames/sec. A computer-controlled cross-hair cursor, controlled by a "mouse," is placed over the new image to digitize points with the aid of a computer and VCR controller unit. An advantage of the frame-grab technique over the overlay method is that a grabbed frame can be computer enhanced to create a clearer image for locating points. It also places a dot on the point digitized to give the user accuracy feedback and a chance to re-digitize (Cheetham and Scheirman, 1988).

Three-dimensional data can be obtained using two or more cameras, as in cinematography. Using the same direct linear transformation (DLT) equations in the computer analysis program, the two-dimensional data from each camera are merged to produce the three-dimensional position data that are used for further calculations of kinematic parameters. When two or more cameras are used for three-dimensional data, the use of a device called a *genlock* to synchronize the cameras in time can be used. Configurations of components can be made to split the screen vertically and horizontally to display the images from both cameras simultaneously. Graphics programs then enable the viewer to choose any view of the subject, to rotate it to be viewed from any direction, and to animate stick figures of the body defined by the digitized joint centers. The screen images can then be printed in black and white or color. Three popular systems at the time of this writing are manufactured by Ariel Dynamics, Inc., Peak Performance Technologies, and NAC (distributed by Instrumentation Marketing Corp.).

The PEAK System has a feature that enables the user to superimpose computer-generated graphics with the original video images, and the composite image may be re-recorded (Figure 17.10). Its use in biomechanical analysis has been reported by Smith, Dillman, and Risenhoover (1988).

The Ariel Performance Analysis System is a system that also provides easy data collection and analysis. In addition to the kinematic data, it can include on the computer screen simultaneous records of force-plate data and muscle activity if instruments for those measurements are used also.

The NAC Model HSV-400 System (Instrumentation Marketing Corp.) includes a specialized color shuttered video camera, which can record 200 full frames/sec and 400 half frames/sec. The exposure times from 500 to 20 microsec produce very clear images on standard 1/2-inch VHS tape (Figure 17.11). The computerized analysis is accomplished also by superimposing crosshairs on the screen to digitize the point coordinates. The data are then sent to an IBM/PC (or a compatible unit) for analysis by means of software included with the system.

These systems have attractive capabilities. Typically, they use the standard color 60-pictures/sec shuttered video cameras, but 200 and 400-pictures/sec NAC cameras can be used also. The PEAK system can be used with video recording systems that record up to 2,000 frames/sec. The digitization process includes a process for predicting the location of each point in the following frame. The software allows the "tracking" of the movement of each point from frame to frame and places the crosshairs on the same point in the next frame; the user can manually adjust the precise location, however. This process is called "semiautomatic digitizing," and it reduces the time necessary for the manual digitization operation. In the PEAK system, reflective markers can be used for automatic digitizing, as is done with the optoelectronic camera systems. With the PEAK system, however, the computer alerts the operator if a potential tracking problem exists, such as two points passing close together, and the operator can intervene to make the correct decision.

The Spin Physics' SP2000 Motion Analysis System (Eastman Kodak), which can record from 60 to 2,000 full pictures/sec, does not use conventional video-imaging tubes; it operates

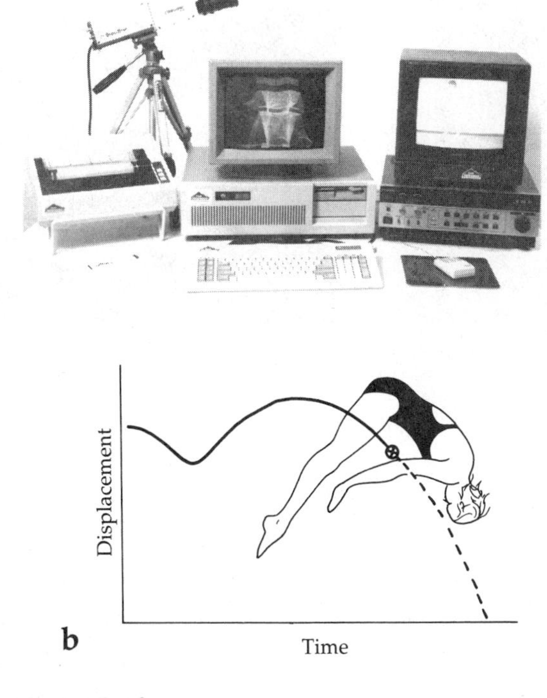

FIGURE 17.10

(a) Components for the Peak Performance Analysis System. (b) A graph of the displacement of a diver's center of gravity (CG) is superimposed on a still video frame. (c) Examples of animated stick figures generated by the analysis software.

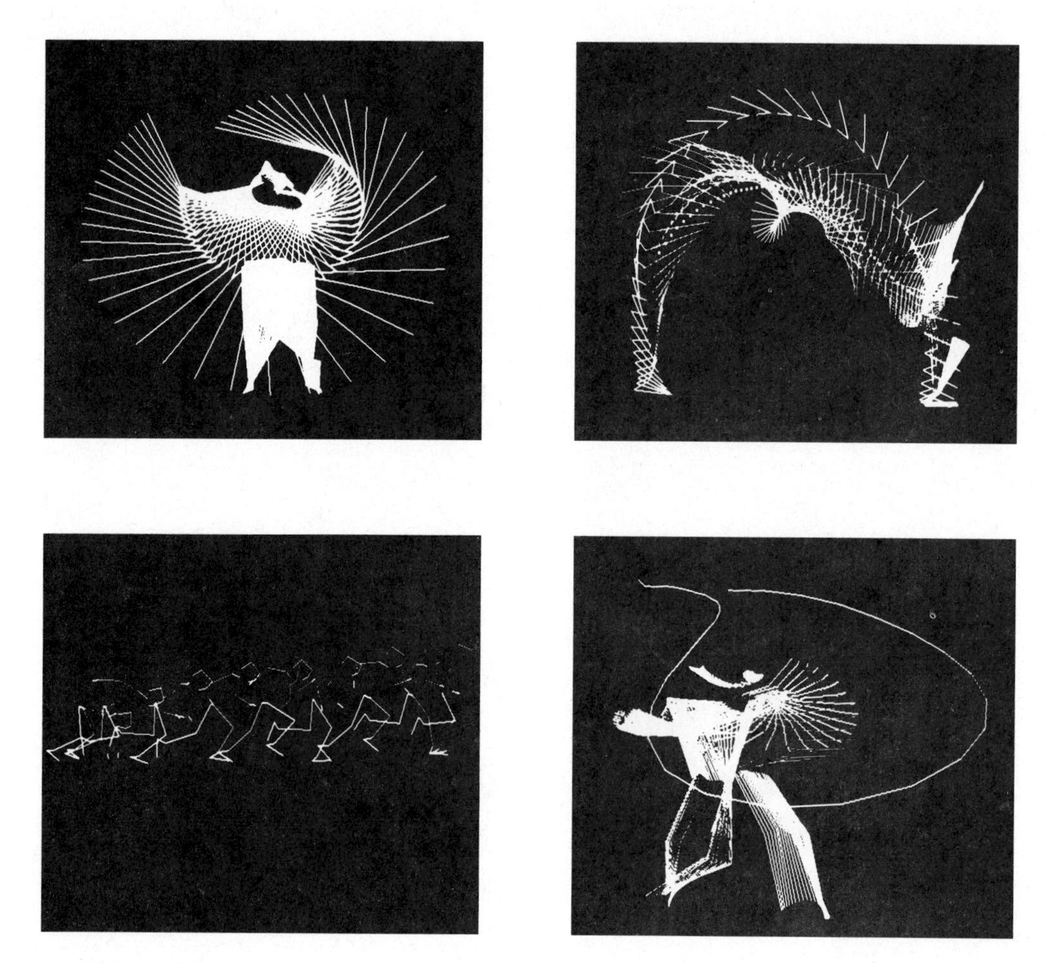

FIGURE 17.10
(continued)

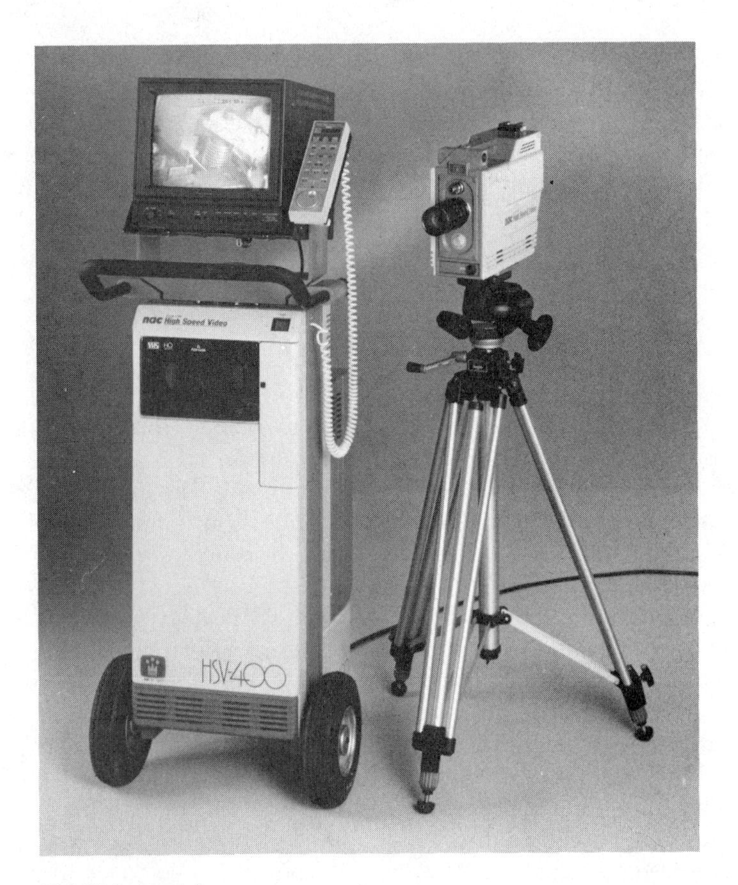

FIGURE 17.11

The NAC HSV-400 Analysis System shown with its high-speed video camera on a stable tripod. (Courtesy of Instrumentation Marketing Corporation).

on solid-state video sensor devices, which enable much faster data recording on special black and white tape. Using the frame-splitting process, each picture can be divided into 2, 3, or 6 horizontal segments, producing up to 12,000 partial pictures/sec. The recorded data are fed as an FM signal to a special recording system and put on tape. The frames can be grabbed and digitized for analysis. Shapiro and Pink (1985) discuss the use of this system in studying a world-record javelin throw. At much less expense is their later model, capable of recording 1,000 full-frame images, or 6,000 partial frames, per second (EKTAPRO 1000 Analyzer).

A variety of other video motion-analysis systems are available at the time of this writing. The VLC-360 Color Motion Analysis System by Video Logic Corporation is another example. With the increasing demand for cost-effective and time-saving devices for analyzing visual data, new and improved systems surely will enter the marketplace.

17.7 Force-Measuring Instruments

Kinetic analyses of movement include the consideration of the forces acting on the system. In the software for film and video analyses, it is possible to calculate forces and joint torques from the kinematic data acquired from the moving body. Because of the error factor in the calculation of forces from mass values and kinematic data, direct measures of forces are preferred. Force- and torque-measuring devices used in biomechanics come in a variety of forms, and the measurements they produce are recorded and analyzed in various ways. When possible, interfacing the device with a microcomputer has become the preferred method. Descriptions of some of the more common devices used in biomechanics follow.

Force Transducers

A **transducer** is any device that transforms one kind of energy or signal to another. A force transducer is one that is sensitive to force and converts the magnitude of force to an electrical signal (voltage). This signal is transmitted to a receiving unit, which may treat the signal in some way so that the recording device can use it. Probably the most versatile and popular force transducer is the *electric resistance strain gauge*. It is a short wire or piece of foil attached to a small piece of metal. As the metal is deformed (visually imperceptibly) by a force, it changes its resistance to electrical current. The change in electrical resistance changes the voltage, which is used as a measure of the changing force. Another type that is also popular is the *piezoelectric crystal*. A solid material is said to be piezoelectric if it generates an electrical charge when its shape is changed by a force, acceleration, or pressure. Quartz crystals are typically used. By attaching electrodes to its surface, the voltage produced by the crystal during deformation is proportional to the magnitude of the applied force. For both strain-gauge and piezoelectric transducers, it is necessary to calibrate the voltage changes with known changes in applied force so that the magnitudes of the unknown forces generated by the subject can be determined.

Force Platforms

A force platform, or force plate, is used for measuring applied forces and torques in three dimensions. In addition, the instantaneous point of force application can be identified. Basically, the force plate is a weighing device with elements that are sensitive to a change in their shape or size. The top plates are interchangeable to provide the surface appropriate for the activity. A 2 × 4-foot plate with a capacity up to 2,000 pounds is common for sports research. Activities that are studied using force plates are many. Impact forces and friction forces are measured for foot strikes in walking, running, jumping, landing, starting, and stopping. With computer software, power measurements can be obtained also. Force plates are very popular devices in rehabilitation studies on weight bearing during stance, posture, and normal and abnormal gait.

The two types of force plates most common in biomechanics are those using strain gauges or piezoelectric crystals. Gagnon, Robertson, and Norman (1987) suggest that the strain-gauge type might be the better type to use for studying postural sway and other quasi-static motions because of their sensitivity properties. The piezoelectric type seems to be more tolerant of temperature changes. Gagnon et al. point out, however, that both types are suitable for most activities that can be studied indoors. The components typically needed are the force plate with voltage amplifiers, which are connected to a computer and/or a display device (monitor screen, printer, or plotter). Some force plates can be used outside the laboratory, and the data can be collected and stored on computer disk or magnetic tape for analysis later. Figure 17.12 shows a computerized biomechanics force platform system, which is designed to record, process, and plot the forces and torques.

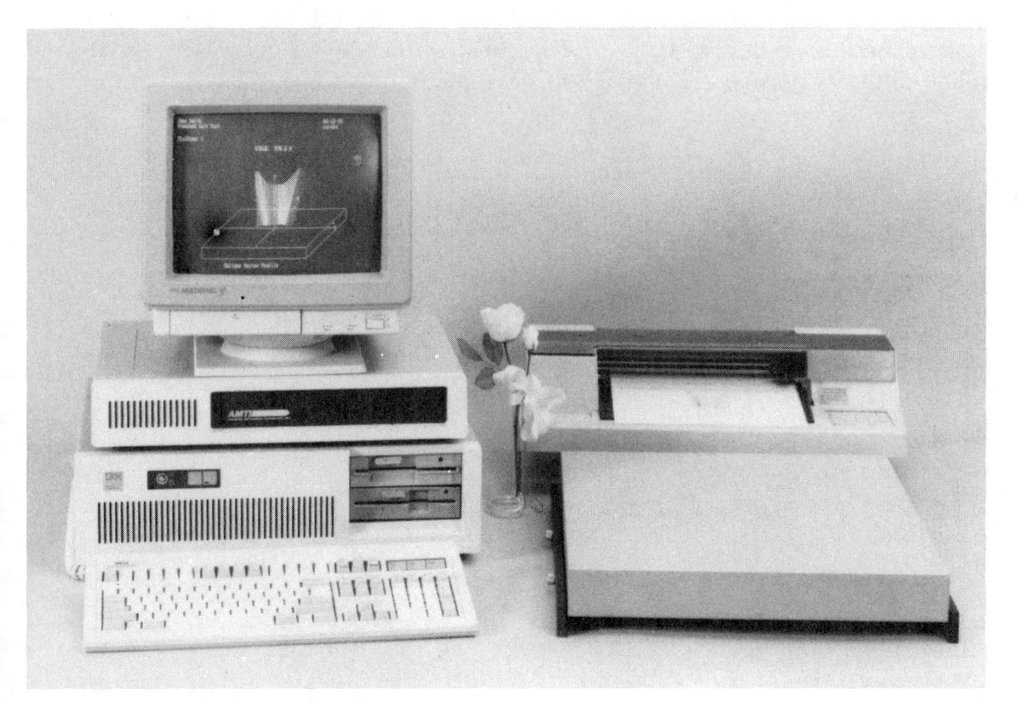

FIGURE 17.12

An AMTI force platform connected to a computer and plotter. (Courtesy of Advanced Mechanical Technology, Inc.).

Because any force must be applied over some amount of time, the data output by a force plate are in the form of impulse (impulse = force × the time during which the force is applied). The force applied over time is not constant but varies from first contact through the contact time. The graphs of these impulses are called *force-time histories,* and they show the changing force (or torque) magnitude over time. If the data are recorded on a chart strip recorder, they must be analyzed manually for the desired information. Manual analysis is time-consuming, so the preferred method is to input the force-plate signals to a microcomputer for storage and analysis by appropriate software (J. G. Richards, 1987). Figure 17.13 shows a typical record of force-plate data collected and analyzed by a microcomputer.

Some type of video or film record of the performance along with the force data is desirable and often necessary for a complete understanding of the performance. With some movement analysis systems, notably the Ariel Performance Analysis System, force-place data can be synchronized and stored simultaneously with the video data to produce composite data pictures of the performance.

Pressure Platforms

A special form of force platform is a pressure platform; it is used to identify the pattern of pressure distribution as the foot first plants and as the force is distributed over the foot area during the contact time. Such data are useful in studying injury mechanisms and in the design of shoes. With force plates, only the location of the point of application of the resultant force is identified (Gagnon et al., 1987). Cavanagh and Michiyoshi (1980) and Draganich et al. (1980) describe mechanisms for this special purpose.

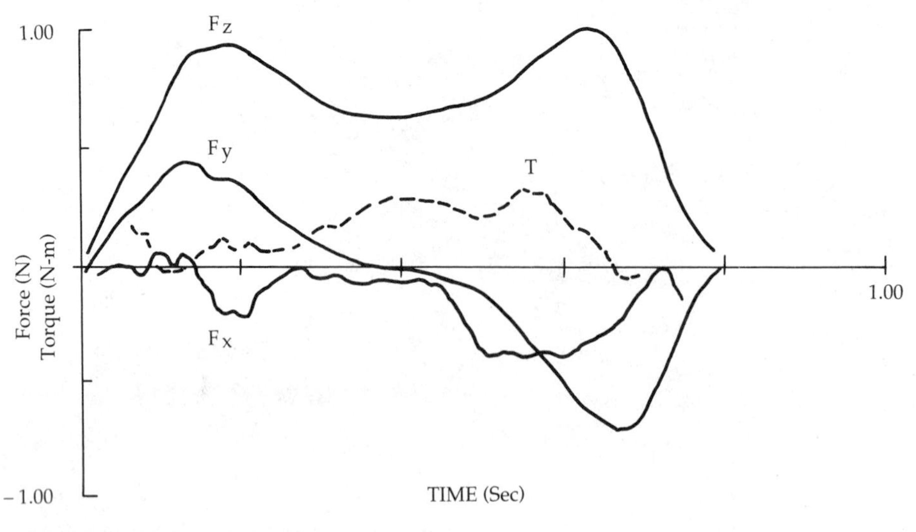

FIGURE 17.13

A graph of force-plate data can be displayed on a computer monitor and printed.

17.8 Accelerometry

An **accelerometer** is a transducer that is sensitive to acceleration and can be used in situations in which film or video techniques may not be appropriate or sufficient. It consists of a specially positioned mass connected to strain gauges or a piezoelectric crystal. When the system is accelerated (or decelerated), the inertia of the mass causes it to lag (or keep moving) and exert a force on the transducers. Accelerometry provides a direct measure of acceleration rather than indirect, such as calculations made from film and video. The piezoelectric crystal is a common type of accelerometer and can be constructed to measure acceleration in one, two, or three directions. Its use is somewhat limited, however. Because of its sensitivity, if it is skin mounted, skin movements cause error. It usually must be mounted on some rigid device or model that can be moved by a subject's limb. It is used in impact and sport equipment design studies where attachment on the equipment or a rigid surface is appropriate (Dainty and Norman, 1987; Norman et al., 1979). The output of the accelerometer produces an acceleration–time graph on an oscilloscope or chart recorder. The manual analysis of the curve is time-consuming, so it is advantageous to collect the data with a microcomputer for easy and more accurate analysis.

17.9 Electrogoniometry

A goniometer is a device used to measure joint angles. Basically, it is two thin rods connected at an axis on a protractor. The center of the protractor is placed over a joint axis, one rod is placed along a bone, and the other movable rod is placed on the connecting bone. The static angle is then read from the protractor. An **electrogoniometer** (ELGON) is one that can measure continuously the changing joint angle during movement. Instead of a protractor, an electrical resistance device called a *potentiometer* connects the two bars, and a wired circuit is formed to a recording device or computer (Figure 17.14).

As joint movement occurs, the contact points in the potentiometer change position, which creates more or less electrical resistance. The voltage fluctuations accompanying these changes are transformed into an angle versus time graph on a chart strip recorder from which measurements can be taken. Hand measurements introduce error as well as increase analysis time; therefore, if possible, the electrical signals are sent to a microcomputer. The position–time data can be used to give angular velocity and acceleration values and graphs.

Another design for an ELGON presented by Nicol (1987) consists of a narrow flexible steel foil band fitted with strain gauges. The low weight and high flexibility of this design has given it broad application possibilities.

The ELGONS described here measure angles in only a single plane. Often a joint movement is biaxial or triaxial, and movement in three dimensions occurs. When feasible, two or three single-plane ELGONS may be attached over a joint, and each one produces a record of its own plane. Such a procedure can often inhibit the real motion of the subject, however. The strain-gauge type has been further developed to provide triaxial measures. The MERU triplanar ELGON described by Hannah (1979) has been used to analyze three-dimensional knee joint movement during running (Smart and Robertson, 1985). The data were found to be reliable,

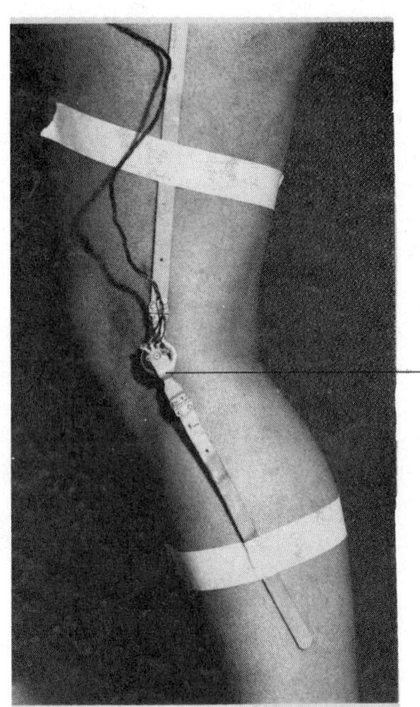

FIGURE 17.14

An electrogoniometer attached to a knee for measuring flexion and extension during movement.

although, during impact, some slippage of the device was noted. Additional information on triaxial goniometry is contained in the functional overview presented by Chao (1980).

As with other measurements taken from localized areas, it is best to have a total view of the performance on film or video, so that the joint-motion data can be put in perspective.

17.10 Electromyography

An *electromyogram* is a record of levels of electrical activity in muscle. **Electromyography** (EMG) is a technique used to identify relative levels of activation of different muscles or parts of a muscle. The activity is detected by skin-mounted electrodes or wire and needle electrodes inserted through the skin into the muscle. Electrical signals resulting from the propagation of the action potentials along the membrane of the active muscle can be amplified, monitored, and recorded. These signals usually are transmitted through wires from the subject to the recorder or computer; however, the use of telemetering systems eliminates the confining conditions presented to the subject by attached wires. Electrical signals are sent to the recorder or computer by means of radio frequencies generated by the level of muscle activity. Both wired and telemetered EMG have been used successfully on land and in underwater activities (Barthels and Adrian, 1971; Clarys, Robeaux, and Delbeke, 1985).

EMG is used to obtain a muscular activity profile during a performance but is made more meaningful if accompanied by a force plate, film, video, or joint-movement record (electrogoniogram) that is synchronized in time with it. Figure 17.15 shows one such synchronized record of vertebral flexor and extensor EMGs with hip, knee, and ankle electrogoniograms.

Simultaneously interfacing EMGs and other data with a computer for processing and analysis has enabled users to retrieve more accurate and meaningful quantitative information about relationships between internal and external forces during a performance. Moreover, EMG data reflect a *neuro*muscular phenomenon, and by using EMG data, along with other tools, researchers in motor learning/control and biomechanists, together making a natural team, can investigate movement behavior at its generational level. The reader is referred to LaGasse (1987) and Dainty and Norman (1987) for detailed information on electromyography and the procedures for its use in biomechanics.

17.11 Using Microcomputers for Collecting and Analyzing Data

The preceding descriptions of measuring devices frequently referred to the desirability of interfacing (connecting) the instrumentation with a microcomputer. Such *on-line* data collection means that the signals from a measuring device are sent directly to a computer instead of to some recorder such as a strip chart recorder or oscilloscope. In an *off-line* operation, these recorded data must be transferred manually by the investigator to a computer for calculations. Accuracy and reliability are improved from the elimination of human handling of some kinds of data, saving much time in the total analysis.

The standard components of an interfacing system include a sensor, which converts a signal to an electrical quantity; an amplifier, which converts the electrical quantity to a voltage suitable for "reading"; a multiplexer, which takes voltages from a number of instruments so that each may be recorded on a different data channel; an analog to digital converter (A/D converter, or ADC), which takes analog (waveform) data and changes them into digital (numbered) form; the microcomputer, which processes the digital information in ways specified by the user's software; and the user input–output (I/O) devices, which consist of the computer's keyboard, monitor, printer, or plotter. An *interactive* system is one that allows the user to have control over the data collection and analysis process by specifying how the computer should handle particular aspects. Most systems are of this type and allow "customization" of the data-gathering capabilities of the system.

Ratzlaff (1987) presents a comprehensive view of computer-assisted experimentation and explains the details of the operation of all the components used in the instruments commonly used in biomechanics and other fields of research. In his justification of computer use with data collecting and analysis, he adds a bit of caution by quoting Albert Einstein as follows: "Everything should be made as simple as possible, but not simpler."

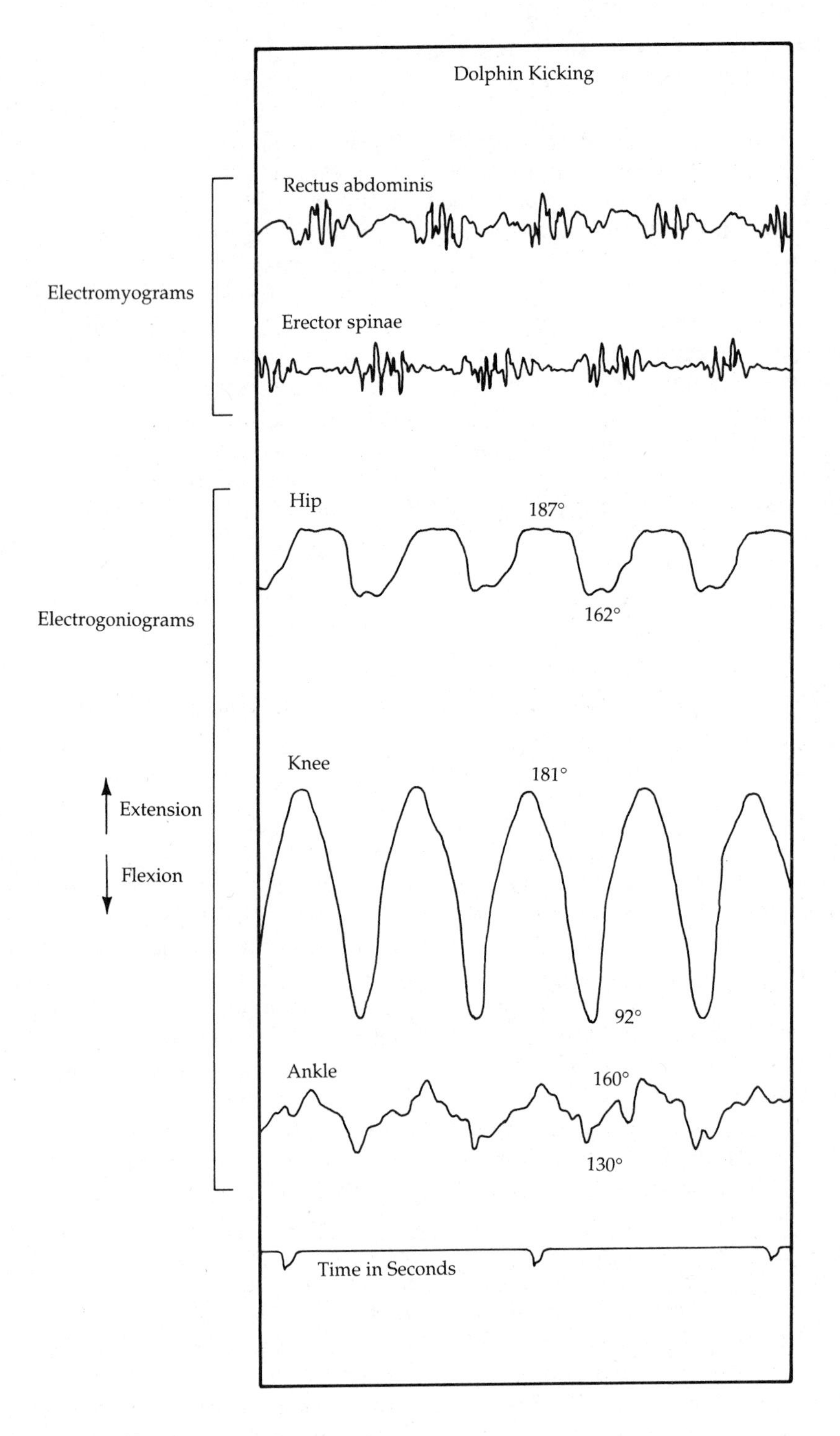

FIGURE 17.15
Electromyograms and electrogoniograms of a butterfly swimmer.

Understanding Biomechanics Instrumentation

1. Why would it be better to analyze the mechanics of a skilled performer during competition rather than practice?
2. What limitation does stroboscopy have as a visual method of analysis?
3. In high-speed cinematography, what factors should be considered when a frame rate is selected?
4. What factors determine exposure time in cinematography?
5. What limitation is inherent in two-dimensional cinematography and videography?
6. What is meant by "digitizing" a film or video image?
7. Compare the advantages and disadvantages of cinematography and videography.
8. Describe two examples of movements that could be studied by using (a) force-measuring devices, (b) electrogoniometry, and (c) electromyography.
9. What does "on-line" data collection mean?

References and Suggested Readings

Abraham, L. D. (1987). An inexpensive technique for digitizing spatial coordinates from videotape. In B. Jonsson (Ed.), *Biomechanics X-B* (pp. 1107–1110). Champaign, IL: Human Kinetics.

Advanced Mechanical Technology, Inc. 151. California Street, Newton, MA 02158.

Ariel Dynamics, Inc. 22000 Plano Trabuco Road, Trabuco Canyon, CA 92679.

Assente, R., Ferrigno, G., Pedotti, A., & Rodano, R. (1987). Application in sports of the "elite": A system for real time processing of TV signals. In J. Terauds, B. Gowitzke, & L. Holt (Eds.), *Biomechanics III & IV* (pp. 324–335). Del Mar, CA: Academic Publishers.

Atha, J. (1984). Current techniques for measuring motion. *Applied Ergonomics, 15*(4), 245–257.

Barr, R. E., & Abraham, L. D. (1987). The punter's profile—A biomechanical analysis. *SOMA, 1*(4), 4–12.

Barthels, K. M. (1974). Three dimensional kinematic analysis of the hand and hip in the butterfly swimming stroke. Unpublished doctoral dissertation, Washington State University, Pullman.

Barthels, K. M., & Adrian, M. J. (1971). Variability in the dolphin kick under four conditions. In L. Lewillie & J. P. Clarys (Eds.), *First International Symposium on Biomechanics in Swimming* (pp. 105–118). Universite Libre de Bruxelles, Brussels, Belgium.

Biomechanics instrumentation. (1984, December). *Journal of Physical Therapy, 64.*

Boucher, J. P., & Flieger, M. S. (1985). Signal characteristics of EMG: Effects of ballistic forearm. In D. A. Winter, R. W. Norman, R. P. Wells, K. C. Hayes, & A. E. Patla (Eds.), *Biomechanics IX-A* (pp. 297–301). Champaign, IL: Human Kinetics.

Cavanagh, P. R., & Michiyoshi, A. (1980). A technique for the display of pressure distribution beneath the foot. *Journal of Biomechanics, 13*(2), 69–75.

Chao, E. Y. (1980). Justification of triaxial goniometer for the measurement of joint rotation. *Journal of Biomechanics, 13*, 989–1006.

Cheetham, P. J., & Scheirman, G. L. (1988, January). The video study of motion. *Advanced Imaging.*

Clarys, J. P., Robeaux, R., & Delbeke, G. (1985).

Telemetered versus conventional EMG in air and water. In D. A. Winter, R. W. Norman, R. P. Wells, K. C. Hayes, & A. E. Patla (Eds.), *Biomechanics IX-B* (pp. 286–290). Champaign, IL: Human Kinetics.

Dabrowska, A., & Kedzior, K. (1985). Investigation and modeling of the relationship between integrated surface EMG and muscle tension. In D. A. Winter, R. W. Norman, R. P. Wells, K. C. Hayes, & A. E. Patla (Eds.), *Biomechanics IX-A* (pp. 308–312). Champaign, IL: Human Kinetics.

Dainty, D. A., & Norman, R. W. (Eds.) (1987). *Standardizing biomechanical testing in sports.* Champaign, IL: Human Kinetics.

Disch, J. G., & Hudson, J. L. (1980). Measurement aspects of biomechanical analysis. In J. M. Cooper & B. Haven (Eds.), *Biomechanics: Proceedings of the 1980 Biomechanics Symposium, Indiana University* (pp. 191–201). Indiana State Board of Health.

Donnelly, J. E. (Ed.) (1987). *Using microcomputers in physical education and sport sciences.* Champaign, IL: Human Kinetics.

Draganich, L. F., Andriacchi, T. P., Strongwater, A. M., & Galante, J. O. (1980). Electronic measurement of instantaneous foot–floor contact patterns during gait. *Journal of Biomechanics, 13,* 875–880.

Ekstrom, H., & Karlsson, J. (1985). Direct force measurement in biomechanical investigations. In D. A. Winter, R. W. Norman, R. P. Wells, K. C. Hayes, & A. E. Patla (Eds.), *Biomechanics IX-B* (pp. 201–205). Champaign, IL: Human Kinetics.

Gagnon, M., Robertson, G., & Norman, R. W. (1987). Kinetics. In D. A. Dainty & R. W. Norman (Eds.), *Standardizing biomechanical testing in sport* (pp. 39–42). Champaign, IL: Human Kinetics.

Garrett, G. E. (1987). A sports analysis system: A biomechanical workstation. In J. Terauds, B. Gowitzke, & L. Holt (Eds.), *Biomechanics III & IV* (pp. 318–321). Del Mar, CA: Academic Press.

Gates, S. C. (1984). Laboratory data collection with an IBM PC. *PC Magazine, 3*(24), 375–377.

Grieve, D. W., Miller, D. I., Mitchelson, D. L., Paul, J. P., & Smith, A. J. (1975). *Techniques for the analysis of human movement.* Princeton, NJ: Princeton Book Company.

Hannah, R. E. (1979). Manual of the C.A.R.S.—University of British Columbia electrogoniometer. Vancouver Arthritis Society. Vancouver, B.C. Canada.

Harrison, J. (1985). The reliability of force-time traces. *Proceedings of the Annual Sport and Science Conference of the British Association of Sports Sciences* (pp. 168–170).

Henning, E. M., Cavanagh, P. R., Macmillan, A., & Macmillan, H. T. (1982). A piezoelectric method of measuring the vertical contact stress beneath the human foot. *Journal of Biomedical Engineering, 4,* 213–222.

Hudson, J. L., Lee, E. J., & Disch, J. G. (1986). The influence of biomechanical measurement systems on performance. In M. Adrian & H. Deutsch, (Eds.), *Biomechanics: The 1984 Olympic Scientific Congress Proceedings* (pp. 347–352). University of Oregon Microform Publications.

Hyzer, W. G. (1981, May). The spin physics SP-2000 motion analysis system. *Photomethods,* pp. 47–48.

Instrumentation Marketing Corp. 820 South Mariposa Street, Burbank, CA 91506.

Kerwin, D. G. (1987). A laboratory interface for use in biomechanical data collection and analysis. *Proceedings of Sports Biomechanics Section of the British Association of Sports Sciences (no. 12).* Abstract in *International Journal of Sport Biomechanics, 4,* 187.

Kistler Instrument Corporation. 75 John Glenn Dr., Amherst, NY, 14120.

Klinger, A., & Adrian, M. (1987). Power output as a function of fencing technique. In B. Jonsson (Ed.), *Biomechanics X-B* (pp. 791–795). Champaign, IL: Human Kinetics.

Kondraske, G. V., et al. (1988). Measuring human performance: Concepts, methods, and applications. *SOMA, 2,* 6–13.

LaGasse, P. (1987). Neuromuscular considerations. In D. A. Dainty & R. W. Norman (Eds.), *Standardizing biomechanical testing in sports* (pp. 59–71). Champaign, IL. Human Kinetics.

Macellari, V., Rossi, M., & Bugarini, M. (1987). Human motion monitoring using the CoSTEL system with reflective markers. In B. Jonsson (Ed.), *Biomechanics X-B* (pp. 260–264). Champaign, IL: Human Kinetics.

Macleod, A., & Morris, J. R. W. (1987). Investigation of inherent experimental noise in kinematic experiments using superficial markers. In B. Jonsson (Ed.), *Biomechanics X-B* (pp. 1035–1039). Champaign, IL: Human Kinetics.

McComb, G. (1985). Pictures to pixels. *Macworld, 2*(4), 68–79.

Medeiros, A. J. M. (1986). Automated clinical motion analysis. In M. Adrian & H. Deutsch (Eds.), *Biomechanics: The 1984 Olympic Scientific Congress Proceedings* (pp. 327–335). University of Oregon Microform Publications.

Merriman, J. S. (1975). Stroboscopic photography as a research instrument. *Research Quarterly, 46,* 256–261.

Miller, C. E. (1981). High speed videography instrumentation and procedures. In J. Terauds (Ed.), *Second International Symposium of Biomechanics Cinematography and High Speed Photography, 291,* (pp. 68–75). Bellingham, WA: International Society for Optical Engineering.

Miller, D. I. (1986). Microcomputers in biomechanics research. *International Journal of Sport Biomechanics, 2,* 63–65.

Miller, D. I. (1989). Microcomputers in biomechanics research & applied settings. In J. S. Skinner, C. B. Corbin, D. M. Landers, P. E. Martin, & C. L. Wells (Eds.), *Future directions in exercise and sport science research* (pp. 209–221). Champaign, IL: Human Kinetics.

Morris, J. R. W. (1973). Accelerometry—A technique for the measurement of human body movements. *Journal of Biomechanics, 6,* 729–736.

Muckenfuss, M. (1988). Building a better mousetrap—Computer-video system. *Swimming Technique, 24*(3), 34–35.

Nakamura, Y., Ohmichi, H., & Miyashita, M. (1985). Tension–EMG relationship during maximum voluntary contraction. In D. A. Winter, R. W. Norman, R. P. Wells, K. C. Hayes, & A. E. Patla (Eds.), *Biomechanics IX-A* (pp. 293–296). Champaign, IL: Human Kinetics.

Nicol, A. C. (1987). A new flexible electrogoniometer with widespread applications. In B. Jonsson (Ed.), *Biomechanics X-B* (pp. 1029–1033). Champaign, IL: Human Kinetics.

Nicol, K., Clasing, D., Thayer, A., El-Din, A. Z. (1987). Time characteristics in swimming. In B. Jonsson (Ed.), *Biomechanics X-B* (pp. 757–760). Champaign, IL: Human Kinetics.

Norman, R. W., Bishop, P. J., Pierrynowski, M. R., & Pezzack, J. C. (1979). Aircrew helmet protection against potential cerebral concussion in low-magnitude impacts. *Aviation Space Environmental Medicine, 50,* 553–561.

Okamoto, T., & Wolf, S. L. (1979). A method for underwater recording of EMG activity using wire electrodes. In J. Terauds & W. Bedingfield (Eds.), *Swimming III* (pp. 160–166). Baltimore: University Park Press.

Otahal, S., Dedik, L., & Karas, V. (1987). Frequency-response characteristics of voluntarily activated and electrically evoked isometric contractions of the quadriceps femoris. In B. Jonsson (Ed.), *Biomechanics X-A* (pp. 325–328). Champaign, IL: Human Kinetics.

Peak Performance Technologies, Inc. 7388 South Revere Parkway, Suite 801, Englewood, CO 80112.

Peeraer, L., Willems, E., Stijns, H., Spaepen, A., & Stijnen, V. (1987). A method to determine segmental moments of inertia, segmental accelerations, and resultant joint forces using accelerometers. In B. Jonsson (Ed.), *Biomechanics X-B* (pp. 1059–1063). Champaign, IL: Human Kinetics.

Pizzimenti, M. (1986). Interfacing a CalComp 9100 series digitizer to an Apple IIe microcomputer. *International Journal of Sport Biomechanics, 2,* 110–116.

Plagenhoef, S. (1971). *Patterns of human motion.* Englewood Cliffs, NJ: Prentice-Hall.

Ramey, M. R. (1975). Force plate designs and applications. In J. H. Wilmore & J. F. Keogh (Eds.), *Exercise and Sport Sciences Reviews* (pp. 303–319). New York: Academic Press.

Ramey, M. R. (1982). The use of force plates for jumping research. In J. Terauds (Ed.), *Biomechanics in sports* (pp. 81–91). Del Mar, CA: Academic Publishers.

Ratzlaff, K. L. (1987). *Introduction to computer-assisted experimentation.* New York: John Wiley & Sons.

Richards, C. L. (1985). EMG activity level comparison in quadriceps and hamstrings in five dynamic activities. In D. A. Winter, R. W. Norman, R. P. Wells, K. C. Hayes, & A. E. Patla (Eds.), *Biomechanics IX-A* (pp. 313–317). Champaign, IL: Human Kinetics.

Richards, J. G. (1987). Microcomputer applications in biomechanics. In J. E. Donnelly (Ed.), *Using microcomputers in physical education and the sport sciences* (pp. 235–262). Champaign, IL: Human Kinetics.

Schieb, D. A. (1987a). The biomechanics piezoelectric force plate. *SOMA, 1*(4), 35–40.

Schieb, D. A. (1987b). The force plate in sports biomechanics research. In J. Terauds, B. Gowitzke, & L. Holt (Eds.), *Biomechanics III & IV* (pp. 337–365). Del Mar, CA: Academic Publishers.

Shapiro, R. (1978). Direct linear transformation method for three-dimensional cinematography. *Research Quarterly, 49*(2), 197–205.

Shapiro, R., Blow, C., & Rash, G. (1987). Video digitizing analysis system. *International Journal of Sport Biomechanics, 3*, 80–86.

Shapiro, R., & Pink, M. (1985). Biomechanical analysis of a world record javelin throw: A case study. *International Journal of Sport Biomechanics, 1*, 73–77.

Smart, G., & Robertson, G. (1985). Triplanar electrogoniometer analysis of running gait. In D. A. Winter, R. W. Norman, R. P. Wells, K. C. Hayes, & A. E. Patla (Eds.), *Biomechanics IX-B* (pp. 144–148). Champaign, IL: Human Kinetics.

Smith, S. L., Dillman, C. J., & Risenhoover, S. G. (1988). Validation of an automatic video analysis system. (Abstract) *Medicine and Science in Sports and Exercise, 20*(2), Supplement: S75.

Sprigings, E. J. (1988). Sport biomechanics: Data collection, modelling, and implementation stages of development. *Canadian Journal of Sport Sciences, 13*, 3–7.

Terauds, J. (1982). Introduction to biomechanics cinematography and video as tools for the researcher and coach. In J. Terauds (Ed.), *Biomechanics in sports* (pp. 71–80). Del Mar, CA: Academic Publishers.

Vaughn, C. L., Smith, D. C., & du Toit, L. L. (1987). Kinematic gait analysis using a digital camera. In B. Jonsson (Ed.), *Biomechanics X-B* (pp. 1041–1047). Champaign, IL: Human Kinetics.

Vaughn, D. L., Smart N. M., Edwards, B. I., & Kwan-Leung, S. (1985). Gait analysis on a microbudget. In D. A Winter, R. W. Norman, R. P. Wells, K. C. Hayes, & A. E. Patla (Eds.), *Biomechanics IX–B* (pp. 265–268). Champaign, IL: Human Kinetics.

Video Logic Corporation, 597 North Mathilda Avenue, Sunnyvale, CA 94086.

Wininger, S., Reese, C. D., & Zebas, C. J. (1985). A portable mounting device for the Kistler force plate. In D. A. Winter, R. W. Norman, R. P. Wells, K. C. Hayes, & A. E. Patla (Eds.), *Biomechanics IX-B* (pp. 191–194). Champaign, IL: Human Kinetics.

Wyss, U. P., & Pollak, V. A. (1985). Surface electromyogram EMG/muscle force: An experimental approach. In D. A. Winter, R. W. Norman, R. P. Wells, K. C. Hayes, & A. E. Patla (Eds.), *Biomechanics IX-A* (pp. 318–324). Champaign, IL: Human Kinetics.

GLOSSARY

Acceleration: The change in velocity (speed or direction or both) per unit of time.

Accelerometer: A transducer that senses acceleration.

Accommodating Resistance Exercise: See ISOKINETIC EXERCISE.

Aerodynamic Drag Force: The fluid force opposing the motion of a body moving through the air.

Aerodynamic Lift Force: The fluid force directed perpendicular to the flow direction past an object with a lift-producing shape or past a spinning ball.

Agonist: A muscle that causes a motion (mover).

Anatomical Position: The upright body position in which all joints are extended and the palms are facing forward.

Angle of Attack: The angle formed between the main plane or longitudinal axis of an object and the direction of fluid flow past it.

Angle of Projection: The angle at which an object or body is projected at release or takeoff; measured in relation to a stated frame of reference, usually the horizontal.

Angular Momentum: The product of a body's rotational inertia and angular velocity.

Antagonist: A muscle that can cause the joint movement opposite to the movement being done by an agonist.

Anthropometrics: The area of study that is concerned with the body's physical composition and quantifiable characteristics.

Appendicular Skeleton: A portion of the human skeleton consisting of the upper and lower extremities.

Anteroposterior (AP) Axis: An imaginary line that runs from the front to the back of the body or through an articulation.

Archimedes' Principle: A body that is partially or totally immersed in a fluid will experience an upward buoyant force that is equal to the weight of the volume of fluid displaced by that body.

Articulation: The junction of two bones.

Attitude Angle: The angle formed between the horizontal and the main plane or longitudinal axis of an airfoil or similar lift-producing object.

Axial Skeleton: A portion of the human skeleton consisting of the skull, thorax, and vertebral column.

Axis of Momentum: The spatial axis about which the resultant angular momentum of a rotating body exists.

Axis of Rotation: The imaginary line or point about which a body or segment rotates.

Base of Support: That region bounded by body parts in contact with some resistive surface that exerts a reaction force against the body.

Bernoulli's Principle: A low-pressure zone is created in a region of high fluid flow velocity, and a high-pressure zone is created in a region of low fluid flow velocity.

Biarticulate Muscle: A muscle that crosses two joints.

Biomechanics: The area of study wherein knowledge and methods of mechanics are applied to the structure and function of the living human system.

Boundary Layer: The invisible thin layer of fluid that is adjacent to the surface of an immersed body.

Buoyant Force: The upward force exerted on an immersed body by the water beneath it.

Cardinal Axis or Plane: See PRINCIPAL AXIS or PLANE.

Center of Buoyancy: The point at which the buoyant force acts on an immersed body's volume; the center of gravity of the volume of water being displaced by the body.

Center of Gravity: The point at which all the body's mass seems to be concentrated; the balance point of a body; the point around which the sum of the torques of the segmental weights is equal to zero. The point of application of gravity's force on a mass; the center of mass.

Center of Pressure: The point of application of a fluid lift force acting on a body moving through the fluid.

Centrifugal Force: The force directed radially outward that is exerted by a rotating body on a structure or mass that exerts a center-directed (centripetal) force on that body.

Centripetal Force: The force directed radially toward the center of rotation that is exerted by a mass on a rotating body and that causes that body to travel in a circular path.

Cinematography: Motion picture photography.

Closed Skill: A skill performed in a predictable environment so that the performer is free to move without having to make quick decisions.

Coefficient of Friction: The ratio of the magnitude of the maximum force of friction to the magnitude of the perpendicular force pressing the two surfaces together (normal force).

Colinear Forces: Forces whose lines of action lie along the same line.

Concentric Tension: The contraction of a muscle during which the muscle shortens and causes movement of one or more attached segments.

Constraint: A restriction to the performance of free human movement patterns; a limiting factor.

Continuous Skill: A skill in which the same movement pattern is performed repeatedly as cycles of the total act.

Crural Index: A ratio of thigh length to leg length.

Curvilinear: Refers to motion along a curved line or path.

Deceleration: The decrease in velocity per unit time.

Density: The mass per unit volume of an object or body (mass density); or weight per unit volume (weight density).

Digitize: To plot or identify the x and y coordinates of a point on an image for quantitative analysis, usually performed with a computerized device.

Discrete Skill: A skill in which a movement pattern is performed as a single act.

Displacement: The change in a body's location in space in a given direction.

Distal: Refers to the end of a segment, bone, or muscle attachment that is farther from the axial skeleton.

Dynamic Equilibrium: The state of a body moving with constant speed and direction, that is, with zero acceleration.

Dynamic (Isotonic) Tension: The contraction of a muscle during which the muscle changes length.

Dynamics: The study of mechanical factors associated with systems in motion.

Eccentric Tension: The contraction of a muscle during which the muscle lengthens and resists segmental motion.

Elastic: That property of a body that causes it to reform following deformation.

Electrogoniometer (ELGON): A device that senses joint angle changes and sends electrical signals to a recording device.

Electromyography (EMG): The technique used to sense and record levels of muscular activity through electrodes placed on the skin or inserted into the muscle.

Endurance: The ability to continue performance of a movement activity.

Equilibrium: The state of a system whose motion is not being changed, accelerated, or decelerated.

External: Outside a defined system.

Exteroceptors: Sensory receptors located in or around the skin that respond to changes in the environment making direct contact with the exterior of the body.

Femoral Torsion: A relative position of the femur and tibia in which the femur is rotated medially around its longitudinal axis relative to the tibia (knees facing inward).

Flexibility: The range of motion permitted between two adjacent segments.

Force: That which causes or tends to cause a change in a body's motion or shape.

Force Arm (Moment Arm, Lever Arm, or **Torque Arm):** The perpendicular distance between the line of action of the force and the axis of rotation.

Form Drag: See PROFILE DRAG.

Friction: The force that resists the sliding of one surface upon another.

Frontal: Refers to a plane that divides a body or segment into front and back portions.

Gyroscopic Stability: The resistance of a rotating body to a change in its plane of rotation.

Impulse: The product of the magnitude of a force or torque and its time of application.

Inertia: The resistance of a body to a change in its state of motion.

Inferior: Refers to a portion of a segment, bone, or muscle closer to the feet.

Instantaneous Velocity: The velocity of a body at a given instant in time.

Internal: Within a defined system.

Interoceptors: Sensory receptors located within the body that respond to changes in the body's internal environment and control visceral functioning.

Isokinetic: Refers to a method of strength training in which the speed of segmental rotation is kept constant throughout the range of muscular contraction.

Isokinetic Exercise: Movements performed on devices designed to match the motive torque applied while maintaining a constant angular speed of joint movement (also referred to as accommodating resistance exercise).

Isometric Tension: See STATIC TENSION.

Isotonic Tension: See DYNAMIC TENSION.

Kinanthropometrics: The area of study that is concerned with the body's physical measurements of size, shape, and proportion as they relate to human movement.

Kinematics: An area of study that is concerned with the time and space factors in the motion of a system.

Kinesiology: The study of human movement from an anatomical or mechanical perspective or both.

Kinesthesis: The perception of segmental and body position and movements.

Kinetic Energy: The ability of a body to do work by virtue of its motion.

Kinetic Friction: The friction that exists between two surfaces sliding past each other.

Kinetic Link Principle: The generation of high end-point velocity accomplished through the use of accelerating and decelerating adjoining links, by the use of internal and external muscle torques, applied to the segments in a sequential manner from proximal to distal, from most massive to least massive, and from most fixed to most free.

Kinetics: An area of study that is concerned with the forces that act on a system.

Lateral: Refers to the side away from the longitudinal midline of the body or segment.

Lever Arm: See FORCE ARM.

Lever System: A mechanism for doing work, consisting of a body with an axis of rotation and eccentrically applied forces.

Linear Momentum: A property of a body in motion; the product of a body's mass and its velocity.

Linear Motion: Motion along a straight (recti-) or curved (curvi-) line.

Longitudinal Axis: An imaginary line that runs along the length of a body or segment.

Mass: The measure of a body's inertia; the amount of matter in a body.

Maximal Voluntary Contraction: The maximal amount of force that is exerted by a muscle during a static contraction against an immovable resistance.

Medial: Refers to the side closer to the longitudinal midline of the body or segment.

Mediolateral (ML) Axis: An imaginary line that runs from the medial to the lateral side of the body or through an articulation.

Mobility: The ease with which an articulation, or a series of articulations, is allowed to move before being restricted by the surrounding structures.

Moment Arm: See FORCE ARM.

Moment of Inertia: See ROTATIONAL INERTIA.

Momentum: A system's resistance to change in its state of motion (inertia) multiplied by its velocity.

Motive Force: A force that causes motion or change in shape.

Motor Unit: A single motor nerve cell and its branches, with associated muscle fibers; the smallest functional unit of muscular contraction.

Movement Pattern: A general series of anatomical movements that have common elements of spatial configuration such as segmental movements occurring in the same plane of motion.

Multiarticulate Muscle: Muscle that crosses more than two joints.

Neutral Equilibrium: The state of a body in which, if the body is displaced, it will remain in the displaced position.

Normal Force: A force directed perpendicular to a surface.

Nutation: Wobbling; the tilting of a body's principal axis of rotation from its original position.

Open Skill: A skill performed in response to an unpredictable, changing environment.

Orthogonal: Refers to the orientation of lines or planes at 90 degrees to one another.

Parabolic Path: A curvilinear path (trajectory) followed by a projectile that is unaffected by air resistance.

Plane of Motion: The two-dimensional space cut by a moving body; the plane along which movement occurs.

Plastic: Refers to the condition of connective tissue (ligaments or tendons) when it has been stretched past its limit and will no longer return to its original shape.

Ponderal Index: A particular ratio of the human body's weight and height that is a measure of stature, or stoutness.

Potential Energy: The ability of a body to do work by virtue of its position above an object (gravitational potential energy) or by virtue of its deformation (elastic potential energy).

Power: The product of an applied force and the speed with which it is applied; the quantity of work done per unit of time.

Pressure: The ratio of force to the area over which the force is applied.

Pressure Drag: See PROFILE DRAG.

Principal (Cardinal) Axis or Plane: An axis or plane that is directed through a body's center of gravity.

Profile Drag (Pressure Drag or **Form Drag):** A net force caused by a difference in pressure from the leading side of the body to the trailing side; the greater the differential, the greater the drag.

Projection Velocity: The speed and direction of an object at the instant of projection.

Proprioceptors: Sensory receptors located in and around joints and muscles that respond to changes in position, length, tension, and acceleration of the host tissues.

Proximal: Refers to that end of a segment, bone, or muscle attachment that is closer to the axial skeleton.

Radius of Gyration: A measure of the distribution of a body's or segment's mass about an axis of rotation.

Radius of Rotation: The linear distance from an axis to a point on a rotating body.

Range of Motion: The total amount of angular displacement through which two adjacent segments may move.

Reaction Force: An equal and opposite force exerted by a second body on the first in response to a force applied by the first on the second.

Recruitment: The process of increasing the number of active motor units in a muscular contraction.

Rectilinear: Refers to motion along a straight line or path.

Repetition Maximum: The maximum amount of resistance that can be moved a designated number of times.

Resistive Force: A force that resists motion or change of shape.

Resultant Vector: The result of the composition (addition) of two or more vector quantities.

Rotary Motion: Motion that describes a circular path about an axis.

Rotational Inertia (Moment of Inertia): The resistance of a body to angular acceleration.

Sagittal: Refers to plane that divides a body or segment into right and left portions.

Scalar: A quantity that has magnitude only; that is, no direction is associated with it.

Sensory Unit: A single nerve cell, with axon, branches, and receptors; such units are located within muscle, connective, and cutaneous tissue.

Skill: A general movement pattern that has been adapted to the constraints of a particular activity or sport.

Skin Friction (Surface Drag): The fluid drag force acting on a body moving through a fluid caused by viscous friction forces.

Spatial: Refers to a set of planes and axes defined in relation to three-dimensional space.

Speed: The magnitude of a body's displacement per unit of time without regard to direction.

Sprain: A tearing of ligamentous tissue.

Stability: The ability of an articulation to absorb shock and withstand motion without injury to the joint; also, the resistance to disturbance of a body's equilibrium.

Stable Equilibrium: The state of a body in which, if the body is displaced, it tends to return to its original position.

Static Equilibrium: The state of a body at rest.

Static Friction: The frictional force generated between two objects tending to slide past each other when no motion is occurring.

Static (Isometric) Tension: Muscular contraction during which no discernible segmental movement is taking place.

Statics: The study of factors associated with nonmoving systems.

Strain: A tearing of muscular tissue.

Strength: The ability of a muscle or group of muscles to exert force against a resistance.

Stroboscopy: A photographic technique used in the dark with a flashing light to create superimposed images on a single photograph.

Style: Individual adaptations—or modifications—to a technique, which are unique to the individual using them.

Superior: Refers to that portion of a segment, bone, or muscle closer to the head.

Surface Drag: See SKIN FRICTION.

Synergy: The cooperative effort of two or more muscles contracting to accomplish a single movement.

System: A body or group of bodies whose state of motion is being examined.

Tangent: A line that meets but does not intersect a curved line or surface and that is perpendicular to the radius of curvature of the arc.

Technique: A particular type, or variation, of the performance of the same skill.

Teleceptors: Sensory mechanisms receptive to smell, sounds, and visual stimuli.

Terminal Velocity: The constant velocity attained by a body moving through a fluid; this is due to the lack of a net force on the body.

Tibial Torsion: A relative position of the femur and the tibia in which the tibia is rotated laterally around its longitudinal axis relative to the femur (knees facing forward).

Torque: A turning or rotary force; the product of a force and the perpendicular distance from the line of action of the force to the axis of rotation.

Torque Arm: See FORCE ARM.

Trajectory: The aerial path followed by a projectile.

Transducer: A device that transforms one kind of energy to another (e.g., a force to an electrical signal).

Transverse: Refers to a plane that divides a body or segment into top and bottom portions.

Uniarticulate Muscle: A muscle that crosses only one joint.

Unstable Equilibrium: The state of a body in which, if the body is displaced, it tends to increase its displacement.

Variable Resistance Exercise: Movement performed on devices designed to vary the resistance torque applied during one excursion of a joint movement.

Vector: A quantity having both magnitude and direction.

Vector Composition (Addition): The process of determining the net or resultant vector from two or more component vectors.

Vector Resolution: The process of determining component vectors from a resultant vector.

Velocity: The speed and direction of a body.

Videography: The process of capturing images on a videotape or videodisc.

Volume: The three-dimensional space occupied by a body (length, width, and height).

Wave Drag: A force caused by and acting against a body moving through water, forming waves at the surface of the water.

Weight: The force of earth's gravitational attraction on a body's mass.

Work: The force applied to a body multiplied by the distance through which that force is applied.

APPENDIX I

Metric and British Units and Conversions

The purpose of this appendix is to provide conversion factors between and within metric and British units of measurement.

TABLE I.1
METRIC PREFIXES

mega	1,000,000	10^6	deci	0.1	10^{-1}
kilo	1,000	10^3	centi	0.01	10^{-2}
hecto	100	10^2	milli	0.001	10^{-3}
deka	10	10^1	micro	0.000001	10^{-6}

TABLE I.2
APPROXIMATIONS

	Metric	Multiply by	British	Multiply by	Metric
DISTANCE	millimeters	0.04	inches	25.40	millimeters
	centimeters	0.39	inches	2.54	centimeters
	meters	3.28	feet	0.30	meters
	meters	1.09	yards	0.91	meters
	kilometers	0.62	miles	1.61	kilometers
MASS	kilograms	0.07	slugs	14.59	kilograms
VOLUME	cubic meters	35.31	cubic feet	0.03	cubic meters
	cubic meters	1.31	cubic yards	0.76	cubic meters

TABLE I.2

APPROXIMATIONS (continued)

	Metric	Multiply by	British	Multiply by	Metric
Area	square centimeters	0.16	square inches	6.45	square centimeters
	square meters	10.76	square feet	0.09	square meters
	square meters	1.20	square yards	0.84	square meters
	square kilometers	0.40	square miles	2.60	square kilometers
Temperature	Celsius	9/5(°C + 32)	Fahrenheit	5/9(°F − 32)	Celsius
Velocity	meters/second	3.28	feet/second	0.30	meters/second
	kilometers/hour	0.62	miles/hour	1.61	kilometers/hour
Acceleration	meters/second^2	3.28	feet/second^2	0.30	meters/second^2
	meters/second^2	2.24	miles/hour/second	0.45	meters/second^2
Force	Newtons	0.22	pounds	4.45	Newtons
Linear Momentum	kilogram-meters/second	0.22	slug-feet/second	4.45	kilogram-meters/second
Impulse	Newton-seconds	0.22	pound-seconds	4.45	Newton-seconds
Angular Momentum	kilogram-meters^2/second	0.74	slug-feet^2/second	1.36	kilogram-meters^2/second
Rotational Inertia	kilogram-meters^2	0.74	slug-feet^2	1.36	kilogram-meters^2
Work	Newton-meters (Joule)	0.74	foot-pounds	1.36	Newton-meters (Joule)
Torque	Newton-meters	0.74	pound-feet	1.36	Newton-meters
Power	Newton-meters/second (watts)	0.74	foot-pounds/second	1.36	Newton-meters/second (watts)
	Newton-meters/second (watts)	0.0013	horsepower	745.7	Newton-meters/second (watts)
	kilocalories/second	5.88	horsepower	0.17	kilocalories/second
	kilocalories/second	3,234.00	foot-pounds/second	0.00031	kilocalories/second

Conversions Within the British System

There are convenient conversions within the British system. The following are examples:

feet/second multiplied by 0.68 = miles/hour
miles/hour multiplied by 1.47 = feet/second

5,280 feet = 1 mile
1,760 yards = 1 mile

APPENDIX II

List of Symbols and Equations

TABLE II.1
SYMBOLS

Symbol	Meaning
A	Area
AP	Anteroposterior
a	Linear acceleration
C_D	Coefficient of drag
C_{fr}	Coefficient of friction
C_L	Coefficient of lift
CG	Center of gravity
D	Distance of projectile (horizontal or vertical)
d	Linear displacement (or distance moved)
$d_\perp$	Perpendicular distance (force arm)
F	Force
F_C	Centripetal and centrifugal force
F_D	Force of drag
F_L	Force of lift
F_{fr}	Friction force
g	Acceleration due to the force of gravity
h	Height
I	Rotational inertia
k	Radius of gyration
KE	Kinetic energy
L	Angular momentum
M	Linear momentum
m	Mass
ML	Mediolateral
P	Power
p	Pressure

TABLE II.1

SYMBOLS (continued)

Symbol	Meaning
PE	Potential energy
PI	Ponderal index
r	Radius of a point on a circle: radius of rotation
RF	Reaction force
ROM	Range of motion
T	Torque
t	Time
V	Volume
v	Speed (velocity when direction is specified)
W	Weight
w	Work
WD	Weight density
α (alpha)	Angular acceleration
θ (theta)	Angular displacement
π (pi)	Ratio of the circumference of a circle to its diameter (3.1416)
ρ (rho)	Mass density
Σ (sigma)	Sum of
ω (omega)	Angular velocity

TABLE II.2

EQUATIONS

Equation Number	Textual Page Reference	Equation	Quantity
B.1	51	$PI = 10^3 \times \frac{\sqrt[3]{W}}{H}$	Ponderal index
C.1	98	radians = $\theta/57.3°$ (θ in degrees)	Radians
C.2	100	$p = F/A$	Pressure
C.3	104	$w = Fd$	Work = force × distance the force is applied
C.4	105	$P = F \times d/t$	Power = force × velocity at which force is applied
C.5	105	$P = Fv$	
C.6	106	$KE = 1/2\ mv^2$	Kinetic energy = 1/2 mass × velocity^2
C.7	106	$PE = Wh$	Potential energy = weight × height the weight is raised
D.1	115	$C_{fr} = F_{fr}/F_{\perp}$	Coefficient of friction

TABLE II.2
EQUATIONS (continued)

Equation Number	Textual Page Reference	Equation	Quantity
D.2	117	$F_C = mv^2/r$	Centripetal and centrifugal force = (mass × tangental velocity²/radius)
D.3	119	$e = \sqrt{\frac{\text{height of rebound}}{\text{height of drop}}}$	Coefficient of elasticity
D.4	129	$c = \sqrt{a^2 + b^2}$	Pythagorean theorem
E.1	139	$T = F(d_\perp)$	Torque = F × force arm
3.1	164	$\Sigma T = 0$	Rotary equilibrium
3.2	174	MA = resistive force/motive force	Mechanical advantage of a muscle lever system
3.3	174	MA = motive force arm/resistive force arm	Mechanical advantage of a muscle lever system
3.4	181	MA = radius of wheel/radius of axle	Mechanical advantage of a wheel–axle system
3.5	182	MA = force on axle/force on wheel	Mechanical advantage of a wheel–axle system
F.1	335	$v = d/t$	Linear velocity (speed)
F.2	336	$a = (v_2 - v_1)/t$	Linear acceleration
F.3	343	$a = F/m$	Newton's second law of motion for linear motion
G.1	353	$M = mv$	Linear momentum
G.2	354	$F(t) = M_2 - M_1$	Linear impulse
10.1	383	$D_{vert} = \frac{v_{vert}^2}{2g}$ or $D_{vert} = \frac{(v \sin \theta)^2}{2g}$	Vertical distance of a projectile with no air resistance
10.2	XXX	$D_{horiz} = v_{horiz} \times t$ or $D_{horiz} = \frac{(v^2 \sin 2\theta)}{g}$	Horizontal distance of a projectile with equal projection and landing heights and no air resistance
H.1	422	$F_D = \frac{1}{2} C_D A \rho v^2$	Fluid drag force
H.2	430	$F_L = \frac{1}{2} C_L A \rho v^2$	Fluid lift force
12.1	466	$WD = W/V$	Weight density
12.2	487	$v = SL \times SF$	Swimming velocity = distance per stroke × stroke frequency
I.1	514	$\omega = \theta/t$	Angular velocity (speed)
I.2	515	$d = r\theta$ (θ in radians)	Linear distance of a rotating point
I.3	515	$v = r\omega$ (ω in radians/sec)	Linear velocity of a rotating point

TABLE II.2

EQUATIONS (continued)

Equation Number	Textual Page Reference	Equation	Quantity
I.4	517	$\alpha = (\omega_2 - \omega_1)/t$	Angular acceleration
I.5	521	$\alpha = T/I$	Newton's second law of motion for angular motion
I.6	521	$I = mr^2$	Rotational inertia of a body with concentrated mass
I.7	523	$I = mk^2$	Rotational inertia of a body with distributed mass
J.1	531	$L = I\omega$	Angular momentum
J.2	532	$T(t) = \Delta L = I\omega_2 - I\omega_1$	Angular impulse

OTHER USEFUL EQUATIONS

Chapter 10

$$X = v_0 t \cos\theta$$

$$Y = v_0 t \sin\theta - \frac{1}{2} g t^2$$

X and Y are the horizontal and vertical locations of a projected object

v_0 is the initial velocity

θ is the projection angle

$$D_{horiz} = \frac{v^2 \sin\theta \cos\theta + v \cos\theta \sqrt{(v \sin\theta)^2 + 2gh}}{g}$$

Horizontal distance of a projectile with unequal projection and landing heights and no air resistance

APPENDIX III

Anthropometric Parameters

TABLE III.1

MEAN SEGMENT WEIGHTS EXPRESSED AS PERCENTAGES OF TOTAL BODY WEIGHT

Segment	Males (N = 35)	Females (N = 100)
Head	8.26	8.20
Whole trunk	55.10	53.20
Thorax	20.10	17.02
Abdomen	13.06	12.24
Pelvis	13.66	15.96
Total arm	5.77	4.97
Upper arm	3.25	2.90
Forearm	1.87	1.57
Hand	0.65	0.50
Forearm and hand	2.52	2.07
Total leg	16.68	18.43
Thigh	10.50	11.75
Leg	4.75	5.35
Foot	1.43	1.33
Leg and foot	6.18	6.68

Adapted from Plagenhoef et al., 1983.

TABLE III.2

MEAN SEGMENT LENGTHS EXPRESSED AS PERCENTAGES OF TOTAL BODY HEIGHT

Segment	Males (N = 35)	Females (N = 73)
Head and neck	10.75	10.75
Trunk (hip to shoulder)	30.00	29.00
Thorax	12.70	12.70
Abdomen	8.10	8.10
Pelvis	9.30	9.30
Upper arm	17.20	17.30
Forearm	15.70	16.00
Hand	5.75	5.75
Thigh	23.20	24.90
Leg	24.70	25.70
Foot	4.25	4.25
Biacromial	24.50	20.00
Bi-iliac	11.30	12.00

Adapted from Plagenhoef et al., 1983.

TABLE III.3

SEGMENTAL CENTER OF GRAVITY LOCATIONS EXPRESSED AS PERCENTAGES OF SEGMENT LENGTHS MEASURED FROM THE PROXIMAL ENDS

Segment	Males (N = 7)	Females (N = 9)
Head and neck	55.0	55.0
Trunk	63.0	56.9
Upper arm	43.6	45.8
Forearm	43.0	43.4
Hand	46.8	46.8
Pelvis	5.0	5.0
Abdomen	46.0	46.0
Thorax	56.7	56.3
Thigh	43.3	42.8
Leg	43.4	41.9
Foot	50.0	50.0
Abdomen and pelvis	44.5	39.0

Adapted from Plagenhoef et al., 1983.

TABLE III.4

SEGMENTAL RADII OF GYRATION EXPRESSED AS PERCENTAGES OF THE SEGMENT LENGTHS MEASURED FROM THE PROXIMAL (P) AND DISTAL (D) ENDS[a]

	Males (N = 36)		Females (N = 100)	
Segment	*P*	*D*	*P*	*D*
Upper arm	54.2	64.5	56.4	62.3
Forearm	52.6	54.7	53.0	64.3
Hand	54.9	54.9	54.9	54.9
Thigh	54.0	65.3	53.5	65.8
Leg	52.9	64.2	51.4	65.7
Foot	69.0	69.0	69.0	69.0
Head and trunk[a]	83.0	60.7	——	——

[a] From Dempster, 1955.
Adapted from Plagenhoef et al., 1983.

TABLE III.5

MEAN BODY SEGMENT VOLUMES EXPRESSED AS PERCENTAGES OF TOTAL BODY VOLUME

Segment	Rotund (%)	Muscular (%)	Thin (%)	Median (%)
Total upper extremity	5.28	5.60	5.20	5.65
Arm	3.32	3.35	2.99	3.46
Forearm plus hand	1.95	2.24	2.23	2.15
Forearm	1.52	1.70	1.63	1.61
Hand	0.42	0.53	0.58	0.54
Total lower extremity	20.27	18.49	19.08	19.55
Thigh	14.78	12.85	12.90	13.65
Leg plus foot	5.52	5.61	6.27	5.97
Leg	4.50	4.35	4.81	4.65
Foot	1.10	1.30	1.46	1.25

From Dempster, 1955.

TABLE III.6

MEAN BODY SEGMENT DENSITIES

Segment	Density (gm/cm^3)
Head and neck	1.11
Trunk	1.03
Thigh	1.05
Leg	1.09
Foot	1.10
Upper arm	1.07
Forearm	1.13
Hand	1.16

From Dempster, 1955.

TABLE III.7

MEANS AND STANDARD DEVIATIONS (SD) OF THE ROTATIONAL INERTIA OF THE BODY IN SELECTED POSTURES

		Rotational Inertia ($lb\text{-}in\text{-}sec^2$)	
Postures	Axis[a]	Mean	SD
Standing	x	115.0	19.3
	y	103.0	17.9
	z	11.3	2.2
Standing	x	152.0	26.1
Arms over head (layout)	y	137.0	25.3
	z	11.1	1.9
Spread eagle	x	151.0	27.1
Shoulders abducted 135°	y	114.0	21.3
Hips abducted 30°	z	36.6	7.9
Sitting	x	61.1	10.3
Elbows flexed 90°	y	66.6	11.6
	z	33.5	5.8
Sitting	x	62.4	9.7
Elbows extended	y	68.1	12.0
	z	33.8	5.9
Sitting (tuck position)	x	39.1	6.0
Hips flexed 145°	y	38.0	5.8
Knees flexed 145°	z	26.3	5.1

TABLE III.7

MEANS AND STANDARD DEVIATIONS (SD) OF THE ROTATIONAL INERTIA OF THE BODY IN SELECTED POSTURES (continued)

Postures	Axis[a]	Rotational Inertia ($lb\text{-}in\text{-}sec^2$)	
		Mean	**SD**
Sitting	x	65.8	10.3
Elbows flexed 80°	y	75.2	14.0
Hips flexed 80°	z	34.2	5.6
Knees flexed 168°			
Relaxed	x	92.2	13.3
(Weightless)	y	88.2	13.3
	z	35.9	5.4

[a] The z axis corresponds to the body's longitudinal axis; the y axis corresponds to the body's mediolateral axis; the x axis corresponds to the body's anteroposterior axis.
From Santschi, Du Bois, and Omotu, 1963.

References and Suggested Readings

Clauser, C. E., McConville, J. T., & Young, J. W. (1969). *Weight, volume, and center of mass of segments of the human body*. (AMRL Technical Documentary Report) (pp. 69–70). Wright-Patterson Air Force Base, Ohio.

Dempster, W. T. (1955). *Space requirements of the seated operator*. (WADC Technical Documentary Report) (pp. 55–159). Wright-Patterson Air Force Base, Ohio.

Kjeldsen, K. (1969). *Body segment weights, limb lengths, and the location of the center of gravity in college women*. Unpublished master's thesis, University of Massachusetts.

Miller, D. M., & Nelson, R. C. (1973). *Biomechanics of sport*. Philadelphia: Lea & Febiger.

Morse, R., & Kjeldsen, K. (1975). *Anthropometric measurements*. Paper presented at the AAHPERD Pre-Convention Workshop, Milwaukee.

Plagenhoef, S., Evans, F. G., & Abdelnour, T. (1983). Anatomical data for analyzing human motion. *Research Quarterly for Exercise and Sport, 54*(2), 169–178.

Santschi, W. R., Du Bois, J., & Omotu, C. (1963). *Moments of inertia and centers of gravity of the living human body*. (AMRL Technical Documentary Report) (pp. 63–66). Wright-Patterson Air Force Base, Ohio.

APPENDIX IV

Methods of Calculating the Center of Gravity

To determine the location of the center of gravity (CG) in any direction, two methods may be used. Both methods use the technique of summing torques about an axis of rotation.

Board-and-Scale Method

The **board-and-scale method** calls for a person to stand on a board that is supported at each end, as seen in Figure IV.1. The sum of the forces downward is equal to the weight of the body plus the weight of the board, and the sum of the forces upward is equal to the sum of the reaction forces acting upward at each end of the board.

When added together, the reaction forces will equal the weight of the body plus the weight of the board. This can be confirmed easily if two scales are placed under the supports of the board. Regardless of where the person stands or lies on the board, the *sum* of the two scale readings will equal the total body weight plus the weight of the board. For convenience, the effect of the weight of the board should be ignored, and this can be accomplished easily if its weight is eliminated from the reaction by setting the readings of the scales to zero after the board has been placed on top of them. In this way only the weight of the body is considered.

To sum the torques on the system, an axis of rotation must be selected. Either end of the board may serve as an axis. If one end of the board is an axis through which one reaction force passes, only two torque-producing forces will be acting on the board about the specified axis: (1) the body weight, and (2) the reaction force acting upward at the end opposite to the axis end. The body weight line of force extends from the CG toward the center of the earth. The reaction force

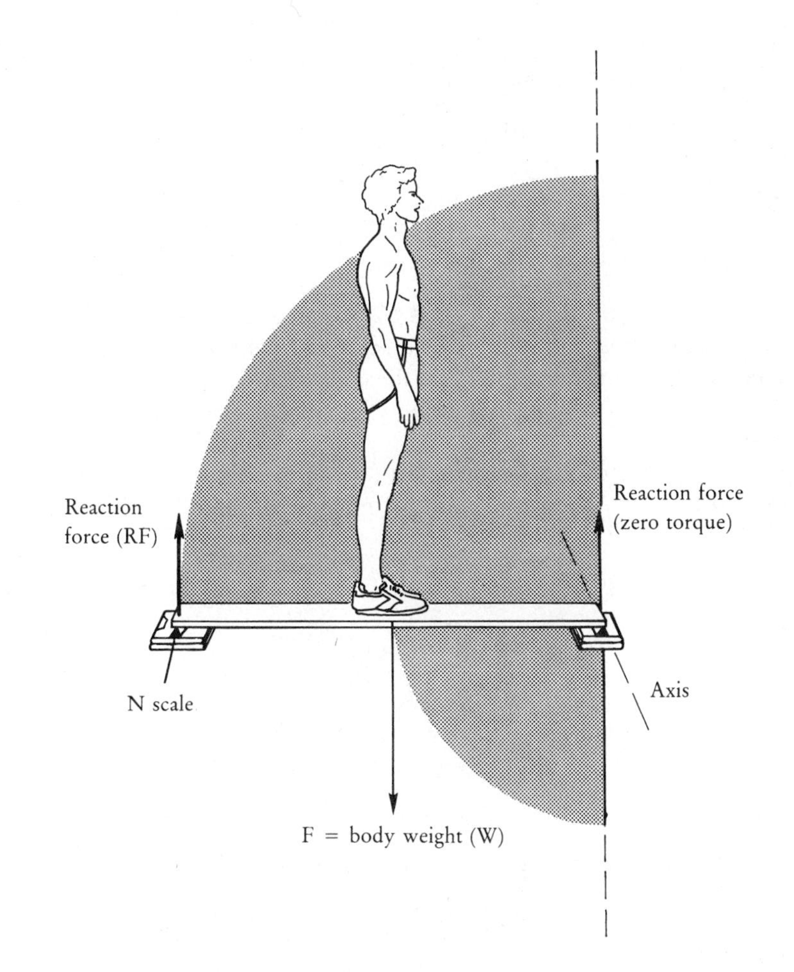

FIGURE IV.1
A body standing on a board supported at each end, for calculating the front-to-back location of the CG.

at the scale acts upward on the board. Because the board is not rotating, the opposing two *torques* about the axis end must be equal. The torque produced by the scale is equal to the reaction force times its perpendicular distance to the axis. This distance will equal the length of the board if the supports are placed as close as possible to the ends of the board. The opposing torque produced by the body is equal to the body weight times the perpendicular distance from the line of gravity to the axis. It is this $d_\perp$ that will be determined.

This is the procedure for calculating the distance of the CG of a 667.5-N body from an axis:

$$\Sigma T = 0$$
$$W(d_\perp) + RF(l) = 0$$

where Σ is sum of, T is torque, W is weight (667.5 N), RF is reaction force (–267 N), and l is length of the board (243.8 cm).

$$(667.5 \text{ N} \times \text{CG distance}) + (-267 \text{ N} \times 243.8 \text{ cm}) = 0$$

$$\text{CG distance} = \frac{65094.6 \text{ N-m}}{667.5 \text{ N}}$$

$$\text{CG distance} = 97.52 \text{ cm}$$

The line of force of the body's CG is 97.53 cm from the axis. To determine where the body's line of gravity is located from front to back within the foot, the distance of the toes from the axis (86.36 cm) must be subtracted from the total (97.53 cm). Thus, the line of gravity in this case is located 11.17 cm into the supporting foot base.

To determine the location of the line of gravity of a body from head to foot and from right to left, the same method may be repeated with the body in a

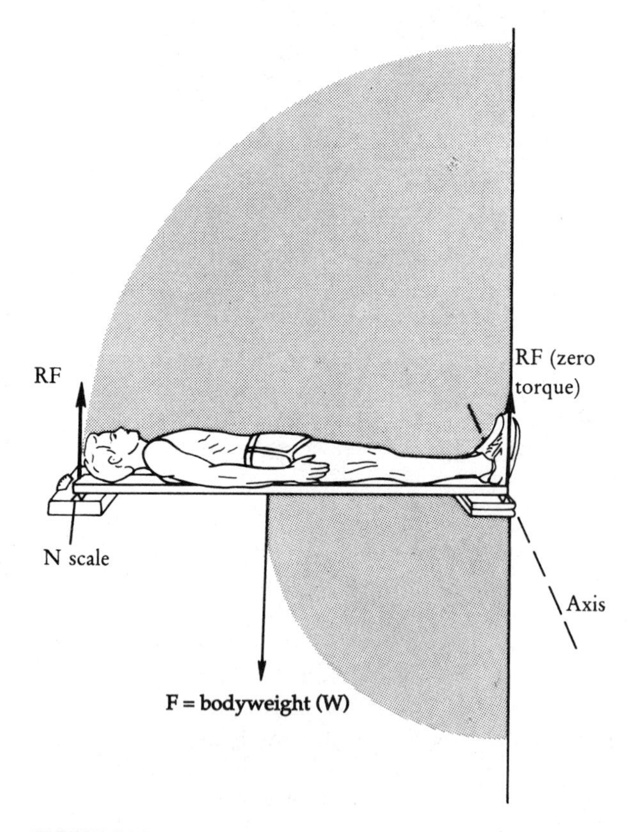

FIGURE IV.2

A body on a board, for calculating the location of the CG from top to bottom.

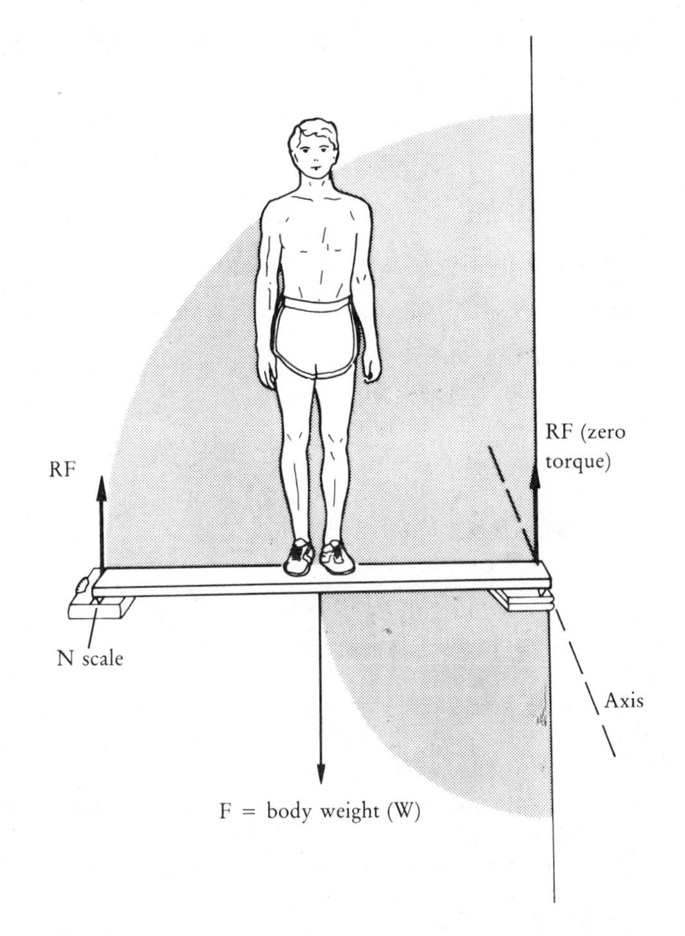

FIGURE IV.3
A body on a board for calculating the location of the CG from right to left.

back lying position (Figure IV.2) and in a standing position (Figure IV.3) with one side toward the axis.

If three scales are used rather than one, a similar method may be used to determine the location of the CG in two directions simultaneously; Figure IV.4 illustrates the setup. An axis is selected through scale 1 and the support (axis A). The reaction torques and the total body torque are calculated and summed, as was done previously with the board and scale, to produce the CG location from axis A. A second axis through the support and the other adjacent scale (scale 3) is selected, and the reaction force torques and body weight torques are summed to produce the CG location from the second axis, B. If a scale were positioned at the support, the sum of the reaction forces indicated on the four scales would equal the total body weight. The calculations about each axis include two torque-producing reaction forces, each with its own $d_{\perp}$ acting in an upward direction (out from the page), and the body weight with its $d_{\perp}$, the CG distance, acting downward about each axis (into the page).

$$\Sigma T = 0$$

Right-to-left location from axis A is calculated by:

$$RF_3(d_{\perp}) + RF_2(d_{\perp}) + W(d_{\perp}) = 0$$

Head-to-foot location from axis B is calculated by:

$$RF_2(d_{\perp}) + RF_1(d_{\perp} + W(d_{\perp}) = 0$$

where RF_1, RF_2, and RF_3 are the respective reaction forces, $d_{\perp}$ is the respective perpendicular distance, and W is body weight.

Segmental Method

The second method of calculating the CG in any direction is called the **segmental method**. This method employs each segmental weight as a separate force acting at some distance from an arbitrary axis.

To use the segmental method, one needs a body outline of the performer taken from a film or photograph. Calculating from a photograph enables location of the CG for bodies in various positions during the execution of a sport skill and does not require the performer to hold a position on the gravity board while the readings and measurements are taken. Another advantage of the segmental method is that the performer is in the exact position used in performance. (A performer lying on the board has difficulty reproducing the exact position of the body during the execution of a sport skill.)

Because the actual segmental weights for every performer are not known, proportional segmental weights listed as percentages of the total body weight may be used. These data are listed for males and females in Appendix III, Table III.1. The proportional weight of each segment is a percentage of the total body weight. The following equation explains arithmetically why percentages may be used:

$$X_tW_t = X_1W_1 + X_2W_2 + X_3W_3 \ldots X_nW_n$$

where X is the distance of the CG from the arbitrary axis, W is the weight of the total body (t) or the weight of each segment (n), t is the subscript used to denote the entire body, and n is the subscript used to denote each segment, 1 to n.

If one divides both sides of the equation by W_t, one finds that the distance of the total body CG, X, from the arbitrary axis is equal to the sum of the weights of each individual segment divided by the total body weight times that segment's distance from the arbitrary axis, or the segmental percentage of total

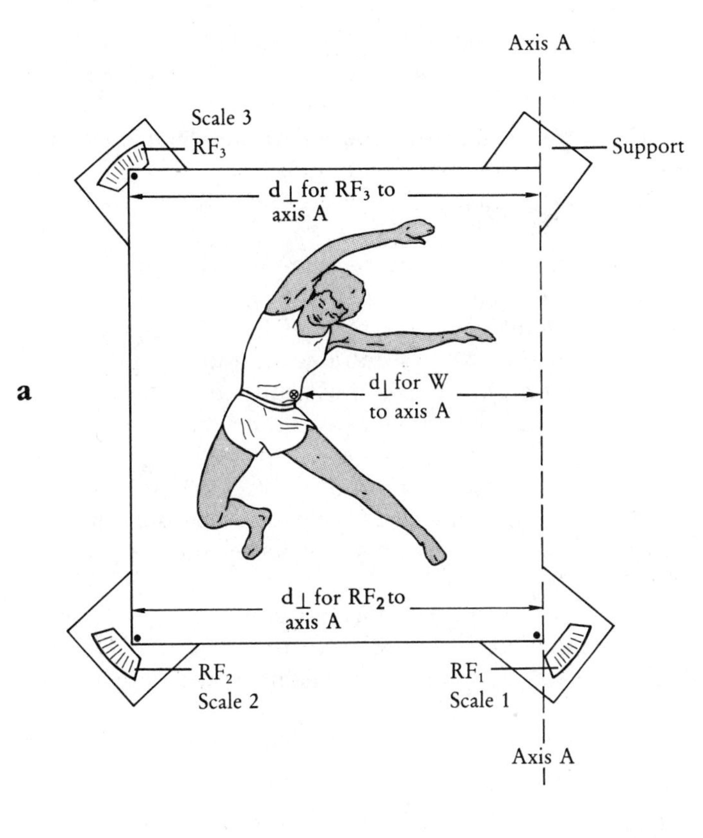

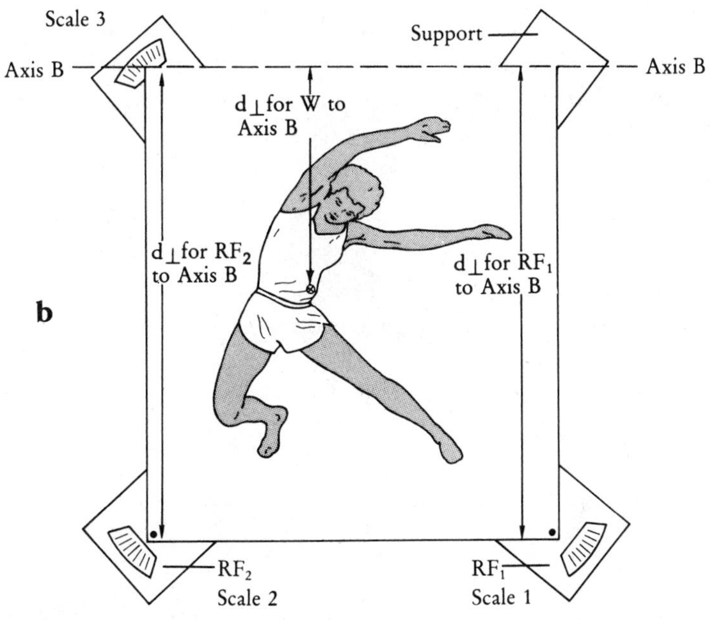

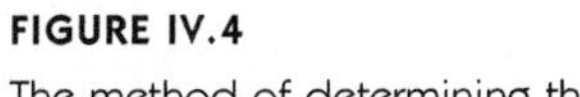

FIGURE IV.4

The method of determining the CG location in two planes, (a) from axis A and (b) from axis B.

body weight times the distance of the individual segment. This relationship may be expressed by the following equation:

$$X = X_1W_1/W_t + X_2W_2/W_t + \ldots X_nW_n/W_n$$

For an example, calculations may be done for the hurdler shown in Figure IV.5.

One must keep in mind that because the CG is that point at which the total body weight is assumed to be concentrated, the torque produced by the total body weight from any arbitrary axis must equal the sum of all the individual segmental weights times each of their perpendicular distances from that same axis.

Therefore, to determine the location of the CG in one direction, one would repeat a procedure similar to the three scales and board method. Two arbitrary axes are drawn on the paper most conveniently outside the body figure. A $d_\perp$ for each segment's weight is drawn from the segmental CG to each of the arbitrary axes. The weight of each segment may be represented by its percentage of the total body weight, and the weight of the total body will be equal to 100%. Knowing the *actual* weight of the body or its segments is unnecessary.

A chart similar to that shown in Table IV.1 should be made for the calculation. Complete the chart in this manner:

1. List all the segments in the left-hand column.
2. From Table III.1 in Appendix III, list each segment's weight (as a percentage) in the % Segmental Weight column.
3. Measure the distance of each segmental CG from each axis and list those distances in the D_{horiz} and D_{vert} columns.

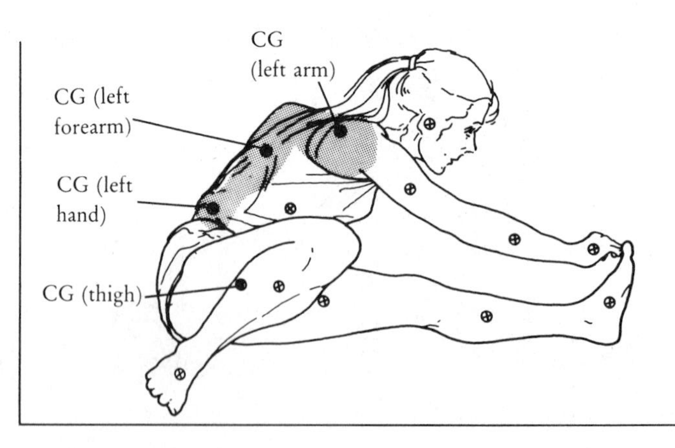

FIGURE IV.5

A hurdler with segmental CGs identified.

TABLE IV.1
CHART FOR CALCULATING THE CENTER OF GRAVITY USING THE SEGMENTAL METHOD

Segment	% Segmental Weight[a]	D_{horiz}	$D_h \times (\%)$	D_{vert}	$D_v \times (\%)$
Right thigh	______	______	______	______	______
Left thigh	______	______	______	______	______
Right leg	______	______	______	______	______
Left leg	______	______	______	______	______
Etc.					
.					
.					
.					
	______		______		______
	100%		Total horizontal distance		Total vertical distance

[a] From Appendix III, Table III.1.

4. Multiply each distance by the segment's weight percentage and list the product in the $D_h \times (\%)$ and $D_v \times (\%)$ columns, respectively. The percentage of segmental weight times the distance of the segment's CG from each axis reflects the torque contribution of that segment to the torque produced by the total body's CG from that axis. Thus, the sum of all the "weighted" segmental distances is equal to the distance that the total body's CG is from that same axis.
5. Sum the head-to-foot (vertical) weighted distances, and sum the front-to-back (horizontal) weighted distances. Place the totals in the spaces provided.
6. Measure each total distance from its respective axis, and draw a line parallel to that axis through that point. The intersection of these two drawn lines represents the total body's CG location. The location of the CG from the performer's right to left side cannot be determined from a single, two-dimensional photograph.

The segmental method is presented here only to allow the user to calculate the CG of a few subjects, enough to let one gain a feel for the contributions of the segments' weights and positions to the total body's CG location. Calculation of the exact location of the CG during the execution of a skill is usually not necessary for the teacher or coach. One should be able to estimate the location and the path of the CG, however, by knowing how the various segments contribute to and affect the CG location during movement.

A precise location and path of the CG may be determined more efficiently and precisely by using sophisticated filming and computer instrumentation (See Chapter 17).

APPENDIX V

Mathematics Review and Trigonometry

I. Decimals can be expressed as positive or negative powers of ten.

Example: $10^0 = 1$

$10^1 = 10$

$10^2 = 100$

$10^{-1} = .1$

$10^{-2} = .01$

$10^{-3} = .001$

$56.24 \times 10^2 = 5,624$

$3.79 \times 10^{-2} = 0.0379$

$2.1 \times 10^{-3} = 0.0021$

II. Proportions, Equations[1]

A. A proportion is a relationship between two numbers or quantities. A proportion is either direct or inverse. There is a direct proportion between two quantities if an exact increase in one causes an exact increase in the other.

Example: Given v = d/t, if d is tripled, v will also triple.

There is an inverse proportion between two quantities if an exact increase in one causes an exact decrease in the other.

[1]Adapted from Barham, J. (1978). *Mechanical kinesiology*. St. Louis: C. V. Mosby Co.

Example: Given v = d/t, if t is doubled, v is halved (decreased by becoming two times smaller).

(a) If two numbers are directly across from = sign or in vertical position, they are directly proportional.

(b) If two numbers are diagonally across from = sign, they are inversely proportional.

Example: $\frac{a}{b} = \frac{c}{d}$

direct: a and b, c and d, a and c, b and d
inverse: a and d, b and c

III. Elementary Trigonometry

A. Triangles, Functions

1. In a right triangle, one angle must be 90 degrees, while the sum of the other two acute angles must be 90 degrees. The sum of the three angles of any triangle is always 180 degrees.
2. In a right triangle, the square of the longest side, or hypotenuse, is equal to the sum of the squares of the other two sides. This yields the following equation: $a^2 + b^2 = c^2$. This is known as the Pythagorean theorem.

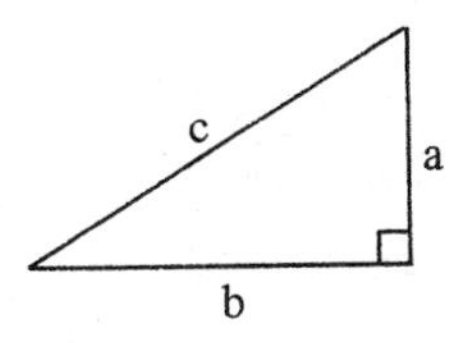

3. In triangle XYZ, side x is opposite angle X and adjacent to angle Y; side y is opposite angle Y and adjacent to angle X; side z (hypotenuse) is opposite the right angle Z.

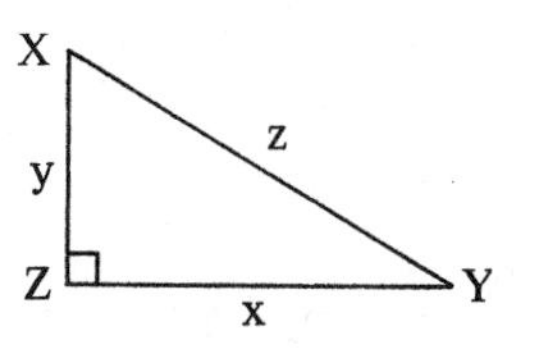

4. Trigonometric functions are ratios between various sides of a right triangle. There are six functions. We are concerned with three of them: sine (sin), cosine (cos), tangent (tan).

From ABC we obtain the following ratios:

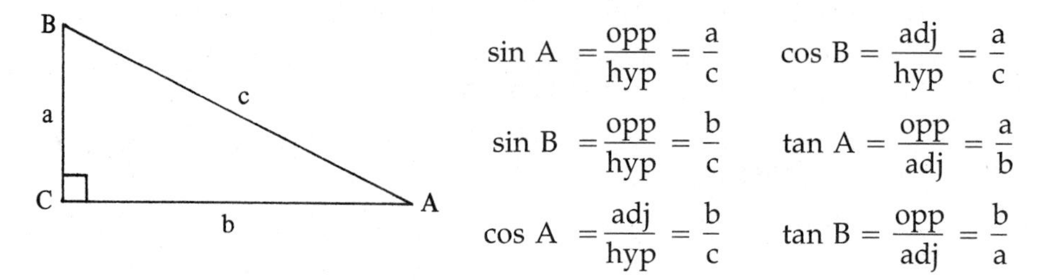

$$\sin A = \frac{\text{opp}}{\text{hyp}} = \frac{a}{c} \qquad \cos B = \frac{\text{adj}}{\text{hyp}} = \frac{a}{c}$$

$$\sin B = \frac{\text{opp}}{\text{hyp}} = \frac{b}{c} \qquad \tan A = \frac{\text{opp}}{\text{adj}} = \frac{a}{b}$$

$$\cos A = \frac{\text{adj}}{\text{hyp}} = \frac{b}{c} \qquad \tan B = \frac{\text{opp}}{\text{adj}} = \frac{b}{a}$$

Note that sin A = cos B, and sin B = cos A.

The value of the sine of a given angle (26 degrees) is a constant value independent of the size of the right triangle. The ratio of the side opposite the 26-degree angle to the hypotenuse will be the same whether the triangle is measured in hectometers or inches. This principle holds true for all the trigonometric functions.

5. A table of trigonometric functions appears at the end of this appendix.

Example: sin 60° = .8660
cos 30° = .8660
tan 22° = .4040

6. Tables of trigonometric functions usually go up to 90 degrees.

a. For functions of angles 91 degrees to 180 degrees, take the function of (180° − ∢). All functions are negative except the sine.

Example: sin 120° = sin 60°
tan 150° = –tan 30°

b. For functions of angles 181 degrees to 270 degrees, take the function of (270° − ∢). All functions are negative except the tangent.

Example: cos 220° = –cos 50°
tan 195° = tan 75°

c. For functions of angles 271 degrees to 360 degrees, take the function of (360° − ∢). All functions are negative except the cosine.

Example: tan 290° = –tan 70°
sin 330° = –sin 30°

7. Trigonometric functions make finding all the components of a right triangle possible if the length of one side and the size of one angle *or* the lengths of two sides are given.

Example: Given triangle LMN with angle L = 42 degrees and hypotenuse n = 20 ft. Find the length of sides 1 and m.

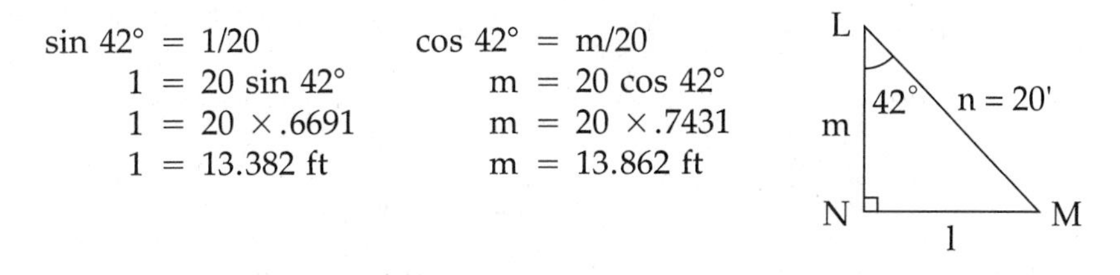

$$\begin{aligned} \sin 42^\circ &= l/20 \\ l &= 20 \sin 42^\circ \\ l &= 20 \times .6691 \\ l &= 13.382 \text{ ft} \end{aligned} \qquad \begin{aligned} \cos 42^\circ &= m/20 \\ m &= 20 \cos 42^\circ \\ m &= 20 \times .7431 \\ m &= 13.862 \text{ ft} \end{aligned}$$

Example: Given triangle QRS with sides of 8 in. and 16 in. Find the length of the hypotenuse and the size of both acute angles.

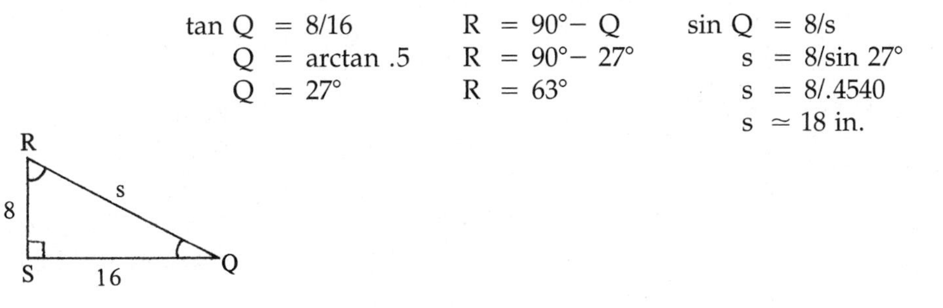

$$\begin{aligned} \tan Q &= 8/16 \\ Q &= \arctan .5 \\ Q &= 27^\circ \end{aligned} \qquad \begin{aligned} R &= 90^\circ - Q \\ R &= 90^\circ - 27^\circ \\ R &= 63^\circ \end{aligned} \qquad \begin{aligned} \sin Q &= 8/s \\ s &= 8/\sin 27^\circ \\ s &= 8/.4540 \\ s &\simeq 18 \text{ in.} \end{aligned}$$

Note: arctan x means "the angle whose tangent is x."

B. Unit Circle, Functions

1. Another way to approach trigonometric functions is by relating them to points on a unit circle (radius = 1, circumference $= 2\pi$) with its center at the origin of a Cartesian coordinate system.

If R is a point on the circle, its coordinates are given as $(\cos\theta, sin\theta)$, where θ is the angle, in radians, through which R has rotated. The x coordinate is $\cos\theta$, the y coordinate is $\sin\theta$.

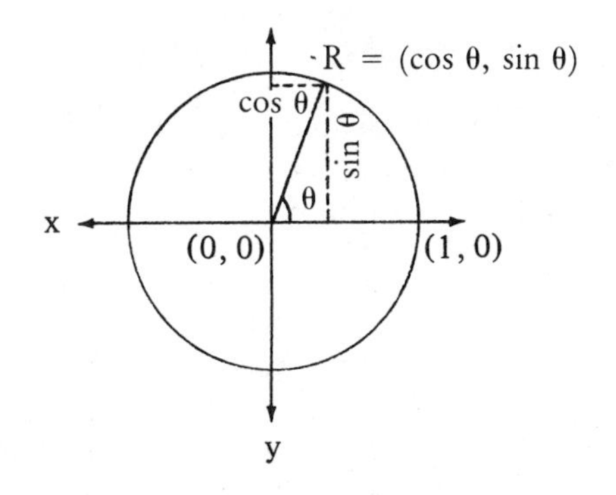

2. Determining the signs (+ or –) of the functions for angles greater than 90 degrees is easy using the coordinate system. Note that a point on the unit circle travels through 360 degrees, or a distance of 2π.

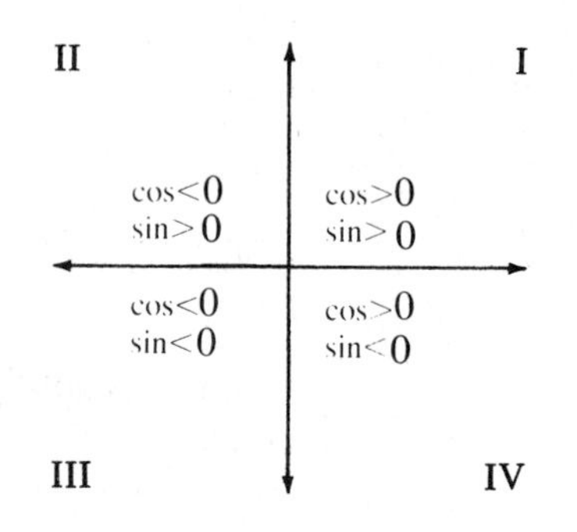

Quadrant I
Contains angles 0 degrees to 90 degrees (0π to $\pi/2$). For point (x,y), x is positive; y is positive; sin, cos, and tan are positive.

Quadrant II
Contains angles 90 degrees to 180 degrees ($\pi/2$ to π). For point (x,y), x is negative; y is positive; sin is positive, cos and tan are negative.

Quadrant III
Contains angles 180 degrees to 270 degrees (π to $3\pi/2$). For point (x,y), x is negative, y is negative, sin and cos are negative, tan is positive.

Quadrant IV
Contains angles 270 degrees to 360 degrees ($3\pi/2$ to 2π). For point (x,y), x is positive, y is negative, cos is positive, sin and tan are negative.

Example: $\sin \pi/2 = \sin 90°$
$\sin 4\pi/3 = \sin 240° = -\sin 30°$
$\tan 2\pi/3 = \tan 120° = -\tan 60°$

C. Geometry of Circles

1. The circumference of a circle is determined by the equation $C = 2\pi r$, where r is the circle's radius and π is the ratio of the circumference of a circle to its diameter, π(pi) is a constant value that is approximated to be 3.1416.

2. When a point on a circle makes one revolution, it travels through 360 degrees or 2π radians. The term *radian* is a geometric expression that is the ratio of the length of a circular arc, p, to the length of the circle's radius, r. When the length of the arc, p, is equal to the length of the circle's radius, r, an angular displacement of one radian has been sectioned off on the arc length. Because a radian is a ratio, one radian always equals 57.3 degrees no matter what the size of the circle's radius.

Example:

Radians	*Degrees*
2π or 6.28	360
π or 3.14	180
$\pi/2$ or 1.57	90

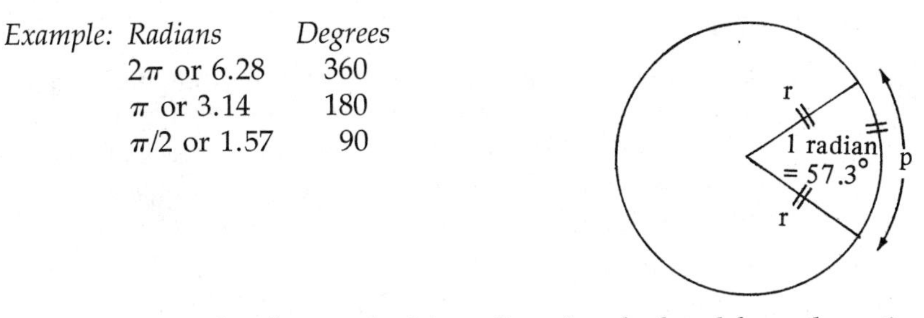

a. The magnitude of an angle θ in radians is calculated from the ratio of the arc length p to the radius r.

Example: θ = p/r radians

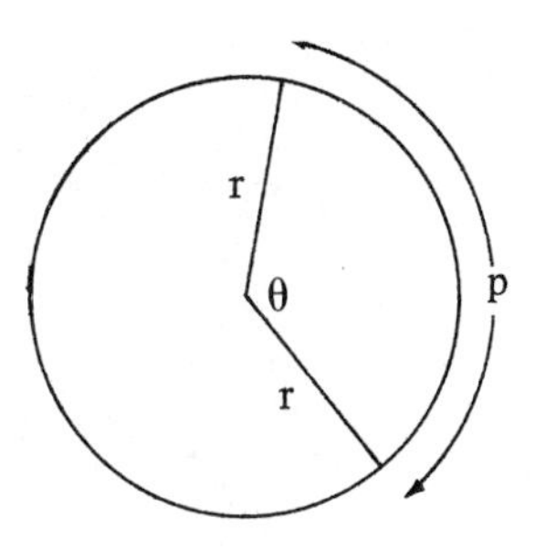

b. The angular measure of a complete circle, measured in radians, is calculated in the same manner.

Example: $\theta = 2\pi r/r$ radians
or 2π radians
or 6.28 radians

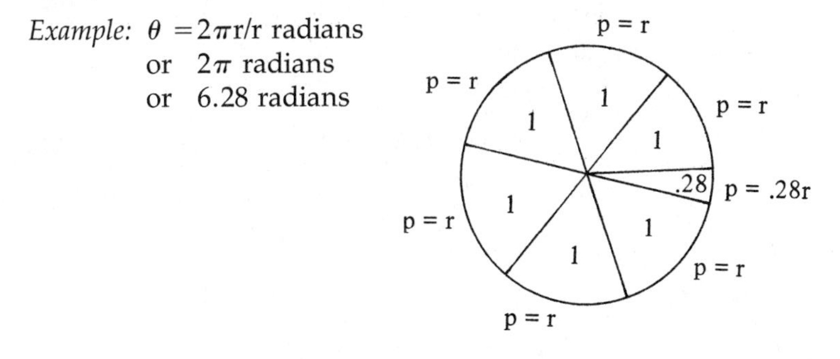

c. Because the units of measurement for r and p are the same, they can be canceled out. Consequently, a radian is a unitless quantity.

Example: number of radians = arc length/radius
= 30 mm/15 mm = 2

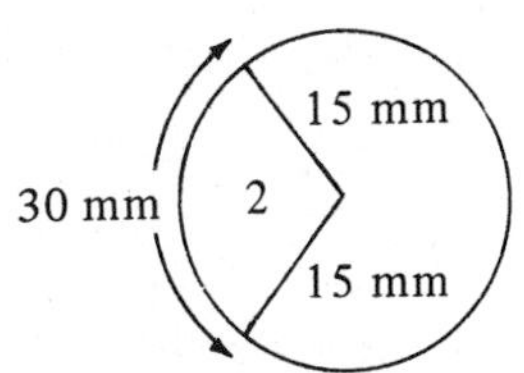

D. Vectors

1. A vector is defined as a line segment that has a magnitude and a direction and represents some quantity (e.g., force, velocity, momentum). The length V of the vector represents the magnitude (size). The direction of the vector is some angle θ which represents the direction. The sum of two (or more) vectors is called the resultant.

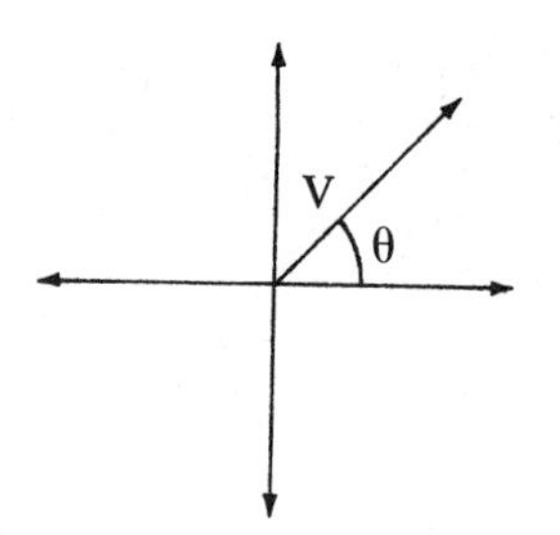

2. Finding resultants

a. A two-dimensional vector can be expressed as an ordered pair of numbers (x,y). The vector V is depicted as a directed line segment whose initial point is the origin (0,0) and whose terminal point is (x,y). The resultant of two vectors can be found by adding the ordered pairs that correspond to the vectors.

Example: Find the resultant of two vectors $V_1 = (3, -2)$ and $V_2 = (4,3)$.

Solution: $V_1 + V_2 = (3, -2) + (4, 3) = |(3 + 4), (-2 + 3)|$
$= (7, 1)$

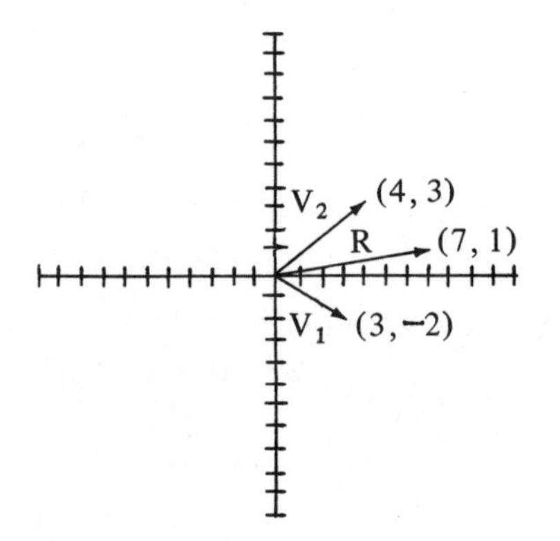

b. Resultants also may be found by simply completing the parallelogram with V_1 and V_2 as sides.

Example: Given that the two forces V_1 and V_2 are applied to an object at A. Find the resultant.

Solution: Draw a vector equivalent to V_2 (same magnitude and direction) with its initial point at the terminal point of V_1. The resultant is V_3.

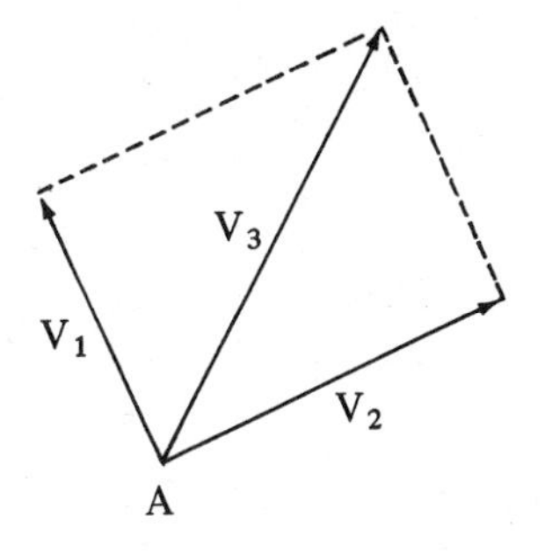

Example: Given three forces 3, 7, and 10 pounds acting on an object at 0 in the indicated directions. Find the resultant.

Solution: The resultant of V_1 and V_2 is V_a, and the resultant of V_3 and V_a is V_b. V_b is the resultant for the system.

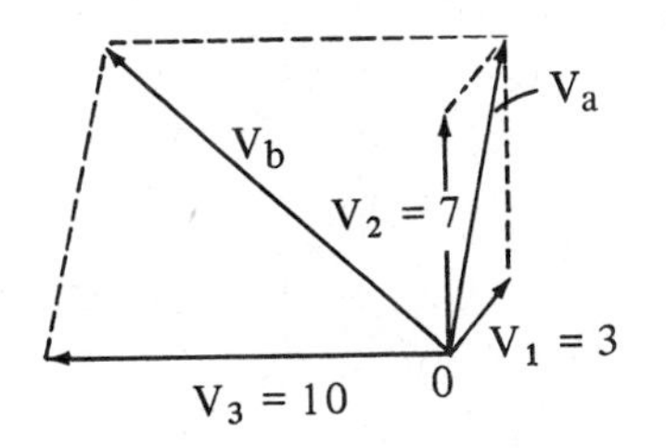

c. A resultant can be determined by placing the vectors head to tail, maintaining their magnitudes and directions. The resultant is the vector drawn from the tail of the first vector to the head of the final vector.

Example: Find the resultant of V_1 and V_2.

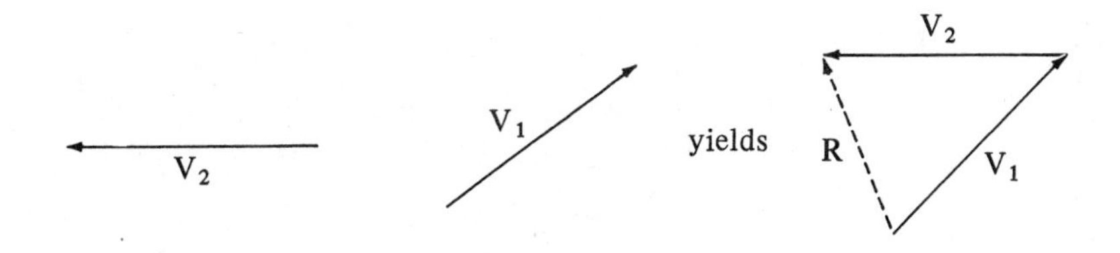

Example: Find the resultant of F_1, F_2, F_3, F_4.

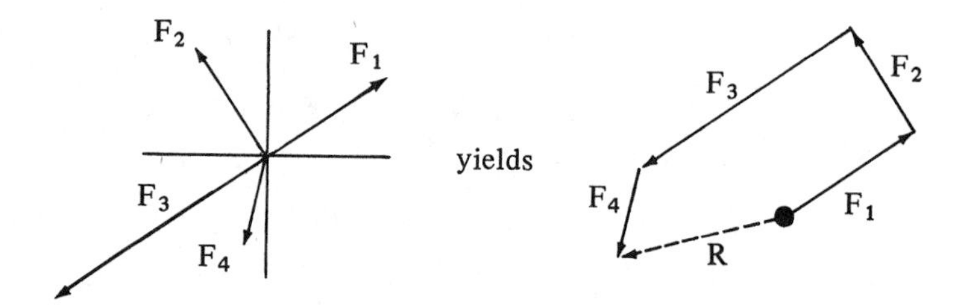

d. To determine trigonometrically the magnitude and direction of the resultant of two vectors that form a nonright angle, the cosine rule can be used.

1. Magnitude of $R^2 = A^2 + B^2 - 2AB \cos \phi$
2. Direction of R: $\theta = \arctan [(A \sin b)/(B + A \cos b)]$

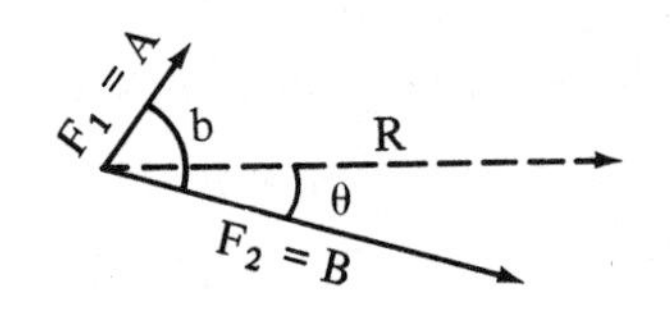

b = angle between vectors A and B
θ = angle between vectors R and B
$\phi = (180° - b)$ = angle between vectors A and B when B is repositioned at the head of A.

Example: Find the resultant magnitude and direction of two vectors, F_1 and F_2 if the angle between them is 40° and F_1 equals 20 lb and F_2 equals 60 lb.

Solution:

a. magnitude of $R^2 = F_1{}^2 + F_2{}^2 - 2F_1F_2 \cos \phi$
$R^2 = (20)^2 + (60)^2 - 2(20)(60)(-.766)$
$R = 76.40$ lb

b. $\theta = \arctan \dfrac{F_1 \sin 40°}{F_2 + F_1 \cos 40°}$

$= \arctan \dfrac{20 \sin 40°}{60 + 20 \cos 40°}$

$= \arctan .17 \quad 9.69°$

$= \arctan .17$

3. Finding horizontal and vertical components

a. Often, determining the horizontal and vertical components of a resultant vector is necessary. This is accomplished by completing the rectangle (right angles) with the resultant as the diagonal.

Example:—

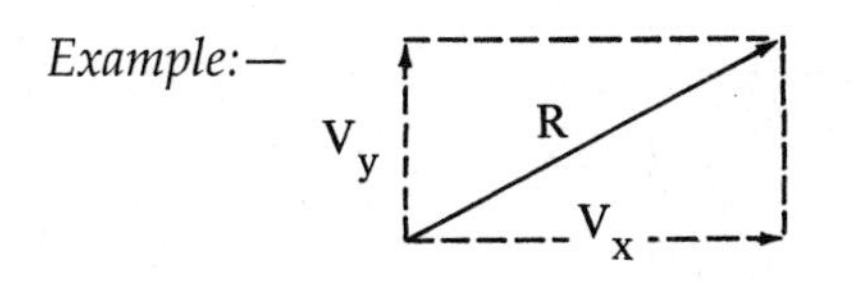

b. Components also may be determined by the use of trigonometric functions.

Example: A resultant of magnitude 80 makes a 30-degree angle with the horizontal. Find its horizontal and vertical components.

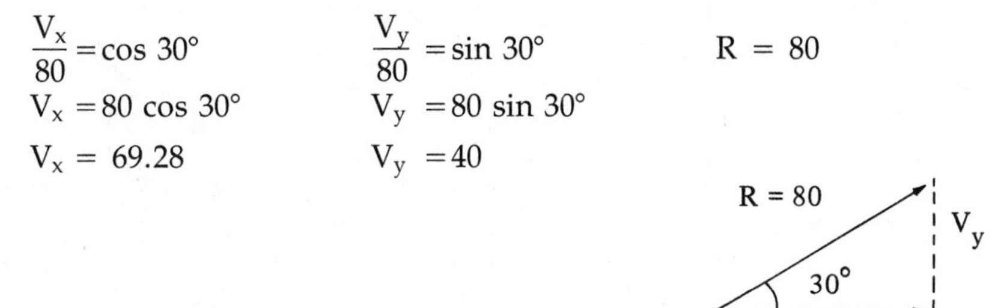

$$\frac{V_x}{80} = \cos 30° \qquad \frac{V_y}{80} = \sin 30° \qquad R = 80$$

$$V_x = 80 \cos 30° \qquad V_y = 80 \sin 30°$$

$$V_x = 69.28 \qquad V_y = 40$$

TABLE V.1

TRIGONOMETRIC FUNCTIONS

Radian	Degree	Sine	Cosine	Tangent	Radian	Degree	Sine	Cosine	Tangent
.000	0	0.0000	1.0000	0.0000	.803	46	0.7193	0.6947	1.0355
.017	1	0.0175	0.9998	0.0175	.820	47	0.7314	0.6820	1.0724
.035	2	0.0349	0.9994	0.0349	.838	48	0.7431	0.6691	1.1106
.052	3	0.0523	0.9986	0.0524	.855	49	0.7547	0.6561	1.1504
.070	4	0.0698	0.9976	0.0699	.873	50	0.7660	0.6428	1.1918
.087	5	0.0872	0.9962	0.0875	.890	51	0.7771	0.6293	1.2349
.105	6	0.1045	0.9945	0.1051	.908	52	0.7880	0.6157	1.2799
.122	7	0.1219	0.9925	0.1228	.925	53	0.7986	0.6018	1.3270
.140	8	0.1392	0.9903	0.1405	.942	54	0.8090	0.5878	1.3764
.157	9	0.1564	0.9877	0.1584	.960	55	0.8192	0.5736	1.4281
.175	10	0.1736	0.9848	0.1763	.977	56	0.8290	0.5592	1.4826
.192	11	0.1908	0.9816	0.1944	.995	57	0.8387	0.5446	1.5399
.209	12	0.2079	0.9781	0.2126	1.012	58	0.8480	0.5299	1.6003
.227	13	0.2250	0.9744	0.2309	1.030	59	0.8572	0.5150	1.6643
.244	14	0.2419	0.9703	0.2493	1.047	60	0.8660	0.5000	1.7321
.262	15	0.2588	0.9659	0.2679	1.065	61	0.8746	0.4848	1.8040
.279	16	0.2756	0.9613	0.2867	1.082	62	0.8829	0.4695	1.8807
.297	17	0.2924	0.9563	0.3057	1.100	63	0.8910	0.4540	1.9626
.314	18	0.3090	0.9511	0.3249	1.117	64	0.8988	0.4384	2.0503
.332	19	0.3256	0.9455	0.3443	1.134	65	0.9063	0.4226	2.1445
.349	20	0.3420	0.9397	0.3640	1.152	66	0.9135	0.4067	2.2460
.367	21	0.3584	0.9336	0.3839	1.169	67	0.9205	0.3907	2.3559
.384	22	0.3746	0.9272	0.4040	1.187	68	0.9272	0.3746	2.4751
.401	23	0.3907	0.9205	0.4245	1.204	69	0.9336	0.3584	2.6051
.419	24	0.4067	0.9135	0.4452	1.222	70	0.9397	0.3420	2.7475
.436	25	0.4226	0.9063	0.4663	1.239	71	0.9455	0.3256	2.9042
.454	26	0.4384	0.8988	0.4877	1.257	72	0.9511	0.3090	3.0777
.471	27	0.4540	0.8910	0.5095	1.274	73	0.9563	0.2924	3.2709
.489	28	0.4695	0.8829	0.5317	1.292	74	0.9613	0.2756	3.4874
.506	29	0.4848	0.8746	0.5543	1.309	75	0.9659	0.2588	3.7321
.524	30	0.5000	0.8660	0.5774	1.326	76	0.9703	0.2419	4.0108
.541	31	0.5150	0.8572	0.6009	1.344	77	0.9744	0.2250	4.3315
.559	32	0.5299	0.8480	0.6249	1.361	78	0.9781	0.2079	4.7046
.576	33	0.5446	0.8387	0.6494	1.379	79	0.9816	0.1908	5.1446
.593	34	0.5592	0.8290	0.6745	1.396	80	0.9848	0.1736	5.6713
.611	35	0.5736	0.8192	0.7002	1.414	81	0.9877	0.1564	6.3138
.628	36	0.5878	0.8090	0.7265	1.431	82	0.9903	0.1392	7.1154
.646	37	0.6018	0.7986	0.7536	1.449	83	0.9925	0.1219	8.1443
.663	38	0.6157	0.7880	0.7813	1.466	84	0.9945	0.1045	9.5144
.681	39	0.6293	0.7771	0.8098	1.484	85	0.9962	0.0872	11.43
.698	40	0.6428	0.7660	0.8391	1.501	86	0.9976	0.0698	14.30
.716	41	0.6561	0.7547	0.8693	1.518	87	0.9986	0.0523	19.08
.733	42	0.6691	0.7431	0.9004	1.536	88	0.9994	0.0349	28.64
.751	43	0.6820	0.7314	0.9325	1.553	89	0.9998	0.0175	57.29
.768	44	0.6947	0.7193	0.9657	1.571	90	1.0000	0.0000	∞
.785	45	0.7071	0.7071	1.0000					

APPENDIX VI:

Muscles and Movements

TABLE VI.1
MAJOR MUSCLES OF THE UPPER EXTREMITY AND THEIR MOVEMENTS[1]

Muscle	Articulations Crossed	Movement Caused
Levator scapulae	Sternoclavicular	Elevation
Subclavius	Sternoclavicular	Depression
Pectoralis minor	Sternoclavicular	Depression Protraction
Pactoralis major	Sternoclavicular	Protraction Depression
	Shoulder	Adduction Transverse adduction Flexion (clavicular portion)
Trapezius	Sternoclavicular	Elevation (upper portion) Retraction (middle portion) Depression (lower portion)
Rhomboid major & minor	Sternoclavicular	Elevation Retraction
Serratus anterior	Sternoclavicular	Depression Protraction
Deltoid	Shoulder	Flexion (anterior) Transverse adduction (anterior) Abduction (middle) Extension (posterior) Transverse abduction (posterior)

TABLE VI.1

MAJOR MUSCLES OF THE UPPER EXTREMITY AND THEIR MOVEMENTS[1] (continued)

Muscle	Articulations Crossed	Movement Caused
Biceps brachii	Shoulder	Flexion (long head)
	Elbow	Flexion
	Radioulnar	Supination
Coracobrachialis	Shoulder	Flexion
		Transverse adduction
[a]Subscapularis	Shoulder	Lateral rotation
[a]Supraspinatus	Shoulder	Abduction
[a]Infraspinatus	Shoulder	Lateral rotation
[a]Teres minor		Transverse abduction
Latissimus dorsi	Shoulder	Medial rotation
Teres major		Extension
Triceps	Shoulder	Extension
		Transverse abduction
	Elbow	Extension
Brachialis	Elbow	Flexion
Brachioradialis	Elbow	Flexion
	Radioulnar	Supination or pronation to midposition
Anconeus	Elbow	Extension
Pronator teres	Elbow	Flexion
	Radioulnar	Pronation
Pronator quadratus	Radioulnar	Pronation
Supinator	Elbow	Extension
	Radioulnar	Supination
Wrist extensors	Elbow	Extension
	Wrist	Extension
Wrist flexors	Elbow	Flexion
	Wrist	Flexion

[a] Rotator cuff muscles.

[1] Adapted from Kreighbaum, E. (1985). Anatomy and kinesiology. In N. Van Gelder (Ed.), *Aerobic dance-exercise instructor manual* (p. 66). San Diego: IDEA Foundation.

TABLE VI.2

SPECIFIC MUSCLES AND MOVEMENTS OF THE JOINTS IN THE UPPER EXTREMITY[1]

Movement	Muscle	Articulation	Action
Shoulder flexion	Biceps brachii	Shoulder	Flexion
		Elbow	Flexion
		Radioulnar	Supination
	Anterior deltoid	Shoulder	Flexion
			Transverse adduction
	Coracobrachialis	Shoulder	Flexion
			Transverse adduction
	Pectoralis major (clavicular)	Shoulder	Flexion
			Transverse adduction
Shoulder extension	Posterior deltoid	Shoulder	Extension
	Triceps	Shoulder	Extension
		Elbow	Extension
	Latissimus dorsi	Shoulder	Extension
			Adduction
	Teres major		Medial rotation
Shoulder abduction	Middle deltoid	Shoulder	Abduction
	Supraspinatus		
Shoulder adduction	Pectoralis major	Shoulder	Adduction
		Sternoclavicular	Protraction
	Latissimus dorsi	Shoulder	Extension
			Adduction
	Teres major		Medial rotation
Shoulder medial rotation	Latissimus dorsi	Shoulder	Medial rotation
	Teres major		
Shoulder lateral rotation	Infraspinatus	Shoulder	Lateral rotation
	Teres minor		
Shoulder transverse adduction	Pectoralis major	Shoulder	Transverse adduction
		Sternoclavicular	Protraction
	Coracobrachialis	Shoulder	Flexion
	Anterior deltoid		Transverse adduction
Shoulder transverse abduction	Triceps	Shoulder	Transverse abduction
	Posterior deltoid		Extension
Elbow flexion	Biceps brachii	Elbow	Flexion
		Shoulder	Flexion
		Radioulnar	Supination
	Brachialis	Elbow	Flexion
	Brachioradialis	Elbow	Flexion
		Radioulnar	Supination and pronation to midposition
Elbow extension	Triceps	Elbow	Extension
		Shoulder	Extension

TABLE VI.2
SPECIFIC MUSCLES AND MOVEMENTS OF THE JOINTS IN THE UPPER EXTREMITY[1] (continued)

Movement	Muscle	Articulation	Action
Radioulnar pronation	Brachioradialis	Radioulnar	Pronation to midposition
		Elbow	Flexion
	Pronator teres	Radioulnar	Pronation
		Elbow	Flexion
	Pronator quadratus	Radioulnar	Pronation
Radioulnar supination	Brachioradialis	Radioulnar	Supination to midposition
	Supinator	Radioulnar	Supination

[1] Adapted from Kreighbaum, E. (1985). Anatomy and kinesiology. In N. Van Gelder (Ed.), *Aerobic dance-exercise instructor manual* (p. 85). San Diego: IDEA Foundation.

TABLE VI.3
MAJOR MUSCLES OF THE LOWER EXTREMITY AND THEIR MOVEMENTS[1]

Muscle	Articulations Crossed	Movement Caused
Iliopsoas	Intervertebral	Flexion
	Hip	Flexion
Gluteus medius Gluteus minimus	Hip	Abduction
Gluteus maximus External rotators	Hip	Extension Lateral rotation Transverse abduction
Tensor fascia lata	Hip	Abduction
Rectus femoris (quadricep)	Hip	Flexion
	Knee	Extension
Pectineus Adductors (Magnus, longus, brevis)	Hip	Adduction Transverse adduction
Gracilis	Hip	Adduction Transverse adduction
Sartorius	Hip	Flexion Lateral rotation
	Knee	Flexion

TABLE VI.3

MAJOR MUSCLES OF THE LOWER EXTREMITY AND THEIR MOVEMENTS[1] (continued)

Muscle	Articulations Crossed	Movement Caused
Biceps femoris (hamstring)	Hip Knee	Extension Flexion Lateral rotation
Semitendinosus (hamstring) Semimembranosus (hamstring)	Hip Knee	Extension Flexion Medial rotation
Popliteus	Knee	Flexion Medial rotation
Vastus lateralis (quadriceps) medialis intermedius	Knee	Extension
Gastrocnemius (Achilles)	Knee Ankle	Flexion Plantar flexion
Soleus (Achilles)	Ankle	Plantar flexion
Tibialis posterior Flexor digitorum longus Flexor hallucis longus	Ankle Subtalar	Plantar flexion Inversion
Peroneus longus Peroneus brevis Peroneus tertius	Ankle Subtalar Ankle Subtalar	Plantar flexion Eversion Dorsiflexion Eversion
Tibialis anterior	Ankle Subtalar	Dorsiflexion Inversion
Extensor digitorum longus	Ankle Subtalar	Dorsiflexion Eversion

[1] Adapted from Kreighbaum, E. (1985). Anatomy and kinesiology. In N. Van Gelder (Ed.), *Aerobic dance-exercise instructor manual* (p. 72). San Diego: IDEA Foundation.

TABLE VI.4

SPECIFIC MUSCLES AND MOVEMENTS OF THE JOINTS IN THE LOWER EXTREMITY[1]

Movement	Muscle	Articulation	Action
Hip flexion	Iliopsoas	Hip	Flexion
		Vertebral column	Flexion
	Rectus femoris	Hip	Flexion
		Knee	Extension
	Sartorius	Hip	Flexion
		Knee	Flexion
Hip extension	Gluteus maximus	Hip	Extension
	Biceps femoris	Hip	Extension
		Knee	Flexion
			Lateral rotation
	Semitendinosus	Hip	Extension
	Semimembranosus		Medial rotation
Hip abduction	Tensor fascia lata	Hip	Abduction
	Gluteus medius		
	Gluteus minimus		
Hip adduction	Pectineus	Hip	Adduction
	Adductor longus		
	Adductor magnus		
	Adductor brevis		
	Gracilis		
Hip lateral rotation	Six external rotators	Hip	Lateral rotation
	Gluteus maximus		
Hip medial rotation	Iliopsoas	Hip	Medial rotation
		Vertebral column	Flexion
	Tensor fascia lata	Hip	Medial rotation
			Flexion
Hip transverse abduction	Gluteus maximus	Hip	Lateral rotation
	Gluteus medius		Extension
	Gluteus minimus		
Hip transverse adduction	Pectineus	Hip	Adduction
	Adductors		Adduction
	Tensor fascia lata		Abduction
			Medial rotation
	Rectus femoris	Hip	Flexion
		Knee	Extension
Knee flexion	Biceps femoris	Knee	Flexion
			Lateral rotation
		Hip	Extension

TABLE VI.4
SPECIFIC MUSCLES AND MOVEMENTS OF THE JOINTS IN THE LOWER EXTREMITY[1] (continued)

Movement	Muscle	Articulation	Action
	Semitendinosus Semimembranosus	Knee	Flexion Medial rotation
		Hip	Extension
	Sartorius	Knee	Flexion
		Hip	Flexion
Knee extension	Rectus femoris	Hip	Flexion
		Knee	Extension
	Vastus lateralis Vastus medialis Vastus intermedius	Knee	Extension
Knee—medial rotation	Semitendinosus	Knee	Medial rotation
		Hip	Extension
		Knee	Medial rotation
	Semimembranosus	Hip	Extension
		Knee	Medial rotation
	Popliteus	Hip	Adduction
Knee—lateral rotation	Biceps femoris	Knee	Lateral rotation
		Hip	Extension
Ankle dorsiflexion	Tibialis anterior	Ankle	Dorsiflexion
		Subtalar	Inversion
	Extensor digitorum longus	Ankle	Dorsiflexion
		Subtalar	Eversion
Ankle plantar-flexion	Gastrocnemius	Ankle	Plantar flexion
		Knee	Flexion
	Soleus	Ankle	Plantar flexion
	Tibialis posterior	Ankle	Plantar flexion
		Subtalar	Inversion
	Flexor digitorum longus	Ankle	Plantar flexion
		Subtalar	Inversion
	Flexor hallucis longus	Ankle	Plantar flexion
		Subtalar	Inversion
	Peroneus longus and brevis	Ankle	Plantar flexion
		Subtalar	Eversion
	Peroneus tertius	Ankle	Dorsiflexion
		Subtalar	Eversion
Subtalar inversion	Tibialis anterior	Subtalar	Inversion
		Ankle	Dorsiflexion
	Flexor digitorum longus	Subtalar	Inversion
		Ankle	Dorsiflexion
	Flexor hallicus longus	Subtalar	Inversion
		Ankle	Dorsiflexion

TABLE VI.4

SPECIFIC MUSCLES AND MOVEMENTS OF THE JOINTS IN THE LOWER EXTREMITY[1] (continued)

Movement	Muscle	Articulation	Action
Subtalar eversion	Tibialis posterior	Subtalar	Inversion
		Ankle	Plantar flexion
	Peroneus longus	Subtalar	Eversion
		Ankle	Plantar flexion
	Peroneus brevis	Subtalar	Eversion
		Ankle	Plantar flexion
	Peroneus tertius	Subtalar	Eversion
		Ankle	Plantar flexion

[1] Adapted from Kreighbaum, E. (1985). Anatomy and kinesiology. In N. Van Gelder (Ed.), *Aerobic dance-exercise instructor manual* (p. 86). San Diego: IDEA Foundation.

Index

Page numbers set in boldface type indicate the page on which an entry is defined.